MODERN ASPECTS OF BULK CRYSTAL AND THIN FILM PREPARATION

Edited by **Nikolai Kolesnikov** and **Elena Borisenko**

Modern Aspects of Bulk Crystal and Thin Film Preparation
Edited by Nikolai Kolesnikov and Elena Borisenko

Published by InTech

Janeza Trdine 9, 51000 Rijeka, Croatia

Publishing Process Manager Sasa Leporic

Technical Editor Teodora Smiljanic

Cover Designer InTech Design Team

Additional hard copies can be obtained from orders@intechopen.com

Modern Aspects of Bulk Crystal and Thin Film Preparation, Edited by Nikolai Kolesnikov and Elena Borisenko
p. cm.
ISBN 978-953-307-610-2

For EU product safety concerns:
IN TECH d.o.o., Prolaz Marije Krucifikse Kozulić 3, 51000 Rijeka, Croatia,
info@intechopen.com or visit our website at intechopen.com.

Meet the editors

Dr. Nikolai N. Kolesnikov graduated from the Physical-Chemical Department of the Moscow Chemical-Technological Institute in 1982. From 1982 to 1993, he worked at the Institute of Solid State Physics (ISSP), Russian Academy of Sciences, and has experienced the journey from Engineer to Senior Researcher. He obtained his PhD degree in Condensed Matter Physics in 1989. Since 1993, Dr. Kolesnikov has been Chief of Laboratory of the Physical-Chemical Basis of Crystallization of ISSP. Nikolai N. Kolesnikov is successfully working in crystal growth from the melt and related fields. His research interests are crystal growth of II-VI compounds (ZnS, ZnSe, ZnTe, CdS, CdSe, CdTe, Cd1-xZnxTe) by zone melting, the Bridgman technique, growth of the nanocrystals of the wide-gap semiconductors, investigation of properties of the refractory compound melts, and the growth of Tl-based HTSC single crystals. Dr. Kolesnikov has published over 100 papers in scientific journals and has received 38 patents.

Dr. Elena Borisenko is currently a Senior Researcher at the Institute of Solid State Physics, at the Russian Academy of Sciences. She graduated from Moscow Institute of Steel and Alloys as a Master of Science in Metal Physics, in 1985 and she did her PhD study, on condensed matter physics, at ISSP RAS, in 1992. Dr. Borisenko is experienced in studying properties and microstructure of advanced materials, especially, inorganic semiconductor compounds and alkali halide crystals. Her research includes phase transitions, physical-chemical aspects of crystal growth, grain growth and textures, structural defects, and testing of mechanical properties of solids.

Contents

Preface

Crystal growth is widely renowned as a sure way to solve a great range of technological tasks, both in the manufacturing of well-known materials and in a search and development of new ones with preset properties. For many technical fields, such as non-linear optics, semiconductor detectors of ionizing radiations, or THz technique, the bulk growth of single crystals often provides a "ley line" to devices with desirable characteristics. This is why the well-known growth methods, like Bridgman, Czochralski, or zone melting, are still in use in production and in research and development. Moreover, new applications for them are found continuously. At the same time, the last decades have revealed high involvement in low-dimensional systems and nanostructures, and here, the crystal growth is a way to prepare new materials both for research purposes and for manufacturing. The concern of crystallization of organic compounds, like proteins, leads to rapid development of this relatively new field. This book is divided into four sections: bulk crystal growth, preparation of thin films, low-dimensional structures, growth of organic crystals, and some theoretical aspects of the field.

The first section contains eleven chapters, and covers the modern act of growing bulk crystals of some silicates, oxides, fluorides, tungstates, nitrides, metals, and intermetallic compounds by means of Czochralski, Bridgman, flux, floating zone, and vapor deposition methods. A few data presented are published for the first time, while other chapters cover the contemporary state of the art, the concrete problems addressed, the materials' characteristics achieved, the characterization methods, crystals' applications, and a bibliography. A wide range of methods is used for the determination of crystal properties, including X-ray diffraction, energy-dispersive spectroscopy, neutron scattering, spectral and thermal analysis, and many others. One of the chapters is devoted to a new class of apparatus for crystal growth, and in other chapters, the growth equipment, as well as the selection of right crucible materials are discussed as well.

The second section includes six chapters on growth of thin films and low-dimensional structures, covering topics on AlN thin films, influence of the substrate temperature on the properties of semiconducting films, ZnO mesostructures preparation, growth of nano-zeolites, green synthesis of metal nanoparticles, and on crystal habit modification via habit modifiers.

The penultimate section includes four chapters focused on crystallization of proteins, covering main aspects of protein nucleation and crystallization, different diagnostic tools, crystallization techniques, and various other strategies. High pressure as a tool for enhancing crystallization of a protein is also discussed. The contemporary knowledge on twinning formation is summarized, and the methods to overcome it are presented. The section also includes the chapter on solution growth of organic crystals for non-linear optics.

The last section contains two chapters describing the simulation of $CaCO_3$ crystal growth through a multiphase reaction, and colloidal crystal formation with focus on capillary growth and its dependence on interparticle interactions, the substrate, and the manipulation of the solvent meniscus.

Acknowledgements

I deeply appreciate the valuable help of the InTech team in editing this book.

Nikolai N. Kolesnikov
Institute of Solid State Physics, Russian Academy of Science, Moscow, Russia

Part 1

Bulk Crystal Growth

1

New Class of Apparatus for Crystal Growth from Melt

Aco Janićijević[1] and Branislav Čabrić[2]
[1]*Faculty of Technology and Metallurgy, Belgrade,*
[2]*Faculty of Siences, Kragujevac,*
Serbia

1. Introduction

In this chapter, we offer original solutions for crystallization devices by presenting a set of cooling devices that are upgraded models of the existing ones used in a well-known apparatus for crystal growth. Many basic ideas from these articles were used as a starting point for the creation of the new, modern multifunctional devices that may be used both as standard school laboratory tool and as industrial equipment. A number of crystal growth devices previously employed were designed to match contemporary technology level and needs for specific monocrystal growth. This led to the additional engagement on the realization of new working conditions, thereby increasing production costs. In light of this problem, while developing new forms od crystal growth apparatus we also have aimed at making the whole process as economical [1], approachable and efficient as possible.

A brief review of twentieth century devices for the crystal growth from the melt [2, 3, 4] reveals widely accepted remarks on some not so good characteristics of specific apparatus components. Let's mention Tamman's test tube and its tip modification which is essential to the crystal germ formation, realization of suitable apparatus geometry, construction of cooler parts in order to have controlled under-cooling, some specific demands for the adequate temperature gradient. Within this chapter, we have defined certain activities conducted (with the set goal in mind) in order to improve existing and to develop new crystal growth devices. We started with a set of simple steps that allowed for the modeling and construction of school type apparatus [5, 6, 7]. Later on, we came up with original solutions and more complex devices with a number of advantages compared to the known crystal growth devices.

Construction of new devices has as its ultimate goal apparatus standardization. Therefore, in a number of papers we have performed calculations that justify the use of newly designed apparatus. As a matter of a fact, in previous research, a standard and widely used approach in technology of crystal growth was to make a specific prototype of apparatus, and then, through a variety of experimentally gained data, to upgrade and improve the characteristics of crystal growth process, depending on the specific demand set for the purpose [2, 3]. That kind of approach was uneconomical regarding time consumption, and a large number of unsuccessful attempts was something one had to count on. For each specific demand a construction of an apparatus almost identical (with a slight modification only) to the one that failed was necessary. In turn, this led to significant material investments for the research, therefore making the crystal growth research a privilege of financially powerful countries that had the opportunity of gathering the top quality researchers from all over the world. Nevertheless, such huge investments had its justification in the fact that some extraordinary results were achieved. This resulted in production of materials of exceptional purity, as well of some new substances and materials whose crystals were realized for the first time in laboratory conditions.

These new materials found its immediate application in the military industry, where high quality materials are imperative, but also in some industrial branches, making these countries top producers of relevant materials (revolutionary novelties in semiconductor technology, telecommunication and optical devices).

Modern apparatus and its modifications presented here have common characteristics of not being financially [1] demanding (starting with the simple to the complex ones). Secondly, it is desirable to have apparatus that will allow for the large number of repetitions of similar processes (with small modifications only and development of new simple parts of equipment for possible improvements of crystal growth conditions). We went even further by developing models and constructing the devices with suitable geometry that allow for the crystallization of a single substance with different crystallization rates and temperature gradients. In addition, it is possible to achieve crystallization of different materials within the single event by employing materials with similar melting points, while having different crystallization rates and temperature gradients.

Along with previously stated advantages of developed apparatuses, we attempted and applied numerical calculations (whenever possible) to get best possible set of parameters in preparation of a new model for crystallization processes. One such analysis takes into account the dimensions of apparatus parts as well as interrelations among the most relevant crystallization factors that will allow for the optimal quality final product - crystal or monocrystal.

In general, intention of the authors is to intertwine these modern devices (large repeatability and multifunctional aspect of crystallization process being the most important advantages) with relevant numerical calculations and existing software. Computer regulated and monitored crystallization would give us more insight on how different parameter variation (such as temperature variations, heat transfer, crystallization rate etc.) and different apparatus dimensions, influences the crystallization process. In other words, there is a tendency to perform all the possible calculations in order to take necessary steps to modify and improve crystallization, so that we would get a crystal of predefined characteristics in a modern and efficient way by using state-of the-art information technologies within the regime of so-called expert systems.

2. New classes of coolers

In accordance with plans based on the variety of possible choices of data on architecture, construction and reconstruction of crystallization apparatus, we came upon a number of creative ideas that are directed towards the adaptations of apparatus shape within the laboratory conditions, the form of coolers and its more efficient role in crystallization apparatus. Long time experience based on the years of the research led author to the conclusion that the heat conduction is one of the essential factors determining the crystallization rate. When, in the conditions of undercooling, the heat is being released, the undercooling will exist only if the heat is being taken away in a proper manner. The rate of heat conduction is a factor quite responsible for the crystallization rate. Crystal growth rate is constant when there is a balance in heat transfer. The heat transfer is quite a complex problem in the sense of regulating the system that has a continuous and controlled operation in accordance with the predefined phases of the crystallization process. From the very start of the germ formation, it is necessary to get a desired temperature drop that defines the initial state of crystallization, and then, by setting an appropriate temperature gradient one can have optimal conditions for obtaining the crystal of specific characteristics.

The temperature aspect of the crystallization that is so significant for the crystal growth and possibility of programming the process parameters through various shapes and positioning of the coolers (which provide cold fluid flow in crystallization apparatus), demands coolers to have multiple roles: firstly, to enable for more precise crystallization, and secondly, to lead to construction of new coolers made of materials of adequate heat conductivity so to have more convenient conditions for crystal growth from the melt.

Besides, suitably designed coolers have such a shape that they may simultaneously serve as ampoule carriers or test tubes with melt. In this way, the crystallization will be easily controlled. When looking back at the devices previously used, it is easy to see that some parts of the devices were burdened by carriers of pots with melt, as well as due to their heating and operating them in and out of the apparatus. Also, realization of adequate temperature gradient and subcooling through complicated pipe constructions or other forms of the coolers of intricate geometries (positioned within the crystallization apparatus), additionally complicated crystallization apparatus, not to mention the other instruments used in the process. Detailed analysis of these problems gave us very useful data that generated a completely new set of ideas, which ultimately resulted in a new, more complex role that coolers have in process. Their multifunctionallity led to significant simplification in apparatus construction in many of the known methods, which, loosely speaking, were reinvented. In some of author's papers, a demand for cooler improvement was set, and it resulted in design of more efficient and modern generation of crystallization devices.

As a basis for the design of novel or significantly improved and modified standard crystallization devices, we have used a series of originally, for the purpose-constructed coolers presented within the chapter. Cooler models presented in Figs 1- 6, whose forms and functionality gained recognition through presentation in few articles, may be divided in several groups, based on its positioning in the apparatus, cooling fluid flow propagation and its intended method application (Tamman, Stober, Czochralski). The general classification, which arises from the position of the cooler within the apparatus, leads us to two types of coolers: vertical and horizontal.

2.1 Vertical air coolers

Coolers where the cooled air is moving along defined (vertical) tube direction, belong to the group of so called vertical air coolers (Fig. 1). Thanks to the different cross sections of the tube, different speeds of airflow are possible. In that way, various crystallization speeds via heat dissipation are established in test tubes that are attached to the body of the cooler in various manners. There is a whole spectrum of coolers based on the positioning of the test tubes: the ones with fixed test tube position, to the ones with mobile rings on mobile coolers. Large number of test tube positions is available (Fig. 3).

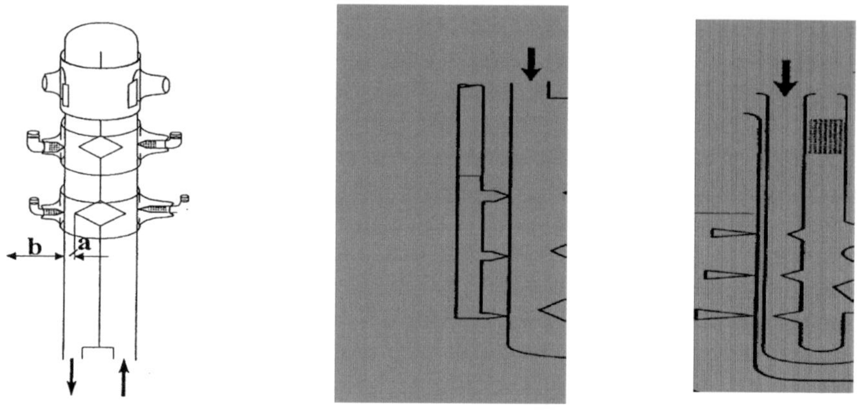

(a) (b) (c)

Fig. 1. Vertical coolers: (a) cold "finger", (b) "cold key" and (c) "cold ear-rings".

For the class of coolers presented in Fig. 2, the line of development was the following one: in certain positions, the tubes were constricted and slightly bended, so to achieve the optimal heat dissipation, and to simultaneously allow for an additional number of test tubes to be positioned. This was followed by coolers where the pipes were ring like bended in a couple of independent levels of crucibles, which allows for an increase in crucible operating capacity.
The operating regime of this class of coolers is such that each ring has a direct fluid flow within it and heat dissipation in the environment. The other opportunity are so called spiral coolers where heat generated during the crystallization process from all the rings is being "collected" and dissipated into environment. Detailed analysis of presented models showed some additional possibilities of vertical coolers. These were used for some novel practical solutions. Depending on the geometry of the space the coolers are in, they may be maneuvered (so called movable vertical air coolers) or be fixed while some of the other pipes (with Tamman test tubes) can be maneuvered on order to get a desired temperature gradient or crystallization rate. Whenever the vertical air coolers are employed, whether its orientation is upside down or vice versa, fluid flow is such that it returns in the opposite direction along the same path.

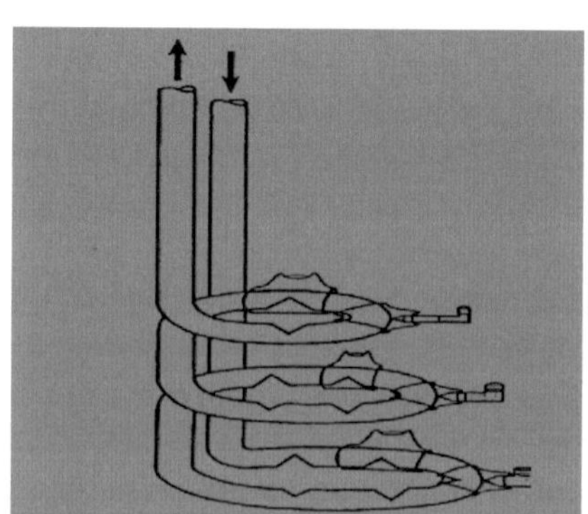

Fig. 2. Air cooler model ("cristallization spiral").

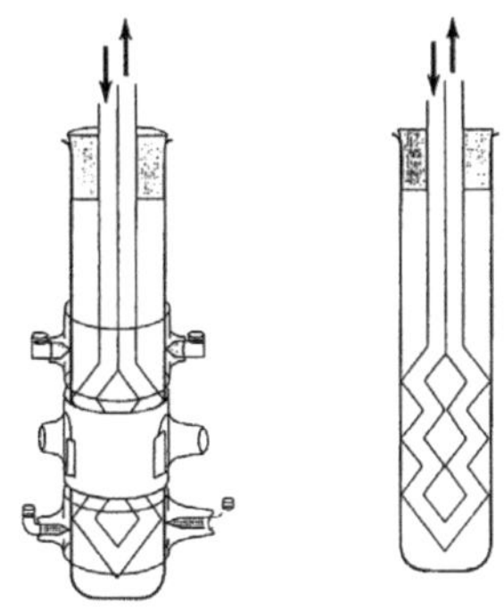

Fig. 3. Air cooler model ("crystallization key").

2.2 Horizontal air coolers

When talking about the horizontal air coolers, there are, basically, two classes with some specific variations:

a. To the first group belong coolers whose fluid flow pipes are horizontal. The cold fluid enters on one side and exits on the other one (single pipe horizontal air cooler, Fig. 4; system may also have two or more horizontal pipes). Couple of horizontal coolers can form an ensemble of instruments in chamber or crucible furnace.
b. Other type of cooler employed in the crystallization purposes, is the one where a horizontal pipe is bended at its end, carrying the fluid in the direction opposite to the initial one, and then the heat is being dissipated into environment (Fig. 4b. and Fig. 5.). If the pipe is bended at 180°, there is a possibility of multiplying initial activities via new conditions and test tube positioning. This allows for a large interval of crystallization rates in direction of the cooler.

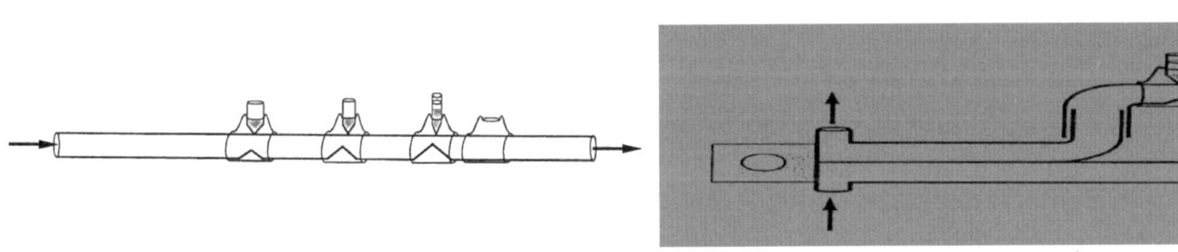

(a) (b)

Fig. 4. Horizontal coolers: (a) pipe, (b) two-pipe (folding)

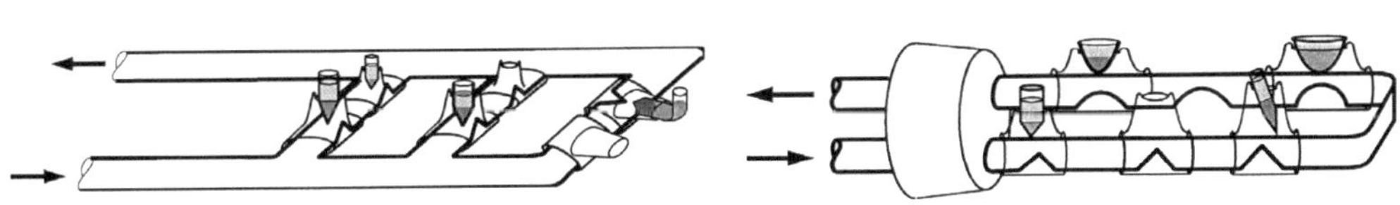

(a) (b)

Fig. 5. Multifunctional horizontal coolers: a) the standard method, b) for the combined methods.

In such cases, we have come up with an original solution. The flow that convects heat below Tamman's test tubes is now being used for cooling the top layer of the melt that is positioned next to the exit pipes of the cooler. In that way, we have assigned it a new role upon bending the initial pipe. It gives us the opportunity of constructing the apparatus with new combined methods (Tamman's and Stober's). In Fig. 5, we present two solutions from a whole family of coolers whose realization is based on previously presented idea that leads to greater operability and more economical functioning in the crystallization process. Solutions presented give a clearly confirm validity of idea of redesigning some parts of cooler as well apparatus as a whole, and undoubtedly pointout their versatile practical purposes.

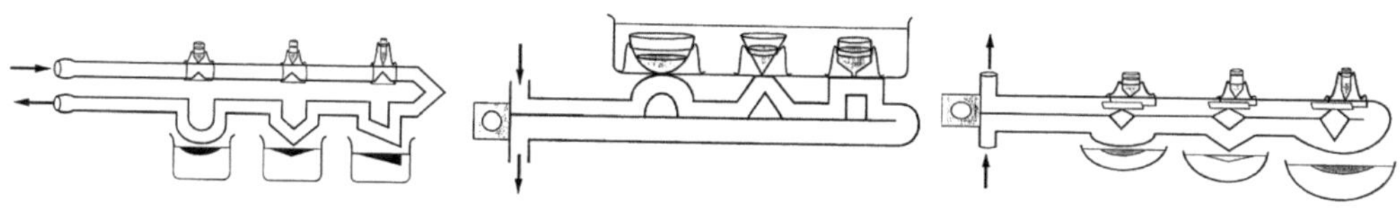

(a) (b) (c)

Fig. 6. Horizontal coolers; modification (a) and (c) combined with multivariate methods (b) variation of a method.

In Fig.6, specific horizontal coolers are given. Some parts of pipes are bended in the outer part of the device (unlike the previous ones where constrictions exist on the inner parts only) having endings of different geometrical shapes that allow different flow velocities. We have therefore met the conditions necessary for Stober method crystallization.

In this way, in the course of a single event, we have enabled crystallization based on the two methods, one during the fluid flow in the one direction, and the other for the opposite direction flow. In one case the cooling fluid flows above the crucibles containing crystallization melt. In the other, the flow goes below the melt where, by under-cooling specific capillary endings of test tubes with melt, a new process of germ creation starts all the way to the final crystallization. A geometrical representation of such coolers reminds of "cold horseshoes" and "cold keys". Fig. 11 demonstrates application of the modified cooler, which comprises two horizontal pipes mutually joined to movable pipe, which is an exceptional improvement compared to former examples in a sense of simplified geometry modification and crystallization conditions.

In some of horizontal coolers with one or more pipes containing cold fluid, another innovation is present. The pipe of cooler is introduced into a pipe of greater diameter, which may consist of one or two parts (Fig. 10) with small openings and slots, in which the position of pots and test tubes with melt may be fixed. Such a solution has clear advantages to the previously described ones, since by simply moving the cylindrical pipe (whose function is to move the cooler pipe and to serve as a test tube carrier all at once) a large number of different crystallization conditions and new crystallization geometries is achieved.

3. Original crystallization apparatus

The installation of the innovated systems for cooling, with the aim to monitor heat removal for the regulations of the processes of crystal growth from melted materials, enabled

obtaining more devices for crystallization. The new classes of cooling devices, with aforementioned advantages linked to the crystallization processes have an additional quality which is that those cooling devices are very adaptive for installation and operative by application in well known laboratories-crucibles, chamber furnaces and tube furnaces. However, more complex cooling systems with the Tamman's test tubes, as a carriers devices, need to create new forms of crystallization apparatuses. The projecting of the new classes of devices for crystal growth of melts, which will be shown in the following text, is the response to the aforementioned need.
From these methods for crystal growth from the melt, it is estimated that in the school laboratory, the Tamman's method is the most convenient one. If we use the advantages of the horizontal single tube aerial cooling system, an original device, the so called "crystallization bench" can be realized [8]. It consists of a tube furnace and a specially adapted cooling system (Fig 7.).

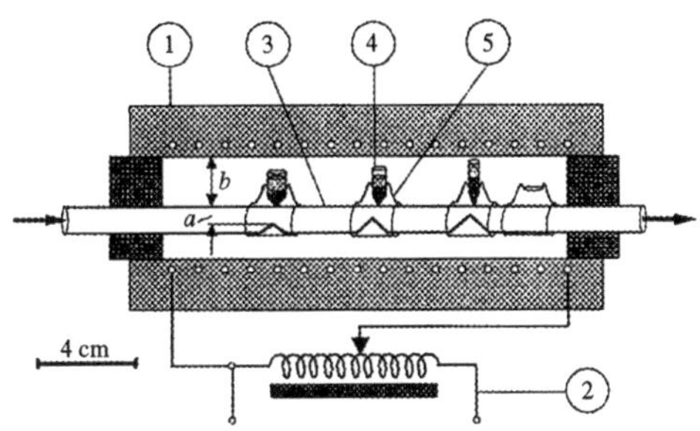

Fig. 7. Crystallization regulation in a tube furnace. (1) electroresistant tube furnace, (2) continuously changeable transformer, (3) air cooler ("cold bench"), (4) Tammann test tubes and (5) rings.

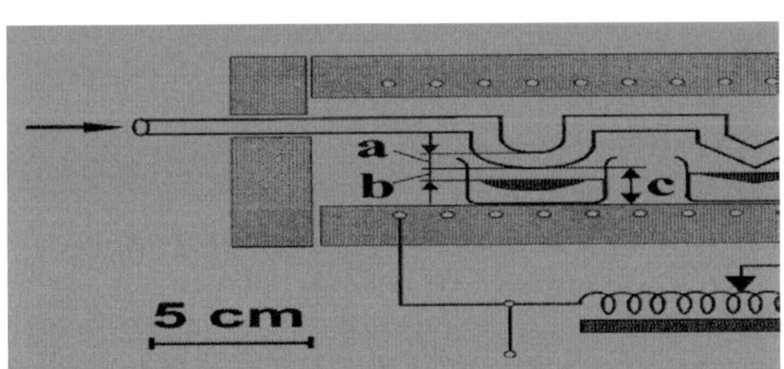

Fig. 8. A chamber furnace for obtaining crystals. (1) laboratory chamber furnace, (2) continuously changeable transformer, (3) air cooler ("cold key"), (4) cold "teeth" (5) crucibles with the floating crystals.

The procedure of choosing the wanted disposition of the test tubes, with the melted material, above the narrowing cooler cross section, is accomplished with moving rings on the cooler tubes. The devices may contain many Tamman's test tubes of various sizes and dispositions. The constructed device enables the simultaneous test of a few various nucleation and crystallization rates. Tamman test tubes of various shapes and dimensions (a family group [2, 3, 4] can be mounted on the test tube rings and thus simultaneously tested).

The variations considering the disposition changes of certain test tubes, as well as simultaneous regulations of some temperature gradients are also possible. The working regime of the devices works as following: at a constant furnace temperature, a weak air flow is turned on through the cooler. There, the crystallization starts on the bottom of the test tube (Fig.9).

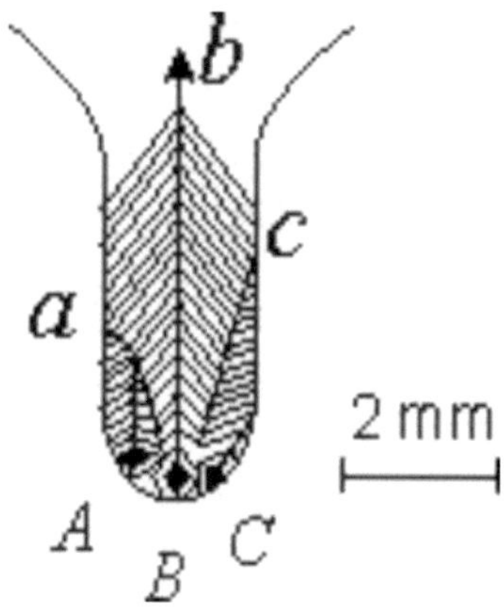

Fig. 9. The beginning of the crystallization at the bottom of the capillary tube.

The bottom of the test tube continues in the capillary, so that in the beginning of the process, only a small amount of the melt is overcooled. Therefore, only certain crystal nucleuses can be formed. The nucleation which grows towards the walls of the capillary, stop growing at a certain time. Only the nucleation which grows towards the axis of the capillary overgrows the other nucleations, and when they exit the capillaries, they expand to the full cross section of the testing tube.

The preparation of crystals of good quality, containing a low concentration of impurities and defects, requires a crystable substance of high purity, test tubes of materials that do not react chemically with the melt, a high degree of temperature stabilization of the furnace, and the absence of shocks [9]. The conditions required to grow crystals of some example substances, wich have low melting temperatures and can be used to obtain single crystals in school laboratory.

The crystallization rate interval [4] in each tube is regulated by the cross section of the air flow (a), i.e. by translation movement of the test tube rings (Fig. 7). The temperature gradient is regulated by distance (b). Different temperature gradients in the tubes can be simultaneously regulated using an inclined cooler, i.e. "inclined cold bench". By varying the internal and external cooler shape and dimensions, a famili of coolers can be modeled for different intervals of temperature gradients and crystallization rates. Different crystallization fronts and rates in crucible columns can also be regulated below the cooler so that crystallization starts on the surface of the melt (Fig. 8). Crystal growth then occurs downward the lower interface on the floating crystal.

By increasing the air flow velocity, the crystallization front spreads to the other end of the testing tube. The interval of the crystallization rates in each of the testing tubes of the devices (Figs 7, 8) is regulated by the air flow, i.e. the cross section *a,* which increases or decreases by relocating the moving rings, along with the cooler tubes. The temperature gradient is regulated by distance regulators *b* (Fig. 7).

Besides the standard case of the "Crystallization bench", other geometric solutions are possible in the design of the part of the devices [10]. Different thermal gradients in test tubes can be simultaneously regulated by the inclined cooler (or some other part of the cooler) relative to the axis of the furnace (Fig. 8). The shape of crystallization fronts and the crystallization rates in the crucibles are regulated by the path and the cross section of the air flow (*a*) of the cooler, as well as by the distance regulator from the surface of the melted material (*b*). By such creations and innovations, considering shapes and cooler functioning, the possibility of the Stober method realization in shamber and tube furnaces is accomplished.

The project of the original developed devices, the so called "the moving crystallization bench" (Fig. 10) contains some of the more complex forms of the aerial cooler which is in the shape of a cyllindric tube, which is located in another tube, which can have one or two parts, with a bearing for the Tamman test tubes [11].

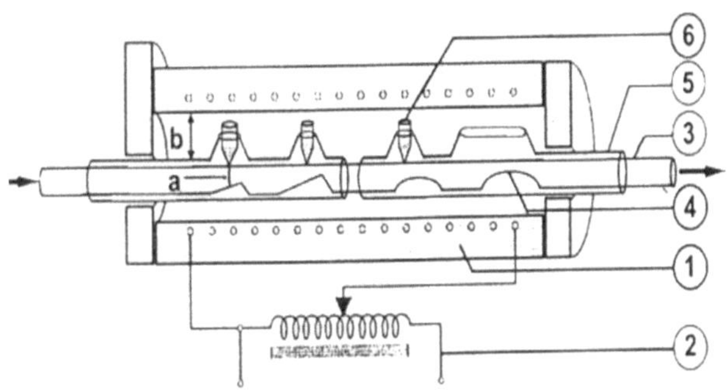

Fig. 10. A tube furnace for obtaining crystals: (1) laboratory tube furnace, (2) continuosly changeable transformer, (3) air coler (telescopic cold bridge), (4) cold "thresholds", (5) cylindrical tube with the mounting holes and grooves (telescopis test sieve) and (6) family group of Tamman test tubes.

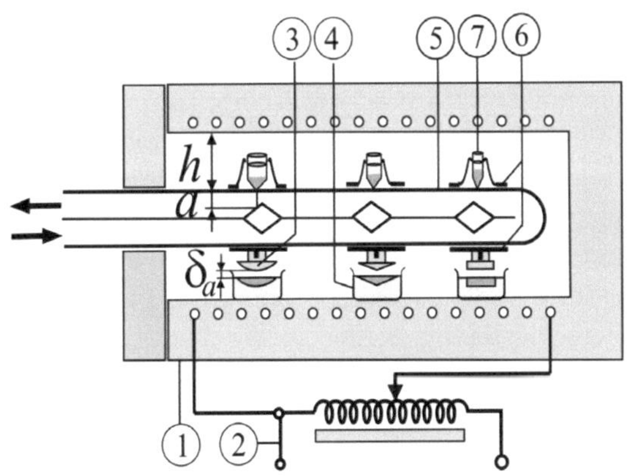

Fig. 11. Apparatus for combining methods: (1) Laboratory chamber furnace, (2) continuosly changeable transformer, (3) movable plugs, (4) columns of crucibles, (5) air cooled toothed tube ("crystallization finger"), (6) movable mounting rings, and (7) Tamman test tubes.

The method of testing crystallization, as well as possible variations of the processes are described. The formula for crystallization rates depending on the parameters of the cooler and the characteristics of the material, as well as respective temperature changes. It creates great possibilities for utilization of various crystallization rates.
Tamman's test tubes of various shapes and sizes can be laid out to the moving cyllindric tube (the so-called ''sieve''). One can accomplish the simultaneous test of the crystallization for a great number of different Tamman's test tubes. They are of various temperature gradients, intervals of crystallization rates, and materials. They can be used for obtaining single crystals from the melt by using cheap and practical modular devices-crystallization apparatus with the moving elements.
The development of the models of one group of apparatuses, whose work is based on single tube horizontal cooler, has developed in several phases. Each one of the phases is characterized by innovations in the series of details, and therefore a very high level has been achieved. That level has gained a special, important confirmation by publishing the paper with newly accomplished results in the professional journal [12].
In the published article [13], the original modification of the devices, which is considerably more sophisticated and efficient than the previous class of the device. Is has been created based on the experience and the series of practical conclusions from the previous models. That article initiated the design of a certain number of devices, which are based on the simultaneous unwinding of the Tamman and Stober methods (Fig. 11). The specially adapted cooler, functional for this purpose, has been installed in the laboratory chamber furnace [13]. The cross section of the fluid current and the distance of the cooler from the surface of the vessel where the melt is located ,define the shape of the fronts and the crystallization rates. Some more demanding and economical variations of these devices contain two tube-coolers, for the arm with Tamman's test tubes. The tube which serves as the test tube carrier can be mobile and can contain more than one series of Tamman's test tubes in telescopic test sieves in the previous paper [13], the possibility of simultaneous realizations of Stober and Tamman's methods has been accomplished. The presented solution and the defined modifications, with the aim of improving the conditions of crystallization by these methods can be applied in the tube furnace in the horizontal positions, too. The most sophisticated devices of so called double-tube horizontal models are achieved by flexing one tube by 180 degrees, or two horizontal tubes linked by vertical linking extensions. This is not only focused on accomplishing simultaneous developments of the crystals utilizing the two methods (Tamman's and Stober's), but it is the invention of the quality forms, the positioning of every single test tube, cooler aperture up to the influence on the front crystallization according to certain calculations [14].
The creation of a certain number of functional vertical coolers, which are previously presented, has made the simultaneous realization of the projects with the crystallization devices with vertical coolers possible.
The model of an air cooler, which is vertically positioned in the laboratory crucible furnace (so-called ''finger'') is presented in the paper [15]. Some bended Tamman's test tubes are positioned on the cooler with the help of rings and sliders of the test tube carrier. The formula of linear crystallization rating in each test tube is derived from using the balance between the latent heat of the solidification and removed heat through the cooler. The possibility of translation of each test tube independently is considered, with the aim of simultaneous probe of the matrices of various crystallization rate intervals.

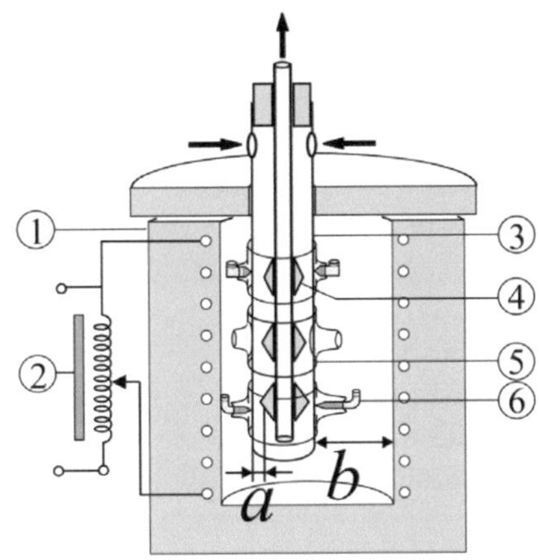

Fig. 12. Multifunction crucibles to obtain crystals: (1) Laboratory crucible furnace, (2) continuosly changeable transformer, (3) air cooler ("cold finger"), (4) movable cold "thresholds", (5) movable mounting rings, and (6) curved Tamman test tubes.

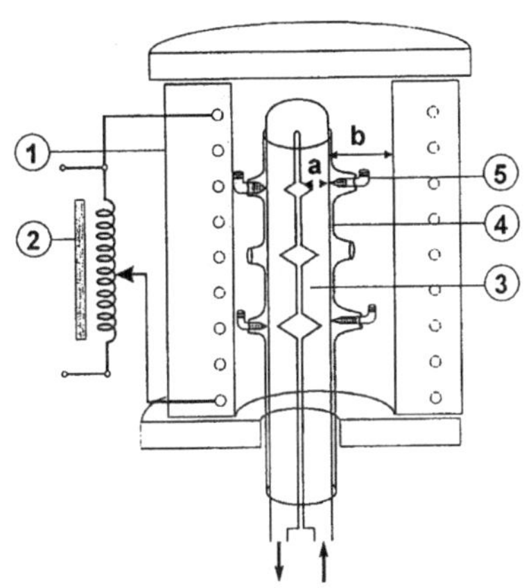

Fig. 13. Apparatus for obtaining crystals: (1) laboratory tube furnace, (2) continuously changeable transformer, (3) air cooler ("cold tree"), (4) movable cylindrical tube with the mouting holes ("test sieve"), (5) family of ("grafted") Tammans test tubes.

The devices in work [16] present the basis for the previous solution of the new apparatus (Fig. 12), but there is a difference in the flexibility of the elements of the devices in the systems as well with the purpose of the geometrical solution of the test tube (with the melt), with many possibilities of the realization of various temperature gradients.
The devices for the crystallizations shown in Fig. 13 presents exactly one level more operative devices [17] than those shown in Fig. 12. These systems usually consist of aerial coolers (with a cold fluid flowing through) and movable cyllindric tubes with placable holes for Tamman's test tubes. The coolers, in this case, are movable, and they give the possibility of definition of certain parameters during the crystallization process. That is how one can very operatively influence the progression of the process and the quality of the obtained crystal.

The mobile test tube carrier, with the melt, can easily enable an adequate position of the test tube tip, depending on the cross section aerial currents through the cooler. It is easier to control the parameters which influence the substantial magnitudes in the crystallization process (crystallization rate, temperature gradient) that way.
An example of an even higher quality of the devices with vertical coolers (Fig. 12), enable us to translate and move the cooler vertically, but also very convenient for quality work due to the various possibilities of the cooler rotation. It enables the regulation of the front crystallization wanted regulation dynamics, which enables the conditions to obtain quality crystals [18]. These devices have emphasized the mobility of elements, and higher potential for the work by choosing the position of the test tube, and the number of the test tube, in the process of crystallization. It is better than that of the devices in Fig. 13, thanks to the fact that it contains a constructive design solution, with a mobile ring and a mobile mechanism of the test tube carrier.
The achieved variety, considering the design, on the crystallization devices, with another cooler class, has brought a new quality in the sense of the possibility ensure a heightened quantity of the melt. It could be in strictly defined and stabilized conditions, which is substantial as an introductory activity, for the crystallization process itself. Such an idea has been realized and has justification in constructive solutions of the devices which are shown in papers [19] and [20].
The original devices for obtaining single crystals from the melt in coherence with the new demands considering the quantity of the melt, as well as obtaining the possibility where more devices can be put in with melted substances with similar melting points. By variation, those crystals are formed in various ways, depending on the conditions. In this case, the idea of applying the Tamman's method with very specific sets of testing tubes, in a laboratory crucible furnace [19] was realized.The regulation and simultaneous crystallization of several substances for a few nucleations of various temperature gradients and crystallization has been made possible.

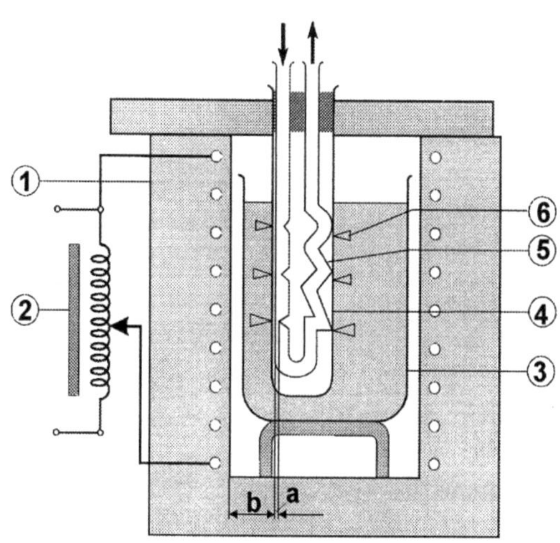

Fig. 14. A crystallization cooler in a crucible furnace. (1) laboratory crucible furnace, (2) continuously changeable transformer, (3) crucible (4) test tube (5) moving air cooler ("cold ear-rings"), and (6) Tamman test tubes.

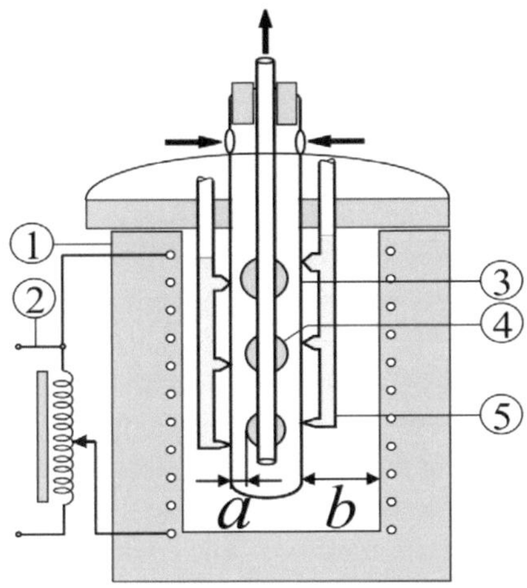

Fig. 15. Crystallization apparatus: (1) laboratory crucible furnace, (2) continuosly changeable transformer, (3) air cooler ("cold key"), (4) movable rings and (5) branched Tamman's test tube ("crystallization test comb").

The combination of several Tamman's test tubes in the form shown in Fig. 15 makes the growth of several crystal from the melt possible, as well as obtaining the conditions for several devices with melted substances who have approximately the same melting points. By variation of the shape and the size of the cooler (inside and out), can model a family of "cold keys" for testing a wider interval of temperature gradients and crystallization rates.

The fusion of the best performance of the devices shown in Figs 12, 13, 15 has benefited the crystallization process in the new devices [14] with a vertical cooler in the crucible furnace, chamber furnace or tube furnace. They also come with a modern form, better functioning and higher economic value, and other important traits which have been greatly improved. Controlled functioning of certain phases during the crystallization process with great reliability for obtaining the crystal's wanted characteristics.

The devices who improve the efficient solution of the form of the apparatus related to the previously described apparatus [21] are shown in Fig. 14 (from the constructional point of view, they are very similar at the first glance). The presence of a larger quantity of the melted substance, but also a bigger number of the test tubes for obtaining adequate crystals with big potential for variation of the conditions, which are very important for the regulation of the crystallization rate, is made possible [22].

3.1 New generation of devices for crystal growth – "Expert systems"

Before mentioned division of coolers on horizontal and vertical ones, and related construction of apparatus could not exist within the given frame. Rich experience in connection with work on crystallization apparatus and vision of development directed towards new possibilities, led to so called "hybrid" solutions for the coolers. In their regime of work, they employ both horizontal and vertical fluid flow.

This, in turn, gives a variety of opportunities for development of original, high quality devices with new possibilities and advantages for crystallization process. A increased efficiency and reduced costs may also be expected.

In [23] a successful realization of combined ("hybrid") device is demonstrated in laboratory chamber furnace. For that purpose, one improved model of crystallization cooler in ladder-like shape on which movable bended Tamman's test tubes are positioned, is presented. By finding the appropriate angle between axes of the test tube and direction of crystallization, defect drainage towards test tube wall may be regulated. To this intermediate group of crystallization devices belongs the apparatus described in [24].

It is quite obvious that fluid current that conducts crystallization heat from a certain level of test tubes, circulates several levels by passing through profiled sections. Therefore, on the remaining semicircular levels, we may, either have the same substance with different cooler cross section on the location of the test tube (in this way we will get the family of crystals of same substance), or, on each level we may have a system of test tubes of different substances from a group of substances having more or less the same crystallization conditions.

In [10], [13] and [25], original modification of devices based on the Stober's method are presented. During the research of methods for crystallization regulation in laboratory chamber furnace for crystal substances with unknown crystallization parameters, we reached the conclusion that combined Tammn-Stober's method can be employed.

Particularly adapted cooler for this purpose was installed in laboratory chamber furnace [26]. The forms of crystallization fronts and crystallization rate in crucibles are regulated via trajectory and the cross section of cooler air flow (d_1), and via the distance of the cooler from the furnace wall (d_2). In more demanding and more economical type of this apparatus, two pipe coolers can contain an array of Tamman's test tubes on one branch [27].

The apparatus presented in [28] is practical realization of combination of more elements from different types of presented devices. It is specific in the sense that it has built in parts of devices that contain several groups of Tamman's test tubes of different shapes, volume and inclination of test tube axes where the formation of crystals is expected. It more complex variant, the apparatus may contain ensembles of coolers and test tubes. In that way, many of the steps can be repeated within single crystallization process [29]. During the research on realization of monocrystals of family of substances with unknown crystallization parameters in laboratory chamber furnace, we have modeled air cooler that enables simultaneous crystallization of several substances at different temperature gradient, shapes of crystallization fronts and crystallization speeds in column of crucibles and test tubes.

By upgrading the existing experiences and improving the characteristics of previously described classes of devices, we have achieved results which give a solid bases for accomplishing the highest goal set during production of new devices for crystal growth from the melt: realization of "smart systems" that control process of crystal growth via computer programming [30].

The first results appeared almost simultaneously in two articles: in Russian journal Instruments and Experimental Techniques [31] (Fig. 16), and another one in American journal American Laboratory [32] (Fig. 17), both published in 2011.

This new class of devices ("superclass") owes its name to its multifunctionality, and ability of its dynamical elements to react almost instantaneously to the tasks regarding regulation and monitoring the crystal growth. It is achieved by establishing permanent connection of devices with computer controlled programs [33]. At this stage of realization, results of process simulation and apparatus conditions are used. Nevertheless, practical realization of establishing direct connection of computer to apparatus and its movable parts (cooler and test tubes with melt) is a matter of time.

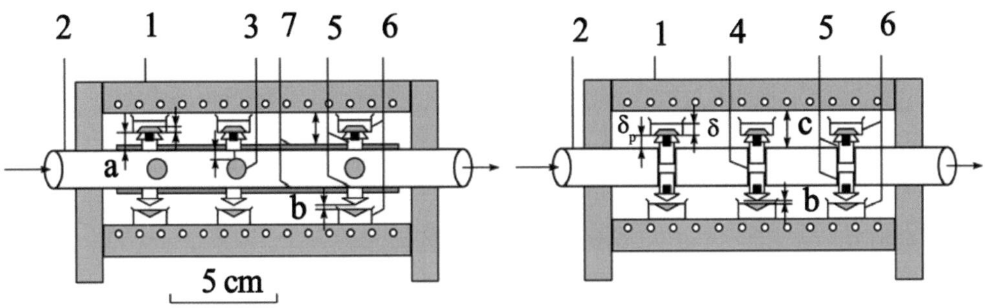

Fig. 16. "Programmed crystallization bench" for crystal growth from melt: (1) laboratory tube furnace, (2) air cooled tube ("crystallization shelf-comb"), (3) radial holes in a horizontal position ("crystallization thresholds"), (4) radial holes in a vertical position ("crystallization sockets"), (5) movable cold plugs, (6) columns of crucibles, and (7) slide bars.

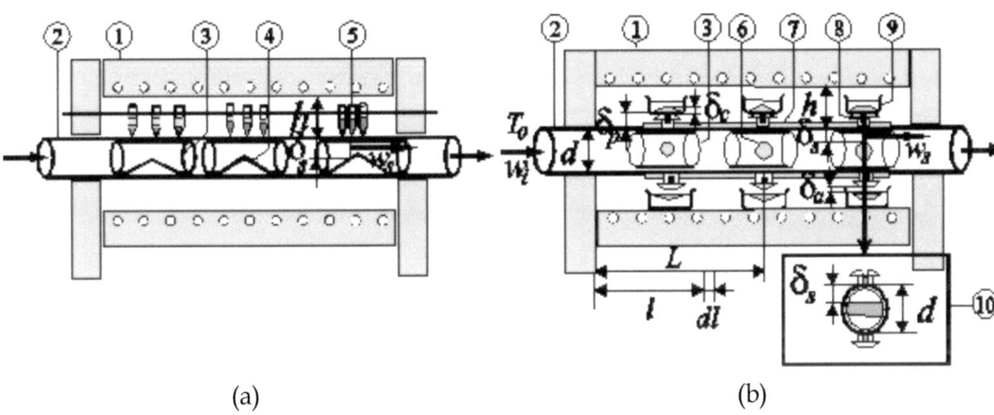

Fig. 17. "Smart" coolers for the combined methods of crystal growth from melt: (modular unilateral (a), and bilateral (b) "crystallization comb" in a tube furnace). (1) laboratory tube furnace, (2) air-cooled tube, (3) modular and movable pipes: (4) "Δ" cold thresholds, (5) string of family group of Tamman test tubes, (6) radial holes in a horizontal (or vertical) position ("O" or "I" cold thresholds), (7) slide bars (or movable mounting rings), (8) movable plugs with modular heads, and (9) column of crucibles; (10) cross section of the cooler.

Programmed conditions will be controlled by computer system that directly define the dynamics of movable systems, such as optimal positioning of cooler within the furnace, controlled heat dissipation in the course of crystallization as well as monitoring the position of test tube carriers where crystallization from the melt is taking place.

4. Results of modeling

The rate of melt solidification depends upon extracting the latent heat of solidification. For a time interval t a crystal layer of thickness δ_c is formed (Fig.17). During the formation of an elementary crystal layer of thickness $d\delta_c$ per unit area, the amount of heat released is

$\lambda\rho d\delta_c$ (λ denotes the latent heat of solidification and ρ the crystal density); the latter is being extracted through the cooler for a time interval dt. On this basis the following equation may be written [34].

$$\lambda\rho\, d\delta_c = \frac{\Delta T(L)}{1/\alpha_s + \delta_p / k_p + \delta_a / k_a + \delta_c / k_c} dt \tag{1}$$

where $\Delta T(L)$ denotes the difference between the temperature of the melt and that of the air stream; α_s is the coefficient of heat transfer from the cooler wall to the air stream; and k_p, k_a and k_c designates the heat conductivity of the plug, air and crystal respectively (Fig. 17b). Transforming equation (1) we obtain

$$R = \frac{d\delta_c}{dt} = \frac{\Delta T(L)}{\lambda\rho(1/\alpha_s + \delta_p / k_p + \delta_a / k_a + \delta_c / k_c)} \tag{2}$$

The quotient $d\delta_c/dt$ denotes the rate of crystal layer growth which is usually represented by the symbol R.
The coefficient of heat transfer from the cooler wall to the air stream can be calculated using the following expression [35]:

$$\alpha_s = \left[4{,}13 + 0{,}23\frac{t}{100} - 0{,}0077\left(\frac{t}{100}\right)^2\right]\left(\frac{U}{4A}\right)^{0.25} (w_{s0})^{0.75} \tag{3}$$

where t is average temperature of the air stream in ^{0}C (up to 1000 ^{0}C), U and A are circumference and area of the cross section of the airstream, respectively – see (10) in Fig. 17b,

$$w_{s0} = w_s\left(\frac{273}{273+t}\right),$$

w_s is average velocity of the airstream (0 ^{0}C, 1.013 bar) in m/s.
On the basis of the continuity of the airstream, and the cross section at the entrance of the tube and on the threshold – see (10) in Fig. 17b, the following expression for the velocity of the air stream on the threshold w_s we have derived:

$$w_s = w_i\left(\frac{\pi}{\varphi - \sin\varphi}\right) \tag{4}$$

where w_i denotes an average velocity of the air stream at the entrance of the tube

$$\varphi = 2\arccos\left(1 - \frac{2\delta_s}{d}\right)$$

in rad, δ_s is the width of the air stream, d is diameter of the tube (Fig. 17b).
Based on the fact that the heat removed from the cooler wall is equal to the heat accepted by the air stream, we have derived the following expression (integral equation) for the difference of the temperature between of the melt and that of the air stream along the cooler $\Delta T(L)$:

$$\Delta T(L) = \Delta T_0 - \frac{4}{d}\int_0^L \frac{\alpha_s \Delta T(l)}{\rho_a c_a} dl \tag{5}$$

where ΔT_0 denotes the difference between the temperature of the melt and that of the air stream at the point $L=0$ (at the entrance of the tube), d is diameter of the tube, α_s is the coefficient of heat transfer from the cooler wall to the air stream - eq. (2), when put

$$\left(\frac{U}{4A}\right) = \left(\frac{1}{d}\right)$$

(i.e. tube without thresholds), $\Delta T(l)$ denotes the difference between the temperature of the melt and that of the air stream at the point l (Fig. 17b), ρ_a designates the air density, c_a is heat capacity of the air.

The crystallization parameters (designed in Fig. 17b) are determined by the numerical analysis of eqs. (2), (3), (4) and (5), in the case of bismuth: T_{melt} = 271 ^{0}C, λ = 52300 J/kg, ρ = 9800 kg/m^3, and k_c = 7.2 W/mK, In all numerical calculation is was taken that: k_p = 0.756 W/mK (pyrex *i.e.* borosilicate glass, softening point ≈ 600 ^{0}C), k_a = 0.0342 W/mK, ρ_a=0.682 kg/m^3,c_a = 1.035 kJ/kgK, ΔT_0 = 251 ^{0}C, T_0 = 22 ^{0}C (the temperature at the entrance of the tube), d = 2cm, δ_p = 5 mm, δ_c = 0 mm (on the bottom or the surface of the melt).

The dependence of the crystallization rate R on the position of the plug along the cooler L is represented on Fig. 18a, when w_s = w_i, eq. (3), *i.e.* the tube without the thresholds. As can be seen from Fig. 18a, the crystallization rate decreases with increasing L, which is the consequence of the fact that ΔT decreases with the increase of the L, eq. (5). Fig. 18b shows the dependence of the crystallization rate R on the with of the air stream δ_s - see (10) in Fig. 17b. As can be seen from Fig. 18b, the crystallization rate increases with decreasing δ_s , which is the consequence of the fact that w_s , eq. (4), and consequently α_s , eq. (3) and R, eq. (2) increases with decrease δ_s.

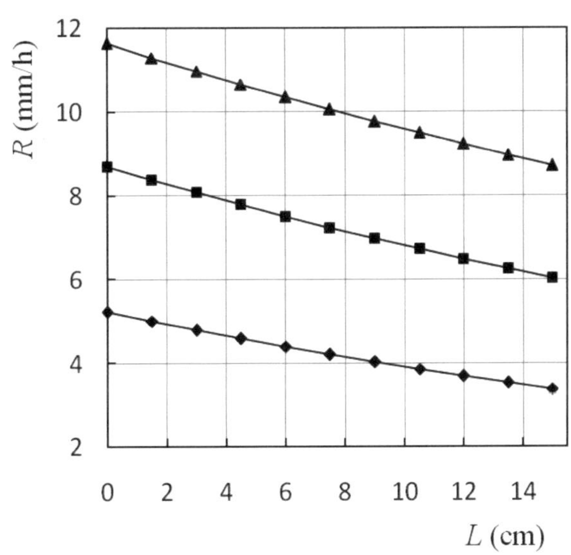

(a)

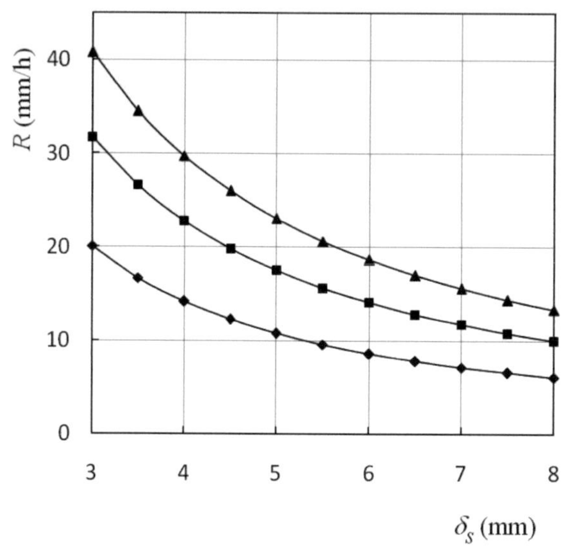

(b)

Fig. 18. Crystallization rate R as a function of the position of the plug along the cooler L, and the width of the air stream δ_s respectively, when: -♦- w_i = 0.3 m/s, -■- w_i = 0.6 m/s, -▲- w_i = 0.9 m/s; $\delta_a = 0$ (crucibles above the plugs) (a) $\delta_s = d$; (b) $L = 9$ cm (Fig. 17).

Fig. 19a represent the possible values of distances of the plug head from the surface of the melt δ_a and the position of the plug along the cooler L, for definite values of crystallization rates. As can be seen, if L is larger, then δ_a must be smaller for the definite crystallization rate, which is the consequence of the fact that ΔT decreases with increasing of the L,- eq. (5) and R increases with decresing δ_a – eqs. (2), (3) and (4).

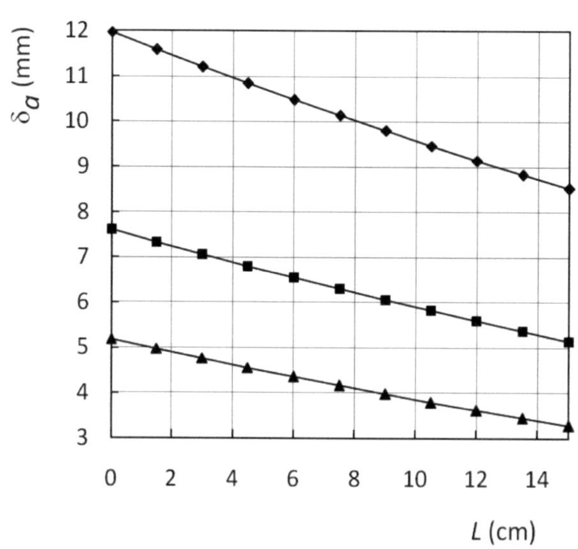

(a)

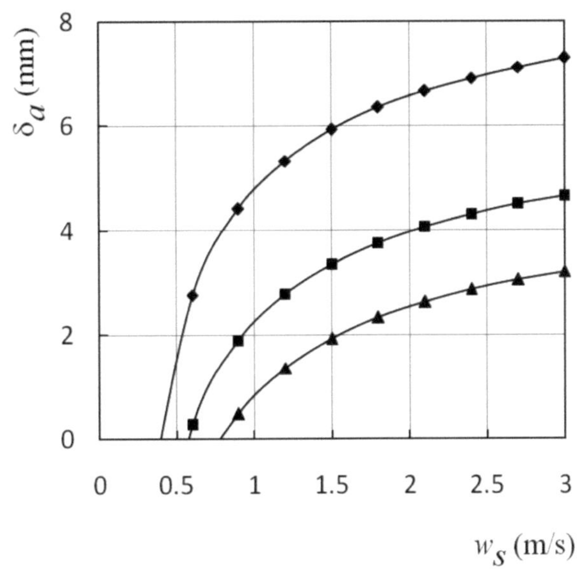

(b)

Fig. 19. Dependence of the distances of the plug head from the surface of the melt δ_a (crucibles below the plugs, Fig. 17b) on the positions of the plug along the cooler *L*, and the velocities of the air stream w_s respectively, for the crystallization rates -♦- R = 5 mm/h, -■- R = 7 mm/h, -▲- R = 9 mm/h, when: (a) w_s = 2.4 m/s; (b) L = 7 cm.

In Fig. 19b the possible values of distances of the plug head from the surface of the melt δ_a and velocities of the airstream $w_{s,}$ are presented, for definite values of crystallization rates. As can be seen from Fig. 19b, if w_s is larger then δ_a must be larger for the same crystallization rate, which is the consequence of the fact that α_s increases with increasing of the w_s - eq. (2), and R decreases with incresing δ_a – eq. (1)

5. Conclusion

The subject of our research belongs to the field of crystal growth from the melt, particularly growth conditions depending on the design and construction of crystallization apparatus, which have significant influence on germ formation conditions and controlled crystal growth. A class of new modern devices for crystal growth from the melt, based on the well-known methods and crystal growth techniques, is presented in the paper. Crystal growth from the melt plays an important role in area of electronic technologies, because it includes a major part of most efficient methods for production of semiconductor and electronic monocrystal materials. In this monograph, we have systematically presented results concerning crystal growth from the melt, from both renowned authors and the author of monograph.

Technology of crystal growth depends on apparatus state-of-the-art and on devices with specific characteristics for particular growth method. However, in order to meet specific demands in their subsequent application, some of the apparatus for production of standard materials have additional peculiarities. Consequently, an upgrade in both constructional and functional sense for a variety of apparatus for crystal growth from the melt was necessary. As presented in the paper, this resulted in development of a class of novel devices with notably improved solutions for both some elements of the device and device as a whole.

Basic settings for the new approach in fulfilling desired crystal growth conditions and flexibility of specific devices while varying some parameters, were obtained through realization of whole set of coolers, starting from elementary specifically positioned to the, so called, mobile coolers of different profiles.

Studying of conditionality of crystallization parameters and physical conditions of the process itself, generated an original idea where modern design coolers gain multifunctional role. On the one side, they have become carriers of the ensemble of test tubes with melt, and on the other side they allow for the positioning of melt in desired spots thereby bringing about the needed temperature gradient. Finally, cooler-melt system has a potential of easily being positioned where necessary by moving it in various directions or by rotating it within the space available in furnace. A particular quality in innovations that we came upon, is that idea of multifunctional coolers triggered an idea of incorporating computer system into the crystallization process. Application of computer systems allows one to define crystallization conditions prior to the crystal growth via simulation process. In addition, it is possible to permanently control and monitor quality of crystallization process. Development of some original programs in MATH LAB only confirmed validity of idea. This gives vast opportunities in presented modern approach to growth of crystals and monocrystals.

All the mentioned innovations in both specific parts of crystal growth apparatus and apparatus as a whole, allowed relatively easy reproducibility of crystallization process. This approach enables, for the predefined conditions, simultaneous growth of a family of crystals of single material in same or different conditions on the one side, and simultaneous growth of different material crystals in, more or less, same crystallization conditions on the other side. Along with the described multifunctionality, a new class of crystal growth devices gained in quality and importance in connection with low cost, efficiency, rationalization and modernization of crystallization process. Usage of computer modeling and development of original computer programs are a good basics for achieving the highest goals of this monograph: to incorporate, via application of information technology in the process of crystal growth from the melt (with the use of latest class of devices designed for crystal growth), in a certain way, the expert systems. In this way, efficiency and accuracy is significantly increased due to a possibility of controlling and simultaneously eliminating undesired effects, in a process that is almost fully automatic and that can be influenced essentially in order to get a crystal of desired quality. Results presented here are of great practical interest for theoretical and applied research in solid state physics, as well as in the area of new materials, all this fulfilling high requirements and standards demanded in both laboratory and industry growth of crystals and single crystals.

6. Acknowledgments

The paper (chapter) was supported by Serbian Ministry of Education and Sciences, grant No. 44002.

7. References

[1] H.J. Scheel and T. Fukuda, *Crystal Growth Technology*, John Wiley & Sons, Ltd. ISBN: 0-471-49059-8, (2003).

[2] K.-T. Vilke, *Virashchivanie kristallov* («Nedra». Leningradskoe odelenie, Leningrad, pp. 258-311), (1977).

[3] K.-Th. Wilke and J. Bohm, *Kristallzüchtung* (Verlag Harri Deutsch: Thun, Frankfurt/Main, , pp. 591-647), (1988).

[4] R. A. Laudise, *The Growth of Single Crystals*, (Prentice-Hall, Englewood Cliffs, NJ, pp 159-172), (1970).

[5] B. Čabrić, B. Žižić, and M. Lj. Napijalo, *An apparatus for crystal growth in the undergraduate laboratory*, Eur. J. Phys., 11, 233, (1990).

[6] B. Čabrić, T. Pavlović, and S. Savović, *A simple programming the crystallization rate from the melt*, Cryst. Res. Technol., 29, No. 7, K96, (1994).

[7] B. Čabrić, T. Pavlović, and B. Žižić, *Crystallization in a laboratory chamber furnace*, J. Appl. Cryst., 27, 199 (1994).

[8] B. Čabrić, T. Pavlović, *Krystallization bench*, J. Appl. Cryst., 33, 387-388, (2000).

[9] I.Tarjan and M. Matrai, *Laboratory Manual on Crystal Growth* (Akademiai Kiado, Budapest, pp. 221-238), (1972).

[10] B. Čabrić, A. Janićijević, *A Method For Obtaining Crystals In A Laboratory Furnace*, Program and Contributed Papers of XVI National Symposium on Condensed Matter Physics SFKM 2004, Ed. Institute of Physics Belgrade - Serbia and Montenegro, str. 325-328, (2004).

[11] (46) B. Čabrić, A. Janićijević, *Obtaining crystals in a laboratory furnace*, J. Cryst. Growth, (Holand.), Vol. 267, str. 362-363, (2004).

[12] B. Čabrić, A. Janićijevic., *Crystallization shelf*, J. Applied Crystallography, 40, 391, 2007).

[13] B. Čabrić, A. Janićijević, *A laboratory furnace for obtaining crystals*, J. Appl. Cryst. Engleska), Vol.37, str. 675, (2004).

[14] [13] (49) B. Čabrić, A. Janićijević, *Crystalization shelves for laboratory furnaces*, Cryst. Res. Technol., 42, No. 4, 342-343, (2007).

[15] B. Čabrić, T. Pavlović, A. Janićijević, *Regulation of the crystallization in a crucible urnace*, J. Cryst. Growth - Holandija, No. 200, str. 339-340, (1999).

[16] B. Čabrić, T. Pavlović, *Obtaining crystals in crucible furnace*, J. Appl. Cryst., 29, 745, 1996).

[17] B. Čabrić, A. Janićijević, *A method for obtaining crystals in a tube furnace*, Extended Abstracts of XII Conference of the Serbian Crystallographic Society, Serbian Crystallographic Society, Belgrade, p. 20-21, (2004).

[18] B. Čabrić, A. Janićijević, *Crystalization regulation in a crucible furnace*, Program and Contributed Papers of XVI National Symposium on Condensed Matter Physics FKM 2001, Arandjelovac, Ed. Institute of Physics Belgrade - Serbia and Montenegro, str. 44, (2001).

[19] B. Čabric, and A. Janicijevic, *Cooler for obtaining crystals in a crucible furnace*, J. Appl. Cryst., 36, 950, (2003).

[20] A. Janićijević, B. Čabrić, R. Simeunović, *Tamman's method for obtaining crystals in a rucible furnace*, VIII Conference of the Serbian Crystallographic Society, Serbian Crystallographic Society, Kragujevac, p. 92-93, (2000).

[21] B. Čabrić, T. Pavlović, and T. Despotović, *A crystallization cooler*, Czech. J. Phys., 49, No. 7, 1115 (1999).

[22] A. Janićijević, and B. Čabrić, *A crystallization coolers*, Extended Abstracts of X Conference of the Serbian Crystallographic Society, Serbian Crystallographic Society, Belgrade, p. 47.(2002)

[23] B. Čabrić, A. Janićijević, T. Despotović, *Regulation of the crystallization in a tube furnace*, VII Conference of the Serbian Crystallographic Society, Serbian Crystallographic Society, Vrnjačka Banja, p. 62-63, (1998).

[24] A. Janićijević, B. Čabrić, *Curcible furnace for obtaing crystals*, Extended Abstracts of X Conference of the Serbian Crystallographic Society, Serbian Crystallographic Society, Belgrade, p. 42, (2001).

[25] B. Čabrić, A. Janićijević, *A method for obtaining crystals in a chamber furnace*, Extended Abstracts of XII Conference of the Serbian Crystallographic Society, Serbian Crystallographic Society, Belgrade, p. 35-36, (2004).

[26] B. Čabrić, A. Janićijević, *Stober's method for obtaining crystals in Tamman a chamber furnace*, Extended Abstracts of X Conference of the Serbian Crystallographic Society Serbian Crystallographic Society, Belgrade, p. 49, (2002).

[27] B. Čabrić, A. Janićijević, *A laboratory furnace for obtaining single crystals*, Program and Contributed Papers of XVI National Symposium on Condensed Matter Physics FKM 2004, Ed. Institute of Physics Belgrade - Serbia and Montenegro, str. 337, 2004).

[28] B. Čabrić, A. Janićijević, *A chamber furnace for obtaing crystals*, Extended Abstracts of X Conference of the Serbian Crystallographic Society, Serbian Crystallographic Society, Belgrade, p. 45, (2001).

[29] A. Janićijević, N. Danilović, and B. Čabrić, *Crystallization Shelves*, Program and Contributed Papers of XVII Symposium on Condensed Matter Physics SFKM 2007, Ed. Institute of Physics, Belgrade - Serbia, p. 240. (2007).

[30] N. Danilović, A. Janicijević, and B. Cabrić, *Crystallization columns in a chamber furnace*, Kragujevac J. Sci., 32, 41 (2010).

[31] B. Čabrić, N. Danilović, and A. Janićijević, *Tube for obtaining crystals in a laboratory furnaces*, Instrum. Exp. Tech., 54, No. 2, 282 (2011).

[32] B. Čabrić, N. Danilović, and A. Janićijević, *Simultaneous crystallization testing in a laboratory furnace*, Am. Lab., 43, No. 7 (2011).

[33] A. Janićijević, " *Methods, techniques and analysis of physical conditions for the growth of crystals from the melt in the laboratory*", doctoral dissertations, PMF, Kragujevac, (2008).

[34] B. Čabrić, N. Danilović, and T. Pavlović, *Cooler for obtaining crystals*, Cryst. Res. Technol., 46, No. 3, 292 (2011).

[35] E.-R. Schramek, *Taschenbuch für Heizung + Klimatechnik* (Oldenbourg Industrieverlag GmbH, München, , p. 152. (2007).

2

Growth and Characterization of Ytterbium Doped Silicate Crystals for Ultra-Fast Laser Applications

Lihe Zheng, Liangbi Su and Jun Xu
Shanghai Institute of Ceramics, Chinese Academy of Sciences, P. R. China

1. Introduction

Diode-pumped solid-state lasers (DPSSL) have predominated over waveguide lasers and fiber lasers when considering the efficiency and operability since the first realization of laser-diode pumped Yb-doped laser at room temperature (Lacovara et al., 1991). As a rule of thumb, DPSSL are preferable for devices operating with high peak power, whereas low-threshold and high-gain operation is much easier to be achieved with waveguide lasers and amplifiers. Besides the application in the fields of double-frequency, remote sensing and biomedical, ultra-fast DPSSL with diversified wavelength and stable system is widely exploited in the fields of mechanics, micro-electrics and ultra-fast photo-communication. DPSSL are composed of laser resonator which is mostly formed with discrete laser mirrors placed around gain medium with an air space in between. Bulk crystals or glasses doped either with rare earth ions or transition-metal ions are adopted as gain medium. With the development of DPSSL industries, the demand for laser crystals with the advantageous physicochemical properties such as efficient energy absorption, high optical uniformity and favorable thermal behavior has dramatically increased over the past few decades (Keller, 2003). With the rapid development of InGaAs laser diodes emitting from 900nm to 980nm, Yb^{3+} doped laser crystals are expected to alternate the traditional Nd^{3+} doped for generating efficient broad tunable and ultra-fast DPSSL in near-IR spectral range (Krupke, 2000).

Yb^{3+} ion with simple quasi-three energy level scheme of $^2F_{7/2}$ and $^2F_{5/2}$ is provided with high quantum efficiency, long lifetime of metastable $^2F_{5/2}$ level and large crystal-field splitting which is beneficial for reducing thermal load and enhancing Yb^{3+} doping level bringing about the realization of compact device without luminescence quenching caused by cross relaxation and excited-state absorption (Giesen & Speiser, 2007; Pelenc et al., 1995). However, the strong re-absorption at emission wavelengths leads to high pump threshold since the thermally populated terminal level of Yb^{3+} lasers at the ground state manifold is contemporary the laser terminal level. To reduce the re-absorption losses at laser emission wavelengths, strong splitting of ground sublevels of $^2F_{7/2}$ in Yb^{3+} ion is required to form a quasi-four-level system as that of Nd^{3+}. Thus, laser crystal hosts with low symmetry structure and strong crystal field splitting are the central issues in exploiting new Yb doped gain media (Du et al., 2006). Crystal hosts such as aluminum, tungsten, oxides, fluorides and

vanadates were explored for diode-pumped ultra-fast lasers (Uemura & Torizuka, 2005; Liu et al., 2001; Griebner et al., 2004; Su et al., 2005; Kisel et al., 2005).
The chapter is devoted to the systematical investigation on Yb doped oxyorthosilicate crystals such as gadollinium silicate (Gd_2SiO_5, GSO), yttrium silicate (Y_2SiO_5, YSO), Lutetium silicate (Lu_2SiO_5, LSO) and scandium silicate (Sc_2SiO_5, SSO) obtained by the Czochralski Crystal Growth System with Automatic Diameter Control, which encompassing distinctive low symmetry monoclinic structure, excellent physicochemical properties and favorable spectroscopic features for DPSSL. Besides, the chapter summarizes the structure properties of the obtained silicate crystals. Afterwards the chapter discusses the optical properties of silicate crystals available for ultra-fast lasers, together with the calculation of spectroscopic parameters such as pump saturation intensities I_{sat}, minimum pump intensities I_{min} and gain spectra of laser medium. Finally, the laser performance of the studied silicates is briefly outlined.

2. Experimental details of silicate crystals

The basic properties of rare earth oxyorthosilicate crystals were presented in Table 1*. Silicate Crystals were obtained by Czochralski method. Structure and spectra method were outlined as well as the laser experiments in the following elaboration.

2.1 Basic properties of rare earth oxyorthosilicate crystals

The monoclinic orthosilicate crystals RE_2SiO_5 could be stably formed according to the binary phase diagram of RE_2O_3-SiO_2, where RE stands for Lu^{3+}, Gd^{3+}, Y^{3+} and Sc^{3+}. RE^{3+} ions occupy two different low symmetry crystallographic sites which Yb^{3+} ion could substitute with selectively. The Rare Earth silicates with larger Rare Earth ion radius from La^{3+} to Tb^{3+} manifest the space group of $P2_1/c$ with typical compounds of La_2SiO_5 and GSO, while those with smaller Rare Earth ion radius from Dy^{3+} to Lu^{3+} as well as Y^{3+} and Sc^{3+} hold the space group of C2/c with typical compound of YSO and LSO.
Researchers show solitude for the crystal growth, structure, opto-electrical properties on Rare Earth doped RE_2SiO_5 crystals (Eijk, 2001; Kuleshov et al., 1997; Melcher & Schweitzer, 1992; Ivanov et al., 2001). Table 1 reveals the comparison of the basic physicochemical properties of monoclinic silicate crystals (Gaume, 2002; Smolin & Tkachev, 1969; Camargo et al., 2002). As seen from Table 1, LSO, YSO and SSO crystals are with Monoclinic C2/c structure occupying two different low symmetric and distorted crystallographic sites which would provide strong crystal field for Yb^{3+} ions with quasi-four-level laser operation scheme. SSO crystal retains the highest thermal conductivity among silicate crystals together with the striking negative calorescence coefficient which is crucial for laser operation regarding laser power resistance (Petit, 2005). The crystallographic sites, coordination and mean distance of RE-O in RE_2SiO_5 crystals are extended for ongoing discussion on energy level splitting as shown in Table 2 (Gaume et al., 2003; Ellens et al., 1997).

2.1.1 Lu_2SiO_5

LSO is a positive biaxial crystal with n_Y axis along the b direction and two other axis a and axis c lying in plane (010) (Wang, 2004). The structure of monoclinic LSO crystal with isolated ionic SiO_4 tetrahedral units and non-Si-bonded O atoms in distorted OLu_4 tetrahedron were determined by neutron diffraction (Gustafsson et al., 2001). The OLu_4 tetrahedron form edge-sharing infinite chains and double O_2Lu_6 tetrahedron along the c axis. The edge-sharing chains are connected to O_2Lu_6 double tetrahedron by isolated SiO_4 units.

Crystal Molecular Formula	LSO	GSO	YSO	SSO	YAG
Space group/simple	C2/c	$P2_1$/c	C2/c	C2/c	Ia3d
Site symmetry	C_1	C_s+C_{3v}	C_1	C_1	D2
Lattice constant (Å)	a=10.2550Å, b=6.6465Å, c=12.3626Å, β=102.422°	a=9.132Å, b=7.063Å, c=6.749Å, β=107.56°	a=10.41Å, b=6.72Å, c=12.49Å, β=102.65°	a=9.961Å, b=6.429Å, c=12.03Å, β=103.8°	a=12.01 Å
Density / g·cm^{-3}	7.4	6.7	4.44	3.52	4.56
Melting point / °C	2100	1950	2000	1920	1950
Mohs hardness	5.8	5.5	5.6	-	8.5
Thermal conductivity undoped@30°C/W·m^{-1}·K^{-1}	4.9	4.5	4.4	7.5	10.7
dn/dT(*10^{-6} K^{-1})	-	-	7.2	-6.3	7.9
Refractive indices	1.82	1.85	1.782(a) 1.785(b) 1.806(c)	n_x=1.82, n_y=1.84, n_z=1.86	1.83

Table 1. Physicochemical property of monoclinic silicate crystals.

Host	Label	Coordination number	Re^{3+} (Å)	Re-O_{mean} (Å)
LSO	Lu(I)	7	1.001	2.277
	Lu(II)	6		2.229
GSO	Gd(I)	7	1.05	2.49
	Gd(II)	9		2.39
YSO	Y(I)	7	0.910	2.3080
	Y(II)	6		2.2520
SSO	Sc(I)	6	0.885	2.2289
	Sc(II)	6		2.2149

Table 2. Crystallographic sites, coordination and mean distance in RE-O in RE_2SiO_5 crystal.

2.1.2 Gd_2SiO_5

As seen from Table 1, Gd^{3+} in monoclinic GSO crystal is coordinated with oxygen atoms of 7 and 9, respectively (Fornasiero et al., 1998). The Gd(I) coordinated with 7 oxygen is linked with three isolated oxygen ions and three [SiO^{4-}] ions. The Gd(II) coordinated with 9 oxygen is bonded with one isolated oxygen and six tetrahedral [SiO^{4-}] ions. As shown in Table 2, the average distance of 2.39 Å in Gd(I)-O is shorter than that of 2.49 Å in Gd(II)-O (Felsche, 1973) The symmetry and intensity of crystal field would be affected by the coordination field after doping with active Rare Earth ions into the host (Suzuki et al., 1992; Cooke et al., 2000).

2.1.3 Y_2SiO_5

The distorted crystallographic sites of Y^{3+} in YSO crystal are with coordination number of 6 and 7 which is similar to those in LSO. Crystallographic site Y(I) forms polyhedron with 5 oxygen from that in tetrahedron $(SiO_4)^{4-}$ and 2 isolated oxygen. Crystallographic site Y(II) comprises reticular formation of pseudo octahedron with 4 oxygen from that in tetrahedron $(SiO_4)^{4-}$ and 2 isolated oxygen. The reticular formation of OY_4 is composed of edge-sharing infinite chains along c axis together with the network structure of OSi_4 (IEM Databases and Datasets).

2.1.4 Sc_2SiO_5

The structure of SSO crystal is analogous to those of YSO and LSO, with monoclinic structure and space group of C2/c. Although the large sized Yb:SSO crystal is difficult to obtain due to crystal diameter controlling, the impressive thermo-mechanical properties in SSO crystal make it an excellent performer in high power laser applications (Ivanov, 2001; Gaume et al., 2003; Campos et al., 2004). As indicated in Table 1, SSO crystal possesses a much stronger thermal conductivity of $7.5W \bullet m^{-1} \bullet K^{-1}$ than that in YSO ($4.4W \bullet m^{-1} \bullet K^{-1}$), LSO ($4.5\ W \bullet m^{-1} \bullet K^{-1}$) and GSO ($4.9\ W \bullet m^{-1} \bullet K^{-1}$). Furthermore, SSO crystal preserves the characterization of minus calorescence coefficient dn/dT of $-6.3\times10^{-6}\ K^{-1}$ which is quite different from that of YSO ($7.2\times10^{-6}\ K^{-1}$) and favorable for high power laser operation.

2.2 Crystal growth

Czochralski method (Cz) is one of the major crystal growth methods of obtaining bulk single crystals with high optical quality and fast growth rates from melt for commercial and technological applications. Cz is named after Polish scientist Jan Czochralski, who discovered the method in 1916 while investigating the crystallization rates of metals.

Yb doped silicate crystals such as Yb:LSO, Yb:GSO, Yb:YSO and Yb:SSO studied in this chapter were grown by the Cz method in inductively heated crucibles under inert atmosphere of 5N nitrogen. The starting materials were Rare-Earth Oxide (Lu_2O_3, Gd_2O_3, Y_2O_3 or Sc_2O_3), Yb_2O_3 and SiO_2 powders with purity higher than 99.995%. The powders were mixed and pressed into tablets followed by sintering at 1400°C for 24 h before loading into the iridium crucible. The chemical reaction is shown in Equation (1). The doping level of Yb ions in the melt was 5at.% with respect to that of Rare-Earth ions in the crystal host. Accordingly the compound formulae of the crystals could be written as $(RE_{0.95}Yb_{0.05})_2SiO_5$.

$$(1-x)RE_2O_3 + SiO_2 + xYb_2O_3 \rightarrow RE_{2(1-x)}Yb_{2x}SiO_5 \quad (1)$$

A precisely and oriented rod-mounted seed crystal with diameter of 4.5mm and length of 40mm was introduced for growth. The seeds for Yb:LSO, Yb:YSO, Yb:SSO were oriented along b-axis while that for Yb:GSO crystal was along [100]. The seed crystal was pulled upwards around 0.8-3mm·h^{-1} and rotated simultaneously at 10-30rpm. To keep convex solid–melt interface was important in growing silicate crystals to eliminate engendered dislocations. Temperature gradients and velocity fields were accurately controlled to gain stable settings (Zheng et al., 2007).

Automatic diameter control (ADC) with computer control system and weight sensor or so-called load cell is applied to detect the weight during crystal growth process which is wholesome for increasing crystal yield as well as reducing thermal stress. Strain gauge weight sensor with sufficient resolution was employed comparing with the total weight of obtained crystals. The input signal of weight sensor via an A/D converter is collected t o calculate the diameter of the generative crystal. A proportional-integral-derivative controller (PID controller) was adopted to better control the loop feedback during temperature regulation.

The difficulty in controlling diameter is originated from encapsulated materials and the high temperature atmosphere which would initiate diameter fluctuation bringing about dislocations besides irregular crystal shape and polycrystallization. The diameter deviation is regulated by controlling the temperature gradients where the double layered zircon cover was designed to optimize the temperature gradients. The heater temperature was increased to reduce the diameter when the measured diameter is larger than expected and vice versa. Crystal boules were finally obtained as shown in Fig. 1.

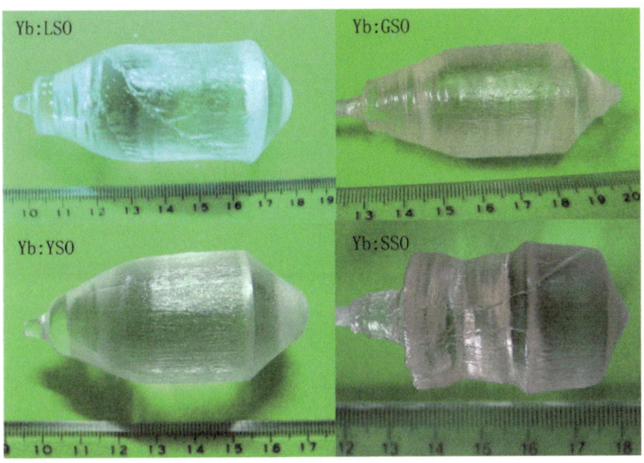

Fig. 1. Bulk crystals of 5at.%Yb doped silicate crystals obtained by Czochralski method.

2.3 Segregation coefficient characterization in silicate crystals

The segregation coefficient of Yb ions in silicate crystal hosts was measured by the inductively coupled plasma atomic emission spectrometer (ICP-AES). Crystal samples adjacent to the seed crystal position were cut and ground into fine powder in an agate mortar. The results of ICP-AES analysis are shown in Table 3.

Sample	seed crystal	Yb^{3+} c_0 (at.%)	Atom content (wt.%)					Yb^{3+} c_{top} (at.%)	k_m
			Yb	Lu	Gd	Y	Sc		
Yb:LSO	LSO	5.0	3.31	70.76	0	0	0	4.515	0.903
Yb:GSO	GSO	5.0	3.06	0	76.22	0	0	3.52	0.704
Yb:YSO	YSO	5.0	5.22	0	0	50.06	0	5.085	1.017
Yb:SSO	SSO	5.0	7.19	0	0	0	36.9	4.82	0.964

Table 3. Segregation coefficients of Yb ions in silicate crystals.

As acquired from Table 3, the segregation coefficient of Yb is 0.903 in Yb:LSO, 0.704 in Yb:GSO, 1.02 in Yb:YSO and 0.964 in Yb:SSO. The solubility of Yb ion in the LSO, YSO and SSO host lattice is higher than that in GSO indicating that Yb ions are liable to incorporate into the crystals with structure of C2/c comparing with that of $P2_1/c$. Meanwhile, the congenial radius of Yb^{3+} (0.868Å) with Lu^{3+}(1.001Å), Y^{3+} (0.910Å) and Sc^{3+} (0.885Å) makes ideal adulteration squaring up that of the Gd^{3+} (1.05Å).

2.4 X-Ray diffraction measurement

The crystal structure of silicate crystals were characterized by powder X-ray diffraction (XRD; Model D/Max 2550V, Rigaku Co., Tokyo, Japan) using Cu Kα radiation (λ = 0.15418 nm). The XRD patterns were inspected using PCPDF software package. Fig. 2 presents the XRD pattern for Yb doped silicate crystals which demonstrated that Yb:LSO, Yb:YSO and Yb:SSO crystals would maintain the primitive monoclinic structure with space group of C2/c, while Yb:GSO maintain that of $P2_1/c$.

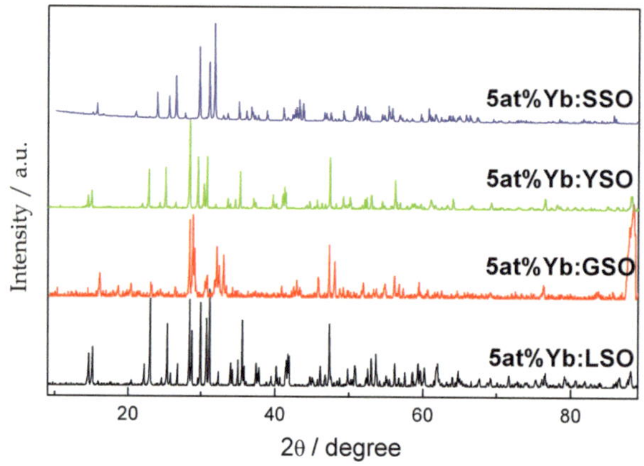

Fig. 2. Comparison of the XRD pattern for Yb doped silicate crystals.

2.5 Absorption and emission spectra

The unpolarized absorption spectra of silicate crystals were measured by JASCO Model V-570 UV/VIS/NIR spectrophotometer at a resolution of 1nm in the range between 860-1100nm with Xe light as pump source.

Fluorescence spectra were measured at a resolution of 1nm from 950nm to 1150nm by TRIAX 550-type spectrophotometer (Jobin-Yvon Company) with 940nm laser pumping source. The fluorescence lifetime with sample thickness of 1mm was pumped with a Xenon lamp and detected with an S-photomultiplier tube, while the data of emission decay curve was collected by a computer-controlled transient digitizer simultaneously.

2.6 Laser experiment

Kerr-lens mode-locked is well-developed for ultrafast pulses in efficient and compact lasers which is initiated by the self-focusing effect yielded from the nonlinear refractive-index variation in laser crystal. Diode-pumped mode-locked Yb:LSO laser with a W-typed cavity was developed based on Kerr-lens effect. The experimental setup for mode-locked Yb:LSO lasers was represented in Fig. 3. The maximum output power of the fiber-coupled diode-pumped semiconductor laser reached 30W around the emitting wavelength of 978 nm. The radius and the numerical aperture of the fiber were with 200 μm and 0.22, respectively. Yb:LSO host was cut into the dimension of 3×3×2 mm^3 and end-coated with antireflection at lasing wavelength of 1030-1080 nm and pump wavelength of 978 nm. Yb:LSO wrapped with indium foil was placed with a small angle to suppress the Fabry-perot etalon effect and mounted in a water-cooled copper block with temperature maintained at 14 °C.

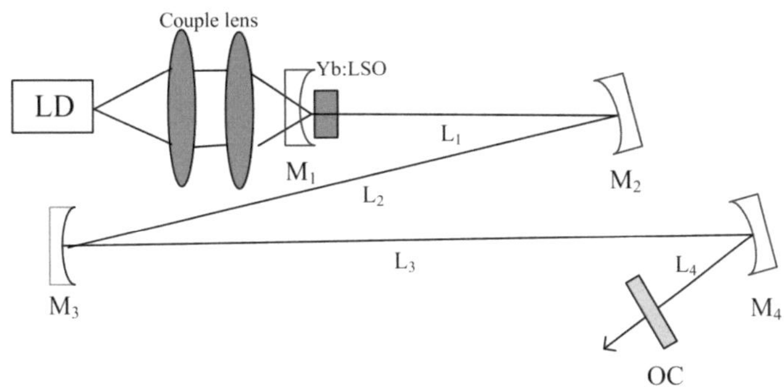

Fig. 3. Schematic of the Kerr-lens Yb: LSO laser.

Laser experiments on Yb:GSO crystal was performed with aperture of 5×6 mm^2 and length of 5 mm. The crystal was end pumped by a 30 W fiber-coupled laser diode with emission centered at 974 nm. The diameter and numerical aperture of the fiber core were 400 μm and 0.22, respectively. The laser cavity consisted of two mirrors M1 and output coupler OC, where dichroic mirror M1 was with anti-reflection coating at 974 nm and high-reflection coating in a broadband of 1020-1120 nm besides OC with various transmission.
The schematic of the Yb:SSO laser is shown in Fig. 4. The b-cut Yb:SSO crystal is with dimensions of 2.2 × 3 × 3 mm^3. The crystal was coated at both the pump and laser wavelengths with high transmission coatings to minimize the Fresnel reflection losses. The crystal was titled to about 6° with respect to the principle axis of the cavity to suppress etalon effects and improve the stability of the mode-locking operation. A single emitter diode with central wavelength at (975 ± 3) nm was used as the pump source.

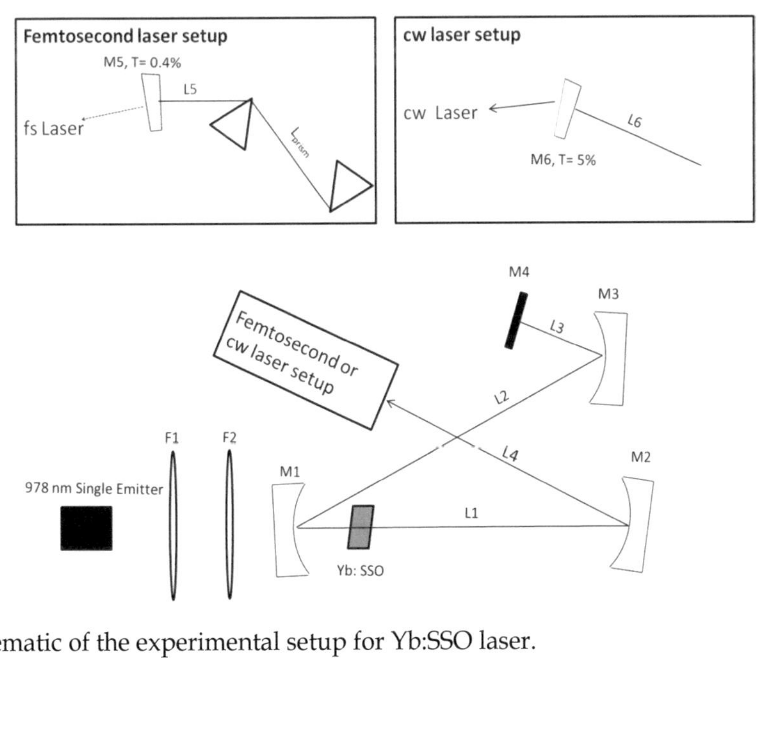

Fig. 4. Schematic of the experimental setup for Yb:SSO laser.

3. Spectra analysis of silicate crystals

Absorption and fluorescence spectra in Yb doped silicate crystals were presented. Based on the above spectra, key parameters were calculated for predicting the laser performance of the studied hosts.

3.1 Absorption and fluorescence spectra

Fig. 5 showed the absorption spectra of Yb doped silicate crystals with main absorption peak located around 920nm and 978 nm which is suitable for InGaAs diode pump.
The absorption bands were corresponding to the typical transitions from ground state $^2F_{7/2}$ to the sublevels of $^2F_{5/2}$. The absorption peak around 978nm belongs to the zero-line transition between the lowest sublevels of $^2F_{7/2}$ and $^2F_{5/2}$ manifolds. Broad absorption bands are beneficial for enhancing the efficiency of diode-pump operation since the laser diodes are typically emitting at a wide spectral range of 5 nm and presenting a thermal shift at the peak wavelength. As manifested in Fig. 5, the broad fluorescence behavior was associated with the typical transitions from the lowest level of $^2F_{5/2}$ to the sublevels of $^2F_{7/2}$ manifold. The emission band above 1080nm was due to transition from the lowest levels of $^2F_{5/2}$ manifold to the highest levels of $^2F_{7/2}$ manifold.

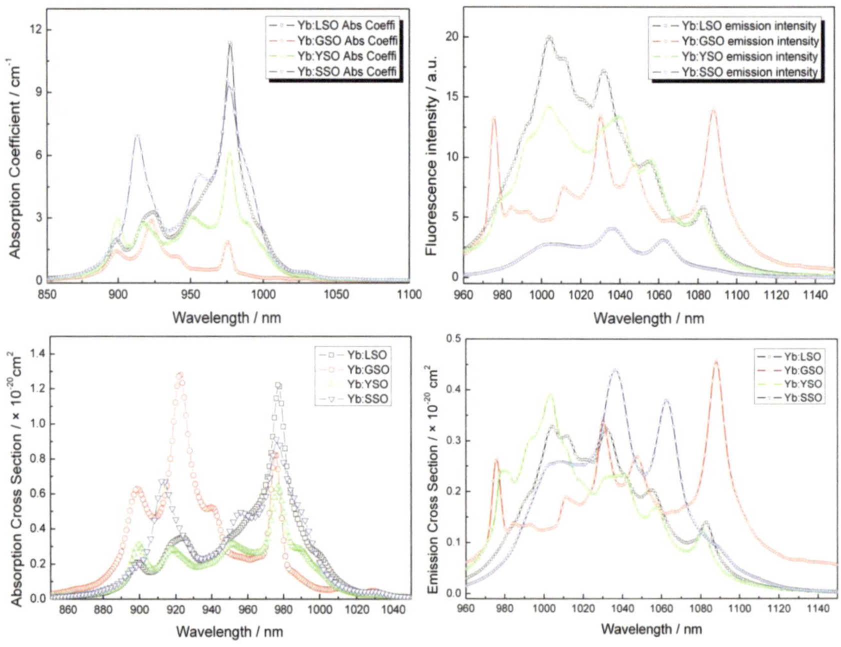

Fig. 5. Absorption coefficients, Fluorescence Intensity, Absorption cross section and emission cross section of Yb doped silicate crystals.

$$\sigma_{abs} = \frac{\alpha}{N} \tag{2}$$

$$\sigma_{em}(\lambda) = \frac{\lambda^5 I(\lambda)}{8\pi n^2 c\tau_{rad} \int \lambda I(\lambda) d\lambda} \tag{3}$$

Parameter	5 at. % Yb:LSO	5 at. % Yb:GSO
Main absorption peaks λ_{abs} (nm)	899, 924, 977	897, 922, 940, 976
Absorption bandwidth (nm)	13, 25, 15	17, 26, 24, 10
Absorption cross-section $\sigma_{abs}(10^{-20}\ cm^2)$	0.21, 0.36, 1.3	0.33, 0.60, 0.39, 0.51
Fluorescence peak (nm)	1004, 1012, 1032,1055, 1083	976, 1011, 1030, 1047, 1088
Emission cross-section σ_{em} $(10^{-20}\ cm^2)$	0.34, 0.32, 0.33, 0.21, 0.14	0.17, 0.45, 0.31, 0.46
Emission bandwidth (nm)	73	72
Fluorescence lifetime (ms)	1.68 @ 1083nm	1.56 @ 1088nm
Parameter	**5 at. % Yb:YSO**	**5 at. % Yb:SSO**
Main absorption peaks λ_{abs} (nm)	899, 917, 950, 977	914, 956, 976
Absorption bandwidth (nm)	15, 24, 31, 13	18, 19, 24
Absorption cross-section $\sigma_{abs}(10^{-20}\ cm^2)$	0.31, 0.28, 0.32, 0.64	0.67, 0.49, 0.91
Fluorescence peak (nm)	980, 1003, 1040, 1056, 1081	1006, 1036, 1062, 1087
Emission cross-section σ_{em} $(10^{-20}\ cm^2)$	0.24, 0.39, 0.23, 0.18, 0.12	0.26, 0.44, 0.38, 0.1
Emission bandwidth (nm)	48	57
Fluorescence lifetime (ms)	1.74 @ 1081nm	1.64 @ 1035nm

Table 4. Spectroscopic parameters of Yb doped silicate crystals.

Fig. 5 also revealed the calculated σ_{abs} of Yb^{3+} according to Equation (2) where α means the absorption coefficient and *N* presents the concentration of Yb^{3+} ions. σ_{em} of Yb^{3+} in silicate hosts explicated in Fig. 5 were calculated by Füchtbauer-Ladenburg formula referring to Equation (3) where τ_{rad} stands for the radiative lifetime of the excited manifold $^2F_{5/2}$ of Yb^{3+}, I(λ) for the emission intensity at wavelength λ, n for refractive index and c for light velocity (Caird et al., 1991; Deloach et al., 1993). σ_{em} is fit for describing available gain at given inversion density. The lifetime of excited manifold $^2F_{5/2}$ was measured to be 1.68ms, 1.56ms, 1.74ms and 1.64ms in Yb:LSO, Yb:GSO, Yb:YSO and Yb:SSO crystals respectively. The lifetime is much longer than that of 0.8ms in Yb:YAG. Table 4 outlined the spectroscopic parameters of Yb^{3+} in silicate hosts based on absorption and emission spectra.

The large overlapping between absorption and emission spectra was observed in Fig. 5 due to the transitions among the multiplets of $^2F_{7/2}$ and $^2F_{5/2}$ manifolds of Yb^{3+}. The emission bandwidth FWHM of Yb:LSO crystal is appropriately 73nm which is much larger than that of Yb:YSO and Yb:GSO crystal. Although Yb:LSO possessed the largest σ_{em} around 1004nm, the reabsorption losses was consequently strong which would detrimentally affect laser action. σ_{em} of the terminal laser level at 1083nm was with 1.4×10^{-21} cm^2 containing the smallest thermal populating as well as the least re-absorption losses. In the case of 1032nm, the emission cross section is large enough to obtain low threshold and high efficient laser operation.

The absorption spectrum in Yb:GSO crystal was mainly composed of four strong bands centered at 897 nm, 922 nm, 940 nm and 976 nm. σ_{abs} at 922 nm and 976 nm were estimated as 6.0×10^{-21} and 5.1×10^{-21}, cm^2 with the absorption bandwidths of 26 nm and 10 nm, respectively. The emission spectrum mainly included four bands centered at 1011 nm, 1030 nm, 1047 nm and 1088 nm with fluorescence lifetime of 1.56 ms at 1088 nm which would be treated as radiative lifetime since the re-absorption loss could be negligible at 1088 nm. The σem at 1088 nm were 4.2×10^{-21} which was with the strongest cross section as well as the lowest pumping threshold. Referring to the spectra in Yb:YSO crystal, the strongest absorption peak was located at 978nm with furthest and weakest emission band centered at 1080 nm. The strongest σem centered at 1003 nm was severely affected by strong re-absorption losses. Yb^{3+} ions in Yb:SSO crystal show an inhomogeneous broad absorption band with peak absorption coefficient of $9.43cm^{-1}$ at 976nm which is higher than that in Yb:GSO crystal. The furthest σem centered at 1087 nm was with 1.0×10^{-21}, cm^2 in Yb:SSO which was at the same level of that in Yb:LSO and Yb:YSO. In conclusion, Yb:SSO with wide spectra bandwidth and long lifetime is favorable for high-efficiency ultra-fast diode-pump lasers.

3.2 Gain cross-section σ_g and minimum pump intensity I_{min}

The pump threshold and energy extraction would affect the application of Yb doped crystals on laser devices. Gain cross section σ_g was calculated according to Equation (4) to predict the laser performance (Deloach et al., 1993).

$$\sigma_g = \beta\, \sigma_{em} - (1-\beta)\, \sigma_{abs} \tag{4}$$

$$I_{sat}=h\upsilon/(\sigma_{abs}\cdot\tau_{em}) \tag{5}$$

The inversion coefficient β is defined as the population ratio on the $^2F_{5/2}$ level over the total Yb^{3+} ions population density. σ_g is equal to the emission cross-section σ_{em} when β is equal to 1. To effectively assemble the diode-pumped Yb-laser systems, a substantial fraction of Yb ions on the ground state should be driven to the excited state in order to achieve adequate gain and to overwhelm the absorption losses. As indicated in Equation (5), Pump saturation intensity I_{sat} would be reduced along with stronger σ_{abs} and longer emission life time τ_{em}. The excitation fraction of Yb ions is given by I_{abs}/I_{sat} where I_{abs} stands for the absorbed pump power. Long τ_{em} would promote accumulation of population inversion at fixed power as a complement of the peak-power-limited InGaAs diode sources.

The minimum excitation fraction β_{min} of Yb^{3+} ions calculated by Equation (6) must be balanced for laser output (Moulton, 1983). The parameter of minimum pump intensity I_{min}, obtained by Equation (7), represents the minimum pumping intensity required to invert population in a quasi-three-level system. The fraction I_{abs}/I_{sat} of excited Yb ions equalling to the required fractional excitation of β_{min} permits the equation of I_{min} and I_{abs}. Hence, I_{min} may be interpreted as the absorbed pump intensity required to surpass threshold in a lossless oscillator or to balance at selected extraction wavelength λ_{ext} in amplifier configuration.

$$\beta_{\min} = \frac{\sigma_{abs}(\lambda_{ext})}{\sigma_{ext}(\lambda_{ext}) + \sigma_{abs}(\lambda_{ext})} \tag{6}$$

$$I_{min}=\beta_{min}\cdot I_{sat} \tag{7}$$

Key spectroscopic parameters are interpreted for designing high power rare-earth lasers. Strong σ_{abs} permits low Yb^{3+} ion doping level which in turn leads to a lower resonance absorption loss and a higher small-signal gain at given pump flux. I_{min} was accounted as a figure-of-merit to evaluate the potential Yb doped gain media for laser application by describing the suitable inversion density. Altogerther, I_{sat} as well as the minimum pump intensities I_{min} would be deduced from the absorption cross sections σ_{abs}.

3.2.1 Spectra parameters on Yb: LSO crystal

Fig. 6 showed the gain cross-section with various inversion ratio β. The emission spectra extending from 1000nm to 1092nm at β values of 0.5 was beneficial to laser output. The minimum pump intensities I_{min} at the selected extraction wavelengths, as well as the saturable pump density I_{sat} and the β_{min} under different laser output were also presented.

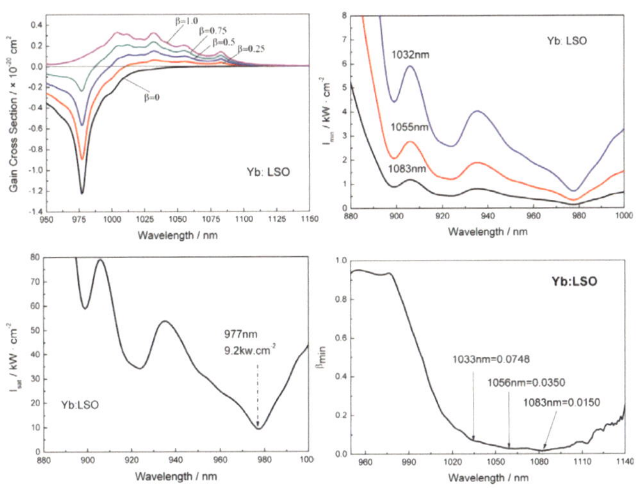

Fig. 6. σ_g, I_{min}, I_{sat} and β_{min} in Yb:LSO crystal.

3.2.2 Spectra parameters on Yb: GSO crystal

Fig. 7 presented the gain cross section σ_g of Yb:GSO crystal at β values of 0, 0.25, 0.5, 0.75 and 1. The emission spectra extending from 999nm to 1150nm at β values of 0.5 was beneficial to ultra-short pulse laser output. The minimum pump intensities I_{min} of 1.9kW/cm^2, 0.71kW/cm^2 and 0.094kW/cm^2 were achieved for selected extraction wavelengths at 1030nm, 1048nm and 1088nm as shown in Fig. 7.

The pump threshold power at the furthest wavelength of 1088nm was much lower than that of 0.13kW/cm^2 at 1083nm in Yb:LSO crystal which predicted low threshold and high efficient laser output in Yb:GSO crystal. Fig. 7 indicated the saturable pump density I_{sat}. The β_{min} at various laser output wavelength in Yb:GSO crystal was shown in Fig. 7 and the comparatively lowest was located at 1088nm with the β_{min} value of 0.0071 which was half less than that of 0.015 at 1082nm in Yb:LSO crystal.

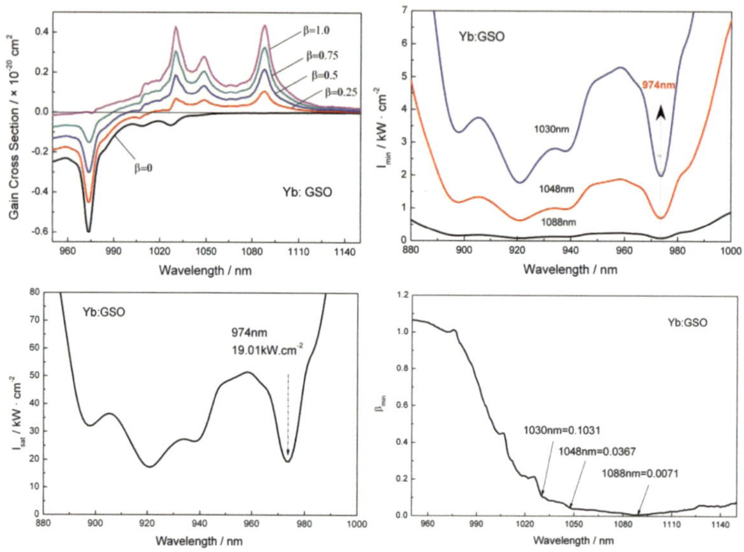

Fig. 7. σ_g, I_{min}, I_{sat} and β_{min} in Yb:GSO crystal.

3.2.3 Spectra parameters on Yb: YSO crystal

Fig. 8 presented the gain cross section of Yb:YSO crystal at β values of 0-1. The emission spectra stretched from 990nm to 1100nm at β values of 0.5. The minimum pump intensities I_{min} of 1.38kW/cm^2, 0.52kW/cm^2 and 0.2kW/cm^2 were achieved for selected extraction wavelengths at 1035nm, 1056nm and 1081nm as shown in Fig. 8. Fig. 8 indicated the saturable pump density I_{sat} in Yb:YSO crystal as well as the β_{min} at various laser output wavelength. The lowest β_{min} was located at 1081nm with the value of 0.015 which was almost the same as that in Yb:LSO crystal.

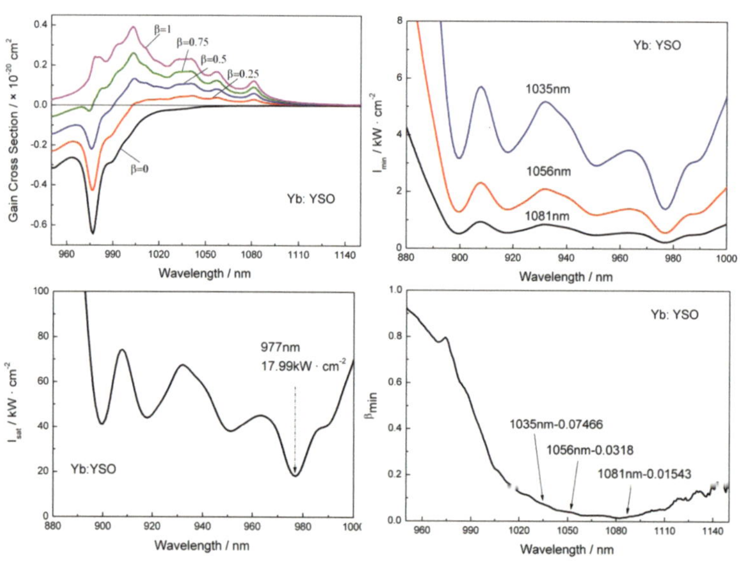

Fig. 8. σ_g, I_{min}, I_{sat} and β_{min} in Yb:YSO crystal.

3.2.4 Spectra parameters on Yb: SSO crystal

Fig. 9 presented the gain cross section of Yb:SSO crystal at β values of 0-1. The emission spectra elongated from 990nm to 1120nm at β values of 0.5. The minimum pump intensities I_{min} of 0.261kW/cm², 0.077kW/cm² and 0.2033kW/cm² were achieved for selected extraction wavelengths at 10356nm, 1062nm and 1087nm as shown in Fig. 9.

Fig. 9 indicated the saturable pump density I_{sat} in Yb:SSO crystal. The β_{min} at various laser output wavelength in Yb:SSO crystal was shown in Fig. 9. The lowest β_{min} was located at 1087nm with the value of 0.007148 which was almost half of that in Yb:LSO crystal. As concluded in Table 5, the laser parameters of Yb doped silicate crystals were listed and compared with that of Yb:YAG crystal (Haumesser et al., 2002).

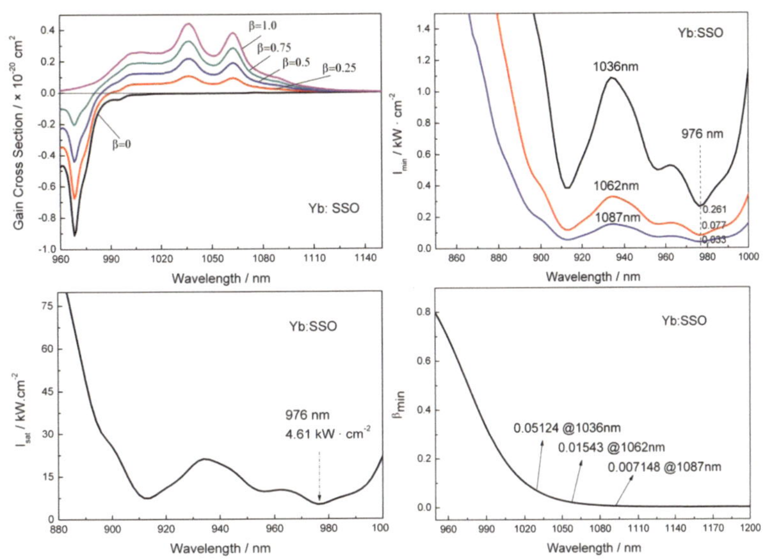

Fig. 9. σ_g, I_{min}, I_{sat} and β_{min} in Yb:SSO crystal.

Host	λ_{pump} (nm)	β_{min}	I_{sat} (kW/cm²)	I_{min} (kW/cm²)	λ_{ext} (nm)	σ_{ext} (10^{-20}cm²)
Yb:LSO	977	0.015	9.2	0.138	1083	0.14
Yb:GSO	974	0.0071	19.01	0.093	1088	0.46
Yb:YSO	977	0.0015	17.99	0.18	1081	0.12
Yb:SSO	976	0.0071	4.61	0.033	1087	0.1
Yb:YAG	942	0.055	28.8	1.53	1030	2.2

Table 5. Calculated Laser parameters of Yb doped Crystals.

3.3 Energy level of Yb in silicate crystal hosts

The energy level schemes for Yb doped silicate crystals were determined by optical spectroscopic analysis and semi-empirical crystal-field calculations using the simple overlap model. The stark levels distributed in the two $^2F_{7/2}$ and $^2F_{5/2}$ manifolds are labeled from one to four in the ground state and five to seven in the excited state. The maximum splitting manifolds of Yb in silicate hosts was calculated according to the spectra shown in Fig. 5.

Fig. 10 explicated the overall energy splitting of $^2F_{7/2}$ manifold of Yb^{3+} in silicate crystal hosts with 1012cm^{-1} in Yb:LSO crystal, 1076cm^{-1} in Yb:GSO crystal, 984 cm^{-1} in Yb:YSO crystal and 1027cm^{-1} in Yb:SSO crystal indicating much stronger crystal-field interaction in silicate hosts than that in Yb:YAG with energy splitting of 785cm^{-1} (Yan et al., 2006).

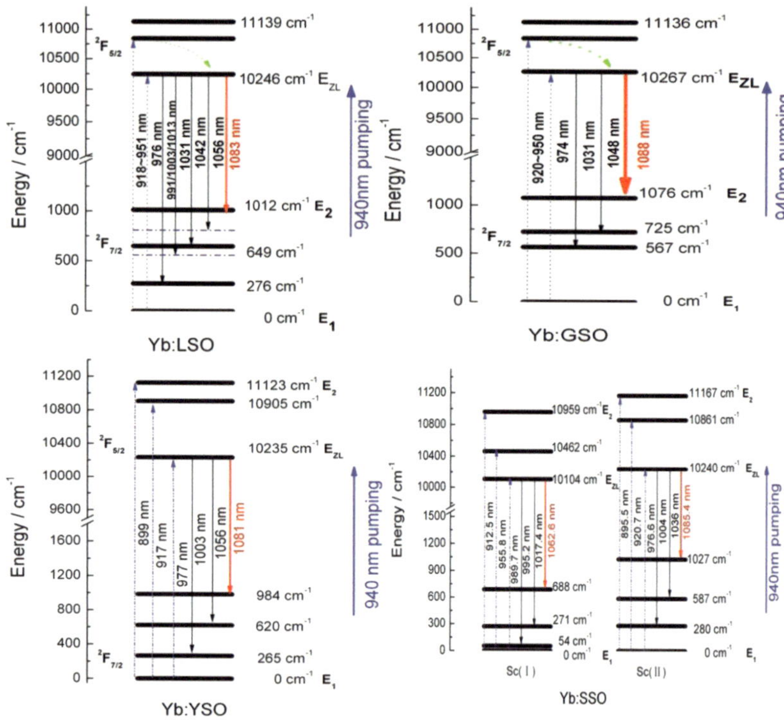

Fig. 10. Energy level of Yb^{3+} in Yb doped silicates.

As indicated in Fig. 10, the manifold splitting for Yb^{3+} in Yb:GSO crystal reached 1076 cm^{-1} which was much larger than that in Yb:LSO, Yb:YSO and Yb:SSO with C2/c structure thanks to the anisotropic and compact structure in GSO crystal. The strong crystal field in Yb:GSO crystal was simultaneously the largest ground-state splitting among Yb doped silicate crystals as obtained by absorption spectra starting from around 897 nm and the emission spectra extending till 1090 nm in Yb:GSO host.

To further elaborate the energy level in Yb doped silicate crystals, we found that the absorption band centered at 976nm in Yb:GSO belongs to the zero-line transition between the lowest level of $^2F_{7/2}$ and $^2F_{5/2}$ manifolds. The emission band at the longest wavelength around 1088 nm corresponds to the transition from the lowest levels of $^2F_{5/2}$ manifold to the highest levels of $^2F_{7/2}$ manifold. The strong electron–phonon coupling among Yb^{3+} ions together with the considerable interaction between Yb^{3+} ions and lattice vibrations give arise to strong vibration sidebands. The supplementary stark level splitting was incited by the resonance between stark levels and phonons which can be easily mistaken for the assignment of electronic transitions. Large crystal-field splitting of the fundamental manifold $^2F_{7/2}$ is critical for limiting the thermal population of the terminal laser level. In other words, less thermal populating of the terminal laser level of $^2F_{7/2}$ ground state would

reduce re-absorption losses and decrease laser threshold as well. The strongest emission cross section was around 1000 nm in either Yb:YSO or Yb:LSO, while that in Yb:GSO was located around 1088 nm suggesting weakest re-absorption losses and least thermal populating of the terminal laser level. Accordingly, Yb:GSO around 1088 nm would produce ultra-fast laser with lowest pumping threshold.
The energy-level diagram of Yb^{3+} ions in crystallographic site Sc(I) and crystallographic site Sc(II) located in Yb:SSO crystal was approximately denoted in Fig. 10. The overall splitting of the $^2F_{7/2}$ manifold of Sc(II) reached about 1027cm^{-1} which was approaching that in Yb:GSO but much larger than 984 cm^{-1} in Yb:YSO and 1012 cm^{-1} in Yb:LSO. Yb ions encountered a stronger crystal-field interaction in SSO host indicating that Yb:SSO crystal would be the suitable solid-state laser gain for high efficient and high power application.

4. Laser peformance of Yb doped silicate crystals

High efficient laser output of 198fs/2.61W, 260fs/2.6W and 343fs/400mW were obtained in Yb:YSO, Yb:LSO and Yb:GSO respectively by Chinese and French researchers (Thibault et al., 2006; Li et al., 2007). Based on laser setup in Fig. 3, diode-pumped Kerr-lens mode-locked Yb:LSO laser was achieved without additional components. The mode-locked laser pulses were obtained in five-mirror cavity with average output power of 2.98 W and repetition rate of 103 MHz under incident pump power of 14.44W. Short pulse of 8.2 ps was realized at wavelength centered at 1059 nm (Wang et al., 2010). With an OC of T=2.5%, the cw Yb:GSO lasers with the slope efficiency of 75% at 1094 nm were achieved. Self-pulsed Yb:GSO lasers were achieved with low pumping threshold from 1091nm to 1105 nm (Li et al., 2006). Efficient diode-pumped laser performance of Yb:SSO was demonstrated with slope efficiency of 45% and output power of 3.55W (Zheng et al., 2008). The passive mode-locking and cw lasing performance of Yb:SSO was carried out in an x-fold cavity end-pumped by a 978 nm single emitter. The laser produced a maximum cw output power of 2.73 W with a slope efficiency of 70%. Preliminary tests regarding the laser operation of the thin-disk Yb:SSO laser were presented with 9.4 W of output power with an optical efficiency of 25.3 % (Wentsch et al., 2011). Passive mode-locking of Yb: SSO was initiated using a semiconductor saturable absorber mirror (SESAM) while dispersion compensation was introduced using a pair of SF10 prisms. The laser mode-locked at 1041 nm, 1060 nm and 1077 nm with near Fourier transformed limited pulse width of 145 fs, 144 fs and 125 fs, and average output power of 40 mW, 52 mW and 102 mW, respectively (Tan et al., 2010).

5. Conclusion

Yb doped oxyorthosilicate single crystals as technologically important laser host family owing to quasi-four level scheme were successfully obtained with high transparency and high quality by Czmethod. The efficient diode pumped ultra-fast laser was achieved with distinguished ground-state splitting up to 1000 cm^{-1}. Key spectroscopic parameters of Gain cross-section σ_g and minimum pump intensity I_{min} for designing high power lasers are specifically interpreted. In future, bulk silicate crystals with favorable thermal properties and multi-crystallographic sites for introducing distorted and broad spectra behavior are among the hot spots in the development of ultra-fast lasers.

6. Acknowledgment

The authors thank the supports from National High Technology Research and Development Program of China(2009AA03Z435), National Natural Science Foundation of China (60938001, 60908030, 61178056, 61177037), Innovation Project of Shanghai Institute of Ceramics (Y04ZC5150G), Hundred Talents Project of Chinese Academy of Sciences and National Science Fund for Distinguished Young Scholars. Special thanks to Dr. Heping Zeng, Dr. Xiaoyan Liang, Dr. Jie Liu, Dr. Weide Tan ,Dr. Guangjun Zhao, Dr. Chengfeng, Yan, Dr. Kejian Yang, Dr. Nengyin Sheng, Ms. Jiao Wang.

7. References

Caird J., Ramponi A. & Staver P. (1991). Quantum efficiency and excited-state relaxation dynamics in neodymium-doped phosphate laser glasses. *Journal of the Optical Society of America B*, Vol.8, No.7, pp. 1391-1403, ISSN 0740-3224

Camargo A., Davolos M. & Nunes L. (2002). Spectroscopic characteristics of Er^{3+} in the two crystallographic sites of Gd_2SiO_5, *Journal of Physics: Condensed Matter*, Vol.14, No.12, pp. 3353-3363, ISSN 0953-8984 (print)

Campos S., Denoyer A., Jandl S., Viana B., Vivien D., Loiseau P. & Ferrand B. (2004). Spectroscopic studies of Yb^{3+}-doped rare earth orthosilicate crystals. *Journal of Physics: Condensed Matter*, Vol.16, No.25 pp. 4579-4590, ISSN 0953-8984 (Print)

Cooke D., Bennett B., Mcclellan K., Roper J., Whittaker M. (2000). Oscillator strengths, Huang-Rhys parameters, and vibrational quantum energies of cerium-doped gadolinium oxyorthosilicate. Journal of Applied Physics, Vol.87, No.11, pp. 7793-7797, ISSN 0021-8979 (Print)

Deloach L., Payne S., Chase L., Smith L., Kway W. & Krupke W. (1993). Evaluation of absorption and emission properties of Yb^{3+} doped crystals for laser applications. *IEEE Journal of Quantum Electronics*, Vol.29, No.4, pp. 1179 - 1191, ISSN 0018-9197

Du J., Liang X., Xu Y., Li R., Xu Z., Yan C., Zhao G., Su L. & Xu J. (2006). Tunable and efficient diode-pumped Yb^{3+}:GYSO laser. *Optics Express*, Vol.14, No.8, pp. 3333-3338, ISSN 1094-4087

Eijk C. (2001). Inorganic-scintillator development. *Nuclear Instruments and Methods in Physics Research Section A*, Vol.460, No.1, pp. 1-14, ISSN 0168-9002

Ellens A., Andres H., Meijerink A. & Blasse G. (1997). Spectral-line-broadening study of the trivalent lanthanide-ion series.I. Line broadening as a probe of the electron-phonon coupling strength. *Physical Review B*, Vol.55, No.1, pp. 173-179, ISSN 1098-0121

Felsche J., (1973). The crystal chemistry of the rare-earth silicates, *Structure and Bonding*, pp. 99-197, ISSN 0081-5993 (Print), Springer-Verlag, New York, 1973

Fornasiero L., Petermann K., Heumann E. & Huber G. (1998). Spectroscopic properties and laser emission of Er^{3+} in scandium silicates near 1.5 μm, *Optical Materials*, Vol.10, No.1, pp. 9-17, ISSN 0925-3467

Gaume M. (2002). Relations structures - propriétés dans les lasers solides de puissance à l'ytterbium. Elaboration et caractérisation de nouveaux matériaux et de cristaux composites soudés par diffusion, These De Doctorat de l'universite pierre et marie curie-paris VI

Gaume R., Viana B., Derouet J. & Vivien D. (2003). Spectroscopic properties of Yb-doped scandium based compounds Yb:$CaSc_2O_4$, Yb:$SrSc_2O_4$ and Yb:Sc_2SiO_5. *Optical Materials*, Vol.22, No.2, pp. 107-115, ISSN 0925-3467

Giesen A. & Speiser J. (2007). Fifteen Years of Work on Thin-Disk Lasers: Results and Scaling Laws. *IEEE Journal of Selected Topics in Quantum Electron*, Vol.13, No.3, pp.598-609, ISSN 1077-260X

Griebner U., Petrov V., Petermann K. & Peters V. (2004). Passively mode-locked Yb:Lu_2O_3 laser. *Optics Express*, Vol.12, No.14, pp. 3125-3130, ISSN 1094-4087

Gustafsson T., Klintenberg M., Derenzo S., Weber M., Thomas J. (2001). Lu_2SiO_5 by single-crystal X-ray and neutron diffraction. *Acta Crystallographica Section C*, Vol.57, No.6, pp. 668-669, ISSN 1600-5759 (Online)

Haumesser P., Gaumé R., Viana B. & Vivien D. (2002). Determination of laser parameters of ytterbium-doped oxide crystalline materials. *Journal of the Optical Society of America B*, Vol.19, No.10, pp. 2365-2375. ISSN 0740-3224

http://database.iem.ac.ru/

Ivanov V., Petrov V., Pustovarov V., Shulgin B., Vorobjov V., Zinevich E., Zinin E., (2001). Electronic excitations and energy transfer in A_2SiO_5-Ce (A=Y, Lu, Gd) and Sc_2SiO_5 single crystals, *Nuclear Instruments and Methods in Physics Research Section A*, Vol.470, No.1-2, pp. 358-362, ISSN 0168-9002

Keller U. (2003). Recent developments in compact ultrafast lasers. *Nature*, Vol.424, No.6950, pp. 831-838, ISSN 0028-0836

Kisel V., Troshin A., Shcherbitsky V., Kuleshov N., Matrosov V., Matrosova T., Kupchenko M., Brunner F., Paschotta R., Morier-Genoud F. & Keller U. (2005). Femtosecond pulse generation with a diodepumped Yb^{3+}:YVO_4 laser. *Optics Letters*, Vol.30, No.10, pp. 1150-1152, ISSN 0146-9592

Krupke W. (2000). Ytterbium solid-state lasers-the first decade. *IEEE Journal of Selected Topics in Quantum Electron*, Vol.6, No.6, pp. 1287-1296, ISSN 1077-260X

Kuleshov N., Shcherbitsky V., Lagatsky A., Mikhailov V., Minkov B., Danger T., Sandrock T. & Huber G. (1997). Spectroscopy, excited-state absorption and stimulated emission in Pr^{3+}-doped Gd_2SiO_5 and Y_2SiO_5 crystals. Journal of Luminescence, Vol.71, No.1, pp. 27-35, ISSN 0022-2313

Lacovara, P., Choi, H., Wang, C., Aggarwal, R. & Fan T. (1991). Room-temperature diode-pumped Yb:YAG laser. *Optics Letters*, Vol.16, No.14, 1089-1091, ISSN 0146-9592 (print)

Li W., Hao Q., Zhai H., Zeng H., Lu W., Zhao G., Zheng L., Su L. & Xu J. (2007). Diode-pumped Yb:GSO femtosecond laser. *Optics Express*, Vol.15, No.5, pp. 2354-2359, ISSN 1094-4087

Li W., Pan H., Ding L., Zeng H., Lu W., Zhao G., Yan C., Su L. & Xu J. (2006). Efficient diode-pumped Yb:Gd_2SiO_5 laser. *Applied Physics Letters*, Vol.88, No.22, pp. 221117(1-3), ISSN 0003-6951 (Print)

Liu H., Nees J. & Mourou G. (2001). Diode-pumped Kerr-lens mode-locked Yb:$KY(WO_4)_2$ laser. *Optics Letters*, Vol.26, No.21, pp. 1723-1725, ISSN 0146-9592

Melcher C. & Schweitzer J. (1992). Cerium-doped lutetium oxyorthsilicate: A fast, efficient new scintillator, *IEEE Transactions on Nuclear Science*, Vol.39, No.4, pp. 502-505, ISSN 0018-9499

Moulton P. (1983). Paramagnetic ion lasers, in: *Handbook of Laser Science and Technology*, M.J. Weber, (Ed.), CRC, ISBN: 978-0-8493-3508-2 , Boca Raton, FL

Pelenc D., Chambaz B., Chartier I., Ferrand B., Wyon C., Shepherd D., Hanna D., Large A. & Tropper A. (1995). High slope efficiency and low threshold in a diode-pumped epitaxially grown Yb:YAG waveguide laser. *Optics Communication*, Vol.115, No.5-6, pp. 491-497, ISSN 0030-4018

Petit J., Goldner P. & Viana B. (2005). Laser emission with low quantum defect in Yb:$CaGdAlO_4$. *Optics Letters*, Vol.30, No.11, pp. 1345-1347, ISSN 0146-9592

Smolin Y. & Tkachev S. (1969). Determination of the structure of gadolinium oxyorthosilicate (Gd_2SiO_5). *Kristallografiya*, Vol.14, pp.22, ISSN 0023-4761

Su L., Xu J., Li H., Yang W., Zhao Z., Si J., Dong Y. & Zhou G. (2005). Codoping Na^+ to modulate the spectroscopy and photo-luminescence properties of Yb^{3+} in CaF_2 laser crystal. *Optics Letters*, Vol.30, No.9, pp. 1003-1005, ISSN 0146-9592

Suzuki H., Tombrello T., Melcher C. & Schweitzer J. (1992). UV and gamma-ray excited luminescence of cerium-doped rare-earth oxyorthosilicate. *Nuclear Instruments and Methods in Physics Research Section A*, Vol.320, No.1-2, pp. 263-272, ISSN 0168-9002

Tan W., Tang D., Xu X., Zhang J., Xu C., Xu F., Zheng L., Su L. & Xu J. (2010). Passive femtosecond mode-locking and cw laser performance of Yb^{3+}:Sc_2SiO_5. *Optics Express*, Vol.18, No.16, pp. 16739-16744, ISSN 1094-4087

Thibault F., Pelenc D., Druon F., Zaouter Y., Jacquemet M. & Georges P. (2006). Efficient diode-pumped Yb^{3+}:Y_2SiO_5 and Yb^{3+}:Lu_2SiO_5 high-power femtosecond laser operation. *Optics Letters*, Vol.31, No.10, pp. 1555-1557, ISSN 0146-9592

Uemura S. & Torizuka K. (2005). Center-wavelength-shifted passively mode-locked diode-pumped ytterbium (Yb):Yttrium aluminum garnet (YAG) laser. *Jpn. J. Appl. Phys.*, Vol.44, No.12-15, pp. L361-L363, ISSN: 0021-8979 (print)

Wang W., Liu J., Chen W., Lu C., Zheng L., Su L., Xu J. & Wang Y. (2010). Diode-Pumped Passively Mode-Locked Yb:LSO/SESAM Laser. *Laser Physics*, Vol.20, No.4, pp. 740-744, ISSN 1054-660X (Print)

Wang X., (2004). *Optical Crystallography*, Nanjing University Press, ISSN 7-305-04088-6

Wentsch K., Weichelt B., Zheng L., Xu J., Abdou-Ahmed M. & Graf T. (2011) Contnious-wave Yb doped Sc_2SiO_5 thin-disk laser. *Optics Letters*, Vol. Accepted 11/08/2011, Doc. ID 153527, ISSN 0146-9592

Yan C., Zhao G., Su L., Xu X., Zhang L. & Xu J. (2006). Growth and spectroscopic characteristics of Yb:GSO single crystal. *Journal of Physics: Condensed Matter*, Vol.18, No.4, pp. 1325-1333, ISSN 0953-8984 (Print)

Zheng L., Xu J., Zhao G., Su L., Wu F., Liang X. (2008). Bulk crystal growth and efficient diode-pumped laser performance of Yb^{3+}:Sc_2SiO_5. *Applied Physics B*, Vol.91, No.3-4, pp. 443-445, ISSN 0946-2171 (Print)

Zheng L., Zhao G., Yan C., Yao G., Xu X., Su L. & Xu J. (2007). Growth and spectroscopic characteristics of Yb:LPS single crystal. *Journal of Crystal Growth*, Vol.304, No.2, pp. 441-447, ISSN 0022-0248

3

Defect Engineering During Czochralski Crystal Growth and Silicon Wafer Manufacturing

Lukáš Válek[1,2] and Jan Šik[1]
[1]ON Semiconductor Czech Republic,
[2]Institute of Physical Engineering,
Brno University of Technology,
Czech Republic

1. Introduction

Single crystal silicon has played the fundamental role in electronic industry since the second half of the 20th century and still remains the most widely used material. Electronic devices and integrated circuits are fabricated on single-crystal silicon wafers which are produced from silicon crystals grown primarily by the Czochralski (CZ) technique. Various defects are formed in the growing crystal as well as in the wafers during their processing. This chapter deals with the topic of engineering of crystal defects in the technology of manufacturing silicon single crystals and silicon wafers for the electronic industry. A basic overview of crystal defects found in semiconductor-grade silicon is provided and mechanisms of their formation are introduced. The impact of crystal defects on the manufacturing and performance of electronic devices is outlined and some of the methods of defect analyses are described. Finally, the most important methods for control of defect formation are summarized.

2. Industrial production of silicon for electronics

Single crystals of silicon for today´s electronic industry are produced primarily by the Czochralski (CZ) method (Czochralski, 1918; Teal & Little, 1950). Only applications with extreme demands on pure bulk material utilize the float zone (FZ) method (Keck & Golay, 1953). The CZ method is based on crystal pulling from the melt, while the FZ method utilizes recrystallization of polysilicon rod which is locally molten by passage of the RF coil. The processes differ mainly in production cost and speed, which favor the CZ method, and in the purity of produced material, which is higher in case of the FZ method. The lower purity, which was seemingly unfavorable, helped CZ silicon become the dominant material as it makes silicon wafers more resistant against thermal stress and metallic contamination. Furthermore the FZ process could not follow the continual increase in crystal diameter. The diameter of 200 mm is the current size limit for FZ crystals while 450 mm capability was demonstrated for CZ process. As the chapter deals mainly with CZ silicon, the CZ crystal growth will be more closely described.

2.1 Czochralski growth of silicon crystals

Today´s Czochralski (CZ) grown silicon single crystals are produced in a mass scale in diameters of up to 300 mm, but the 150 mm and 200 mm processes are still considered as standard. A typical CZ puller is shown in Fig. 1. The puller consists of an upper and lower chamber formed by steel water-cooled shells. The lower chamber contains a graphite hot zone with active central part and thermally insulating outer parts. A silica crucible is placed in the heart of the hot zone which is supported by a graphite susceptor on the pedestal attached to the lower shaft. The seed holder is fastened onto the upper shaft (or affixed to a cable). The heater is a meandering-coil element heated by high electric current. Both chambers are piped to a vacuum system. The puller is typically purged with inert gas (usually argon). In the beginning of the process the quartz crucible is loaded with a charge of polysilicon chunks and the single crystalline seed is fitted into the seed holder. After closing the puller the chambers are evacuated and re-filled by inert gas to the desired process pressure. The process starts with melting the polysilicon charge by applying high power to the heater. Once the charge is molten, the melt flow is stabilized under steady conditions and the seed is lowered towards the melt. After the seed is dipped into the melt, the system is adjusted to achieve a stable interface between the melt and the seed crystal. Pulling the seed upwards crystallizes the melt at the solid-liquid interface and the crystal proceeds to grow.

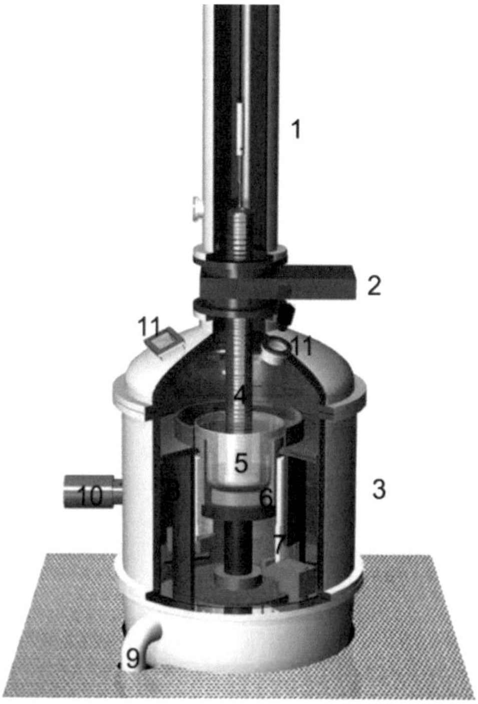

Fig. 1. A typical configuration of a CZ silicon puller. (1) Upper chamber, (2) isolation valve, (3) lower chamber, (4) grown crystal, (5) silica crucible, (6) graphite susceptor, (7) heater, (8) insulation, (9) vacuum pipe, (10) heater pyrometer window, (11) operator and camera windows.

Once the seed touches the melt surface, it is subjected to a huge thermal shock leading to generation of numerous dislocations. In order to achieve dislocation-free growth, "necking" is performed. To achieve this condition, the crystal pulling rate is increased to about 3 to 6 mm per minute, and the crystal diameter is reduced to about 5 to 2 mm, which allows dislocations to partially freeze in the neck and partially move to the crystal surface. Dislocation-free growth is usually achieved after several centimeters of the neck growth. Then the pulling rate is significantly decreased and the diameter is slowly increased. The crystal grows into the form of a cone called the crown. As the diameter increases to the desired crystal diameter the pull rate is gradually increased (the so called "shouldering stage") until the crystal grows with the desired diameter and the proper growth rate. Then, the cylindrical portion of the crystal, the "crystal body", is grown.

The melt and the crystal are in intimate contact at the solid-liquid interface. The melt surface forms a meniscus to the crystal which reflects the light from the hot crucible to the chamber windows. This results in the appearance of a shiny ring on the melt surface around the crystal. As the meniscus height increases with crystal radius, changes in the meniscus height can be sensed and used for crystal diameter control during the growth of the crystal body. Another option for crystal diameter control is a measurement of the meniscus diameter with a CCD camera. The crystal diameter is controlled by the pulling rate and simultaneously the pulling rate is adjusted by the heater power to be within the empirically-determined process window for dislocation-free growth, typically at or below 1 mm per minute. Solidification heat is conducted to the crystal surface and radiated to the chamber. For longer ingots the heat conduction is reduced and therefore the pulling rate has to be reduced. The crystal is rotated to homogenize the distribution of impurities and to suppress inhomogeneities in the temperature field. The crucible is rotated in the opposite sense to the crystal to stabilize the melt flow and control the oxygen concentration in the crystal.

The final stage of the crystal growth is the tail growth where the diameter is slowly decreased and a conical shape is achieved. The diameter of the solidification interface is reduced and dislocation formation is suppressed due to minimization of the thermal shock. Once the crystal has detached from the melt the power to the puller is decreased and the crystal is cooled down while being lifted into the upper chamber. At the end of the process, the crystal is removed from the puller for further processing.

2.2 Manufacturing of silicon wafers

Manufacturing of silicon wafers involves a series of mechanical, physical and chemical processes, all optimized to provide superior properties of the final wafer. A simplified manufacturing flow of the silicon wafer is shown in Fig. 2.

Wafer manufacturing follows the crystal growth process. First, the crystal crown and tail are cut off and the crystal body is divided into several pieces. Then the crystal quality (resistivity, oxygen and carbon content, dislocation-free state) is assessed on test wafers. Each section of the crystal is then surface-ground to the desired diameter, the crystal is oriented, and the flat is ground onto the cylindrical ingot. The flat identifies the orientation of the silicon wafers with respect to specific crystallographic directions; usually it corresponds to the (1 1; $\bar{0}$) plane. Silicon wafers are sliced from the crystal sections using wire-saws or the inner-diameter (ID) saws. After edge grinding the wafer is lapped, etched and polished. Finally, the polished silicon wafer may have an epitaxial layer deposited on the prime surface by silicon epitaxy methods. Optionally the wafer backside can be coated with a polysilicon layer and/or a protective layer of silicon oxide. After final cleaning and inspection the silicon wafers are suitable for device or integrated circuit (IC) manufacturing.

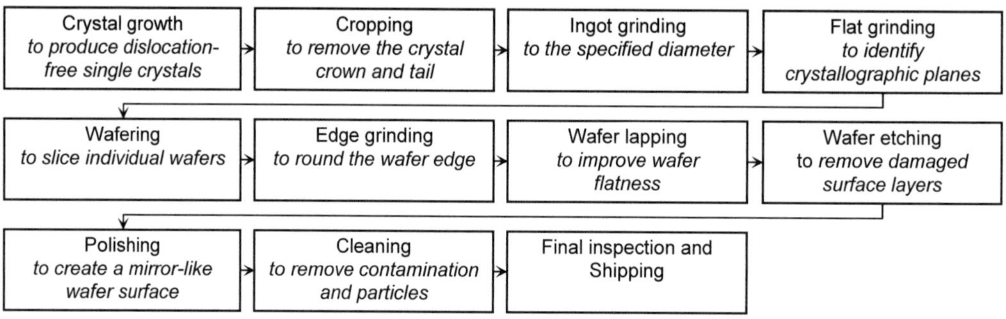

Fig. 2. Schematic manufacturing flow of the polished silicon wafer.

3. Defects in CZ silicon

Single crystalline CZ silicon wafer is a well-defined material of a very high quality. Nevertheless, it still may contain various defects which are formed either during the crystal growth or during processing of the silicon wafer. Defects which arise most frequently will be briefly introduced in the following sections. A detailed overview of defects in silicon can be found e.g. in (O´Mara, 1990).

3.1 Introduction

Soon after the start-up of mass production of dislocation-free silicon in the 1960's it was realized that the highly pure silicon wafers suffered from enhanced formation of slip lines and degraded device yields as compared to formerly-used dislocated silicon. Later it was found that the issues were actually related to impurities and crystal defects.

Swirl-like distributions of agglomerates of silicon self-interstitials (so called A- and B-defects) were described first (Föll & Kolbesen, 1975), and vacancy agglomerates (so called C- and D- defects) were observed soon after (Roksnoer & van den Boom, 1981). Although the initial investigations were focused on FZ silicon, the defects were observed in CZ silicon as well, and obeyed essentially the same rules. The intensive research in the field revealed that formation of crystal defects depended on the crystal growth conditions. Several attempts were made to explain the occurrence of vacancy- and interstitial-type defects in silicon, but a unifying and generally accepted theory was not provided until the work of Voronkov (Voronkov, 1982).

The complete picture of defects in CZ silicon is even more complicated due to influence of impurity inherently tied to CZ silicon, namely oxygen. Oxygen preferentially occupies interstitial sites in the silicon lattice, but below about 1200°C it becomes supersaturated and tends to cluster and precipitate. Moderate oxygen precipitation significantly enhances the mechanical strength of the silicon wafer, whereas too strong precipitation can degrade the mechanical properties. Oxygen precipitates in the active region of an electronic device usually results in degradation of the device performance. On the other hand, oxygen precipitates outside of the active region may have a highly beneficial effect through intrinsic gettering (Rozgonyi, 1976). As there are both negative and positive effects closely related to oxygen precipitation, it is a phenomenon which has to be carefully engineered in the silicon manufacturing technology. Oxygen precipitation depends on the concentration of interstitial oxygen, annealing conditions and oxygen interactions with intrinsic point defects. Hence, a lot of effort was directed to the investigation of defect formation in silicon during past decades.

3.2 Silicon crystal structure

Silicon crystallizes in the diamond cubic lattice (Bullis, 1991), which structure can be constructed as two interpenetrating face-centered cubic (fcc) lattices displaced along the unit cell body diagonal by the quarter of its length. The lattice constant of pure silicon is representing the length of the side of the fcc cube and has the value of 0.5431 nm at room temperature. Each silicon atom from one fcc lattice is covalently bound to its four nearest neighbors which belong to the other fcc lattice. In principle silicon wafer can be manufactured with the silicon lattice being oriented in any arrangement with respect to wafer surface. The common crystallographic orientations of wafer surface, described by Miller indices, are (100), (111) and (110). The crystal planes belonging to the families of these low-index crystal planes, i.e. {100}, {111} and {110} family, are the most important planes in the silicon lattice with the largest differences in various properties among these planes.
The closely spaced planes of the {100} family are the {400} planes with the interplanar distance of 0.1358 nm. The {220} planes with spacing of 0.1920 nm are the closely spaced planes of the {110} system. While spacing of all the {400} and {220} planes is equidistant, the situation is more complicated in case of the {111} system. The {111} planes of the fcc lattice are stacked in an ...ABCABC... manner, where none of the atoms in each layer is aligned with atoms in other layers when viewed perpendicularly to the stack. The {400} planes follow the ...ABCDABCD... scheme and the {220} planes follow the ...ABAB... scheme. Since the diamond lattice consists of two fcc lattices displaced along the direction perpendicular to the {111} system, stacking along <111> follows the ...AaBbCcAaBbCc... scheme. Here atoms in planes of the same letter overlap each other when viewed in <111> direction and capitalization distinguishes planes belonging to the two interpenetrating fcc lattices. The distance between adjacent planes of the same set (A-B, B-C and C-A) is 0.3135 nm; the distance between the closely spaced planes of the system a-B, b-C and c-A is 0.0784 nm and 0.2352 nm for the system A-a, B-b, C-c, which is the length of the covalent bond silicon-silicon. The closely spaced planes are bound strongly and often appear as one double plane. Different spacing of the crystal planes corresponds to different surface atom densities of 6.78 $\times 10^{14}$ cm^{-2} for {100} planes, 7.83 $\times 10^{14}$ cm^{-2} for {111} planes and 9.59 $\times 10^{14}$ cm^{-2} for {110} planes. The {111} plane shows the lowest surface energy and crystallizing silicon therefore tends to be bound by the {111} planes. Also the growth rate of the epitaxial layer is the lowest for the {111} oriented surface.

3.3 Classification and overview

Crystal defects are usually classified according to their shape and dimension. Silicon crystals and wafers may contain:

a. point defects- silicon self-interstitials, vacancies, interstitial impurities such as oxygen, substitutional impurities such as dopants and carbon,
b. line defects- edge and screw dislocations, dislocation loops,
c. planar defects- stacking faults,
d. bulk defects- agglomerates of point defects.

Some of the crystal defects are shown schematically in Fig. 3.

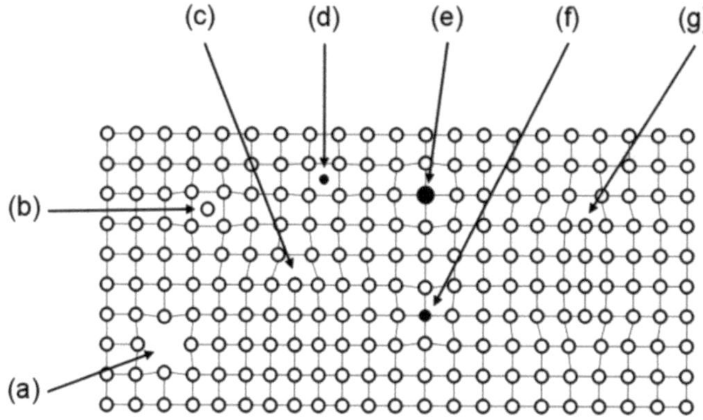

Fig. 3. Schematic 2D representation of crystal defects in silicon. (a) Vacancy, (b) self-interstitial atom, (c) edge dislocation, (d) interstitial impurity atom, (e) substitutional impurity atom of larger atomic radius, (f) substitutional impurity atom of smaller atomic radius, (g) extrinsic stacking fault.

3.4 Intrinsic point defects

Vacancies and silicon self-interstitials (Pichler, 2004) are intrinsic point defects inherent to the material whose occurrence in the lattice arises from thermodynamic equilibrium. Intrinsic point defects are incorporated into the growing crystal at the melt-crystal interface in their equilibrium concentrations. Point defects in the crystal are further created or annihilated by Shottky and Frenkel mechanisms, recombination, interaction with bulk defects, and injection during thermal wafer processing.

A silicon atom removed from its lattice site leaves behind four broken covalent bonds which can be arranged in several configurations. When the dangling bonds reform into molecular orbitals a neutral vacancy is formed. These orbitals are strained and can be relatively easily broken. Interaction with charge carriers results in formation of singly and doubly positively- or negatively-charged vacancies. The silicon self-interstitial is considered as an additional, more or less free atom occupying tetrahedral or hexagonal interstitial sites. Small clusters of vacancies and interstitials may be considered as a point defect. The precise atomic arrangement of vacancies and interstitials in silicon still remains somewhat uncertain, yet their effective thermodynamical properties such as concentration and diffusivity have been experimentally determined.

3.5 Extrinsic point defects

Dopant atoms are the primary extrinsic point defects in CZ silicon. The common dopant species - boron, arsenic, antimony and phosphorus - are introduced deliberately into the silicon melt during the crystal growth. They occupy substitutional sites in the crystal lattice and their concentrations are typically in the range of 10^{15} - 10^{19} cm^{-3}. The effect of dopant atoms on the properties of the crystalline silicon (besides the primary electric effect) originates from different atomic volumes, from interactions with intrinsic point defects, the formation of clusters, and from the effects on the diffusion and redistribution of impurities (Pichler, 2004).

Extremely undesirable point defects are *atoms of metallic elements* in the middle of the periodic table (Graff, 2000). Deep energy levels near the middle of the silicon forbidden band arising from metallic species act as effective generation-recombination centers and pose serious issues for device performance and process yield. The transition metals such as copper, nickel and iron are the most abundant contaminants. At high temperatures the metal atoms are usually distributed in the lattice occupying both interstitial and substitutional sites. As the temperature is reduced the metal atoms can form complexes with dopant atoms, and may eventually precipitate into bulk defects. The state of metals in the lattice depends on their solubility in silicon and on the imposed thermal cycles. The level of metallic contamination in semiconductor silicon for today's processes is typically below 10^{12} cm^{-3}. The steep decrease in solubility with decreasing temperature combined with a high diffusivity at low temperatures allows relaxation-type gettering techniques to remove the metals from the active device regions.
An unavoidable impurity in CZ silicon is *carbon* (O´Mara, 1990) introduced into the crystal growth system mainly from the graphite elements of the hot zone. Typical concentration of carbon in semiconductor silicon is below about 0.1 ppma ($5x10^{15}$ cm^{-3}). Carbon predominantly occupies substitutional sites in the silicon lattice. Due to its four valence electrons substitutional carbon is electrically inactive. Substitutional carbon forms complexes with oxygen. Interactions with silicon self-interstitials may lead to a displacement of the carbon into an interstitial position (Pichler, 2004). Interstitial carbon readily forms complexes with intrinsic point defects, substitutional carbon or dopants, which may become electrically active. Carbon was reported to promote oxygen precipitation. However, due to the low content of carbon in today's silicon, its the role in defect formation is of minor importance.
After dopant elements, *oxygen* is perhaps the most important extrinsic defect in silicon (O´Mara, 1990; Shimura, 1994). Oxygen is incorporated into the CZ silicon during the crystal growth. The silica crucible wall slowly dissolves into the silicon melt and oxygen is transported by diffusion and convection to the melt-crystal interface. Much of the oxygen evaporates from the melt free surface and only about 1% is incorporated into the growing crystal. The solubility of oxygen in silicon is a monotonically increasing function of temperature; its value at the melting temperature is about $2x10^{18}$ cm^{-3}. Oxygen concentration in the CZ silicon crystal is typically on the order of 10^{17} cm^{-3}. Oxygen atoms occupy interstitial positions in the silicon lattice where they are covalently bound to the two nearest silicon atoms. Interstitial oxygen may form electrically active chains known as "thermal donors". Oxygen also forms complexes with intrinsic point defects (Pichler, 2004) which may play a role during the formation of bulk defects and promote oxygen diffusion.

3.6 Line defects

Today's CZ silicon crystals are grown in a dislocation-free mode (A defects are not taken into account). Dislocations appear in silicon mainly due to stress generated by high temperature operations during the manufacturing of the wafer and devices on it when the critical resolved shear stress (CRSS) is exceeded. Edge and screw dislocations and dislocation loops are created under the high stress conditions, and have been observed in silicon mainly accompanying oxygen precipitates, self-interstitial agglomerates, dislocation arrays and tangles, and slip.

3.7 Planar defects

Two types of *stacking faults* (SFs) are generally distinguished, namely intrinsic and extrinsic faults. Intrinsic SF is formed by several missing atomic planes while the extrinsic SF is formed by excess atomic planes. SFs in silicon are always of the extrinsic nature (O´Mara, 1990) formed by discs of double planes inserted into the regular AaBbCc order of silicon {111} atomic planes. As the layers in the SF have to be added in pairs, one can drop the double index notation and describe the packing of the sequence with ...ABCABC... In the SF region the order of planes is changed, such as ...ABCACABC.... SFs are considered to be low-energy faults because they involve no change in the covalent bonds of the four nearest neighbors in the lattice (Hirth & Lothe, 1967). On the other hand, faults which disturb the nearest-neighbor covalent bonds are called high-energy faults (these are, for example, dislocations). The SF is bound by a Frank-type partial dislocation. A schematic drawing of such a SF is shown in Fig. 3. Other types of planar defects such as twin planes or grain boundaries are not present in properly-grown single crystal silicon.

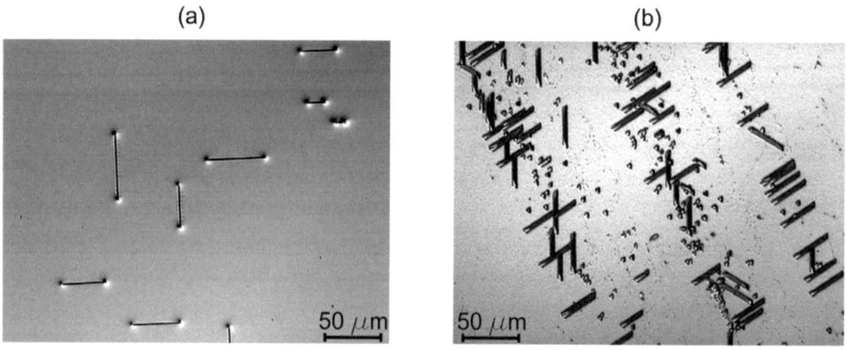

Fig. 4. (a) Bulk stacking faults on a (100) silicon surface. The varying SF length originates from the various depths of nucleation sites below the etched surface. (b) Surface stacking faults and dislocations on a (111) silicon surface. Predominantly equal SFs length confirms nucleation on the surface.

SFs originate from the condensation of silicon self-interstitials (Ravi & Varker, 1974), preferentially on suitable nucleation sites such as oxide precipitates, metal precipitates or mechanically damaged locations in the lattice. Surface and bulk SFs are distinguished according to the location of their nucleation site. Stacking faults can be delineated by selective etching of the sample surface and studied by optical microscopy or other inspection methods. The crystallographic origin determines the appearance of the SFs on the surface of silicon wafers of different orientations (see Fig. 4).

Bulk SFs in the silicon wafer may grow in the regions of strong oxygen precipitation (O´Mara, 1990), which is accompanied by a strong ejection of silicon self-interstitials. Growth of surface SFs can be also enhanced by an increased population of self-interstitials created by surface oxidation. SFs observed on the surface of polished silicon wafer after oxidation are referred to as "oxidation induced stacking faults" (OISFs).

3.8 Bulk defects

Vacancy-type defects. Supersaturated free vacancies present in the silicon crystal at high temperatures can agglomerate into voids (Itsumi, 2002). Voids (Fig. 5) take on an octahedral shape bounded by {111} planes. The octahedral shape may be incomplete, truncated by {100} planes. Under favorable conditions, double- or triple-voids may appear. The inner void surface is covered by an oxide layer typically 2 - 4 nm thick. A typical void dimension is around 100 nm and a typical density in the crystal is about 10^6 cm^{-3}.

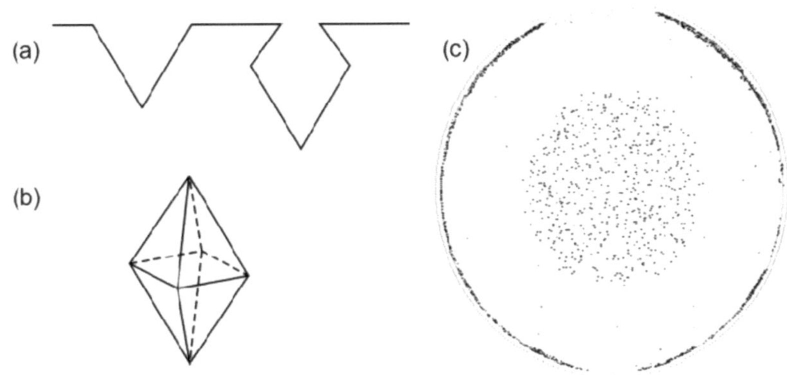

Fig. 5. Schematic representation of (a) COP defects on the wafer surface, and (b) a void defect in wafer bulk. (c) COP distribution observed by laser-type particle counter after dedicated SC1 treatment of the wafer with a vacancy-type core.

A void intersected by the wafer surface creates a pit referred to as the "crystal-originated particle" (COP) (Ryuta et al., 1990). The relation of a COP to a void is demonstrated in Fig. 5. The typical dimension of a COP does not allow direct observation on the polished wafer surface by common industrial equipment; a method such as atomic force microscopy (AFM) has to be used for the analyses. It was found that COPs can be observed after repetitive SC1 cleaning by commercial laser-type particle counters. COPs on the wafer surface can be delineated by other methods such as Secco etching where they appear as the wedge-shaped "flow pattern defects" (FPDs) (Yamagishi et al., 1992). Voids detected by infrared light scattering tomography are denoted as "LSTD" (Vanhellemont et al., 1997). Being vacancy agglomerates, COPs appear in the vacancy-type silicon (Fig. 5c).

Interstitial-type defects. Supersaturated silicon self-interstitials may coalesce into so called A- and B- defects (swirls). The B-defects are considered to be coherent globular clusters; the A-defects are large dislocation loops possibly formed by the collapse of the B-defects (Föll & Kolbesen, 1975). Large dislocation loops appear in size ranging from roughly a micrometer up to a few tens of micrometers with a density typically around 10^8 cm^{-3}.

Oxide precipitates. Due to the rapid decrease of oxygen solubility with temperature and the rather high oxygen concentrations in CZ silicon crystals, oxygen in silicon is usually present in a supersaturated state at most common process temperatures. The resulting precipitation of oxygen interstitials leads to the formation of oxide particles referred to as oxide precipitates or oxygen precipitates (Shimura, 1994). Oxide precipitates are formed of amorphous SiO_x, where x ranges from 1 to 2. Oxide precipitates of various morphologies have been observed: rod-like, square platelet, truncated octahedral, and polyhedral to spherical. The morphology of oxide precipitates depends on the formation conditions, mainly temperature and the degree of supersaturation. In principle it is determined by the stress energies associated with the precipitate growth (Borghesi et al., 1995). Growth of oxide precipitates is usually accompanied by the formation of extended defects such as stacking faults and punched-out dislocation loops.

4. Impact of crystal defects on device yield

While the vacancy- and interstitial-type crystal defects have a detrimental effect on yield and the performance of electrical devices, oxide precipitates and associated extended defects can have both a positive and negative effect.

COP defects were shown to degrade gate oxide integrity (GOI) (Yamagishi et al., 1992) causing extrinsic dielectric breakdown of MOS capacitors and failures of the DRAM modules. Voids below the surface can result in excessive leakage currents in power devices. The gate oxide grown across a COP may be locally thinned or stressed on the COP edges where the electric field is locally greater; both these effects may lead to dielectric breakdowns, shifted device thresholds and leakage.

The presence of interstitial-type dislocation loops in the active regions of the devices may cause degradation of the minority carrier lifetime, alter diffusion profiles, and shift p-n junction characteristics (Abe et al., 1983). Due to their relatively large size they are especially harmful for today's processes characteristic by a high degree of device integration. The dislocations act as fast diffusion paths for impurities and dopant species. In particular, when decorated by metals, dislocation cores are sources of shorts and leakage currents.

Oxygen thermal donors can induce resistivity shifts in very lightly-doped silicon which affects numerous resistivity-dependent device parameters. The thermal donors are formed by rather weak binding energies and can be easily dissolved at temperatures above 600°C. As oxygen thermal donors are formed at a peak rate around 450°C the cooling rates during device manufacturing processes should be optimized in order to control the resistivity of the sensitive materials. On the other hand, the oxide precipitates are quite stable defects. Oxide precipitates in the active region of the device result in increased leakage currents in the p-n junctions, reduced refresh times in DRAM memories, reduced breakdown voltages in bipolar devices, degraded minority carrier generation/recombination lifetimes, and other serious failure mechanisms (see Schröder, 1989, and references therein). The root-causes of the failures are the precipitates themselves as well as the associated extended defects. On the other hand, oxide precipitates may have a positive effect through mechanical strengthening through impurity hardening (Sumino, 1980) and intrinsic gettering (Rozgonyi, 1976). However, oxygen precipitation has to be controlled in an optimal range for a given wafer processing flow, and a defect-free near surface region (so called "denuded zone") should be guaranteed.

The effect of crystal defects on manufacturing processes and the properties of silicon wafers and electronic devices is quite varied and the impact is often fundamental. Hence, engineering of defects is inseparable across crystal growth and wafer processing. Due to the continuous evolution of silicon devices it is also a never-ending part of the technology of silicon crystal growth and wafer manufacturing.

5. Formation of grown-in defects in CZ silicon

In order to control formation of crystal defects in silicon wafers one has to understand the mechanisms of their formation from the very early stages of crystal growth. The basic mechanisms of defect formation during silicon crystal growth are described by the Voronkov theory (Voronkov, 1982) which will be briefly introduced in following sections.

5.1 Point defects - A pathway to crystal defects

Voronkov theory describes the formation of crystal defects consisting of several stages:

a. the incorporation of silicon self-interstitials and vacancies into the crystal,
b. transport, diffusion and recombination of the point defects,
c. nucleation and growth of defect clusters.

It is assumed that both silicon self-interstitials and vacancies are incorporated into the growing crystal at the melt-crystal interface. High diffusivities near the melting temperature and the proximity of the crystal surface acting as an effective source/sink allows the intrinsic point defects to exist in their equilibrium concentrations.

It is further assumed that the recombination rate of the point defects is high enough to maintain the product C_IC_V near its equilibrium value $C_I^{eq}C_V^{eq}$. As the equilibrium concentrations of the point defects decrease very rapidly with temperature, the $C_I^{eq}C_V^{eq}$ product quickly drops below the initial concentration of one of the point defects. This defect essentially vanishes and only the second specie survives. The type of the surviving defect is determined by the concentration of vacancies and interstitials prior to recombination, which is governed by the flux of the defects towards the region of effective recombination.

Vacancies and interstitials in the crystal diffuse by Fickian diffusion and thermodiffusion, and are drifted by the growing crystal. Voronkov assumed that the equilibrium concentrations of interstitials $C_I^{eq}(T_m)$, and vacancies $C_V^{eq}(T_m)$, at the melting temperature T_m are comparable, but $C_V^{eq}(T_m)$ is somewhat higher than $C_I^{eq}(T_m)$. Therefore, the drift flux of vacancies should be larger than that of interstitials. The drift (convection) is thus responsible for the supply of vacancies into the crystal, and the flux of vacancies into the crystal is proportional to the crystal growth rate, v. The diffusion flux is proportional to the concentration gradient which arises from the defect recombination processes. This, in turn, is determined by the axial temperature profile above the melt-crystal interface. The diffusion flux is therefore scaled in proportion with the temperature gradient, G. It is assumed that the diffusion coefficient of the self-interstitials is larger relative to that of vacancies. Hence, diffusion is responsible for the supply of interstitials into the crystal and the flux of interstitials into the crystal is proportional to the temperature gradient, G.

The concentration of the point defects above the melt-crystal interface is determined by the competition of the drift flux supplying vacancies and the diffusion flux supplying interstitials. As these fluxes are proportional to the crystal growth rate v, and to the temperature gradient G, respectively, the v/G ratio determines the type and concentration of point defects which survive the recombination stage. Point defect recombination proceeds at a significant rate around 1300°C. A typical temperature profile in the crystals grown in modern CZ crystal pullers results in a recombination length of about 2 - 3 cm from the melt-crystal interface. The v/G ratio in fact determines whether the silicon crystal a few centimeters above the melt-crystal interface (or at about 1300°C) contains vacancies or interstitials.

There is a critical value of the v/G parameter corresponding to the state when the drift and diffusion fluxes (or the vacancy and interstitial fluxes) are roughly balanced. In such a case the concentrations of vacancies and interstitials before the recombination are comparable and recombination leaves essentially defect-free crystal. The critical value of v/G, also referred to as ξ_t, is given by the point defects properties near the melting temperature. The critical ratio ξ_t separates two cases: (a) when the crystal growth process results in $v/G < \xi_t$, defect recombination results in a crystal populated by excess self-interstitials; or (b) when $v/G > \xi_t$, the crystal contains excess vacancies. The two cases are known as the interstitial-type crystal, and the vacancy-type crystal, respectively.

The surviving point defects become further supersaturated during cooling of the growing crystal which results in formation of bulk defects – vacancy-type defects in the vacancy-type crystal, and interstitial-type defects in the interstitial-type crystal. Also other phenomena such as oxygen precipitation may differ substantially between the vacancy-type and interstitial-type crystals. A silicon matrix rich in vacancies can easily absorb silicon interstitials emitted during the growth of oxide precipitates, while emission of interstitials into a matrix already rich in interstitials is energetically unfavorable. Oxygen precipitation in a vacancy-type silicon is generally stronger as compared to oxygen precipitation in an interstitial-type material. Taking into consideration the influence of crystal defects on wafer properties and device performance one realizes the importance of the crystal growth process, which determines the value of v/G.

5.2 Spatial distribution of defects

Principal distribution of defects in silicon crystal is determined by the relation of the v/G value to the critical value ξ_t. The crystal growth rate at a given crystal length can be considered constant across the melt-crystal interface from a macroscopic viewpoint. The temperature gradient, however, strongly varies across the radius. Due to cooling of the crystal surface by radiation and convective heat loss to the atmosphere the axial temperature gradient at the melt-crystal interface increases from the crystal center to the perimeter. The v/G parameter thus decreases from the center to the perimeter (see Fig. 6). The relation of the ξ_t value to the v/G curve determines whether an interstitial-type, a vacancy-type, or a mixed-type crystal is grown, as illustrated in Fig. 6. The transition between the vacancy-type and interstitial-type portion of the crystal is called the vacancy–interstitial boundary (V-I boundary). As the process conditions vary during the crystal growth, the v/G curve changes also with the crystal length.

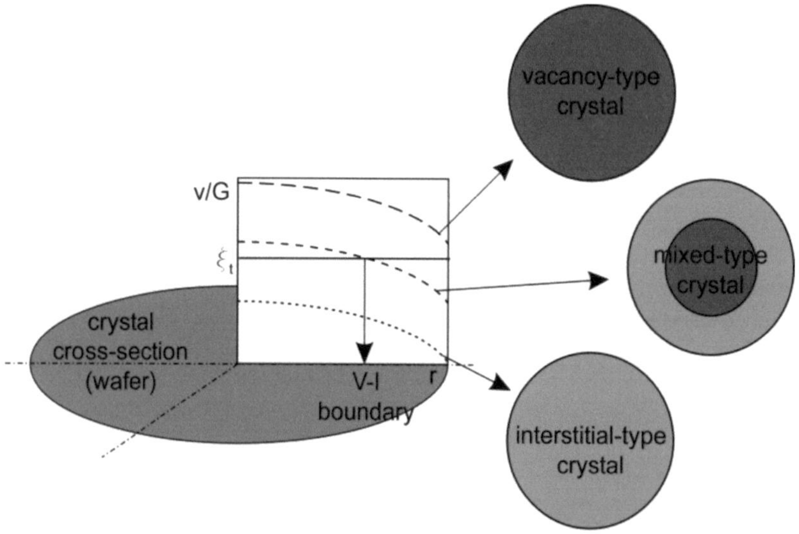

Fig. 6. Left: a typical radial dependence of the v/G ratio and three possible relations of the $v/G(r)$ curve to the critical parameter ξ_t. Right: crystal (wafer) types determined by the relation of $v/G(r)$ to ξ_t. The cross-section image corresponds to a wafer sliced from the crystal.

As noted above, formation of bulk defects in silicon is strongly influenced by the type of the material. The nature of the crystal from the defect point of view (distribution of the vacancy-type and interstitial-type regions) in principle determines the presence and distribution of bulk defects in the silicon wafers. Today's FZ silicon crystals are typically of fully vacancy-type due to high growth rate which is typically 2 - 3 times higher compared to CZ silicon crystals. On the other hand, CZ silicon crystals of all three types shown in Fig. 6 can be found. First CZ silicon crystals were of the interstitial-type, due to a relatively high value of G in the crystals of small diameter. Today's 6" and 8" crystals produced at a rather high pull rates are usually of vacancy-type. The 300 mm crystals may be of the mixed-type and even larger crystals may become fully interstitial-type because of the pull rate which has to be decreased when increasing the crystal diameter up to these levels (von Ammon et al., 1999). Doping elements and their concentration in the crystal can significantly influence formation of crystal defects and shift the process in favor of either vacancy-type or interstitial-type (Borionetti et al., 2002).

5.3 Oxygen precipitation

Oxygen precipitation, i.e., the formation and growth of oxide precipitates from the supersaturated solid solution of oxygen in silicon can be characterized as a two-step process consisting of a nucleation stage and a growth stage. A comprehensive introduction on oxygen precipitation theory can be found in (Shimura, 1994; Borghesi, 1995).

Nucleation is the process of formation of aggregates of a small number of oxygen atoms - the nuclei (or precipitate embryos). Due to relatively low annealing temperatures the supersaturation is high but the diffusivity of oxygen is low. The change in the concentration of interstitial oxygen in the silicon matrix is usually negligible and mainly the precipitate density is established. Precipitate growth (often referred to as the precipitation stage) is caused by diffusion of oxygen atoms and attachment to the existing nuclei at higher temperatures. While the generation of new nuclei during the precipitation stage is negligible the existing nuclei grow substantially while forming oxide particles of sizes reaching up to the micrometer scale. The supersaturation in this stage is lower but the diffusivity of oxygen is much higher than in the nucleation step. During precipitation the reduction in the concentration of interstitial oxygen may be dramatic.

Annealing Step		Precipitated oxygen [%]
750°C/10h	-	1.0 ± 0.5
750°C/40h	-	46 ± 3
750°C/10h	1050°C/20h	91.4 ± 0.5

Table 1. Influence of annealing temperature and time on the amount of precipitated oxygen in a lightly boron doped wafer with an interstitial oxygen content of 9.7×10^{17} cm^{-3}.

The influence of various annealing steps on the reduction of interstitial oxygen is shown in Table 1. Although shorter annealing at low temperature (750°C/10h) reduces the interstitial oxygen concentration only by about 1%, the nuclei have sufficient size for further growth during a precipitation step at 1050°C. It is also shown that even low temperature annealing can reduce interstitial oxygen significantly if the annealing time is extremely long and the initial interstitial oxygen concentration is sufficiently high.

Capturing the nucleation stage has been the most difficult problem over the decades of oxygen precipitation studies. The precipitate nuclei are too small to be observed directly therefore only indirect observations have been used for model validations. Further complications resulted from the uncertainties in the thermo-physical properties of the various defects involved in the nucleation phase. The topic has been treated by numerous authors who assumed both homogeneous and heterogeneous nucleation mechanisms (e.g., Borghesi, 1995, and references therein). Regardless of the particular nucleation mechanism there is a parameter crucial for all theoretical considerations – the critical radius of the nucleus which determines the temperature stability of the nucleus. Nuclei of radius smaller than the critical radius corresponding to the actual temperature tend to dissolve while the larger ones continue to grow.

In order to overcome the lack of analytical methods for identifying the nucleation stage the experimental studies often use various thermal treatments for growing the oxide precipitates to a detectable size and deducing the nucleation processes from the precipitation behavior. The early work resulted in the establishment of procedures for optimization of oxygen precipitation, e.g. for intrinsic gettering. Classical precipitation treatment of silicon wafer consists of a low temperature nucleation step (~700°C) followed by a high temperature precipitation step (~1000°C).

While the nucleation stage of oxygen precipitation remained rather mysterious, the growth of oxide precipitates became experimentally well-described. As shown by many authors oxide precipitates formed during wafer thermal treatments usually show the diffusion-limited growth according to the theory of Ham (Ham, 1958). The precipitate growth controlled by oxygen diffusion results in a square-root dependence of the precipitate size on annealing time at a given temperature. However, due to several assumptions, Ham's theory is not always applicable. The early works usually considered silicon wafer as "clean" input material. However, it has been gradually recognized that some nuclei are formed already during the silicon crystal growth process, and the importance of the thermal history and intrinsic point defects has been recognized.

State-of-the-art investigations of defect formation in silicon thus start from the early beginnings of the crystal growth process, and the interplay of all intrinsic and extrinsic point defects present in the crystal must be considered.

5.4 OISF ring

An interesting feature related to oxygen precipitation during crystal growth is observed on silicon wafers of the mixed-type after oxidation of the surface. Rapid oxidation of the wafer surface (wet oxidation) results in the injection of silicon interstitials below the oxidized surface. These interstitials can condense into stacking faults on grown-in oxide precipitates in the near-surface region. A ring of relatively large oxygen precipitates can be formed near the vacancy-interstitial boundary during the crystal growth process (Voronkov, 2008). This phenomenon results in a ring-like distribution of stacking faults (Hasebe et al., 1989) during subsequent wet oxidations. The observed feature (see Fig. 7) is referred to as the "oxidation-induced stacking fault ring" (OISF ring).

The OISF ring can be easily visualized on an oxidized wafer by stripping the oxide and selectively etching the surface. The etched OISFs are visible by optical microscopy and the OISF ring is detectable even by the naked eye. The OISF ring is located on the edge of the vacancy-type core in mixed-type crystals close to the V-I boundary. Due to the ease of detection it has been widely used for delineating the V-I boundary and investigating the defect distribution in silicon crystals.

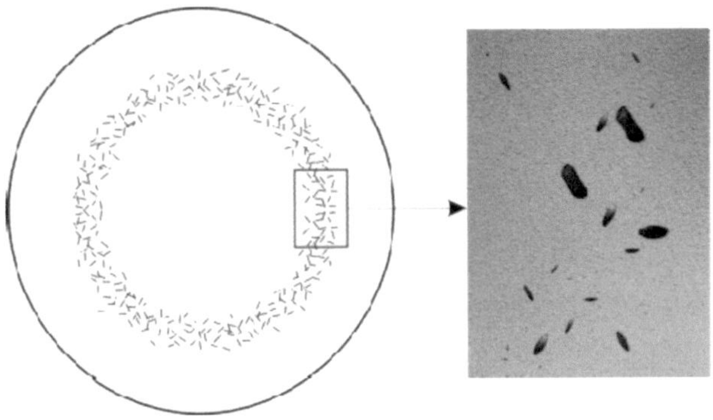

Fig. 7. OISF ring on the surface of a silicon wafer.

6. Methods of studying crystal defects

In order to capture the distinctive nature of crystal defects and the wide range of their size and density, various methods are used for defect analyses. The most common techniques are briefly introduced in following section.

6.1 Preferential etching

Analytical methods based on preferential etching of silicon are frequently used for studying defects in silicon technology. All modern defect etchants for silicon are based on the chemistry of formation and dissolution of silicon dioxide, and therefore all etchants include hydrofluoric acid. The dissolution process is enhanced at and around defects due to weakened bonds and coalescence of impurities around them. The electronic properties may play a role, too. Etching anisotropy with respect to various crystallographic planes (atom density) also has to be taken into account. In general, the etch rate is lowest for the {111} crystal planes. The most common etching solutions are those of Wright, Sirtl, Schimmel, Yang, Secco, MEMC, and Dash. Individual solutions are optimized for various surface orientations, conductivity types or resistivity levels. For more details on use and application see (ASTM standard F 1809–02, 2003).

Preferential etching allows for rather simple detection of dislocations, stacking faults, slip lines, and oxide precipitates as shown in Fig. 8. Individual defects on the etched surface may be observed by optical microscopy or SEM, while defect distributions across a wafer can be observed by the naked eye in most cases.

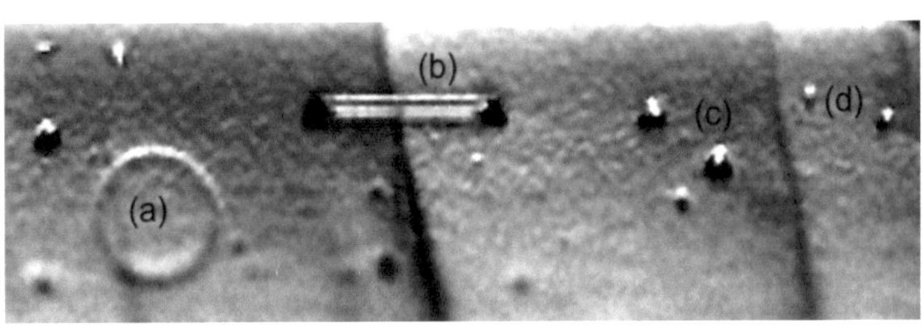

Fig. 8. Defects in silicon revealed by preferential etching of a cleaved surface. (a) A stacking fault in the cleavage plane, (b) a stacking fault tilted with respect to the cleavage plane (note the bounding dislocation), (c) dislocations, (d) oxide precipitates.

6.2 X-ray topography

X-ray topography (Lang, 1978) is based on Bragg diffraction. A monochromatic X-ray beam incident on a surface is reflected by the atomic planes of the crystalline sample. Constructive interference occurs when the Bragg condition is fulfilled. The diffraction pattern is recorded on the detector either in reflection geometry (Bragg geometry) or transmission geometry (Laue geometry).

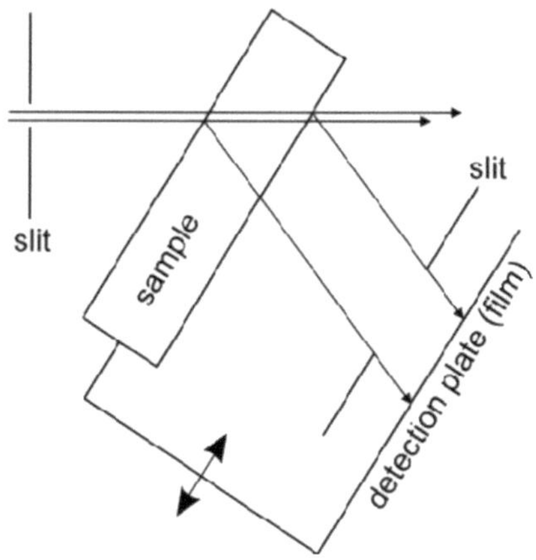

Fig. 9. Schematic layout of the X-ray section and projection topography in Laue geometry.

In the case of section topography, a very narrow beam of dimensions in the order of 10 μm is used. The section topographs therefore investigate only a small volume (section) of the sample. The beam is diffracted at different depths, each one contributing to the image on a different location on the detector. Section topography is therefore used for depth-resolved defect analysis. When the sample and detector are simultaneously moved with respect to the incident beam, the sample is scanned over a larger volume and a projection topograph is formed. The schematic configuration of X-ray topography in Laue geometry is shown in Fig. 9. A homogeneous crystal lattice generally results in a homogeneous distribution of intensity across the topograph. Section topography, however, results in spatial variations in the intensity of diffracted rays even in a perfect crystal (so called Pendelösung fringes) due to the dynamic nature of diffraction and a small investigated volume. Irregularities of the crystal lattice, such as defects and strain, are captured as a distortion of the image of the perfect crystal. X-ray topography can detect irregularities such as phase boundaries, defective areas, cracks, scratches, growth striations, and most of the common crystal defects such as dislocations, oxide precipitates, stacking faults, and interstitial-type defects. An example of X-ray topography results is shown in Fig. 16.

6.3 Fourier transform infrared spectroscopy

Fourier transform infrared spectroscopy (FTIR) is the most common technique for determining the concentration of oxygen in silicon materials. The polychromatic source radiates infrared (IR) light through a Michelson interferometer. The radiation is transmitted through the silicon wafer and directed to a detector. Undoped silicon is transparent to the IR radiation. Impurities cause localized absorption from lattice vibrations. The absorption due to the anti-symmetric vibrational mode at 1107 cm^{-1} has been assigned to interstitial oxygen and it is thus used for determining the oxygen concentration. In order to suppress the influence of the silicon lattice vibrations and to allow the quantitative analysis, a reference spectrum is subtracted from the measured spectrum. The reference spectrum is measured on float-zone silicon prepared with very low oxygen content (less than 1×10^{16} cm^{-3}). After signal processing and taking the wafer thickness into account, the intensity of the oxygen absorption peak is calculated, and thus, the concentration of oxygen may be determined as the product of the peak intensity and a tabulated conversion factor. The selected conversion factor has to be reported together with the concentration value. The oxygen concentrations presented in this work are given in accordance with the ASTM standard (ASTM Standard F1188-02, 2003) commonly referred to as "new ASTM". The FTIR method is not applicable for heavily-doped silicon as this material is not transparent to IR radiation due to the high concentration of free carriers.

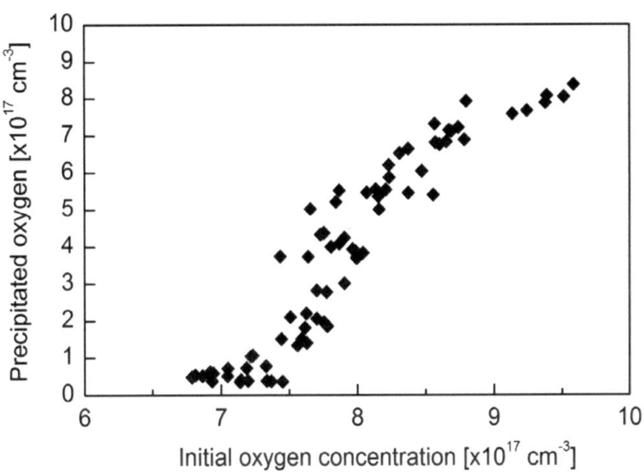

Fig. 10. An example of an "S-curve" characterizing oxygen precipitation in the crystal grown by a particular crystal growth process.

As the contribution of precipitated oxygen atoms to the absorption peak of interstitial oxygen is small (absorption energies are shifted), the FTIR measurements of interstitial oxygen concentration can be used for characterization of oxygen precipitation. The difference in the content of interstitial oxygen prior to, and after, the heat treatment is the measure of oxygen precipitation, relevant for the specific sample material and thermal cycles. For overall characterization of a crystal grown by a particular growth process the dependence of the precipitated oxygen on the initial oxygen concentration is constructed for a standardized thermal cycle. The dependence is called the "S-curve" because of its typical sigmoidal shape (Fig. 10). Another application of the S-curves is the comparison of different thermal cycles (e.g., in IC production) with respect to oxygen precipitation.

6.4 Precipitation test

This test is aimed at assessing the precipitation behavior in the wafers. Oxide precipitates formed in the crystal during growth (grown-in precipitates) are usually too small to be detected directly as noted previously. Oxygen precipitation is therefore evaluated in silicon wafers subjected to a standardized thermal treatment sequence.

Generally, two types of the heat treatment are used: (a) single step annealing at high temperature when the grown-in oxide precipitates are allowed to grow, and (b) two-step annealing which includes also a low-temperature nucleation step. Typical thermal cycles for such heat treatments are the 1050°C/16h annealing, and 750°C/4h + 1050°C/16h annealing. The extent of oxygen precipitation is evaluated as the difference in oxygen concentration before, and after, the thermal treatment as measured by FTIR, or by "cleave-and-etch" analyses. In such cases, the wafer cross-section is prepared by cleaving or grinding, the sample is preferentially etched, and the defect distribution and density is evaluated under the microscope. An example of the cleave-and-etch analysis is shown in Fig. 11.

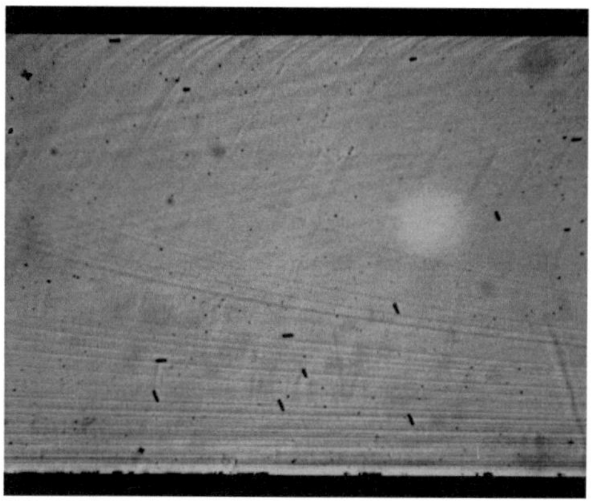

Fig. 11. A wafer cross-section prepared by the cleave-and-etch method. The wafer front surface (polished) is at the top of the figure. The analysis reveals oxide precipitates (dots) and stacking faults (short lines) in the bulk of the wafer.

6.5 OISF test

The OISF test uses the fact that silicon interstitials injected below the wafer surface during oxidation under usual conditions needs a suitable nucleation site for formation of OISFs (Ravi & Varker, 1974). The test wafer is subjected to an oxidation cycle (e.g., 1000°C/1hr in a "wet" atmosphere), then the grown oxide is stripped with HF acid and the surface is preferentially etched. The stacking faults observed on the surface decorate the nucleation centers in the near-surface wafer region. The nucleation sites can arise from the sub-surface damage caused by the wafer manufacturing processes, handling, etc., from wafer contamination, and defects such as dislocations and oxygen precipitates. The OISF test is therefore used as a general method for assessing the quality of the wafer surface. If the manufacturing process is optimized and wafer contamination is avoided, the OISF test can be used for delineation of crystal defects in the wafer as in the case of the OISF ring.

6.6 COP test

Delineation of COPs is based on SC1 cleaning and "particle" inspection methods (Ryuta et al., 1990). As the typical COP dimensions do not allow direct observation on the polished wafer surface by common industrial equipment, the SC1 cleaning chemistry is used to etch the COPs and enlarge them to dimensions greater than the detection limit of the instruments. The size of COP defects after the SC1 treatment is typically 100 - 300 nm and the surface density is of the order of 10 to 100 cm^{-2}. An example of the measurement output is shown in Fig. 5c.

7. Engineering of crystal defects in silicon

Defect engineering in production manufacture of silicon wafers comprises modifications of the crystal growth process and dedicated heat treatment of the wafers. The primary objective is usually the control of oxygen precipitation and the associated defect structures. The specific requirements on the spatial distribution, size and density of oxide precipitates may differ for various devices, but the basic features are essentially the same. It is desirable to achieve a defect-free region near the surface of the wafer so that the devices fabricated there do not suffer from the electrical effects resulting from the presence of the defects. On the other hand, a high density of oxide precipitates is usually required in the bulk of the wafer in order to provide effective intrinsic gettering. The near-surface region free of oxide precipitates is called a "denuded zone".

7.1 Denuding and precipitation of oxygen

The classical procedure for forming the desired denuded zone and achieving intrinsic gettering utilizes a three-step thermal treatment (Nasagawa et al., 1980) demonstrated in Fig. 12a. The first step is the denudation at high-temperature which results in dissolution of the grown-in oxide precipitates and subsequent out-diffusion of oxygen from the wafer surface. Oxygen diffusivity above 1100°C is sufficiently high to result in a depleted zone of several tens of micrometers depth, while the oxygen concentration in the bulk far from the wafer surface retains its initial value. The shape of the transition from the solubility limit value at the surface to the bulk value is determined by the annealing time (Fig. 12b, $C_0 = 8\times10^{17}$ cm^{-3}).

The second step causes the nucleation of new oxide precipitates at temperatures around 700°C. During the third step, called the precipitation stage, nucleated oxide precipitates grow in the regions of sufficiently high oxygen concentration, while no oxide precipitates grow in the depleted region near the surface and the denuded zone is formed. The experimentally observed limit of oxygen concentration necessary for appreciable precipitation is around 6.6 - 7 $\times10^{17}$ cm^{-3} (see Fig. 12b) and varies depending on the type and concentration of dopant species.

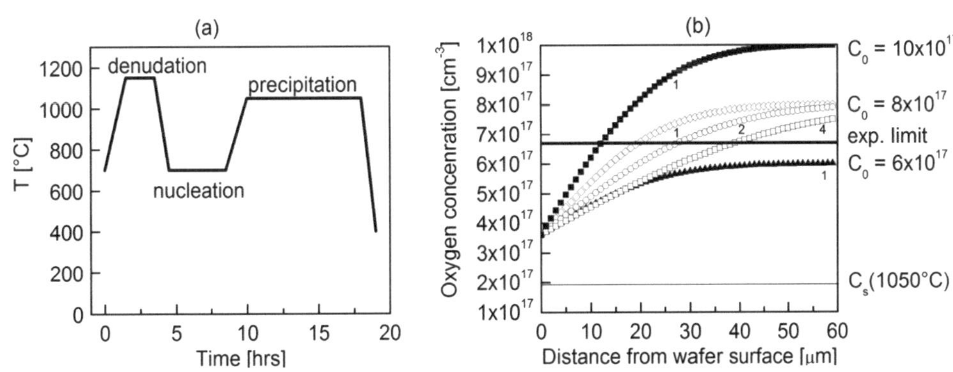

Fig. 12. (a) An example of a thermal cycle for formation of the denuded zone and effective intrinsic gettering. (b) The concentration of oxygen in the wafer with initial oxygen concentration C_0 denuded at 1150°C. Denuding duration in hours is noted in the plot below the curves. The thin horizontal line in (b) represents the equilibrium concentration of oxygen at the temperature of the precipitation step (1050°C); the thick solid line represents the experimental limit of oxygen precipitation. Out-diffusion profiles in panel (b) were calculated after (Andrews, 1983).

The depth profile of the precipitated oxygen in the silicon wafer can be calculated on the basis of the interstitial oxygen depth profile (Borghesi et al., 1995). As shown in Fig. 12b, the oxygen concentration after out-diffusion provides certain supersaturation even in the denuded zone, but the amount of precipitated oxygen depends more strongly on the density of nuclei than on the level of oxygen supersaturation during precipitation. This latter point is critical in making material stable to large thermal budget wafer processing. The nucleation rate during the second step is very strongly reduced for low oxygen concentrations in denuded zone, e.g., the rate was reduced by more than ten times for oxygen concentration reduced by 2×10^{17} cm^{-3} (Borghesi et al., 1995). As a result the region near the surface is precipitate-free and the bulk region is precipitate-rich. The transition between the bulk and the surface is not sharp; it follows the initial oxygen profile (see the dependence in Fig. 14 below).

7.2 Wafer annealing

It was found that high temperature annealing can help control and manage also other types of crystal defects, not simply oxide precipitates. The COP defects on the wafer surface can be annihilated during high temperature annealing in hydrogen (Nadahara et al., 1997), or in argon ambient (Adachi et al., 1998). As shown in Fig. 13, proper treatment of the polished wafer can significantly reduce or even eliminate COP defects on the surface of polished silicon wafers. However, attention has to be paid to oxygen precipitation in the bulk of the wafer, since during the high temperature annealing applied for annihilation of COPs the grown-in oxide precipitates tend to dissolve. Subsequent oxygen precipitation in the annealed wafer may become suppressed and the wafer can loose its gettering capability.

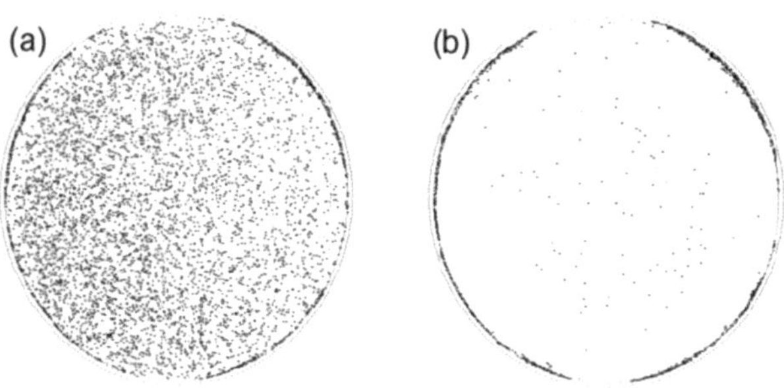

Fig. 13. COP distributions measured by a particle detector (lower dimensional detection limit is 0.13 μm) on vacancy-type wafers (a) before and (b) after hydrogen annealing (1180°C/2hr).

7.3 Optimization of oxygen concentration in the crystal

The denuded zone and intrinsic gettering are closely related to oxygen precipitation, which in turn strongly depends on the oxygen concentration in the wafer. Fig. 12 demonstrates the thickness of the denuded zone, and Fig. 14 shows the density of oxide precipitates, both as the function of oxygen concentration. As the oxygen concentration in silicon is determined during the crystal growth process, control of the oxygen concentration in the growing crystal is one of the primary tasks for crystal growers.

As was previously discussed, oxygen is incorporated into the crystal at the melt-crystal interface, while its concentration in the crystal relates to concentration in the melt. The concentration of oxygen in the melt below the crystal is determined by the source (crucible materials), by the effectiveness of the transport in the melt, and by the strength of the sink, i.e., by the dissolution rate of the silica crucible, by the melt flow, and by the evaporation from the melt surface. Critical process parameters driving these phenomena are the hot zone design, the crucible rotation rate, and gas atmosphere parameters. All of the parameters are usually optimized to reach a desired oxygen concentration in the whole crystal with limited variation and good reproducibility. Typical concentrations of oxygen in CZ silicon crystals range from about 11×10^{17} cm^{-3} down to about 6×10^{17} cm^{-3}; special designs of the hot zone (usually utilizing shields for gas flow control) allow decreasing oxygen concentration slightly below 5×10^{17} cm^{-3}.

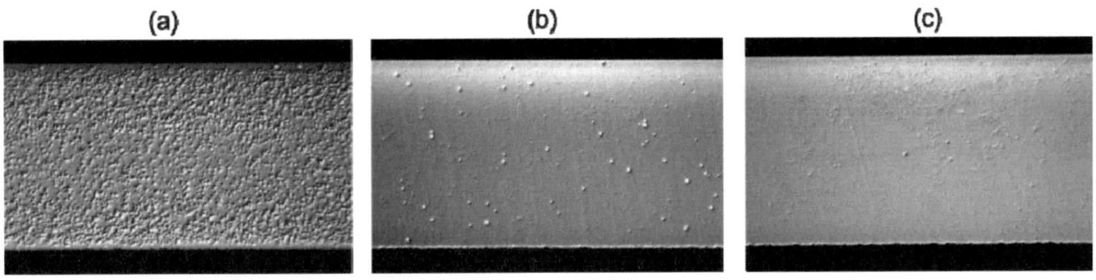

Fig. 14. Oxide precipitates delineated by preferential etching after 750°C/4hr + 1050°C/16hr annealing in lightly phosphorus-doped silicon wafers with oxygen content of (a) 8×10^{17} cm^{-3}, (b) 7×10^{17} cm^{-3}, (c) 6×10^{17} cm^{-3}.

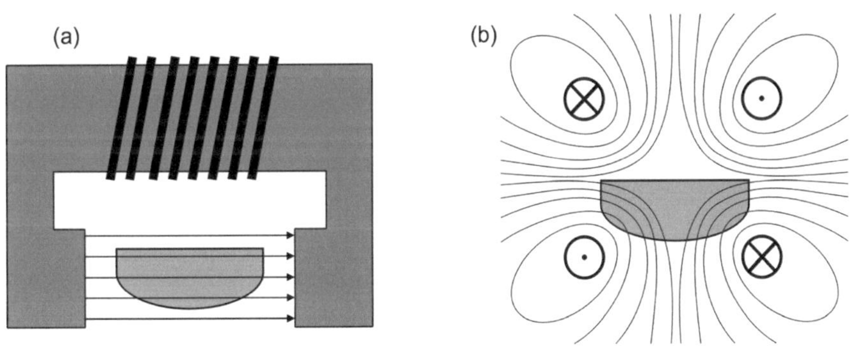

Fig. 15. Schematic drawing of (a) transverse and (b) cusp magnetic fields for control of the melt convection during crystal growth.

There are applications which require extremely low oxygen concentrations or very uniform distributions of oxygen in the crystal. These requirements are beyond the capability of most standard Czochralski crystal growth processes. Magnetic Czochralski process (MCZ) has to be used to achieve reduced concentrations of oxygen in the silicon crystals. The principal aspect of MCZ lies in the influence of the Lorentz force on the convection in the melt, which arises from the interaction of magnetic field with the fluid flow of the highly electrically conductive silicon melt. The effect of various kinds of magnetic fields has been studied extensively (Hurle, 1993). Based on many investigations, only transverse (horizontal) and cusp magnetic fields are used in industrial applications. The two configurations are schematically shown in Fig. 15. The horizontal magnetic field is used for growth of large diameter crystals, while for smaller diameter crystals the cusp field is applied. The two most important effects of the magnetic field generally consist in damping of micro-scale growth rate instabilities at the melt-crystal interface through damping of melt temperature fluctuations, and in the possibility of tuning the incorporation of impurities into the growing crystal. Application of the magnetic field adds additional degrees of freedom to the crystal growth process which widens the capability of the process, but simultaneously brings complexity related to the higher technological level of the process.

7.4 Nitrogen doping

Extremely low oxygen concentration allows for the formation of very deep denuded zones, but also results in negligible oxygen precipitation in the bulk. Oxygen precipitation is generally low also in heavily-doped n-type wafers due to enhanced evaporation of dopant atoms in the form of oxides. For applications requiring effective intrinsic gettering this drawback can be solved by nitrogen doping. Introducing nitrogen in silicon crystal results in enhanced precipitation of oxygen during subsequent wafer annealing due to stable nuclei formed during the crystal growth, an increased concentration of free vacancies available for oxygen precipitation and consumption of the interstitials emitted during the precipitate growth (von Ammon et al., 2001). Effective intrinsic gettering then can be achieved even at low oxygen concentrations or under conditions of unfavorable thermal treatments, such as argon or hydrogen annealing (Ikari et al.). The effect of nitrogen doping on oxygen precipitation is demonstrated in Fig. 16.

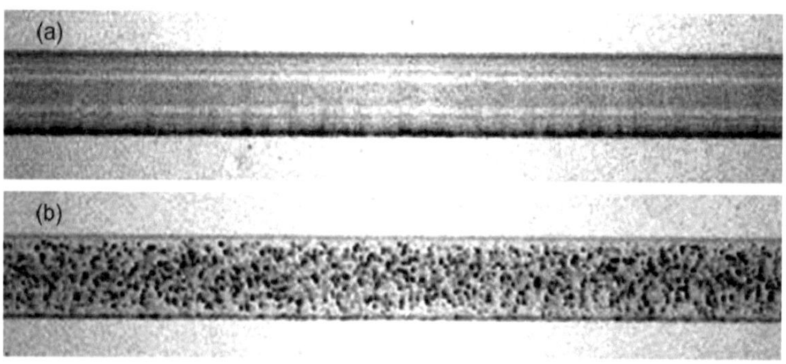

Fig. 16. An X-ray section topograph showing the effect of nitrogen doping on oxygen precipitation in heavily antimony-doped wafers with low oxygen concentration. (a) Standard wafer. (b) A wafer co-doped with nitrogen at the level of 5×10^{14} cm^{-3}. Wafer (a) is "defect-free". The horizontal Pendelösung fringes are clearly visible (see Section 6.2). Strain related to the polysilicon layer on the wafer backside (bottom) results in dark features in the topograph. A section topograph of wafer (b) shows oxide precipitates (dark spots) and residual damage on the wafer backside (bottom). The wafer thickness is 625 μm.

7.5 Vacancy-controlled denuded zone

An elegant solution to overcome the dependence of denuded zone parameters on the oxygen concentration was developed by Falster et al. (Falster et al., 1998). The concept of so-called "Magic Denuded Zone" (MDZ) utilizes the importance of vacancies for oxygen precipitation. The concentration profile of vacancies in the wafer is modified via rapid thermal annealing (RTA). Annealing at temperatures of about 1200°C forms a population of vacancies and interstitials in equal concentrations through the generation of Frenkel pairs. Point defects reach their equilibrium concentrations in a short period (a few tens of seconds) throughout the thickness of the thin wafer by diffusion to and from the surface. During cooling of the wafer to low temperatures, the equilibrium concentrations of point defects decreases, providing a driving force for enhanced diffusion and recombination. This favors diffusion to the surface, where equilibrium is maintained. However, when the cooling rate is fast enough, the diffusion is effective only in the near-surface region. In the bulk of the wafer, only point defect recombination takes place leaving behind a population of excess vacancies at a concentration given by $C_v^{eq}(T_a)$ - $C_i^{eq}(T_a)$, where T_a is the annealing temperature. The MDZ wafer finally contains an essentially vacancy-free surface and a transition region below the surface with steadily increasing vacancy concentrations up to the bulk level.

When an MDZ wafer passes through a high temperature anneal oxide precipitates nucleate and grow only in the region with sufficient vacancy concentrations which results in the denuded zone at the wafer surface. The depth of the Magic Denuded Zone can reach even deeper into the bulk than the classical denuded zone, but the solution has another major advantage- the result is practically independent of oxygen concentration in the wafer. Although the MDZ comprises an extra treatment of the wafer it suppresses the need for complicated oxygen control in CZ silicon crystals.

7.6 Optimization of v/G

Sometimes annealing of the wafer cannot meet the requirements of the wafer processing thermal budget. In such cases, formation of defects in the wafer can be adjusted by controlling the distribution of point defects in the growing crystal. As described above, vacancy-type or interstitial-type regions are formed in the crystal depending on the v/G parameter. The different regions within the crystal contain different types of defects and also substantially differ in oxygen precipitation characteristics. Tuning of the crystal growth rate v, and/or the temperature gradient G, can shift the V-I boundary and so influence defect formation.

There are often issues with excessive oxygen precipitation in heavily-boron doped wafers which results in electronic devices with excessive leakage currents. Analyses showed that the strong precipitation is constrained within the vacancy-type region in the portion of the crystal with the highest oxygen concentration (Valek et al., 2007, 2008). If the oxygen profile of the particular crystal growth process cannot be altered, then the crystal growth rate may be optimized to eliminate the vacancy-type region from the crystal (Fig. 17). The low growth rate shifts the v/G curve below the critical value ξ_t (see Fig. 6), the material becomes interstitial-type in full cross-section and oxygen precipitation is suppressed.

7.7 Computer simulations of crystal growth

Distribution of crystal defects in silicon wafers is significantly predetermined by the formation of defects during the crystal growth. Incorporation of point defects into the crystal and formation of crystal defects are driven by the time evolution of the temperature field, which governs the solubility and the diffusivity of the various species. However, measurement of the temperature field (necessary for its control) in the growing crystal is practically impossible. The analysis of grown-in defects after the crystal growth is very difficult due to their small size which is often below the detection limit of common techniques. Therefore, detailed numerical models have been developed to overcome these obstacles by computer simulations which help to describe and predict defect formation from the crystal growth up to the point of wafer processing and device manufacturing (Brown et al., 2001; Dupret & Van den Bogaert, 1994; Kalaev et al., 2003).

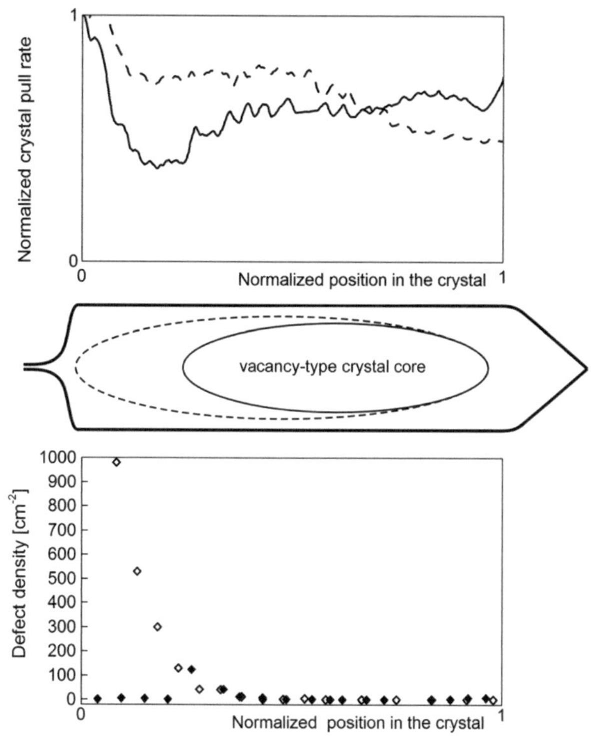

Fig. 17. An example of defect engineering through *v/G* control. The axial profile of the crystal pull rate (top), the distribution of point defects (middle), and the corresponding defect density (bottom) of the original (dashed lines, open symbols) and optimized (solid lines, solid symbols) crystal growth process.

Simulations of *v/G* are the simplest method for prediction of defect formation in the crystal. Comparing the simulated *v/G* values with the critical *v/G* value allows the identification of the vacancy-type and interstitial-type regions of the crystal separated by the vacancy-interstitial boundary. For example, a complicated distribution of OISFs on the oxidized surface of heavily boron-doped wafers was identified (Valek et al., 2007, 2008). The OISFs formed a pattern of two concentric rings (radii R_1 and R_2 in Fig. 18) on wafers from the beginning of the crystal. With increasing distance from the seed end of the crystal the OISF density decreased and the rings gradually disappeared.

In order to identify origin of the rings the temperature gradient, *G*, at the melt-crystal interface was modeled at several positions in the crystal using FEMAG code (Dupret & Van den Bogaert, 1994) and the corresponding radial *v/G* curves were calculated. Using the ξ_t values the V-I boundary was constructed. In Fig. 18 the V-I boundary was found to go through the whole crystal. It was concluded that also the OISF pattern should be formed throughout the whole crystal. This conclusion was supported by a special OISF test with pre-annealing of the wafers. This annealing was performed to enlarge the grown-in oxide precipitates to a supercritical size for the OISF test. Finally, the origin of the "strange" OISF pattern was explained in terms of oxygen concentration and thermal history, see (Valek et al., 2007, 2008), for more details. This work demonstrated the value of computer simulations of crystal growth as a necessary tool for defect engineering.

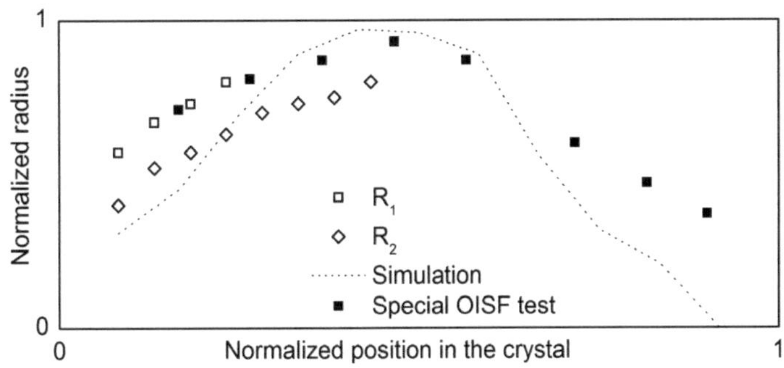

Fig. 18. Plot of the radius of the rings forming the OISF pattern, the simulated V-I boundary, and the radius of the OISF pattern delineated by a special OISF test (with pre-annealing).

8. Conclusion

Crystal defects are formed in silicon during the growth of the single crystals and processing of the wafers. Depending on their nature, density and size, these defects may substantially influence silicon material properties and consequently strongly impact process of device manufacturing. We provided the reader with a brief outlook of crystal defects in silicon and introduced the mechanisms of their formation. Knowledge of these mechanisms is the fundamental requirement for the art of defect engineering.

Crystal defects can be controlled from the early beginning by influencing their formation during the crystal growth. We discussed several aspects of controlling oxygen concentration in Czochralski silicon crystals, which is the basic method for control of oxygen precipitation in silicon wafers. Defect formation can be adjusted by optimization of the crystal growth process with respect to the v/G parameter. Tuning the crystal pull rate v and the hot zone design, which determines the temperature gradient G, one can produce either silicon containing vacancy-type defects and showing enhanced oxygen precipitation or silicon containing interstitial-type defects and showing suppressed oxygen precipitation. Sometimes it can be advantageous to modify the processes of defect formation by doping of the silicon crystal with an extra element like carbon or nitrogen to promote oxygen precipitation. Since the material properties obtained by application of the above mentioned methods may not be invariable or sufficient, it may be necessary to treat also the silicon wafer.

Another reason for treatment of the wafer can be an effort to modify properties of wafer surface. For example oxygen precipitation and also appearance of the vacancy-type defects, the COPs, can be controlled by high temperature annealing of polished silicon wafers in optimized ambient. In this manner a wafer which contains a defect-free near-surface denuded zone and an optimized defect density in the bulk of the wafer providing effective gettering of metal impurities can be obtained.

We showed that various methods of defect engineering can be utilized during the growth of silicon crystals and manufacturing of wafers to produce silicon wafers of desired quality. Properties of the wafers have to be optimized for specific requirements of individual types of electronic devices and integrated circuits. Following the development of electronic industry, engineering of crystal defects remains an inseparable and never-ending part of the silicon manufacturing technology.

9. Acknowledgment

This work was partially supported by grant FR-TI3/031 awarded by the Ministry of Industry and Trade of the Czech Republic. Critical reading of the manuscript by Dr. John M. Parsey, Jr., is highly acknowledged.

10. References

Abe, T.; Harada, H. & Chikawa, J. (1983). Swirl Defects in Float-Zoned Silicon Crystals, *Physica B+C*, Vol. 116, No. 1-3, (February 1983), pp. 139-147, ISSN 0378-4363

Adachi, N.; Hisatomi, T.; Sano, M.; & Tsuya, H. (1998). Reduction of Grown-in Defects By High Temperature Annealing, In: *Semiconductor Silicon 1998*, Huff, H.R.; Gösele, U.; Tsuya, H. (Eds.), pp. 698-706, The Electrochemical Society, ISBN 978-1-56677-193-1698, Penington, NJ, USA

Andrews, J. (1983). Oxygen Out-Diffusion Model for Denuded Zone Formation in Czochralski-Grown Silicon with High Intersitital Oxygen Content, In: Defects in Silicon, Murray Bullis, W. & Kimerling, L. C. (Eds.), pp. 133-141, The Electrochemical Society, Pennington, NJ

ASTM Standard F1188-02 (2003). *Standard Test Method for Interstitial Atomic Oxygen Content of Silicon by Infrared Absorption with Short Baseline*, ASTM International, West Conshohocken, PA, USA

ASTM standard F 1809–02 (2003), *Standard Guide for Selection and Use of Etching Solutions to Delineate Structural Defects in Silicon*, ASTM International, West Conshohocken, PA, USA

Borghesi, A.; Pivac, B.; Sassella, A. & Stella, A. (1995). Oxygen Precipitation in Silicon, *Journal of Applied Physics*, Vol. 77, No. 9, (May 1995), pp. 4169-4244. ISSN 0021-8979

Borionetti, G.; Gambaro, D.; Porrini, M. & Voronkov, V.V. (2002). In: *Semiconductor Silicon 2002*, Huff, H.R.; Fabry, L. & Kishino S. (Ed.), pp. 505-516, The Electrochemical Society, ISBN 1-56677-374-1, Penington, NJ, USA

Bullis, W.M. (1990). Silicon Material Properties, In: *Handbook of Silicon Technology*, O´Mara, W.C.; Herring R.P.; & Hunt, L.P., (Eds.), 347-450, Noyes Publications, ISBN 0-8155-1237-6, Park Ridge, NJ, USA

Brown, R.A.; Wang, Z. & Mori, T. (2001). Engineering Analysis of Microdefect Formation During Silicon Crystal Growth, *Journal of Crystal Growth*, Vol. 225, No. 2-4, (May 2001), pp. 97-109, ISSN 0022-0248

Czochralski, J. (1918). Ein neues Verfahren zur Messung des Kristallisationsgeschwindigkeit der Metalle, *Zeitschrift für Physikalische Chemie*, Vol. 92, (1918), pp. 219-221

Dupret, F. & Van den Bogaert, N. (1994). Modelling Bridgman and Czochralski growth, In: *Handbook of Crystal Growth*, Vol. 2B, Chapter 15, Hurle, D.T.J. (Ed.), 875-1010, North Holland / Elsevier, ISBN 0-444-81554-6, Amsterodam, Holand

Falster, R.; Gambaro, D.; Olmo, M.; Cornara, M. & Korb, H. (1998). The Engineering of Silicon Wafer Material Properties Through Vacancy Concentration Profile Control and the Achievement of Ideal Oxygen Precipitation Behavior, *Material Research Society Syposium Proceedings*, Vol. 510, pp. 27-36, ISSSN 0272-9172

Föll, H. & Kolbesen, B.O. (1975). Formation and Nature of Swirl Defects in Silicon. *APPLIED PHYSICS A: MATERIAL SCIENCE AND PROCESSING*, Vol. 8, No. 4, (December 1975), pp. 319- 331, ISSN 0947-8396

Graff, K. (2000). *Metal Impurities in Silicon-device Fabrication* (Second, Revised Edition), Springer-Verlag Berlin Heidelberg New York, ISBN 3-540-64213-7, Berlin, Germany

Ham, F.S. (1958). Theory of Diffusion-Limited Precipitation, *Journal of Physics and Chemistry of Solids*, Vol. 6, No. 4, (September 1958), pp. 335-351, ISSN 0022-3697

Hasebe, M. Takeoka, Y. Shinoyama S. & Naito S. (1989), Formation Process of Stacking Faults with Ringlike Distribution in CZ-Si Wafers, *Japanese Journal of Applied Physics*, Vol. 28, No. 11, (January 1980), pp. L1999-L2002, ISSN 0021-4922

Hirth, J.P. & Lothe, J. (1967). *Theory of dislocations*, McGraw-Hill, ISBN 0-521-86436-4, New-York, US

Hurle, D.T.J. (1993). *Crystal Pulling from the Melt*, Springer-Verlag Berlin-Heidelberg New York, ISBN 3-540-56676-7, Berlin, Germany

Ikari, A.; Nakai, K.; Tachikawa, Y.; Deai, H.; Hideki, Y.; Ohta, Y.; Masahashi, N.; Hayashi, S.; Hoshino, T. & Ohashi, W. (1999). Defect Control in Nitrogen Doped Czochralski Silicon Crystals, *Solid State Phenomena*, Vol. 69-70 (1999), pp. 161-166, ISSN ISSN 1662-9779

Itsumi, M. (2002). Octahedral Void Defects in Czochralski Silicon, *Journal of Crystal Growth*, Vol. 237-239, No. 3 (April 2002), pp. 1773-1778, ISSN 0022-0248

Kalaev, V.V.; Lukanin, D.P.; Zabelin, V.A.; Makarov, Yu.N.; Virbulis, J.; Dornberger, E. & von Ammon, W. (2003), *Materials Science in Semiconductor Processing*, Vol. 5, No. 4-5, (August-October 2003), pp. 369–373, ISSN 1369-8001

Keck, P.H. & Golay, M.J.E. (1953), Crystallization of Silicon from a Floating Liquid Zone, *Physical Review*, Vol. 89, (March 1953), pp. 1297

Lang, A. R. (1978), Techniques and interpretation in X-ray topography. In: *Diffraction and Imaging Techniques in Materials Science*, Amelinckx, S.; Gevers, R. & Van Landuyt, J. (Eds.), 2nd ed. rev. (1978), pp 623–714, North-Holland, ISBN 9780444851307, Amsterdam, Holand

Nadahara, S.; Kubota, H. & Samata, S. (1997). Hydrogen Annealed Silicon Wafer, *Solid State Phenomena*, Vol. 57-58 (1997), pp. 19-26, ISSN 1662-9779

Nagasawa, K.; Matsushita, Y. & Kishino, S. (1980). A New Intrinsic Gettering Technique Using Microdefects in Czochralski Silicon Crystal: A New Double Preannealing Technique, *Applied Physics Letters*, Vol. 37, No. 7, (October 1980), pp. 622-624, ISSN 0003-6951

O´Mara, W.C. (1990). Oxygen, Carbon and Nitrogen in Silicon, In: *Handbook of Silicon Technology*, O´Mara, W.C.; Herring R.P.; & Hunt, L.P., (Eds.), pp. 451-549, Noyes Publications, ISBN 0-8155-1237-6, Park Ridge, NJ, USA

Pichler, P. (2004). *Intrinsic Point Defects, Impurities, and Their diffusion in Silicon*, Springer-Verlag Wien New York, ISBN 3-211-20687-6, Wien, Austria

Ravi, K.V. & Varker, C.J. (1974). Oxidation-Induced Stacking Faults in Silicon. I. Nucleation phenomenon, *Journal of Applied Physics*, Vol. 45, No. 1, (January 1974), pp. 263-271, ISSN 0021-8979

Roksnoer, P.J. & van den Boom, M.M.B. (1981). Microdefects in a Non-Striated Distribution in Floating-Zone Silicon Crystals. *Journal of Crystal Growth.* Vol. 53, No. 3, (June 1981), pp. 563- 573, ISSN 0022-0248.

Rozgonyi, G.A.; Deysher, R.P. & Pearce, C.W. (1976). The Identification, Annihilation, and Suppression of Nucleation Sites Responsible for Silicon Epitaxial Stacking Faults, *Journal of The Electrochemical Society*, Vol. 123, No. 12, (December 1976), pp. 1910-1915, ISSN 0013-4651.

Ryuta, J.; Morita, E.; Tanaka, T. & Shimanuki, Y. (1990). Crystal-Originated Singularities on Si Wafer Surface after SC1 Cleaning, *Japanese Journal of Applied Physics*, Vol. 29, No. 10, (October 1990), pp. L1947-L1949, ISSN 0021-4922

Schröder, D.K. (1989). Lifetime in Silicon, In: *Gettering and Defect Engineering in the Semiconductor Technology*, Kittler, M. (Ed.), pp. 383-394, Sci-Tech, ISBN 3-908044-04-9, Vaduz, FL

Shimura, F. (1994). *Oxygen in Silicon*, Academic Press, ISBN 0-127-52142-9, London, UK

Sumino, K.; Harada, H. & Yonenaga, I. (1980). The Origin of the Difference in the Mechanical Strengths of Czochralski-Grown Silicon and Float-Zone-Grown Silicon, *Japanese Journal of Applied Physics*, Vol. 19, No. 1, (January 1980), pp. L49-L52, ISSN 0021-4922

Teal, G.K. & Little, J.B. (1950). Growth of Germanium Single Crystals, *Physical Review*, Vol. 78, (1950), pp 647

Válek, L.; Lysáček, D. & Šik, J. (2007). OISF Pattern and Grown-in Precipitates in Heavily Boron Doped Silicon, *Journal of The Electrochemical Society*, Vol. 154, No. 10 (August 2007), pp H904-H909, ISSN 0013-4651

Válek, L.; Šik, J. & Lysáček, D. (2008). Enhanced Oxygen Precipitation during the Czochralski Crystal Growth, *Solid State Phenomena*, Vol. 131-133, (2008), pp. 167-174, ISSN 1662-9779

Vanhellemont, J.; Senkader, S.; Kissinger, G.; Higgs, V.; Trauwaert, M.-A.; Graef, D.; Lambert, U. & Wagner, P. (1997). Measurement, Modelling and Simulation of Defects in As-grown Czochralski Silicon, *Journal of Crystal Growth*, Vol. 180, No. 3-4 (October 1997), pp. 353-362, ISSN 0022-0248

von Ammon, W.; Dornberger, E. & Hansson, P.O. (1999). Bulk properties of very large diameter silicon single crystals, *Journal of Crystal Growth*, Vol. 198/199, No. 1, (March 1999), pp. 390-398, ISSN 0022-0248

von Ammon, W.; Holzl, R.; Virbulis, J.; Dornberger, E.; Schmolke, R.; Graf, D. (2001). The Impact of Nitrogen on the Defect Aggregation in Silicon, *Journal of Crystal Growth*, Vol. 226, No. 1, (June 2001), pp. 19-30, ISSN 0022-0248

Voronkov, V.V. (1982). The Mechanism of Swirl Defects Formation in Silicon, *Journal of Crystal Growth.* Vol. 59, No. 3, (October 1982), pp. 625- 643, ISSN 0022-0248.

Voronkov, V.V. (2008). Grown-in defects in silicon produced by agglomeration of vacancies and self-interstitials, *Journal of Crystal Growth*, Vol. 310, No. 7-9, (April 2008), pp. 1307-1314, ISSN 0022-0248

Yamagishi, H.; Fusegawa, I.; Fujimaki, N. & Katayama, M. (1992). Recognition of D Defects in Silicon Single Crystals by Preferential Etching and Effect on Gate Oxide Integrity, *Semiconductor Science and Technology*, Vol. 7, No. 1A, (January 1992), pp A135-A140, ISSN 0268-1242

4

Growth and Characterization of Doped CaF_2 Crystals

Irina Nicoara and Marius Stef
West University of Timisoara, Timisoara, Romania

1. Introduction

The alkaline-earth fluorides crystallize in the cubic structure and constitute an important class of relatively simple ionic crystals whose optical and lattice-dynamical properties have theoretical and experimental interest. The CaF_2 crystals have been used for long time in many optical components due to its exceptional transparency in the UV as well as in the IR spectral domain. CaF_2, SrF_3 and BaF_2 have been among the first solid-state laser hosts and they were lased at the beginning of the 1960s doped with RE^{3+} ions; these rare-earth doped crystals, however, have been abandoned as laser systems during a long time. The reason resides in the charge compensation which is required to maintain the electrical neutrality of crystals. This process gives rise to a rich multisite structure including so-called isolated centers and more or less complex centers [Petit et al., 2008], which leads to broad absorption and emission bands comparable with those of glasses. Rare earth doped CaF_2 recently have a new interest firstly because it was found that clustering of ions in these materials could be favorable to produce some infrared laser emission. Yb^{3+} -doped CaF_2 has been proved in recent years to be one of the most attractive Yb^{3+} laser materials for different reasons. It is also proved that the proportions of the different luminescent centers vary with the considered RE^{3+} ion and with the nature of the substituted divalent cation. It is expected, indeed that the transition strengths associated with various centers are different; for example, the transition strengths associated with tetragonal centers will be greater than those of the trigonal ones. At high dopant concentrations (about 1 at%) which become interesting for the laser application, the ions generally aggregate and form more or less complex centers that in turn can weaken the emission transitions. Such a detrimental pairing effect can be decreased and some improvement can be obtained by co-doping the crystals with charge compensating buffer ions, such as monovalent ions or non-optically active rare-earth trivalent ions. For example, after Na^+ ions were introduced as charge compensators the IR emission intensity of the YbF_3-doped CaF_2 crystal was enhanced several times [Su et al., 2007].

Taking into account the interesting properties of CaF_2 crystals doped with various impurities, in this work we describe our investigations about the optical properties of Pb^{2+}, Yb^{3+} doped CaF_2 crystals, as well as the influence of Na^+ ions on the formation of various charge compensating defects.

2. Crystal growth

Crystallization process is essentially a phenomenon of periodic arrangement of the constituent ions of a given material obtaining a crystalline state. There are many methods of crystal growth. Controlled solidification in a crucible of a molten material is one of the most used method. First used by Bridgman [Bridgman, 1925], improved by Stockbarger [Stockbarger, 1949] this technique is used successfully (about 40% of production of artificial crystals) because it is a simple technology and does not require complicated control systems. In the design of vertical Bridgman-type technique (VB) it is important to predict the thermal profiles in the growing crystal. The position and shape of the solidification interface and the axial and radial thermal gradients are of particular interest in controlling the growth process. The relatively high axial temperature gradient necessary in order to obtain crystals with large diameters ($\Phi > 20$mm) cannot be obtained by means of a single heating element; two or more independently controlled heaters are necessary. The temperature gradient is directly dependent on the "isolated" zone between the hot and cold zones [Fu&Wilcox, 1980; Mikkelsen, 1980]. In order to obtain crystals with melting temperature higher than 1300^{0}C, graphite, molybdenum, silicon carbide, etc heaters must be used [Stockbarger, 1949; Gault et al., 1986; Jones et al., 1966]; the high temperature gradient necessary at the crystallization interface is obtained either by supplementary heaters [Stockbarger, 1949; Jones et al., 1966], by thermal screen translation or by using two heaters with a moving temperature gradient and a stationary crucible [Gault et al., 1986]. Taking into account the analysis of the heat transfer in VB technique [Chang&Wilcox, 1974; Fu&Wilcox, 1980; Naumann, 1982; Jasinski et al., 1983], we designed [D.Nicoara, 1975; D. Nicoara et al., 1985a; D. Nicoara et al., 1985b; I. Nicoara et al., 1987; D. Nicoara&I. Nicoara, 1988] several types of graphite heaters whose characteristics would satisfy the growth conditions to obtain crystals with melting point up to 2000^{0}C. In this paragraph we will focus on the growth particularities in order to obtain pure and various concentrations of PbF_2, YbF_3- doped and NaF-codoped CaF_2 crystals using Bridgman technique [D. Nicoara&I. Nicoara, 1988; Nicoara et al., 2008a, 2008b; Pruna et al., 2009; Paraschiva et al., 2010]

2.1 Shaped graphite heaters

The general view of the crystal growth set-up is shown in figure 1a [D.Nicoara 1975]. Figure 1b illustrates the longitudinal section of three types of heaters and the temperature distribution along them. The model contains an adiabatic zone and a booster heater zone that can be used to increase the thermal gradient near the solidification interface. Built of graphite (see fig 7), the heater produces an almost constant temperature zone (the B-C hot zone) necessary to melt the charge and a high gradient temperature zone (D-E). The C zone (the booster heater) ~ 20 mm long in the case of type I heater is designed as a meander-type resistance with wall thickness of 1.5 mm, by means of which an overheating is obtained in the lower level of the upper zone. The D zone is inserted for the improvement of heat transfer upwards the upper part of the heater, the E zone ~ 10 mm long, which appears like a wall thickening (up to 5 mm), has the role of an „isolated" zone and allows to reach a high temperature gradient. The F zone -the cold zone - is used as the lower cooling chamber. The wall thickness of this zone increases from 1.5 mm to 4 mm in the lower part. The wall thickness of the II type heater in the B zone decreases from 3mm to 2mm; the C zone is ~ 15 mm long with wall thickness of 1.5 mm. The III type heater produces a relatively flat temperature profile in the hot zone B-C. The wall thickness decreases gradually from 3 mm in the upper part of the B zone to 1.5 mm in the C-D zone, then increases to 4mm in the lower part of the F zone. Curve I represents the temperature distribution in I type heater and the curve I_a is the temperature recorded during the growth of a 30 mm diameter CaF_2 crystal. Comparing these two distributions one can see that the furnace parameters were well chosen, the axial temperature distribution remaining approximately the same. In order to avoid thermal loss, the graphite heater is surrounded by a set of concentric screens made of molybdenum, graphite, and stainless steel.

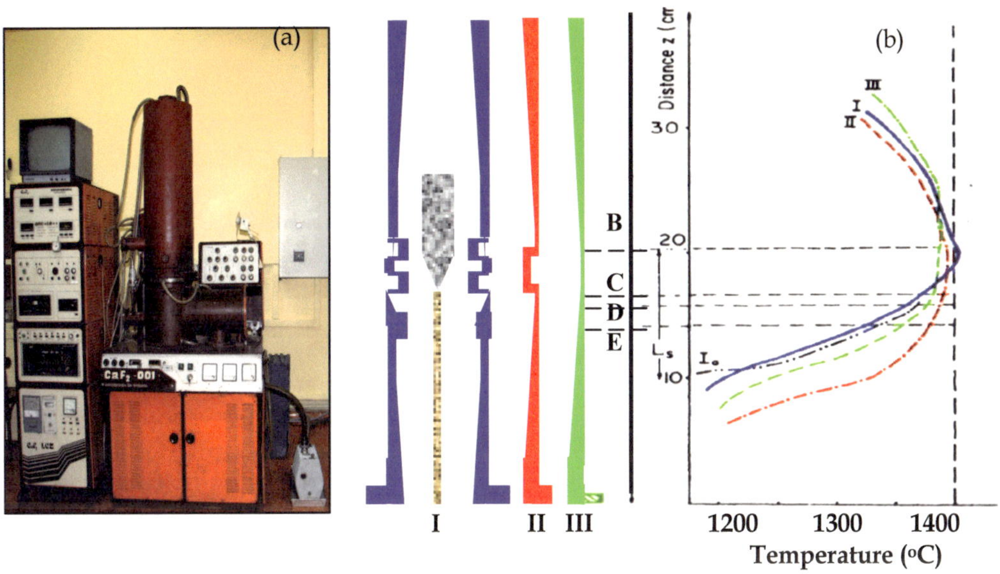

Fig. 1. (a) The crystal growth set-up; (b) Longitudinal sections of three types of heaters and temperature distribution along them, [D. Nicoara, 1975; D. Nicoara et al., 1985a,b].

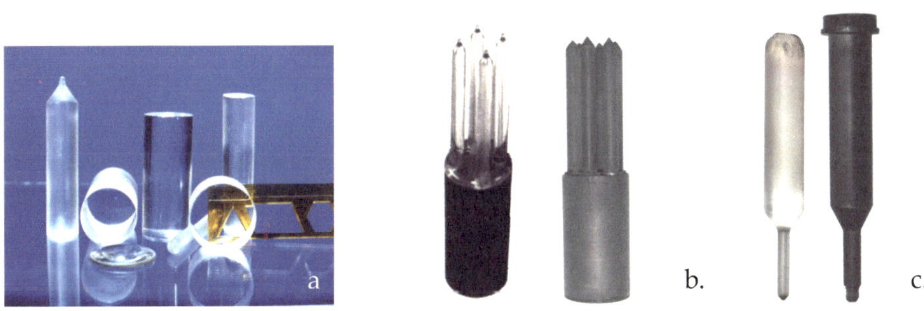

Fig. 2. (a) Calcium fluoride crystals; (b) multiple type crucible; (c) crucible with seed.

In order to obtain crystals, with diameter up to 10 mm, the C zone can be eliminated, the necessary temperature gradient (7-10^0 C/cm) can be obtained only by the heater wall thickness adjustments, like III type heater. For crystals with diameter less than 20 mm, type II and III heaters could be employed. Using these types of heaters fluoride single crystals with diameter of up to 50 mm and 150 mm in length were obtained (fig. 2a). With a multiple type crucible [D. Nicoara et al., 1985a; 1985b] several up to 12 mm diameter crystals can be obtained during a single growth process (fig. 2b). Oriented crystal can be grown using a seed (Fig. 2 c) [D. Nicoara et al., 1983].

The thermal stresses can be reduced by gradual cooling rate to room temperature. After the crystal pulling is finished, the crucible is raised in the B-C zone and by using an automatic system the power supply is gradually lowered.

From our experiments it can be concluded that the use of these types of heaters in the VB technique in order to obtain high quality fluoride crystals is advantageous because the same heater can be used for about hundred growth processes, the growth conditions being reproducible and easy to control.

2.2 Growth of PbF_2-doped CaF_2 crystals

PbF_2-doped CaF_2 crystals were grown in our crystal research laboratory using vertical Bridgman method. Suprapure grade (Merck) calcium fluoride and PbF_2 were used as the starting materials. It is known [Stockbarger, 1949; Yonezawa et al., 2002] that Pb^{2+} ions hardly remain in fluoride crystals if the usual VB technique is used for the growth of CaF_2 crystals. In order to obtain PbF_2-doped CaF_2 crystals the following procedure was used. First, pure, oxygen-free CaF_2 crystals were grown using the usual growth conditions, namely adding to the starting material an amount of 4 wt% PbF_2 as oxygen scavenger [Stockbarger, 1949]. The obtained CaF_2 crystals do not contain any undesired amount of lead ions or other impurities, as results from the optical absorption spectrum; then the PbF_2-doped crystals were grown from the crushed pure fluoride crystals doped with the desired amount of PbF_2. To prevent the evaporation of the PbF_2, a thin floating graphite lid was put on the charge in a sealed graphite crucible. Transparent colorless crystals of about 10 mm in diameter over 6–7 cm long were obtained in spectral pure graphite crucible in vacuum (~ 10^{-1} Pa) using a shaped graphite furnace [D. Nicoara&I. Nicoara, 1988]. The pulling rate was 4 mm/h. The crystals were cooled to room temperature using an established procedure. The as-grown single crystals are shown in Fig. 3.

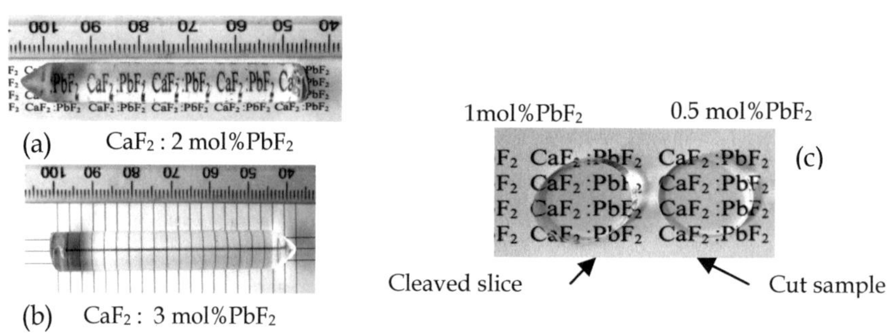

Fig. 3. As-grown x mol%PbF_2-doped CaF_2 crystals: (a) $x = 2$; (b) $x = 3$; (c) cleaved slice of 1 mol%PbF_2 doped CaF_2 crystal and cut sample 0f 0.5 mol% PbF_2 doped CaF_2 crystal.

CaF_2 crystallizes in cubic structure with a typical fluorite lattice. In order to understand the optical properties and the dopant distribution in CaF_2 crystals it is better to see this structure as consisting of a simple cubic lattice of fluorine ions in which every other body center position is occupied by a Ca^{2+} ion. When various ions, such as alkaline metals, rare-earths or heavy metal ions, like Pb^{2+} are introduced into the lattice they usually occupy Ca^{2+} sites. If the introduced impurity ions have other valence than the Ca^{2+} ion, the valence mismatch is compensated in a variety of ways: by vacancy formation, by interstitial fluorine ion, etc. The Pb^{2+} ion has the same valence as Ca^{2+} but with a larger geometric size (0.143 nm) than the Ca^{2+} ion (0.126 nm), and for high dopant concentrations this will lead to distortion of the crystal lattice; this is the reason why the crystals doped with more than 3 mol%PbF_2 reveal structural defects, like blocks with different crystallographic orientations (Figure 4).

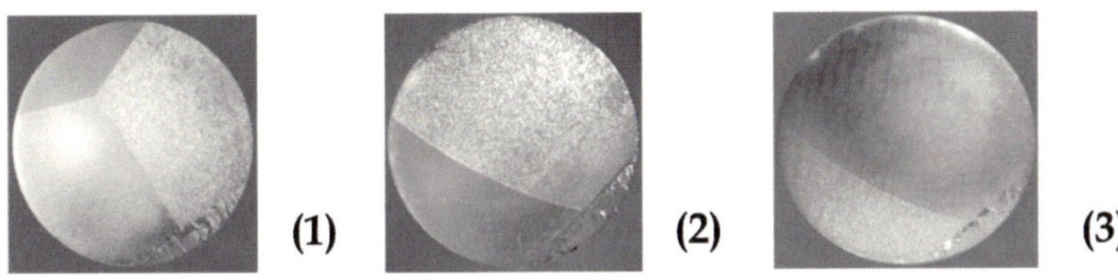

Fig. 4. Misoriented high PbF_2 concentration sample; (1) the 2mm thick sample was cut from the bottom of the cystal, (2) after 4mm only two blocks remain and (3) after other 2 mm a single oriented crystal appears.

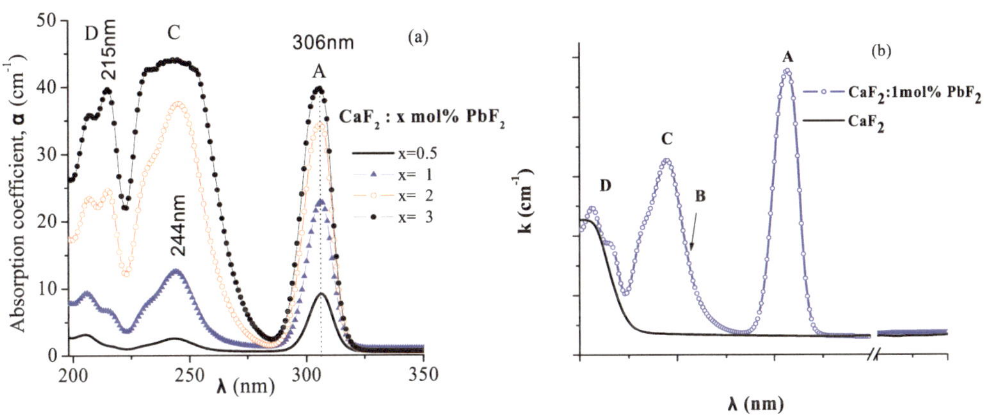

Fig. 5. (a) Room temperature absorption spectra of PbF_2 doped CaF_2 crystals; (b) absorption spectra of pure and 1 mol%PbF_2-doped CaF_2 crystals.

The presence of impurity ions in CaF_2 lattice with the ns^2 ground state configuration (like Pb^{2+} ions) induces absorption bands both in vacuum ultraviolet (VUV) and in UV domain [Jacobs, 1991]. Four characteristic bands are located in UV domain, denoted by A, B, C, and D [Jacobs, 1991]. The absorption spectra of various concentrations CaF_2:PbF_2 samples reveal the four characteristic absorption bands of Pb^{2+} ions (Fig. 5). As Pb^{2+} -ions concentration increases, the shape of the C and D absorption bands modifies due to the overlap of many new bands that appear as a result of the energy levels splitting, only the A band conserves his sharp shape.

2.3 Rare-earth ions-doped CaF_2 crystals

Rare-earth (RE) ions-doped CaF_2 crystals are used as laser active media due to the well-known good optical, mechanical and thermal properties of the CaF_2 host and due to the broad band transitions of the various RE ions. The optical properties of the Yb^{3+} ions in the CaF_2 host are well known and the luminescence properties have been intensively studied due the strong near-IR emission that can be easily pumped with laser diodes. It is known that several of RE ions, such as Sm, Eu, Ho, Tm and Yb, can be stabilized in the divalent state in alkaline earth halide lattices, besides the trivalent state, with which can coexist. The spectroscopic properties of the Yb^{2+} ions have been less investigated, mainly only for its intense and broad yellow-green (540–560 nm) luminescence [Feofilov, 1956; Kaplyanskii&Feofilov, 1962; Kaplyanskii et al., 1976]. When YbF_3 is dissolved in CaF_2, normally the ytterbium ions are in the trivalent state, but it is known that a certain fraction of any of RE^{3+} ions can be reduced to divalent state by various methods.

The change of the valence can be attained by exposing the crystals to ionizing radiation, baking them in a suitable atmosphere [Kirton&McLaughlan, 1967; Kaczmarek et al., 2005] or by electrolytic reduction [Fong, 1964]. The observed emission is weak for the crystals containing Yb^{2+} ions obtained using one of these methods. There are a few reported results [Feofilov, 1956; Kaplyanskii&Feofilov, 1962; Kaplyanskii&Smolyanskii 1976] about the properties of the Yb^{2+} ions, with high concentration in the as-grown crystals. The intense broad yellow-green luminescence has been obtained only for low temperature. Near-UV luminescence at room temperature has not been reported. This was one of the reason why we studied the spectroscopic properties of YbF_3-doped CaF_2 crystals with high divalent ytterbium content in the as-grown crystals. In order to obtain high Yb^{3+} -Yb^{2+} conversion in the as grown crystals, we have developed a special growth procedure.

Rare-earth ions -doped and PbF_2-codoped CaF_2 crystals were grown using vertical Bridgman method [Nicoara et al., 2006a, 2006b, 2008a, 2008b, 2008c; Pruna et al.,2009; Paraschiva, 2010]. As starting materials suprapure grade calcium fluoride, YbF_3 , ErF_3 and PbF_2 have been used. The Yb:CaF_2 and Er:CaF_2 crystals were obtained by adding YbF_3 and ErF_3 to the melts in molar concentrations varying between 0.07 and 2 mol%. First, pure CaF_2 crystals were grown using the standard growth conditions. In all cases, 0.4wt% PbF_2 was used as oxygen scavenger. Fig. 5a shows the time dependence of the power (in arbitrary units) and temperature in the furnace in order to determine the growth conditions.

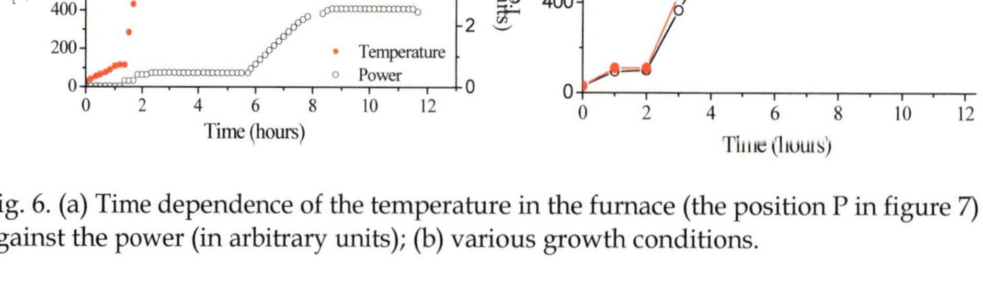

Fig. 6. (a) Time dependence of the temperature in the furnace (the position P in figure 7) against the power (in arbitrary units); (b) various growth conditions.

If the crystal growth process is not correctly established (for example, the T_0 temperature is not reached and then keep the raw material for 3 h at a temperature lower than T_0), the obtained crystals may contain some amount of undesired lead ions. This is why we study the influence of the Pb^{2+} ions on the optical properties of the YbF_3-doped CaF_2 crystals [Nicoara et al., 2006b; 2008b]. The growth procedure started to run after the whole furnace chamber was vacuumed to 10^{-1} Pa; the temperature was kept at 100^0 C for 3 h to remove water from the raw materials and then 3 h at $T_0=800^0$ C to eliminate the PbO obtained after the reaction $CaO+PbF_2 \rightarrow CaF_2+PbO$.

Pure CaF_2 crystals were grown using the conditions shown in the Fig. 6b, alternative 1. The obtained CaF_2 crystals do not contain any undesired lead ions amount, as seen in the optical absorption spectrum (Fig. 5. the inset).

The YbF_3 (or ErF_3)-doped CaF_2 crystals were grown from crushed pure CaF_2 crystals obtained using a correct growth process. No oxygen scavenger has been added in order to obtain the doped crystals; in this case the alternative 2 shown in figure 6b was used.

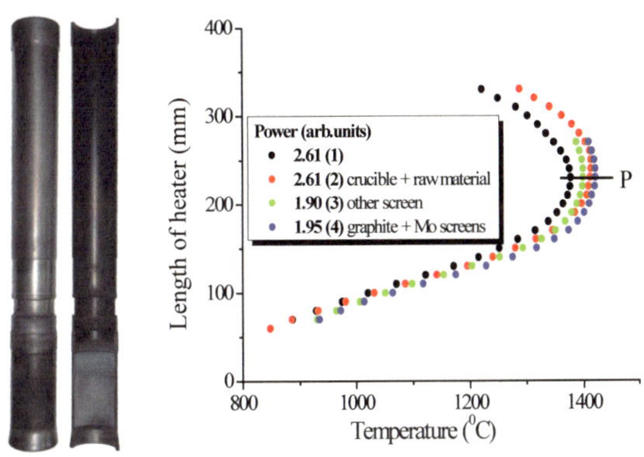

Fig. 7. Axial temperature distribution in the furnace for various powers and growth conditions: (1) without crucible and charge, (2) with crucible and charge, supplementary screen, (3) other positions of the screen, (4) double screen: graphite and molybdenum.

In order to obtain Pb^{2+} codoped crystals, 1 mol% PbF_2 was added in the starting mixture in melt. To prevent the evaporation of the PbF_2, a thin graphite lid was put on the charge in the sealed crucible. In order to obtain Pb^{2+} codoped $YbF_3 : CaF_2$ crystals, alternative 2 (fig. 6b) was used. The crystal growth begins by lowering the crucible in the temperature distribution in the furnace shown in Fig. 7. For alternative 1 the growth process starts to run at position R_0G_0 (the Fig. 6b) and for alternative 2 at RG.

Transparent colorless crystals (CaF_2, $CaF_2 : YbF_3$, $CaF_2 : YbF_3 +PbF_2$; $CaF_2 : YbF_3 +NaF$; $CaF_2 : PbF_2$) of about 10mm in diameter over 6–7 cm long were obtained. The crystals were cooled to room temperature using an established procedure. The as-grown single crystals are shown in Fig. 8. The ErF_3 doped crystals are transparent pink color.

The optical absorption spectra (190–1090 nm) reveal the characteristic UV absorption bands of the Yb^{2+} ions with more than 10 times higher absorption coefficient than the one of the Yb^{3+} ions (see Fig. 9). We assigned the high Yb^{2+} ions content in the as-grown crystals to the reducing conditions during the growth process due to the presence of the graphite components of the growth set-up and the lack of the oxygen scavenger during the growth of the doped crystals.

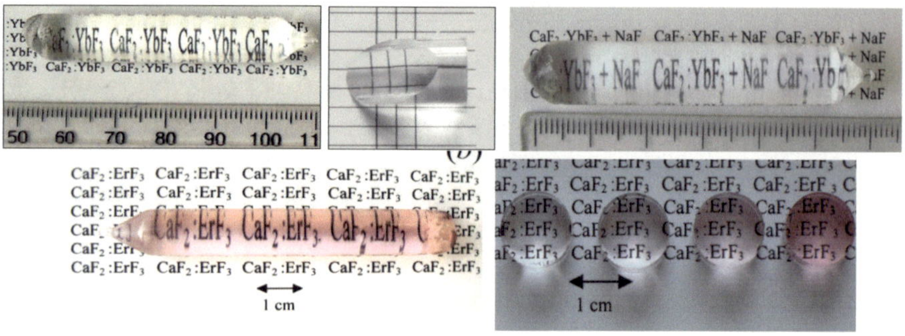

Fig. 8. As-grown crystals: (a) CaF_2: 0.17 mol%YbF_3; (b) CaF_2: 0.7 mol%YbF_3+2.5 mol% NaF; (c) CaF_2:2mol%ErF_3; (d) CaF_2 : x mol%ErF_3, x= 0.69, 0.8, 1.1, 5.

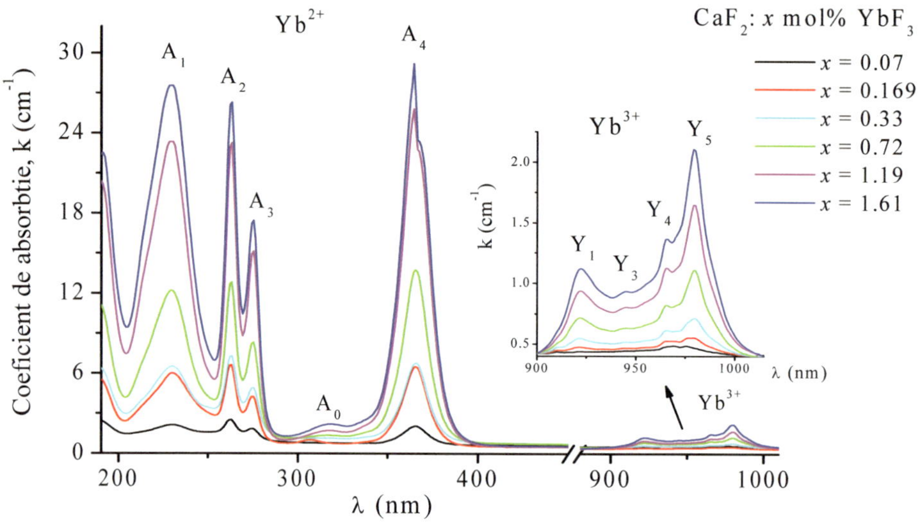

Fig. 9. Absorption spectra of YbF_3 doped CaF_2 crystals.

In order to obtain YbF_3-doped CaF_2 and NaF codoped crystals the following procedure was used. First, pure, oxygen-free CaF_2 crystals were grown using the usual growth procedure, namely adding to the starting material an amount of 4 wt% PbF_2 as an oxygen scavenger. The obtained CaF_2 crystals do not contain any undesired lead ions amount or other impurities, as results from the optical absorption spectrum; these crystals were then crushed and the powder was doped with the desired amount of YbF_3 and NaF codopant. To prevent the evaporation of NaF, a thin floating graphite lid was put on the charge in a sealed graphite crucible.

3. Characterization

Taking into account that these crystals are used as optical materials for various purposes it is important to know how the impurities influence the optical properties of the crystals. Various structural defects appear, from the so-called isolated centers to more or less complex centers. In order to study the structural defects in crystals we used various methods as are described below.

3.1 Dislocations distribution in crystals

The quality of the crystals was studied by examining the dislocations distribution using the chemical etching method. For this purpose the fresh cleavage surface (111) of the crystal were subjected to etching in aqueous solution of 2N-8N HCl.

The effect of temperature and concentration of HCl aqeuos solutions on the etching behavior on the cleavge surface of CaF_2 crystals was investigated [Nicoara et al., 1986]. It has been observed that the shape and the evolution of etch pits and the values of dissolution rate depend on the etching conditions. This behavior suggests that the dissolution rate along a certain direction changes with the temperature and the etchant concentrations. Taking into the ionic arrangement in the (111) plan of the CaF_2, the experimental observations [Sangval&Arora, 1978] and the theoretical considerations [Benson & Dempsey, 1962] on the repulsive energy between the two fluoride ions, we suppose that the activation energy depends on the crystallographic directions on the (111) surface and on the etching parameters. These explain the shape of the etch pits. Various etch pits shapes were also observed in doped calcium fluoride crystals.

The formation, multiplication and high mobility of the dislocations in ionic crystals lead not only to high densities of individual dislocations, but also cause arrangements of dislocations into well-developed grain sub-boundaries, which are stable and nearly immobile in contrast to individual dislocations (fig 10). Sub-boundaries cannot be easily removed by annealing. Some sub-boundaries appear at the beginning of the growth process.

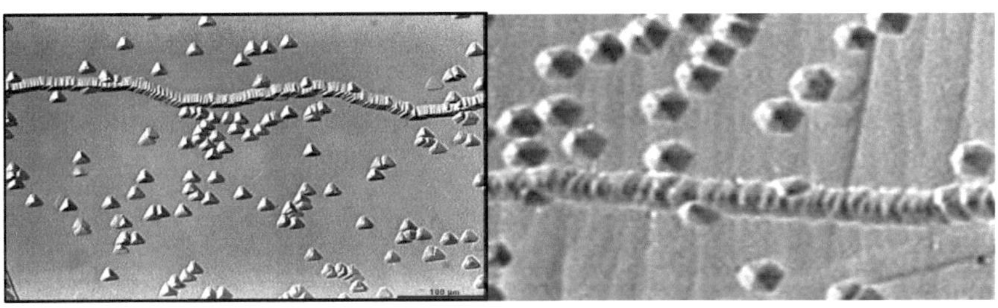

Fig. 10. Individual etch pits and grain sub-boundaries.

An interesting feature is shown in the cross-section of the crystals that are grown in multiple crucibles: the dislocation density is greater in that part of the crystal which is oriented towards the axis support of the crucible [Nicoara et al., 1987]. This fact is explained by the existence of an inadequate radial temperature gradient, the crucible being asymmetrically placed in the thermal field of the furnace, one of its parts near the hot wall. This disadvantage is eliminated by attaching a reflective molybdenum cylindrical screen to the axis, which ensures uniform thermal field for the crucible assembly.

The dislocation density along the sample and for a given cross-section in radial direction was examined. The density of the etch pits and sub-boundaries is quite low and fairly uniform through the sample for crystals with diameter up to 12 mm which were grown in a I type heater with a rate of 6 mm/h and temperature gradient of 7^0 C/ cm. The dislocation density is about 10^3-10^4 dis/cm^2. For higher pulling rate the dislocation density increases [Nicoara et al., 1986; 1987].

The influence of the various dopant and dopant concentration on the dislocations density and etch pits morphologies were studied. The dislocation density and the etch pits shape observed in various crystals are summarized in figure 11and table 1 and 2.

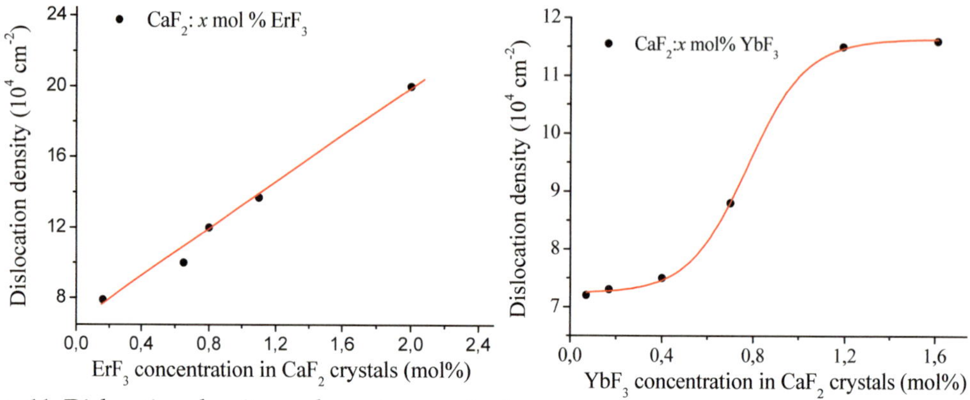

Fig. 11. Dislocation density vs dopant concentration.

CaF_2 : x mol% ErF_3	x=0.17	x=0.69	x= 0.8	x=1.1
Etch pit shape			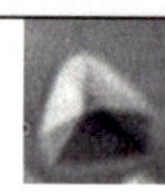	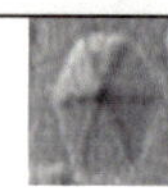
Dislocation density	$7.9 \cdot 10^4$	$10 \cdot 10^4$	$11.9 \cdot 10^4$	$13.7 \cdot 10^4$

Table 1. Dislocation density (dis/cm^2) and etch pit shape in ErF_3 doped CaF_2 crystals.

Crystal	Etch pit shape	Dislocation density (dis/cm^2)	Etch pit dimensions (μm)
CaF_2		$5.5 \cdot 10^4$	18; 4.6; 15.7; 3.7; 16.4; 3.7;
CaF_2: 1 mol% PbF_2	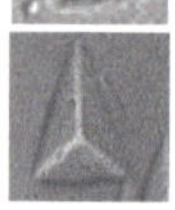	$6.0 \cdot 10^4$	38.9; 33.8; 40.9
CaF_2: 2 mol% PbF_2	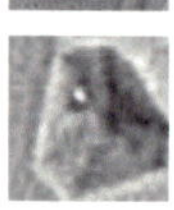	$6.6 \cdot 10^4$	5.1; 15.3; 5.0 12.3; 4.5; 16.2
CaF_2: 0.169 mol% YbF_3		$7.3 \cdot 10^4$	47.6; 37.9; 42;
CaF_2: 0.7 mol% YbF_3	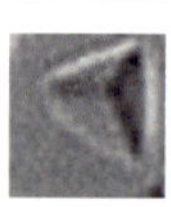	$8.8 \cdot 10^4$	22; 22.7; 22.3;
CaF_2:1,2mol%YbF_3 + 2.5 mol% NaF	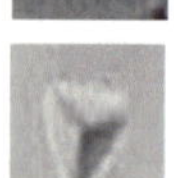	$12.3 \cdot 10^4$	12; 3; 11.5; 3; 11.3; 3;

Table 2. Etch pit shape and dislocation density in doped CaF_2 crystals.

3.2 Dopant distribution alnog the crystals

The homogeneous distribution of the dopant in laser crystals is important because this affects the efficiency of the laser.
The effective segregation coefficient (k_{eff}) at a given growth rate is defined by

$$k_{eff} = C_S / C_L^0, \tag{1}$$

where C_S is the dopant concentration at the bottom of the as-grown crystal and $C^0{}_L$ is the dopant concentration in the initial melt. The value of C_S can be measured by various methods, or estimated from optical absorption measurements [Kuwano, 1982; Sun et al., 2005].
The effective segregation coefficient determination of the Er^{3+}, Yb^{3+}, Yb^{2+} and Pb^{2+} ions in CaF_2 host by optical absorption method is based on the following two laws.
(a) According to the Beer–Lambert law the absorption coefficient is proportional to the sample concentration (C), $\alpha = aC$ where a is the absorption coefficient for unit ion concentration and unit light path length; a may be recognized as constant for the investigated ions concentration [Kuwano, 1982; Sun et al., 2005]. The various ions concentration can be estimated from the measured optical density,

$$O.D. = \log(I_0 / I) \tag{2}$$

where I_0 is the light intensity incident on the sample, I is the transmitted light intensity.
Taking into account the relation

$$I = I_0 \exp(-\alpha \cdot d) \tag{3}$$

where α is the absorption coefficient and d is the sample thickness, the dopant concentration of a slice can be estimated using the relation [Kuwano, 1982]:

$$C = \frac{\alpha}{a} = \frac{O.D. \cdot 2.30258}{a \cdot d} \tag{4}$$

(b) The dopant concentration (along the growth axis) at the distance z from the origin of the crystal can be obtained by using the classical relation [Hurle, 1993]:

$$C_S(z) = C_L^0 k_{eff} \left[1 - g(z)\right]^{k_{eff} - 1} \tag{5}$$

where $g(z)$ is the crystallized fraction of the melt given by $g(z) = V\,t/L = z/L$, V is the crystal growth rate, t is the growth time and L is the crystal length, so Vt is the grown crystal length, z, at the moment t. The more the k_{eff} differs from unity, the larger is the concentration gradient in the crystal.
Taking into account the Beer–Lambert law, the dopant distribution along the crystal length can be estimated using the optical absorption method. In order to determine the effective segregation coefficient, we cut the crystal into i slices with the same thickness (Fig. 12) and calculated the absorption coefficient for a particular absorption peak, from the optical absorption spectrum of every slices ($\alpha(z)$). Using the relations (4) and (5), the following expression is obtained in order to determine the effective segregation coefficient:

$$\lg \alpha(z) = (k_{\mathrm{eff}} - 1)\lg(1 - z / L) + \lg(a k_{\mathrm{eff}} C_L^0) \quad (6)$$

The effective segregation coefficient can be calculated from the slope $m = k_{eff} - 1$, of the fitting line of log $\alpha(z)$ versus log(1–z/L). This was the method used to calculate the segregation coefficient.

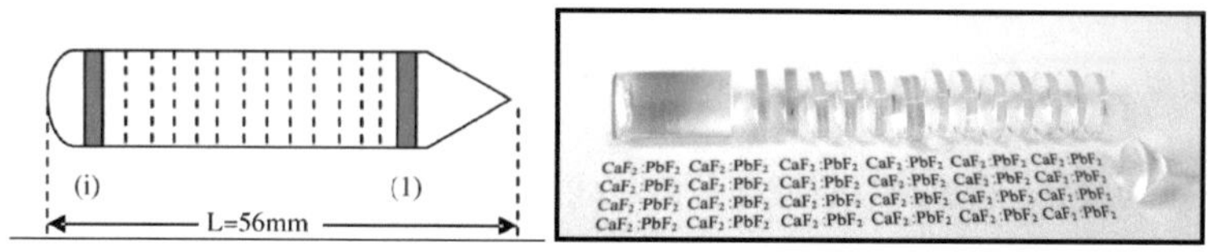

Fig. 12. Crystal cut for segregation coefficient measurement.

Segregation coefficient of Yb^{3+} and Yb^{2+} ions

Room-temperature optical absorption spectra were recorded using a Shimadzu 1650PC spectrophotometer. The absorption coefficient increases as the YbF_3 concentration in the CaF_2 host increases. Fig. 13 shows the influence of the YbF_3 concentration on the absorption coefficient of the peaks at 979 nm and 365 nm.

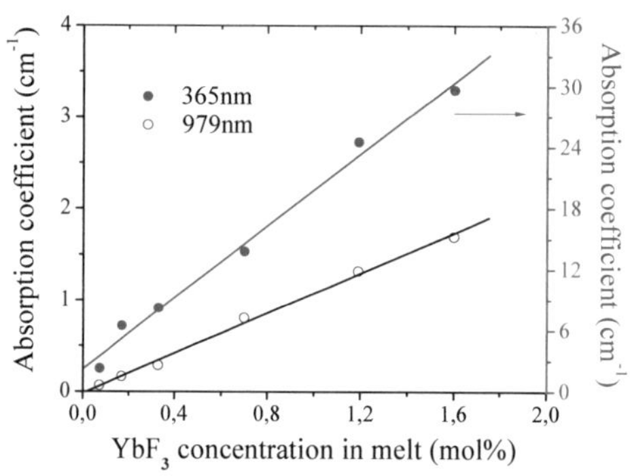

Fig. 13. Dependence of the absorption coefficient of 979 nm and 365 nm on the initial YbF_3 concentration in the melt.

The variation of Yb^{2+} and Yb^{3+} ions' concentration distribution along the as-grown crystals has been studied only for two concentrations, namely with YbF_3 added to the melt in molar concentrations equal to 0.7 mol% and 1.6 mol%.

The absorption coefficient α(z) of every slice was calculated from the absorption spectra for two characteristic absorption bands: 979 nm for Yb^{3+} ions and 365 nm for Yb^{2+} ions [Nicoara et al., 2008a; Nicoara et al., 2008b]. Fig. 14a shows the Yb^{3+} ions' concentration (characterized by the absorption coefficient, α) distribution along the crystals. For both investigated crystals the absorption coefficient of the slides is almost constant along the crystals; this means that the trivalent Yb ions are distributed homogeneously along the crystals. The slope of the fitting line of the Yb^{3+} ions concentration distribution along the crystals, namely of lg $\alpha(z)$ vs. lg(1-z/L) (Fig. 14b), have been used to calculate the segregation coefficient of Yb^{3+} ions in YbF_3 : CaF_2 crystals. The obtained values of the segregation coefficient are close to the unity for both investigated crystals: k ≈ 1.00 for the CaF_2:1.61 mol% YbF_3 sample and k ≈ 0.98 for the CaF_2 : 0.72mol% YbF_3 sample; this indicates a rather homogeneous distribution of Yb^{3+} ions in YbF_3:CaF_2 crystals [Nicoara et al., 2008b]. These values are in good agreement with those reported, for example k ≈ 1.07 for CaF_2:1.96 at% YbF_3 [Su et al., 2005].

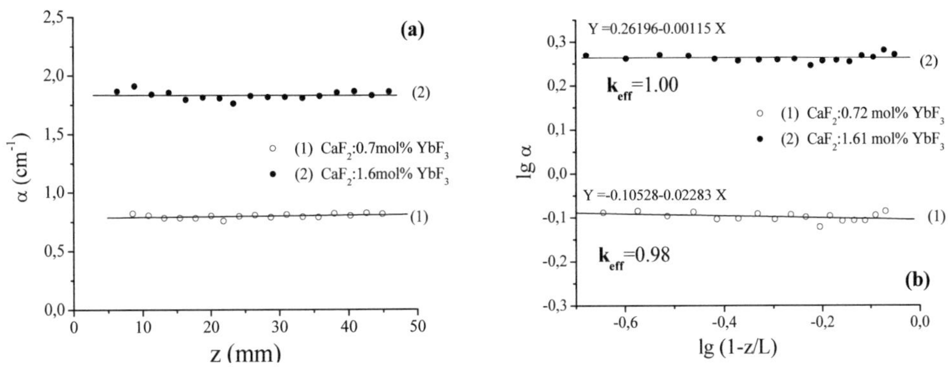

Fig. 14. (a) Yb^{3+} ions distribution along the crystal; (b) fitting lines of lgα vs lg (1-g).

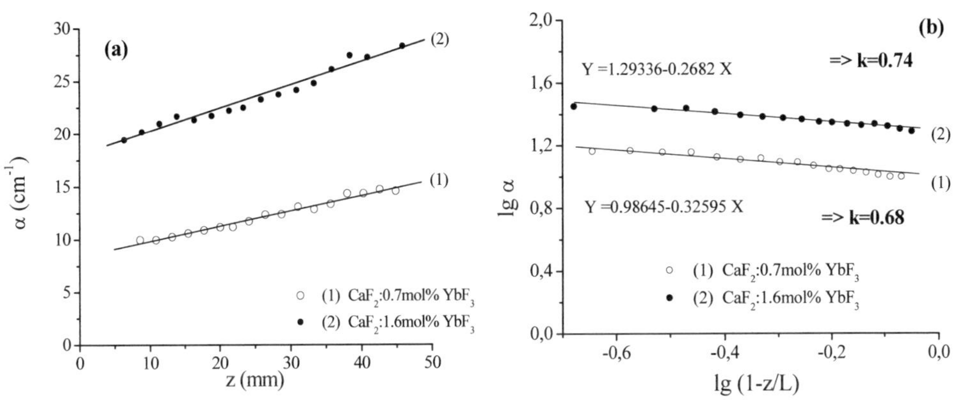

Fig. 15. (a) Yb^{2+} ions distribution along the crystal; (b) fitting line of lg α vs lg (1-g).

Fig. 15a shows the Yb^{2+} ions concentration (characterized by the absorption coefficient, α) distribution along the crystals. The absorption coefficient increases along the crystals. and as the concentration of YbF_3 in the initial melt increases the absorbtion coefficient, α, increases too. The fitting lines of *lg a (z) vs. lg (1-g)* for the two studied Yb:CaF_2 crystals are shown in Fig. 6b. The segregation coefficient is less than unity for both crystals: k ≈ 0.68 for the CaF_2:1.6mol% YbF_3 sample and k ≈ 0.74 for the CaF_2:0.72mol% YbF_3 sample. The more *k* differs from unity, the larger the dopant concentration gradient along the crystal is and hence, no homogeneous distribution of Yb^{2+} ions in YbF_3 : CaF_2 crystals can be obtained.

Segregation coefficient of Er^{3+}

The segregation coefficient of the $Er^{3=}$ ion in CaF_2 host was determined for two ErF_3 concentrations: 0.8mol% ErF_3 and 2 mol%ErF_3 using the method described above.

The absorption coefficient *a* $(z)_i$ of every slice *i* was calculated from the absorption spectra for the following absorption peaks: 406nm, 968nm and 979nm. The dopant distribution along the crystals shows some oscillations of the Er^{3+}-ions concentration. These types of oscillations have also been observed for other crystals grown by Bridgman technique; this behavior was not explained yet [Barat, 1995; Mitric et al., 2006]. The strongest oscillatory behavior of the dopant distribution along the crystal has been observed for the CaF_2: 2mol% ErF_3 sample.

The obtained values of effective the segregation coefficient are: k_{eff} = 1.01-0.99 for CaF_2: 0.8mol% ErF_3 sample and 1.03-1.04 for CaF_2: 2mol% ErF_3 crystal [Munteanu et al., 2010]. As the concentration of the ErF_3 in the initial melt increases the effective segregation coefficient, k_{eff}, increases too; the dependence of the effective segregation coefficient on the dopant concentration was also observed in other host [Lifante et al., 1999; Barraldi et al., 2005]. Because the effective segregation coefficient is almost equal with unity, the Er^{3+} ions distribution in CaF_2 crystals is approximately uniform along the crystal and from this point of view, ErF_3- doped CaF_2 crystals could be a good laser material.

Segregation coefficient of Pb^{2+}

Figure 16 shows the Pb^{2+} ions distribution (characterized by the absorption coefficient) along the four investigated crystals, with various PbF_2 concentrations in the initial melt. The dopant distribution along the crystals shows some oscillations of the Pb^{2+} ions concentration. For CaF_2 : 2 mol% PbF_2 sample the dopant distribution along the crystal has the strongest oscillatory behavior. The calculated effective segregation coefficient for the studied samples is: 0.85 for CaF_2 : 0.5 mol%PbF_2 crystal, 0.925 for CaF_2 : 1mol%PbF_2 crystal, 1.002 for CaF_2:2 mol%PbF_2 crystal and 1.15 for CaF_2:3 mol%PbF_2 crystal.

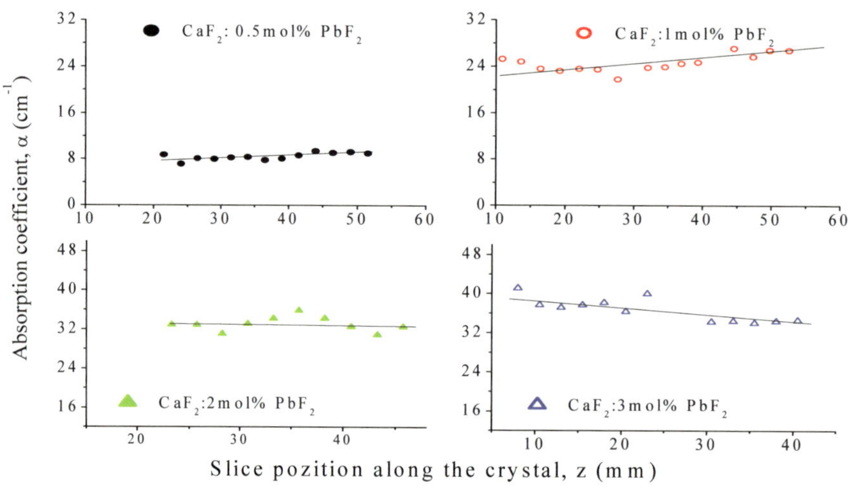

Fig. 16. Variation of the absorption coefficient along the PbF_2 doped CaF_2 crystals

As the concentration of the Pb^{2+} ions in the initial melt increases, the effective segregation coefficient increases, too [Paraschiva et al., 2010]. Dependence of the segregation coefficient of different ions in the CaF_2 host on the dopant concentration is shown in fig. 17.

3.3 Dielectric relaxation

Information on impurity-defect aggregates can be obtained from spectroscopic studies and dielectric relaxation technique. The optical absorption provides information about the nature and site symmetry of the defects. Trivalent RE ions in CaF_2 tend to form pairs of adjacent ions, for charge compensation, even at low RE concentrations. The extra positive charge is usually compensated by interstitial F^- ions. Besides the tetragonal (C_{4v}) symmetry of the predominant dipolar complex, many other simple or cluster configurations appear. The resultant dipolar complexes can reorient by "jumps" of one of the charges to other lattice sites. In order to use the laser properties of the crystals it is necessary to study the influence of the various type of defects introduced by various impurities, such as the RE activator ions or other ions on the properties of the crystals. Information on impurity-defect aggregates can be obtained from spectroscopic and dielectric relaxation techniques, the last being sensitive to aggregates with a dipole moment which can reorient through migration of the anions. Temperature and frequency dependence of the complex dielectric constant give information about the relaxation processes and permits the determination of the activation energy and the reciprocal frequency factor, τ_0, of the relaxation time, τ, and the number of dipoles that contribute to the relaxation process. [Fontanella& Andeen, 1976; Andeean et al., 1977; Fontanella&Tracy, 1980; Andeen et al., 1979; Andeen et al., 1981].

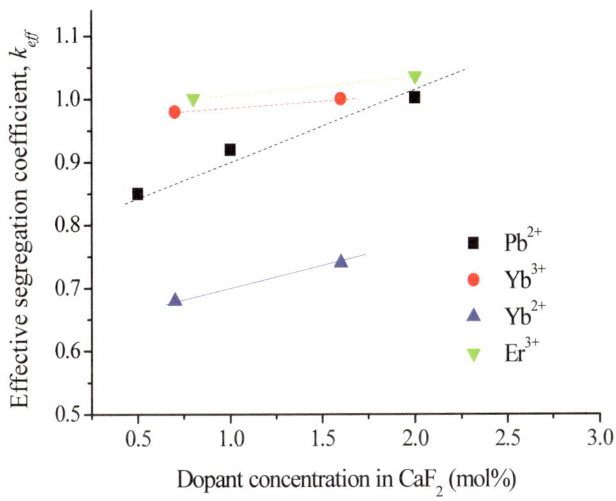

Fig. 17. Dependence of the k_{eff} on the dopant concentration for Pb^{2+}, Yb^{2+}, Yb^{3+} şi Er^{3+} ions.

The impurity-defects in CaF_2 crystals doped with various ions-Er^{3+}, Yb^{3+}, Pb^{2+} and Na^+ - were studied using the dielectric relaxation method [Nicoara et al., 2006; Nicoara et al., 2006b; Nicoara et al., 2008; Pruna et al., 2009].
Capacitance (C) and dielectric loss ($D=\tan\delta$) measurements were performed on the samples using a RLC Meter type ZM2355, NF Corporation, Japan over the temperature range of 150–320 K at seven audio-frequencies. The real part of the complex dielectric constant, ε_1, was calculated from the measured capacitance C. The imaginary part of the complex dielectric constant, ε_2 was then calculated from $D=\varepsilon_2/\varepsilon_1$. Linear heating rates of 2 K/min were employed from liquid nitrogen to room temperature. The dielectric properties have been measured on the 10 mm diameter and 0.6 mm thick disks using an Ag (Leitsilber) contacts. The measurements have been performed on as the (111) cleavage plane as on perpendicular to growth direction cut samples. Using this method we obtained information about the influence of dopant concentration on the formation of various charge compensating defects in doped CaF_2 crystals.

After doping some dipolar complex appear which can reorient (relax) by "jumps" of one of the charges to other lattice sites. Such dipoles are usually characterized by a relaxation time, τ given by

$$\tau = \tau_0 \exp(E/kT) \tag{7}$$

In order to determine the activation energy for reorientation, E, and the reciprocal frequency factor τ_0, the complex dielectric constant $\varepsilon(\omega, T) = \varepsilon_1(\omega, T) + i\,\varepsilon_2(\omega, T)$ has to be determined. The real and imaginary parts of the complex dielectric constant are given by the Debye equations. Since ε_2 has a maximum for $\omega\tau = 1$, it follows from Eq. (1) that: $\ln\omega = -(E/kT_{max}) - \ln\tau_0$, where T_{max} is the temperature at which ε_2 has a maximum at a given frequency and E and τ_0 can be determined from the plot of T^{-1}_{max} vs $\ln\omega$.
Dependence of the real and the imaginary part of the complex dielectric constant of CaF_2: 0.17 mol% ErF_3 sample on the temperature and on the frequency is plotted in Fig. 18.

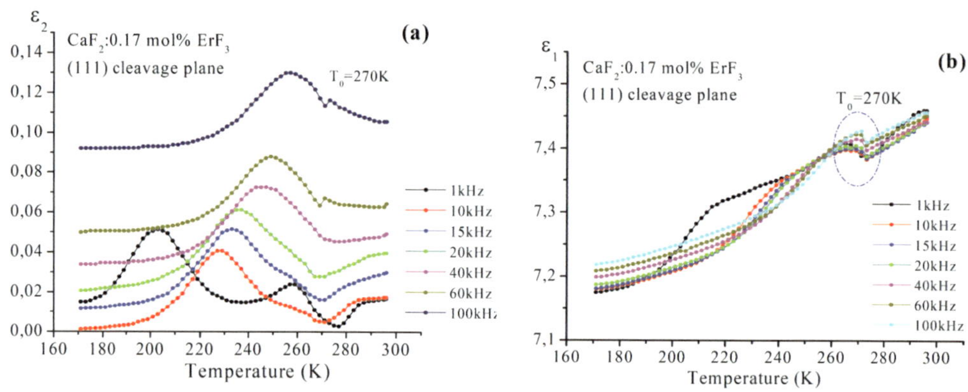

Fig. 18. Temperature and frequency dependence of (a) the imaginary part and (b) real part of the complex dielectric constant. The measurements have been made on (111) cleavage plane.

The peaks occurring at higher temperature as the frequency increases is associated to the R_I, Debye-like relaxation [Fontanella&Andeen, 1976] due to the relaxation of Er^{3+} - F^-_i pair in which the F^- occupies the nearest-neighbor (NN) interstitial position with respect to the Er^{3+} ion. The R_{IV} relaxation [Fontanella&Andeen, 1976] (around 240-260 K, with activation energy > 0.5 eV) increases as the concentration increases and is very clear only for Pb^{2+}-codoped samples [Nicoara et al., 2008]. The values of the relaxation parameters for the observed relaxations are shown in the Table 3.
The value of ε_1 and ε_2 increases as the rare-earth ions concentration increases; around $T_0 = 269K$ an anomaly of ε_1 and ε_2 behavior has been observed which can be assigned with a phase transition of the order-disorder type. The temperature dependence of the dielectric constant and loss tangent for this sample is shown in Fig. 19a. The loss tangent has a maximum at a slightly lower temperature than the ε_1 maximum and a sharp minimum at a slightly higher temperature. The temperature dependence of the reciprocal of ε_1 against $T-T_0$ is shown in Fig. 19b. We can observe that the slopes of the $\varepsilon_1^{-1}(T)$ plots are different on both sides of T_0. These types of anomalies in the dielectric properties have been observed for some perovskite-type compounds [Smolenskii et al., 1958] and are assigned with an order-disorder type phase transition [Strukov&Levanyuk, 1998].

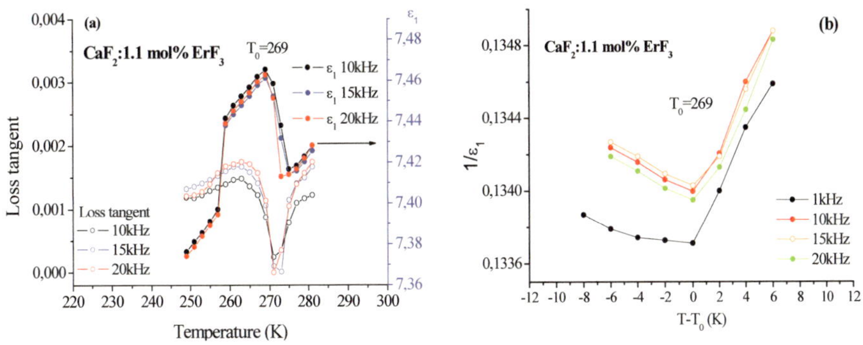

Fig. 19. (a)Temperature dependence of ε_1 and of the loss tangent. (b) Temperature dependence of the reciprocal dielectric constant.

The dielectric spectra of YbF_3 doped CaF_2 crystals has same behavior as described below, the obtained relaxation parameters are summarized in Table 3.

It has been observed [Doualan et al., 2010;] that the detrimentral pairing effect of the RE ions [Corish et al., 1982; Petit et al., 2007; Petit et al., 2008] can be decreased by co-doping the crystals with charge compensating buffer ions, such as monovalent ions, like Na^+ ions [Doualan et al., 2010; Su et al., 2005, 2007] or non-optically active rare-earth trivalent ions. In the case of the double doped (with Yb^{3+} and Na^+ ions) CaF_2 crystals, the Na^+ can work also as charge compensator for Yb^{3+} ions, entering in interstitial (or substitutional) positions near the Yb^{3+} ion and leading to C_{3v} (or C_{2v}) symmetry, sites without dipolar properties. In conclusion, codoping with Na^+ leads to new (Na^+ -V_F) and (Yb^{3+}-Na^+) centers. These defects can be studied using the dielectric relaxation. Five YbF_3 –doped crystals were investigated with the following amount of YbF_3 added in the starting mixture in the melt: 0.07, 0.17, 0.72, 1.19 and 1.6mol%YbF_3; six NaF co-doped YbF_3:CaF_2 crystals with different Na:Yb ratios of R=2, 4, 16, 28, 36 have been also grown.

A typical temperature and frequency dependence of the real and imaginary parts of the complex dielectric constant is plotted in Fig. 20 for CaF_2: 0.16mol%YbF_3 +2.5mol % NaF sample (R=16). The value of ε_1 and ε_2 increases as the YbF_3 concentration increases. In the range of the investigated temperatures, for YbF_3 doped CaF_2 crystals there is only one maximum in the temperature dependence of ε_2, at a given frequency, that corresponds to relaxation of NN (C_{4v}) dipoles [Fontanella&Andeen, 1976]. An "anomaly" of ε_1 and ε_2 behavior, like a maximum has been observed around the temperature T_0 = 270-279K for YbF_3 doped samples and 275-276K for NaF codoped samples, depending on the sample concentration (Fig. 20 and 21a). The value of the temperature T_0 does not depend on the frequency, so is not a relaxation. This behavior is assigned to an order-disorder type phase transition.

The temperature and frequency dependence of ε_2 for Na^+ ions co-doped samples, for ratio R<20 is characterized by three peaks (Fig. 20b, 21a). For all samples the peak occurring at around 195K at 1kHz is associated to the R_I (NN) center relaxation. The second peak is associated to the order-disorder transition and the third peak, that appears only for Na^+ ions codoped crystals, is associated to (Na^+-V_F) dipoles relaxation, formed by a substitutional Na^+ ion and a F^- vacancy created to maintain the electrical neutrality of the crystal Johnson et al., 1969; Shelley&Miller, 1970].

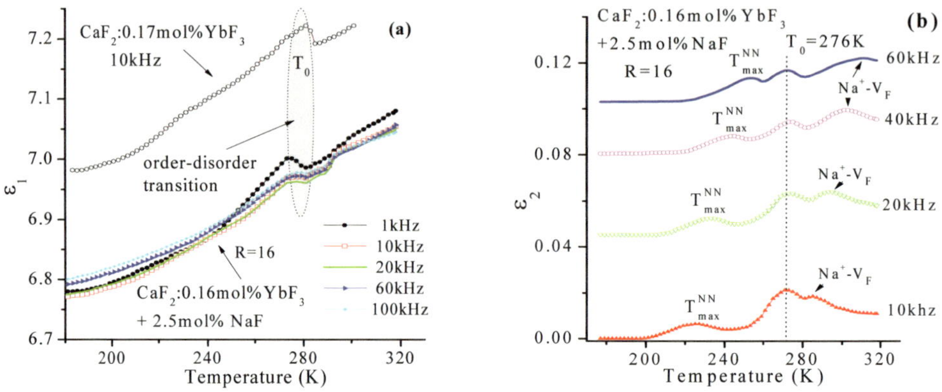

Fig. 20. Temperature and frequency dependence of (a) the real part and (b) the imaginary part of the complex dielectric constant

For heavier NaF- codoped samples, namely for ratio R>20, only one relaxation is observed, associated to the (Na^+-V_F) dipole (see Fig.21b). This confirms the observed suppression of the peaks of optical absorption spectra, corresponding to centers with C_{4v} (NN dipole) symmetry.

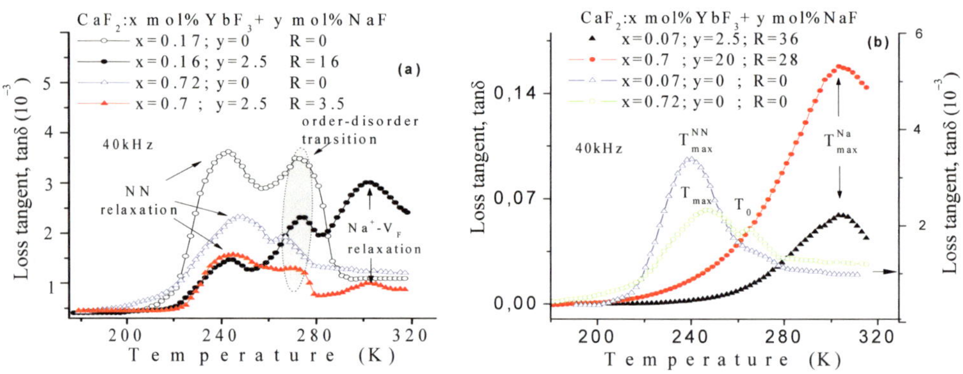

Fig. 21. Influence of Na^+ ions on the loss tangent for (a) low and (b) for high R ratios.

The number of dipoles N_D that contribute to the dielectric relaxation peak can be calculated from the dielectric spectra using the methods described in [Fontanella&Andeen, 1976; Campos&Ferreira, 1974].

Figure 22a shows the variation of the number of NN dipoles (N_{NN}) with the YbF_3 concentration for un-codoped and 2.5mol%NaF codoped samples. The number of NN dipoles that contributes to the dielectric loss peak decreases as the Yb^{3+} ions concentration increases. This effect was also observed by Fontanella [Fontanella&Andeen, 1976] for CaF_2: ErF_3 samples doped with concentration higher than 0.1mol%. This indicates that as the YbF_3 concentration increases the predominant dipoles are NNN and/or clusters types (that relax at lower and/or higher temperature then we have investigated), diminishing in this way the number of dipoles that contribute to the dielectric relaxation of NN (C_{4v} symmetry) type dipoles.

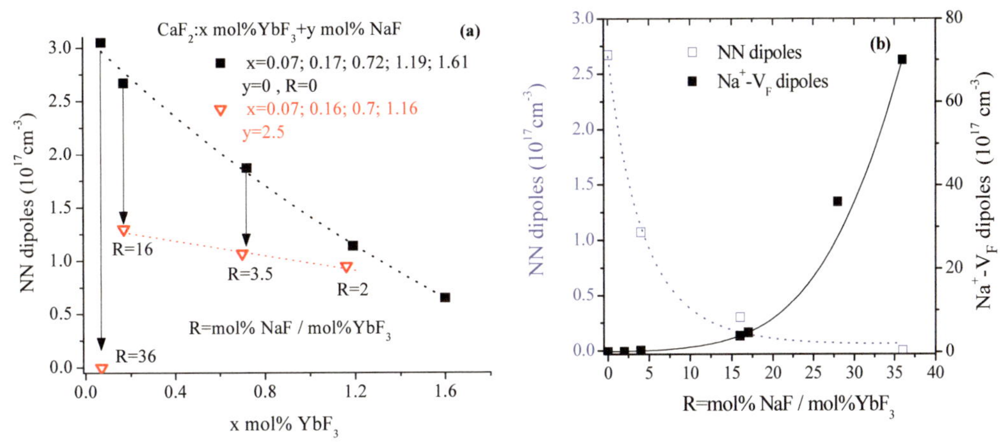

Fig. 22. (a) Influence of YbF_3 concentration and ratio R on the NN dipoles concentration for samples doped with 2.5mol%NaF (open symbols); (b) Influence of ratio R on the NN and Na^+-V_F dipoles concentration

The Na^+- ions co-doped YbF_3:CaF_2 crystals reveal a decrease of the NN dipoles in comparison with the YbF_3- doped samples, indicating that the Na^+ ions reduce the formation of NN type dipoles with C_{4v} symmetry; this is confirmed by the absorption spectra. This effect depends on the ratio R= ymol%NaF/ xmol%YbF_3, as is shown in Figs.22a,b.; as the ratio R increases, the number of NN (C_{4v}) centers decreases and the number of (Na^+-V_F) dipoles increases (fig.22b); for ratio R=28 and 36 no NN type relaxation was detected (fig.21b), this means a very small concentration of C_{4v} sites.

CaF_2:	0.17 mol% ErF_3	0.69 mol% ErF_3	1.1 mol% ErF_3	0.5 mol% PbF_2	1 mol% PbF_2
dipole	(Er^{3+}- F^-)NN	(Er^{3+} - F^-) NN	(Er^{3+}- F^-)NN	(Pb^{2+}-F_{vac})	(Pb^{2+}-F_{vac})
E(eV)	0.383	0.335	0.345	0.438	0.358
τ_0 (s)	$5.1*10^{-14}$	$26.2*10^{-14}$	$52*10^{-14}$	$1*10^{-14}$	$3*10^{-14}$
CaF_2:	0.07 mol% YbF_3	0.17 mol% YbF_3	0.72 mol% YbF_3	CaF: 0.17 mol%YbF_3+2.5mol% NaF	
dipole	(Yb^{3+}- F^-)NN	(Yb^{3+}-F)NN	(Yb^{3+}-F)NN	(Yb^{3+}- F^-) NN	(Na^+-F_{vac})
E(eV)	0.35	0.42	0.34	0.305	0.515
τ_0 (s)	$2.1*10^{-13}$	$0.055*10^{-13}$	$4.4*10^{-13}$	$2.1*10^{-12}$	$7.45*10^{-15}$

Table 3. Relaxation parameters.

3.4 Tailoring the charge compensating defects in YbF_3 doped CaF_2 crystals

When YbF_3 is dissolved in CaF_2, normally the Yb ions are in a trivalent state. It is also known that a certain fraction of any of RE^{3+} ions can be reduced to divalent state depending on the growth conditions. The Yb^{3+}-ions usually occupy a cation substitutional position, but charge compensation is required to maintain the electrical neutrality of the system; the extra positive charge is compensated by an interstitial fluorine ion (F_i^-). For low YbF_3 concentration (< 0.1mol%), besides the O_h cubic symmetry with no local charge compensation, the so-called isolated dipolar centers are predominant with tetragonal symmetry (C_{4v}) in which the F_i^- ion occupies a nearest-neighbor (NN) interstitial site and trigonal (C_{3v}) NNN site [Kirton&McLaughlan, 1967; Corish et al., 1982; Petit et al., 2007; Petit et al., 2008]. At higher Yb^{3+} ion concentrations, the dopant ions aggregate and form more or less complex clusters [Corish et al., 1982; Petit et al., 2007; Petit et al., 2008]. Such a complicated structure leads to broad optical absorption bands. The detrimental pairing effect can be decreased by co-doping the crystals with charge compensating buffer ions, such as monovalent ions [Doualan et al., 2010; Su et al., 2005, 2007] or non-optically active rare-earth trivalent ions. After Na^+ ions were introduced as charge compensators the IR emission intensity of the YbF_3-doped CaF_2 crystal was enhanced several times [Su et al., 2005, 2007].

When a Na^+ ion is introduced into the CaF_2 lattice, this enters substitutionally and is compensated by a fluorine vacancy (Na^+ -V_F) [Johnson et al. 1969; Shelley&Miller, 1970; Fontanella et al., 1980]. In the case of the double doped (with Yb^{3+} and Na^+ ions) CaF_2 crystals, the Na^+ can work also as charge compensator for Yb^{3+} ions, entering in interstitial (or substitutional) positions near the Yb^{3+} ion and leading to C_{3v} (or C_{2v}) symmetry, sites without dipolar properties. In conclusion, codoping with Na^+ leads to new (Na^+ -V_F) and (Yb^{3+}-Na^+) centers.

In order to study the varieties of Yb^{3+} sites in CaF_2 host, several YbF_3 doped and NaF-codoped CaF_2 crystals with different Na:Yb ratios were grown by Bridgman method. Six NaF co-doped YbF_3:CaF_2 crystals with different Na:Yb ratios of 2, 4, 16, 28, 36 have been also grown. Room temperature absorption spectra and dielectric spectra were measured to study the effect of Na^+ ions on the charge compensating defects formation [Pruna et al., 2009].

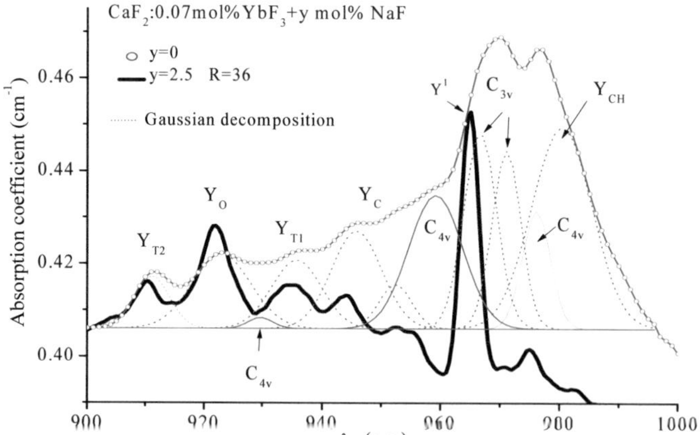

Fig. 23. Absorption spectra of (0.07mol%YbF_3+ymol%NaF):CaF_2 crystals; the decomposition by Gaussian multi-peak fits is shown for CaF_2:0.07mol%YbF_3 sample.

Figures 23 and 24 show the absorption spectra of NaF-codoped $YbF_3:CaF_2$ crystals in comparison with the YbF_3 doped samples; the influence of the Na^+ ions on the absorption spectra depends on the ratio R= y mol%NaF/ x mol% YbF_3. The absorption spectra show that the codoping with Na^+ ions in different R ratios can modulate the spectroscopic properties of Yb^{3+} ions in CaF_2 lattice in a large scale. The differences among the spectra corresponding to ratio R<10 and those R>15 are distinct (see Figs. 23 and 24). The absorption spectra of the crystals with R<10 and without NaF are broad and non-structured.

The absorption bands of the samples with R>15 became narrower and clearly resolved into six peaks, (see Figs.23 and 24) corresponding to trigonal T_2 (Y_{T2}=910nm), cubic O_h (Y_O=922nm), trigonal T_1 (Y_{T1}=936nm), clusters sites (Y_C at 945.6 and 955nm) and trigonal C_{3v} site (Y^1=965nm) with Na^+ ion as charge compensator. As it results from the analysis of these spectra, the presence of Na^+ ions, at ratio R>20 suppresses the peak Y_{CH}=980nm that is attributed to 1-5 transition of hexamer type clusters, and reduces also the intensity of the transitions corresponding to "small clusters", Y_C. Another effect of the Na^+ ions is to reduce the tetragonal C_{4v} (NN) sites; for samples with R>20 this site is drastically reduced, and this effect is confirmed by the dielectric relaxation measurements described in III.3.

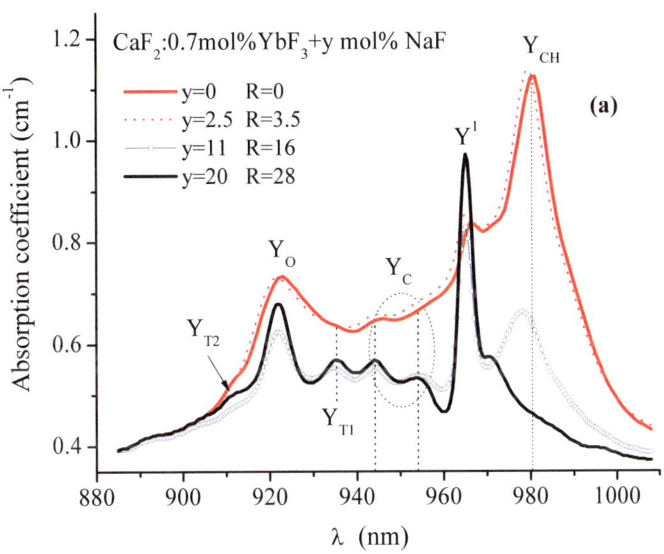

Fig. 24. Optical absorption spectra of 0.7mol%YbF_3 doped CaF_2 crystals codoped with different amount of NaF.

The peak corresponding to hexamer cluster (Y_{CH}=980nm) is completely suppressed as well as the C_{4v} site (see Fig. 23) for the sample with R=36 (0.07mol% YbF_3 + 2.5 mol%NaF) : CaF_2. The reducing effect depends on the R ratio. Comparing the absorption spectra of Yb doped and Na^+ ions codoped crystals it is clear that we can "tailor" the type of the compensating defects by choosing the suitable ratio R of the Na:Yb ions.

The dielectric spectra reveal two peaks that correspond to relaxation of two type of dipoles: NN (Yb^{3+}- F^-_i centers with C_{4v} simmetry) and (Na^+-V_F) dipoles (see III.3).The temperature and frequency dependence of ε_2 for Na^+ ions co-doped samples, for ratio R<20 is characterized by three peaks. For all samples the peak occurring at around 195 K at 1kHz is associated to the R_I (NN) center relaxation [Fontanella&Andeen, 1976; Andeen et al., 1981].

The second peak is associated to the order-disorder transition [Smolenskii et al., 1959; Nicoara, 2008] and the third peak, that appears only for Na^+ ions codoped crystals, is associated to (Na^+-V_F) dipoles relaxation, formed by a substitutional Na^+ ion and a F^- vacancy (V_F) created to maintain the electrical neutrality of the crystal [Johnson et al., 1969; Shelley&Miller, 1970].

For heavier NaF- codoped samples, namely for ratio R>20, only one relaxation is observed, associated to the (Na^+-V_F) dipole. This confirms the observed suppression of the peaks of optical absorption spectra, corresponding to centers with C_{4v} (NN dipole) symmetry (see the optical absorption spectra Figs. 23 and 24).

The number of dipoles N_D that contribute to the dielectric relaxation peak can be calculated from the dielectric spectra using the methods described in [Fontanella&Andeen, 1976; Campos&Ferreira, 1974].

The Na^+- ions co-doped YbF_3:CaF_2 crystals reveal a decrease of the NN dipoles in comparison with the YbF_3- doped samples, indicating that the Na^+ ions reduce the formation of NN type dipoles with C_{4v} symmetry; this is confirmed by the absorption spectra, see figures 23 and 24. This effect depends on the ratio R= ymol%NaF/ xmol%YbF_3, as is shown in Fig. 22a,b.; as the ratio R increases, the number of NN (C_{4v}) centers decreases and the number of (Na^+-V_F) dipoles increases (fig.22b); for ratio R=28 and 36 no NN type relaxation was detected (fig.21b), this means a very small concentration of C_{4v} sites.

Figure 25 illustrates the variation of the (Na^+-V_F) dipoles concentration on the YbF_3 concentration for all the studied NaF-codoped crystals, the ratio R for every sample is specified. The insert shows the dependence of the NN dipoles concentration on the NaF concentration for a given YbF_3 concentration, namely for 0.7mol% YbF_3.

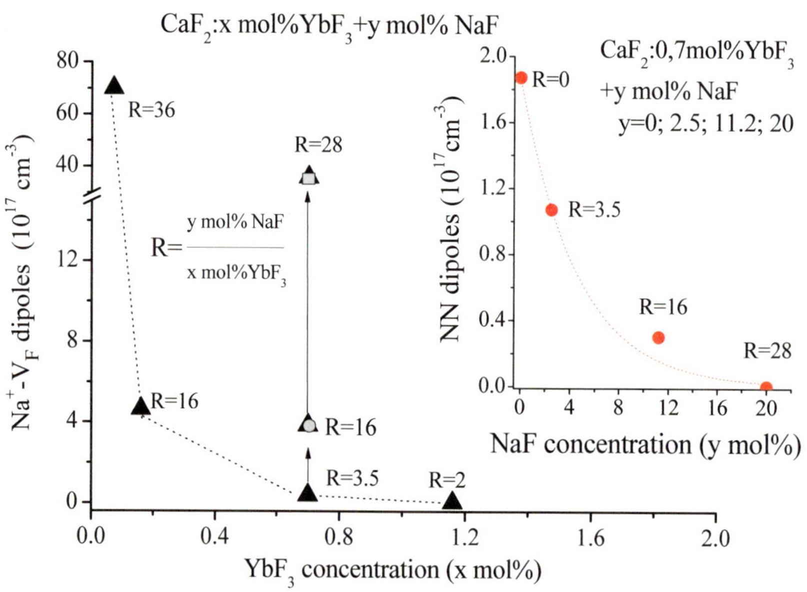

Fig. 25. Influence of YbF_3 concentration and of ratio R on the Na^+-V_F dipoles concentration. The inset shows the influence of NaF concentration on the NN dipoles concentration for CaF_2: 0.7 mol%YbF_3 + y mol%NaF samples.

Influence of Na^+ ions on the charge compensating defects

From the dielectric spectra results that Na^+ ions lead to a decrease of the NN dipoles (see figs.22b and 25), the charge compensating defects with tetragonal C_{4v} symmetry and this behavior is confirmed by the absorption spectra of the samples.

Figure 26 shows the dependence on NaF concentration of the line intensity corresponding to the C_{4v} center and of the NN dipoles concentration for samples doped with 0.7 mol%YbF_3. The inset shows the dependence on the NaF concentration of the line intensity corresponding to C_{3v} site with Na^+ ion as charge compensator, respectively of the line intensity corresponding to clusters. Taking into account that the peak intensity is proportional with the absorbant centers, we observe that as the Na^+ ions concentration increases the charge compensating defects corresponding to C_{4v} site decrease and those corresponding to C_{3v} site increase; the decrease of the cluster type defects is also clear. The decreases of the concentration of the defects corresponding to C_{4v} site is confirmed by the decrease of the NN (C_{4v}) dipoles concentration calculated from dielectric spectra.

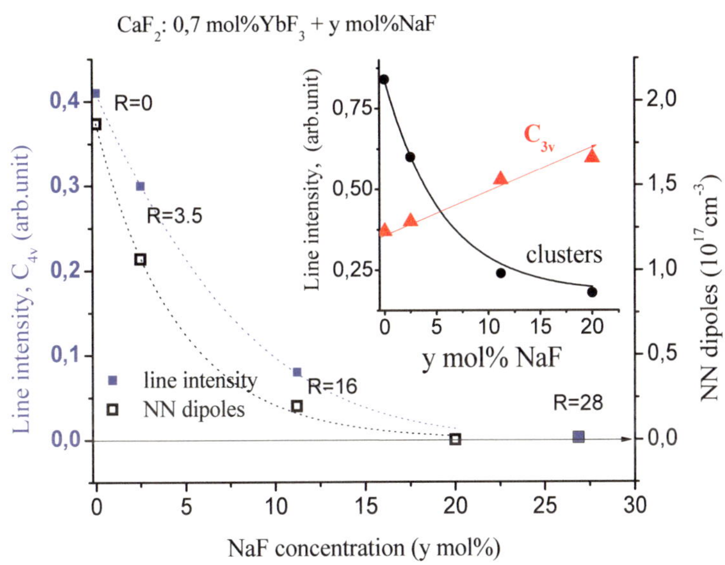

Fig. 26. Influence of NaF concentration on the line intensity of C_{4v} site and on the NN dipoles concentration. The inset shows the influence of NaF concentration on the line intensity corresponding to defects with C_{3v} symmetry with Na^+ ion as charge compensator and on the line intensity corresponding to the clusters.

The effect of Na^+ ions on the formation of the compensating defects can be explained this way. It is known that for concentration above about 0.05at% the rare-earth ions aggregate to form pairs, to form more or less complex centers, clusters. The most common pairs are the dimmers that consist by two Yb^{3+} ions and two F_i^- ions and the hexametric clusters [Petit et al., 2008]. The Na^+ ions substitute one of the partners of the pairs and lead to an increased number of pseudo-isolated centers, like C_{3v} site with Na^+ ion as charge compensator, without dipolar properties. This pair-breaking effect is more effective on the hexameric clusters than on the smaller clusters; for R= 28 and 36 the optical absorption peak (at 980nm) associated to hexamer clusters is suppressed, as is also illustrated in figures 23 and 24. As results from our investigation, codoping with Na^+ the YbF_3 doped CaF_2 crystals it is possible to "tailor" the type of the compensating defects by choosing the suitable ratio R of the Na:Yb ions.

The experimental results showed that codoping with Na^+ ions in different Na:Yb ratios we can modulate the spectroscopic properties of Yb^{3+} ions in CaF_2 host. The influence of Na^+ ions on the defects formation is discussed taking into account the optical absorption spectra and the calculated number of NN dipoles whose relaxation are observed. By choosing the suitable ratio of the Na:Yb ions it is possible to "tailor" the type of the compensating defects in YbF_3 doped CaF_2 crystals.

4. References

Andeen, C.; Link, D. & Fontanella, J. (1977). Cluster-associated relaxations in rare-earth-doped calcium fluoride. *Phys Rev. B*, Vol. 16, pp. 3762-3767

Andeen, C.; Matthews, G. E.; Smith, M.K. & Fontanella, J. (1979). Electric dipole relaxation of mixed clusters in double-doped CaF_2. *Phys. Rev. B*, Vol. 19, pp. 5293-5298

Barat, C. (1995), Origines thermiques et convectives des segregations solutales dans des alliages semiconducteurs. PhD. Thesis, University of Rennes

Barraldi, A.; Capelleti, R.; Mazzera, M.; Ponzoni, A.; Amoretti, G.; Magnani, N.; Toncelli, A. & Tonelli, M. (2005). Role of Er^{3+} concentration in high-resolution spectra of BaY_2F_8 single crystals. *Phy. Rev. B*, Vol. 72, pp. 075132

Benson, K.K. & Dempsey, E. (1962). The cohesive and surface energies of some crystals possessing the fluorite structure. *Proc. Roy. Soc. Lond. A*, Vol. 366, pp. 344-358

Bridgman, P.W. (1925). Crystal growth apparatus. *Proc. Am. Acad. Arts Sci.*, Vol. 60, pp. 305

Campos, V.B. & Leal Ferreira, G.F. (1974). Dipolar studies in CaF_2 with Ce^{3+}. *J. Phys. Chem. Solids*, Vol. 35, pp. 905-910

Chang, C.E. & Wilcox, W.R. (1974). Control of interface shape in the vertical bridgman-stockbarger technique. *J. Cryst. Growth*, Vol. 21, pp. 135-140

Corish, J.; Catlow, C.R.A.; Jacobs, P.W.M. & Ong, S.H. (1982). Defect aggregation in anion-excess fluorites. Dopant monomers and dimers. *Phys. Rev. B*, Vol. 25, pp. 6425

Doualan, J.L.; Comy, P.; Brnayad, A.; Minard, V.; Moncorge, R.; Boudeile, J.; Druon, F.; Georges, P. (2010), Yb^{3+} doped (Ca,Sr,Ba)F_2 for high power laser applications, In: *Laser Physics*, Pashemin, P., pp.533-537, Springer Publishing, ISSN 1054-660X

Feofilov, P.P. (1956). Absorption and luminescence of divalent rare-earth ions in fluoride crystals. *Opt. Spectrosc.*, Vol. 1, pp. 992-1000

Fong, F.K. (1964). Electrolytic reduction of trivalent rare-earth ions in alkalin-earth halides. *J. Chem. Phys.*, Vol. 41, pp. 2291-2296

Fontanella, J. & Andeen, C. (1976). The dielectric spectrum of erbium doped CaF_2. *J. Phys. C*, Vol. 9, pp. 1055

Fontanella, J. & Treacy, D.J. (1980). The effect of quenching on the defect structure of calcium fluoride doped with erbium. *J. Chem. Phys.*, Vol. 72, pp. 2235-2244

Fontanella, J.; Chadwick, A.V.; Carr, V.M.; Wintersgill, M.C. & Andeen, C.G. (1980). Dielectric relaxation studies of alkali-metal-doped calcium fluoride. *J. Phys. C: Solid State Phys.*, Vol. 13, pp. 3457-3466

Fu, T.W. & Wilcox, W.T. (1980). Influence of insulation on stability of interface shape and position in the vertical Bridgman-Stockbarger technique. *J. Cryst. Growth*, Vol. 48, pp. 416-424

Gault, W.A.; Monberg, E.M. & Clemans, J.E. (1986). A novel application of the vertical gradient freeze method to the growth of high quality III–V crystals. *J. Cryst. Growth*, Vol. 74, pp. 491-506

Hurle, D.T.J. (1993), Crystal pulling from the melt, Springer-Verlag, Berlin, ISBN 3540566767

Jasinski, T.; Rohsenow, W.M. & Witt, A.F. (1983). Heat transfer analysis of the Bridgman-Stockbarger configuration for crystal growth: I. Analytical treatment of the axial temperature profile. *J. Cryst. Growth*, Vol. 61, pp. 339-354

Johnson, H.B.; Tolar, N.J.; Miller, G.R. & Cutler, I.B. (1969). Electrical and mechanical relaxation in CaF_2 doped with NaF. *J. Phys. Chem. Solids*, Vol. 30, pp. 31-42

Kaczmarek, S.M.; Tsuboi, T.; Ito, M.; Boulon, G. & Leniec, G. (2005). Optical study of Yb^{3+}/Yb^{2+} conversion in CaF_2 crystals. *J. Phys.:Condens. Mater.*, Vol. 17, pp. 3771-3786

Kaplyanskii, A.A. & Feofilov, P.P. (1962). Absorption spectra of divalent rare-earth ions in fluoride crystals. *Opt. Spectrosc.*, Vol. 13, pp. 235-241

Kaplyanskii, A.A. & Smolyanskii, P.L. (1976), Effect of external fields on optical absorption spectra of $SrCl_2 : Yb^{2+}$. *Opt. Spectrosc.*, Vol. 40, pp. 528-540

Kirton, J. & McLaughlan, S.D. (1967). Correlation of electron paramagnetic resonance and optical absorption spectra of CaF_2:Yb^{3+}. *Phys. Rev.*, Vol. 155, pp. 279-284

Kuwano, Y. (1982). Effective distribution coefficient of neodymium in Nd:$Gd_3Ga_5O_{12}$ crystals grown by the Czochralski method. *J. Crystal Growth*, Vol. 57, pp. 353-361

Lifante, G.; Cantelar, E.; Munoz, J.A.; Nevado, R.; Sanz-Garcia, J.A. & Cusso, F. (1999). Zn-diffused $LiNbO_3$:Er^{3+}/Yb^{3+} as a waveguide laser material. *Opt. Mat.*, Vol. 13, pp. 181-186

Mikkelsen, I.C. Jr. (1980). Three-zone Bridgman–Stockbarger crystal growth furnace. *Rev. Sci. Instrum.*, Vol.51, pp. 1564-1566

Mitric, A.; Duffar, T.; Diaz-Guerra, C.; Corregidor, V.; Alves, L.C.; Garnier, C. & Vian, C. (2006). Growth of $Ga_{(1-x)}In_xSb$ alloys by Vertical Bridgman technique under alternating magnetic field. *J. Cryst. Growth*, Vol. 287, pp. 224-229

Naumann, R.I. (1982). An analytical approach to thermal modeling of Bridgman-type crystal growth : I. One-dimensional analysis. *J. Cryst. Growth*, Vol. 58, pp. 554-568

Nicoara, D. (1975). Metal single crystals growth set-up. Romanian Patent nr.62842

Nicoara, D.; Schlett, Z. & Nicoara, I. (1983). Optical crystals growth set-up. Romanian Patent nr. 82663

Nicoara, D. & Nicoara, I. (1984). Crucibles for optical crystals growth. Romanian Patent nr. 85993

Nicoara, D.; Nicoara, I. & Schlett, Z. (1985). Heater for large diameter crystals growth. Romanian Patent nr. 88497

Nicoara, I.; Aczel, O.F.G.; Nicoara, D. & Schlett, Z. (1986). Dissolution kinetics and etch pit morphology of CaF_2 single crystals. *Crys. Res. Technol.*, Vol. 21, pp. 647-652

Nicoara, I.; Nicoara, D. & Aczel, O.F.G. (1987). Crystal growth in an improved bridgman variable shape graphite furnace. *Crys. Res. Technol.*, Vol. 22, pp. 1139-1144

Nicoara, D. & Nicoara, I. (1988). An improved Bridgman-Stockbarger crystal-growth system. *Mat. Science and Engineering A*,Vol. 102, L1-L4

Nicoara, I; Munteanu, M.; Pecingina-Garjoaba, N.; Stef, M. & Lighezan, L. (2006a). Dielectric spectrum of rare-earth-doped calcium fluoride crystals. *J.Crystal Growth*, Vol. 287, pp. 234-238

Nicoara, I; Munteanu, M.; Pecingina-Garjoaba, N.; Stef, M. & Lighezan, L. (2006b). Dielectric relaxation in PbF_2–doped and X-ray irradiated CaF_2 crystals. *ECS Transactions*, Vol. 3, pp. 51-58

Nicoara, I.; Pecingina-Garjoaba, N. & Bunoiu, O. (2008a). *J. Crystal Growth*, Vol. 310, pp. 1476-1480

Nicoara, I.; Lighezan, L.; Enculescu, M. & Enculescu, I. (2008b). Optical spectroscopy of Yb^{2+} ions in YbF_3-doped CaF_2 crystals. *J. Crystal Growth*, Vol. 310, pp. 2026-2032

Nicoara, I.; Stef, M. & Pruna, A. (2008c). Growth of YbF_3-doped CaF_2 crystals and characterization of Yb^{3+}/Yb^{2+} conversion. *J. Crystal Growth*,Vol. 310, pp. 1470-1475

Nicoara, I.; Munteanu, M.; Preda, E. & Stef, M. (2008d). Some dielectric and optical properties of ErF_3-doped CaF_2 crystals. *J. Crystal Growth*, Vol. 310, pp. 2020-2025

O'Connor, J.R. & Bostick, H.A. (1962). Radiation effects in CaF_2:Sm. *J. Appl. Phys.*, Vol. 33, pp. 1868-1870

Paraschiva, M.; Nicoara, I.; Stef, M. & Bunoiu, M. (2010). Distribution of Pb^{2+} ions in PbF_2-doped CaF_2 crystals. *Acta Physica Polonica A*, Vol. 117, pp. 466-470

Petit, V.; Camy, P.; Doualan, J.L. & Moncorge, R. (2007). Refined analysis of the luminescent centers in the Yb^{3+}:CaF_2 laser crystal. *J. of Luminesc.*, Vol. 122–123, pp. 5-7

Petit, V.; Camy,P.; Doualan, J.L.; Portier, J. & Moncorge, R. (2008). Spectroscopy of Yb^{3+}:CaF_2: From isolated centers to clusters. *Phys. Rev. B*, Vol. 78, pp. 085131

Pruna, A.; Stef, M. & Nicoara, I. (2009). Dielectric spectra of Li^+ (Na^+) codoped CaF_2:YbF_3 crystals. *Phys. Stat. Sol. A*, Vol. 206, pp. 738-744

Sangwal, K. & Arora, S.K. (1978). Etching of MgO crystals in acids: kinetics and mechanism of dissolution. *J. Mater. Sci.*, Vol. 13, pp. 1977-1985

Shelley, R.D. & Miller, G.R. (1970). Ionic thermocurrent study of the dipole-dipole interaction in CaF_2 doped with NaF. *J. Solis State Chemistry*, Vol. 1, pp. 218-226

Stockbarger, D.C. (1949). Artificial fluorite. *J. Opt. Soc. Am.*,Vol. 39, pp. 731-740

Strukov, B.A. & Levanyuk, A.P. (1998), Ferroelectric Phenomena in crystals, Springer, Berlin, ISBN 3540631321

Su, L.; Xu, J.; Li, H.; Wei, L.; Yang, W.; Zhao, Z.; Si, J.; Dong, Y. & Zhou, G. (2005). Crystal growth and spectroscopic characterization of Yb-doped and Yb, Na-codoped CaF_2 laser crystals by TGT. *J. Crystal Growth*, Vol. 277, pp. 264-268

Su, L. & all. (2005). Sites structure and spectroscopic properties of Yb-doped and Yb, Na-codoped CaF_2 laser crystals. *Chem. Phys. Lett.*, Vol. 406, pp. 254-258

Su, L. & all. (2007). Quaternary doping to improve 1.5 μm quantum efficiency of Er^{3+} in CaF_2 single crystal. *J. of Luminesc.*, Vol. 122-123, pp. 17-20

Sun, D.; Zhang, Q.; Wang, Z.; Su, J.; Go, C.; Wang, A. & Yin, S. (2005). Concentration distribution of Nd^{3+} in Nd:$Gd_3Ga_5O_{12}$ crystals studied by optical absorption method. *Cryst. Res. Technol.*, Vol. 40, pp. 698-702

Yonezawa, T.; Nakayama, J.; Tsukuma, K. & Kawamoto, Y. (2002). Behaviors of trace amounts of metal-oxide impurities in CaF_2 crystal grown by Stockbarger's method. *J. Cryst. Growth*, Vol. 244, pp. 63-69

5

The Growth and Properties of Rare Earth-Doped $NaY(WO_4)_2$ Large Size Crystals

Chaoyang Tu*, ZhenYu You, Jianfu Li, Yan Wang and Zhaojie Zhu
Key Laboratory of Photoelectric Materials Chemistry and Physics of CAS,
Fujian Institute of Research on the Structure of Matter,
Chinese Academy of Sciences,
P.R. China

1. Introduction

Recently, strong attention has been focused on development of a new-advanced material for optoelectronics applications. $MRe(WO_4)_2$ [M=alkali metal, Re=rare earth] single crystals is noticed as an interesting self-frequency conversion solid-state laser host material because of stimulated Raman scattering[1]. $NaY(WO_4)_2$ crystal is classified among the disorder crystalline host for lasing rare-earth ions[2]. Because of the disorder structure, the optical features in the absorption and emission spectrum even at low temperature are broadened.

The lattice parameters of $NaY(WO_4)_2$ crystal are a=b=5.205 Å and c=11.251 Å respectively with the space group of $I4_1/a$ [3]. This crystal is a typical tetragonal scheelite-type crystal with a formula $MT(WO_4)_2$, where M is a monovalent alkali cation and T a trivalent cation . In these materials the M and T cations are randomly distributed in the 2b and 2d sites [4], which can be replaced by rare earth ions, such as Nd^{3+}, Yb^{3+}, Tm^{3+}, Ho^{3+} and Ce^{3+}. As a consequence, the optical absorption and emission lines of rare earth doping ions become broadened, which allow some laser tunability as well as a better match with the available diode laser emissions used for pumping. As it melts congruently, large size single crystal can be easily obtained by the Czochralski (CZ) method. Furthermore, the higher concentration of rare earth ions can be accepted in the crystal because of the higher covalent characteristic results in the lower luminescent quenching efficiency. Compared to the other laser host crystals such as YAG and YVO_4 crystal, $NaY(WO_4)_2$ crystal has lower melting point and its raw materials for crystal growth is in-nocuity. As a result, $NaY(WO_4)_2$ crystal can serve as an excellent laser host. In this chapter, the crystal growth, thermal characteristic, optical and spectrum and laser properties of rare earth doped- $NaY(WO_4)_2$ crystals are presented.

2. The growth of large size crystals

Rare earth-doped $NaY(WO_4)_2$ crystals were grown in air along <001> direction by using Czochralski method[1~3]. The chemicals used were analytical grade Na_2CO_3, WO_3, Y_2O_3 and spectral grade Re_2O_3 (Re=Yb, Tm, Ho,Ce, Nd, Er). The starting materials were prepared by mixing Y_2O_3, Na_2CO_3, WO_3 and Re_2O_3 powders according to reaction formula:

$$Na_2CO_3+(1-x)Y_2O_3+x\ Re_2O_3+4WO_3 \rightarrow 2NaY_{1-x}Re_x(WO_4)_2+CO_2\uparrow$$

The weighed materials with doping 6 at% Re^{3+} were thoroughly mixed and pressed and put into a platinum crucible with Φ50×50 mm³, then heated to 750 °C and kept for 18 h to decompose the Na_2CO_3, and ground, mixed again, and then reheated to 800 °C, kept for 24 h. The obtained sample was very hard ceramics.

The synthesized material melted congruently at 1210 °C. The platinum crucible was heat by conventional RF-heating method. Crucible size is 50 mm in diameter and 50 mm in height. The pulling rate was 1-1.5 mm/h and the crystal rotation rate at 12-20 r.p.m. To release the stress produced in the temperature-lowering process, the crystals were annealed at 1200 °C for 5-6 h and then cooled down to room temperature at a rate of 20 K/h.

The earlier grown crystals as shown in Fig.2.1 occur screwy crack during anneal process. In order to avoid the cracking of the crystal, the designed after-heater should be used and the above crystal must be taken to anneal again in O_2 atmosphere according to a special temperature-controlled procedure.

As a result, high-quality (Tm^{3+},Ho^{3+},Nd^{3+},Yb^{3+},Er^{3+}/Yb^{3+}) rare earth doped-NYW cylinder crystal with dimension of Φ25 mm×100 mm (shown in Fig.2.2~2.5). The result shows that its optical homogeneity is 4×10^{-5}, as shown in Fig.2.5. It shows that it has excellent quality.

In order to obtain the large-sized rare earth doped-$NY(WO_4)_2$ crystals with high optical homogeneity, the control of growing processes and conditions are very important firstly. Then the used raw materials must be highly pure. Furthermore, to get the defined composition of the melt, the preparation of chemicals was found to be important. Thirdly, the seed surface must be melted to remove the defect in the seed before growing. And the growth point temperature must be a little higher than the saturation point. The control of the pulling rate, rotating rate and annealing rate is also very important. Finally, a designed after-heater should be used to avoid the crack of large crystal.

The concentration of rare earth ions in the $NaY(WO_4)_2$ crystal has been measured to by the inductively coupled plasma-atomic emission spectrometry (ICP-AES) method. A sample for the experiment has been cut from the top to eh boule. The concentration of Yb^{3+} ions has been 1.73 wt%. The distribution coefficient (K_0) of Yb^{3+} ions in the Yb:$NaY(WO_4)_2$ crystal has been calculated using the following relation:

$$K_0=C_A\,/\,C_0\,;$$

Where C_A is the Yb concentration at the top of the grown crystal and C_0 is the initial concentration of the admixture. The result indicates that the segregation coefficient of Yb^{3+} ions in Yb: $NaY(WO_4)_2$ crystal is approximately 1.02.

3. The thermal characteristic

The a and c axes were obtained by the YX-2 X-ray Crystal Oridentation Unit (produced by Dandong Radiative Instrument Co,Ltd). Two pieces of square samples with the size 5×5×5 mm³ having polished faces perpendicular to the a and c crystallophysical directions were used to carry out the measurements. The thermal expansion of as-grown Yb^{3+}:$NaY(WO_4)_2$ crystal was measured by using Diatometer 402 PC instrument from 300 K to 1273 K[1]. Because of the relatively lower reliability of the room temperature cell parameter arising out of presence of water in the sample chamber, only the data from 473 to 1273 K is considered in calculating the expansion coefficients. The thermal expansion pattern was obtained (shown in the Fig.3.1). The thermal expansion coefficients of the Yb^{3+}:$NaY(WO_4)_2$ crystal were calculated over different temperature ranges. In this case, the linear thermal expansion coefficients for different crystallographic direction c- and a-axes are , 1.83×10^{-5} K^{-1}, 0.85×10^{-5} K^{-1}, respectively.

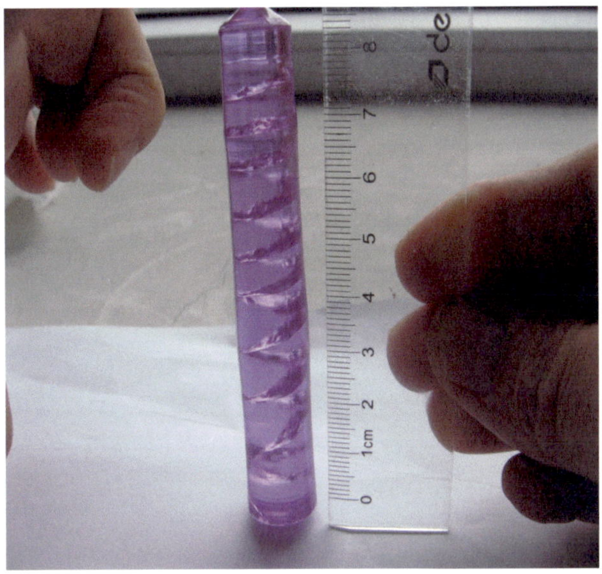

Fig. 2.1 The cracking Nd^{3+}:$NaY(WO_4)_2$ crystal.

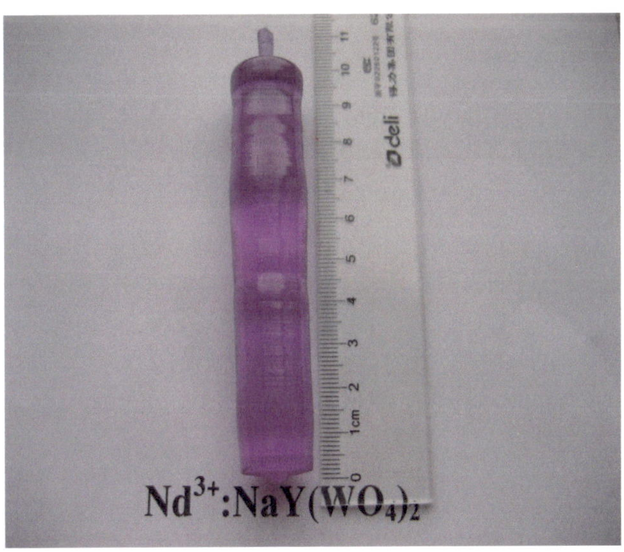

Fig. 2.2 The grown Nd^{3+}:$NaY(WO_4)_2$ crystal.

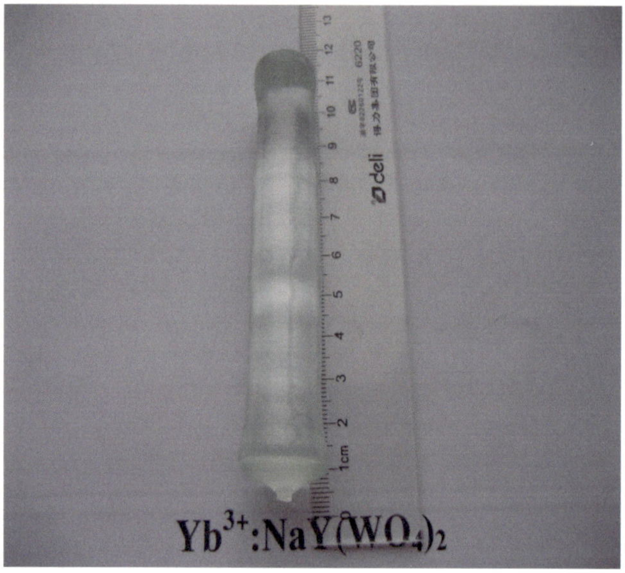

Fig. 2.3 The grown $Yb^{3+}:NaY(WO_4)_2$ crystal.

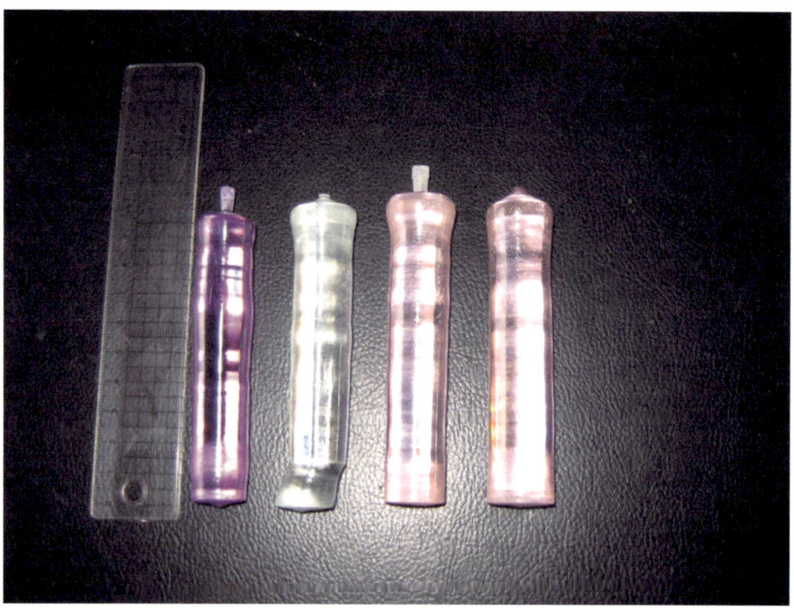

Fig. 2.4 The grown $Re^{3+}:NaY(WO_4)_2$ crystals (Re=Yb, Nd,Er/Yb).

Fig. 2.5 The grown Re^{3+}:$NaY(WO_4)_2$ crystals (Re=Tm/Ho).

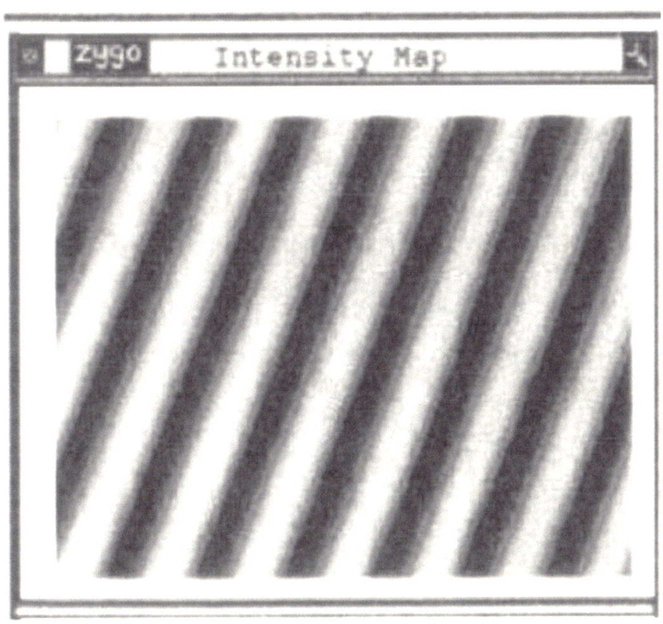

Fig. 2.6 Interference fringe of crystals.

The thermal-expansion coefficient $\left[\alpha_{ij}\right]$ of a crystal is a symmetrical second-rank tensor and it can be described by the representation quadric. The $NaY(WO_4)_2$ crystal belongs to the tetragonal system and 4/m point group. The unique symmetry axis is a fourfold axis along the crystallographic c-axis; the axes of the crystallographic and crystallophysical coordinate systems in $NaY(WO_4)_2$ have the same direction. In this case the value of thermal expansion along a- and b-axis are comparable and the values of α_1 and α_3 can be obtained by measuring the thermal expansion of the a- and c-oriented crystal.
The expansion coefficient in the [001] is about two times larger than that of the [100] direction according to our experimental results, which means that the $NaY(WO_4)_2$ crystal has anisotropic thermal expansion. The reason for the themal expansion coefficient along the c-axis being larger than that along the a- or b-axis can be explained by the structure of the $NaY(WO_4)_2$ crystal. The $NaY(WO_4)_2$ crystal has a scheelite structure according to the XRPD experiment results. According to Fig.3.1, it can be seen that there are five layers and three layers perpendicular to the c- and the a- or b-axis, respectively. According to the XRPD experiment results, the distance of the interlayer of five layers and three layers are c/4 and a/2 (or b/2),which is equal to 2.813×10^{-10} and 2.603×10^{-10} m, respectively. The larger the distance of the interlayer is, the weaker the chemical bonds of the interlayer will be according to the crystal lattice vibration dynamics. It can be seen that the interaction force along the c-axis is weaker than that along the a- or b-axis, and there are more layers in the c-direction than in the a-direction. Thus when the crystal is heated, the thermal expansion of the Yb^{3+}:$NaY(WO_4)_2$ crystal along the c-axis is larger than that along the a- or b-axis.

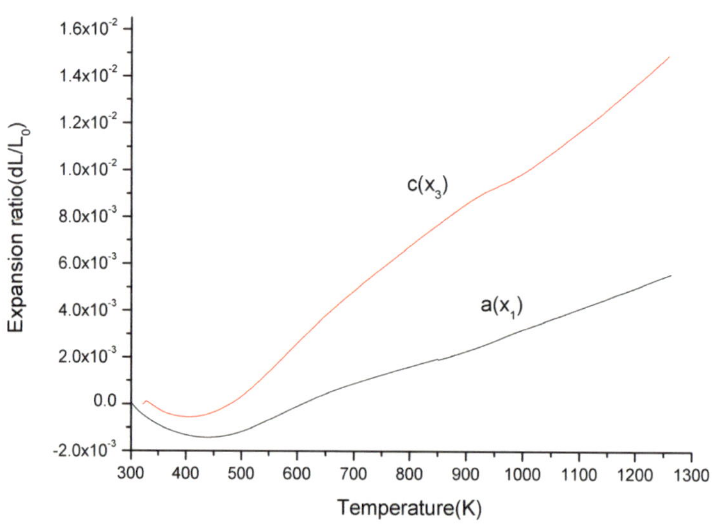

Fig. 3.1 The curve of thermal expansivity of Yb:$NaY(WO_4)_2$.

4. The spectroscopic characteristics

4.1 The spectroscopic characteristic of Nd^{3+}: $NaY(WO_4)_2$ crystal

Fig.4.1 shows the RT absorption spectrum of Nd^{3+}:$NaY(WO_4)_2$ Crystal. Owing to the disordered structure and the high Nd-doping concentration, the strong absorption intensity and broad FWHM of every band are shown, especially for the 806 nm[1]. Its FWHM is about 16 nm and the cross-section is about 2.8×10^{-20} cm^2 at 806 nm, which is benefit to the pumping of commercial laser diode. Fig.4.2 shows the RT emission spectrum with the pumping perpendicular to (001) planes. There are six emission peaks at follows wavelength: 894, 917,1063,1087,1339 and 1389 nm. The value of emission cross-section at 1063 nm is about 4.6×10^{-20} cm^2. Fig.4.3 shows the fluorescence decay of $^4F_{3/2}$ level of Nd^{3+} in NYW crystal at RT and the lifetime of $^4F_{3/2}$ level is about 85 μs and relative luminescent quantum efficiency is about 47% . Tab.4.1 presents the integrated absorbance, the line strengths, the experimental and calculated oscillator strengths. Table 4.2 shows the calculated radiative probabilities, radiative branching ratios and radiative time for the emissions from the $^4F_{3/2}$ level of Nd^{3+}:NYW crystal. Table 4.3-4 give the comparison of spectrum parameters in Nd:NYW and other Nd-doped crystals.

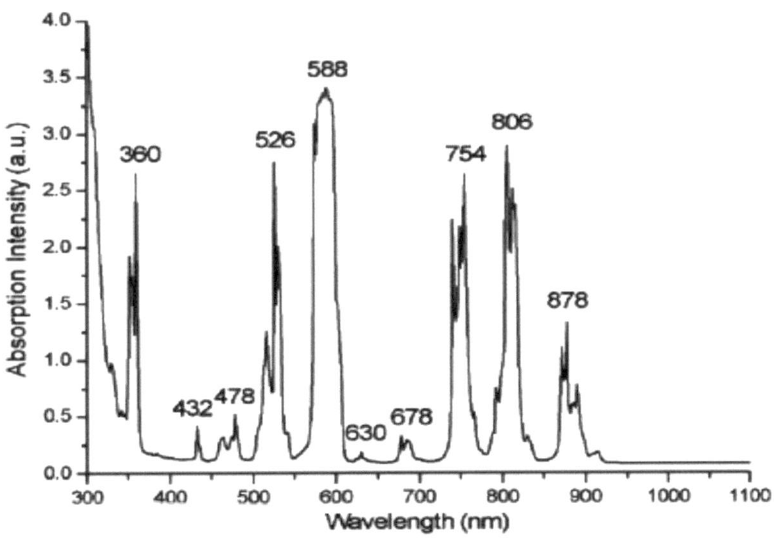

Fig. 4.1 Absorption spectra of Nd:$NaY(WO_4)_2$ crystal.

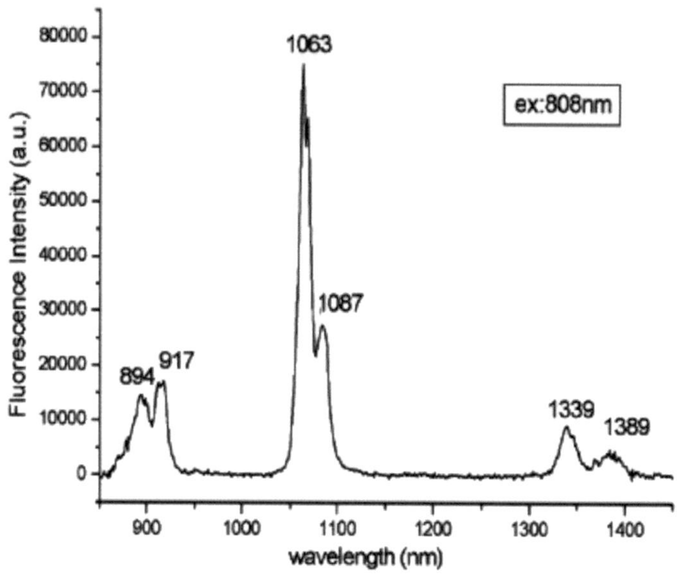

Fig. 4.2 Fluorescence spectra of $Nd:NaY(WO_4)_2$ crystal.

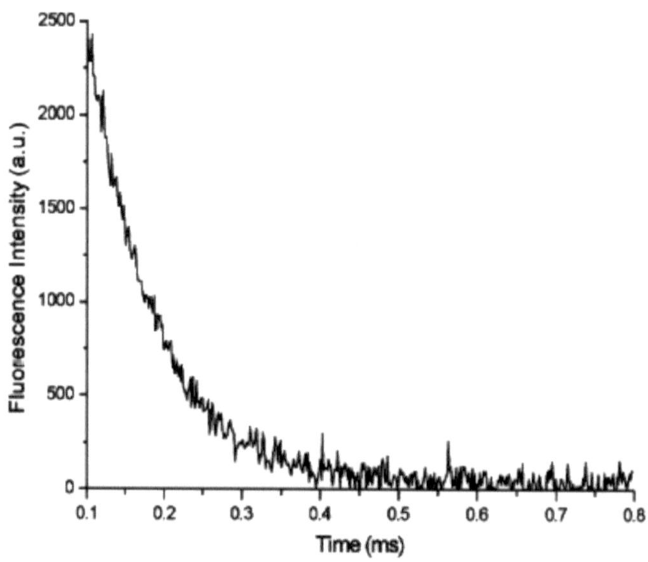

Fig. 4.3 Fluorescence decay of Nd:NYW crystal at RT.

Excited state	Wavelength (nm)	Γ (nm/cm)	S_{mea} ($10^{-29}cm^2$)	S_{cal} ($10^{-29}cm^2$)	f_{exp} (10^{-6})	f_{cal} (10^{-6})
$^4D_{1/2}+^4D_{3/2}+^5D_{5/2}$	360	71.2	2.862	3.127	15.874	17.342
$^2P_{1/2}$	432	4.57	0.153	0.221	0.708	1.021
$^2K_{13/2},,^2G_{9/2},,^2P_{3/2},,^2G_{11/2}$	478	18.42	0.558	0.416	2.329	1.739
$^4G_{7/2}+^4G_{9/2}+^2K_{13/2},$	526	130.86	3.600	2.186	13.666	8.296
$^4G_{5/2}+^2G_{7/2}$	588	375.88	9.251	9.345	31.412	31.732
$^2H_{11/2}$	630	1.47	0.034	0.061	0.107	0.193
$^4F_{9/2}$	678	10.2	0.218	0.238	0.641	0.702
$^4F_{7/2}+^4S_{3/2}$	754	159.94	3.070	3.115	8.129	8.249
$^4F_{5/2}+^2H_{9/2}$	806	191.84	3.444	3.681	8.532	9.118
$^4F_{3/2}$	878	78.2	1.289	1.554	2.931	3.533

Table 4.1 The integrated absorbance, the line strengths, the experimental and calculated oscillator strengths of Nd:NYW crystal.

Start levels	Wavelength(nm)	A(S^{-1})	β	τ (μs)
$^4I_{9/2}$	894	2612	0.473	180
$^4I_{11/2}$	1063	2434	0.441	
$^4I_{13/2}$	1339	450	0.082	
$^4I_{15/2}$	1852	23	0.004	

Table 4.2 Calculated radiative probabilities, radiative branching ratios and radiative time for the emissions from the $^4F_{3/2}$ level of Nd^{3+}:NYW crystal.

Crystal	Ω_2 (10^{-20} cm^2)	Ω_4 (10^{-20} cm^2)	Ω_6 (10^{-20} cm^2)	FWHM (nm)	σ_a at 808 nm (10^{-29} cm^2)	*Ref.*
Nd:NYW	5.8	5.74	4.32	16	2.80	[1]
Nd:NBW	30.9	12	9.3	10	2.6	[2]
Nd:KYW	8.80	3.11	3.16	8	3.13	[3]
Nd:KGW	12.67	10.15	7.48	12	26	[4,5]
Nd:YVO_4	5.88	4.08	5.11	8	27	[6]
Nd:$GdVO_4$	12.629	4.828	8.125	4.7	9.396	[7]
Nd:YAG	0.2	2.7	5.0	0.9	7.0	[8,9]
Nd:YAP	0.69	3.69	4.56	3	10.2	[10,11]
Nd:GAB	3.118	2.676	5.343	8.7	4.3	[12]

Table 4.3 Comparison of spectral values in Nd:NYW and other Nd-doped crystals.

Crystal	σ_e at 1064nm (10^{-20} cm^2)	τ_f (μs)	η (%)	References
Nd:NYW	4.56	85	47	[1]
Nd:NBW	16	122	85	[2]
Nd:KYW	4.5	154	78.6	[3]
Nd:KGW	38	110	92.4	[4,5]
Nd:YVO_4	100	98	46.8	[6]
Nd:$GdVO_4$	7.6	90	45.5	[7]
Nd:YAG	34	240	91	[8,9]
Nd:YAP	24.4	180.4	88	[10,11]
Nd:GAB	30	55.6	19	[12]

Table 4.4 Comparison of the emission spectroscopic parameters of some Nd-doped laser crystal.

4.2 The spectroscopic characteristic of Yb^{3+}: $NaY(WO_4)_2$ crystal

Fig.4.4 shows the RT Polarized absorption spectrum of Yb^{3+}:$NaY(WO_4)_2$ Crystal. The largest absorption cross-section is located at 976 nm in the $\sigma-$ and $\pi-$ spectrum, which is the main pump wavelength of the possible Yb^{3+} laser in $NaY(WO_4)_2$ crystal using the InGaAs LD, and the value is 1.81×10^{-20} cm^{-2}. This value is smaller than that of Yb^{3+}:$KY(WO_4)_2$ (13.3×10^{-20}) crystals at 981 nm [13], but larger than those of Yb^{3+}:YAG crystal (0.8×10^{-20} cm^2) [14] at 942 nm and Yb^{3+}:YCOB crystal (1.0×10^{-20} cm^2) [15] at 976nm. Fig.4.5 shows the RT Polarized emission spectrum of Yb^{3+}:$NaY(WO_4)_2$ Crystal. The emission cross-sections of crystal calculated from the fluorescence spectra by the reciprocity method and the Füchtbauer-Ladengurg formula are shown in Fig.4.6 [16~18]. The radiative lifetime τ_r of the $^2F_{3/2}$ manifold is measured to be 0.902 ms. The gain coefficient was calculated for several values of population inversion P (P=0, 0.1, 0.2…) and is shown in Fig.4.7 (a) and Fig.4.7 (b). Positive gain coefficient for P values larger than 0.5, which are encountered in a free-running laser operation, implies a tuning range from 990 to 1070 nm.

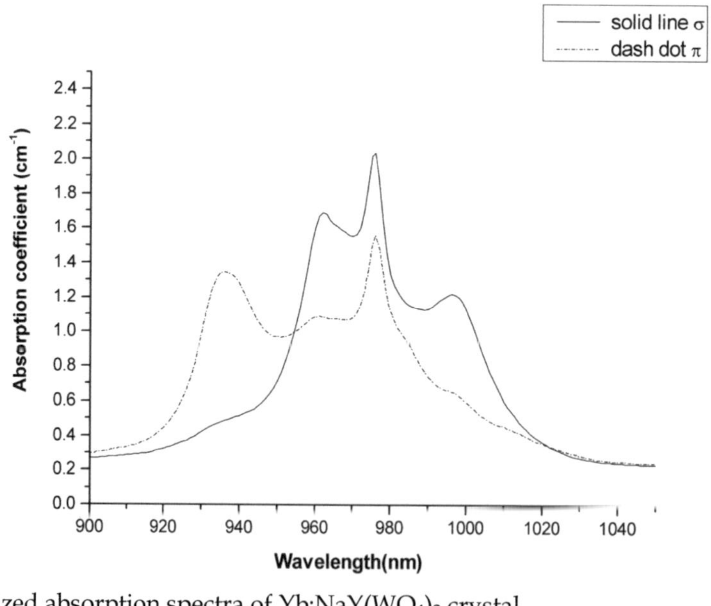

Fig. 4.4 Polarized absorption spectra of Yb:$NaY(WO_4)_2$ crystal.

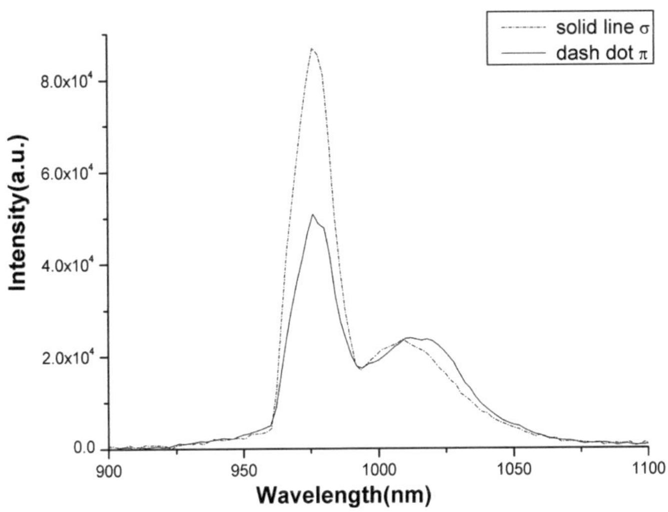

Fig. 4.5 Polarized fluorescence spectra of $Yb:NaY(WO_4)_2$ crystal.

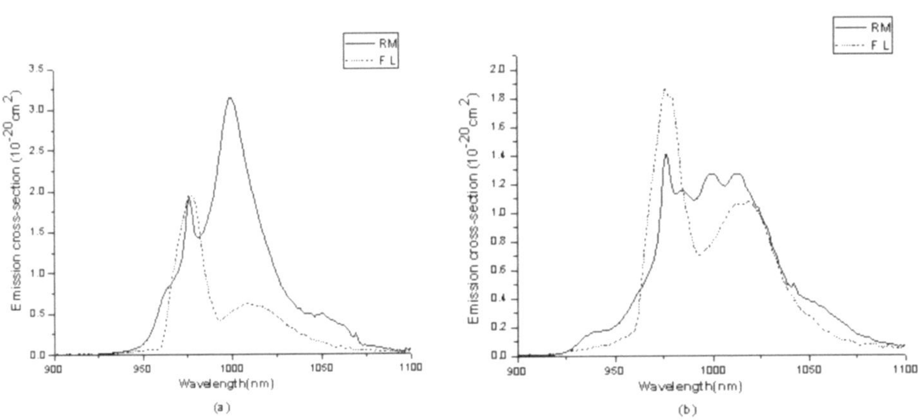

Fig. 4.6 The emission cross-sections of crystal calculated from the fluorescence spectra by the reciprocity method and the Füchtbauer-Ladengurg formula.

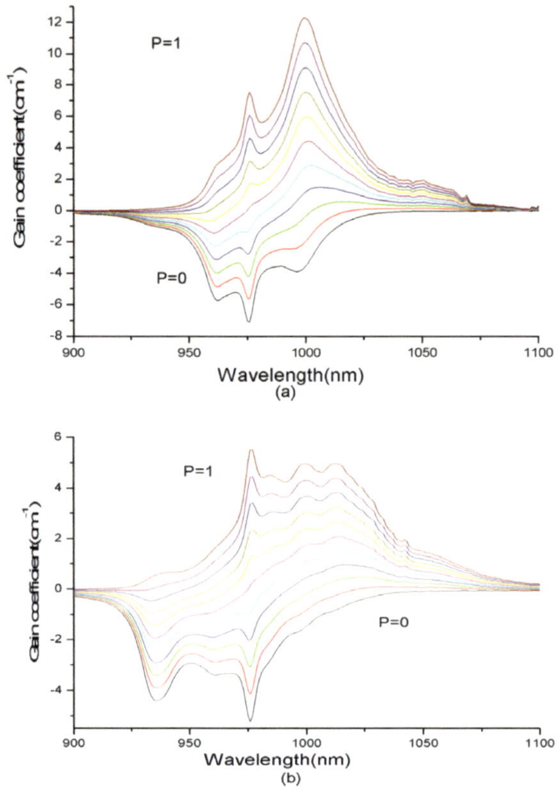

Fig. 4.7 The gain coefficient was calculated for several values of population inversion P (P =0, 0.1, 0.2...).

4.3 The spectroscopic properties of Tm^{3+}, Ho^{3+}: $NaY(WO_4)_2$ crystal

Fig.4.8 shows the Room temperature absorption spectra of Tm^{3+}-, Ho^{3+}-doped and Tm^{3+}/Ho^{3+} co-doped $NaY(WO_4)_2$ crystals (a) in the range 300-850 nm and (b) in the range 1100-2100 nm. The spectrum of Tm^{3+}: $NaY(WO_4)_2$ crystal consists of six resolved bands associated with the transitions from the ${}^3H_6$ ground state to the ${}^3F_4$, ${}^3H_5$, ${}^3H_4$, ${}^3F_{2,3}$, ${}^1G_4$ and ${}^1D_2$ excited states. It can be seen that the absorption band of the σ polarization is narrower and has a larger peak cross section than the π absorption band. The spectrum of Ho^{3+}: $NaY(WO_4)_2$ crystal consists of ten resolved bands associated with the transitions from the ${}^5I_8$ ground state to the ${}^5I_7$, ${}^5I_6$, ${}^5F_5$, ${}^5F_4$+${}^5S_2$, ${}^5F_3$, ${}^3K_8$+${}^5F_2$, ${}^5F_1$+${}^5G_6$, ${}^5G_5({}^3G_5)$,, ${}^3H_6$+${}^5F_2$+${}^3H_5$ and ${}^3K_6$+${}^3F_4$+${}^3H_4$+${}^3G_4$ excited states[19]. Some absorption bands of Tm^{3+} and Ho^{3+} ions overlap in the Tm^{3+}/Ho^{3+}:$NaY(WO_4)_2$ crystal. Compared to Ho^{3+} ions concentration in Ho^{3+}: $NaY(WO_4)_2$ and Tm^{3+} ions concentration in Tm^{3+}/Ho^{3+}:$NaY(WO_4)_2$ crystal, the concentration of Ho^{3+} ions in Tm^{3+}/Ho^{3+}:$NaY(WO_4)_2$ crystal is very low; the ${}^5I_8 \rightarrow {}^5I_7$ (Ho^{3+}) transition of Tm^{3+}/Ho^{3+}:$NaY(WO_4)_2$ crystal is extremely weak.

Fig.4.9 shows the absorption cross sections and polarized stimulated emission cross sections associated with the (a) ${}^3F_4 \rightarrow {}^3H_6$ transition for the Tm^{3+}:$NaY(WO_4)_2$ and (b) ${}^5I_7 \rightarrow {}^5I_8$ for Ho^{3+}:$NaY(WO_4)_2$ crystal derived by the reciprocity method. The maximum values of σ_{em}

are 1.399x10^{-20} cm^2 for σ polarization at 2044 nm and 1.426x10^{-20} cm^2 for π polarization at 2047 nm. For comparison, the σ_{em} obtained for Tm^{3+} in NLuW are 2.0(±0.1)x10^{-20} cm^2 at 1798 nm and 1.9(±0.1)x10^{-20} cm^2 at 1830nm, respectively[20] . The FWHMs of the emission bands for σ and π polarizations are 161 and 130 nm, respectively.

Fig.4.10 presents the gain cross-section calculated for different values of P (P=0.1~0.5) for (a) the $^3F_4 \rightarrow ^3H_6$ transition of Tm^{3+} in NaY(WO$_4$)$_2$ crystal and (b) the $^5I_7 \rightarrow ^5I_8$ transition of Ho^{3+} in NaY(WO$_4$)$_2$ crystal. The gain curves at a wavelength longer than 1900 nm are obscure due to the low signal-to-noise ratio of the absorption spectrum. The positive gain cross-section can be obtained at about 2.0 μm when P exceeds 0.2. The positive gains for P=0.5 are in a range from 1758 to about 1954 nm for σ polarization and from 1758 to about 1977 nm for π polarization, respectively.

Fig.4.11 presents the room temperature fluorescence spectra of Tm^{3+}-, Ho^{3+}-doped and Tm^{3+}/Ho^{3+} co-doped NaY(WO$_4$)$_2$ crystals.

Fig.4.12 gives the decay curves of $^3F_4$ manifold in the samples of bulk and powder in the Tm^{3+} doped NaY(WO$_4$)$_2$ crystals. Fig.4.13 also gives the decay curves of Ho: $^5I_7$ level in the (a) samples of bulk and powder in the Ho^{3+}:NaY(WO$_4$)$_2$ and (b) Tm, Ho:NaY(WO$_4$)$_2$ crystals.

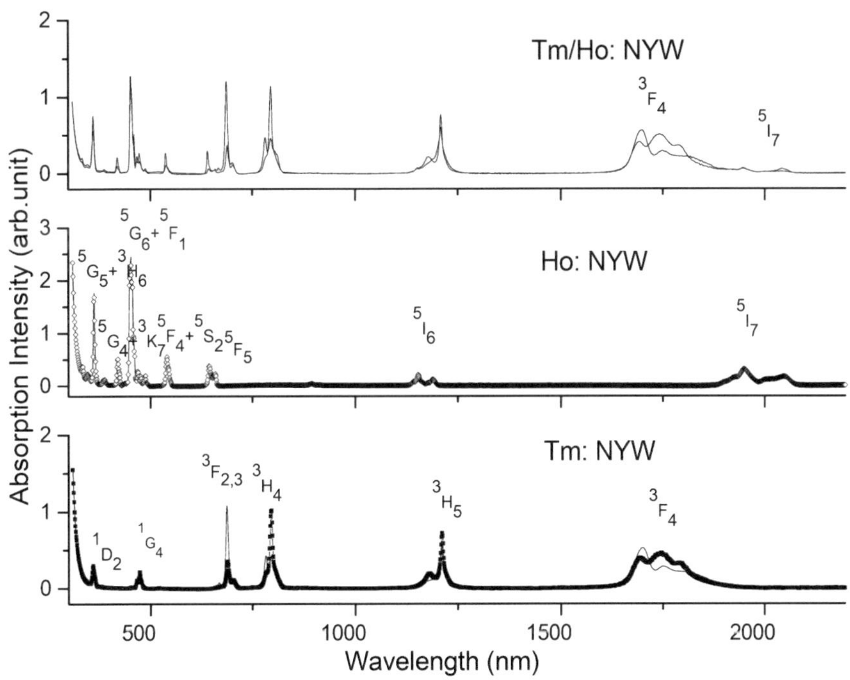

Fig. 4.8 Room temperature absorption spectra of Tm^{3+}-, Ho^{3+}-doped and Tm^{3+}/Ho^{3+} co-doped NaY(WO$_4$)$_2$ crystals.

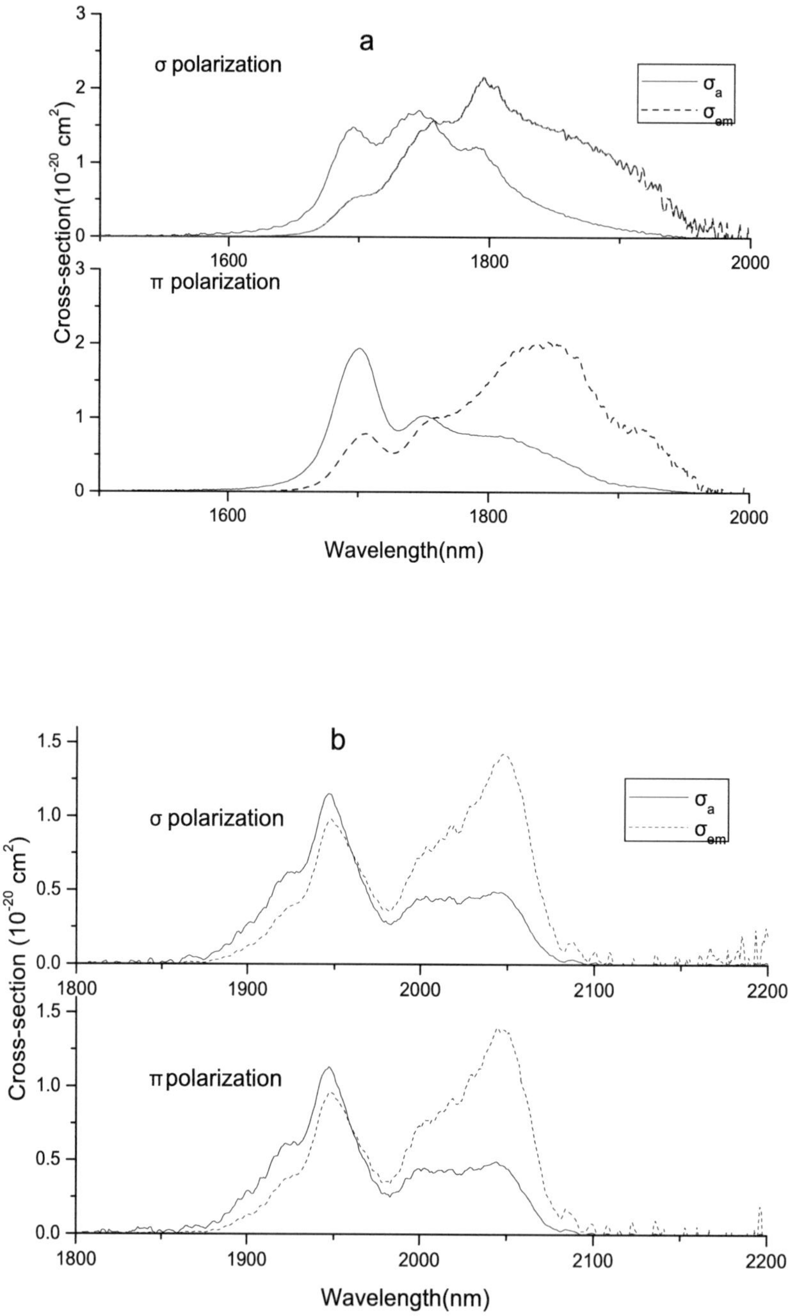

Fig. 4.9 Absorption cross sections and polarized stimulated emission cross sections associated with the (a) $^3F_4 \rightarrow ^3H_6$ transition for the Tm^{3+}:$NaY(WO_4)_2$ and (b) $^5I_7 \rightarrow ^5I_8$ for Ho^{3+}:$NaY(WO_4)_2$ crystal derived by the reciprocity method.

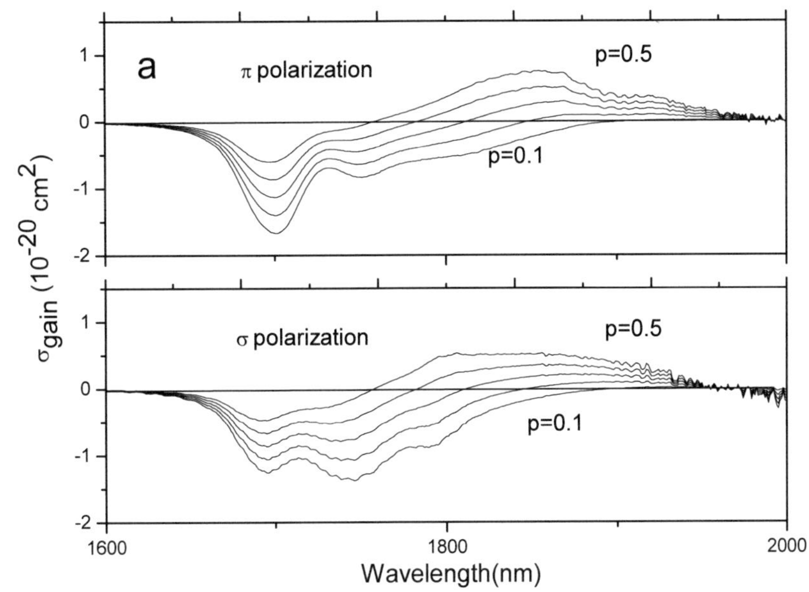

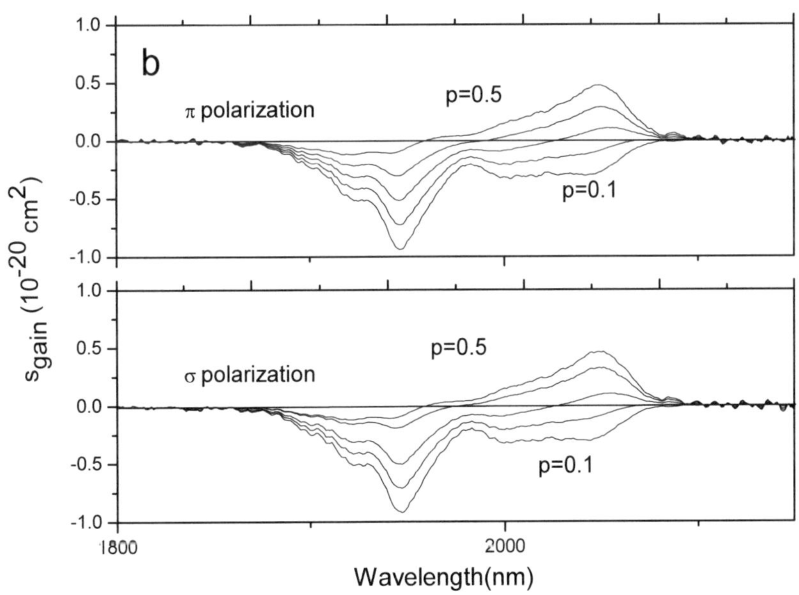

Fig. 4.10 Gain cross-section calculated for different values of P (P=0.1~0.5) for (a) the $^3F_4 \rightarrow ^3H_6$ transition of Tm^{3+} in $NaY(WO_4)_2$ crystal ,(b) the $^5I_7 \rightarrow ^5I_8$ transition of Ho^{3+} in $NaY(WO_4)_2$ crystal.

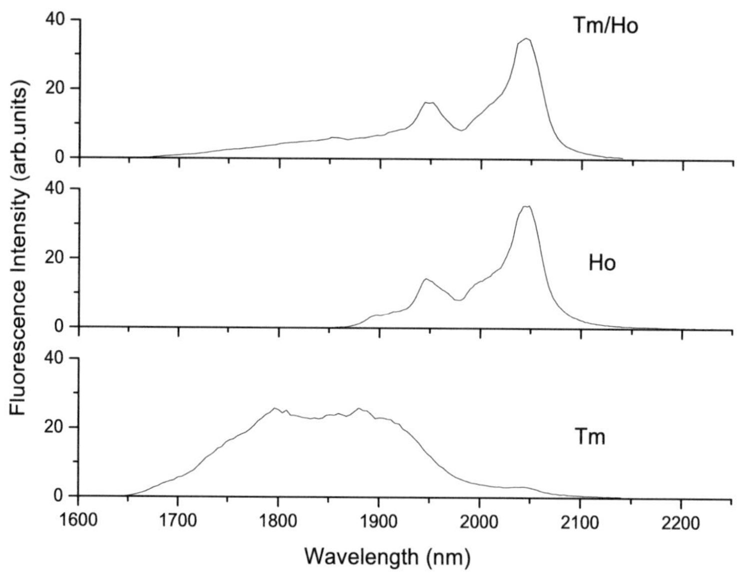

Fig. 4.11 Room temperature fluorescence spectra of Tm^{3+}-, Ho^{3+}-doped and Tm^{3+}/Ho^{3+} co-doped $NaY(WO_4)_2$ crystals.

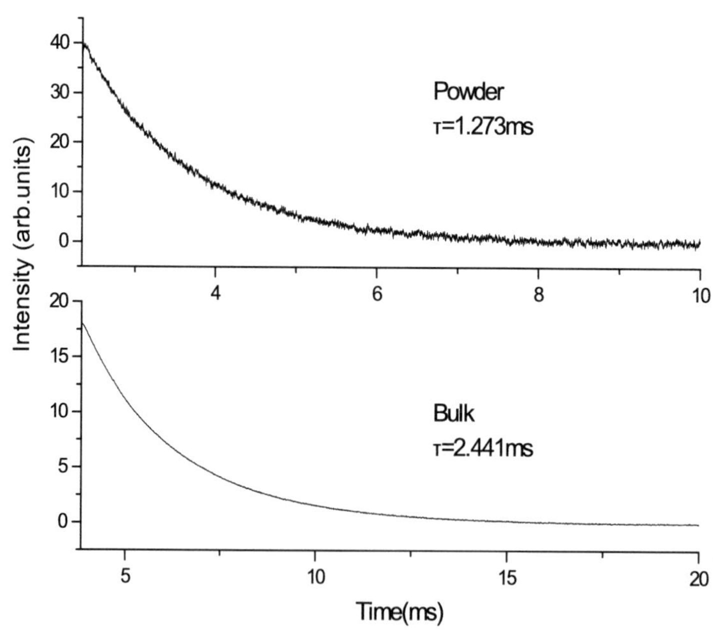

Fig. 4.12 Decay curves of $^3F_4$ manifold in the samples of bulk and powder in the Tm^{3+} doped $NaY(WO_4)_2$ crystals.

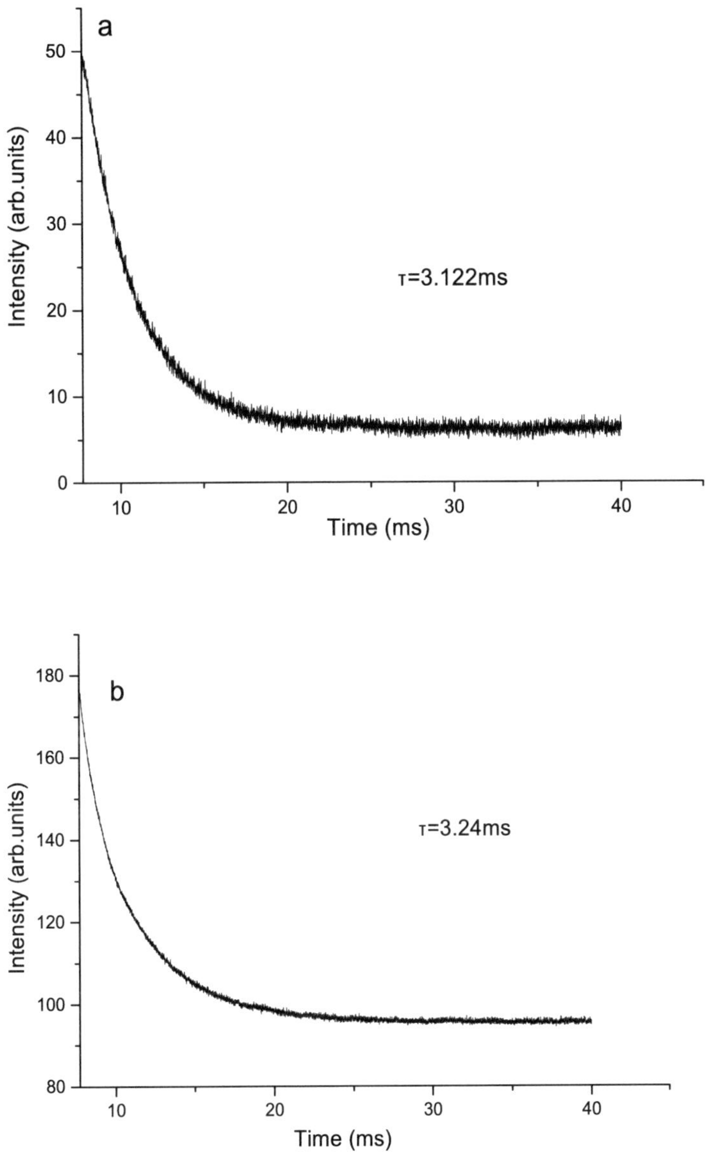

Fig. 4.13 Decay curves of Ho: $^5I_7$ level in the (a) samples of bulk and powder in the Ho^{3+}: $NaY(WO_4)_2$ and (b) Tm, Ho: $NaY(WO_4)_2$ crystals.

5. The laser characteristics

5.1 The laser characteristics of Nd^{3+}: $NaY(WO_4)_2$ crystal

The Nd^{3+}:$NaY(WO_4)_2$ crystal was made into laser stick and the laser experiment was performed using a xenon flash lamp as a pump source[1]. Maximum pulse energy of 786 mJ with a repetition rate of 1 Hz has been obtained. A maximum output power of 87 mW at 532 nm has been obtained and the double-frequency conversion efficiency is more than 25% when a LBO optical crystal was used as the frequency-doubling crystal. Table 5.1 shows the data of input and output energy and Fig.5.1 presents the relationship between the Iuput energy and output energy. Table 5.2 presents the Comparison of laser properties of Nd:NYW crystal and Nd:YAG crystal. It can be found that the Nd:NYW crystal has the higher laser efficiency than Nd:YAG crystal. Table.5.3 shows the frequency-doubling laser output power and conversion efficiency and Fig.5.2 presents the relationship between the pump power and output power of SH generation. Fig.5.3 shows the laser facula of the SH generation.

Input V (J)	static state output (mJ)	Wavelength (μm)
440 (9.68)	7	1.063
480 (11.5)	33	
520 (13.5)	67	
560 (15.7)	106	
600 (18.0)	173	
640 (20.5)	215	
680 (23.1)	275	
720 (25.9)	346	
760 (28.9)	376	
800 (32.0)	443	
840 (35.3)	512	
880 (38.7)	572	
920 (42.3)	671	
960 (46.1)	735	
1000 (50)	786	

Table 5.1 Data table of pumping energy and output energy.

Crystal	Size	Input	Output
Nd:NYW Crystal	Φ4.5 mmx52 mm	50 J	786 mJ
Nd:YAG Crystal	Φ8 mmx120 mm	50 J	~800 mJ

Table 5.2 Comparison of laser properties of Nd:NYW and Nd:YAG.

Input	Output/530nm	Threshold	Optical to optical efficiency for SHG	Slope efficiency	Optical to optical efficiency for ground frequency
1500 mW	87 mW	410 mW	5.80%	7.98%	> 25%

Table 5.3 SH generation power and conversion efficiency.

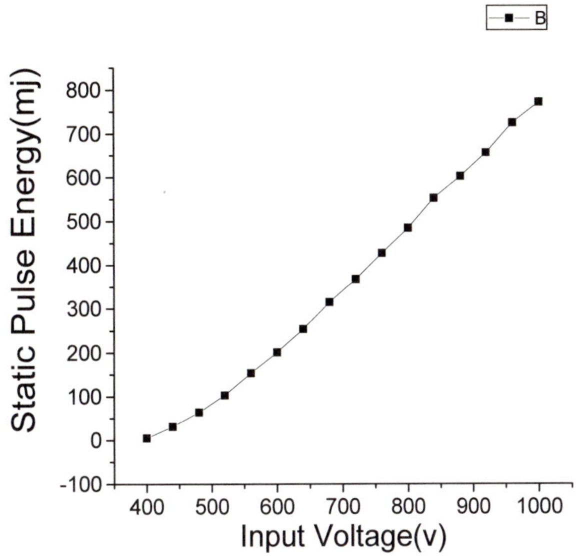

Fig. 5.1 The relationship between the Iuput energy and output energy at 1.062 um.

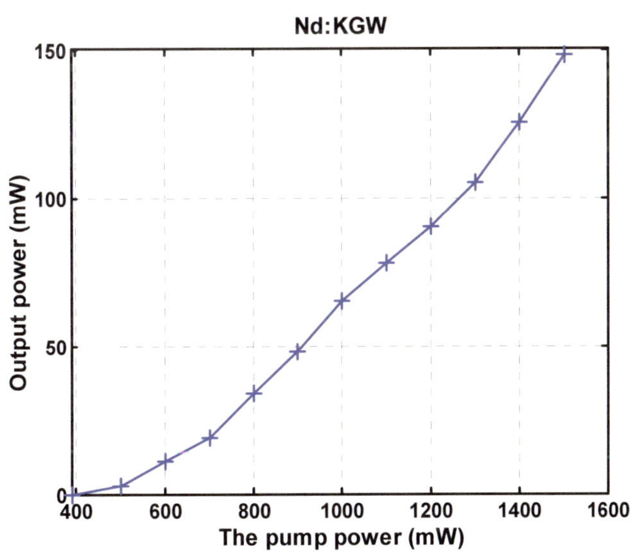

Fig. 5.2 The relationship between the pump power and output power of SH generation.

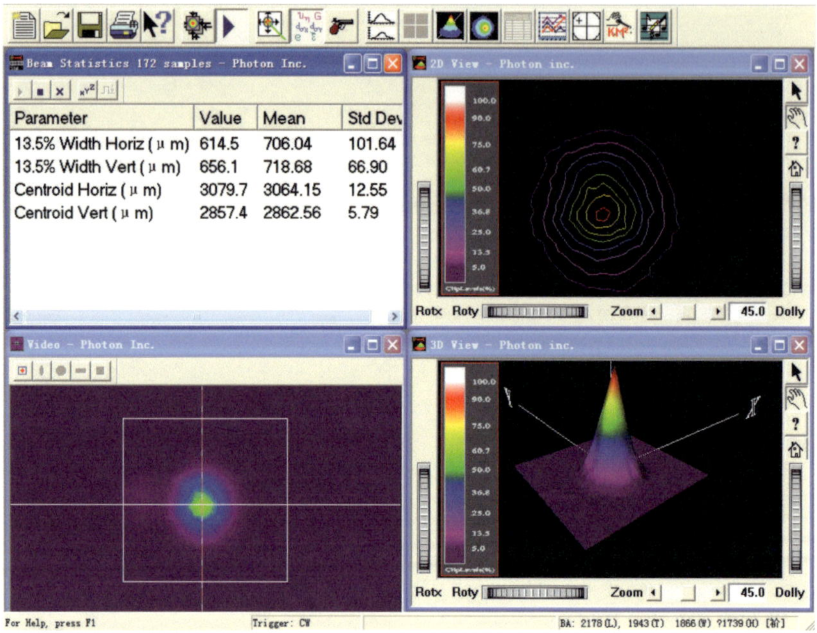

Fig. 5.3 The laser facula of the SH generation.

5.2 The laser characteristics of Tm, Ho: $NaY(WO_4)_2$ crystal

An infrared laser output at 2.07 μm with Tm, Ho:$NaY(WO_4)_2$ crystal end-pumped by 795 nm laser diode at room temperature is reached[2,3]. Fig.5.4 shows the experimental configuration of the LD-end-pumping Tm, Ho: NYW laser. The crystal used with the concentrations of 5 at% Tm^{3+} and 1 at% $Ho^{3+}$was grown by the Czochralski method. The highest output power was up to 2.7 W corresponding to the crystal temperature being controlled at 283 K. Fig.5.5 presents the output power versus pumping power at different temperatures. The overall optical conversion efficiency was 5.4% and the slop efficiency was 26%. The output characteristics and the laser threshold affected by the pulse duration and temperature have been studied. It can be found that the stability of the output power was correlative with the crystal temperature heavily. In addition, the wider pulse duration of pump could promote the output power efficiently as shown in Fig.5.6, which presents the output power versus pulse duration.

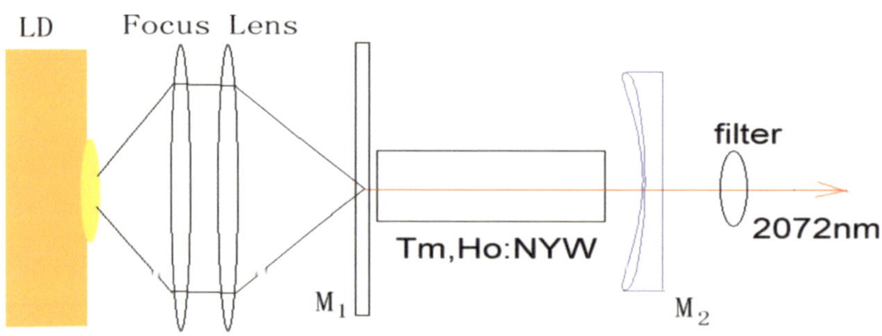

Fig. 5.4 The experimental configuration of the LD-end-pumping Tm, Ho: NYW laser.

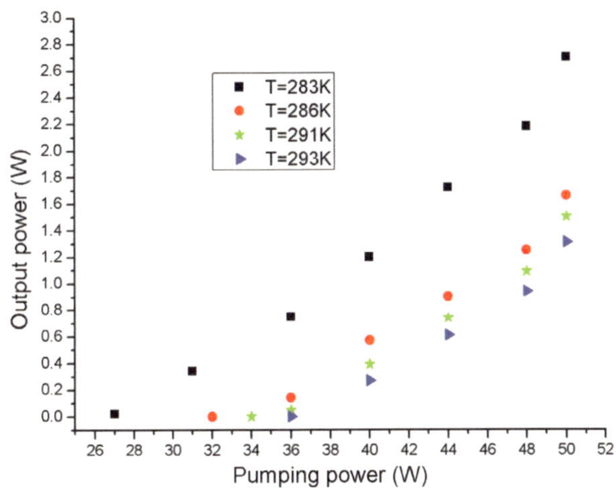

Fig. 5.5 The output power versus pumping power at different temperatures.

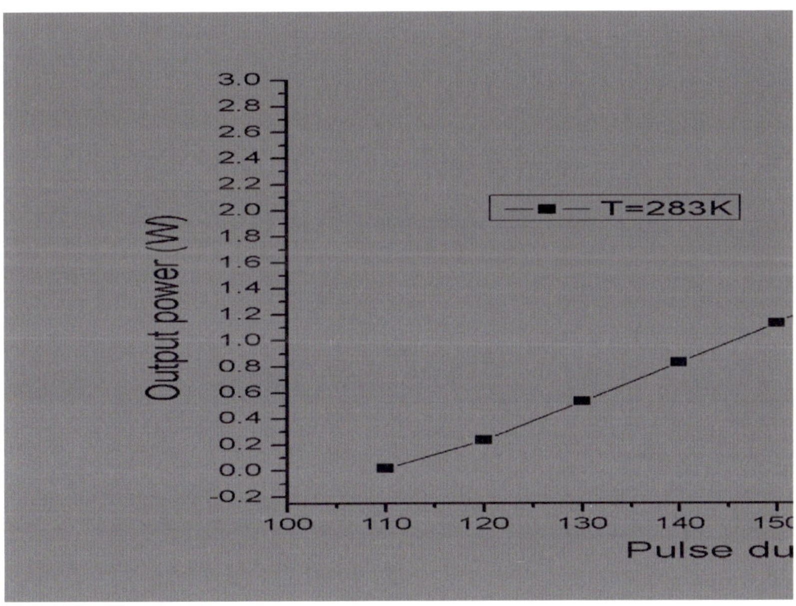

Fig. 5.6 The output power versus pulse duration.

With Ti:sapphire laser pumping at 795 nm, a slope efficiency and a maximum output power as high as 48% and 265 mW, respectively, have been obtained at 2050 nm from a Tm:Ho: $NaY(WO_4)_2$ crystal by Prof.C.Zaldo[4]. Tuning from 1830 nm to 2080 nm has also been achieved using an intracavity Lyot filter. Fig.5.7 shows Cw laser performance of Tm(5 at%), Ho(0.25 at%): $NaY(WO_4)_2$ crystal. Fig.5.8 shows Cw and quasi Cw laser performance of Tm(5 at%), Ho(0.5 at%): $NaY(WO_4)_2$ crystal.

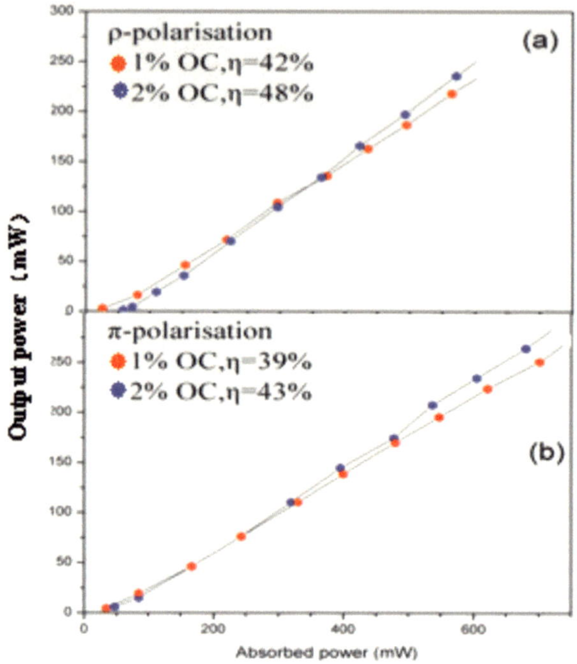

Fig. 5.7 Cw laser performance of Tm(5 at%),Ho(0.25 at%): $NaY(WO_4)_2$ crystal. (a) σ -pol (1.55-mm-long crystal).(b) π -pol (3.4-mm-long crystal).

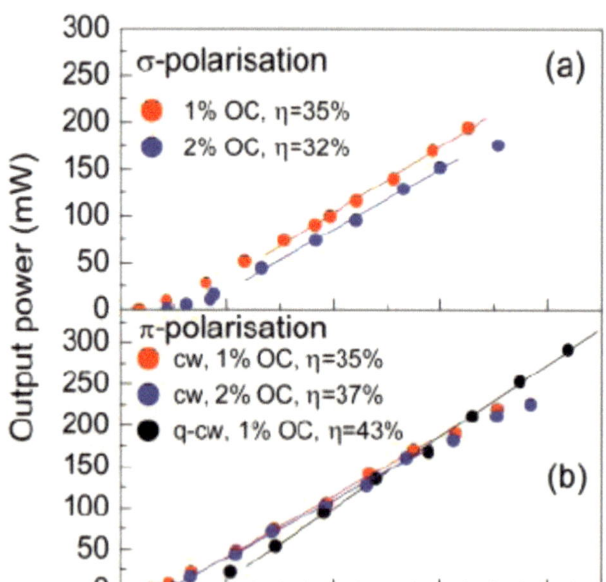

Fig. 5.8 Cw and quasi Cw laser performance of Tm(5 at%),Ho(0.5 at%): $NaY(WO_4)_2$ crystal. (a) σ -pol (1.75-mm-long crystal).(b) π -pol (3.5-mm-long crystal).

Prfo.A.A.Lagatsky and C.Zaldo [5] also reported the femtosecond-pulse operation of a Tm:Ho:$NaY(WO_4)_2$ laser at around 2060 nm by using an ion-implated InGaAsSb quantum-well-based semiconductor saturable absorber mirror for passive mode-locking maintenance for the first time. Transform-limited 191fs pulses are produced with an average output power of 82 mW at a 144 MHz pulse repetition frequency. Maximum output power of up to 155 mW is generated with a corresponding pulse duration of 258 fs. Fig.5.9 presents the Input-output characteristics of the mode-locked Tm:Ho:$NaY(WO_4)_2$ laser.

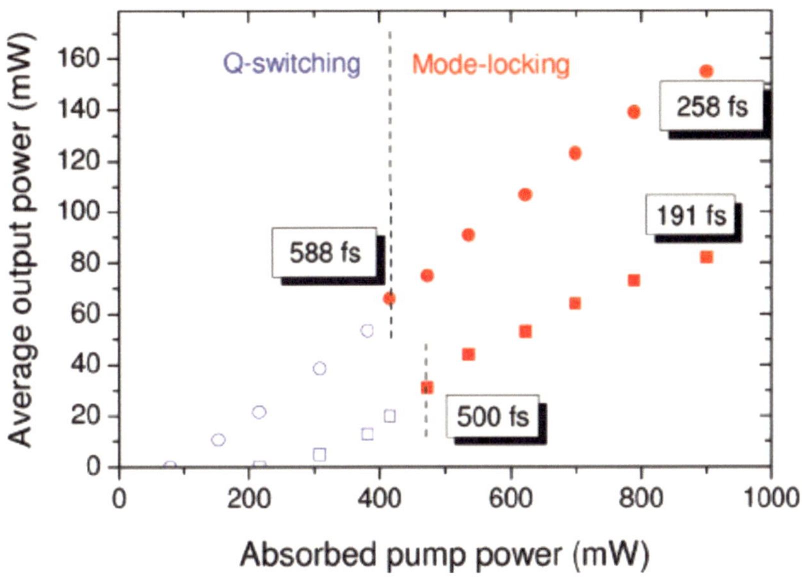

Fig. 5.9 Presents the Input-output characteristics of the mode-locked Tm:Ho:$NaY(WO_4)_2$ laser.

Two different operation regimes, shorter-pulse and higher-power, are indicated by squares and circles, respectively. Q-switching and mode-locking regimes are represented by open and closed symbols, respectively.

5.3 The laser characteristics of Tm, Ho, Ce: $NaY(WO_4)_2$ crystal

An infrared laser output at 2.07 μm with Tm,Ho,Ce:$NaY(WO_4)_2$ single crystal end-pumped by 795 nm laser diode at room temperature[3,6]. The crystal used with the concentrations of 5 at% Tm^{3+}, 1 at% Ho^{3+} and 30 at% Ce^{3+} was grown by the Czochralski method. The highest output power was up to 0.2 W corresponding to the pumping power of 50 W and the threshold was about 40 W at 293 K. Figure 5.10 shows the output power versus the pump power. The introduction of Ce^{3+} brought about a novel phenomenon. End-pumping with the 795 nm LD, it was found the up-conversion was repressed heavily and the green emission disappeared thoroughly in Tm,Ho,Ce:$NaY(WO_4)_2$ crystal, which was particularly different from the crystal Tm,Ho:$NaY(WO_4)_2$, where the green emission was obvious and weakened the sensitized transition energy.

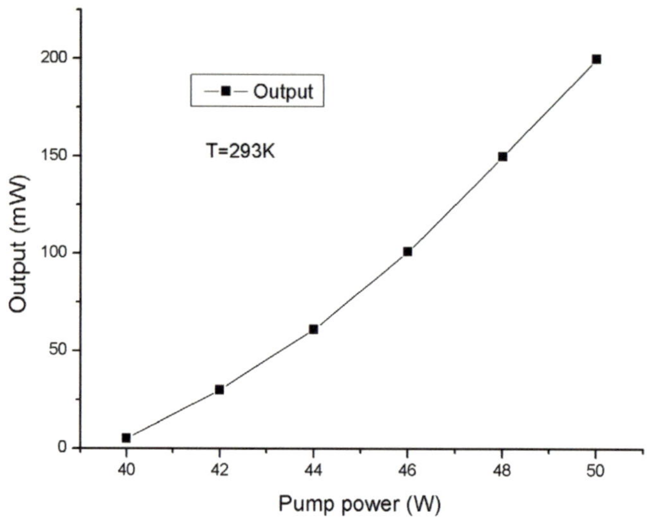

Fig. 5.10 The output power versus the pump power.

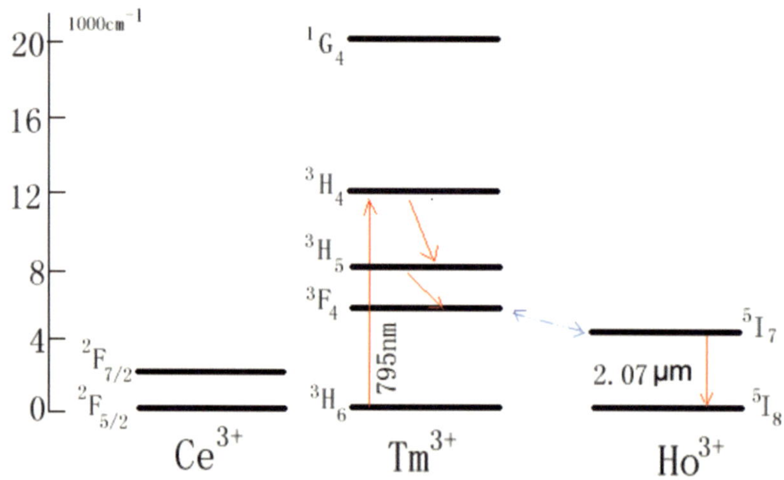

Fig. 5.11 The illustration of Ce^{3+}, Tm^{3+}, Ho^{3+} energy levels.

The original intention of selecting the Ce^{3+} was to compensate the up-conversion loss. The compensation mechanisms of the Ce^{3+} lie in its transition energy. As shown in Fig.5.11, the transition energy of ${}^2F_{7/2}\rightarrow{}^2F_{5/2}$ (Ce^{3+}) is close to that of ${}^3H_5\rightarrow{}^3F_4$ (Tm^{3+}) and half of the

$^3H_4 \rightarrow ^3H_5$ (Tm^{3+}). Pumped with 795 nm, the electrons will transit from $^3H_6$ to $^3H_4$, and jump to $^3H_5$, $^3F_4$ depending on the radiationless transition. Because the energy level $^3F_4$ (Tm^{3+}) was close to $^5I_7$ (Ho^{3+}), the electrons will transit from $^3F_4$ (Tm^{3+}) to $^5I_7$ (Ho^{3+}), which is just the sensitized process. At last, the transition $^5I_7 \rightarrow ^5I_8$ (Ho^{3+}) generates the2.07 μm laser. In the complex sensitized process, only few of the electrons will transit from the upper pumping energy level $^3H_4$ (Tm^{3+}) into the $^5I_7$ (Ho^{3+}), which is the reason of the lower laser efficiency. By virtue of the Ce^{3+}, the electrons of the $^3H_4$ (Tm^{3+}) can transit fast into the energy level $^3F_4$ (Tm^{3+}). More important, the multiple transition energy can guide the electrons towards $^3F_4$ (Tm^{3+}) instead of irregular radiationless transition. That is to say, in the shorter time, there are more electrons gathering into the energy level $^5I_7$ (Ho^{3+}), which is just the demand of the high laser efficiency. Here, in our experiment, the disappeared green emission is the certification of the function of the Ce^{3+}, which contributes to the improvement of the 2 μm laser.

6. Conclusion

In this review, the growth of rare earth (Tm^{3+},Ho^{3+},Nd^{3+},Yb^{3+}, Er^{3+}/Yb^{3+})-doped $NaY(WO_4)_2$ large crystal with the dimensions of Φ25 mm×100 mm is reported. The thermal, optical and spectrum characteristics of these crystals are presented. The laser characteristics of Nd^{3+};Tm^{3+}/Ho^{3+}:$NaY(WO_4)_2$ laser crystals are also covered. Maximum pulse energy of 786 mJ with a repetition rate of 1Hz has been obtained from Nd^{3+}-doped $NaY(WO_4)_2$ crystal pumped by xenon flash lamp. It can be found that the Nd:NYW crystal has the higher laser efficiency than Nd:YAG crystal. An infrared laser output of 2.7 W at 2.07 μm with Tm,Ho:$NaY(WO_4)_2$ crystal end-pumped by 795 nm laser diode at room temperature is also reached. Furthermore, the femtosecond-pulse operation of a Tm:Ho:$NaY(WO_4)_2$ laser at around 2060 nm is obtained for the first time. Transform-limited 191fs pulses are produced with an average output power of 82 mW at a 144MHz pulse repetition frequency. Maximum output power of up to 155 mW is generated with a corresponding pulse duration of 258 fs. Also, it is found that the co-doped Ce^{3+} can depress the green up-conversion emission of Tm^{3+} and thus improves the 2 μm laser. All the above performances demonstrate that $NaY(WO_4)_2$ crystal can serve as an excellent laser host.

7. Acknowledgements

Some works of this chapter were supported by National Nature Science Foundation of China (No.50902129, 61078076), Major Projects from FJIRSM (SZD09001), the Knowledge Innovation Program of the Chinese Academy of Sciences (Grant No. KJCX2-EW-H03), Science and Technology Plan Major Project of Fujian Province of China (Grant No. 2010I0015).

8. References

1.
[1] Z.X. Cheng, S.J. Zhang, J.R. Han, et al.,, Cryst. Res. Technol. 36 (2001) 135;
[2] A.A. Kaminskii, H.J. Eichler, K. Ueda,, et al., Appl. Opt. 38 (1999) 4533;
[3] Z. L. Zhu, Y. N. Qian, J. H. Liu,et al., J. Cera. Soci. 35(2007) 991-994;
[4] C. Cascales, M. D. Serrano, F. Esteban-Betegón, et al.,, Phys. Rev. B. 74(2006) 174114;

2.

[1] Yan Wang,Chaoyang Tu, Zhenyu You, et al.,Journal of Crystal Growth, 285(1-2)2006:123-129;

[2] Chengli Sun, Fugui Yang, Ting Cao, et al., Journal of Alloys and Compounds, 509(25)(2011): 6987-6993;

[3] Zhenyu You, Master dissertation, The study of Nd^{3+} and Yb^{3+} doped in $NaY(WO_4)_2$ and Cr^{3+}doped in $LiNbO_3$ crystals,Graduated School of Chinese Academy of Sciences,2010.

3.

[1] Zhenyu You, Master dissertation, The study of Nd^{3+} and Yb^{3+} doped in $NaY(WO_4)_2$ and Cr^{3+}doped in $LiNbO_3$ crystals,Graduated School of Chinese Academy of Sciences,2010.

4.

[1] Yan Wang,Chaoyang Tu, Zhenyu You, et al.,Journal of Crystal Growth, 285(1-2)2006:123-129;

[2] A. Mendez-Blas, M. Rico, V. Volkov, et al., Mol. Phys. 101 (2003): 941.

[3] X. Han and G. Wang, J. Crystal Growth 247 (2003):551.

[4] Y. Kalisky, L. Kravchik and C. Labbe, Opt. Commun. 189 (2001): 113.

[5] R. Moncorge, B. Chambon, J.Y. Rivorie,et al., Opt. Mater. 8 (1997):109.

[6] M.H. Randles, J.E. Creamer and R.F. Belt, OSA technical digest ser. Opt. Soc. Am. 10 (1998): 289.

[7] H.D. Jiang, H.J. Zhang, J.Y. Wang, et al., Opt. Commun. 198 (2001):447.

[8] W.F. Krupke, IEEE J. Quantum Electron. QE-7 (1991):153.

[9] W. Koechner, Solid state laser engineering, Optical Science (third ed.), Springer Verlag, New York (1992).

[10] H. Zhang, Z. Luo, A. Zhen, C.M. Wu and G.S. Li, Chin. J. Infrared Res. A 7 (1998): 297.

[11] A.A. Kaminskii, Laser crystals. In: H.F. Ivey, Editor, Springer, Berlin, Heidelberg, New York (1981).

[12] C.Y. Tu, Master dissertation, Study on the blue–green laser crystal Nd3+:GdxY1−xAl3 $(BO_3)_4$, Graduated School of Chinese Academy of Sciences, 2001.

[13] N.V. Kuleshov, A.A. Lagatsky, A.V. Podlipensky, et al., Opt.lett.22(1997)1317.

[14] K.I. Schaffers, L.D. Deloach, S.A. Payne, IEEE J. Quantum electron. 32 (1996) 741.

[15] H.Zhang, X.Meng, P. Wang, et al., Appl. Phys. B 68 (1999) 1147.

[16] K. Ohta, H. Saito, M. Obara, J. Appl. Phys. 73,(1993) 3149.

[17] D.E. McCumber, Phys. Rev. A 134(1964) 299.

[18] S.A. Payne, L.L. Chase, L.K. Smith, et al., IEEE J. Quantum Electron. QE-28 (1992) 2619.

[19] W.T. Carnall, P.R. Fields, K. Rajnak, J Chem Phys, 49 (1968) 4424.

[20] X. M. Han, J. M. Cano-Torres, M. Rico,et al., J. Appl. Phys. 103(2008) 083110.

5.

[1] Zhenyu You, Master dissertation, The study of Nd^{3+} and Yb^{3+} doped in $NaY(WO_4)_2$ and Cr^{3+}doped in $LiNbO_3$ crystals,Graduated School of Chinese Academy of Sciences,2010.

[2] F. G. Yang, C. L. Sun, Z. Y. You, et al., Laser Phys.20(2010)1695;

[3] Fugui Yang, Doctoral dissertation, The study of LD pumped yellow and 2μm laser properties, Graduated School of Chinese Academy of Sciences,2011.

[4] X.Han, F.Fusari, M.D.Serrano, et al. OPTICS EXPRESS 18(6)(2010)5413;

[5] A.A.Lagatsky, X.Han, M.D.Serrano, et al. OPTICS LETTERS 35(18)(2010)3027;

[6] F.G. Yang, F.P. Yan, Z.Y. You, et al., Laser Phys. Lett. 7(2010)867.

6

The Influence of Atmosphere on Oxides Crystal Growth

Morteza Asadian
Iranian National Center of Laser Science and Technology, Tehran, Iran

1. Introduction

Oxide single crystals are known for wide application in electronics and optical industries, e.g. lasers, substrates, scintillators, nonlinear and passive optical devices (Brandle, 2004). The Czochralski technique has become an important method of choice for the growth and production of many bulk oxide materials owing to the possibility of growing crystals with large size, core free with good optical quality, and a high concentration of dopant, e.g., Nd^{3+}, Yb^{+3} with better homogeneity (Zhaobing et al., 2007; Albrecht et al., 1998).

Numerous researchers (Dupret and Bogaert, 1994; Cockayne et al., 1976; Valentino and Brandle, 1974; Brandle, and Barns, 1974; Jacobs et al., 2007; Mateika et al., 1982; Piekarczyk and Pajaczkowska, 1979) have studied the effect of different parameters on the quality of oxide crystals grown by Czochralski technique. They found that Crystal growth atmosphere composition and melt stoichiometry are strongly influenced on the quality of crystals as well as solid-liquid interface shape.

However, the growth of oxides such as $Gd_3Ga_5O_{12}$ [Brandle et al., 1972], ZnO [Klimm et al., 2008a], SrLaGaO4 [Pajaczkowska et al., 2001], single crystals by standard melt growth techniques are difficult due to their high melting point and thermal decomposition upon heating. Otherwise, if the composition of melt would be serious, the chemical composition of these oxides melt will depart from the congruent point and superfluous components would become the harmful impurities for crystal growth.

Due to the volatilization of oxide melts and the necessity of an oxidizing atmosphere, the crucible that contains the melt should consist of a noble metal. The only materials that have been found to tolerate these violent conditions are the platinum group of metals and their alloys (Day, 1963; Cockayne, 1974). Although the growth from platinum crucible can be performed in presence of oxygen, unfortunately, platinum with T_m=2042 K and its alloys cannot use for mostly oxides because of their high melting point (Darling et al., 1970). Very often platinum or alloys of platinum are used only for oxides with low melting point (T<1850 K) (Uitert, 1970). Actually, it is necessary to choose materials of higher melting point, such as the rhodium-platinum alloys, or rhodium or iridium metals (Cockayne, 1968). Owing to a high melting point and high mechanical strength even at T> 2250 K° (Handley,1986), iridium crucible are widely used for growing high melting oxides such as GGG, Al_2O_3 and ZnO. It is important to note that, iridium is more sensitive to oxidation than platinum or rhodium (Weiland et al., 2006), especially in the medium temperature (~870-1000 K°) as the metal would oxidize to IrO_2 (s) (Asadian et al., 2010). Consequently, the addition of oxygen to the furnace chamber reacts with iridium crucible at medium temperatures and it complicates the process of crystallization (Lipinska et al., 2009).

The ceramic oxide crucibles are usually utilized for melting metals (Day, 1963). Since they are all chargeable to chemical react by the molten oxides, unfortunately, they cannot apply as a part to contact directly to the oxides melt (Klimm and Schroder, 1999). The high melting point metal such as molybdenum (T_m=2896 K) and tungsten (T_m= 3680 K) are oxidized rapidly by oxygen at high temperatures and can only be used in reducing or neutral atmospheres (Klimm et al., 2008b).

This chapter is organized as follows. An outline of thermodynamic background is illustrated in section 2. Section 3 contains a detailed discussion of the thermodynamic behaviour of construction materials (crucible, seed holder, after heater). Section 4 describes partial pressures effect of gases atmosphere. The short review of growth atmosphere effect on crystal growth process including two examples is presented in section 5. Section 6 concludes the chapter.

2. Thermodynamic background

Consider the general redox equilibrium reaction

$$aM_xO_y \leftrightarrow bM_zO_w + 1/2O_2 \qquad (1)$$

With

$$\mathrm{Log}\left(K_p\right) = \frac{1}{2.30R}(\Delta S^o - \frac{\Delta H^o}{T})(J.K^{-1}.mol^{-1}) \qquad (2)$$

The equilibrium constant (K_p) gives the required combination of the activities of the reactants as

$$K_p = (\frac{a^b_{M_zO_w} \times P^{1/2}_{O_2}}{a^a_{M_xO_y}}) .$$

R is the universal gas constant, ΔH° is the standard enthalpy change and ΔS° is the standard entropy change for the reaction. The choice of the pure material a standard states makes the activity of material has values of unity in equation (2). In which case for material oxides, if the oxide pure is solid oxides, the activity of solid oxides will be unit. Moreover, when the material oxides would be gas forms the activity of the oxides replace to their pressures (Gaskell, 2003).

For the oxidation of (a)dn mole material oxide from the m valent state to the m- s valent state proceeds under the consumption of (s/2)dn mole oxygen. Herein m=2y/x and s=2/ax. m valent gives a number between 1-8 and always $m \geq s$. The Table 1 indicates four possible conditions of the redox equilibrium reactions which are obtained with simplifying the equation (1).

m	s	The redox equilibrium reaction
even	even	$2MO_{(m/2)} \leftrightarrow 2MO_{(m-s)/2} + \frac{s}{2}O_2$
even	odd	$2MO_{(m/2)} \leftrightarrow M_2O_{(m-s)} + \frac{s}{2}O_2$
odd	even	$M_2O_m \leftrightarrow M_2O_{(m-s)} + \frac{s}{2}O_2$
odd	odd	$M_2O_m \leftrightarrow 2MO_{(m-s)/2} + \frac{s}{2}O_2$

Table 1. The different conditions of redox equilibrium reaction.

The reaction equilibrium between pure material M, its oxide and oxygen gas where m=s would take place. In which case ΔG° is a linear function of temperature and a plot of ΔG°=RT Log (PO_2) versus temperature gives the lines for each oxidation reaction on an Ellingham diagram (Elingham, 1944). These lines separate phase regions where one oxidation state overcomes and the whole graph represents a predominance phase diagram for the material M and its oxides.

3. Construction materials

Platinum group metals and their alloys can usually be safely heated for long periods in contact with the more refractory oxides without serious risk of contamination. Due to their excellent chemical stability, oxidation resistance, and resistance to the action of many molten oxides, the platinum group metals: iridium, platinum and rhodium are widely used for high-temperature applications involving simultaneous chemical attack and mechanical strength. Important applications of platinum group metals are as crucibles for oxides crystal growth from melt. Oxides of major optical interest and their most suitable crucible materials are listed in Table 2.

Material	Oxide melting point (K)	Crucible	Material use
Lithium Niobate ($LiNbO_3$)	1523	Platinum or rhodium-platinum	Electro-optic
Calcium Tungstate ($CaWO_4$)	1839	Rhodium	Laser host
Gadolinium Gallium Garnet ($Gd_3Ga_5O_{12}$)	2023	Iridium	Laser host
Zinc Oxide (ZnO)	2248	Iridium	Green laser-UV light emitters
Sapphire (Al_2O_3)	2327	Iridium	Laser host

Table 2. Application of oxides single crystal and appropriate crucibles.

3.1 Platinum

The only materials that have been found to withstand high temperature chemical stability, oxidation resistance and resistance to react with many molten oxides are the platinum group of metals and their alloys. Although growth from Platinum crucibles can be performed in air, platinum does not appear to be widely used for the growth of oxide materials, probably due to its relatively low melting point (2042 K) in comparison to that for most refractory oxides or mixtures of oxides. The low melting point of lithium niobate ($LiNbO_3$), T_m=1523 K, (Day, 1963) and the chemical inertness of platinum suggested that this metal might be a useful crucible material, while platinum also offered some advantage in cost over the platinum group metals.

3.2 Rhodium

In a few instances, unalloyed platinum crucibles have been found satisfactory, but generally, it is necessary to choose materials of higher melting point, such as the rhodium platinum alloys, or the metals rhodium or iridium.

In the past, there has been some hesitation in the use of rhodium and iridium crucibles in oxidizing atmospheres at high temperatures through fear of high losses from the formation of volatile rhodium and iridium oxides. Based on this concept, Nassau and Broyer (1962) have used rhodium and iridium crucibles successfully for growing barium titanate single crystals. They were grown at about 1920 K from rhodium crucibles, when 0.36 weight percent of rhodium was detected in the crystals and from iridium crucibles when 0.02 weight percent of iridium was observed.

It is known that Rh reacts with oxygen above 1000 K (Chaston, 1965) forming Rh_2O_3 (s). Fig. 1 shows the stability regions phase in the Rh-O system from available thermodynamic data (Binnewies and Milke, 2008). As illustrated in Fig. 1, Rh_2O_3 (s) is stable up to 1412 K where the oxygen pressure reaches 1.0 atm. Above this temperature, Rh_2O_3 (s) decomposes to gaseous phase according to the equation

$$Rh_2O_3\,(s) = RhO_2\,(g) + 1/2O_2 \tag{3}$$

As pointed out, the dissociation of Rh_2O_3 at very low oxygen pressure $PO_2 < 10^{-35}$ (atm) and at room temperature, it maybe possible to take place this reaction

$$Rh_2O_3\,(s) = Rh\,(s) + 3/2O_2 \tag{4}$$

From Fig. 1, it is limited to apply rhodium at high oxygen pressure as crucible because of evaporate to RhO_2. Therefore, the oxygen partial pressure must be low to avoid oxidation of crucible especially at low temperature.

3.3 Iridium

Owing to have a high melting point and high mechanical strength even at T> 2250 K°, iridium is a particularly suitable material for applications such as the stress-rupture strength, creep behavior and thermal shock which preclude the use of platinum alloys or rhodium (Weiland et al., 2006). Important applications of iridium is as crucible for pulling refractory oxides crystals such as GGG (T_m=2023 K) and ZnO (T_m=2248 K).

It is noticeable that iridium crucible are sensitive to oxygen, especially at the medium temperature (~870-1000 K°) as the metal would oxidize to solid iridium oxide. IrO_2 (s) decomposes to iridium metal at temperature higher than 1370 K° according to the equation

$$IrO_2 (s) = Ir (s) + O_2 \quad (5)$$

As illustrated in Fig. 2, IrO_2 (s) is stable at the temperature less than 1370 K° at standard condition. If oxygen were applied at higher temperature than 1370 K°, iridium parts would not be oxidized.

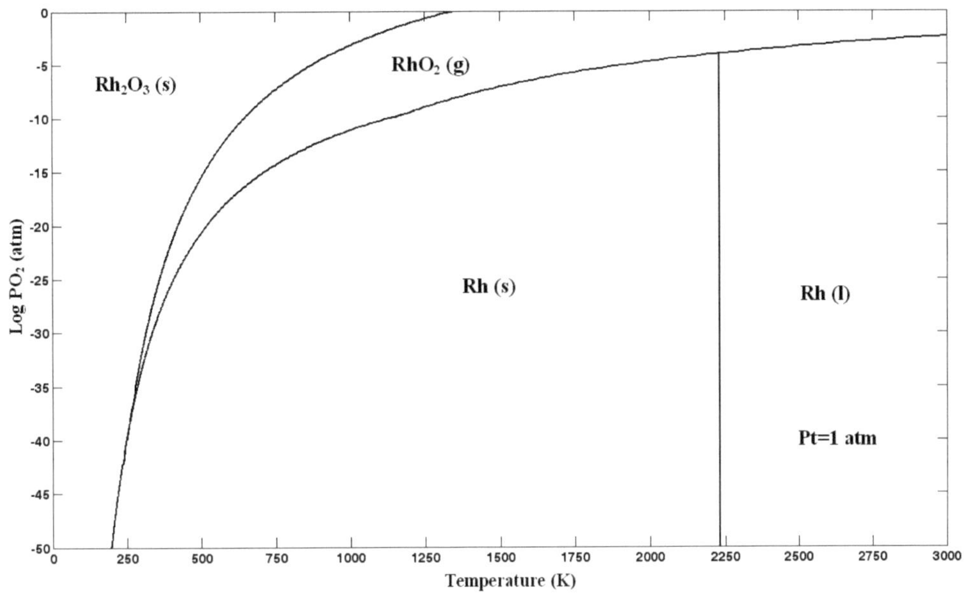

Fig. 1. Predominance diagram for Rh-O_2 system.

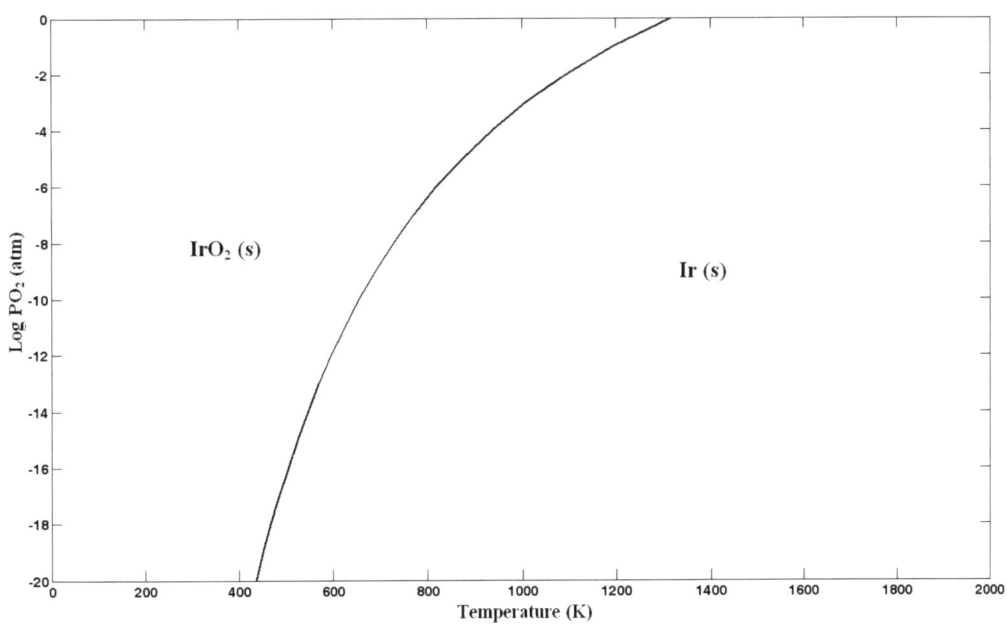

Fig. 2. Predominance diagram for Ir and IrO_2 (s) in the temperature- PO2 plane.

Several studies (Cordfunke and Mayer, 1962; Schafer and Heitland, 1960) have been shown that the oxidation of iridium at high oxygen pressure forms IrO_3 (g). As mentioned before, the decomposition temperature of IrO_2 (s) in oxygen at 1 atm is about 1370 K. Hence, above this temperature the equilibrium forming IrO_3 (g) follows

$$2Ir\ (s) + 3O_2 = 2IrO_3\ (g) \tag{6}$$

Below 1370 K the volatile oxide dissociates to IrO_2 (s) and oxygen by the reaction (Cordfunke and Meyer, 1962)

$$2IrO_3\ (g) = 2IrO_2\ (s) + O_2 \tag{7}$$

The major volatile species in the iridium-oxygen system is IrO_2 (g) (Chandrasekharaiah et al., 1981; Carpenter, 1989). For the reaction

$$Ir\ (s) + O_2 = IrO_2\ (g) \tag{8}$$

The results of these thermodynamics calculation are plotted in Fig. 3 as Log PO_2 against temperature. It can be seen that IrO_3 (g) is identified as the major gaseous species at low temperatures (800 K- 1700 K). At 2033 K, Norman et al. (1965) determined the IrO_2 (g) pressure to be 1.9×10^{-8} atm and the IrO_3 (g) pressure to be 3.1×10^{-9} atm. This is indicated that at elevated temperatures IrO_2 (g) is predominant gaseous species (seen Fig. 3).

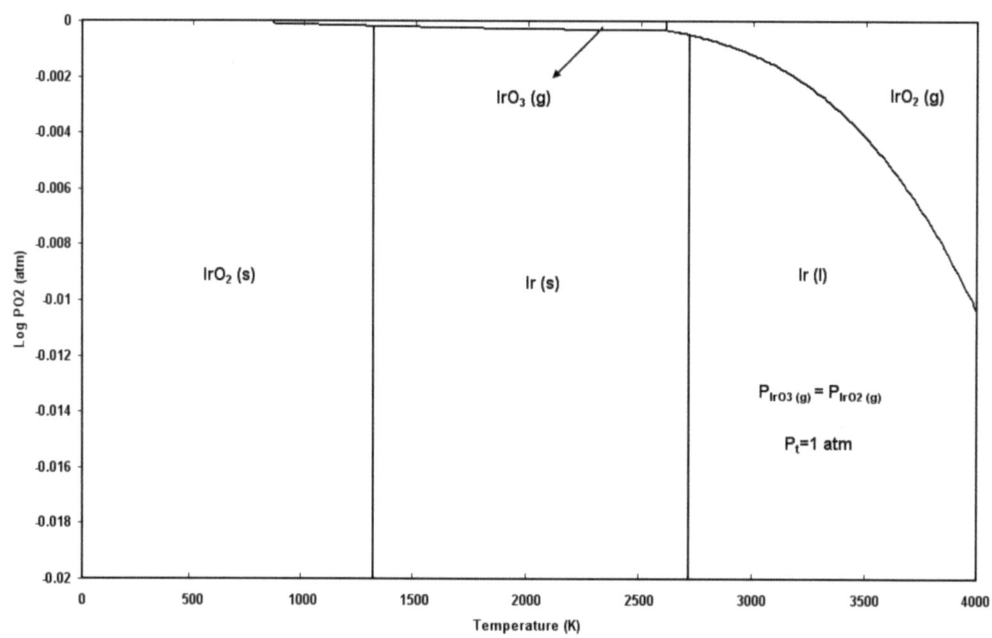

Fig. 3. Predominance diagram for Ir-O_2 system.

To sum up, although iridium is more sensitive to oxidation than platinum (E= -0.1474 V for Ir and E= -0.4422 V for Pt (Klimm et al., 2008c), it is the most chemically resistant of all metals. The high melting point of iridium (2716 K) and its resistance to attack by stable oxide melts is maintained up to temperatures above 2250 K make it a particularly suitable material for oxides crystal growth from hot melt which preclude the use of platinum alloys or rhodium.

4. Growth atmosphere

The oxides crystal growth atmosphere should have two conditions
1) Consider a type of oxide material Me, in Table 1, i. e.,

$$Me_2O_m(s,l) \leftrightarrow Me_2O_{(m-s)}(s,l,g) + \frac{s}{2}O_2 \tag{9}$$

The oxygen partial pressure must be in that range where the favored valency of the oxide material (Me_2O_m) is stable. Always PO_2 system should be more than $(PO_2)_{min}$ to suppress the decomposition of oxide. Herein $(PO_2)_{min}$ is the minimum partial pressure of O_2 where the reaction (9) moves to the left side for the given growth temperature $((PO_2)_{eq} > (PO_2)_{min})$.
2) For the reaction equilibrium between a pure solid construction material M, its pure oxide and oxygen gas

$$2M(s) + \frac{m}{2}O_2 \leftrightarrow 2MO_{(m/2)}(s, g) \tag{10}$$

The oxygen partial pressure should be low enough to avoid oxidation of construction parts (crucible, seed holder) being in contact with the melt. Herein PO_2 system should be less than $(PO_2)_{max}$ with $(PO_2)_{max}$ - the maximum oxygen partial pressure where the construction material is equilibrium with its oxide in the m valency state $((PO_2)_{max} > (PO_2)_{eq})$. Because the oxide construction parts are often stable at lower temperatures and oxide crystal almost decomposed at higher temperatures, both conditions are accomplished, if $(PO_2)_{min} < (PO_2)_{eq} < (PO_2)_{max}$. Commonly, amounts of oxygen, e.g. for the growth of $Gd_3Ga_5O_{12}$ (GGG) is about 1-2 vol% (Ganschow, 2010), add to the growth atmosphere to prevent volatile oxide. However, this value (PO_2) often is more than $(PO_2)_{max}$ at lower temperatures. Therefore, it causes that the construction material oxidized.
A solution to this problem is that mixtures containing an oxygen bearing gas like CO_2 or H_2O would be utilized in the growth atmosphere. They produce a temperature dependent oxygen partial pressure in this manner that the O_2 partial pressure increases with temperature. Hence, oxidation of construction parts decrease.
Another solution is that the protective gas (N_2, Ar) atmosphere would be charged into the furnace chamber in ambient temperature before heating the raw material and beginning the crystal growth process. The atmospheric pressure of the chamber should be more than ambient pressure. To prevent the evaporation of oxide, amount of oxygen should add into the during the crystal growth process at temperatures higher than temperature $MO_{m/2}$ decomposition. Growth experiments for Nd:GGG (Asadian, 2011) prove that the iridium crucible is not oxidized when the required oxygen was charged at T>1370 K°.

4.1 CO_2-CO system

CO_2 decomposes into CO and oxygen with increasing temperature according to

$$2CO_2 = 2CO + O_2 \tag{11}$$

If a moles of CO and 1 mole of CO_2 are mixed (a=CO/CO_2), then from stochiometry of Eq. (11), 2x moles of CO_2 would decompose to form 2x moles of CO and x moles of O_2 such that, at total pressure P_t, the PO_2 in CO-CO_2 system in various temperatures would be

$$Log\frac{x(a+2x)^2}{(1+a+x)(1-2x)^2}P_t = 9.04 - \frac{29613.35}{T} \tag{12}$$

Which,

$$PO_2 = \frac{x}{1+a+x}P_t \tag{13}$$

The partial pressure of oxygen can be ''automatically'' increased in the heating-up phase of the growth process by the thermal decomposition of carbon dioxide Eq. (11). Temperature dependence of PO_2 with comparing different gases for the above reaction is shown in Fig. 4.

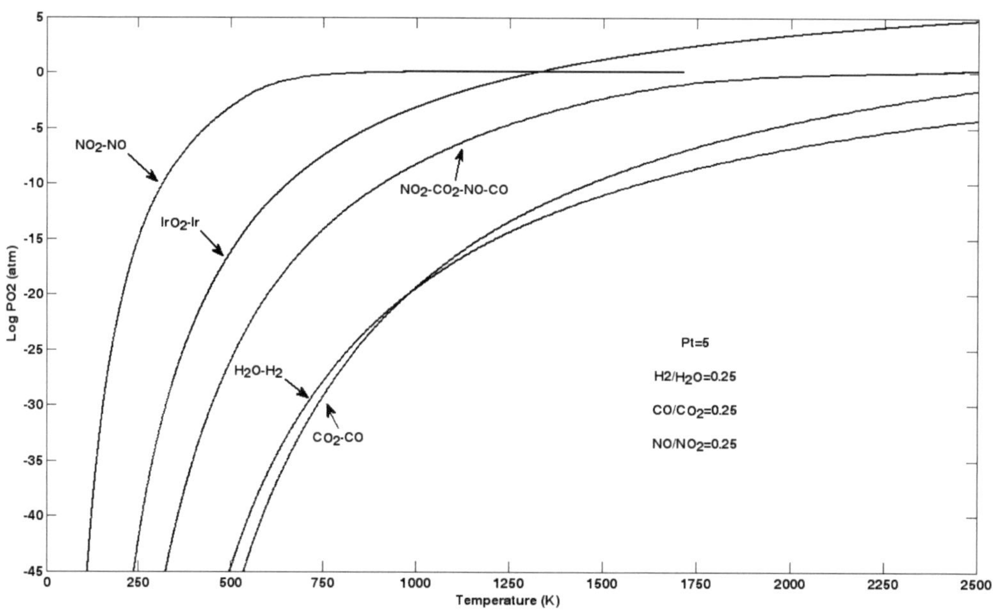

Fig. 4. Temperature dependence of PO_2 within different gases and gas mixtures comparing with stability regions for iridium (iridium metal and its solid oxide).

Thermodynamic calculations (Tomm et al.,2000) of the chemical equilibria between Ga_2O_3 (solid and liquid phase) and the surrounding gas-phase showed that an Ar+10% CO_2 atmosphere delivers an oxygen partial pressure that increases in such a manner with temperature, that Ga_2O_3 crystals can be grown from the melt. This amount of liberated oxygen is just sufficient to decrease the evaporation of molten Ga_2O_3 as well as low enough to allow the use of iridium crucibles, i.e., allowing Czochralski growth of Ga_2O_3.

4.2 H_2O-H_2 system

H_2O-H_2 system are used when it is required that the partial pressure of oxygen in a gas phase be fixed at a very low value at low temperatures. For example, if it were required to have a gaseous atmosphere containing an oxygen partial pressure of 10^{-30} atm at 705 K, then such an oxygen potential pressure can be obtained with simple relation by establishing the equilibrium

$$2H_2O = 2H_2 + O_2 \tag{14}$$

The partial pressure of oxygen in H_2-H_2O gaseous atmosphere in various temperatures from Eq. (14)

$$Log\frac{x(c+2x)^2}{(1+c+x)(1-2x)^2}P_t = 4.56 - \frac{25302.15}{T} \tag{15}$$

Which,

$$C=H_2/H_2O$$

$$PO_2 = \frac{x}{1+c+x}P_t \tag{16}$$

The oxygen pressure of H_2-H_2O is shown in Fig. 4. Comparing the H_2-H_2O system with CO-CO_2 indicate that H_2-H_2O system products more oxygen at lower temperatures. However, at higher temperatures CO-CO_2 system is the more efficient oxidant agent.

4.3 NO_2-NO system

Similarly, the partial pressure of oxygen in NO_2-NO system can be determined by establishing the reaction

$$2NO_2 = 2NO + O_2 \tag{17}$$

Thus

$$Log\frac{x(b+2x)^2}{(1+b+x)(1-2x)^2}P_t = 7.67 - \frac{5985.45}{T} \tag{18}$$

Which,

$$b=NO/NO_2$$

$$PO_2 = \frac{x}{1+b+x}P_t \tag{19}$$

The oxygen partial pressure in NO_2-NO gaseous atmosphere in various temperatures is drawn in Fig. 4. Comparing with others gaseous components, at lower temperatures the released oxygen is too much in NO_2-NO system. Therefore, the iridium crucible indeed would be oxidized.

4.4 NO_2-CO_2-CO-NO system

Consider the reaction CO_2 and NO_2 are mixed in the molar ratio 1:1 to form NO, CO and O_2 according to

$$CO_2 + NO_2 = CO + NO + O_2 \tag{20}$$

To obtain this gas mixture, CO_2 and CO, NO_2 and NO at total pressure P_t, would be mixed in the molar ratio $CO/CO_2=a$, $NO/NO_2=b$ and allowed to equilibrate.
Which,

$$Log\frac{x(a+x)(b+x)}{(2+a+b+x)(1-x)^2}P_t = 8.36 - \frac{17799.40}{T} \tag{21}$$

$$PO_2 = \frac{x}{2+a+b+x}P_t \tag{22}$$

According Fig. 4, NO_2-NO and CO_2-CO gas mixtures can be used when it is required that the partial pressure of oxygen very low at lowers temperature to suppress oxidation of iridium crucible. Also they can be utilized when it is required that the high partial pressure of oxygen at elevated temperature. In this case, ZnO crystal growth is a good example. For the growth of ZnO in iridium crucible by Czochralski method, the oxygen partial pressure of chamber at ZnO melting point (2248 K) should be more than $(PO_2)_{min}$=0.35 atm (Klimm et al. , 2009) to have ZnO melt. In order to suppress the oxidation of iridium crucible, the PO_2 of system should be less than $(PO_2)_{max}$. According Table 3, NO_2-NO and CO_2-CO gas mixtures can be used for ZnO crystal growth.

PO_2 (atm) / Temperature	$(PO_2)_{min}$ (Ga_2O_3)	$(PO_2)_{min}$ (ZnO)	H_2-H_2O $(PO_2)_{eq}$	CO-CO_2 $(PO_2)_{eq}$	NO-NO_2 $(PO_2)_{eq}$	CO_2-NO_2-CO-NO $(PO_2)_{eq}$	$(PO_2)_{max}$ (iridium)
900 K	7.94×10^{-38}	9.52×10^{-14}	5.6×10^{-23}	2.0×10^{-23}	1.06	6.3×10^{-11}	2.57×10^{-5}
(T_m(G_2O_3)) 2080 K	1.12×10^{-5}	1.53×10^{-2}	4.89×10^{-7}	1.01×10^{-4}	1.33	0.63	10^2 <
(T_m(ZnO)) 2248 K	------	7.76×10^{-2}	4.0×10^{-6}	1.20×10^{-3}	1.41	0.89	10^2 <

Table 3. Maximum, minimum and required PO_2 that is supplied by different gas mixtures at different temperatures.

5. Oxides crystal growth

Oxides crystal growth are often performed from melts contained in crucibles, e.g. by Czochralski or Bridgman technique. For oxide components, oxygen partial pressure during growth is one of the most important parameters to decide about success or failure of crystal growth process. Many oxides may be easily decomposed at high temperatures and low oxygen partial pressure. Therefore, oxides crystal growth is often carried out in a protective gas atmosphere such as Ar and N_2 to avoid the oxidization of crucibles and to minimize dissociation of oxide. An amount of oxygen partial pressure is always necessary to remain the stability of the oxide.

5.1 Ga_2O_3

The growth of GGG single crystal by the Czochralski method, in high-quality and large-sized is hard because of the dissociation of Ga_2O_3 (Luo et al. ,2005; Li et al. , 2007). At the T> 1500 K° (Klimm et al., 2008), reduction to evaporate Ga_2O may occur, whereas, the serious evaporation of Ga_2O_3 takes place at high temperatures. Fig. 5 represents a schematic of the experimental setup employed for the Nd:GGG crystal growth.

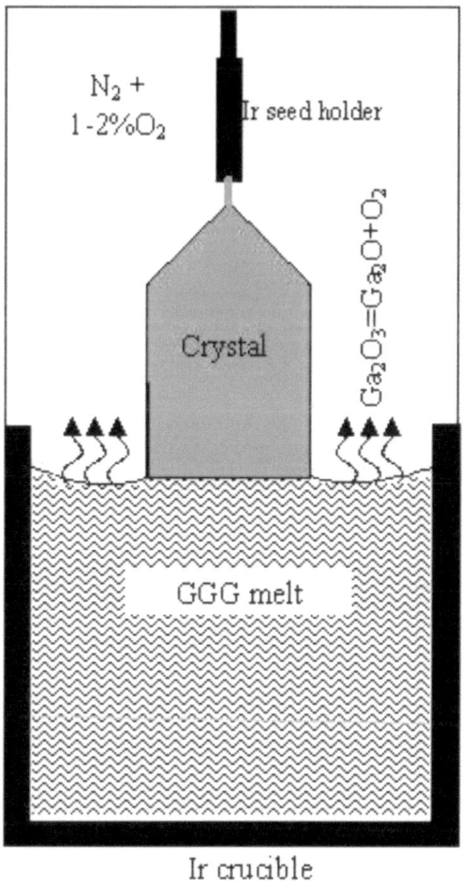

Fig. 5. Sketch diagram of GGG crystal growth setup.

Based on thermodynamic analysis, Ga_2O (g) and Ga (l) is main decomposed constituent via heating Ga_2O_3. During single crystal growth process by the Czochralski method, Ga_2O_3 sublimes by dissociation:

$$Ga_2O_3\ (s,l) = Ga_2O\ (g) + O_2 \tag{23}$$

The diagram Log P_{O_2} (T) for gallium oxides is shown in Fig. 6. At above 1500 K°, the evaporation of Ga_2O_3 may take place but the most serious Ga_2O_3 volatile will be happened at temperatures than higher 1950 K°. According to the chemical equilibria law, charging amount of oxygen into system decreases the volatilization of Ga_2O_3. For GGG crystal growth by Czochralski method, the pressure of oxygen (PO_2) should be more than 10^{-4} atm to preserve stability of growth process.
Additionally, it would be extracted from Fig. 6, at lower pressures ($PO_2<10^{-11}$ atm), the decomposition of Ga_2O_3 to liquid gallium may occur.

$$Ga_2O_3\ (s) = 2Ga\ (l) + 1.5O_2 \tag{24}$$

The liquid gallium will rapidly make to alloy with iridium crucible. Unfortunately, the formation of iridium-gallium alloys can definitely destroy the iridium crucible. Moreover, the Ga^+ ions decomposed from the Ga_2O_3 enter the lattice of Nd:GGG crystal ultimately cause the spiral growth.

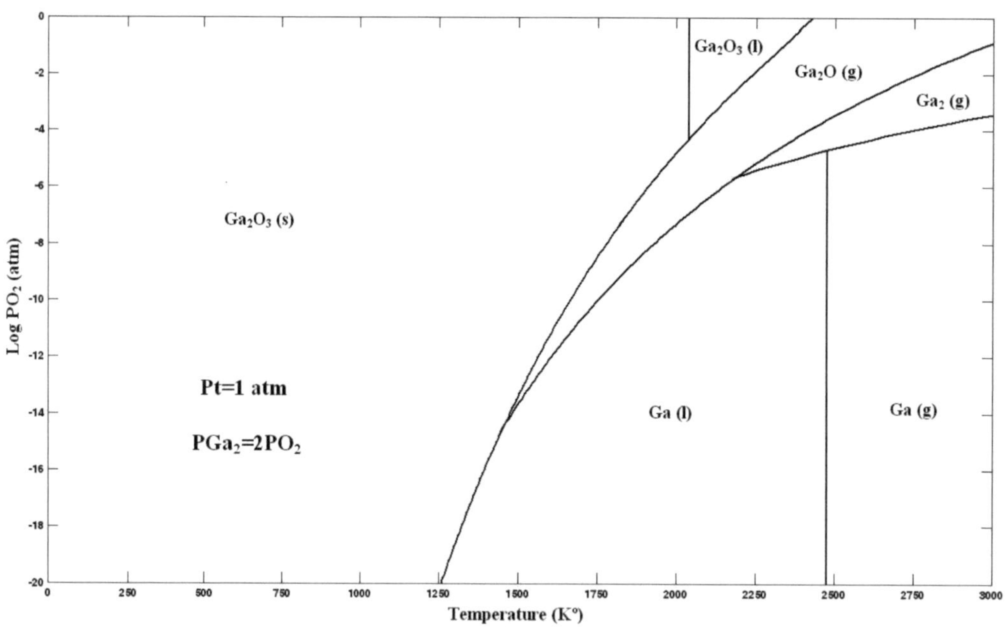

Fig. 6. Predominance diagram for Ga-O_2 system in dependence on T and the oxygen pressure PO_2 at total pressure 1 atm.

A solution to this problem is that the addition of oxygen would be replaced by carbon oxide (CO_2) in the growth atmosphere. The O_2 partial pressure increases with temperature; hence, oxidation of iridium crucible decreases (see Fig. 6). Matika et al. (1982) used a gas mixture of 50% CO_2 and 50% N_2 to reduce the dissociation of Ga_2O_3 and to suppress the formation of iridium within crystals. They obtained the densities of dislocations and inclusions for single crystals were below 5 cm^{-2}. Tomm et al. (2000) reported the first successful growth of Ga_2O_3 single crystal by Czochralski method. They have decreased evaporation of molten Ga_2O_3 by means of Ar plus 10% CO_2 instead of O_2 in the atmosphere.

5.2 ZnO

Zinc oxide is a group II-VI semiconductor that most commonly exhibits wurtzite structure in its unstrained phase (Park et al., 2006). Wurtzite crystals can be cleaved cleanly along several different planes. The c-plane (0001) intercepts only the z axis and in the case of zinc oxide produces polar surfaces as a result of the lack of inversion symmetry along the c axis (Miller, 2008). Furthermore, because of its wurtzite crystal structure and lattice parameter (a=3.25A° in the a-direction) it is isomorphous with GaN and has potential as a substrate material for GaN-based epitaxial devices (Pearton, et al., 2005). It has a large exciton binding energy of 60 meV and, as a result, stimulated emission and lasing have been observed at room temperature (Bagnall et al., 1998)]. This means that ZnO excitons will remain bound at much higher temperatures than GaN and in doing so, offer the potential for greatly improved efficiency over GaN. This indicates that ZnO can be the preferred choice for future opto-electronic devices.

Attempts are made to grow ZnO a large number of different methods. Bulk ZnO crystals can be grown using melt (Klimm et al.,2008a), hydrothermal (Suscavage et al.,1999) and vapor phase growth techniques (Look et al., 2002).

Unfortunately, traces of the solvent are always incorporated in such hydrothermal crystals and hydrogen and lithium are found typically in concentrations of several 10^{18} cm^{-3}. In addition, the suitability of ZnO crystals for epitaxy is highly dependent on surface preparation and subsequent characterization.Therefore; Attempts have recently been made to grow single crystal bulk from the melt.

Single crystals of ZnO can be grown from the melt at high oxygen pressure and high temperature. The Czochralski (Klimm et al., 2008a) and Bridgman (Jacobs et al., 2009) methods are the most common melt-growth techniques used for production of the bulk single-crystal. The advantages of the Czochralski and Bridgman growth methods are relatively high growth rates (in the range of several millimeters per hour) and the nearly thermodynamically equilibrium conditions are capable to generate bulk crystals of high structural perfection. In addition, the doping techniques are well established in these two methods.

There are technical obstacles to the growth of single crystal ZnO from the melt. The triple point of ZnO (the temperature and pressure at which all three phases of that substance, gas, liquid and solid phases, coexist in thermodynamic equilibrium) is 2248 K under the total (Zn and O_2) vapor pressure of 1.06 atm.

Figs. 7 and 8 show the predominance diagram in the relevant temperature range of the Zn-O system at total pressures 1 and 5 atm, respectively. As it can be seen from Fig. 7, before the melting point, ZnO evaporates under dissociation

$$ZnO\ (s) = Zn\ (g) + 1/2O_2 \tag{25}$$

It has strongly impeded the development of bulk crystal growth from the melt. To maintain the ZnO melt stable, the total pressure of oxygen-containing atmosphere in the growth chamber must be considerably larger than 1.06 atm.

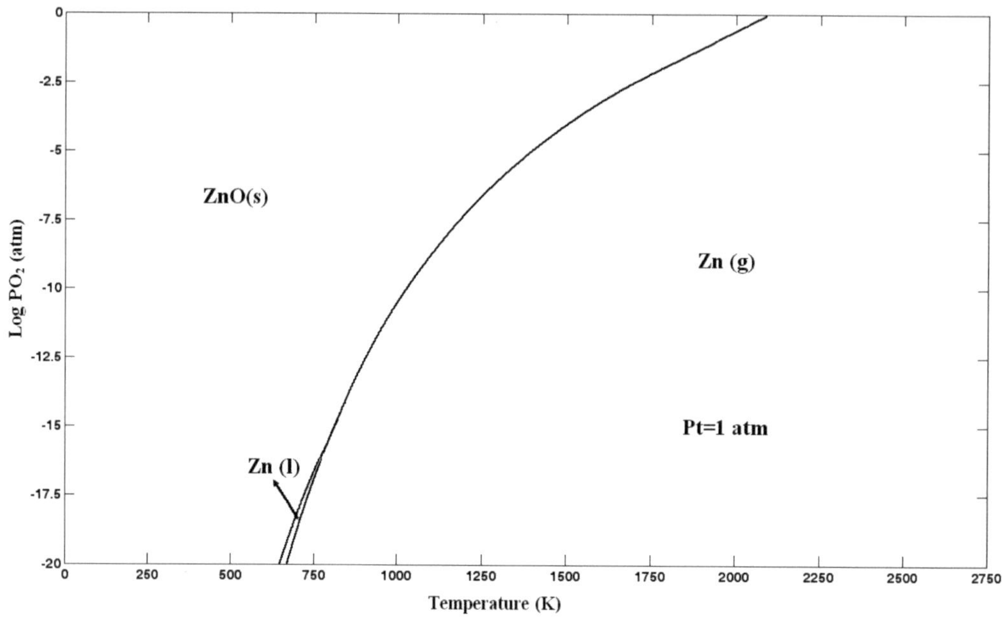

Fig. 7. Predominance diagram for Zn-O_2 system in dependence on temperature and the oxygen pressure PO_2 at total pressure 1 atm.

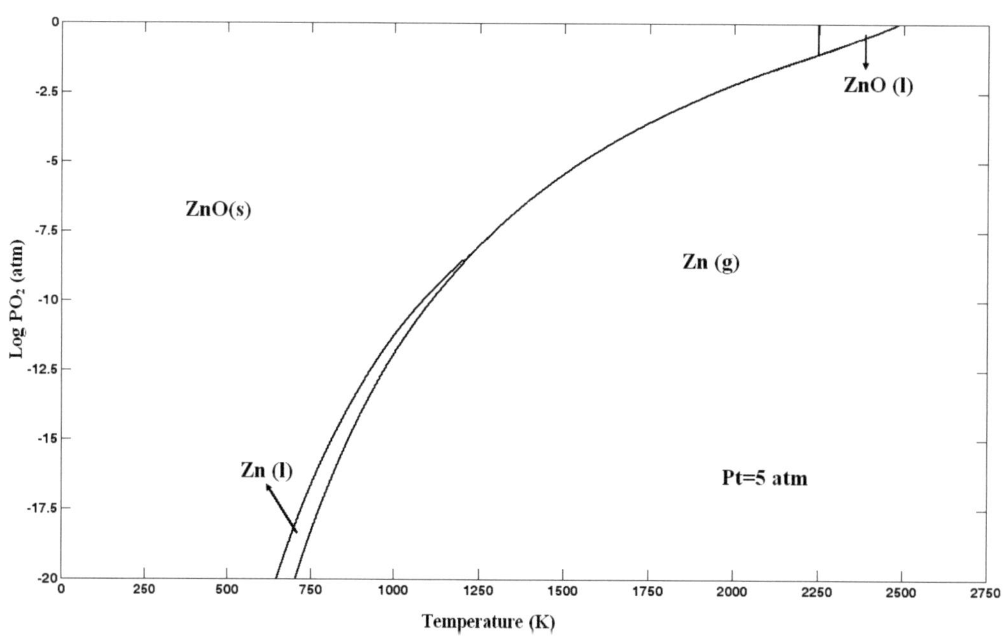

Fig. 8. Predominance diagram for Zn-O_2 system in dependence on temperature and the oxygen pressure PO_2 at total pressure 5 atm.

However, an essential difficulty to overcome is the lack of a perfect crucible material. The only metal withstanding high temperatures and an oxidizing atmosphere appears to be iridium. The design of a crucible containing molten ZnO capable of withstanding highly oxidizing conditions (temperatures of about 2300 K in oxygen containing atmosphere) is a serious challenge. The growth of ZnO crystals from a melt contained in an iridium crucible is continuously investigated by several references (Klimm et al., 2008a; Jacobs et al., 2009).
Indeed, it is well known that an atmosphere where iridium parts are to be heated must not contain more than 1-2% O_2 to avoid oxidizing of the metal. A solution could be try to heat the growth set up in the protective gas (N_2, Ar) –like GGG crystal growth- with an oxygen free atmosphere to 1000-1400 K and to add O_2 later, but practically this is not a solution for the problem. Although iridium crucible would not oxidize when the required oxygen was charged at T>1370 K°, crystal growth process would not stable growth because of sublimation of zinc oxide.
Similar problems during the melt growth of Ga_2O_3 were solved by working in a CO/CO_2 gas mixture. CO_2 yields an oxygen partial pressure well inside that corridor. At low temperatures, equilibrium reaction (11) is far on the left side and the oxygen partial pressure is very low. With increasing temperature, the equilibrium is shifted more and more to the right side and the resulting oxygen partial pressure is represented by curve in Fig. 9. Using CO_2 at a total pressure of approximately 10 bars, the authors (Klimm et al., 2008a; Klimm et al., 2008c; Jacobs et al., 2009) have successfully grown ZnO crystals from the melt in a Bridgman-like configuration.

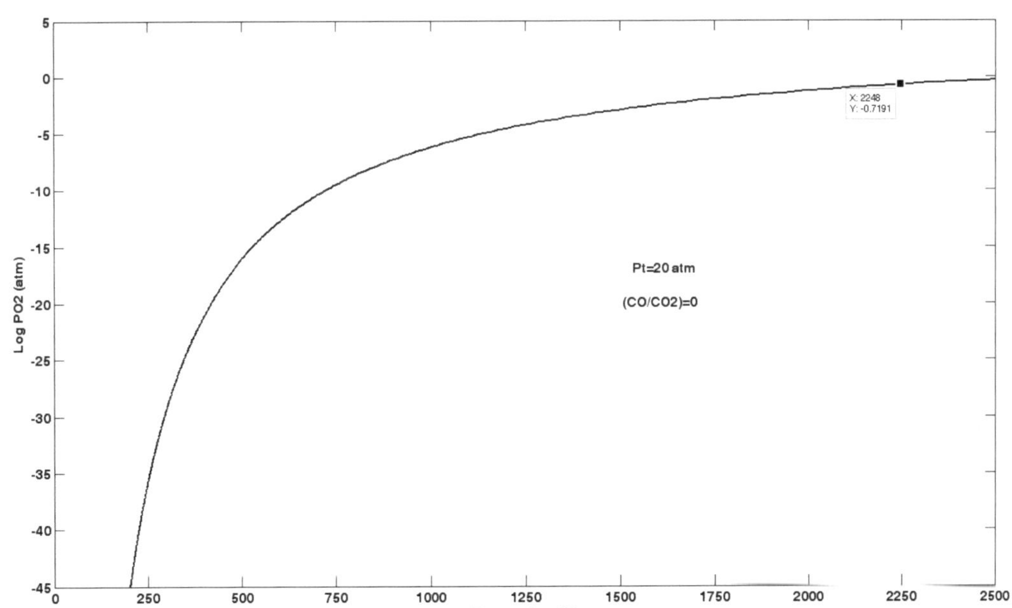

Fig. 9. The Oxygen partial pressure of pure CO_2 is against temperature at total pressure 20 atm.

The researches on growth of ZnO single crystals by the Czochralski method were carried on until now. Unfortunately, all attempts to obtain crystallization of bulk ZnO failed. Only polycrystalline ZnO solidified on the iridium seed rod was reported (Klimm et al., 2008a).

The control of the Czochralski experiments failed as the evaporation rate of ZnO at the melting point is high, even under the pressures up to 20 atm that were used. However, the partial pressure of O_2 at total pressure 20 atm that is supplied by pure CO_2 barely reaches to 1.90×10^{-1} atm at the melting point of ZnO crystal (seen Fig. 9). As shown in Fig. 10, the evaporated material forms white fume laying above the melts surface making optical control of the seeding process almost impossible (Klimm et al., 2008a). Moreover, evaporated ZnO condenses partially on the iridium seed rod, especially where it is lead through the thermal insulation. The sublimate creates mechanical contact between both parts, thus hindering mass control of the crystallizing ZnO by the balance on top of the seed rod that would otherwise allow automatic diameter control of the Czochralski growth process.

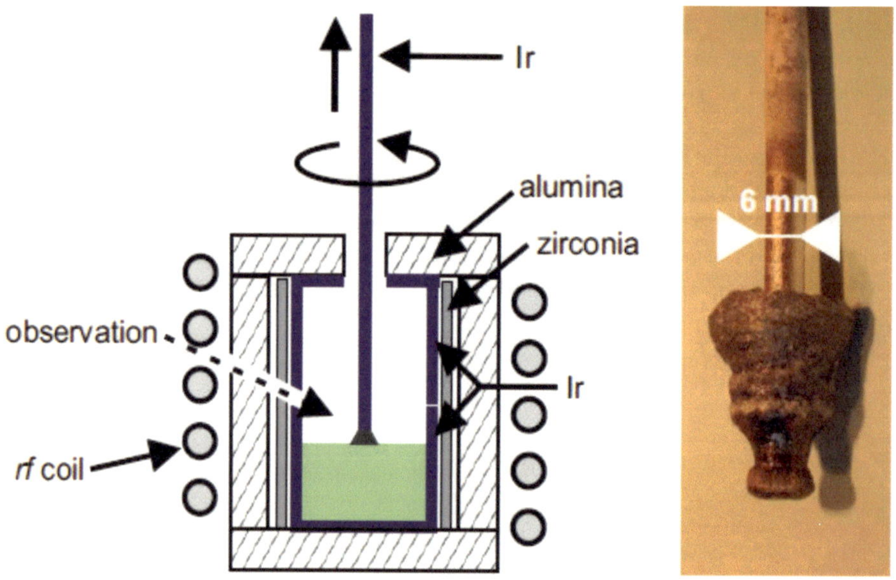

Fig. 10. Left: A schematic illustration of the ZnO crystal growth apparatus,
Right: polycrystalline ZnO solidified on the iridium seed rod. (Klimm et al., 2008a)

The Table 3 shows that at T_m=2248 K (melting point of ZnO), the pressure of oxygen at total pressure 5 atm should be between $(PO_2)_{min}$ and $(PO_2)_{max}$ points ($-1.11<\text{Log } PO_2<2$) in order to have growth stability and suppress the burning of the construction parts (crucible, seed holder). Based on Table 3, the oxygen partial pressure of gas mixtures containing (CO-NO-CO_2-NO_2) with $(CO/CO_2)=0.25$ and $(NO/NO_2)=0.25$ is reached to Log $PO_2=-0.05$ atm. Compared with NO-NO_2 system, It is not only placed among of $(PO_2)_{min}$ and $(PO_2)_{max}$ at the melting point of ZnO but also it is less than the partial pressure of oxygen at lower temperatures of 1370 K. Best of all, the ZnO crystal growth can be performed at lower total pressures (< 5 atm) compared with CO-CO_2 system (>20 atm).

6. Conclusion

During the growth of refractory oxides crystal from melt, appropriate partial pressure of O_2 in the chamber is needed to keep the growth process stable. On the other hand, the partial pressure of O_2 should be kept low adequate in lower temperatures to suppress the oxidation of constructive parts specially crucible. The best solution to this problem is that the PO_2 that is supplied by–depends on thermodynamic behavior of oxide- pure CO_2 or by a mixture of CO-CO_2 or NO-CO-CO_2-NO_2 would be utilized in the growth atmosphere. They produce a temperature dependent oxygen partial pressure in this manner that the O_2 partial pressure increases with temperature. This amount of liberated oxygen is just sufficient to decrease the evaporation of molten oxides as well as low enough to allow the use of iridium crucibles, i.e., allowing Czochralski growth of Ga_2O_3 or ZnO.

7. References

Albrecht G.F., Sutton S.B., George E.V., Sooy W.R., Krupke W.F. (1998), *Laser Part. Beams*, vol.16, pp.605

Asadian M., Hajiesmaeilbaigi F., Mirzaei N., Saeedi H., Khodaei Y., Enayati Sh. (2010), *Journal of Crystal Growth*, vol.312 pp.1645

Asadian M., Mirzaei N., Saeedi H., Najafi M., Mashayekhi Asl I. (2011), *Solid State Sciences*, In Press, Corrected Proof

Bagnall D. M., Chen Y. F., Zhu Z., Yao T., Shen M. Y., Goto T. (1998), Appleid Physics Letter vol.73 pp.1038

Binnewies M. and Milke E. (2008), *Thermochemical Data of Elements and Compounds*, Wiley-VCH Verlag GmbH, Weinheim, thierd Edittion

Brandle C. D., Miller D. C., Nielsen J. W. (1972), *Journal of Crystal Growth*, vol.12 pp.195

Brandle C.D. and Barns R.L. (1974), *Journal of Crystal Growth*, vol.26 pp.169

Brandle, C.D. (2004), *Journal of Crystal Growth*, Vol.264, pp. 593-604

Carpenter J. H. (1989), *Journal of the Less Common Metals*, vol.152 pp.35-45

Chandrasekharaiah M. S., Karkhanavala M. D., Tripathi S. N. (1981), *Journal of the Less Common Metals*, vol.80 pp.9-17

Chaston J. C. (1965), *Platinum Metals Review*, vol.9 pp.126

Cockayne B. (1968), *Platinum Metals Review*, vol.12, pp.16

Cockayne B., Lent B., Roslington J.M. (1976), *Journal Material Science*, vol.11 pp.259

Cockaynen B. (1974), *Platinum Metals Review*, vol.18 pp. 86

Cordfunke E. H. P. and Meyer G. (1962), *Rec. Trav. Chimestry*, vol.81 pp.680

Cordfunke E.H.P. and Mayer G. (1962), *Rec. Trav. Chemistry*, vol.81 pp.495

Darling A. S., Selman G. L. Rushforth R. (1970), *Platinum Metals Review*, pp.14, 54-60

Day J. G. (1963), *Platinum Metals Review*, vol. 7 pp.50

Dupret F., Van Den Bogaert N. (1994), *Handbook of Crystal Growth*, Vol. 2, North-Holland, Amsterdam,

Elingham H. J. T. (1944), Reducibility of Oxides and Sulfides in Metallurgical Processes, *Journal Society Chemistry*, vol.63 pp.125

Ganschow S., Schulz D., Klimm D., Bertram R., Uecker R. (2010), *Crystal Research Technology*, vol.45 pp.1219

Gaskell D. R. (2003), *Introduction to the thermodynamics of materials*, 4th Edition, Taylor and Francis, ISBN 1-56032-992-0

Handley J. R. (1986), *Platinum Metals Review*, vol.30, pp.12-13

Jacobs K., Schulz D., Klimm D., Ganschow S. (2009), *Solid State Science*, vol.12 pp.307

Jia Z., Tao X., Dong C., Cheng X., Zhang W., Xu F., Jiang M. (2006), *Journal of Crystal Growth*, vol.292 pp.386

Klimm D. and Schroder W. (1999), *Journal Korean Association of crystal Growth*, vol.9 pp.360

Klimm D., Ganschow S., Schulz D., Bertram R., Uecker R., Reiche P., Fornari R. (2008b), Preprint submitted to Journal of Crystal Growth, for CGCT4 Sendai

Klimm D., Ganschow S., Schulz D., Fornari R. (2008a), *Journal of Crystal Growth*, vol.310 pp.3009

Klimm D., Ganschow S., Schulz D., Fornari R. (2008c), Preprint submitted to Journal of Crystal Growth, for CGCT4 Sendai;

Klimm D., Ganschow S., Schulz D., Fornari R. (2009), *Journal of Crystal Growth*, vol.310 pp.534- 536

Li X., Hu Z. G., Li J. (2007), *Optical Materials*, vol.29 pp.854

Lipinska L., Ryba-Romanowski W., Rzepka A., Ganschow S., Lisiecki R., Diduszko R., Pajaczkowska A. (2009), *Crystal Research and Technology*, vol.44 pp.477

Look D. C., Reynolds D. C., Litton C. W., Jones R. L., Eason D.B., Cantwell G. (2002), *Applied Physics Letters*, vol. 81pp.1830

Luo Z., Lu M., Bao J., Liu W., Gao Ch., *Materials Letters*, vol.59 pp.1188

Mateika D., Laurien R., Rusche Ch. (1982), *Journal of Crystal Growth*, vol.56 pp.677

Miller P. (2008), Zinc Oxide: *a sperctroscopic Investigation of Bulk Crystals and Thin Films*, Degree of Doctor of Philosophy in Physics, University of Canterbury New Zealand

Nassau K., Broyer A. M. (1962), *Journal American Ceramic Society*, vol.45 pp.474

Norman J. H., Staley H. G. Bell W. E (1965), *Journal Chemistry Physics*, vol.42 pp.1123

Pajaczkowska A., Novosselov A. V., Zimina G. V. (2001), *Journal of Crystal Growth*, vol.223 pp.169

Park J. H., Jang S. J., Kim S., Lee T. (2006), *Applied Physics Letters*, vol.89 pp.121108

Pearton S. J., Norton D.P., Ip K., Heo Y.W., Steiner T. (2005), Applied Physics Letter, vol.50 pp. 293.

Piekarczyk W., Pajaczkowska A. (1979), *Journal of Crystal Growth*, vol.46 pp.483

Schafer H. and Heitland H. J. (1960), *ZAAC*, vol.304 pp.249-265

Suscavage M., Harris M., Bliss D., Yip P., Wang S. Q., Schwall D., Bouthillette L., Bailey J. , Callahan M., Look D.C., Reynolds D.C., Jones R.L., Litton C.W. (1999), *Journal Nitride Semiconductor Research*, vol. 4 pp.40

Tomm Y., Reiche P., Klimm D., Fukuda T. (2000), *Journal of Crystal Growth*, vol.220 pp.510;

Valentino A.J. and Brandle C.D. (1974), *Journal of Crystal Growth*, vol.26 pp.1

Van Uitert L. G. (1970), *Platinum Metals Review*, vol.14 pp.118

Weiland R., Lupton D. F., Fischer B., Merker J. Scheckenbach C., Witte J. (2006), *Platinum Metals Review*, vol.50 pp. 158

Zhaobing W., Qingli Z., Dunlu S., Shaotang Y. (2007), *Journal Rare Earths*, vol. 25, pp. 244

7

Controlling the Morphology and Distribution of an Intermetallic $Zn_{16}Ti$ Phase in Single Crystals of Zn-Ti-Cu

Grzegorz Boczkal
AGH-University of Science and Technology, Faculty of Non-Ferrous Metals, Cracow, Poland

1. Introduction

The work discusses the relatively poorly investigated area of various phenomena accompanying monocrystallization of hexagonal metal alloys, containing second phase inclusions in the structure. Metals of hcp structure form a, not very numerous but having a fundamental industrial importance, group.

Altogether about 27 metals of hcp structure are known, including 11 actinides [1,2,3]. The industrial applications of hexagonal metals cover many different sectors, from zinc sheets used for roofing up to light titanium-based alloys for parts of planes. Single crystal layers based on zinc are also used in electronics [4].

Hexagonal metals are characterised by features unprecedented for other metals. These include the value of c/a ratio (Table 1), which largely determines the metal properties [1,2], as well as very low mutual solubility observed in alloys composed of two metals with hexagonal structure each (Zn-Ti, Cd, Zn, Mg-Zn , Mg-Zr, Mg-Sc and others) [5,6].

The value of the c/a ratio and the related disorder in an ideal structural model are inherently related with the physics of interatomic bonds. In the case of metals with regular face centred lattice and regular body centred lattice, the bonding is of a purely metallic (non-directional) character and there is no phenomenon of interpenetration of the wave functions originating from lower orbitals. In hexagonal metals, the metallic bond predominates, but there is also a significant share of atomic bonds originating from an interaction that is said to exist between the wave functions of the neighbouring atoms [1,2,7,8].

Metals such as Zn, Cd, Tl were classified by Wyatt as a class of intermediate elements, i.e. the elements of a coordination number low enough to enable the occurrence of covalent interactions [1]. The directionality typical of an atomic bonding disturbs the perfect sequence of atomic layers and, consequently, the c/a ratio assumes values different than 1.633, which is perfect quantity when a rigid sphere model is used. Since the interatomic interaction in metals with c/a <> 1.633 is a combination of metallic and atomic bonds, any change in temperature must significantly affect the properties of metals with a hexagonal structure. Lowering the temperature reduces the distance between the atomic cores and, consequently, increases penetration of the lower orbital wave functions, which leads to an increase in the share of atomic bonds. Additionally, the structure where the atoms do not fill in an ideal space is characterised by low energy required to produce and promote migration of point defects (vacancies), and also by strong variation of diffusion coefficients for different crystallographic directions [7,8,9]. Both the temperature and the degree of deformation also strongly influence the number of vacancies in the material. These phenomena are responsible for the specific properties of hexagonal metals at low temperatures, such as the anomalies of a hardening coefficient and yield strength [9].

Metal	a [Å]	c [Å]	c/a
Na-α	3.767	6.154	1.6336
Be-α	2.287	3.583	1.5666
Mg	3.209	5.210	1.6235
Sc-α	3.309	5.273	1.5935
Y-α	3.647	5.731	1.5714
La-α	3.770	12.159	3.2251
Pr-α	3.673	11.835	3.2221
Nd-α	3.658	11.779	3.2200
Ti-α	2.951	4.684	1.5872
Zr-α	3.232	5.148	1.5928
Hf-α	3.195	5.051	1.5809
Tc	2.743	4.400	1.6040
Re	2.761	4.458	1.6146
Ru	2.706	4.281	1.5820
Os	2.735	4.319	1.5791
Zn	2.664	4.947	1.8569
Cd	2.979	5.619	1.8862
Tl-α	3.456	5.525	1.5986

Table 1. Some of the hexagonal metals and their c/a coefficients [2,3].

Hexagonal metals are characterised by large stress variations in the individual slip systems. With proper orientation respective of the stress applied, these metals can be deformed in one system, operating as a primary system, obtaining a wide range of the deformation values [10-14]. It is typical, in particular, of metals having the c/a ratio > 1.633, such as zinc and cadmium. Zinc single crystals of "soft" orientation are deformed to more than 100% (elongation) within the range of an easy slip in a (0001) <11-20> system. Only a very serious change of orientation caused by deformation and the strong strain hardening effect in the basal system of (0001) <11-20> are capable of activating a different slip arrangement [15].

Conducting research on the phenomena and processes occurring in metallic materials requires samples with clearly defined structure of both matrix and lattice obstacles in the form of phases precipitated or introduced from the outside. Metals and alloys in the polycrystalline form are not suitable for studies aiming at the identification of the mechanism of deformation because of the need to examine the phenomena which occur in different and separated areas characterised by different crystal orientations (grains). The use of materials with single crystal structure of the matrix eliminates this problem, as it becomes possible to analyse various phenomena within the whole sample volume, in a well-defined research environment. The ability to control the structure of the lattice obstacles while maintaining the same single crystal matrix enables, moreover, designing of technological processes by means of which products of the desired structure can be made in a reduced number of the necessary technological steps.

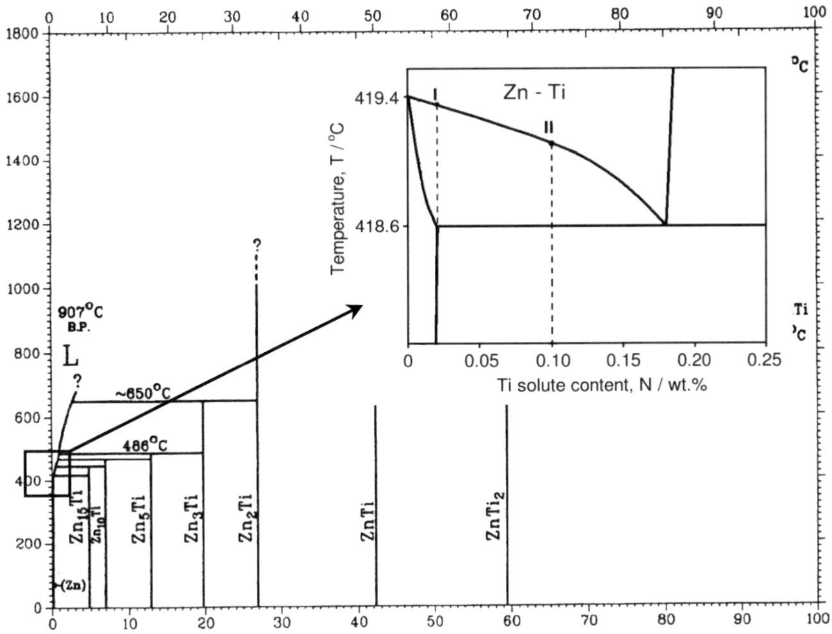

Fig. 1. The Zn-Ti phase diagram [5,6,7].

The specific crystallography of metals with a hexagonal structure, combined with the strong anisotropy of diffusion coefficients and elasticity constants has a strong influence on morphology of the obtained structures.

For example the elastic constants c of the pure zinc are property [16,17]:

- $c_{||}[0001] = 13.1$ cm² dyne⁻¹ x 10⁻¹³
- $c_{\perp}[0001] = 1.93$ cm² dyne⁻¹ x 10⁻¹³

The results of other authors suggest moreover, that elasticity of zinc is very sensitive to small variations in amount and kinds of impurities [15,17].

The additional special feature is also a strong influence of even small amounts of the alloying elements on the surface tension of molten zinc, which is a critical parameter for the process of monocrystallization [16,23,24,26].

Another important factor is the existence of a limited number of directions characterised by a high coefficient of diffusion and self-diffusion. At a temperature of 293K, the zinc coefficients of self-diffusion D_0 are, respectively [9]:

- $D_{0\,[0001]} = 0.13 \times 10^{-4}\ m^2s^{-1}$
- $D_{0\,<11\text{-}20>} = 0.58 \times 10^{-4}\ m^2s^{-1}$

In hexagonal metals there are three directions preferred by the diffusion; these are the directions of the most dense packing <11-20> located in a hexagonal column on the plane (0001). These are also the privileged directions for the growth of a single crystal structure. This tendency is the stronger, the higher is the imposed from the outside speed of crystallisation and the higher is the content of alloying elements. In the case of single crystals grown without a nucleus, this means a constant axial orientation consistent with the direction [11-20]. This effect is so strong that all attempts at obtaining an orientation different than the one preferred by the growth mechanism require, beside the presence of a nucleus of the desired orientation, also the crucible of special design [18].

Alloys based on hexagonal metals offer low solubility to other metals in the solid state and a tendency to create numerous intermetallic phases [5,6,19-22].

The structural anisotropy strongly influences the mode of nucleation and growth of the secondary phases in single crystals with a hexagonal structure. The large difference in elastic constants at different crystallographic directions of the hcp lattice [16,17] enforces some permanent relationships between the crystal lattice of the newly emerging intermetallic phases and matrix.

2. Methodology

Single crystals of Zn-Ti and Zn-Cu-Ti used in this study were grown by Bridgman method with sliding temperature gradient.

Bridgman's original method [23], which consists in lowering a crucible with the charge inside through the zone of strong temperature gradient, was modified by application of the sliding motion of the furnace, while the nucleus and the charge were left immobile. The growth process was carried out in crucibles made of spectrally pure graphite. To prevent oxidation and also a reaction between the charge and the crucible material, an argon protective atmosphere was used. Permanent purging of the furnace chamber allowed the removal of oxygen and unwanted gaseous products formed during melting of charge. Compositions of the investigated single crystals and crystallisation speeds are summarised in Table 2. The resulting single crystals were oriented with a Bruker D8 Advance X-ray diffractometer, and were cut next into specimens with orientation of the observation planes (0001) and (11-20), respectively. The surfaces of the specimens were pre-polished with abrasive papers and diamond paste, first, and etched next with a chromium reagent. To reveal the shape of precipitates and determine the crystallographic relationship with the matrix, deep selective etching was used. Structural examinations were performed under Hitachi 3300 scanning microscopes with an EDS and EBSD attachments, and under the TESLA-302 microscope.

Single crystal chemical composition	Growth rate [mm/h]
Zn-Ti0.02 wt.%	3
Zn-Ti0.10 wt.%	3
Zn-Ti0.10 wt.% - Cu0.1 wt.%	1.8 6 10 16
Zn-Ti0.2 wt.% - Cu0.15 wt.%	1.8

Table 2. The investigated single crystals and growth rates.

3. Characteristics of phases present in the examined Zn-Ti and Zn-Ti-Cu single crystals

3.1 Types of phase lattice

In the case of Zn-Ti and Zn-Ti-Cu single crystals analysed in this study, the only stable intermetallic phase observed in an around-eutectic range (titanium content in zinc of about 0.2 wt.%) is the $Zn_{16}Ti$ phase [10-14,25]. It is a tetragonal phase of Cmcm structure containing 6.33 at.% of titanium. A single cell of this phase is built of 51 atoms. Each Ti atom is surrounded by 15 Zn atoms, with the additional 16th atom located in a space between the "chain" elements. The $Zn_{16}Ti$ phase has the following lattice parameters [25]:

a. 772.0 pm
b. 1144.9 pm
c. 1177.5 pm

For pure zinc elemental cell has a dimensions [2,3]:

a. 266.49 pm
b. 266.49 pm
c. 494.68 pm

Interplanar distances for the planes (0001) normal to the 'c' Zn direction are $d_{(0001)}=½c$ = 247.34 pm. On the other hand, the 'a' parameter of the $Zn_{16}Ti$ phase is 772 pm which is the value roughly three times higher (3 x 247.34 pm = 742.02 pm). Hence, the edge length ratio $a_{Ti}/½c_{Zn}$ is 772pm/247.34pm = 3.12.

Other combinations of ratios between the lattice constants of the $Zn_{16}Ti$ phase and the 'a' constant of a unit cell of zinc are:

a. $Zn_{16}Ti$ / a_{Zn} ~ 2.9
b. $Zn_{16}Ti$ / a_{Zn} ~ 4.3
c. $Zn_{16}Ti$ / a_{Zn} ~ 4.42

The edge ratio of a_{Zn16Ti}/a_{Zn}~2.9 is also close to an ideal multiple, but creating an interface of this configuration is energetically less favourable, as it does not affect the planes with the highest packed coefficient.

In this situation, one should expect a strong tendency to the formation of a crystallographic configuration, in which a_{Zn16Ti} edges of the phase cells and $c_{\alpha\text{-}Zn}$ edges of the matrix will be mutually parallel to each other.

3.2 Chemical composition and hardness of phases

Hardness of the $Zn_{16}Ti$ phase and of the α-Zn matrix was measured on the (0001) plane of a Zn-Ti0.2-Cu0.15 single crystal.

The α-Zn matrix of these single crystals contains 0.15 wt.% Cu and ~ 0.02 wt.% Ti. The measurements were taken using STRUERS microhardness tester, applying a load of 0.1 kg. To eliminate the error due to possible inhomogeneity of properties and small dimensions of the measured object, mapping was performed on the surface of 1.6 x 1.8 mm in the region containing a single exposed particle of $Zn_{16}Ti$ phase and pure matrix. Altogether, 80 measurements were taken. The results are shown in Figure 2 in the form of a 3D chart.

The obtained results showed that an average hardness of the $Zn_{16}Ti$ phase reaches 296 μHV as compared to 58 μHV of the matrix, with standard deviation not exceeding 8%. Previous studies performed on Zn-Ti and Zn-Ti-Cu alloys have proved that the $Zn_{16}Ti$ phase has little influence on hardening at the initial stage of deformation [11]. What predominates at that stage is the effect of precipitation hardening derived from copper, which all enters into the solution, and from small amounts of titanium (at the level of hundredths of a weight percent). Its main role in the hardening effect, the $Zn_{16}Ti$ phase starts playing at higher deformations when, owing to their shape and dimensions, the needle-like particles of this phase are acting in a way analogous to fibres in composite materials. Partially coherent combination of the $Zn_{16}Ti$ phase with matrix confirmed by the results of EBSD [10,11,12] suggests good transfer of stresses from the matrix, while high hardness of this phase will increase the properties of the material taken as a whole.

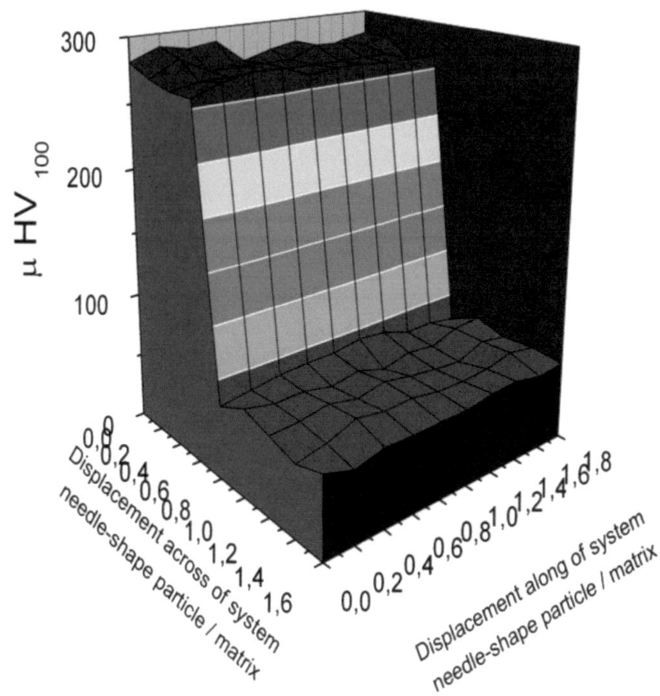

Fig. 2. Microhardness results for system $Zn_{16}Ti$ / α_{Zn}. Displacement in [mm].

4. Crystallographic relations in a matrix/precipitate system

The results of tests performed on Zn-Ti and Zn-Ti-Cu single crystals by TEM and EBSD techniques confirmed the existence of a close crystallographic relationship between the hexagonal crystal lattice of the matrix and the lattice of an intermetallic $Zn_{16}Ti$ phase. The a_{Zn16Ti} edges of the cells of the $Zn_{16}Ti$ phase and the $c_{\alpha\text{-}Zn}$ edges of the matrix are parallel to each other. At the same time, it has been observed that the direction of the b_{Zn16T} edges of the crystal cells in the $Zn_{16}Ti$ phase is in the majority of cases parallel to the $a_{\alpha\text{-}Zn}$ $\langle 11\,20 \rangle$ direction of the hexagonal lattice of the matrix. This type of relationship proves an interrelation that exists between the phase lattice and the matrix lattice and, consequently, a partial coherence between the precipitates and the matrix.

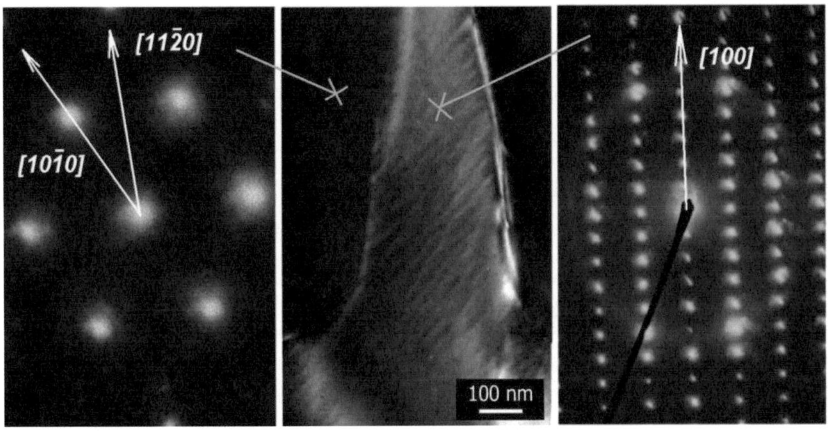

Fig. 3. TEM diffraction results from matrix and the $Zn_{16}Ti$ particle. Plane (0001) of the Zn-Ti0.1wt.% single crystal [11].

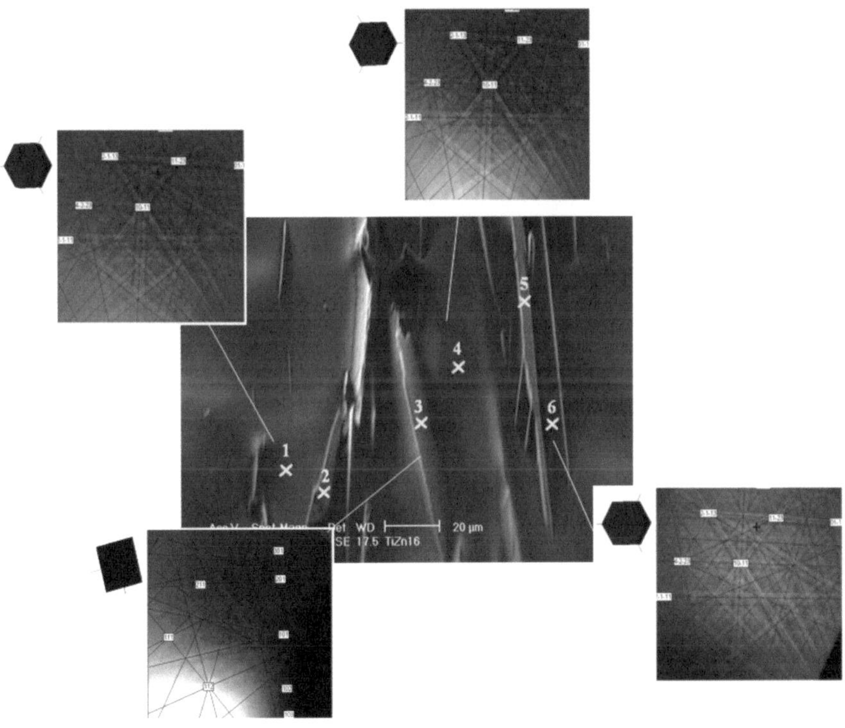

Fig. 4. EBSD results. Plane (0001) of the Zn-Ti0.1wt.% single crystal [26].

5. Control of the Zn16Ti phase morphology

The intermetallic $Zn_{16}Ti$ phase commonly occurs also in industrial zinc alloys. Due to its chemical composition, even alloys with a low content of titanium contain a large volume amount of this phase. This applies, first of all, to the hypoeutectic alloys with titanium content below 0.1 wt,%, which are used as a covering sheet metal in building constructions [26].

Previous studies have shown considerable potential for controlling the $Zn_{16}Ti$ phase morphology through the choice of alloy composition and crystallisation speed. In the case of hypoeutectic binary Zn-Ti alloys, the capabilities of forming different structures are limited by the low stability of the crystallisation front which, in turn, translates into intensive nucleation and growth of intermetallic phases in the form of rods of complex cross-sections. This applies to single crystals obtained by Bridgman method, where the limit growth rate for Zn-Ti0.02wt.% alloys and Zn-Ti0.1 wt% alloys with single crystal matrix does not exceed 3 mm/h [26]. With the rate so low, the factor deciding about the morphology of the $Zn_{16}Ti$ phase in a single crystal is the chemical composition. In the case of Zn-Ti0.02wt.% single crystals, the formation of needle-like crystallites is observed. The reason for this is a relatively low concentration of titanium, which forces predominant growth in the directions defined by diffusion and elastic constants of the crystal matrix. The result is an acicular form of precipitates of the $Zn_{16}Ti$ phase.
Much higher content of titanium in Zn-Ti0.1wt.% single crystals enables the growth of crystallites of an intermetallic phase with larger and more complex cross-sections than in the previous case. The low growth rate combined with a high concentration of titanium allows for efficient growth, also on less preferred planes. The result is the presence of the intermetallic phase particles in the form of rods of irregular cross-sections (Figure 5).

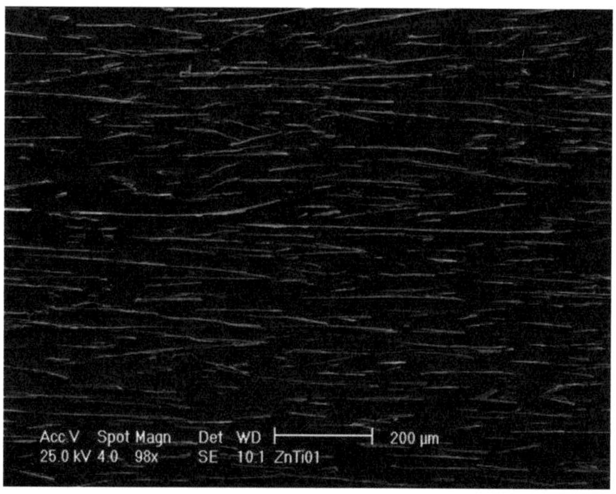

Fig. 5. The needle-shape form of the intermetallic phase $Zn_{16}Ti$ observed on (0001) plane in Zn-Ti0.1wt.% single crystals obtained at growth rate of 3 mm/h.

Enhancing the capability of generating the particles with different morphologies is possible through stabilisation of the crystallisation front with an alloying addition increasing the surface tension of the front and thereby significantly limiting the amplitude of thermal fluctuations on its surface (Figure 6). In zinc alloys, very effective has proved to be copper, used in an amount of 0.075wt.% .... 0.15wt.% as a component of industrial Zn-Ti-Cu alloys. Copper addition to Zn-Ti0.1wt.%-Cu0.1wt.% single crystals allows increasing the crystallisation rate up to 16mm/h, which brings a wide range of changes to the morphology of particles produced in an intermetallic phase precipitating at the crystallisation front (morphology changing from columnar through lamellar to acicular) [18].

For a low rate of single crystal growth in the Zn-Ti0.1wt.%-Cu0.1wt.% alloy, i.e. 1.8mm/h, the intermetallic phase occurs in the form of rods of irregular cross-sections (Figure 7). This situation is analogous to the previously discussed single crystals of binary Zn-Ti alloys. The morphology of this phase is shown in Figure 5. Increasing the growth rate of single crystals in Zn-Ti0.1wt.%-Cu0.1wt.% alloy up to 6mm/h changes the $Zn_{16}Ti$ phase morphology from columnar to lamellar (Figure 8).

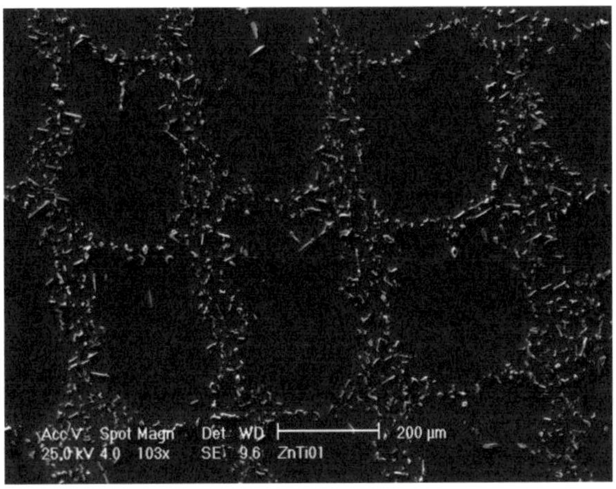

Fig. 6. Structure observed on (11-20) plane in Zn-Ti0.1wt.% single crystals obtained at growth rate of 3 mm/h.

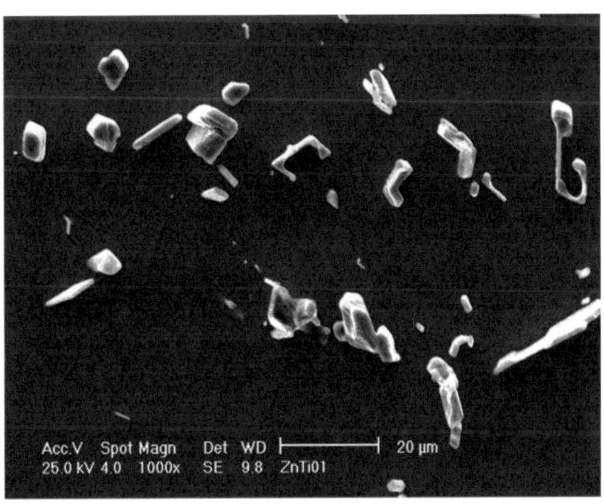

Fig. 7. The $Zn_{16}Ti$ intermetallic phase in a form of rods with irregular crossections, elongated on the [11-20] direction, observed in Zn-Ti0.1wt.%-Cu0.1wt.% single crystals obtained at growth rate of 1.8 mm/h.

Fig. 8. The lamellar form of the $Zn_{16}Ti$ intermetallic phase observed in Zn-Ti0.1wt.%-Cu0.1wt.% single crystals obtained at growth rate of 6 mm/h. The (11-20) plane [18].

The resulting structure is characterised by a high degree of ordering. Constant crystallographic relationships have been observed between the arrangement of the $Zn_{16}Ti$ phase lamellae and matrix orientation. The precipitates of the $Zn_{16}Ti$ phase in the form of lamellae prefer for growth the pairs of planes from the family {10-11} which, in the case of zinc crystal lattice with the ratio of c/a = 1.856, are oriented at angles of 51° (Figure 9). The choice of a specific pair of planes is determined by the direction of growth of a single crystal belonging to the <11-20> family, which also forms an axis for the band of the selected pair of planes (Figure 10).

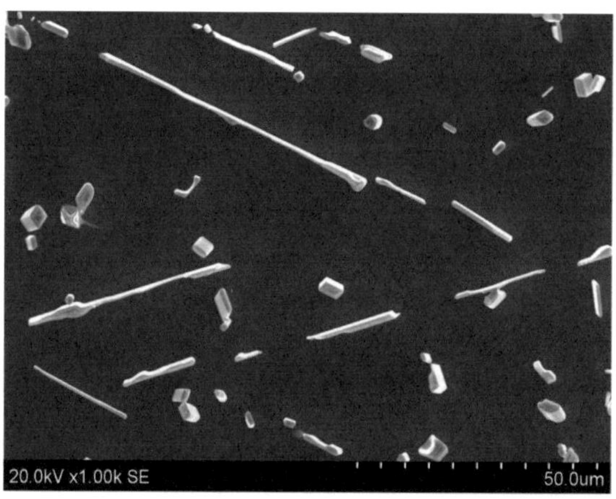

Fig. 9. The lamellar form of the $Zn_{16}Ti$ intermetallic phase observed in Zn-Ti0.1wt.%-Cu0.1wt.% single crystals obtained at growth rate of 6 mm/h. The (11-20) plane [14].

Another change in the morphology of the $Zn_{16}Ti$ intermetallic phase precipitates in single crystals of the Zn-Ti0.1wt.%-Cu0.1wt.% alloy is observed at the growth rate of 10mm/h. In the case of the examined alloy, this is the speed limit at which the crystallisation front is no longer a smooth surface because of the appearance of thermal fluctuations near the axis of growth. A consequence of this phenomenon is the locally varied growth rate of the single crystal matrix structure, changing the kinetics of growth of the intermetallic $Zn_{16}Ti$ phase precipitates at the crystallisation front. This situation is illustrated in Figure 11. The structure visualised here was observed on the plane (11-20) perpendicular to the axis of growth. A clear difference is observed between the morphology of precipitates in the central part of the single crystal cross-section and outer zone close to the crystal faces. The $Zn_{16}Ti$ phase in the outer zone has a lamellar morphology, qualitatively identical with that occurring in single crystals at a growth speed of 6mm/h, while central zone is characterised by a morphology close to the acicular one. The reason accounting for this phenomenon is the structure growth velocity locally increased due to thermal fluctuations in the 'empty' oval areas free from the precipitates, and a small radius of the front curvature in these areas. Under such conditions, the nucleation and growth of an intermetallic phase takes place at the inflection points, which best serve this purpose because of the energy expenditure needed to create a nucleus. This situation is shown in Figure 12.

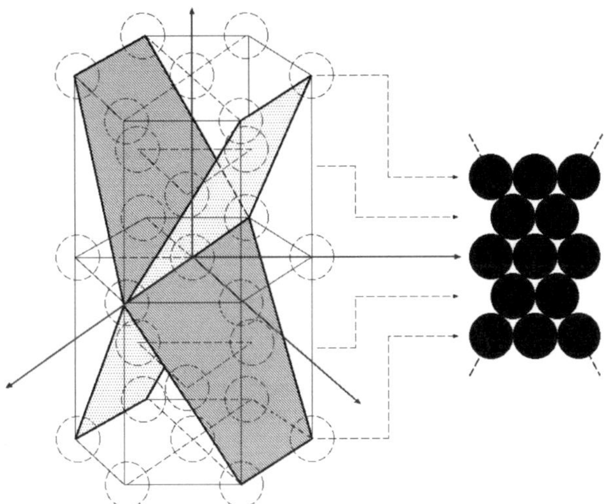

Fig. 10. The planes of preferred growth from {10-11} family, characteristic for lamellar shape of the $Zn_{16}Ti$ intermetallic phase.

With the speed of single crystals growth in the Zn-Ti0.1wt.%-Cu0.1wt.% alloy equal to 16 mm/h, the thermal fluctuations now cover the entire surface of the crystallisation front. This is shown in Figure 13. Within the whole examined area, the precipitates of purely acicular morphology arise. Compared to the previously discussed variant of a lower growth rate, the needles of the $Zn_{16}Ti$ phase now have round cross-sections, devoid of branches. It has also been observed that the oval areas devoid of precipitates show some degree of elongation on direction which is a trace of the plane (0001), characterised by the closest packing in a hexagonal structure (Figure 11, 12).

In the case of the Zn-Ti0.1wt.%-Cu0.1wt.% alloy, the speed of 16mm/h was the highest one for which a single crystal structure could be obtained (Figure 13,14). Therefore the next change to the morphology of precipitates was initiated by modification of the chemical composition. Considering the amount of the $Zn_{16}Ti$ phase formed in alloys with 0.1wt.% of titanium, single crystals of the Zn-Ti0.2wt.%-Cu0.15wt.% alloy were obtained. It is a composition similar to the eutectic point which, on account of the Zn-Ti alloys belonging to the group of "off-eutectic" alloys, has not been at this rate accurately determined.

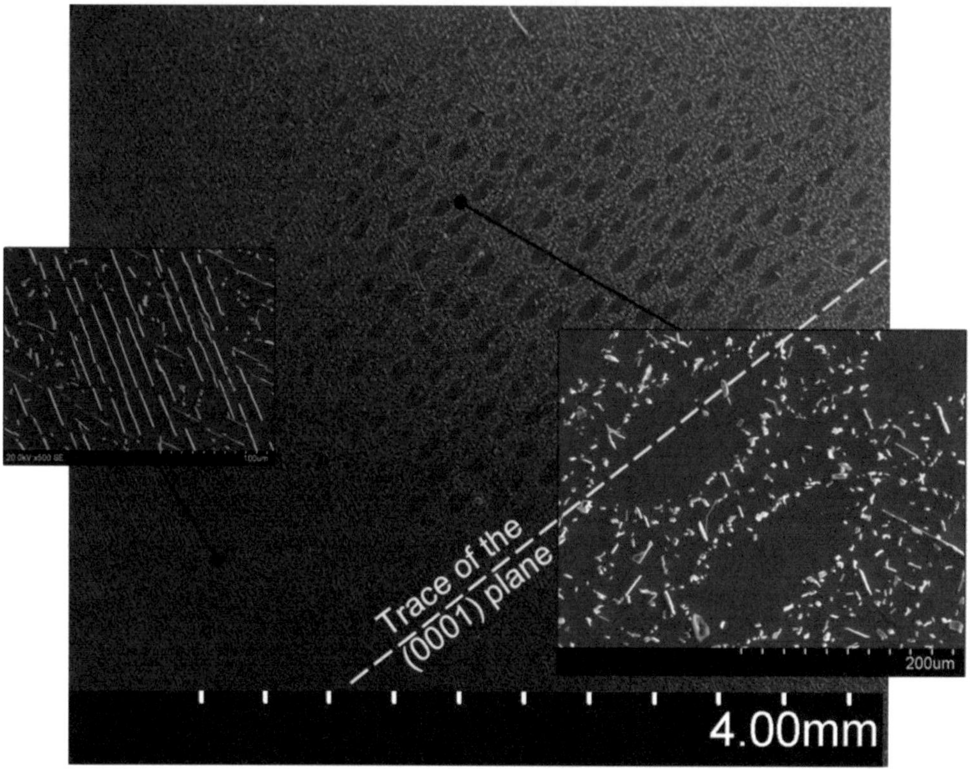

Fig. 11. Structure observed on (11-20) plane in Zn-Ti0.1wt.%-Cu0.1wt.% single crystals obtained at growth rate of 10 mm/h. Morphology of the precipitates in a core of crystals is different from morphology observed near outside walls of the crystal.

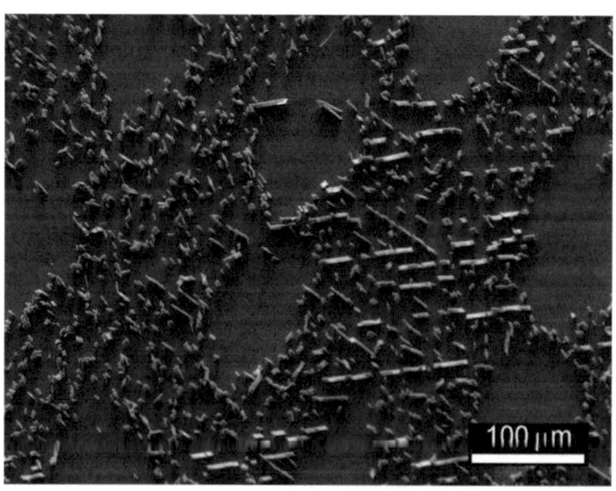

Fig. 12. The beginning of lamellar → fibre transformation observed in a core of the crystal.

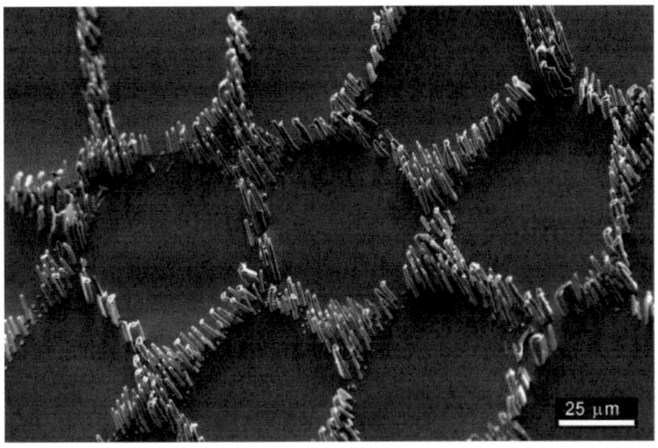

Fig. 13. Structure observed on (11-20) plane in Zn-Ti0.1wt.%-Cu0.1wt.% single crystals obtained at growth rate of 16 mm/h [18].

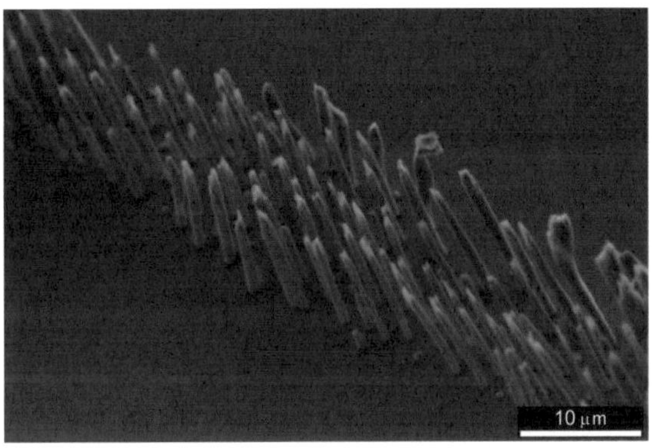

Fig. 14. Structure observed on (11-20) plane in Zn-Ti0.1wt.%-Cu0.1wt.% single crystals obtained at growth rate of 16 mm/h.

Single crystals in the Zn-Ti0.2wt.%-Cu0.15wt.% alloy were obtained at a speed of 1.8mm/h. Structural analysis revealed in the structure the presence of particles of very complex morphology (Figure 15). Observations revealed the existence of precipitates of an acicular morphology and macroparticles shaped like discs and polygons. The performed chemical analysis has shown that all the observed types of precipitates are composed of one intermetallic phase only, i.e. $Zn_{16}Ti$ [21]. Until now it has not been possible to identify clearly the reasons which have made one phase assume so many different types of morphologies. One of the reasons can be different mechanism of nucleation of the individual types of particles. The constitution of Zn-Ti0.2wt.%-Cu0.15wt.% macroparticles observed in the structure of Zn single crystals significantly deviates from previous observations of the $Zn_{16}Ti$ phase. In the case of macroparticles (Figure 16), the precipitates take the form of a eutectic system, in which there are alternately arranged lamellae of $Zn_{16}Ti$ / α / $Zn_{16}Ti$ / α / .../ α /. The nucleation of such a system is energetically privileged, compared to the process by which isolated particles of one phase only are expected to precipitate from the solution. This follows from the mechanism of the coupled growth of eutectic phases [27].

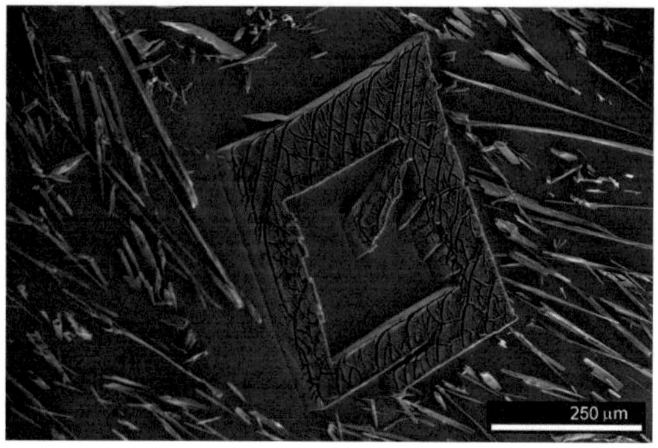

Fig. 15. Multishape morphology of the $Zn_{16}Ti$ intermetallic phase observed on (0001) plane in Zn-Ti0.2wt.%-Cu0.15wt.% single crystals obtained at growth rate of 1.8 mm/h.

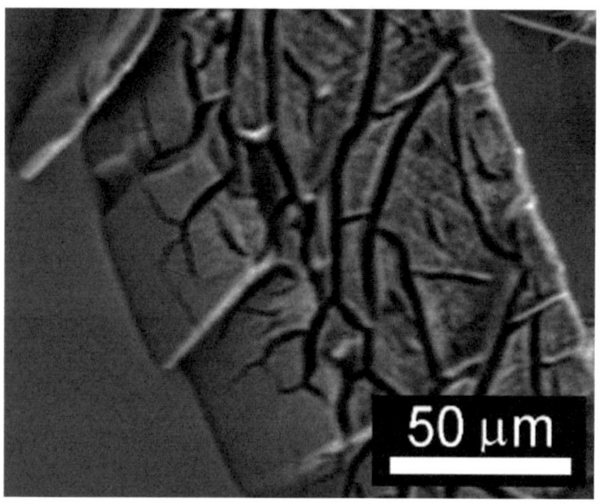

Fig. 16. The macroparticleas a eutectic form $Zn_{16}Ti/\alpha/$ $Zn_{16}Ti/\alpha/$... observed on (0001) plane in Zn-Ti0.2wt.%-Cu0.15wt.% single crystals obtained at growth rate of 1.8 mm/h.

6. The distribution of $Zn_{16}Ti$ phase in function of growth conditions

6.1 Oscillatory structures

The conditions of growth and chemical composition affect not only the morphology of the $Zn_{16}Ti$ phase precipitates, but also distribution of this phase. A special case of the structure formed at low growth rates of the Zn-Ti and Zn-Ti-Cu single crystals is oscillatory structure [18,26,28]. It is characterised by the cyclic occurrence of typical elements, such as areas rich in precipitates interwoven with areas of pure matrix totally devoid of precipitates.

Oscillatory structures were observed for the first time in the peritectic Sn-Cd alloys [28]. They are formed as a result of changes in the concentration of an alloying constituent ahead of the moving crystallisation front, which affects the conditions necessary for nucleation of individual phases. With proper velocity of the crystallisation front movement, the preferred growth is alternately exhibited by phases a and b (Figure 17).

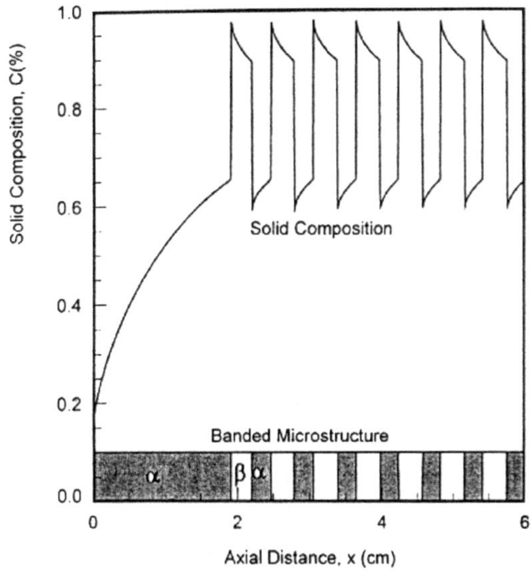

Fig. 17. Growth mechanism of the oscillatory structure in peritectic [28].

Massive macroparticles use in building of their structure a significant amount of titanium, thus reducing the concentration of this element in solution. Then, from the rest of the titanium-depleted solution, the precipitation of fine-grained phases occurs; on account of the low concentration of titanium, these phases do not form a eutectic system in the solution.

A similar phenomenon as described previously for the peritectic Sn-Cd alloys was also obtained in the eutectic Zn-Ti and Zn-Ti-Cu alloys. Studies showed that the condition necessary for the formation of such structures in single crystals based on zinc is low travel speed of the cooling zone, and hence low rate of the crystal growth. In cases under discussion, it did not exceed 6 mm/h. Additional important factors are the chemical composition of the alloy and differences in temperature between the melting point of pure metal (in this case zinc) and the eutectic point. The lower is the difference in temperatures, the stronger tendency the alloy shows to the formation of an oscillatory structure. Depending on the degree to which these conditions are satisfied, the following types of oscillatory structures can be obtained:

a. purely oscillatory, i.e. a sequence of alternately occurring regions rich in inclusions of the second phase and layers of pure matrix (alpha phase) free from the precipitates [18],

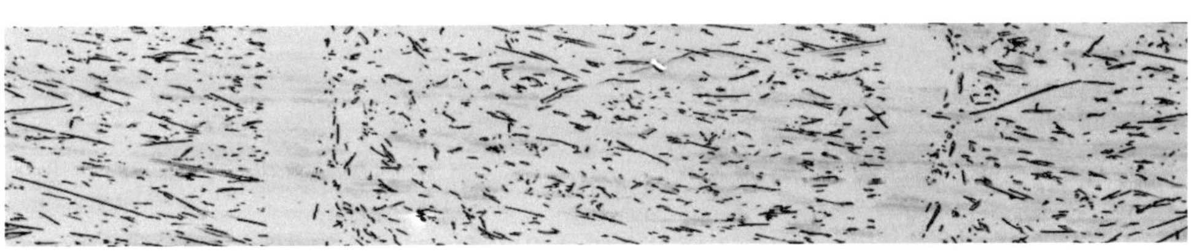

Fig. 18. Oscillatory structure of "a-type" observed in Zn-Ti-Cu single crystals.

b. oscillatory, in the case of which areas totally free from the precipitates are not observed any longer, while the degree of filling the alternately occurring layers with inclusions of the intermetallic phases is periodically changing.

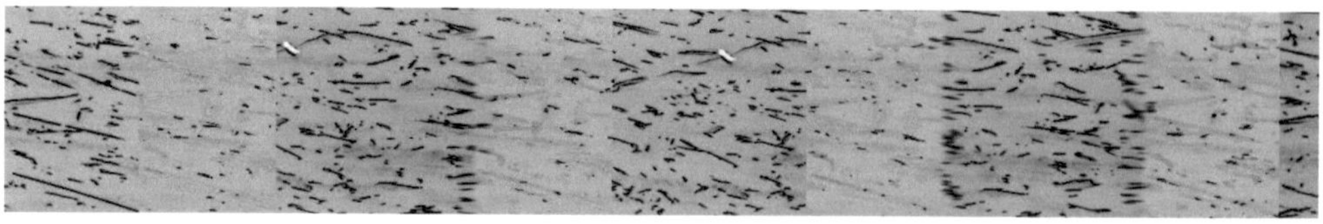

Fig. 19. Model of oscillatory structure of "b-type" observed in eutectic Zn-Ti-Cu single crystals.

Case (a) occurs in titanium alloys with 0.1 wt.% Ti and less. In alloys with the titanium content so low, the zonal segregation occurring during single crystal growth by Bridgman method is sufficiently strong to form areas with the concentration of alloying elements so low that the process of the second phase precipitation is inhibited.
Another phenomenon observed in the case of the oscillatory structures of type (a) is the, linearly changing along the growth direction, density of the inclusions of the second phase in areas rich in precipitates. This phenomenon is particularly evident in single crystals growing at the lowest rates (1.8mm/h). This is presented in Figure 20a. Increasing the growth rate to 6 mm/h makes the structure preserve its oscillatory character but with a uniform distribution of precipitates within the layer (Figure 20b). The reason for this is lower concentration of titanium ahead of the crystallisation front caused by a higher rate of the cooling zone transfer, which weakens the effect of zonal segregations, and consequently changes the conditions for nucleation of an intermetallic phase.

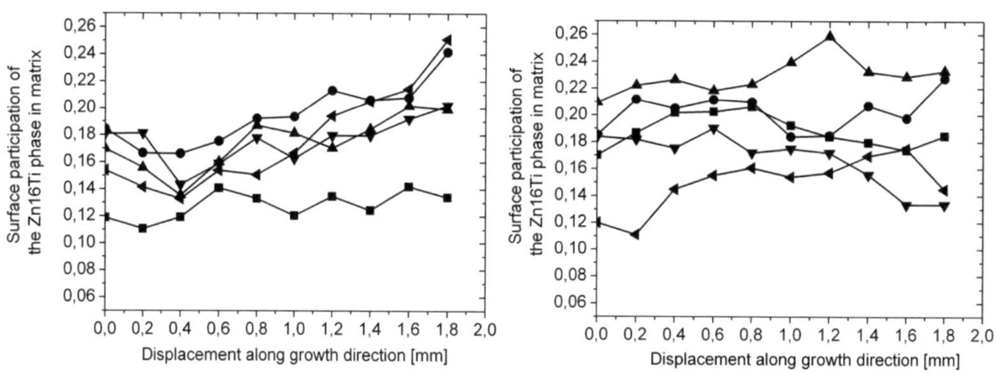

Fig. 20. The $Zn_{16}Ti$ phase distribution along of a precipitation-rich layer: a) oscillatory structure obtained at 1.8 mm/h growth rate; b) oscillatory structure obtained at 6 mm/h growth rate.

On the other hand, structures of type (b) have been observed in Zn-Ti-Cu alloys with around-eutectic titanium content. In this case, the concentration of titanium in the entire volume of the solution is so high that, regardless of the speed of crystal growth, there are conditions for the nucleation and growth of the $Zn_{16}Ti$ phase. What changes is only the intensity of this phenomenon, which manifests itself in an alternate occurrence in the structure of the layers of higher and lower concentration of the precipitates.

6.2 Model analysis of oscillatory structure

To better understand the growth mechanism of oscillatory structures, a mathematical model developed by Wolczynski has been used [29]. It allows us to estimate the impact of growth conditions and chemical composition on the forming sequence of layers rich in an intermetallic phase and free from the precipitates [11,18,26,29,30]. This model has been developed for binary alloys and is based on the diffusion-related changes in alloying constituent concentration. It includes the three major processes occurring during alloy crystallisation:

1. changes in the concentration of alloying element depleting the liquid phase with progressing crystallisation along the liquidus line,

$$N^L(x;\alpha) = N_0\left(1+\alpha\, k\, x - x\right)^{(k-1)/(1-\alpha k)}$$

2. microsegregations at the crystallisation front, resulting from the constituent segregation,

$$N^S(x;\alpha) = k N_0\left(1+\alpha\, k\, x - x\right)^{(k-1)/(1-\alpha k)}$$

3. redistribution of alloying constituents in grains after the occurrence of reverse diffusion.

$$N^B\left(x;X^0,\alpha\right) = \left[k + \beta^{ex}\left(x;X^0\right)\beta^{in}\left(X^0,\alpha\right)\right] N^L(x;\alpha)$$

where:

- k - partition ratio, x - amount of growing crystal,
- N_0 - nominal concentration of the alloy, α - back-diffusion parameter,
- β^{ex}; β^{in} - coefficients of the redistribution extension and intensity.

$$i_E(N_0) = i_K(1,N_0)$$

$$i_D(\alpha,N_0) = i_K(\alpha,N_0) - i_E(N_0)$$

$$i_K(\alpha,N_0) = 1 - x_K(\alpha,N_0)$$

$$x_K(\alpha,N_0) = \frac{1}{1-\alpha k}\left[1-\left(\frac{N_E}{N_0}\right)^p\right]$$

This method allows precise determination of the amount of precipitates of both equilibrium i_E and non-equilibrium i_D character, where x_K is the amount of pure single crystals between the fringes.

$$p = \frac{1-\alpha k}{k-1}$$

If α is equal to zero, then the model is diffusionless. Such simplification makes the model unrealistic, although it may serve as a benchmark or reference.

Two binary alloys, i.e. Zn-Ti0.02wt.% and Zn-Ti0.1wt.%, were selected for tests. The structures of single crystals of these alloys are of an oscillatory character, but differ considerably in respect of volume fraction of the $Zn_{16}Ti$ phase and its distribution. Layers rich in precipitates observed in the Zn-Ti0.02wt.% alloy are much smaller in thickness than the layers without precipitates.

Single crystal	Titanium contents [wt.%]	Thickness of precipitation-rich layer [μm]	Thickness of precipitation-free layer [μm]
Zn-Ti0.02wt.%	0.023	10	120.4
Zn-Ti0.1wt.%	0.10	56.6	127.3

Table 3. Structural parameters of the Zn-Ti0.02wt.% and Zn-Ti0.1wt.% single crystals.

For single crystals of Zn-Ti.01wt.% alloy, the layers rich in precipitates are much thicker, while spacing between them is similar to that observed in single crystals of lower titanium content.

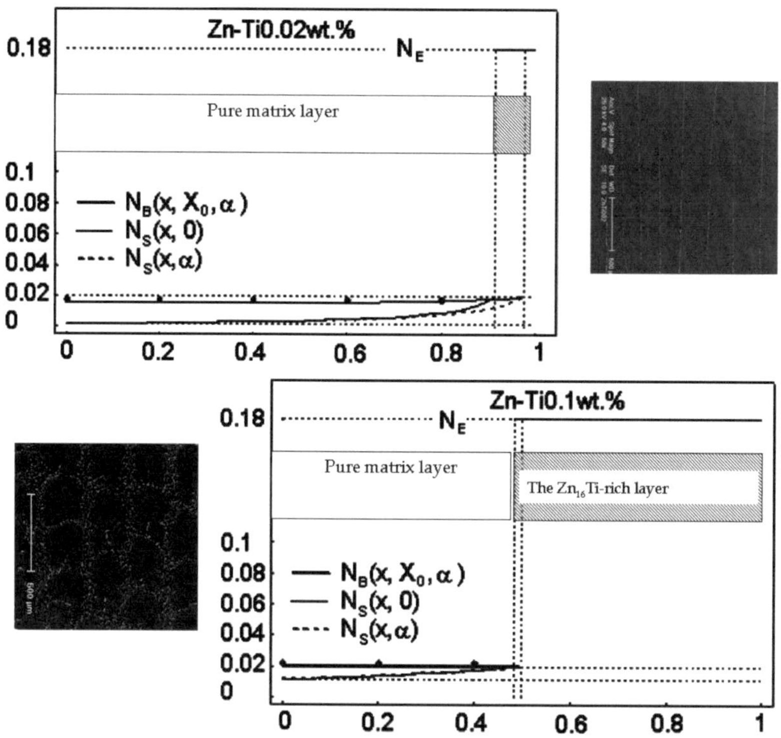

Fig. 21. Calculatin results obtained for the model.

Designations used in Figure 21:

- N_E - eutectic concentration of the Ti-solute [wt.%]
- N_B - solute redistribution after back-diffusion [wt.%]
- N_S - "historical" solute concentration on the solid/liquid interface
- α - back-diffusion parameter [wt.%]

The proposed model based on calculations of the diffusion of the alloying elements fairly well reflects the growth of an oscillatory structure in the binary Zn-Ti alloys. The compliance is particularly strong for the structure of single crystals of Zn-Ti0.02wt.%, where there is clear division between individual elements of the structure. In the case of single crystals of Zn-Ti0.1wt.%, a large amount of the alloying constituent disturbs the arrangement of layers because of "bridges" that are formed in between them. However, in the case of a model based solely on theoretical data, the results obtained can be considered correct. Additionally, Figure 22 presents a hypothetical path of crystallisation for the Zn-Ti0.1wt.% alloy [30].

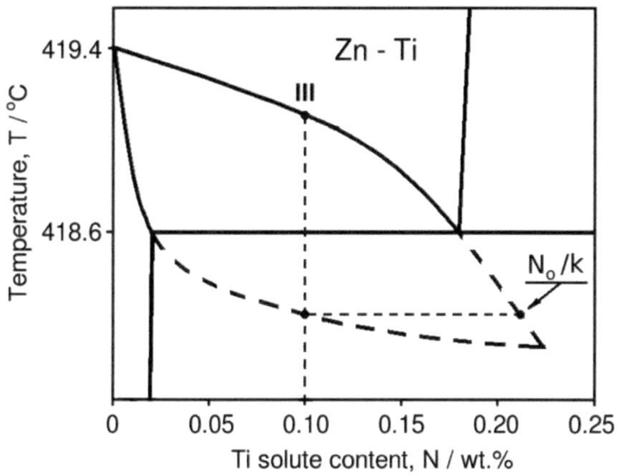

Fig. 22. Solidification $III \rightarrow N_0 / k$ for an equilibrium solidification of the single crystal with $N_0 = III = 0.1\,[wt.\%Ti]$ [30].

6.3 Continuous and cellular structure in single crystals of Zn-Ti0.1-Cu0.1

Increasing the growth rate of Zn-Ti-Cu single crystals leads to disappearance of oscillatory structure and formation of structures with a uniform distribution of inclusions of the second phase, accompanied by simultaneous change in the morphology of an intermetallic phase.

In the case of an oscillatory structure, the observed phase has the form of rods with heterogeneous cross-sections, while with the crystallisation rates of 6.10 mm/h a lamellar phase appears [18].

The morphological transformation at 10mm/h is a consequence of the crystallisation front changing from flat to cellular. The changing character of the crystallisation front can affect the morphological transformation of the second phase. The phase transformation from lamellar to acicular takes place due to the change of nucleation conditions. According to the thermodynamic rules, the lamellae/rods transformation takes place if and when the minimum free energy for lamellae ΔG_L is equal to the minimum free energy for rods ΔG_R [24,29].

Further increase in the front travel rate brings next change to the distribution and morphology of an intermetallic phase. The distribution of precipitates starts changing from continuous to cellular. This transformation is also accompanied by the appearance of the $Zn_{16}Ti$ phase in the form of needles. The beginning of this transformation can be observed in the core of a single crystal, while near the outer walls, the continuous structure with lamellar morphology of the precipitates still prevails. This situation occurs at the rate of growth equal to 10 mm/h.

Thermal fluctuations at the crystallisation front are responsible for different growth rates in microregions, causing the appearance of pseudo-cellular structure. It has the appearance of a lattice with regularly spaced meshes free from the precipitates and surrounded by an acicular phase. The lattice meshes seen on the plane (11-20) are of an oval shape, elongated in the direction indicated by traces of the plane (0001). Observed on the plane (0001), this structure gives a picture of axial sections of the strongly elongated "rods" of pure matrix, enclosed from the outside by acicular precipitates.

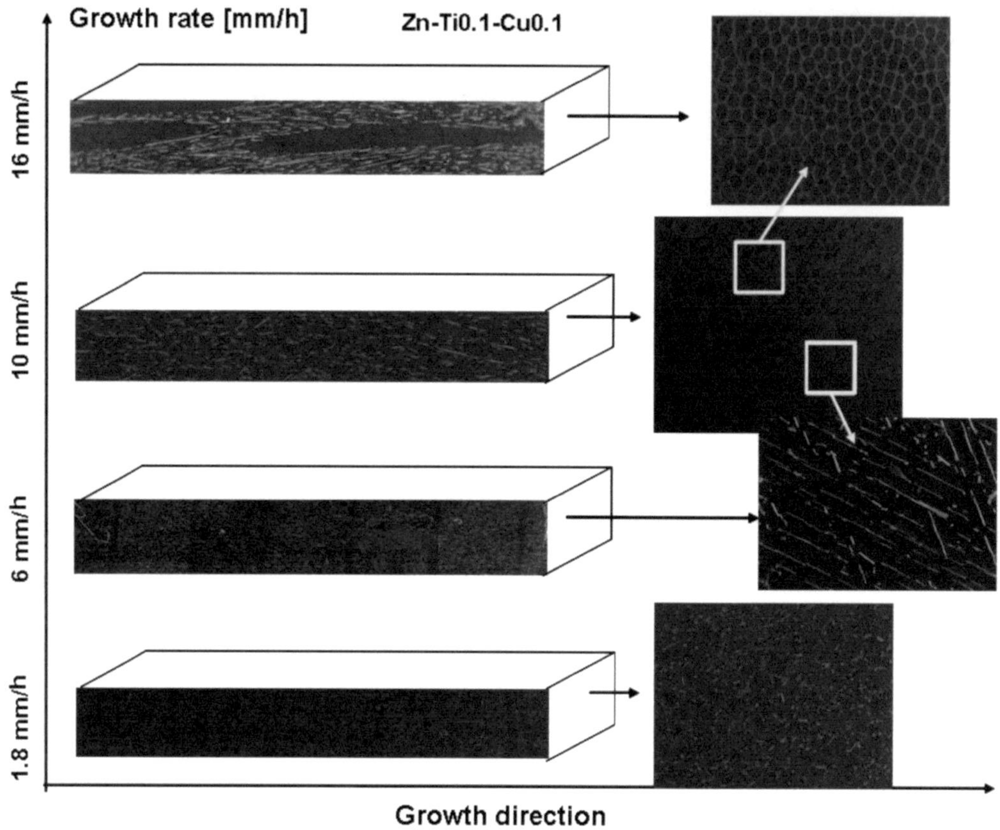

Fig. 23. Comparison of morphology and distribution of the $Zn_{16}Ti$ intermetallic phase in relationship to growth rate of the Zn-Ti0.1wt.%-Cu0.1wt.% single crystals.

An increase in the travel speed of the cooling zone to 16mm/h makes the whole crystal volume assume the form of a pseudo-cellular structure. The observed precipitates assume the form of strongly elongated needles with circular cross-sections, while areas surrounded by them are free from the precipitates and assume an equiaxial shape. Further increase in the crystallisation rate makes the matrix lose its crystallographic continuity and starts the formation of a structure with strongly elongated grains [17].

7. Summary

Regardless of the applied variant of the Zn-Cu-Ti Ti single crystals growth containing up to 0.2wt.% Ti, only one type of the intermetallic $Zn_{16}Ti$ phase was observed to occur. Changing the growth conditions and the chemical composition strongly affected the morphology and distribution of the precipitates of this phase in a single crystal matrix.

It was found that low growth rates of zinc-based single crystals led to the formation of an oscillatory structure. At the same time, in the case of binary Zn-Ti alloys, it was the only available type of structure. Generating other types of structure required higher rates of the single crystals growth. On the other hand, the acceleration of the growth process required stabilised surface of the crystallisation front obtained by reduced thermal fluctuations. This was achieved by adding copper to the Zn-Ti alloy, which resulted in increased surface tension at the crystallisation front and enabled generation of structures characterised by a single crystal matrix at rates in the range of 1.8 - 16 mm/h.

	Shape of the solid/liquid interface	**Type of the structure**	**The intermetallic compound morphology**
1.8 mm/h	almost plane	oscillatory	irregular bands
6 mm/h	concave, without instabilities	oscillatory	regular bands
10 mm/h	instabilities appear in the core	continous	bands/needles (fibres)
16 mm/h	instabilities on a whole surface	cellular/pipe	rod-like needles (fibres)

Table 4. Structure types and the intermetallic phase morphology obtained at different growth rates.

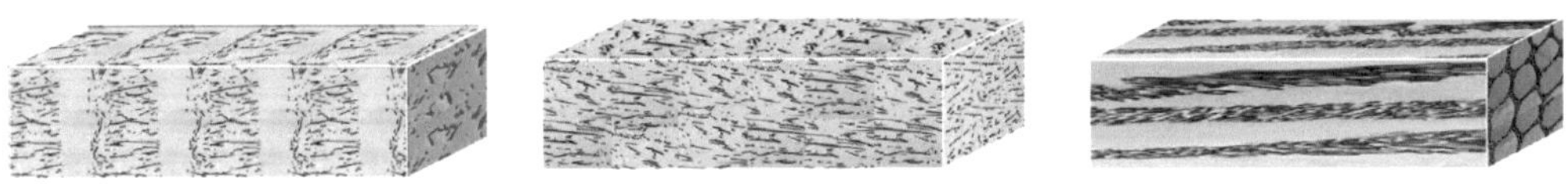

Fig. 24. Change of the intermetallic phase distribution in relationship to growth rate observed in the Zn-Ti0.1wt.%-Cu0.1wt.% single crystals.

Changing the speed of Zn-Ti0.1-Cu0.1 single crystals growth strongly influences the distribution of the $Zn_{16}Ti$ phase in a single crystal matrix, as shown in Figures 23 and 24. In the case of oscillatory structures, the following relationships were observed:

a. precipitates-free layers are located at the same distance, λ, for the growth rates 1.8 mm/h and 6 mm/h
b. thickness of these layers, δ, decreases with the growth rate
c. precipitates-rich layer thickness, λ -δ, increases with the growth rate
d. oscillatory structure vanishes, cells structure appears in the core of the single crystal, above the growth rate 10 mm/h.

The mathematical model proposed for description of the formation of an oscillatory structure operates well in the case of binary alloys and can be used as a tool helpful in the selection of crystallisation conditions when shaping the desired structure.

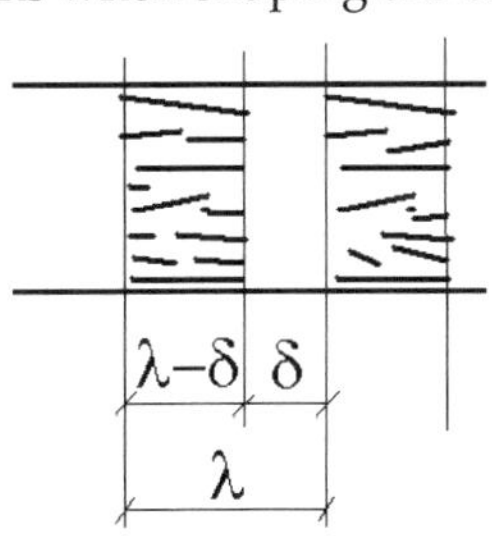

8. Acknowledgment

The authors acknowledge the support of the Polish Committee for Scientific Research, Grant N N508 4800 38 and Grant No. 11.11.180.255

9. References

[1] Oliver H.Wyatt, D.Dew-Hughes: "Metals, Ceramics and Polymers - an introduction to the structure and properties of enginnering materials", Cambridge University Press (1974)

[2] H.Ibach, H.Luth: "Solid-State Physics", Springer-Verlag, (1991)

[3] Landolt-Bornstein, New Series Vol.III, b "structure Data of Elements and Intermetallic Phases", Springer, Berlin, Heidelberg, (1971)

[4] E. Fortunato et al., Materials Science Forum, 514-516, 3, (2006)

[5] J.L.Murray; "Phase Diagrams of Binary Titanium Alloys", J.L.Murray, ed.,ASM International, Metals Park, OH, 336-339 (1987)

[6] T.B.Massalski,: "Binary Alloy Phase Diagrams", editor: H.Okamoto, ASM International (1996)

[7] V. Heine, "Group Theory in Quantum Mechanics", Pergamon, London, (1960)

[8] M. Tinkham, "Group Theory and Quantum Mechanics", McGraw-Hill, New York, (1964)

[9] Gustav E.R. Schulze, Metallphysik, Akademie-Verlag, Berlin (1974)

[10] G.Boczkal, B.Mikulowski, I.Hunsche, C-G.Oertel, W.Skrotzki, Cryst.Res.Technol.,No.2, pp.135-140 (2008)

[11] B.Mikulowski,G.Boczkal, Arch. Metall. Mater.,. 54,: p. 197-203, (2009)

[12] G.Boczkal, B.Mikulowski, I.Hunsche, C-G.Oertel, W.Skrotzki, Cryst. Res. Technol. 45, 111 (2010)

[13] G.Boczkal, B.Mikułowski, Journ.of Alloys and Compound 378, pp.135-139, (2004)

[14] G.Boczkal, Materials Science Forum Vol. 674, pp 245-249,(2011)

[15] B.Mikulowski, Strain Hardening of Zinc Monocrystals with Additions of Silver or Gallium, Metallurgy and Foundry Practice, Scientific Bulletin of Univ. of Mining and Metallurgy 96, Cracow (1982).

[16] G.A.Alers, J.R. Neighbours, The elastic constants of zinc between 4.2° and 670°K ,J.of Phys and Chem of Solids, Vol.7, Iss.1, pp. 58-64, (1958)

[17] C.W.Garland, R.Dalven, Elastic Constants of Zinc from 4.2K to 77.6K, Phys.Rev., vol.111, n.5, Sept. 1, (1958)

[18] G.Boczkal, B.Mikulowski, W.Wolczynski, Materials Science Forum Vol. 649, pp 113-118,(2010)

[19] E.A.Anderson, E.J.Boyle and P.W.Ramsey, Trans. AIME 156, p.278, (1944)

[20] J.A.Spittle, The Effect of Composition and Cooling Rate on the as Cast Microstructure of ZnTi Alloys, Metallography 5, pp.423-447, (1972)

[21] J.A.Spittle, Metallography, Vol.6, pp.115-121 (1973)

[22] Von W.Heine, U.Zwicker, Bd. 53, H.6, (1962)

[23] W. D. Lawson, S. Nielsen: Preparation of Single Crystals. Butterworths Scientific Pub. London. (1958)

[24] B. Chalmers: Principles of Solidification. New York, John Wiley & Sons Inc. (1964)

[25] M.Saillard, G.Develey, C.Becle, J.M.Moreau, D.Paccard, The Structure of ZnTi16, Act.Cryst. 37B, 224-226, (1981)

[26] G.Boczkal, Ph.D. Thesis, AGH University of Science and Technology (2006)

[27] E.Fraś, *Crystallization of metals*, WNT, Warsaw, (2003)

[28] P.Mazumder, R.Trivedi, A.Karma, Metall.and Mat.Trans.A, Vol.31A, pp.1233-1246, (2000)

[29] W. Wołczyński: Modelling of Transport Phenomena in Crystal Growth. Ed.:Szmyd J.S., Suzuki K., Southhampton, Boston, WIT Press (2000)

[30] W.Wolczynski, B.Mikulowski, G.Boczkal, Materials Science Forum Vol. 649, pp 125-130,(2010)

8

High Quality $In_xGa_{1-x}As$ (x: 0.08 – 0.13) Crystal Growth for Substrates of λ= 1.3 μm Laser Diodes by the Travelling Liquidus-Zone Method

Kyoichi Kinoshita and Shinichi Yoda
Japan Aerospace Exploration Agency
Japan

1. Introduction

The growth of compositionally uniform alloy crystals is promising for variety of applications because lattice parameters as well as electrical and optical properties can be controlled by composition. Among them, $In_xGa_{1-x}As$ bulk crystals are expected as substrates of laser diodes with emitting wavelength of 1.3 μm. High optical gain with small temperature dependence was demonstrated for strained quantum well grown on $In_xGa_{1-x}As$ substrates (Ishikawa, 1993; Ishikawa & Suemune, 1994). However, growth of homogeneous $In_xGa_{1-x}As$ bulk single crystals is very difficult due to large separation of liquidus and solidus lines in the pseudobinary phase diagram (Bublik & Leikin, 1978). $In_xGa_{1-x}As$ bulk crystals were grown by the liquid encapsulated Czochralski (LEC) method with supplying GaAs (Nakajima et al., 1991), zone levelling method (Sell, 1991) and multicomponent zone melting method (Nishijima et al. 2005). InAs mole fraction was limited to 0.2 in LEC method and zone levelling method due to temperature fluctuation in the melt and large separation of liquidus and solidus lines. Multicomponent zone growth (MCZG) using a seed with graded InAs concentration produced a single crystal with InAs mole fraction of 0.3 and length of about 17 mm (Nishijima et al. 2005). However, MCZG requires complicated growth technique and no good reproducible results were obtained. In all of the methods, the most difficult point is to keep the freezing interface temperature constant for growing compositionally uniform crystals since interface travelling rate depends on temperature gradient, mass transport in a melt and solute concentration gradient ahead the interface. As a result, no device quality $In_xGa_{1-x}As$ crystals with uniform composition have been obtained so far.

The travelling liquidus-zone (TLZ) method was invented for keeping the interface temperature constant and growing compositionally uniform alloy crystals (Kinoshita et al., 2001, Kinoshita et al., 2004). In the TLZ method, the interface travelling rate can be determined exactly if temperature gradient in the zone is known and then the interface position can be fixed in relation to the heater position by translating a sample device in accordance with the interface travelling rate (Kinoshita et al. 2002). Principle of the TLZ method is proven by the precise measurement of temperature gradient in the melt zone, growth rate and the composition of grown crystals (Nakamura et al. 2003). Since the TLZ method requires diffusion limited mass transport, diameter or thickness of the melt is limited to small values on the ground for suppressing convection in a melt. 2 mm thick platy $In_xGa_{1-x}As$ (x: 0.1 - 0.13) crystals were grown by the TLZ method as substrates of laser diodes with λ = 1.3 μm (Kinoshita et al. 2008). Quality of grown crystals was characterized by electron probe micro analyzer (EPMA), electron back scattering pattern (EBSP), X-ray diffraction (XRD) and by fabricating laser diodes. High temperature stability of output power of laser diodes was proven (Arai et al. 2009).

In this chapter, principle of the TLZ method, growth of high quality $In_xGa_{1-x}As$ platy crystals, fabrication of laser diodes with λ = 1.3 μm and characterization of laser diodes are reviewed.

2. TLZ method

The TLZ method was originally invented for compositionally uniform alloy crystal growth in microgravity. Principle of the TLZ method is studied by growing small diameter $In_xGa_{1-x}As$ crystals on the ground (Kinoshita et al., 2002, 2003, Nakamura et al., 2003, Kinoshita and Yoda, 2010) because convection in a melt can be suppressed in capillary tubes even on the ground. In this section, principle of the TLZ method, one dimensional model and evaluation of a model are reported.

2.1 Principle of a TLZ method

It is well known that compositionally uniform alloy crystals can be grown if diffusion controlled steady-state growth is achieved (Tiller et al., 1953). However, even if such growth is realized during directional solidification, initial transient region is inevitable and long crystal length is required to reach the steady-state. Moreover, growth rate should be sufficiently high because solute in the boundary layer diffuses out and solute should be supplied to keep constant solute concentration at the interface. Segregation at solidification supplies solute and amount of supplied solute is determined by growth rate. This is the reason why sufficiently high growth rate is required for the steady-state growth. High growth rate, however, tends to cause the constitutional supercooling (Tiller et al., 1953). Many investigators tried to grow compositionally uniform alloy crystals by diffusion controlled steady-state growth but none of them obtained satisfactory results. Microgravity experiments are also expected to realize diffusion controlled steady-state growth by suppressing convection in a melt but no one succeeded in growing homogeneous alloy crystals having expected composition. Residual gravity in space crafts on the order of 10^{-4} g hinders diffusion controlled growth. Therefore, a new growth method for obtaining compositionally uniform alloy crystals is anticipated.

When we consider diffusion boundary layer in the steady-state growth, we noticed that zone melting method under temperature gradient is easier to maintain constant solute concentration at the interface because solute is saturated at freezing and dissolving interfaces. Moreover, solute is almost saturated throughout the zone if zone thickness is thin enough. As described later in more detail, in such situation solute concentration gradient in the zone is controlled by temperature gradient and compositionally uniform alloy crystal growth becomes much easier than other growth method. We named this zone melting method under temperature gradient a travelling liquidus-zone (TLZ) method since crystal growth is carried out by travelling almost saturated liquid-zone. Sample configuration, temperature profile, solute profile in a sample and its relation to a phase diagram in the TLZ method is depicted in Fig. 1 as referring to the growth of $In_{0.3}Ga_{0.7}As$. It is needless to say that convection in a thin melt zone is suppressed more effectively than in a long melt even in microgravity.

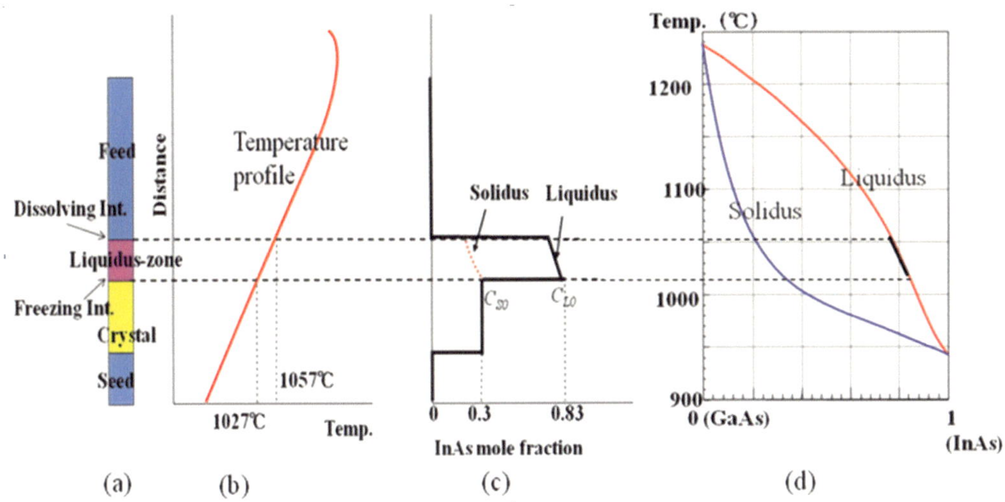

Fig. 1. Schematic drawing of a TLZ method as referring to the growth of $In_{0.3}Ga_{0.7}As$. (a) sample configuration, (b) temperature profile, (c) InAs concentration profile, (d) GaAs-InAs pseudobinary phase diagram.

In Fig. 2, solute concentration profile in the diffusion boundary layer in the steady-state growth is compared between directional solidification method and TLZ method (Kinoshita and Yoda, 2011). When temperature gradient is low and zone thickness is thin, solute concentration profile is linearly approximated and it can be related to the liquidus line in the pseudo binary phase diagram as shown in Fig. 1. Liquidus line is also linearly approximated in a narrow temperature range. Then, it should be noticed that solute concentration gradient in a zone can be controlled by temperature gradient in such case as is given by eq. (1), where C_L is solute concentration in the zone, T is temperature, Z is a distance measured from the freezing interface and $\partial C_L/\partial T$ is reciprocal of the slope of the liquidus line in the phase diagram. Then, freezing rate V is given by eq. (3) if we suppose the diffusion controlled steady-state growth is realized as given by eq. (2) and V is determined by temperature gradient in the zone, where D is interdiffusion coefficient between solute and solvent, C_{L0} and C_{S0} are liquidus and solidus concentrations at the freezing interface.

$$\left(\frac{\partial C_L}{\partial Z}\right)_{Z=0} = \left(\frac{\partial C_L}{\partial T}\right)\left(\frac{\partial T}{\partial Z}\right)_{Z=0} \tag{1}$$

$$V(C_{L0} - C_{S0}) = -D\left(\frac{\partial C_L}{\partial Z}\right)_{Z=0} \tag{2}$$

$$V = -\frac{D}{(C_{L0} - C_{S0})}\left(\frac{\partial C_L}{\partial Z}\right)_{Z=0} = -\frac{D}{(C_{L0} - C_{S0})}\left(\frac{\partial C_L}{\partial T}\right)\left(\frac{\partial T}{\partial Z}\right)_{Z=0} \tag{3}$$

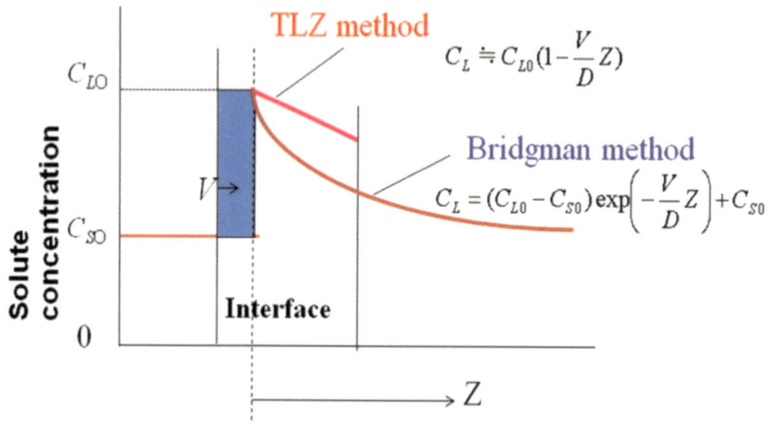

Fig. 2. Comparison of solute concentration profile between directional solidification (Bridgman method) and zone melting method under temperature gradient (TLZ method).

If we know interdiffusion coefficient D and phase diagram date, we can calculate the freezing rate V using eq. (3). Then, it is very easy to fix the interface position as shown in Fig. 3. If a sample device is translated toward opposite direction at the same rate of interface travel rate as calculated by eq. (3), the interface position is fixed and the interface temperature is kept constant, resulting in a compositionally uniform alloy crystals. Thus, compositionally uniform alloy crystals are grown by simply translating a sample device at constant rate in the TLZ method. According to our numerical analysis, 1 to 2 hours are required for establishing an almost saturated solute concentration profile in a zone and spontaneous crystal growth starts.

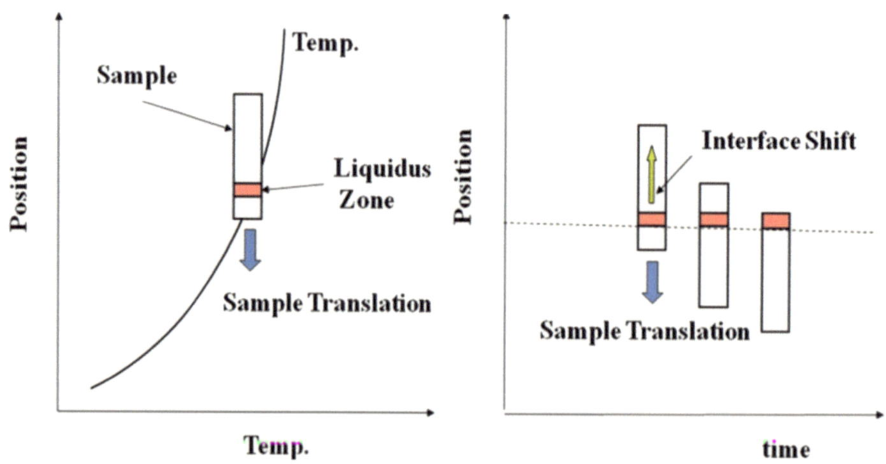

Fig. 3. Schematic view of a liquidus-zone shift in the TLZ method and a way to keep the freezing interface at a fixed position.

2.2 Evaluation of a TLZ growth model

One dimensional TLZ model equation (3) is evaluated by growth experiments. A seed, a zone forming material and a feed were inserted into a boron nitride crucible and the crucible was vacuum sealed in a quartz ampoule at about 1×10^{-4} Pa. In the growth of $In_xGa_{1-x}As$, the zone forming material is InAs or InAs rich solid solution with GaAs. The ampoule was inserted into a temperature gradient furnace and was heated to about 1100℃ at temperature gradients of 10 - 20℃/cm. InAs-GaAs interdiffusion coefficient was measured in microgravity using a sounding rocket (Kinoshita et al. 2000). It was measured to be $1.5 \pm 0.2 \times 10^{-8}$ m²/s at 1070℃. For $In_{0.3}Ga_{0.7}As$ growth, C_{L0} is given by 0.83 from phase diagram. $\partial C_L / \partial T$ is also given by phase diagram as -465 K/mol. Then, V is calculated to be 0.22 mm/h for the temperature gradient of 10℃/cm (Nakamura et al., 2003). Crystal growth experiments were performed for crystals with 2 mm diameter in order to suppress convection in a melt on the ground. Concentration profiles were measured by EPMA. Results are shown in Fig. 4. In the figure, crystal growth started at the length of about 8 mm and the sample was quenched at the length of about 32 mm. Therefore, about 24 mm long crystal was grown. It should be noted that expected InAs concentration (15 at%) is realized without initial transient region. This is one unique point typical to the TLZ method when compared with diffusion controlled steady-state growth in the directional solidification. This is because steady-state solute concentration profile can be established without transient region in the case of zone melting method under temperature gradient. Compositional homogeneity is excellent; In concentration 15±0.5 at% is achieved for the length of more than 20 mm. Scattered concentration region was liquidus-zone before quenching. Dendrite crystals were grown during quenching and dendrite growth resulted in concentration scattering.

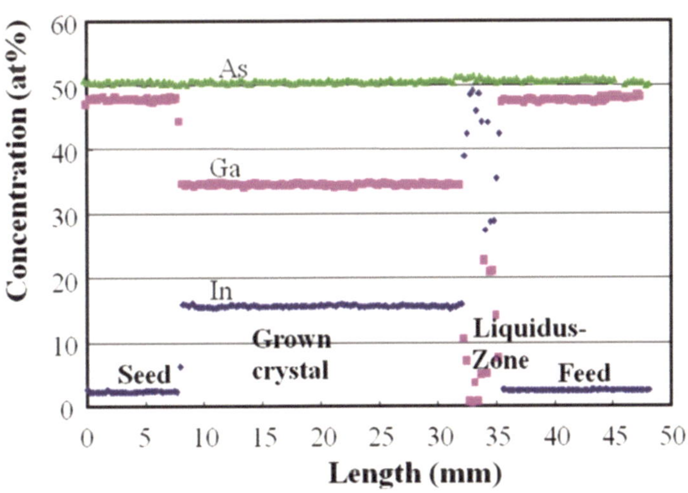

Fig. 4. Axial In, Ga and As concentration profiles for a TLZ-grown 2 mm diameter crystal at a sample translation rate R = 0.22 mm/h.

It is astonishing that excellent compositional homogeneity was achieved at just calculated sample translation rate using eq. (3). Since temperature gradient in the zone was measured precisely by sophisticated method namely by measuring solidus compositions at two interfaces (freezing and dissolving interfaces), model evaluation was possible. Solidus compositions measured at two interfaces are shown in Fig. 5 together with a determined temperature gradient in the zone (Nakamura et al., 2003, Kinoshita et al., 2005). At the freezing interface, In concentration was 15.1 at% and this concentration gives solidus temperature of 1083.1°C from the phase diagram, while In concentration at dissolving interface 12.3 at% gives solidus temperature of 1099.8°C. Distance between two interfaces is 16.8 mm and temperature gradient 10°C/cm was determined. In the temperature gradient measurement, a special sample whose In concentration is gradually decreasing towards a feed and solid-liquid coexistence at the dissolving interface was used. This enabled us to measure solidus composition at the dissolving interface.

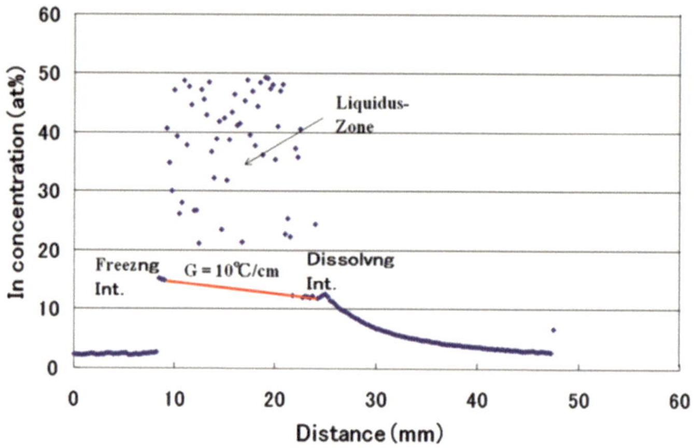

Fig. 5. Measured solidus In concentrations at freezing and dissolving interfaces and determination of temperature gradient (G) in the melt zone.

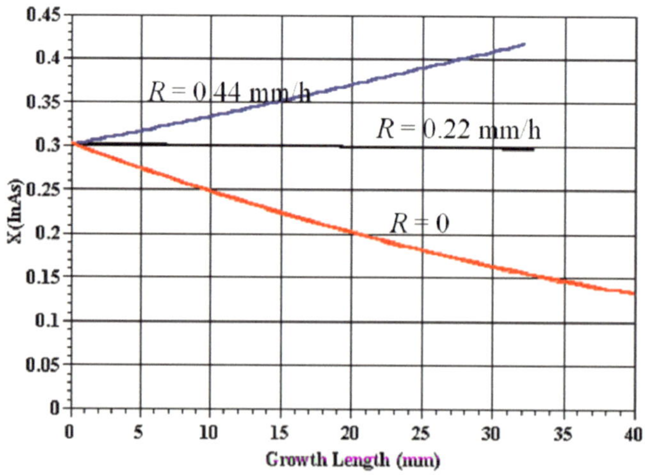

Fig. 6. Results of numerical analysis of axial InAs concentration in mole concentration as a parameter of sample translation rate R.

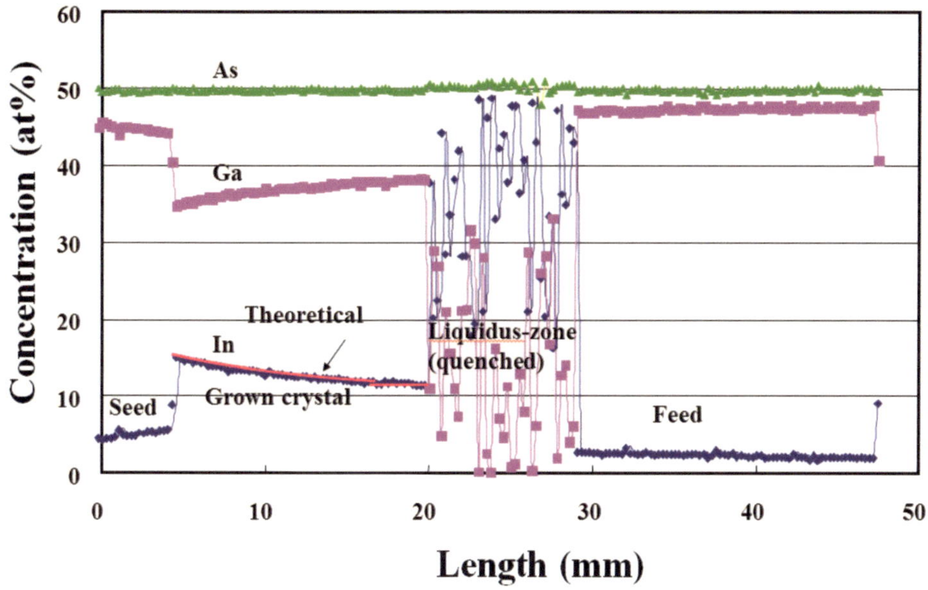

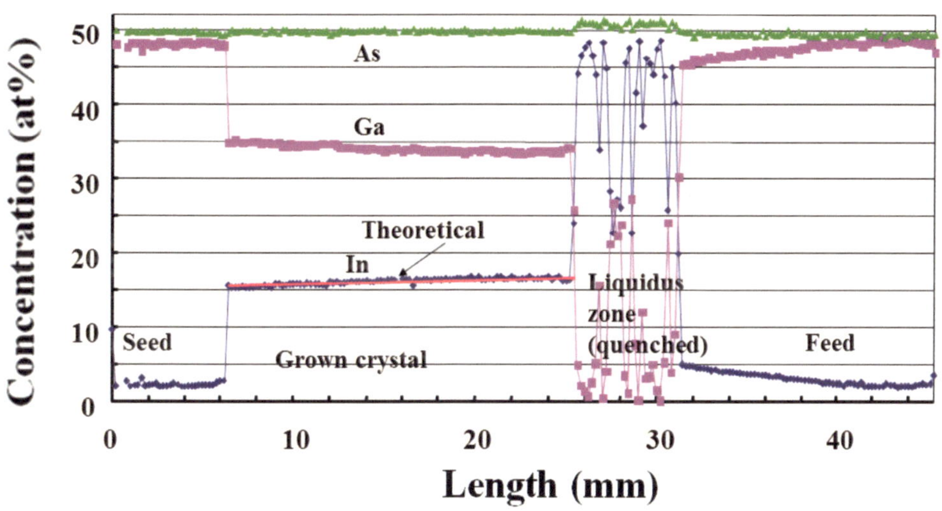

Fig. 7. Axial compositional profiles of TLZ-grown $In_xGa_{1-x}As$ crystals at a sample translation rate (a) $R = 0$ and (b) $R = 0.27$ mm/h.

Numerical analyses show that InAs concentration increases when sample translation rate is faster than the predicted one from eq. (3), while it decreases when sample translation rate is slower than the predicted one (Fig. 6). It should be noted that spontaneous growth occurs even if sample is stationary. This is because solute concentration difference is established between the two interfaces and interdiffusion occurs in a melt zone. At the freezing interface, In concentration is higher and In is transported towards the dissolving interface by diffusion. Indium concentration decrease causes solidification at the freezing interface because equilibrium temperature gets higher. However solidification segregates In at the interface. The segregated In is diffused away towards the dissolving interface. At the dissolving interface, a feed is dissolved by transported In because In concentration gets higher than the equilibrium one. Thus, crystal grows at the freezing interface and a feed is dissolved at the dissolving interface; crystal growth proceeds spontaneously towards lower In concentration side (dissolving interface side, which is a higher temperature side). Sample translation towards a lower temperature side plays a role of keeping the interface temperature constant and it is not a driving force for crystal growth. The driving force in the TLZ growth is interdiffusion between solute and solvent in a zone and segregation on solidification. This should be noted clearly as one of features of the TLZ growth. Figures 7 (a) and 7 (b) show experimental results when sample translation rates do not match the freezing rate of 0.22 mm/h at a temperature gradient of 10°C/cm (Kinoshita et al., 2005). Fig. 7 (a) is a case of $R = 0$ (stationary) and Fig. 7 (b) is a case where R is faster than 0.22 mm/h. Both results agree well with the numerical analyses. As described later, higher sample translation rate than 0.3 mm/h caused constitutional supercooling and resulted in fluctuation of composition. Therefore, in Fig. 7 (b) an experimental result at $R = 0.27$ mm/h is compared with a numerical analysis. Thus, TLZ growth model was confirmed by growth experiments at various sample translation rates and comparison of compositional profiles with those obtained by numerical analyses. Latent heat effect on growth rate was also investigated by numerical analysis and the effect is shown to be negligibly small except for end menber region (Adachi et al., 2004b).

2.3 Limitation in the TLZ method

TLZ growth model is established when solute concentration gradient in the melt zone is linearly approximated as shown in Fig. 2. When temperature gradient is low and zone thickness is thin, such approximation is valid. However, freezing rates deviate from those predicted by eq. (3) at high temperature gradient and for long zone length. Amount of solute in a zone also limits the length of a grown crystal. Moreover, TLZ growth model is based on the diffusion controlled steady-state growth and compositional uniformity is affected by convection in a melt. The TLZ method also lies side-by-side with constitutional supercooling since almost saturated melt is utilized. In this section, such limitations will be described in more detail.

2.3.1 Limitation by temperature gradient

TLZ method is classified as diffusion controlled steady-state growth method as well as a zone method. Solute concentration profile in the diffusion boundary layer is expressed as eq. (4) (Tiller et al., 1953). When VZ/D is small, concentration gradient is approximated by linear relation as eq. (5). Such situation has already been described in Fig. 2. Low temperature gradient gives small value of V and gives a small error in growth rate calculation using eq. (3). However, when temperature gradient is high, V gets large and an error in growth rate calculation becomes large (Adachi et al. 2004a). Figure 8 shows deviation of growth rate from linear relation to temperature gradient. From Fig. 8, it can be said that TLZ growth rate defined by eq. (3) is valid within 10% error when temperature gradient is smaller than 40°C/cm for the zone thickness of 20 mm. Therefore, TLZ growth should be carried out at low temperature gradient like 10 - 20°C/cm at zone thickness of 20 mm.

$$C_L = (C_{L0} - C_{S0})\exp\left(-\frac{V}{D}Z\right) + C_{S0} \tag{4}$$

$$C_L = C_{L0}\left(1 - \frac{V}{D}Z\right) \tag{5}$$

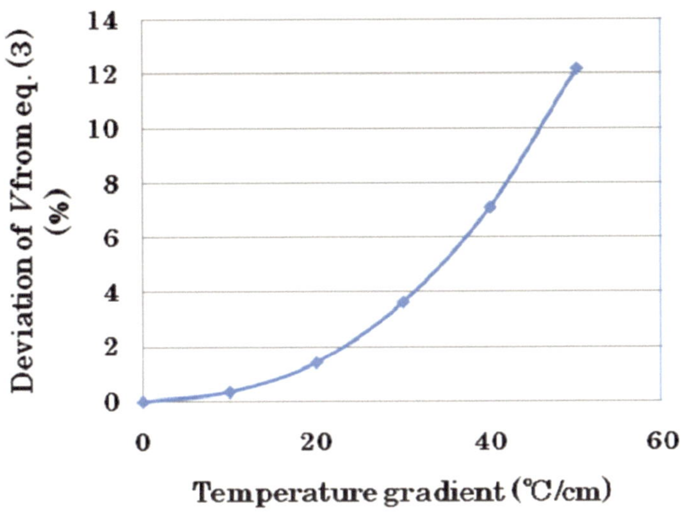

Fig. 8. Deviation of *V* from eq. (3) as a function of temperature gradient.

2.3.2 Limitation by zone length

Distance *Z* is also a variable in eq. (4). Therefore, zone length gives similar effect in growth rate calculation. When the temperature gradient is 20°C/cm, an error in the growth rate calculation using eq. (3) is about 7 % for the zone length of 40 mm but it beyonds 12 % when the zone length is 50 mm. If growth rate deviation from linearity is desired less than 10%, zone length should be less than 45 mm. Limitation of zone length is natural since long zone length cannot be distinguished from directional solidification.

2.3.3 Limitation of growth length

In the $In_xGa_{1-x}As$ crystal growth, we usually start from a combination of a GaAs seed, an InAs zone forming material and a GaAs feed. GaAs saturated InAs rich zone is formed and this zone travel towards a GaAs feed. Therefore, InAs is consumed according as crystal growth proceeds. If an InAs length is 10 mm, length of a grown crystal is limited to 10/3 mm for an $In_{0.3}Ga_{0.7}As$ crystal. In order to increase the length of grown crystal longer zone or InAs contained feed should be used. However, zone length is limited from the point of zone growth as described above and in an InAs contained feed InAs content is restricted to low for preventing melting of a feed. In this meaning, the length of a TLZ-grown crystal is limited.

2.3.4 Limitation by convection in a melt

The TLZ method is possible in the diffusion limited regime. If convection occurs in a melt, solute concentration profile ahead the interface is disturbed and the growth rate does not obey eq. (3). We grew 2 mm diameter $In_xGa_{1-x}As$ crystals by suppressing convection in a melt and confirmed the TLZ growth model. However, 2 mm diameter crystals cannot be used for opt-electronic devices and we tried increase of crystal diameter. Result of a growth experiment for 10 mm diameter crystal is shown in Fig. 9. Axial compositional homogeneity is achieved for the length of more than 30 mm but radial concentration inhomogeneity exists in a 10 mm diameter crystal. According to numerical analysis, convective vortices occur at both freezing and dissolving interfaces as shown in Fig. 10. When such convective vortices are taken into consideration, radial concentration difference is understood; In-rich melt is transported towards the centre by convection and InAs concentration increases in the centre of a grown crystal. As for the strength of convective driving force, Grashof number G defined by eq. (6) gives order of magnitude where g is gravity acceleration, β is volume expansion coefficient of a melt, ΔT is a temperature difference between the two ends of a melt, L is characteristic length of a melt and v is kinematic viscosity of a melt. Comparison of convective driving force between 2 and 10 mm diameter melts, convection in 10 mm diameter melt gives 2 orders of magnitude higher than that in 2 mm diameter melt. Result in Fig. 9 shows importance of suppressing convection in a melt for compositionally homogeneous alloy crystals in radial direction as well as in axial direction (Kinoshita et al., 2006).

$$G_r = \frac{g\beta\Delta TL^3}{\nu^2} \qquad (6)$$

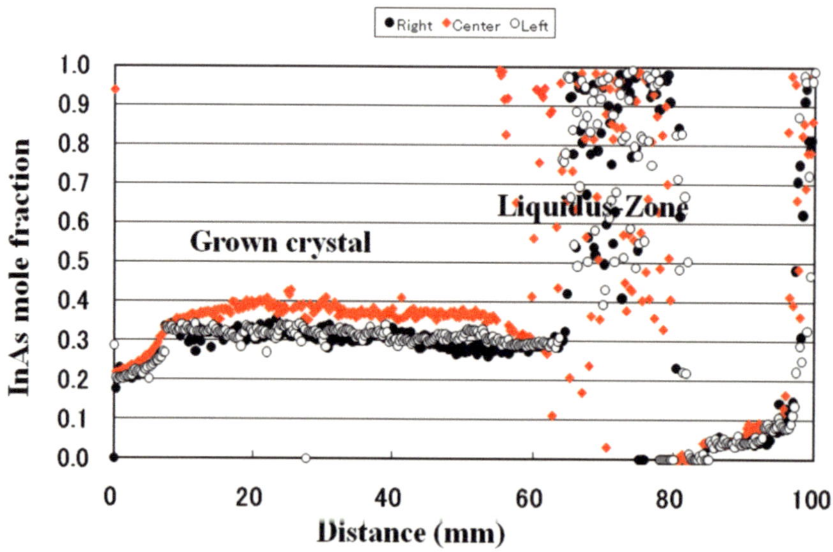

Fig. 9. Axial InAs concentration profiles compared among centre line, right and left peripheral regions.

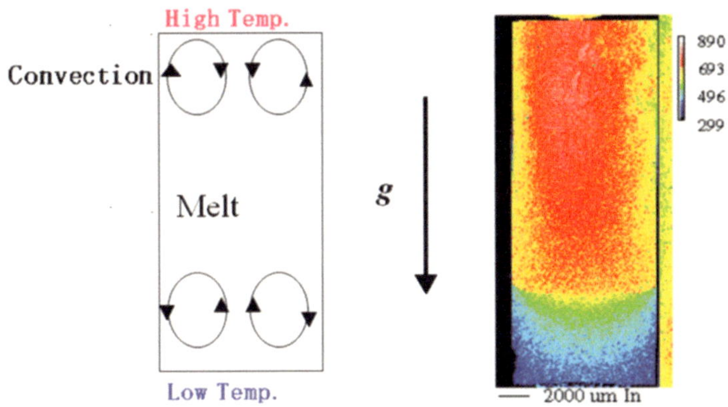

Fig. 10. Convective vortices in a melt and two dimensional In concentration map for a 10 mm diameter crystal.

2.3.5 Limitation by constitutional supercooling

The TLZ growth utilizes almost saturated melt. In this point of view, constitutional supercooling tends to occur. Constitutional supercooling in the directional solidification of alloy crystals in the diffusion limited growth was studied extensively by Tiller et al. (1953). According to them, the growth rate should be high and should fulfil conditions defined by eq. (7) where G is temperature gradient, R is growth rate, m is the slope of the liquidus line. The TLZ method utilizes solute concentration gradient in diffusion boundary layer ahead the freezing interface. Therefore, growth rate should fulfil the conditions proposed by Tiller et al. When we change notation of eq. (3) to similar ones given by Tiller et al. for comparison, equation (8) is given. Equation (8) is further modified into eq. (9). When we compare eq. (7) with eq. (9), we notice that the growth rate for realizing compositional uniformity in the TLZ method is the critical growth rate for avoiding constitutional supercooling. If growth rate is higher than this, constitutional supercooling will occur. From this discussion, it is clearly shown that homogeneous crystal growth in the TLZ method is side-by-side the constitutional supercooling.

$$\frac{G}{R} \geq \frac{m(C_{L0} - C_{S0})}{D} \tag{7}$$

$$R = \frac{D}{(C_{L0} - C_{S0})} mG \tag{8}$$

$$\frac{G}{R} = \frac{m(C_{L0} - C_{S0})}{D} \tag{9}$$

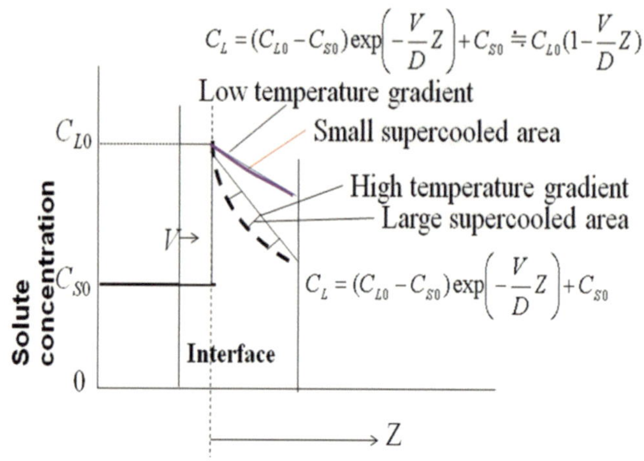

Fig. 11a. Comparison of constitutionally supercooled area during TLZ growth at low and high temperature gradients. At high temperature gradient, non-linearity of solute concentration profile aheat the freezing interface is large and gives larger supercooled area

Constitutional supercooling is inevitable at high temperature gradient (Kinoshita and Yoda, 2011). This may be peculiar from the requirement of high *G/R* ratio shown by eq. (7) but this requirement comes from constant *G/R* for compositional uniformity and solute saturation at two interfaces (freezing and dissolving interfaces) in the TLZ method. At high temperature gradient, non linearity of solute concentration profile increases as shown in Fig. 11(a). In the figure, the straight line shows equilibrium concentration at the given temperature gradient and dashed region is constitutionally supercooled region. It should be noted that constitutionally supercooled area is larger at higher temperature gradient. This is unique point in the TLZ method. It is obvious that degree of constitutional supercooling is greater at higher temperature gradient because deviation from linearity in solute concentration profile is greater at higher temperature gradient (Kinoshita et al., 2011). From this point of view, temperature gradient is limited and growth rate in turn is also limited. In the $In_xGa_{1-x}As$ crystal growth experiments, it is shown that temperature gradient that exceeds 30°C/cm resulted in constitutional supercooling. This means that growth rate is limited to 0.66 mm/h. In general, compositional fluctuation increases in constitutionally supercooled region. An example of compositional profile of constitutionally supercooled sample is shown in platy crystal growth in later section (an $In_{0.15}Ga_{0.85}As$ platy crystal grown at a temperature gradient of 37°C/cm showed large compositional fluctuation as shown in 3.2.3).

3. Platy crystal growth

For device application, large surface area is required. However, increase in crystal size increases convection in a melt and compositional uniformity of TLZ-grown crystals is deteriorated. Suppressing convection in a melt and large surface area is fulfilled by the growth of platy crystals if thickness of the platy crystal is sufficiently thin (Kinoshita et al., 2008, 2010). Such situation is schematically depicted in Fig. 11(b). We have experimentally confirmed that convection in a 2 mm diameter melt is suppressed and TLZ mode crystal growth is possible. Then, we determined thickness of platy crystals as 2 mm. In this section, growth and characterization of platy crystals for substrates of laser diodes with emitting wavelength λ = 1.3 μm are described.

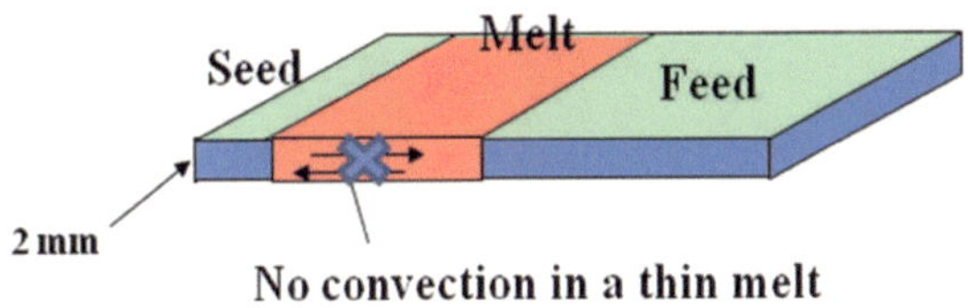

Fig. 11b. Schematic view of a platy crystal growth.

3.1 Growth procedures and characterization methods

Growth method for platy crystals is similar to that of the cylindrical crystal. A GaAs seed, an InAs zone forming material, and a GaAs feed were cut into plates with 2 mm thickness and were inserted into a boron nitride crucible with a rectangular bore and were sealed in vacuum in a quartz ampoule and then heated in a gradient heating furnace. Growth interface temperature was adjusted around 1100℃ so that $In_xGa_{1-x}As$ (x: 0.1 - 0.15) crystals were grown. Temperature gradient was set around 20℃ so that constitutional supercooling at high temperature gradient was avoided. Samples were translated at the rate calculated using eq. (3). Seed orientation was <100> or <110>. Sn was doped as n-type dopant. At the beginning of crystal growth, we thought that $In_{0.3}Ga_{0.7}As$ composition was required as substrates for λ = 1.3 µm laser diodes. However, it turned out that composition around $In_{0.1}Ga_{0.9}As$ can be used as substrates since thin film growth technology has been developed and large lattice mismatch between the substrate is conquered. As for crystal size, we started from the growth of 10 mm width and increased crystal width. At the present, we are successful in growing 50 mm wide platy crystals. Grown crystals were polished and crystalline nature was investigated. Compositional profiles of grown crystals were measured by EPMA on polished surfaces. Two-dimensional semi-quantitative mapping analysis was performed for measuring overall compositional distribution. Quantitative analysis was performed along growth axis with precision of 0.1 at% for each constituent element. Crystal quality was evaluated by measuring X-ray rocking curves. Electrical properties of crystals were measured by Hall measurements. Etch pit density was measured after KOH etching.

3.2 Results and discussion

Experimental results for platy crystal growth are summarized and factors that effect on crystal quality are discussed in this section. Factors for affecting crystal quality are common for the TLZ method except for seeding in rectangular crucible.

3.2.1 Single crystalline nature

Typical two examples of roughly polished surface of two platy crystals are shown in Fig. 12. Composition of Fig. 12 (a) is $In_{0.10}Ga_{0.90}As$ and that of Fig. 12 (b) is $In_{0.13}Ga_{0.87}As$. When InAs mole fraction in a grown crystal is less than 0.1, a single crystal that takes over the seed orientation was grown. However, when InAs mole fraction is more than 0.13, single crystal growth that takes over the seed orientation became very difficult. This may be due to increase in lattice mismatch between a GaAs seed and a grown crystal. Difficulty in single crystallization was settled by using a feed having the same orientation of a seed. In Fig. 12 (b), polycrystallization is observed at the seed and grown crystal interface. However, a single crystal was grown again as crystal growth proceeded.

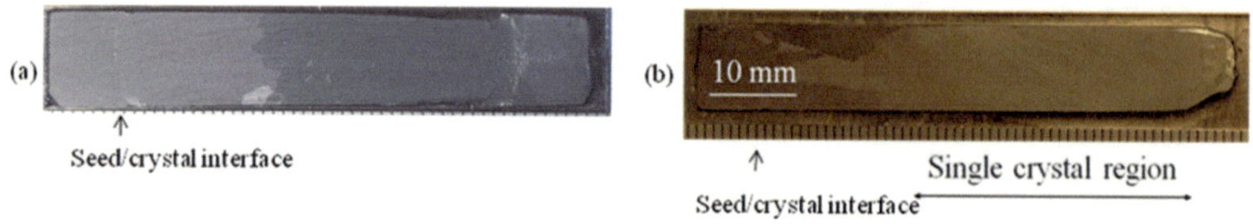

Fig. 12. Roughly polished surface of two platy crystals, (a) $In_{0.10}Ga_{0.90}As$ and (b) $In_{0.13}Ga_{0.87}As$.

The mechanism of single crystal growth is not clear but it may be related to a crucible shape having rectangular bore and nucleation of the same orientation as a feed might occur in a narrow gap between the crucible wall and feed surface. In wider crystal growth too, single crystals were grown by the same mechanism. Reproducibility of single crystal growth was very good and single crystal length more than 50 mm was obtained.

3.2.2 Growth temperature

Freezing interface temperature determines composition of grown crystals. In this point of view, growth temperature is very important. In the TLZ method, solute in a melt is almost saturated and it is very easy to control interface temperature. We set 8 thermocouples around a growth ampoule and one of them is set at seed/zone interface. Only procedure is to monitor temperatures measured by these thermocouples and to change the interface temperature. It is usual that surface temperatures of a quartz ampoule are not equal to inner temperatures of a crucible but difference between the two was less than 5°C and minor adjustment gave target compositions. Figure 13(a) is an example of $In_{0.2}Ga_{0.8}As$ crystal and Fig. 13(b) is an example of $In_{0.13}Ga_{0.87}As$ crystal. In $In_{0.2}Ga_{0.8}As$, interface temperature was set at 1057°C and in $In_{0.13}Ga_{0.87}As$ it was set at 1095°C. In both cases, target compositions were obtained at the seed/crystal interface and no initial transient regions were found as is the case of $In_{0.3}Ga_{0.7}As$ (Fig. 4). Compositional uniformity was excellent for $In_{0.13}Ga_{0.87}As$; InAs mole fraction was 0.13±0.005, but uniformity was degraded for $In_{0.2}Ga_{0.8}As$. Single crystallinity was also better for lower In concentration crystals. Since we succeeded in laser operation at the wavelength of 1.3 μm for laser diodes on $In_{0.13}Ga_{0.87}As$ substrates (Arai et al, 2007), we improved crystal quality for the composition of around $In_{0.13}Ga_{0.87}As$.

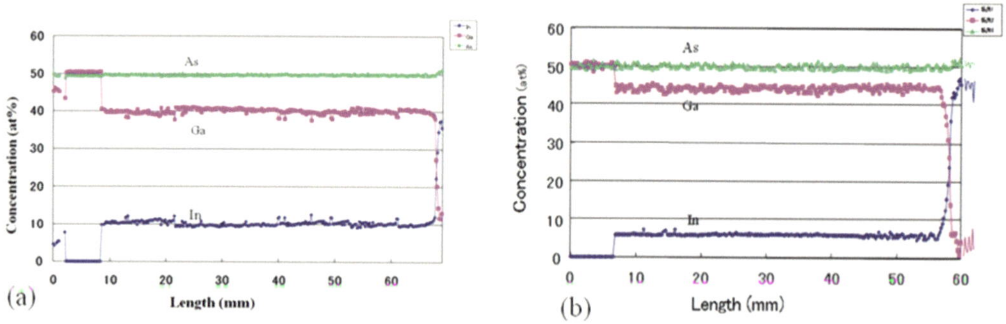

Fig. 13. Axial concentration profiles of platy crystals; (a) for $In_{0.2}Ga_{0.8}As$ and (b) for $In_{0.13}Ga_{0.87}As$.

3.2.3 Temperature gradient

Temperature gradient determines spontaneous growth rate in the TLZ method as given by eq. (3). So long as the sample translation rate matches this spontaneous growth rate, compositionally uniform alloy crystals can be grown. However, high temperature gradient causes constitutional supercooling and results in compositional fluctuation. Figure 14 compares compositional profiles of crystals grown at different temperature gradient (a) 25°C/m and (b) 37°C/cm. Note that the crystal grown at lower temperature gradient shows lower compositional fluctuation. As described in the section 2.3.5, maximum temperature gradient for suppressing constitutional supercooling may be 30°C/cm. It is concluded that temperature gradient should be as low as possible for suppressing compositional fluctuation but the lowest limit of temperature gradient should be determined from the point of mass productivity of crystals.

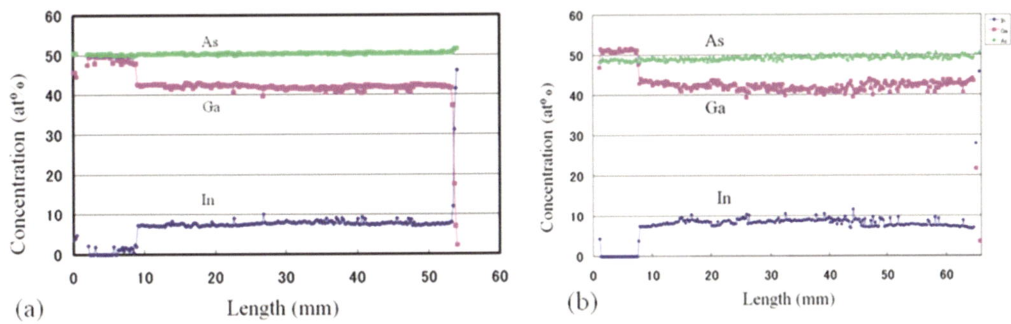

Fig. 14. Comparison of axial concentration profiles of crystals grown at different temperature gradient (a) 25℃/m and (b) 37℃/cm. Average composition of both crystals is about $In_{0.15}Ga_{0.85}As$.

3.2.4 Temperature stability

Effect of temperature stability on compositional uniformity was investigated with average composition of $In_{0.13}Ga_{0.87}As$ (Kinoshita et al., 2008). The most stable temperature was achieved when air flow in a furnace tube was shut. Temperature stability was ±0.1°C as shown in Fig. 15 (a). With air flow conditions, temperature stability got worse to ±0.2°C as shown in Fig. 15 (b). The stability difference was small but this small difference gave a considerable effect on the compositional uniformity as shown in Fig. 16. When the temperature stability was ±0.1°C, InAs mole fraction was 0.13 with σ of 0.0005 where σ is the standard deviation in the distance between 20 and 40mm. When the temperature stability was ±0.2°C, InAs concentration uniformity was 0.13 with σ of 0.006 in the distance between 20 and 40mm. The reason why the temperature fluctuation influence on the compositional uniformity may be related to the crystallization from the almost saturated melt, namely, temperature fluctuation gives rise to the formation of constitutionally supercooled region in a melt and such region crystallizes earlier, then it is not strange that earlier crystallized region has lower In concentration. Improvement in compositional stability resulted in higher crystal quality. X-ray rocking curve measurements showed that full width at half maximum (FWHM) of rocking curve ranges from 0.03 to 0.04 degrees for crystals grown at temperature stability ±0.1°C. Such small FWHM shows good crystallineity which can be used as a substrate of laser diodes.

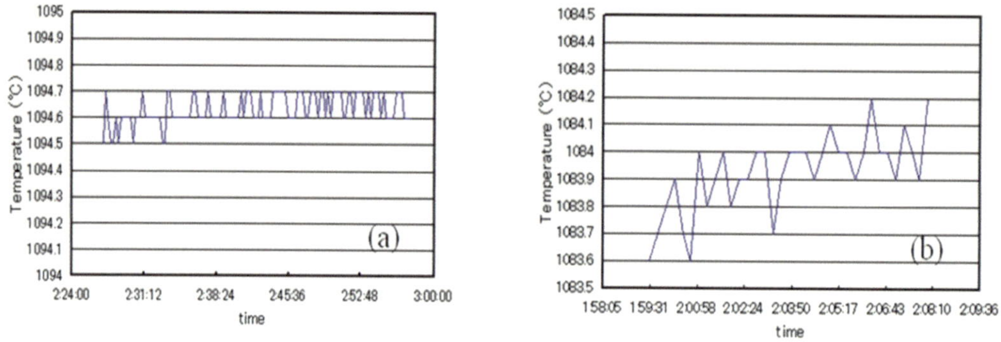

Fig. 15. Comparison of temperature stability.

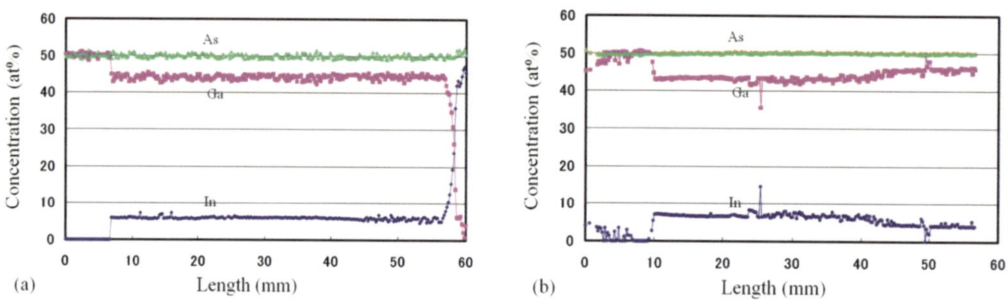

Fig. 16. Comparison of concentration profiles for two crystals grown at different temperature stability (a)±0.1℃ and (b)±0.2℃.

3.2.5 Characterization as substrates of laser diodes

Improvement in compositional stability resulted in higher crystal quality. Figure 17 shows X-ray rocking curves in the distance between 26 mm and 56 mm at an interval of 5mm for the crystal grown at the temperature stability of ±0.1°C. The ω-scan around the (004) diffraction peak was detected by an open detector. Cu $k_{\alpha 1}$ radiation from an X-ray tube (loaded power; 40 kV, 30 mA) was monochromated by a four crystal Ge(220) monochromator and was incident on a crystal. The illuminated area was limited to about 1×1 mm^2 by a divergence and a scatter slit. Full width at half maximum (FWHM) of rocking curve ranges from 0.03 to 0.04 degrees (from 108 to 144 arc seconds). Such small FWHM shows good crystallineity which can be used as a substrate of laser diodes. High quality region extended to 10×30 mm^2. When InAs mole fraction increases, FWHM increased.

This phenomenon is usual in many alloy crystals. When solute concentration increases, compositional uniformity is deteriorated and strain increases in the grown crystal. The increased strain degrades crystalline quality. Therefore, the lower InAs mole fraction, the higher is the quality of grown crystals. In the course of our study, we found that $In_{0.3}Ga_{0.7}As$ could be used as substrates of λ = 1.3 μm laser diodes instead of $In0._{3}Ga_{0.7}As$. This is beneficial for ternary crystal growth since higher quality can be expected. Sn was doped as n-type dopant for substrate use. Carrier concentration was measured to be 5 – 8×10^{18} cm^{-3}. Etch pit densities were in the range between 1×10^3 and 3×10^4 cm^{-2}, which is sufficiently low for a substrate.

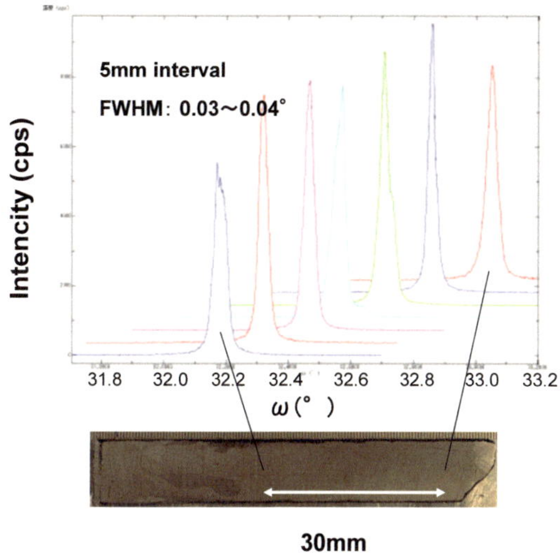

Fig. 17. FWHMs of X-ray rocking curves at seven positions. They range from 0.03 to 0.04° for the length of 30 mm.

3.3 Increase of a crystal width

Convection in a melt is suppressed in a thin melt. In platy crystal growth, width is considered to be independent of convection strength when thickness of crystals is limited. From this point of view, width of platy crystals was increased step by step from 10 mm to 20, 30, and 50 mm (Kinoshita et al., 2010). In all cases, thickness of platy crystals was set to be 2 mm. The most difficult point was to keep high temperatures in a feed area as growth ampoule was translated towards low temperature side. When width of platy crystals was increased, diameter of boron nitride (BN) crucible increased. BN has higher thermal conductivity than quartz and large diameter BN crucible transferred more heat to lower temperature side. For keeping high temperatures in a feed area, length of BN crucible was increased to receive more radiation from a heater.

Figure 18 shows an example of 30 mm wide platy crystal. Outlook of roughly polished surface is shown in Fig. 18 (a) and InAs mole fractions measured along a centre line and 10 mm away from the centre line are shown in Fig, 18 (b). Single crystalline area larger than 30×30 mm^2 was obtained. InAs mole fractions were constant and 0.10 ± 0.01 was achieved for the distance of about 40 mm. Good crystallinity with the FWHM of less than 0.04° in X-ray rocking curves was obtained. Such excellent compositional uniformity and crystal quality owe to matching of sample translation rate and freezing rate and resulting in a fixed freezing interface position relative to heater segments. In this point of view, constant temperature gradient during translation of a sample was important for obtaining compositional uniformity. Such increase in surface area in platy crystals without deteriorating crystal quality also shows that convection in a melt is suppressed by the limitation of melt thickness (2 mm) and increase in melt width did not cause convection. If convection occurs in a melt, local inhomogeneity will be resulted due to stirring of a melt by convection. Good homogeneity shows that convection was suppressed in the course of crystal growth.

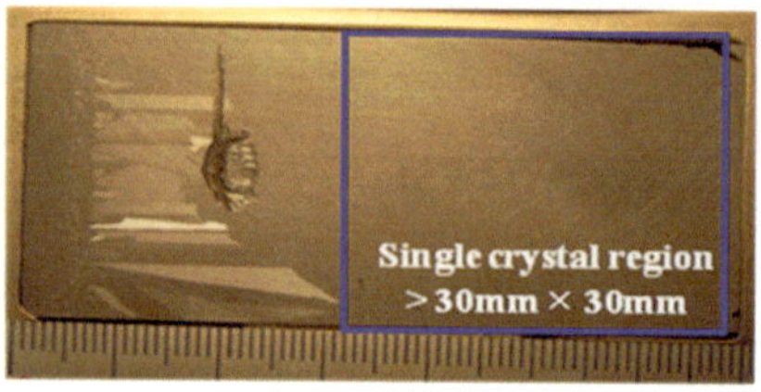

(a)

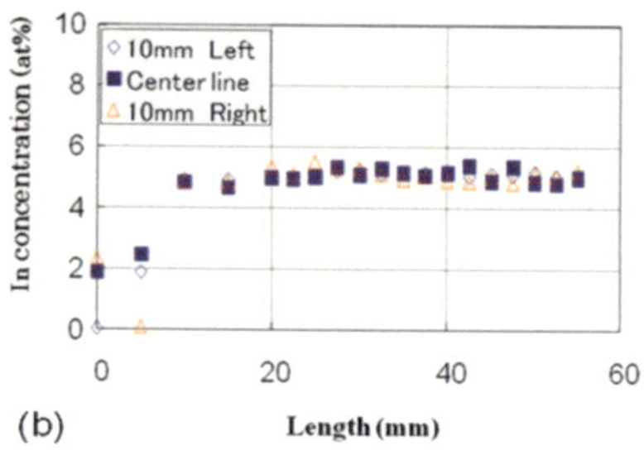

(b)

Fig. 18. An example of 30 mm wide platy crystal; (a) roughly polished surface of a crystal and (b) InAs mole fractions along growth axis.

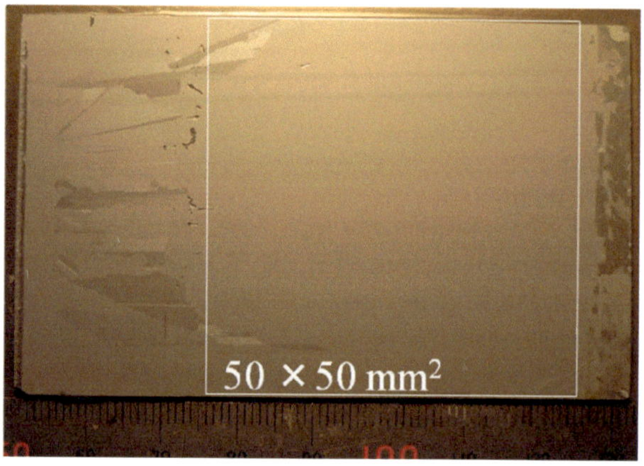

Fig. 19. An example of 50 mm wide crystal (Single crystalline area of almost 50×50 mm² is obtained).

Now we have succeeded in growing 50 mm wide platy crystals. Figure 19 shows an example of 50 mm wide crystal. Single crystalline area of almost 50 ×50 mm² is obtained. For mass production of laser diodes, large platy crystals with good reproducibility are required. As described earlier, crystals do not take over seed orientation but take over feed orientation. In this case, similar single crystallization mechanism works, too. Reproducibility is now more than 90%.

4. Laser diodes fabrication

Laser diodes with emitting wavelength λ = 1.3 μm were fabricated on $In_xGa_{1-x}As$ (x: 0.1 - 0.13) substrates. Ishikawa (1993) demonstrated merits of a ternary substrate for high optical gain with small temperature dependence. In the course of study, we found that $In_xGa_{1-x}As$ (x: 0.1 - 0.13) crystals can be used as substrates owing to development of thin film growth technology. Low InAs content crystals are beneficial to substrates because they have better thermal conductivity than high InAs content crystals.

4.1 Fabrication procedures

$In_{0.13}Ga_{0.87}As$ platy crystals were polished mechano-chemically to 0.5 mm thickness. Surface roughness of mirror polished substrates was measured to be less than several nano-meters. Strained multiple-quantum-wells (MQWs) having the combination of $In_{0.12}Ga_{0.88}As$ and $In_{0.38}Ga_{0.62}As$ as shown in Fig. 20 (a) were prepared by metal-organic vapor phase epitaxy (MOVPE) on the cleaned substrate (Arai et al., 2009). The active region was consisted of four $In_{0.38}Ga_{0.62}As$ wells and five $In_{0.12}Ga_{0.88}As$ barriers sandwitched by two seperate-confinement heterostructure (SCH) layers. The stripe mesa type laser was fabricated by chemical etching. Cross sectional view of a fabricated mesa stripe is shown in Fig. 20 (b). The stripe mesa was in the [011] direction, forming a reverse mesa. Bottom ridge width of 1.7 μm enabled single mode lasing. Selective wet etching was utilized for such narrow ridge formation.

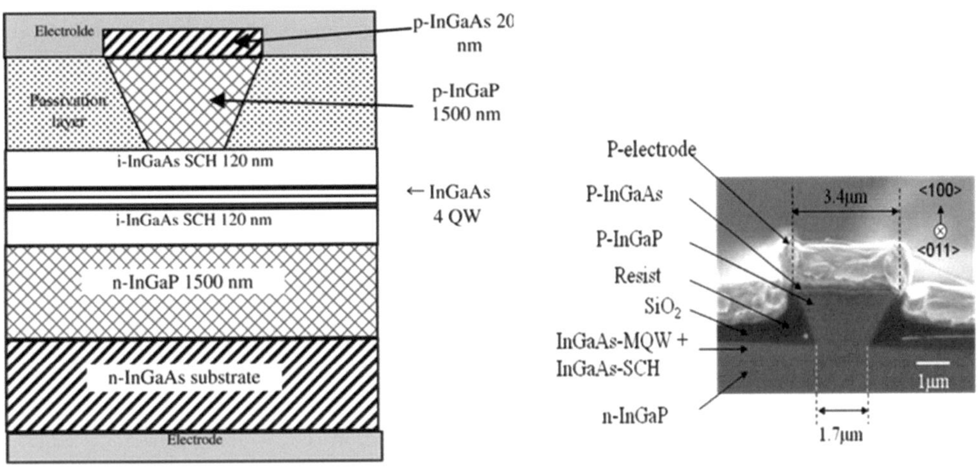

(a) (b)

Fig. 20. Cross sectional view of a fabricated laser diode, (a) schematic drawing and (b) electron micrograph.

4.2 Characterization of laser diodes

Figure 21 (a) shows the bias current versus output power (*I-L*) characteristics of a fabricated laser diode The threshold current is 7.2 mA. Figure 21 (b) is the lasing spectrum at a bias of 20 mA. The lasing wavelength is 1.31μm. The laser exhibits a maximum operating temperature of 80 and 150°C under CW and pulsed operation, respectively. Therefore, the limiting factor was self-heating. An effective way of overcoming this thermal problem is to reduce the indium content of the $In_xGa_{1-x}As$ substrate and to introduce InAlGaAs barrier layers. We therefore tried to realize 10 Gbps modulation over 85°C by eliminating the thermal problems and we prepared an $In_{0.08}Ga_{0.92}As$ substrate.

Laser diodes on an $In_{0.08}Ga_{0.92}As$ substrate showed improved lasing characteristics. Self heating problem due to bad thermal conductivity of $In_{0.13}Ga_{0.87}As$ substrates was settled by reducing InAs content in the substrate. However, emitting wavelength was 1.26 μm due allowance limit of InAs content in active layers. Figure 22 shows continuous wave (CW) lasing characteristics of this laser diode at various temperatures. CW operation was possible up to 150°C and was much improved. Characteristic temperature of the threshold current density between 25 and 125°C was about 90 K and showed a higher value compared with that of 1.31 μm laser (about 70 K). Such high temperature stability in output power is also expected in 1.31 μm laser diodes by optimizing heat sink structure and heat sink materials.

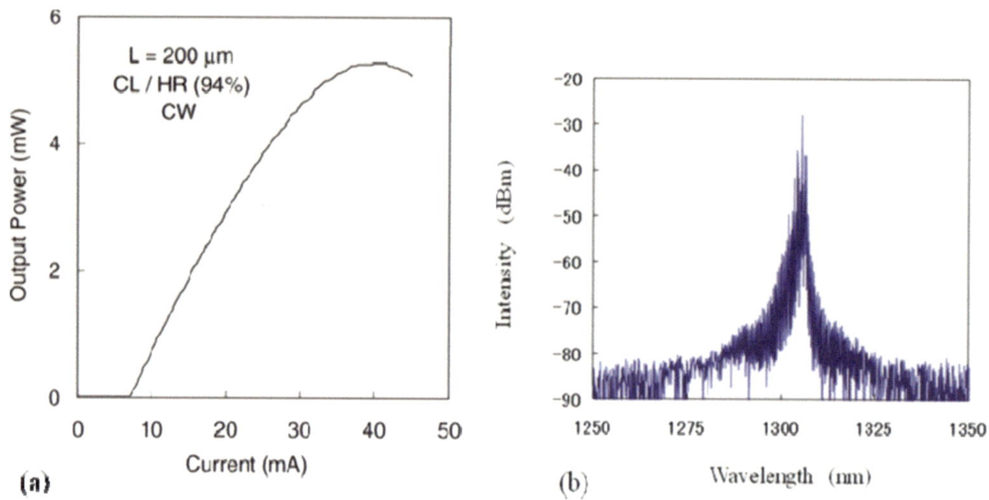

Fig. 21. Lasing characteristics of a fabricated laser diode. (b) is the lasing spectrum at a bias of 20 mA.

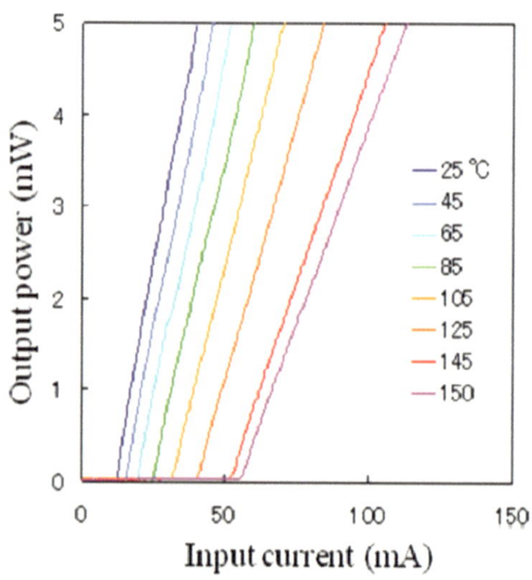

Fig. 22. CW lasing characteristics at various temperatures.

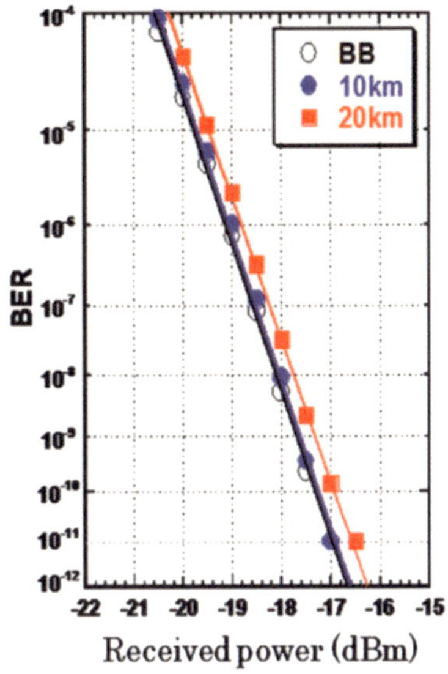

Fig. 23. Bit error rate after 20 km transmission.

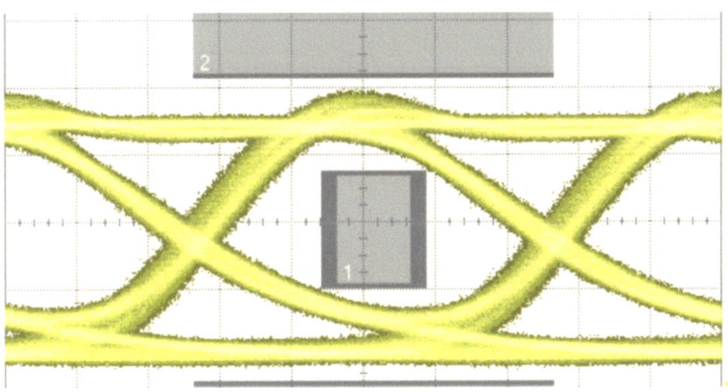

Fig. 24. Eye diagram at the modulation of 10 Gbps at 85°C.

A 10-Gbps direct modulation and transmission tests through a single mode fiber up to 20 km at 25°C were successfully performed using a fabricated laser diode with λ = 1.31 μm as shown in Fig. 23. Bit error rate (BER) for back-to-back configuration after 20 km transmission was less than 10^{-7} at -18 dBm. We also tested 1.26 μm wavelength laser diodes. In this laser diodes, 10 Gbps modulation at 85°C was confirmed by an electrically filtered back-to-back (BB) eye diagram as shown in Fig. 24. The extinction ratio and the mask margin of the synchronous optical network (SONET) mask were 7 dB and 9%, respectively. These results show feasibility of fabricated laser diodes.

5. Conclusions

The TLZ method has been invented for growing compositionally uniform alloy crystals. The solute concentration gradient is controlled by applied temperature gradient utilizing saturation of solute in a zone. Owing to this merit, the spontaneous growth rate is calculated from the diffusion controlled steady-state growth conditions. Compositionally uniform alloy crystals are obtained merely by translating samples at the calculated growth rate relatively to a heater. Principle of the TLZ method was proved by the growth of 2 mm diameter crystals since convection in a melt is suppressed and diffusion limited growth is possible in capillary tubes even on the ground. As predicted, compositionally uniform crystals were grown without initial transient region which is typical to directional solidification method. Although the TLZ method is a superior method, it has limitations and they were also studied theoretically and experimentally. Limitation by temperature gradient, limitation by zone length, limitation by constitutional supercooling, and limitation by convection in a melt were made clear. In microgravity, TLZ growth is free from limitation by convection in a melt. In this point, microgravity is beneficial for TLZ growth. We grew platy $In_xGa_{1-x}As$ (x: 0.08 – 0.13) crystals by the TLZ method on the ground for substrates of 1.3 μm wavelength laser diodes. In platy crystal growth, convection in a melt was suppressed by limiting thickness of platy crystals to 2 mm and large surface area was obtained by increasing width of platy crystals. Laser diodes on $In_xGa_{1-x}As$ (x: 0.08 – 0.13) substrates showed excellent temperature stability in output power as predicted by Ishikawa et al. (1993). Now 50×50 mm^2 platy crystals were reproducibly grown and mass production of laser diodes with emitting wavelength of 1.3 μm is expected.

6. Acknowledgments

We are grateful to Dr. Adachi for the numerical analysis of the TLZ growth, to Dr. Y. Kondo, Dr. M. Arai, Dr. Y. Kawaguchi and Dr. F. Kano for laser fabrication and characterization, to Dr. H. Aoki, Dr. T. Hosokawa, Dr. S. Yamamoto and Dr. M. Matsushima for wide platy crystal growth. This work was supported by the New Energy and Industrial Technology Development Organization (NEDO).

7. References

Adachi, S.; Ogata, Y.; Koshikawa, N.; Matsumoto, S.; Kinoshita, K.; Yoshizaki, I.; Takayanagi, M.; Yoda, S.; Kadowaki, A.; Tsuru, T.; Miyata, H. & Muramatsu, Y. (2004a). Numerical analysis of growth rates in the traveling liquidus-zone method. J. Crystal Growth, Vol.270, No.1-2 (September 2004), pp. 42-49, ISSN 0022-0248

Adachi, S.; Ogata, Y.; Koshikawa, N.; Matsumoto, S.; Kinoshita, K.; Yoshizaki, I.; Takayanagi, M. & Yoda, S. (2004b). Latent heat effect on growth rate in the traveling liquidus-zone method. J. Crystal Growth, Vol.271, No. 1-2 (October 2004), pp. 22-28, ISSN 0022-0248

Arai, M.; Watanabe, T.; Yuda, M.; Kinoshita, K.; Yoda, S. & Kondo, Y. (2007). High-Characteristic-Temperature 1.3μm Band Laser on an InGaAs Ternary Substrate Grown by the Traveling Liquidus-Zone Method. IEEE J. Selected Topics in Quantum Electronics, Vol.13 (2007), pp. 1295-1300, ISSN 1077-260X

Arai, M.; Kobayashi, W.; Fujisawa, T.; Yuda, M.; Tadokoro, T.; Kinoshita, K.; Yoda, S. & Kondo, Y. (2009). 10-Gbps Direct Modulation Using a 1.31μm Ridge Waveguide Laser on an InGaAs Ternary Substrate. Appl. Phys. Express, Vol.2, No.1 (January 2009) pp. 022101-1-3, ISSN 1882-0778

Bublik, V.T. & Leikin, V.N. (1978). Calculation of pseudobinary alloy semiconductor phase diagrams. physica status solidi (a), Vol.46, No.1 (March 1978), pp. 365-372, ISSN 1862-6319

Ishikawa, H. (1993). Theoretical gain of strained quantum well on an InGaAs ternary substrate. Appl. Phys. Lett., Vol.63, No.6, (August 1993), pp. 712-714, ISSN 0003-6951

Ishikawa, H. & Suemune, I (1994). Analysis of Temperature Dependent Optical Gain of Strained Quantum Well Taking Account of Carriers in the SCH Layer. IEEE Photonics Technology Lett., Vol.6, No.3, (March 1994), pp. 344-347, ISSN 0003-6951

Kinoshita, K.; Kato, H.; Matsumoto, S.; Yoda, S.; Yu, J.; Natsuisaka, M.; Masaki, T.; Koshikawa, N.; Nakamura, Y.; Nakamura, T.; Ogiso, A.; Amano, S.; Goto, K.; Arai, Y.; Fukazawa, T.; Kaneko, M. & Itami, T. (2000). InAs-GaAs interdiffusion measurements. J. Jpn. Soc. Microgravity Appl., Vol.17, No.2 (April 2000), pp.57-63, ISSN 0915-3616

Kinoshita, K.; Kato, H.; Iwai, M.; Tsuru, T.; Muramatsu, Y. & Yoda, S. (2001). Homogeneous In0.3Ga0.7As crystal growth by the traveling liquidus zone method. J. Crystal Growth, Vol.225, No.1 (May 2001), pp.59-66, ISSN 0022-0248

Kinoshita, K.; Hanaue, Y.; Nakamura, H.; Yoda, S.; Iwai, M.; Tsuru, T. & Muramatsu, Y. (2002). Growth of homogeneous mixed crystals of In0.3Ga0.7As by the traveling liquidus-zone method, J. Crystal Growth, Vol.237-239, Part 3 (April 2002), pp.1859-1863, ISSN 0022-0248

Kinoshita, K.; Ogata, Y.; Adachi, S.; Koshikawa, N.; Yoda, S.; Miyata, H & Muramatsu, Y. (2004). A new crystal growth method for growing homogeneous mixed crystals of In0.3Ga0.7As: the traveling liquidus-zone (TLZ) method. Adv. in Astronautical Sciences, Vol.117 (2004), pp.865-872, ISSN 0065-3438

Kinoshita, K.; Ogata, Y.; Koshikawa, N.; Adachi, S.; Yoda, S.; Iwai, M.; Tsuru, T. & Muramatsu, Y. (2005). Improvement of compositional homogeneity in In1−xGaxAs bulk crystals grown by the travelling liquidus-zone method. Int. J. Materials and Product Technology, Vol.22, No.1-3 (January 2005) pp. 95-104, ISSN 0268-1900

Kinoshita, K.; Ogata, Y.; Adachi, S.; Yoda, S.; Tsuru, T.;, Miyata, H. & Muramatsu, Y. (2006). Convection effects on crystallinity in the growth of In0.3Ga0.7As crystals by the traveling liquidus-zone method, Ann. New York Acad. Sci., Vol.1077 (September 2006), pp. 161-171, ISSN 0077-8923

Kinoshita, K.; Ueda, T.; Yoda, S.; Arai, M.; Kawaguchi, Y.; Kondo, Y.; Aoki, H.; Hosokawa, T.; Yamamoto, S. & Matsushima, M. (2008). High Quality InxGa1-xAs (x: 0.1 - 0.13) Platy Crystal Growth for Substrates of 1.3 μm Laser Diodes. Proceedings of 20th International Conference on Indium Phosphide and Related Materials, TuB1.2, ISBN 1424422582,Versailles, France, May 25-29 , 2008,

Kinoshita, K.; Yoda, S.; Aoki, H.; Hosokawa, T.; Yamamoto, S.; Matsushima, M.; Arai, M.; Kawaguchi, Y.; Kondo, Y. &Kano, F. (2010). Growth of large platy InGaAs crystals and fabrication of semiconductor laser diodes. Proceedings of 22nd International Conference on Indium Phosphide and Related Materials, TuB3-3, ISBN 9781424459193, Takamatsu, Japan, May 31-June 4, 2010

Kinoshita, K. & Yoda, S. (2011). Growth of homogeneous semiconductor mixed crystals by traveling liquidus-zone method. J. Crystal Growth, Vol.318, No. 1(March 2011), pp. 1026-1029, ISSN 0022-0248

Nakajima, K.; Kusunoki, T. & Takenaka, C. (1991). Growth of ternary InxGa1-xAs bulk crystals with a uniform composition through supply of GaAs, J. Crystal Growth, Vol.113, No.3-4 (September 1991), pp. 485-490, ISSN 0022-0248

Nakamura, H.; Hanaue, Y.; Kato, H.; Kinoshita, K. & Yoda, S. (2003). A one-dimensional model to predict the growth conditions of InxGa1-xAs alloy crystals grown by the traveling liquidus-zone method. J. Crystal Growth, Vol.258, No.1-2 (October 2003), pp. 49 – 57, ISSN 0022-0248

Nishijima, Y.; Tezuka, H. & Nakajima, K. (2005). A modified zone growth method for an InGaAs single crystal. J. Crystal Growth, Vol.280, No.3-4 (July 2005), pp. 364-371, ISSN 0022-0248

Sell, H.J. (1991). Growth of GaInAs bulk mixed crystals as a substrate with a tailored lattice parameter. J. Crystal Growth, Vol.107, No.1-4 (January 1991), pp. 396-402, ISSN 0022-0248

Tiller, W.A.; Jackson, K.A.; Rutter, J.W. & Chalmers, B. (1953). The redistribution of solute atoms during the solidification of metals. Acta Metallurgica, Vol.1, No.7 (July 1953), pp. 428-437. ISSN 0001-6160

9

Pattern Selection in Crystal Growth

Waldemar Wołczyński
Institute of Metallurgy and Materials Science,
Polish Academy of Sciences,
Poland

1. Introduction

Ebeling & Ashby, 1966 were able to prove that the Cu - single crystal can be strengthened by spherical particles of the SiO_2 - phase. Zarubova & Sestak, 1975 studied the analogous phenomenon in the Fe single crystals strengthened by the Si - addition. In the current study, the phenomenon of strengthening is observed in the hexagonal (Zn) - single crystal equipped with regular stripes which contain the Zn-16Ti intermetallic compound.

The studied (Zn) - single crystal was doped by the small amount of titanium and copper. The addition of copper modifies the specific surface free energy at the solid/liquid (s/l) interface. The specific surface free energies are to be determined in the current description for the triple point of the s/l interface to ensure the mechanical equilibrium. Copper does not form an intermetallic compound with the zinc but is localized in the zinc/titanium solid solution, (Zn). The titanium forms, additionally, the intermetallic compound with the zinc, $Zn_{16}Ti$. The (Zn)-$Zn_{16}Ti$ system is the pseudo-binary eutectic system. It exists an opportunity to control the growth of the (Zn) - hexagonal single crystal, and first of all to control the width of the (Zn)-$Zn_{16}Ti$ - stripes which appear cyclically in the single crystal.

Some experiments were performed by means of the *Bridgman* system with the moving thermal field. The system was equipped with the graphite crucible of the sophisticated geometry. It allowed to localize a crystal seed of the desired crystallographic orientation just at the crucible bottom.

The full thermodynamic description of the (Zn) - single crystal growth with periodic formation of the (Zn)-$Zn_{16}Ti$ - stripes requires to consider the model for solute redistribution (Wołczyński, 2000). Next, a steady-state solution to the diffusion equation which yields the solute micro-field in the liquid at the s/l interface is given, (Wołczyński, 2007). The above solution involves a proper localization of the thermodynamic equilibrium at the inter-phase boundary (together with the mechanical equilibrium). This solution also allows for calculating the entropy production in such a system and additionally for describing the transition from lamellar into rod-like structure, (Wołczyński, 2010).

2. Solute redistribution within the (Zn) - Single crystal growing cyclically

Solute redistribution along the single crystal changes significantly due to localization of the *liquidus* and *solidus* lines in the phase diagram. In the case of the (Zn) - single crystal which contains strengthening stripes, Fig. 1a, the Ti - solute redistribution should be measured along the part of the (Zn) - single crystal formed between two neighbouring stripes, Fig. 1b.

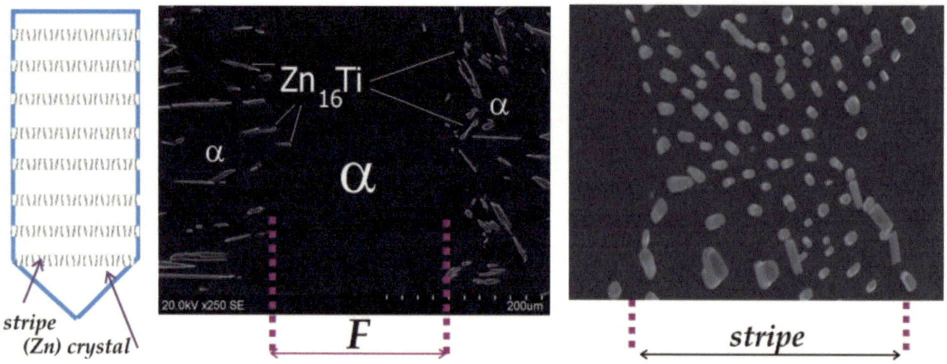

Fig. 1. (Zn) - single crystal structure; a. stripes localized cyclically in a (Zn) - single crystal (scheme), b. the F - distance between stripes; $(Zn) \equiv \alpha$, c. rod-like morphology of a stripe.

2.1 Use of the Zn-Ti phase diagram for the control of the (Zn) - Single crystal growth

According to the Zn-Ti phase diagram the solubility of the Ti in the Zn is equal to 0.000546 at.% at the ambient temperature, Fig. 2. Thus, the strengthening by a solubility is to be neglected. However, an experiment was made to study a strengthening by the eutectic precipitates, for three (I, II, III) solute contents.

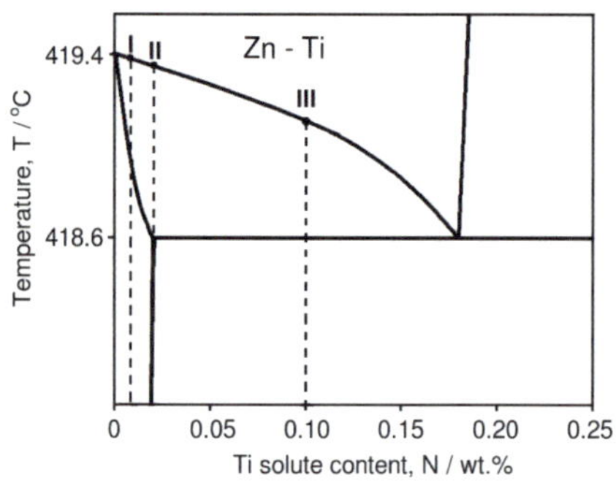

Fig. 2. Zn-Ti phase diagram with the localization of the eutectic point, $N_E = 0.18\,[wt.\%Ti]$, (Murray, 1990), and all studied nominal solute concentration of the single crystal: I, II, III.

Since the difference between melting point of the Zn and eutectic temperature is small it involves a possibility of the single crystal strengthening by the periodic precipitates of eutectic phase ((Zn) + $Zn_{16}Ti$). The (Zn) – eutectic phase, Fig, 1b, Fig. 1c, is the same as in the bulk single crystal. Therefore, it is to be supposed that the $Zn_{16}Ti$ eutectic phase (intermetallic compound) is only responsible for the strengthening of the (Zn) – single crystal.

A peculiar construction of the Zn-Ti phase diagram, Fig. 2, involves a possibility in the control of the precipitates amount. However, a proper model for the solute segregation / redistribution is necessary to predict the amount of the eutectic or intermetallic phase precipitation within a given single crystal.

In the current study the *Bridgman* furnace was working as a closed system and the control of strengthening could be done, to some extent, by an adequate choose of a nominal concentration of titanium in the alloy used for single crystal growth, Fig. 2, and by applying a proper growth rate, v , on which the $\alpha(v)$ back-diffusion parameter depends.

The selection of the nominal solute concentration decides on the stripe thickness as visible in Fig. 2. The nearest is the distance between nominal solute concentration on the *liquidus* line and the eutectic point the wider are the stripes.
Moreover, the applied growth rate decides on the stripes morphology. Three ranges of growth rates were distinguished in the discussed (Zn) – single crystal formation: a. at some low growth rates, the L – shape irregular rod-like structure appears in stripes, b. at some middle growth rates the regular lamellar structure is observed in stripes, c/ at the elevated growth rates the regular rod-like structure exists as an exclusive morphology of stripes.
It is evident that two transformations of morphology were recorded. At the first threshold growth rate the L-shape irregular rods were transformed into the regular lamellae. This transformation cannot occur immediately since the L-shape rod-like into lamellar structure transition is accompanied by the irregular into regular morphology transition. Thus, this transformation occurred continuously within a certain range of rates. However, the lamella into rod transition occurred just at the second threshold rate, immediately.
The control of the stripe thickness can be explained while applying the equilibrium solidification to the *Bridgman* system, Fig. 3.

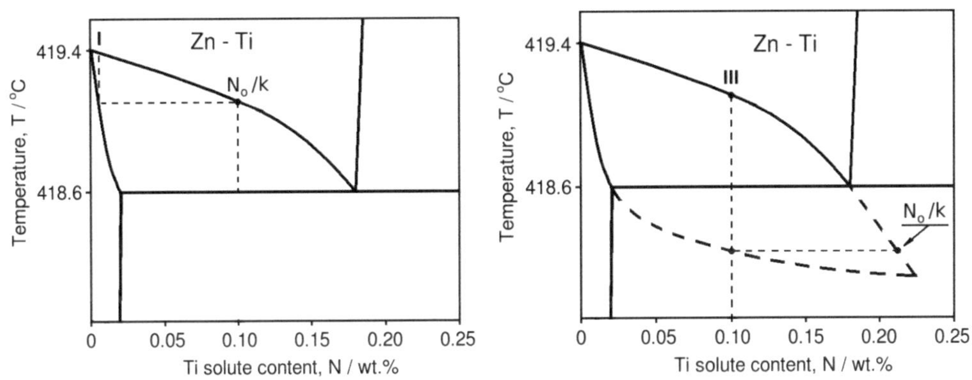

Fig. 3. Solidification paths ($I \rightarrow N_0 / k$) and ($III \rightarrow N_0 / k$) for an equilibrium solidification of the (Zn) - single crystal with: a. $N_0 = I = 0.01\,[wt.\%Ti]$, and b. $N_0 = III = 0.1\,[wt.\%Ti]$.

The precipitation is impossible for the equilibrium solidification with the nominal solute concentration: $N_0 = I = 0.01$, since $N_0 / k < N_E$, Fig. 3a. The precipitation is always ensured for the nominal solute concentration: $N_0 = III = 0.1\,[wt.\%Ti]$, since $N_0 / k > N_E$, Fig. 3b.
Moreover, the solidification occurs along the solidification path cyclically, and precipitates are rejected at the end of each cycle to form the stripes.

2.2 Model for the solute microsegregation / redistribution after the back-diffusion

A theory for solute microsegregation accompanied by the back-diffusion which is defined by the - α parameter, has already been delivered, (Brody & Flemings, 1966), with:

$$\alpha = D_S\, t_l\, L^{-2} \tag{1}$$

D_S - diffusion coefficient into the solid, $[m^2 / s]$; t_l - local growth time, [s]; L - half the crystal spacing, [m]. This theory is not able to describe the solute redistribution since the mass balance is not satisfied, there. Thus, an improved model, (Wołczyński, 2000), based on the α - back-diffusion parameter, Eq. (1) has been delivered. The model describes the solute redistribution after back-diffusion and can be used to calculate an amount of precipitates. The equation which describes the solute redistribution after back-diffusion is as follows:

$$N^B\left(x;X^0,\alpha\right)=\left[k+\beta^{ex}\left(x;X^0\right)\beta^{in}\left(X^0,\alpha\right)\right]N^L\left(x;\alpha\right) \tag{2}$$

k - partition ratio, $[mole\ fr./\ mole\ fr.]$; x - crystal amount, [dimensionless]; $x = X^0$ - crystal amount at arrested growth; β^{ex} - coefficient of redistribution extent, [dimensionless]; β^{in} - coefficient of redistribution intensity, [dimensionless]; N^L - liquid content, $[mole\ fr.]$; Eq. (3).

$$N^L\left(x;\alpha\right)=N_0\left(1+\alpha\,k\,x-x\right)^{(k-1)/(1-\alpha k)} \tag{3}$$

The above equation results from the differential formula:

$$\left[1+\alpha\,k\,x-x\right]dN^L\left(x;\alpha\right)=\left(1-k\right)N^L\left(x;\alpha\right)dx \tag{4}$$

Both, β^{ex} - coefficient of the redistribution extent and β^{in} - coefficient of the redistribution intensity are defined due to the mass balance consideration (Wołczyński, 2000). Their product $\beta\left(x;X^0,\alpha\right)=\beta^{ex}\left(x,X^0\right)\beta^{in}\left(X^0;\alpha\right)$ is equal to zero: $\beta\left(x;X^0,0\right)=0$ for the non-equilibrium solidification (when $\alpha=0$). The product is equal to $\left(1-k\right)\left(1-x\right)$ for the equilibrium solidification (when $\alpha=1$ with $X^0=1$) that is: $\beta\left(x;1,1\right)=\left(1-k\right)\left(1-x\right)$. The single crystal growth occurs slowly, thus its growth is performed under condition close to the equilibrium state. The physical limitation for the the α - back-diffusion parameter has also been determined while assuming: $t_D=L^2/D_S$. Then, $\alpha=t_l/t_D$ and finally, $0\le\alpha\le1$. with t_D - diffusion time necessary for the full homogeneity of the crystal. Thus, the mass balance is satisfied at each step of the crystal growth. Therefore, Eq. (2) can perfectly fit some measurement points obtained by the EDS technique. On the other side, the solute microsegregation is observable at the moving s/l interface during the experiment. Thus, the redistribution is the only concentration which can be measured after the crystal growth.

Usually, the crystal growth is accompanied by some precipitates. The growth rate, imposed in the *Bridgman* system for a single crystal growth (with a seed) $v>0$, does not follow the equilibrium solidification. Thus, the solidification path is longer than that shown in Fig. 2. Therefore, it was possible to obtain precipitates even for the nominal solute concentrations: $N_0=I=0.01\,[wt.\%Ti]$, $N_0=II=0.02\,[wt.\%Ti]$, when the imposed growth rate, v, involves an elongation of the solidification path beyond the eutectic point ($N_E=0.18\,[wt.\%Ti]$) till the N_K - point on the *liquidus* line. N_K is also the solute content in the liquid at the end of solidification, $[mole\ fr.]$ (N_0, N_E are to be expressed in $[mole\ fr.]$ while calculating N_K).

The ratio between an amount of the crystal, x_K, and an amount of precipitate, i_K, depends on the mentioned elongation beyond the eutectic point by the solidification path:

$$x_K/i_K=\left(N_E-N_0\right)/\left(N_K\left(\alpha,N_0\right)-N_E\right) \tag{5}$$

The end of solidification path is: $N_K\left(\alpha,N_0\right)=N^L\left(x_K,\alpha\right)$, and x_K is defined as follows:

$$x_K(\alpha,N_0)=\frac{1}{1-\alpha k}\left[1-(N_E/N_0)^{\frac{1-\alpha k}{k-1}}\right] \text{ when } 0\le\alpha\le\alpha_E(N_0) \tag{6a}$$

$$x_K(\alpha,N_0)=1 \quad \text{when} \quad \alpha_E(N_0)<\alpha\le 1 \tag{6b}$$

with $(\alpha_E k)^{\frac{k-1}{1-\alpha_E k}}=N_E/N_0$, (Wołczyński, 2000).

According to the above considerations, $i_K(\alpha,N_0)=1-x_K(\alpha,N_0)$, [dimensionless]. The precipitate is divided into the so-called equilibrium precipitate: $i_E(N_0)=i_K(1,N_0)$ and non-equilibrium precipitate: $i_D(\alpha,N_0)=i_K(\alpha,N_0)-i_E(N_0)$.
The precipitate visible in the crystal reproduces the s/l interface shape which existed "historically" at a given stage of the (Zn) - single crystal growth, Fig. 4.

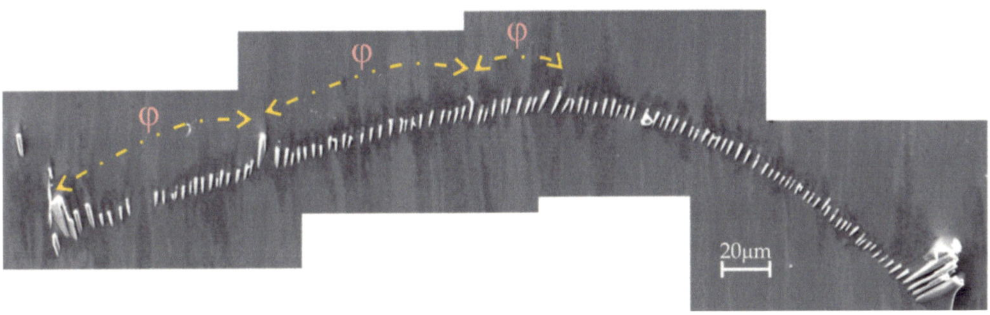

Fig. 4. A stripe deflected by the perturbation wave created at the solid / liquid interface during the (Zn) - single crystal growth; the stripe (precipitate) contains the (Zn) - phase coherent with the bulk single crystal and rods of the $Zn_{16}Ti$ - intermetallic compound; the stripe follows the pattern which seems to be the result of a solitary wave, (φ).

2.3 Measurement / simulation of the Ti - Solute redistribution along the single crystal

The solute redistribution model, discussed above, allows to make some simulations of the Ti - solute redistribution for the (Zn) - single crystal growth. The simulation is developed to fit the measurement points obtained by the EDS technique, Fig. 5. The Ti - solute redistribution was measured along the bulk (Zn) - single crystal just between two neighbouring stripes and for two different samples.
The theoretical solute microsegregation is calculated twice: a. $N^S(x,0)=kN^L(x,0)$ - without the presence of back diffusion, (Scheil, 1942), b. $N^S(x,\alpha)=kN^L(x,\alpha)$ - with the presence of back-diffusion phenomenon, (Wołczyński, 2000), Fig. 5. The theoretical amount of the bulk crystal, $x_K(\alpha)$ and corresponding amount of the ((Zn) + $Zn_{16}Ti$) - precipitate, $i_K(\alpha)$ (red lines) are also shown. The theoretical solute redistribution is calculated for a selected value of the α - back-diffusion parameter, ($0\le\alpha\le1$), [dimensionless], and additionally for $X^0=1$. The partition ratio is applied due to the Zn-Ti phase diagram for the stable equilibrium as equal to $k=0.11[wt.\%/wt.\%]$, (Murray, 1990).

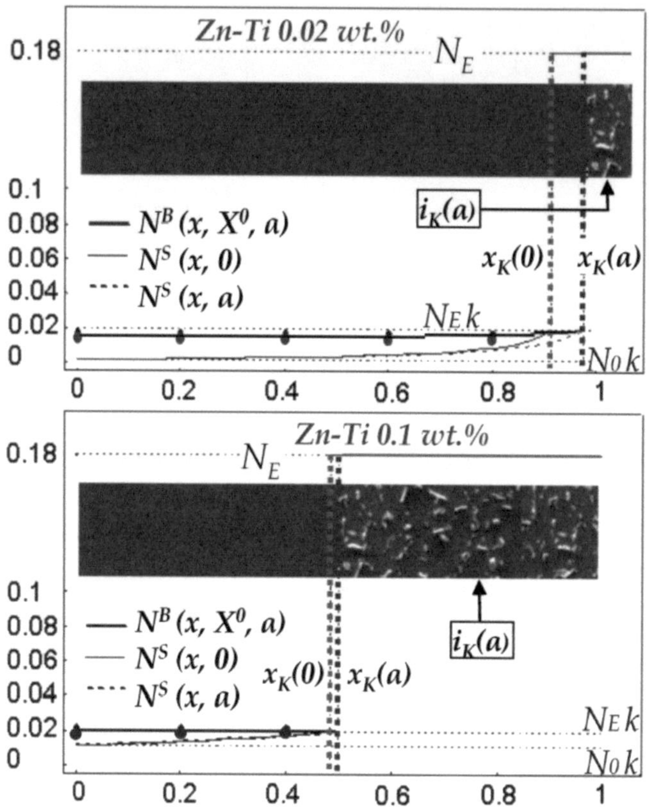

Fig. 5. Ti – solute redistribution as calculated for two alloys of the nominal Ti – solute concentration equal to: 0.02 *wt.*% and 0.1 *wt.*%, respectively; (Boczkal et al., 2010).

3. Solute micro-field in the liquid and the strengthening phase protrusion

The eutectic precipitates are formed under stationary state from the non-homogeneous liquid phase, (Jackson & Hunt, 1966). The solution to the diffusion equation gives a description of the solute concentration micro-field ahead of the s/l interface. The recent solution to the diffusion equation supposes the existence of the mechanical equilibrium at the triple point of the s/l interface and the thermodynamic equilibrium along the inter-phase boundary, (Wołczyński, 2007). The steady-state solution to the diffusion equation is given separately for each eutectic lamella: α - phase lamella, and β - phase lamella. So,

a. for the α - eutectic phase formation (the (Zn) – phase in the Zn – Ti system),

$$\delta C(x,z)=\sum_{n=1}^{\infty}A_{2n-1}\cos\left(\frac{(2n-1)\pi x}{2S_\alpha}\right)\exp\left(-\frac{(2n-1)\pi}{2S_\alpha}z\right) \tag{7}$$

$$A_{2n-1}=-\frac{4}{(2n-1)\pi}\int_0^{S_\alpha} f_\alpha(x)\cos\left(\frac{(2n-1)\pi x}{2S_\alpha}\right)dx \quad n=1,2,\ldots \tag{7a}$$

b. for the β - eutectic phase formation (the $Zn_{16}Ti$ – compound in the Zn – Ti system),

$$\delta C(x,z)=\sum_{n=1}^{\infty}B_{2n-1}\cos\left(\frac{(2n-1)\pi(x-S_\alpha+S_\beta)}{2S_\beta}\right)\exp\left(-\frac{(2n-1)\pi}{2S_\beta}z\right) \tag{8}$$

$$B_{2n-1}=-\frac{4}{(2n-1)\pi}\int_{S_\alpha-S_\beta}^{S_\alpha}f_\beta(x)\cos\left(\frac{(2n-1)\pi(x-S_\alpha+S_\beta)}{2S_\beta}\right)dx \quad n=1,2,... \tag{8a}$$

C - solute concentration within the micro-field formed in the liquid, $[at.\%]$; S_j - half the width of the eutectic phase lamellae, $[m]$, ($j=\alpha,\beta$), respectively; f_j - function used in formulation of the boundary condition for the α and β - eutectic phases formation, $[at.\%]$, ($j=\alpha,\beta$), respectively; x,z - geometrical coordinates, $[m]$.
The total mass balance is satisfied within the solute concentration micro-field, Eq. (9), Fig. 6.
D - diffusion coefficient in the liquid, $[m^2/s]$; v - crystal growth rate, $[m/s]$, identical to the thermal field movement rate in the *Bridgman* system; moreover,

$$B_{2n-1}=A_{2n-1}\left(S_\alpha/S_\beta\right)^2, \quad n=1,2,... \tag{9a}$$

The satisfaction of the total mass balance is shown in Fig. 6. The local mass balance is also satisfied, Fig. 7. But, in the case of the local mass balance the phase protrusion is to be considered, Eq. (10).

$$\int_0^{S_\alpha}\delta C(x,0)dx+\int_{S_\alpha}^{S_\alpha+S_\beta}\delta C(x,d)dx=0 \tag{10}$$

d - protrusion of the β - leading eutectic phase over the α - wetting eutectic phase, $[m]$.

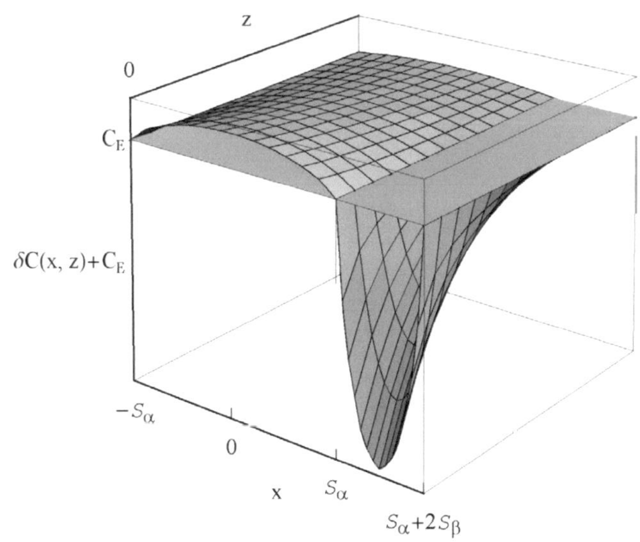

Fig. 6. Total mass balance within the solute concentration field for an eutectic system growing under stationary condition; C_E - eutectic concentration of the solute, $[at.\%]$.

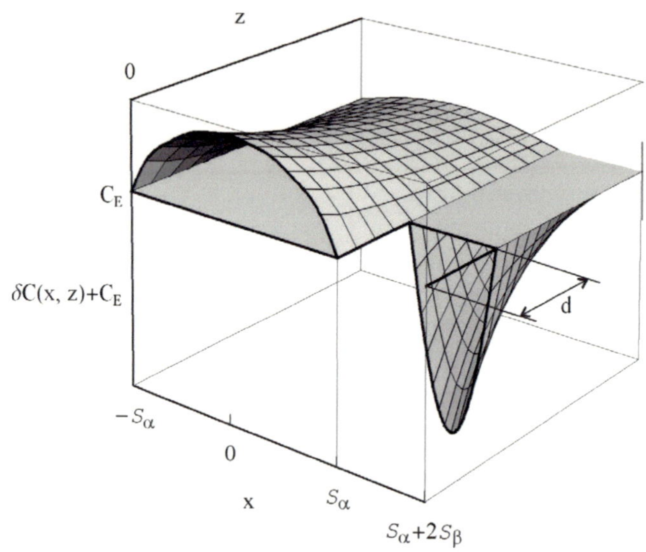

Fig. 7. Local mass balance within the solute concentration field for an eutectic system growing under stationary condition.

The protrusion of the eutectic leading phase, predicted theoretically, Fig. 7, and formerly observed experimentally, (Davies, 1964) has also been revealed within the stripes which contain the ((Zn) + $Zn_{16}Ti$) eutectic, Fig. 8.

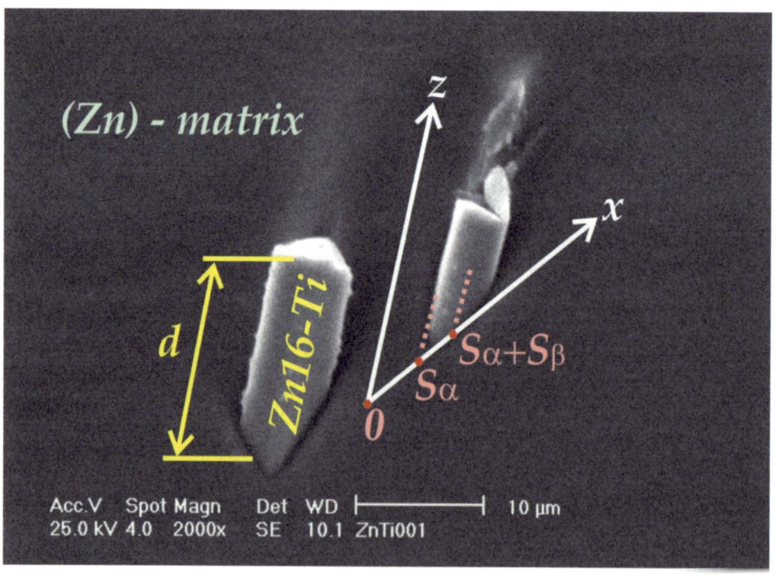

Fig. 8. ((Zn) + $Zn_{16}Ti$) regular rod-like eutectic revealed in the stripes (the EDS observation); the x,z - coordinate system presents the localization of the Ti – solute concentration field; the d - leading phase protrusion is defined due to the (Zn) – single crystal growth arresting.

Since the (Zn) - single crystal growth was arrested and the s/l interface was frozen, it is also possible to reveal the s/l interface shape of the non-faceted (Zn) - eutectic phase, Fig. 9.

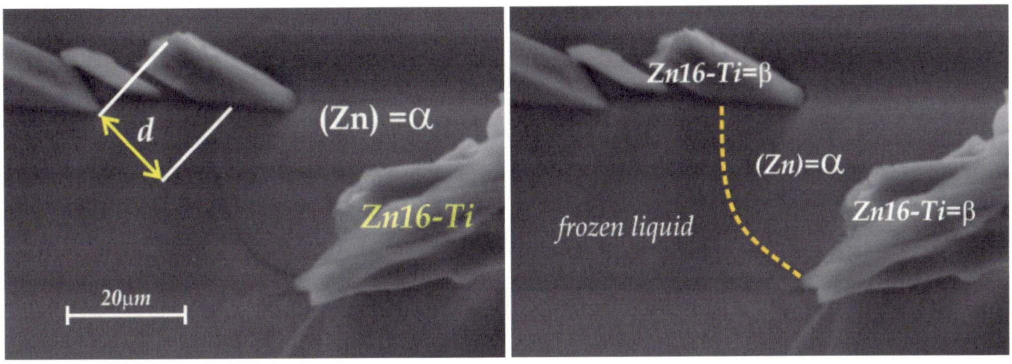

Fig. 9. Frozen solid / liquid interface of the ((Zn) + $Zn_{16}Ti$) regular eutectic revealed within the strengthening stripes; a. the EDS observation, b. identification of the eutectic phases: the $Zn_{16}Ti$ faceted phase, the (Zn) - non-faceted phase, the frozen liquid, the parabolic envelope (yellow dashed line) superposed onto the s/l interface of the (Zn) - phase.

Since the revealed solid/liquid interface has a parabolic shape, Fig. 9, and no interface destabilization is observed, it is evident that the obtained structure is the regular structure.

4. Thermodynamic selection of the pattern in the growing eutectic structure

(Zn) - single crystal growth occurs under stationary state in the *Bridgman* system with an imposed constant growth rate. Neither, the thermal field does not vary during solidification. Eventually, the $Zn_{16}Ti$ eutectic phase (compound) has the same size, Fig. 10, and an inter-phase spacing should not vary during crystal growth.
It should be emphasized that the only condition which defines the stationary state is the criterion of minimum entropy production, (Prigogine, 1968).

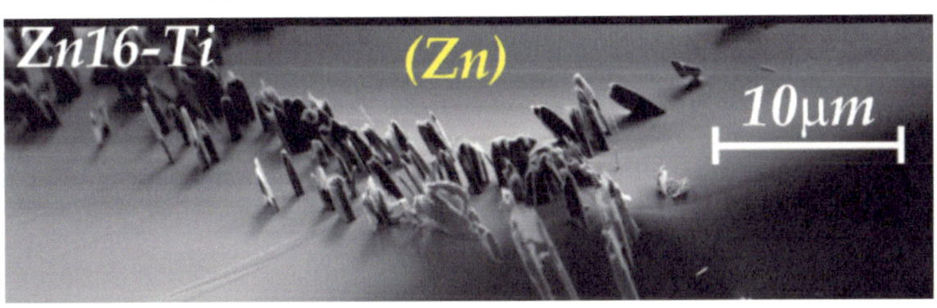

Fig. 10. A strengthening stripe ((Zn) + $Zn_{16}Ti$) within the (Zn) - single crystal frozen during its growth in the *Bridgman* system; the $Zn_{16}Ti$ - rods are all of the same size, mainly.

Many eutectic systems exhibit either a lamellar or rod-like structure depending on solidification conditions, (Elliott, 1977). It is well visible in the case of directional growth of the Al-Si eutectic alloy, (Toloui & Hellawell, 1976) and (Atasoy, 1984). Moreover, a transition from the lamellar structure into the rod-like structure is observable at a the threshold growth rate typical for a given eutectic alloy, (Cupryś et al., 2000).
Especially, growth rate plays a crucial role in the lamella into rod transformation. Some impurities also involve the transition, (Steen & Hellawell, 1975). The impurities change the specific surface free energies and finally modify a mechanical equilibrium at the triple point

of the solid/liquid interface. However, according to the current model assumptions, the mechanical equilibrium varies (rotates) at the s/l interface of the (Zn) – single crystal not only with the copper addition but with solidification conditions (growth rate) as well. An imposed growth rate results in the crystal orientation. Some changes of the orientation from an initial state into a final one give also an effect on the lamella.rod transition.

In the case of the (Zn) – single crystal growth two threshold rates were revealed for the changes in the stripes morphology, as mentioned above. First threshold growth rate, v_1, is connected with the transition from the L – shape irregular (with branches) rod-like structure into the regular lamellar structure. Second threshold growth rate, v_2, is associated with the transition from the regular lamellar structure into the regular rod-like structure.

4.1 Promotion of the rod-like structure or lamellar structure formation

The formerly developed theory (Jackson & Hunt, 1966) has tried to predict the threshold rate at which a structural transition should occur. The theory (Jackson & Hunt, 1966) is based on the description of the s/l interface undercooling with the undercooling defining both lamellar growth and rod-like growth, Eq. (11).

$$\left(\Delta T_L^*\right)^2 = 4 v m^2 a^L Q^L ; \quad \left(\Delta T_R^*\right)^2 = 4 v m^2 a^R Q^R \tag{11}$$

ΔT_L^* - s/l interface undercooling for the lamellar structure formation, $[K]$; ΔT_R^* - s/l interface undercooling for the rod-like structure formation, $[K]$; v - growth rate, $[m/s]$; $1/m = 1/m_\alpha + 1/m_\beta$, with: m_α - slope of the α - *liquidus* line, $[K/at.\%]$; m_β - slope of the β - *liquidus* line, $[K/at.\%]$, $a^L;a^R;Q^L;Q^R$ - parameters associated with the eutectic system capillarity and with a given phase diagram, (Jackson & Hunt, 1966).

This description yields a certain inequality, (Jackson & Hunt, 1966), according to which the rods or lamellae are formed, Fig. 11.

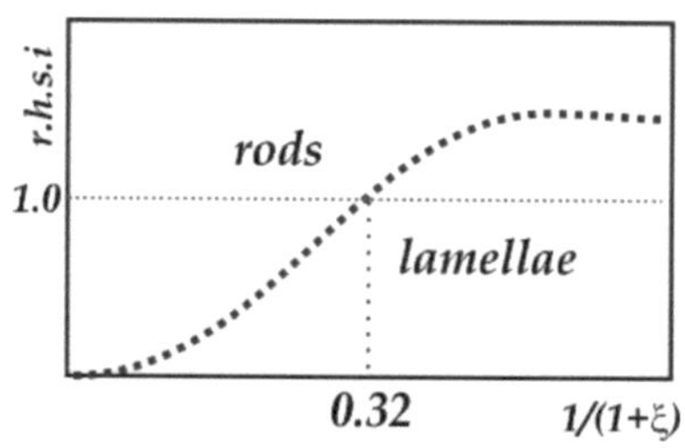

Fig. 11. Prediction of rod-like or lamellar structure formation, (Jackson & Hunt, 1966).

For the isotropic s/l interfacial free energies the r. h. s. of the considered inequality is equal to one, Fig. 11. The discussed inequality, shown in Fig. 11, yields a result, according to which, when the following inequality: $1/(1+\xi) < 0.32$, is satisfied (for the assumed isotropy) then a rod-like structure is stable form, (Jackson & Hunt, 1966). However, this parameter is equal to 0.114, ($1/(1+\xi) = 0.114$), for the Al-Si system. Thus, the rod-like structure should be the stable form. Meanwhile, according to experimental observations a lamellar structure is formed instead of a rod-like structure. Moreover, the lamellar structure transforms into the rod-like structure at a threshold growth rate. It is evident that the inequality, illustrated in Fig. 11, is completely misleading. The discerning analysis shows that the above inequality is able to predict whether an eutectic alloy will manifest lamellar or rod-like structure, only. Thus, the discussed inequality characterizes a given phase diagram. The inequality cannot be applied to describe the lamella into rod transition.

4.2 Thermodynamics of the eutectic s/l interface formation in the pattern selection

Thermodynamics of the eutectic s/l interface formation has already been discussed to some extent, (Wołczyński, 2010). The consideration was focused on the Al-Si eutectic alloy which manifests the irregular structure (with branches) due to directional solidification. It is known that the regular structure areas exist among the generally irregular structure, Fig. 12. It results from the experimental observations of lamellar and rod-like structures formation that the threshold growth rate at which the transformation lamella into rod begins, is equal to 400 $[\mu m / s]$. The transition is completed at the growth rate equal to about 700 $[\mu m / s]$. Both structures coexist within the operating range for the lamella into rod transition, ($400 \mu m / s \div 700 \mu m / s$), as visible in Fig. 13b, but the lamellar structure is obtainable below the threshold solidification rate, exclusively, as it is shown in Fig. 13a.

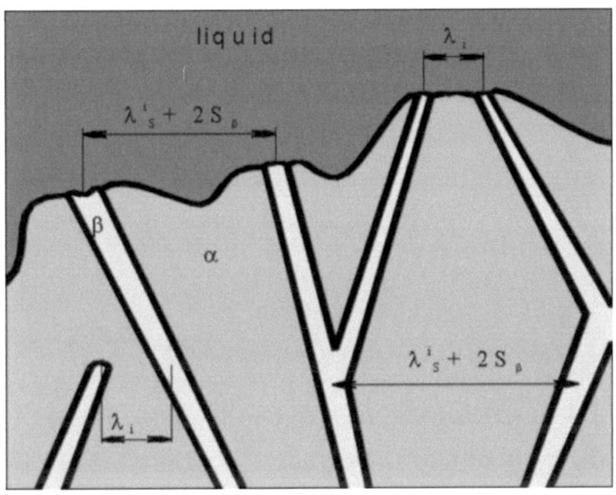

λ_i - inter-lamellar spacing for the regular structure, $[m]$; $\lambda_s^i + 2S_\beta$ - inter-lamellar spacing for the maximal destabilization of the s/l interface of the α - non-faceted phase, $\alpha \equiv (Al)$, $[m]$

Fig. 12. Generally irregular eutectic structure formation; scheme, (Fisher & Kurz, 1980).

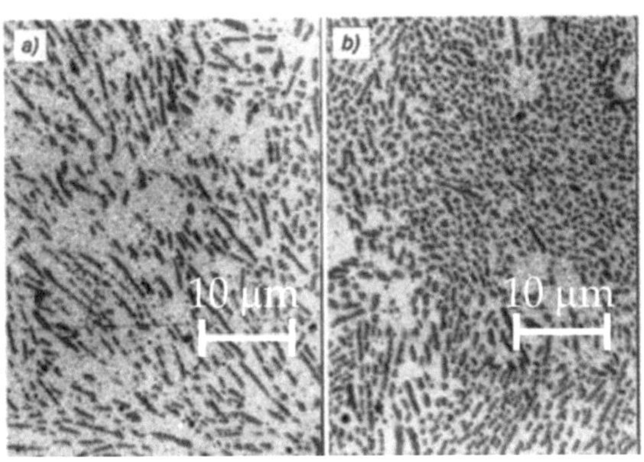

Fig. 13. A cross-sectional morphology of a directionally solidifying Al-Si irregular eutectic; a. for $v = 370 \mu m / s$, and thermal gradient equal to $G = 100 K / cm$, (lamellae exclusively); b. for $v = 500 \mu m / s$, and thermal gradient equal to $G = 40 K / cm$, (lamellae + rods).

The thermodynamics of the s/l interface involves the calculation of the *Gibbs'* free energy for the considered eutectic system. The *Gibbs'* free energy formulas which are concerning the s/l interface formation are as follows, for the regular lamellar and regular rod-like structure:

$$\Delta G_L^* = v\lambda Q_{CW}^L - a_{CW}^L \lambda^{-1} \quad \Delta G_R^* = vRQ_{CW}^R - a_{CW}^R R^{-1} \tag{12}$$

Eq. (12) was developed from Eq. (11) by means of the transformation of the undercooling into the *Gibbs'* free energy.
$\Delta G_L^*;\Delta G_R^*$ - *Gibbs'* free energy for regular lamellar or rod-like growth, respectively, $\left[J/m^3\right]$; $\lambda \equiv \lambda_i$ - inter-lamellar spacing, $[m]$; R - inter-rod spacing, $[m]$.

$$Q_{CW}^L = \left[m\left(L_\alpha\xi + L_\beta\right)/T_E\right]\left[P^*(1+\xi)N_0/(\xi D)\right] \tag{13a}$$

$$Q_{CW}^R = \left[m\left(L_\alpha\xi + L_\beta\right)/T_E\right]\left[4EN_0/(\xi D)\right] \tag{13b}$$

$$a_{CW}^L = \left\{m(1+\xi)\left[\sigma_\alpha^L \sin\theta_\alpha^L / m_\alpha + \left(\sigma_\beta^L \sin\theta_\beta^L\right)/\left(\xi m_\beta\right)\right] + \sigma_{\alpha\beta}^L\right\} \tag{14a}$$

$$a_{CW}^R = 2\left\{m\sqrt{1+\xi}\left[\sigma_\alpha^R \sin\theta_\alpha^R / m_\alpha + \left(\sigma_\beta^R \sin\theta_\beta^R\right)/\left(\xi m_\beta\right)\right] + \sigma_{\alpha\beta}^R / \sqrt{1+\xi}\right\} \tag{14b}$$

$$\xi = S_\beta / S_\alpha \quad \text{with } \lambda = 2\left(S_\alpha + S_\beta\right) \tag{15a}$$

$$r_\alpha = 1/\sqrt{1+\xi} \quad \text{with } R = r_\alpha + r_\beta \tag{15b}$$

$L_\alpha;L_\beta$ - heat of fusion per unit volume of a given eutectic phase, $[J/m^3]$; T_E - eutectic melting point, $[K]$; P^*, E - parameters referred to lamellae width, (Jackson & Hunt, 1966); N_0 - difference of the solubility of the B – eutectic element in the A – element and solubility of the A – element in the B – element, according to a given eutectic phase diagram, $[at.\%]$; D - diffusion coefficient in the liquid, $[m^2/s]$; σ_j^L - specific surface free energy for the lamellar structure, $j=\alpha,\beta$, $[J/m^2]$; σ_j^R - specific surface free energy for the rod-like structure, $j=\alpha,\beta$, $[J/m^2]$; $\sigma_{\alpha\beta}^L$ - α/β phase boundary free energy for lamellar structure, $[J/m^2]$; $\sigma_{\alpha\beta}^R$ - α/β phase boundary free energy for the rod-like structure, $[J/m^2]$.
Geometry for both types of eutectic morphology is shown in Fig. 14.

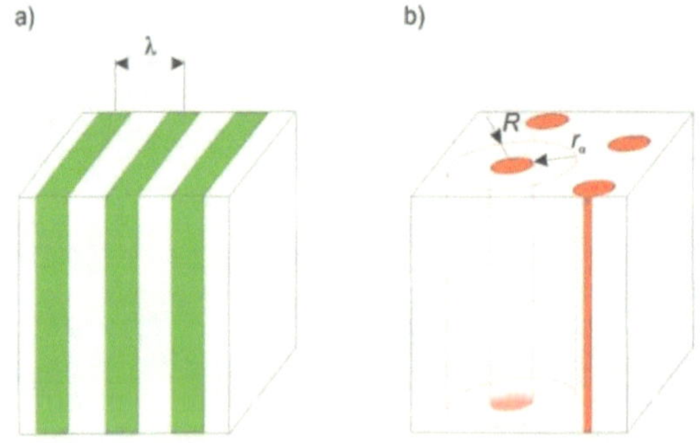

Fig. 14. Geometrical parameters defining both morphologies; a. inter-lamellar spacing, λ in the regular structure; b. rod/(rod + matrix) radiuses in the regular structure, Eq. (15).

Some rearrangements of Eq. (12) allows to formulate the following inequality:

$$\left\{m\left[a_{\alpha}^{L} / m_{\alpha} + a_{\beta}^{L} / \xi m_{\beta}\right] + \sigma_{\alpha\beta}^{L} / (1+\xi)\right\} / \left\{m\left[a_{\alpha}^{R} / m_{\alpha} + a_{\beta}^{R} / \xi m_{\beta}\right] + \sigma_{\alpha\beta}^{R} / (1+\xi)\right\}$$
$$> 4\left(E / P^{*}\right)\left(1/(1+\xi)\right)^{1.5} \tag{16}$$

The obtained inequality was developed in the same manner as the inequality illustrated in Fig. 11. The developed inequality cannot predict the lamella into rod transformation in the eutectic system. Analogously to the mentioned theory, (Jackson & Hunt, 1966), this inequality can be applied, Eq. (16), to some extent, in order to predict, only, whether a given eutectic system (phase diagram) promotes the lamellae or rods formation.
Therefore, the *Gibbs'* free energy, Eq. (12), was calculated with applying the evolution (with the growth rate) of a mechanical equilibrium at the triple point of the s/l interface, Fig. 15.

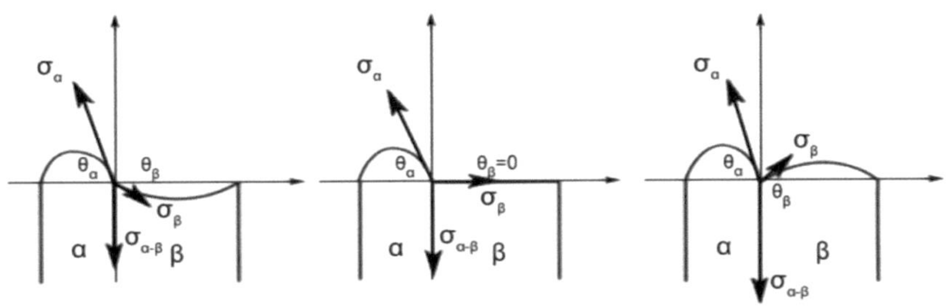

Fig. 15. Evolution (with varying growth rate) of the s/l interface curvature for the regular eutectic growth and adequate mechanical equilibrium situated at the triple point, (scheme).

$\alpha \equiv Si$, and $\beta \equiv (Al)$ - eutectic phases; $\sigma_{\alpha} \equiv \sigma_{\alpha}^{L}$ or $\sigma_{\alpha} \equiv \sigma_{\alpha}^{R}$; $\sigma_{\beta} \equiv \sigma_{\beta}^{L}$ or $\sigma_{\beta} \equiv \sigma_{\beta}^{R}$; $\sigma_{\alpha-\beta} \equiv \sigma_{\alpha\beta}^{L}$ or $\sigma_{\alpha-\beta} \equiv \sigma_{\alpha\beta}^{R}$; $\theta_{\alpha} \equiv \theta_{\alpha}^{L}$ or $\theta_{\alpha} \equiv \theta_{\alpha}^{R}$; $\theta_{\beta} \equiv \theta_{\beta}^{L}$ or $\theta_{\beta} \equiv \theta_{\beta}^{R}$, Fig. 15.
The mechanical equilibrium for the both structure types is:

$$\sigma_{\alpha}^{L} \sin\theta_{\alpha}^{L} \mp \sigma_{\beta}^{L} \sin\theta_{\beta}^{L} - \sigma_{\alpha\beta}^{L} = 0\,; \quad \sigma_{\alpha}^{R} \sin\theta_{\alpha}^{R} \mp \sigma_{\beta}^{R} \sin\theta_{\beta}^{R} - \sigma_{\alpha\beta}^{R} = 0 \tag{17}$$

The *Gibbs'* free energy, Eq. (12) calculated for the threshold growth rate is shown in Fig. 16. The threshold growth rate in calculation equals $v = 400\,\mu m / s$; $\Delta G^{*} \equiv \Delta G_{L}^{*}$ or $\Delta G^{*} \equiv \Delta G_{R}^{*}$, respectively; dot L - denotes localization of the average inter-lamellar spacing, $\overline{\lambda}$ and dot R - denotes localization of the average inter-rod spacing, $\overline{R}$; $\theta_{\alpha}^{L};\theta_{\beta}^{L};\theta_{\alpha}^{R};\theta_{\beta}^{R}$ - angles, Fig. 15.
All the minima of the *Gibbs'* free energy for considered growth rates are gathered in Fig. 17. According to the result of the *Gibbs'* free energy calculation this structure is stable form which has its minimum situated lower, Fig. 17. Thus, the calculation allows to determine the threshold growth rate, when minima are at the same level, Fig. 16. The operating range for transition cannot be described, alas. The present simulation, Fig. 17, was possible since the evolution of the s/l interface curvature, Fig. 18, was implemented into the calculation.
The evolution involves some changes of crystallographic orientation of the s/l interface and specific surface free energies together with inter-phase boundary free energy, Fig. 15.

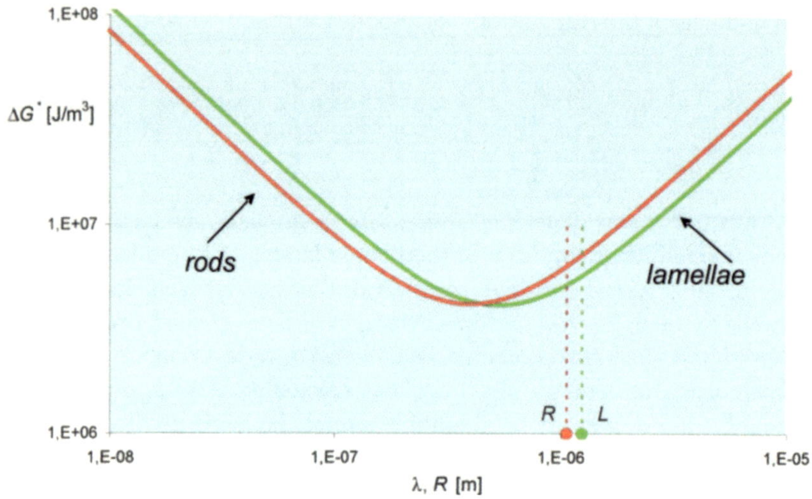

Fig. 16. *Gibbs'* free energy simulated for the both structures formation at the threshold rate.

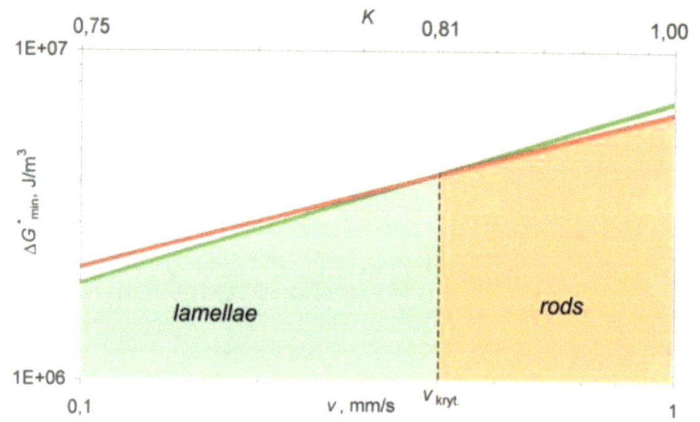

Fig. 17. *Gibbs'* free energy minima calculated for lamellar and rod-like structures formation; the critical growth rate, $v_{kryt.}$, for lamella into rod transition results from the trajectories intersection; $v_{kryt.}$ is equal to the experimental threshold growth rate: $v_{thr.}$.

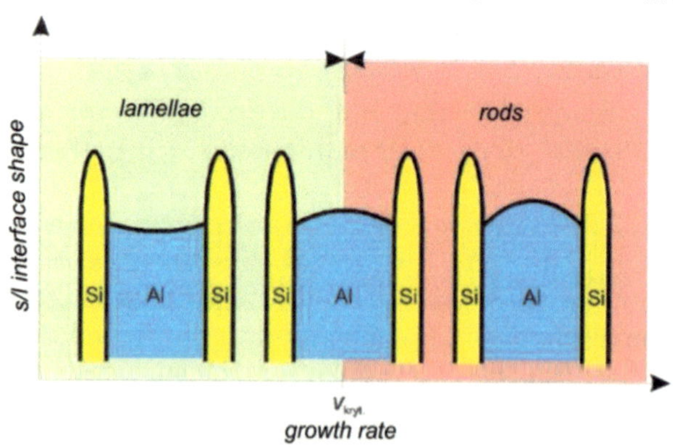

Fig. 18. A model of the s/l interface curvature evolution with the growth rates; $v_{kryt.} \equiv v_{thr.}$.

4.3 Thermodynamics of the whole solidification of regular eutectic structure

Thermodynamics of the whole solidification process supposes a calculation of the entropy production for the regular structure growth, (Lesoult & Turpin, 1969). The regular structure areas can be easily distinguished among generally irregular structure, (Fisher & Kurz, 1980). The regular structure selection under stationary state with the assumption of an isothermal s/l interface is to be described by the criterion of minimum entropy production, (Kjelstrup & Bedeaux, 2008). The entropy production per unit time, P_S^D, $S = R, L$, is as follows:

$$P_S^D = \int_V \sigma \, dV \tag{18}$$

V is the so-called „thermodynamic macroscopic point" (volume) inside of which all essential fluxes are observed, (Glansdorff & Prigogine, 1971). V - volume is shown in Fig. 19.

$$\sigma = R^* \varepsilon C^{-1} (1-C)^{-1} D \nabla^2 C \tag{19}$$

R^* - gas constant, $[J/(mole\,K)]$; ε - thermodynamic factor, [dimensionless]; C - solute content, $[at.\%]$; σ - entropy production per unit time and unit volume, $\left[mole\,fr.^2/(m^3 s)\right]$.

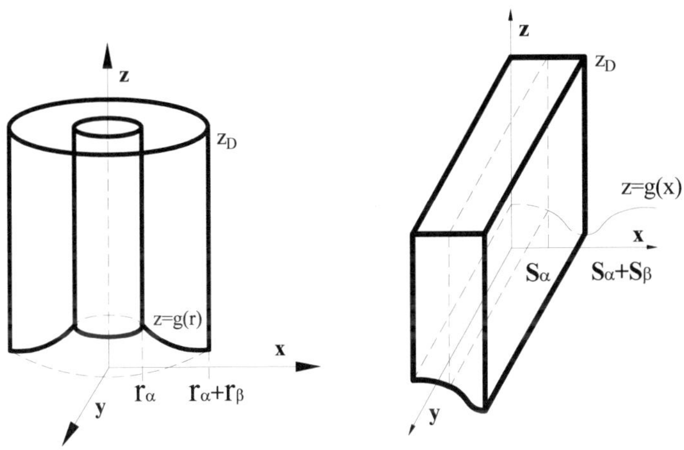

Fig. 19. V - volume, applied in calculation of the entropy production per unit time, Eq. (18); a. for the rod-like structure, $z = g(r)$ - function which describes the s/l interface curvature; z_D - thickness of the diffusion zone, $[m]$; b. for the lamellar structure, $z = g(x)$ - function which describes the s/l interface curvature; z_D - thickness of the diffusion zone, $[m]$.

The solution to Eq. (18) gives the following result:

a. entropy production (associated with the diffusion) for a rod-like structure formation:

$$P_R^D = V_1\, v\left(r_\alpha + r_\beta\right)^{-1} + V_2\, v\left(r_\alpha + r_\beta\right)^{-2} + V_3\, v^2 + V_4\, v^2\left(r_\alpha + r_\beta\right) + V_5\, v^3\left(r_\alpha + r_\beta\right)^2 \tag{20a}$$

b. entropy production (associated with diffusion) for a lamellar structure formation:

$$P_L^D = W_1\, v\left(S_\alpha + S_\beta\right)^{-1} + W_2\, v\left(S_\alpha + S_\beta\right)^{-2} + W_3\, v^2 + W_4\, v^2\left(S_\alpha + S_\beta\right) + W_5\, v^3\left(S_\alpha + S_\beta\right)^2 \tag{20b}$$

$V_n; W_n \quad n = 1,...,5$ - constants contain material parameters, (Wołczyński & Billia, 1996).
The calculations of the entropy production applied to the Zn-Ti system are connected with some experimental observations. According to the observations, for:

- $0 < v < v_1$ irregular L-shape rods (equipped with branches) appear, Fig. 20,
- $v_1 < v < v_1^{'}$ L-shape rods with disappearing branches and regular lamellae coexist, Fig. 21
- $v_1^{'} < v < v_2$ regular lamellae are formed exclusively within the range of growth rates, Fig. 22,
- $v > v_2$ regular rods are stable form within the range of growth rates, Fig. 23.

The characteristic crystal growth rates have the following experimentally determined value: $v_1 \approx 5\,mm/h$; $v_1^{'} \approx 5.8\,mm/h$; $v_2 \approx 10\,mm/h$.

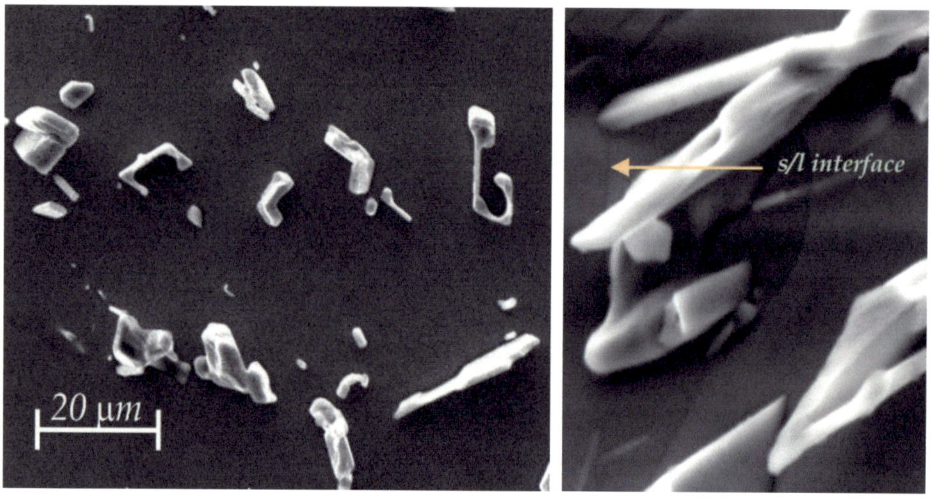

Fig. 20. ((Zn) + $Zn_{16}Ti$) - eutectic morphology in stripes obtained for the growth rate $v < v_1$; a. L-shape irregular rods; b. branching phenomenon with the frozen planar s/l interface.

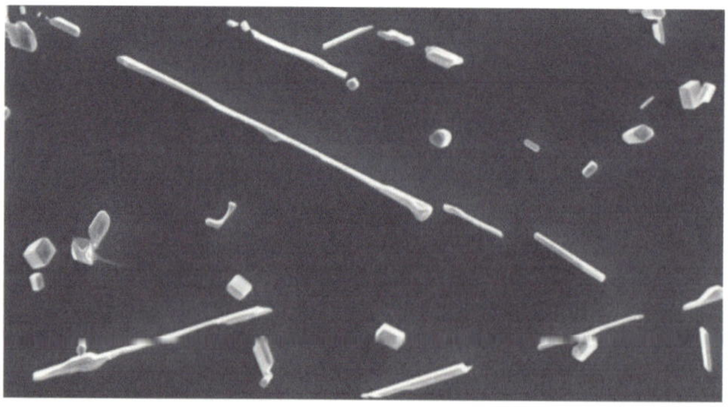

Fig. 21. Coexistence of L-shape rods (disappearing branches) and regular lamellae, $v < v_1^{'}$.

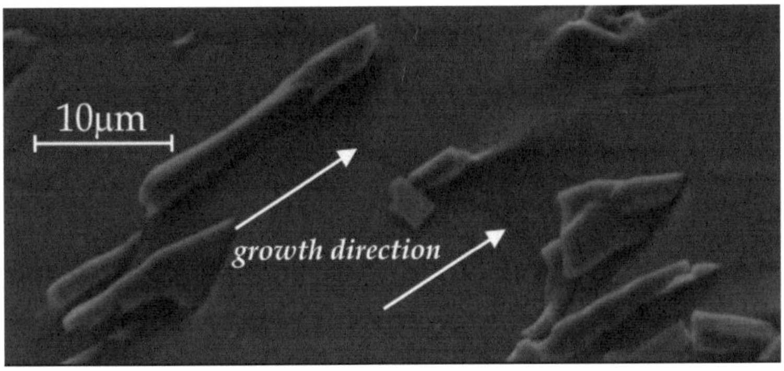

Fig. 22. Lamellar ((Zn) + $Zn_{16}Ti$) - eutectic morphology for the growth rate $v_1^{'} < v < v_2$.

Arrows juxtaposed onto the morphology, Fig. 22, show the parabolic shape of the s/l interface of the (Zn) – non-faceted phase and emphasize the regularity of eutectic structure.

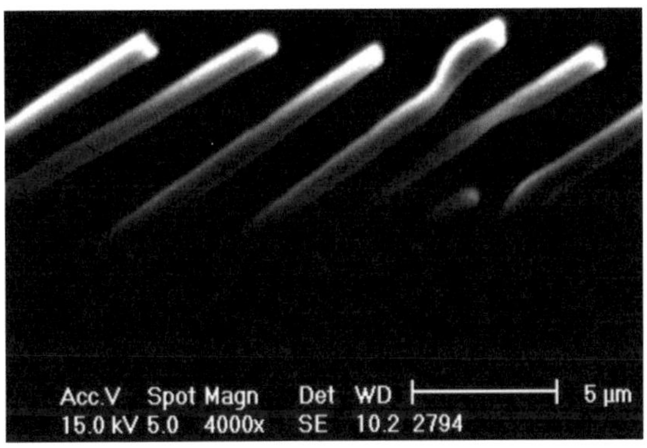

Fig. 23. Regular rods of the $Zn_{16}Ti$ – intermetallic compound obtained for $v > v_2$.

The calculation of the entropy production, P_D, applied to the Zn-Ti system is made with the implementation of the model for rotation of mechanical equilibrium around the triple point of the s/l interface, Fig. 24. The rotation is performed with the increasing growth rate and causes some changes of the s/l interface curvature of the non-faceted (Zn) – phase, Fig. 24.

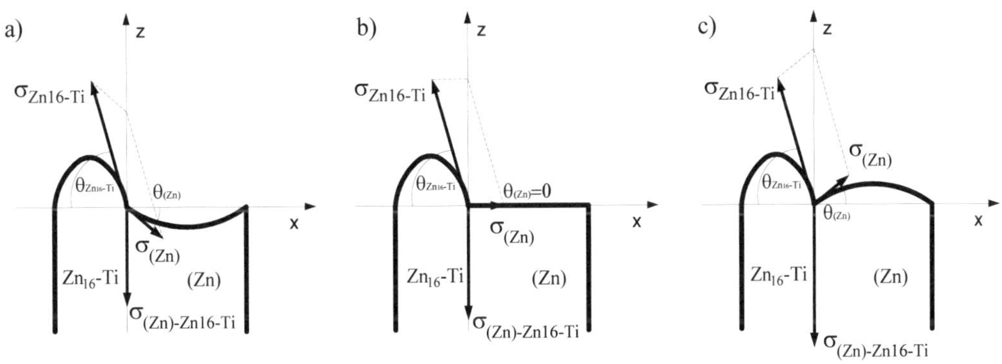

Fig. 24. Rotation of the mechanical equilibrium around the triple point of the s/l interface.

The mechanical equilibrium is established by the parallelogram of the anisotropic specific surface free energies together with the α / β - inter-phase boundary free energy, Fig. 24. As a result of the varying growth rate, the s/l interface curvature changes due to rotating crystallographic orientation. $\sigma_{(Zn)} \equiv \sigma^{L}_{(Zn)}$ and $\sigma_{Zn16-Ti} \equiv \sigma^{L}_{Zn16-Ti}$ - specific surface free energies for (Zn) - phase and $Zn_{16}Ti$ - compound, respectively, $[J / m^2]$; $\sigma_{(Zn)-Zn16Ti}$ - inter-phase boundary free energy, $[J / m^2]$; θ_j - angles, $[^0]$, $j = (Zn), Zn_{16}Ti \equiv \alpha, \beta$
The rotation of the mechanical equilibrium yields some changes of capillarity parameters. This is shown in function of the increasing growth rate, Fig. 25.

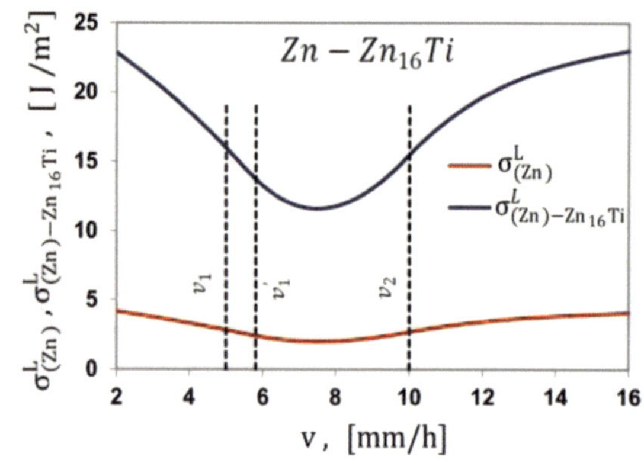

Fig. 25. Changes of the specific surface free energy, $\sigma^{L}_{(Zn)}$, and inter-phase boundary free energy, $\sigma_{(Zn)-Zn_{16}Ti}$; as applied to calculation of the entropy production.

The appearance of the observed structure is accompanied by a proper entropy production, as postulated. The considered solidification occurs under stationary state, therefore, the entropy production, P_D, manifests its minimum. The calculations are made to show the competition between two eutectic regular structures: rod-like and lamellar structure, Fig. 26.

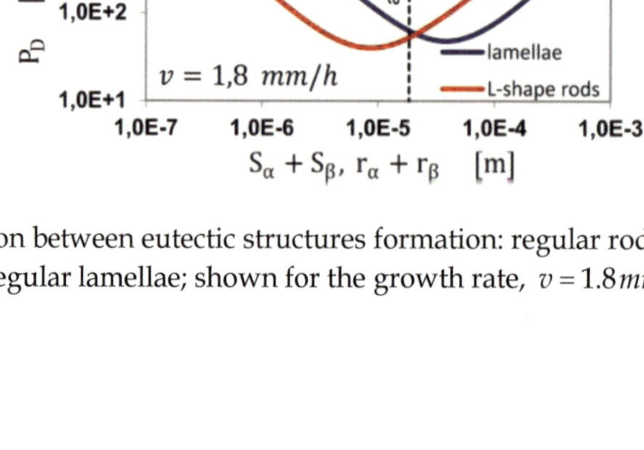

Fig. 26a. Competition between eutectic structures formation: regular rods among irregular L-shape rods and regular lamellae; shown for the growth rate, $v = 1.8 mm / h$; $0 < v < v_1$.

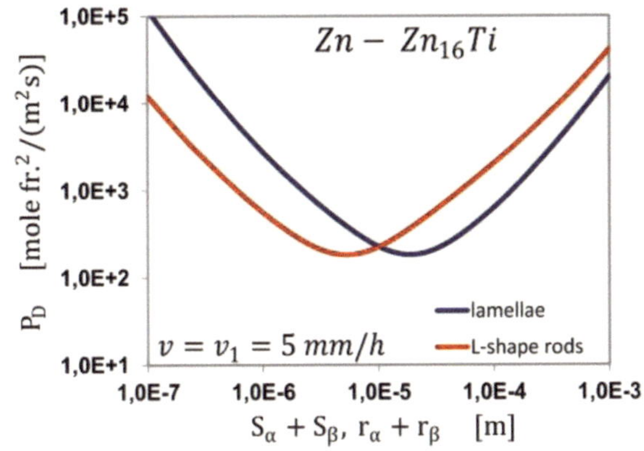

Fig. 26b. Competition between eutectic structures formation: regular rods among irregular L-shape rods and regular lamellae; shown for the first threshold growth rate, $v_1 \approx 5\,mm/h$.

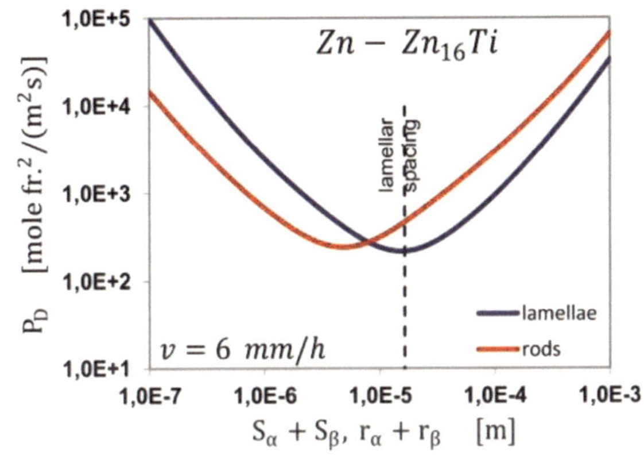

Fig. 26c. Competition between eutectic structures formation: regular lamellae and regular rods; shown for the growth rate, $v = 6mm/h$; $v_1^{'} < v < v_2$.

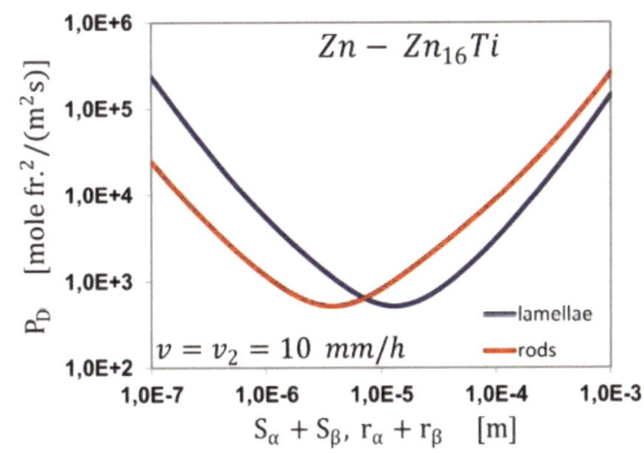

Fig. 26d. Competition between eutectic structures formation: regular lamellae and regular rods; shown for the second threshold growth rate, $v_2 \approx 10mm/h$.

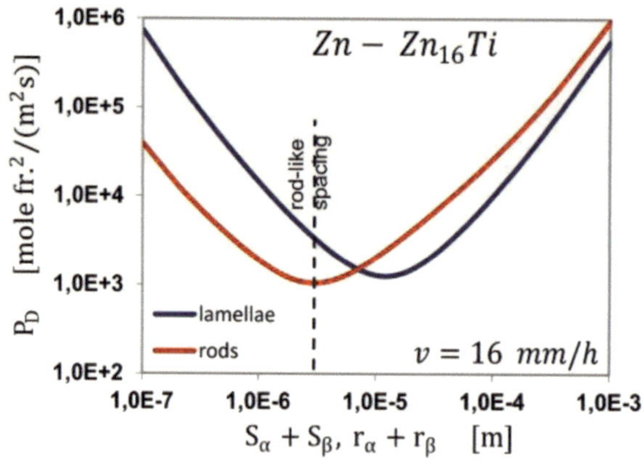

Fig. 26e. Competition between eutectic structures formation: regular lamellae and regular rods; shown for the growth rate; $v = 16 mm/h$, $v > v_2$.

It is visible that the entropy production calculated for the observed structure formation within a proper range of growth rates, locates its minimum lower than the entropy production determined for a competitive structure, Fig. 26.

Next, the calculated entropy productions were minimized in order to compare their minimal values for all studied growth rate ranges: a. $0 < v < v_1$, where the L-shape rods equipped with branches, (irregular eutectic structure) appear, Fig. 20, b. $v_1 < v < v_1^{'}$, where the L-shape rods with disappearing branches and regular lamellae coexist, Fig. 21, c/ $v_1^{'} < v < v_2$, where the regular lamellae are exclusive form, Fig. 22, and d/ $v > v_2$, where the regular rods are created, Fig. 23. The results of calculation are shown in Fig. 27.

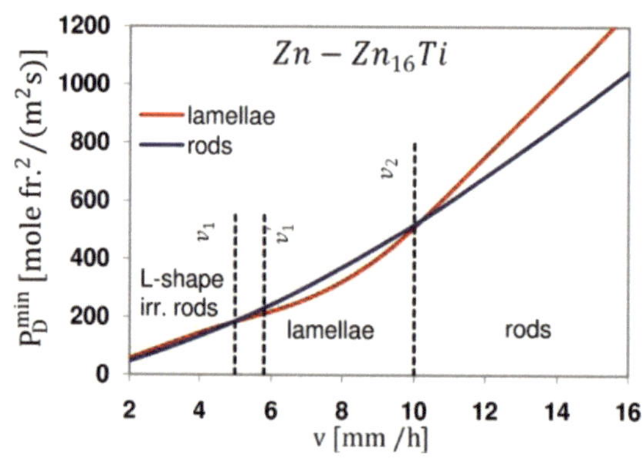

Fig. 27. Comparison of minimal values of the entropy production calculated for both regular rod-like structure formation (formed twice) and regular lamellar structure formation.

The estimation of the average inter-rod spacing for the first rates range: $0 < v < v_1$ requires to apply a definition of the mean value of the investigated spacing.

$$\bar{R} = 0.5R + R_\beta^i + 0.5\lambda_S^i \tag{21}$$

$R = r_\alpha + r_\beta$ - defined within the areas of the regular structure, Fig. 14b, which are formed among the generally irregular structure, $[m]$; R^i_β - radius of the non-coupled rod which manifests an excess protrusion over the s/l interface, $[m]$, (Wołczyński, 1996).

$$\lambda^i_s = 2\pi\left(\Gamma_{(Zn)} / \left(G - m_{(Zn)} G_C\right)\right) \tag{22}$$

where λ^i_s - a wavelength of perturbation which appears at the s/l interface of the (Zn) – non-faceted phase, (as defined in Fig. 12.) $[m]$, with $\Gamma_{(Zn)}$ - *Gibbs-Thomson* capillarity parameter; G - thermal gradient at the s/l interface, $[K/m]$; $m_{(Zn)}$ - slope of the *liquidus* line of the (Zn) – phase, $[K/at.\%]$; G_C - solute concentration gradient at the s/l interface of the (Zn) – phase, $[at.\%/m]$. In the range: $v_1 < v < v_1^{'}$, the following transformation occurs:

$$\lambda^i_s \to R; \quad R^i_\beta \to 0 \quad \text{and finally} \quad \bar{R} \to R \tag{23}$$

It means that irregular rod-like structure transforms into the regular rod-like structure. This transition leads to the local disappearing of branches. Eventually, the regular rods transform into the regular lamellae within the range, $v_1 < v < v_1^{'}$, Fig. 21.

5. Concluding remarks

A pattern selection among the eutectic morphologies is performed through the competition. In the case of the eutectic morphologies which were created within the stripes strengthening the (Zn) single crystal the L-shape irregular rods, regular lamellae or regular rods were revealed. According to the current model, this eutectic structure appeared which was the winner just in the thermodynamic competition. Therefore, it was necessary to determine the minimum entropy production for the appearance of the all studied structures. The formed structure had its minimum entropy production situated lower than the other participant of the competition. So, it has been proved that the regular L-shape rods (existing among the generally irregular L-shape rods) have their minimum entropy production situated lower than the minimum entropy production determined for the regular lamellae growth (when $0 < v < v_1$). However, the regular lamellae are the winner within the growth rates range $v_1^{'} < v < v_2$. Eventually, the regular rods exist exclusively when the growth rate fulfils the following inequality $v > v_2$ since the minimum entropy production calculated for their formation is localized lower than the minimum entropy production determined for the regular lamellae growth, Fig. 27. Both minima being in competition are at the same level for the threshold growth rates, v_1, and v_2, Fig. 26b, Fig. 26d, respectively, (also Fig. 27).

It has been revealed that the leading eutectic phase (the $Zn_{16}Ti$ intermetallic compound) was growing with the d - protrusion, Fig. 8, Fig. 9 and Fig. 10. The protrusion can be determined based on the solute concentration micro-field in the liquid ahead of the s/l interface, Eq. (7), and Eq. (8), but the local mass balance is to be considered, Eq. (10).

In fact, the theoretical definition for the protrusion results from the analysis of the local mass balance (Wołczyński, 2007) and is formulated for the slow solidification as follows:

$$\sum_{n=1}^{\infty} A_{2n-1} \frac{(-1)^{n-1}}{(2n-1)} \left[1 - \frac{S_{\alpha}}{S_{\beta}} \exp\left(-\frac{(2n-1)\pi}{2S_{\beta}} d \right) \right] = 0 \qquad (24)$$

A_{2n-1} - constant in the *Fourier* series. According to the observation of the leading phase protrusion, d, Fig. 8, Fig. 9 and Fig. 10, it is suggested to introduce the revealed protrusion into the structural scheme shown in Fig. 12.

Also, the operating range for the irregular eutectic growth (as defined in Fig. 12) can be justified by some structural observations. These observations are connected with the L-shape rod-like structure revealed for the growth rates range, $0 < v < v_1$. A typical spacing is to be distinguished within the operating range for the irregular eutectic growth:

a. the λ_i - spacing which is associated with the regular lamellar structure formation among generally irregular structure, (or the R - spacing which is associated with the regular rod-like structure formation, Fig. 28a); this spacing is referred to the criterion of the minimum entropy production,

b. the λ_s^i - spacing which corresponds to the maximal destabilization of the s/l interface of the (Zn) - non-faceted phase, Fig. 28b; this spacing (treated as perturbation wave, λ_s^i), is referred to the state of marginal stability which is created at that very moment at the s/l interface.

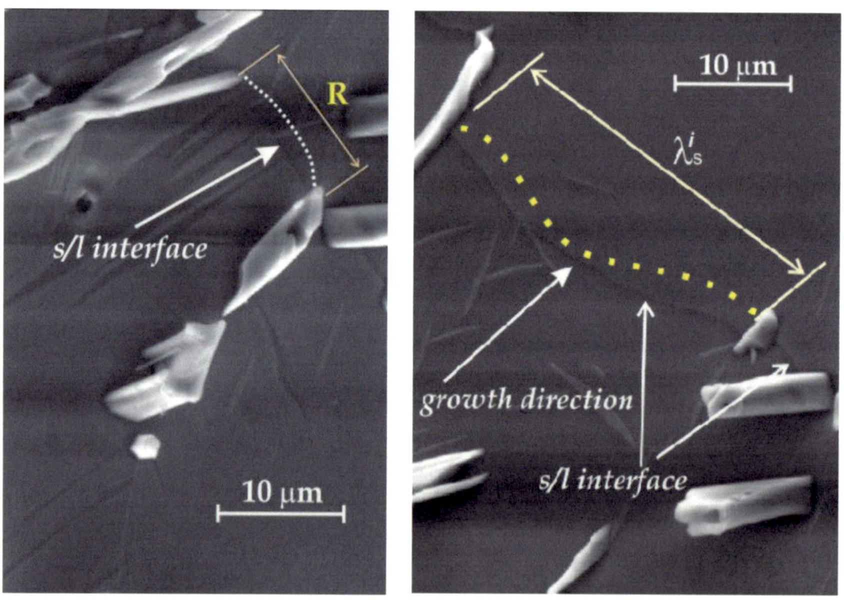

Fig. 28. Selection of the spacing in the irregular eutectic growth; a. λ_i - spacing selected by the minimum entropy production, and b. λ_s^i - spacing selected by the marginal stability.

It is evident that the oscillation of the interlamellar or interrod spacing occurrs in the (Zn) - $Zn_{16}Ti$ eutetcic system. The analyzed structural oscillation is shown schematically in Fig. 29.

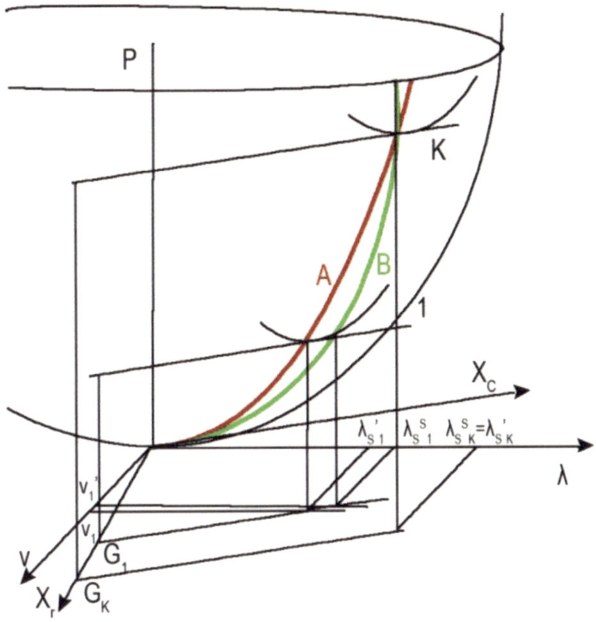

Fig. 29. A paraboloid of the entropy production drawn in function of two thermodynamic forces (X_C, X_r), with the added so-called "technological" coordinate system, (v, λ).

X_C - generalized thermodynamic force associated with the mass transport, $[at.\% / m]$; X_r - generalized thermodynamic force associated with the heat transfer; $[K / m]$; v - crystal growth rate, $[m / s]$; λ - inter-lamellar spacing within the regular eutectic structure, $[m]$
The paraboloid of the entropy production is drawn schematically in the "thermodynamic" coordinate system, that is, in function of two thermodynamic forces. Next, it is assumed (for the simplicity) that the paraboloid does not change its shape when entropy production, P, is calculated for the crystal growth in function of v, λ, Eq. (20b) or v, R, Eq. (20a).
Two trajectories are superposed onto the paraboloid, Fig. 29: a. A - trajectory of local minima of the paraboloid for the regular structure formation, Fig. 28a, /B - trajectory of the marginal stability referred to the maximal destabilization of the s/l interface, Fig. 28b. Both trajectories intersect each other (at K - point) for the critical thermal gradient, G_K, at which an oscillation disappears and irregular structure is completely transformed into regular eutectic structure. However, for a given condition of solidification (crystal growth rate, v_1, and thermal gradient, G_1), the destabilized s/l interface moves slower, with the $v_1^{'}$ - rate, ($v_1^{'} < v_1$), and the perturbation wave, λ_s^{i}, is equal to λ_{s1}^{s}. On the other side, the regular structure appears at the v_1 - rate and involves the $\lambda_i = \lambda_{s1}^{'}$ - spacing formation, Fig. 29. Thus, the discussed oscillation occurs between both trajectories as shown for v_1, G_1, Fig. 29.
The oscillation of the eutectic spacing, Fig. 28, is observed in the case of the irregular eutectic structure formation formed within the stripes strengthening the (Zn) - single crystal for the growth rates range equal to $0 < v < v_1$, as mentioned.

The discussed oscillation can be illustrated on the parabola of entropy production. Such a parabola is to be created by an intersection of the paraboloid, Fig. 29, by the plane given for the imposed thermal gradient, $G = const.$, (Prigogine, 1980), as shown in Fig. 30.

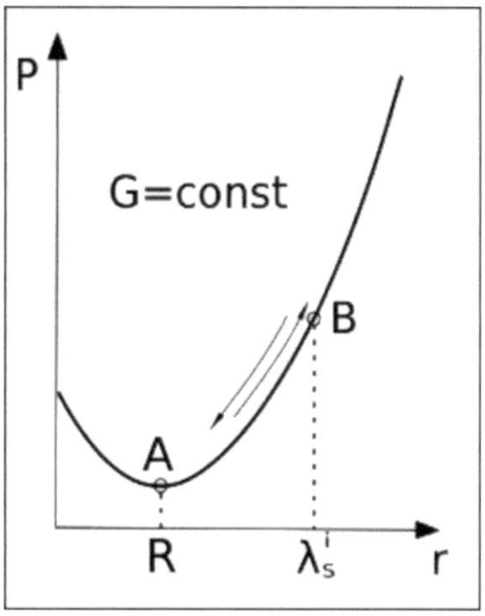

Fig. 30. A parabola of the entropy production for the eutectic rod-like structure formation.

It results from the current model, Fig. 30, that:

a. the regular part of the generally irregular eutectic structure is formed at the minimum entropy production localized at the A - parabola minimum which performs the role of attractor for the eutectic system,
b. the maximal destabilization at the s/l interface of the non-faceted eutectic phase is observed for the B - marginal stability of the eutectic system which corresponds to the branching phenomenon (as visible in Fig. 12) and to maximal deviation from the attractor.

6. Acknowledgement

The financial support from the Polish Ministry of Science and Higher Education (MNiSW) under contract N N 508 480038 is gratefully acknowledged.
Sincere gratitude is expressed to Dr Grzegorz Boczkal. Without his constant support and great patience with experimental observations this chapter would never have been written.

7. References

Atasoy, O. (1984). Effects of Unidirectional Solidification Rate and Composition on Inter-particle Spacing in Al-Si Eutectic Alloy, *Aluminium*, Vol. 60, pp.109-116

Boczkal, G.; Mikułowski, B.: & Wołczyński, W. (2010). Oscillatory Structure of the Zn-Cu-Ti Single Crystals, *Materials Science Forum*, Vol. 649, pp. 113-118

Brody, H.D.; & Flemings, M.C. (1966). Solute Redistribution in Dendritic Solidification, *Transactions of the Metallurgical Society of the AIME*, Vol. 236, pp. 615-624

Cupryś, R.; Major, B.; & Wołczyński, W. (2000). Transition of Flake into Fibre Structure in Eutectic Al-Si, *Materials Science Forum*, Vol. 329/330, pp.161-166

Davies, V.L. (1964). Mechanisms of Crystallization in Binary Eutectic Systems, *Journal of the Institute of Metals*, Vol. 93, pp. 10-14

Ebeling, R.; & Ashby, M.F. (1966). Dispersion Hardening of the Copper Single Crystal, *Philosopher Magazine*, Vol. 13, pp. 805-834

Elliott, R. (1977). Eutectic Solidification, *International Metals Reviews*, Vol. 219, pp. 161-186

Fisher, D.J.; & Kurz, W. (1980). A Theory of Branching Limited Growth of Irregular Eutectics, *Acta Metallurgica*, Vol. 28, pp. 777-794

Glansdorff, P.; & Prigogine, I. (1971). *Thermodynamic Theory of Structure, Stability and Fluctuations*, Wiley - Interscience, 306 pages, John Wiley & Sons, (Ed.), ISBN: 0 471 30280 5, London, UK - New York, USA - Sydney, Australia - Toronto, Canada

Jackson, K.A.; & Hunt, J.D. (1966). Lamellar and Rod Eutectic Growth, *Transactions of the Metallurgical Society of the AIME*, Vol. 236, pp. 1129-1142

Kjelstrup, S.; & Bedeaux, D. (2008). *Non-Equilibrium Thermodynamics of Heterogeneous Systems*, World Scientific Pulishing Co. Ltd., 434 pages, M. Rasetti, (Ed.), ISBN: 13-978-981-277-913-7 and ISBN: 10 981-277-913-2, New Jersey, USA - London, UK - Singapore, - Beijing, China - Shanghai, China - Hong Kong - Taipei, Taiwan- Chennai, India

Lesoult, G.; & Turpin, M. (1969). Etude Theorique sur la Croissance des Eutectiques Lamellaires, *Revue Scientifique de la Revue de Metallurgie*, Vol. 66, pp. 619-631

Murray, J. (1990). Zn-Ti Phase Diagram for the Stable Equilibrium, In: *Binary Alloy Phase Diagrams*, T. Massalski, (Ed.), ASM International, pp. 3500-3502, ISBN: 0-87170-406-4, Park Ohio, USA

Prigogine, I. (1968). *Introduction a la Thermodynamique des Processus Irreversibles*, (Monographies DUNOD), 160 pages, John Wiley & Sons, (Ed.), Paris, France

Prigogine, I. (1980). *From Being to Becoming: Time and Complexity in the Physical Sciences*, The Maple-Vail Book Manufacturing Group, 286 pages, W.H. Freeman and Company, (Ed.), ISBN: 0-7167-1107-9 and ISBN: 0-7167-1108-7, USA

Scheil, E. (1942). Über die Eutektische Kristallisation, *Zeitschrift für Metallkunde*, Vol. 34, pp. 70-80

Steen, H.A.H.; & Hellawell, A. (1975). The Growth of Eutectic Silicon - Contributions to Undercooling", *Acta Metallurgica*, Vol. 23, pp. 529-536

Toloui, B.; & Hellawell, A. (1976), Phase Separtion and Undercooling in Al-Si Eutectic Alloy - The Influence of Freezing Rate and Temperature Gradient, *Acta Metallurgica*, Vol. 24, pp. 565-572

Wołczyński, W. (1996), Thermodynamics of Irregular Eutectic Growth, *Materials Science Forum*, Vol. 215/216, pp. 303-312

Wołczyński, W. (2000). Back-Diffusion Phenomenon during the Crystal Growth by the Bridgman Method, In: *Modelling of Transport Phenomena in Crystal Growth*, J.S. Szmyd & K. Suzuki, (Ed.), pp. 19-59, WIT *PRESS* ISBN: 1-85312-735-3, Ashurst Lodge, Southampton, UK - Boston, USA

Wołczyński, W. (2007). Concentration Micro-Field for Lamellar Eutectic Growth, *Defect and Diffusion Forum*, Vol. 272, pp. 123-138

Wołczyński, W. (2010). Lamella / Rod Transformation as described by the Criterion of the Minimum Entropy Production, *International Journal of Thermodynamics*, Vol. 13, pp. 35-42

Wołczyński, W.; & Billia, B. (1996). Influence of Control and Material Parameters on Regular Eutectic Growth and Inter-Lamellar Spacing Selection, *Materials Science Forum*, Vol. 215/216, pp. 313-322

Zarubova, N.; & Sestak, B. (1975). Plastic Deformation of Fe-3 wt.% Si Single Crystals in the Range from 113 to 473 K, *Physica Status Solidi*, Vol. 30, pp. 365-374.

10

Development of 2″ AlN Substrates Using SiC Seeds

O.V. Avdeev et al.*
Nitride Crystals Ltd., Saint-Petersburg,, 194156
Russia

1. Introduction

The unique properties of the group III-nitrides (Edgar et al., 1999; Jain et al., 2000; Kasap & Capper, 2006) make them the best semiconductor material for

- optoelectronic devices emitting light in the visible and UV spectral ranges (Orton & Foxon, 1998), including sources for general illumination (Craford, 2005; Liu, 2009; Miyajima et al., 2001; Nakamura et al., 2000; Schubert & Kim, 2005; Schubert, 2006; Taguchi, 2003; Zukauskas et al., 2002),
- photodetectors for these spectral ranges, including solar-blind UV detectors,
- high power/high frequency electronic devices capable of operating at high temperatures and in harsh environment (Bennett et al., 2004; Morkoc, 1998; Pearton et al., 2000; Shur, 1998; Skierbiszewski, 2005; Xing et al., 2001).

To fully exploit the potential of the group III-nitrides in optoelectronics and communication technology, two problems are to be solved:

1. difficulties of doping group III-nitrides, especially attaining the high-level p-doping - in 1999 AlN was even called an "undopable" material (Fara et al., 1999) (p-doping is also a problem for other wide bandgap semiconductors - oxides such as ZnO and chalcogenides such as ZnSe;) and
2. the lack of large high crystalline quality native substrates with required electrical properties.

1.1 Substrates for III-nitride devices

The requirements for a good substrate are:

- lateral and vertical[1] lattice matching,
- thermal expansion coefficient (TEC) matching,

---

*T.Yu. Chemekova, E.N. Mokhov, S.S. Nagalyuk (Nitride Crystals Ltd., Saint-Petersburg,, 194156, Russia), H. Helava, M.G. Ramm (Helava Systems Inc., 181 E Industry Court, Site B, Deer Park, NY 11729, USA), A.S. Segal (STR Group - Soft-Impact Ltd., Saint-Petersburg, 194156, Russia), A.I. Zhmakin (A.F. Ioffe Physical Technical Institute, RAS, Saint-Petersburg, 194021, Russia) and Yu.N. Makarov (Nitride Crystals, Inc., Richmond VA 23238, USA)

[1] A vertical lattice mismatch generates additional crystalline defects by upsetting the epitaxial layers, including inversion domain boundaries and stacking faults (Kasap & Capper, 2006).

- chemical compatibility,
- large size ("a size agreeable to an industry" (Schujman & Schowalter, 2010) - at least 2″),
- affordable price.

Sometimes additional features are desirable, e.g., the cleavability for laser diodes or the high electrical resistivity for field effect transistors (FETs).
A substrate determines the crystal orientation, polarity, polytype, surface morphology; the difference in the chemical composition of the substrate and the epitaxial layer could lead to the contamination of the layer with substrate elements.
The lattice mismatch for the semiconductor layer with the lattice constant a is measured by the misfit parameter $f_m = (a - a_s)/a_s$, where a_s is the lattice constant of the substrate. It leads to misfit strain in the grown layer that should be accommodated if the layer thickness exceeds the critical one[2] (according to either the energy minimization theory of Frank-Van der Merwe (Frank & van der Merwe, 1949) or the force balance theory of Matthews-Blakeslee (Matthews & Blakeslee, 1974), see, for example, (Jain et al., 1997; Jain & Hayes, 1991; Zhmakin, 2011b)) via the introduction of misfit dislocations, the modulation of the free surface profile (waviness (Freund, 1995)), the composition modulation (Seol et al., 2003) or the interdiffusion between the layer and the substrate (Lim et al., 2000); formation of the V-grooves and the random stacking faults (Cho et al., 2000) and the mosaic structure of slightly misoriented subgrains in the epitaxial films (Srikant et al., 1997) are also observed. These mechanisms could be cooperative as well as competitive (Hull et al., 2002). In addition to the misfit strain, surface free energies and interface energies are the factors that determine the growth mode (Frank-van der Merve, Stranski-Kristanov, Volmer-Weber, polycrystalline, columnar) (Gilmer et al., 1998; Wadley et al., 2001). Nitride layers grown on the substrates with the large lattice mismatch can also include twin crystals (Matsuoka, 2005; Rojas et al., 1998).
Homoepitaxy provides a better control over the surface morphology and the defect density compared to the heteroepitaxial growth (Cao et al., 2007; Kirchner et al., 1999). The best substrates for GaN-based devices are GaN and AlN. AlN ones are superior for the structures with high Al content, e.g., UV devices (AlN is transparent down to about 200 nm), FETs and both vertical and lateral Schottky and p-i-n diodes (Luo et al., 2002) (unintentionally doped (UID) AlN crystals are insulators[3] in contrast to GaN ones that have a high background

[2] The critical thickness is the equilibrium entity, thus metastable metamorphic layers (Hull & Stach, 1996) of higher thickness could exist at low temperatures: e.g., recently an n-type AlGaN layer on the bulk AlN substrate with a thickness more than an order of magnitude greater than the critical value was reported (Grandusky et al., 2009).

[3] The concentration of the residual oxygen is high for the sublimation grown AlN crystals (Herro et al., 2006); however, most of oxygen is incorporated not as a substitutional shallow donor, but as impurity-forming complexes with point and extended defects located deep in the gap (Freitas, 2005; Schultz, 2010; Slack et al., 2002). Below a critical concentration, the presence of oxygen leads to the formation of Al vacancies; at higher concentrations a defect based on octahedrally coordinated Al atoms is formed (Kasap & Capper, 2006).
The effect of the oxygen contamination is more severe on the thermal conductivity of AlN due to the phonon scattering on oxygen defects whose nature does depend on the oxygen concentration (Kasap & Capper, 2006; Kazan et al. , 2006a). For example, the substitutional defect associated with aluminum vacancy serves as a very efficient center for the phonon scattering (Rojo et al., 2001).

concentration of free electrons (Freitas, 2005))[4]. AlN substrates are also the best choice for for nitride-based power devices due to the high thermal conductivity.

The high thermal conductivity of AlN is attributed to the simplicity of the atomic structure, it's low atomic mass, the existence of strong covalent bonding, and low anharmonicity within the lattice (Dinwiddie et al., 1989). Note that the thermal conductivity of AlGaN alloys is smaller than that of pure GaN, to say nothing of AlN, due to the high degree of the disorder in the system (Liu & Balandin, 2005).

The cooling of devices on AlN substrates could be effectively better than of that on SiC substrates that have higher thermal conductivity since the former do not require a buffer layer between the device structure and the substrate that is highly defective and thus presents considerable thermal resistance (Schowalter et al., 2004).

AlN crystals, possessing the highest surface and bulk acoustic wave velocity known for piezoelectric materials, a small temperature coefficient of delay, and a large piezoelectric coupling coefficient, are needed for the surface acoustic wave (SAW) devices[5]. Aluminum nitride for such applications has an evident advantage - AlN crystals could be used up to 1150 °C (Fritze, 2011) (conventional piezoelectric materials are not suited for high-temperature applications, for example, α-quartz αSiO_2 undergoes a phase transformation at 573 °C while lithium niobate $LiNbO_3$ and lithium tetraborate $Li_2B_4O_7$ decompose at 300 °C). AlN crystals are also considered as perspective material for the integration of semiconductor electronic and SAW devices (Cleland et al., 2001).

1.2 Foreign substrates

In the lack of native GaN or AlN substrates, a number of foreign ones are used in practice leading to low quality of the grown epitaxial layers in terms of large threading dislocation density, inversion domain boundaries, partial dislocations bounding stacking faults, bowing of the structure, cracks (Cherns, 2000; Speck, 2001; Speck & Rosner, 1999; Wu et al., 1996; 1998). The near-interfacial region of the film could contain a mixture of cubic and hexagonal GaN (Wu et al., 1996).

The island coalescence is considered as the main mechanism, at least during MBE growth, of the dislocation appearance (Waltereit et al., 2002). Similar conclusion is reached in the study of the early stages of HVPE epitaxy - all the edge and most of the mixed threading dislocation are originated from the island junctions of the high temperature (HT) buffer layer (Golan et al., 1998). An increase of the smoothness was also observed and attributed to the subsequent surface and bulk reconstruction. A strong correlation between the final threading dislocation

---

[4] For example, in GaN crystals grown by the hydride (halide) vapour-phase epitaxy (HVPE) an unintentional n-type conductivity originates from the background doping by silicon and oxygen from the quartz elements of the reactor or from the process gases (Paskova et al., 2010). Fe compensating doping allows one to achieve semi-insulating properties of the layer in this growth method (Vaudo et al., 2003) with the the lowest free carrier concentration reported so far $5 \cdot 10^{13}$ cm^{-3} (Paskova et al., 2009).
However, the performance parameters of AlGaN/GaN heterostructure field-effect transistor (HFET) structure formed by depositing a layer of AlGaN on a relatively thick semi-insulating GaN epitaxial layer are greatly improved by replacing a GaN epitaxial layer with a highly resistive AlN epitaxial layer in the device structure, in particularity, parasitic conduction in the GaN epilayer, leakage current through the GaN epilayer, and the channel electrons spillover into the GaN epilayer have been completely eliminated and the drain current collapse has been reduced (Fan et al., 2006).

[5] Even polycrystal AlN films obtained, for example, by low-cost sputtering are of great interest since the corresponding SAW devices could operate at frequencies above 1GHz (Epelbaum et al., 2004). However, the properties of elasic waves in polycrystal depend on the grain size and orientation.

density in the thick films and the initial island density in the high temperature buffer layer was registered in the MOCVD grown GaN layers on the sapphire substrate (Fini et al., 1998). Early stages of the film growth frequently occur via coherent (dislocation-free) island formation due to the existence of the energy barrier to the introduction of dislocations; dislocations arise at the island edges for large enough islands since those edges are characterized by the large stress (Eisenberg & Kandel, 2002).

A review of the study of the morphology of the heteroepitaxially grown GaN layers using the scanning tunnel microscopy could be found in Ref. (Bakhtizin et al., 2004).

Threading dislocation are electrically active with the charge density of approximately 2 elementary charges per nanometer dislocation length (Müller et al., 2006); surface depressions caused by the high strain-energy density near dislocations are observed on the surface of GaN films (Heying et al., 1999). Edge and mixed (screw/edge) dislocations in nitride materials act as nonradiative recombination centers and as conduits for charge transport resulting in leakage currents and breakdown (Amano et al., 2003; Davis et al., 2002).

While high threading dislocation density is acceptable for light emitting diodes (LEDs)[6] — usually their small effect on the performance is attributed to the potential fluctuations related to the indium composition fluctuations (resulting from the poor In incorporation in the epitaxial layer during the growth or from the phase separation due to the large miscibility gap (I-hsiu & Stringfellow, 1996; Karpov, 1998; Korcak et al., 2007)) observed in the active layers made from the ternary solid solution InGaN and in the corresponding QWs (Christen et al., 2003; Limb, 2007; Mukai et al., 2001) that provide the localization of carriers and reduce their in-plane diffusion to the non-radiative recombination centers — the high dislocation density, however, is fatal for laser diodes and power transistors; even for the LEDs including indium-free violet ones (Usikov et al., 2003) it leads to the drop in the efficiency and is the main factor of the device failure (Karpov, 2009; Roycroft et al., 2004).

The most important foreign substrates are sapphire Al_2O_3 and silicon carbide (6H-SiC or 4H-SiC). The list of other substrates used for group III nitride epitaxy includes GaAs, AlAs, GaP, ZnO, MgO, Mg(Zn) Fe_2O_4, $LiGaO_2$, $LiAlO_2$, SiO_2, $NdGaO_3$, $ScAlMgO_4$, ZrO_2, Mo, glass, quartz glass (Jain et al., 2000; Kukushkin et al., 2008; Miskys et al., 2003).

Sapphire is a good choice for the nitride layers from the crystallographic point of view: crystal orientations of sapphire and gallium nitride grown on c-plane (0001) are parallel, with the unit cell of GaN being rotated by 30° about c axis with respect to the unit cell of sapphire; the $(1\bar{1}00)$ axis of GaN is parallel to the $(1\bar{2}\bar{1}0)$ sapphire axis (Jain et al., 2000).

However, sapphire is inferior (due to the high TEC mismatch and the lattice mismatch being about 15% that corresponds to a critical thickness of less than a monolayer (Jain et al., 2000) compared to 3.5 % for 6H SiC) in the quality of the grown nitride layers leading to higher threading dislocation density.

The insulating nature of sapphire restricts the device architecture, excluding the vertical die design. Bipolar devices fabricated on the sapphire substrates should employ mesa structures with the lateral geometry of the anode and cathode electrodes (with both contacts placed in the same plane) and are especially prone to the current crowding effect - a nonhomogenous in-plane distribution of the current density, especially at the vicinity of the edge of the metal electrodes.

This effect is one of the limiting factors of the efficiency of light emitting diodes, but it is also of concern in other semiconductor devices, e.g., bipolar transistors and Schottky diodes (Chen

[6] Threading dislocations with vacancies at Ga sites (Hovakimian, 2009) acting as non-radiative recombination centers (Choi et al., 2004) present more serious problem for GaN/AlGaN system than for GaN/InGaN one (Akasaki & Amano, 2006).

et al., 2007; Paskova et al., 2010). The current crowding can lead to the localized overheating and the formation of thermal hot spots, lowering the internal quantum efficiency of LED and affecting the series resistance of the diode as well as can result in a premature device failure (Bogdanov et al., 2008; Bulashevich et al., 2007; Evstratov et al., 2006).

Another weak point of sapphire as a substrate for the light emitting diodes is a low refractive index of sapphire in comparison to III-nitrides that leads to waveguiding of light emitted in the active region of the LED and thus to the decrease of the light extraction efficiency (Karpov, 2009).

Nevertheless, at present time over 80 % of LEDs are fabricated on sapphire 2″ substrates (Russel, 2006) due to the low cost and availability.

A relatively new approach to the effective use of the sapphire substrates that has been developed for the nitrides light-emitting diodes (LEDs) growth is the nano-patterning of the surface of the substrate: it is expected that the strain induced by the lattice misfits between the GaN epilayers and the sapphire substrates can be effectively accommodated via the nano-trenches (Yan et al., 2007) (the patterning also enhances light extraction from the device (Zhmakin, 2011a)).

A few years ago a possibility to directly obtain aluminum nitride layers on the sapphire by nitriding α-SiO_2 surface on large (2 inch) substrate by chemical reaction with N_2 - CO gas mixture under carbon-saturation conditions was demonstrated (Fukuyama et al., 2006).

The drawback of SiC substrates is the high cost and the surface roughness on the scale of a few atomic bilayers, leading to a stacking-mismatch defects at the interface between SiC and III-nitride layers in the form of a 2H-polytype (Karpov, 2009; Potin et al., 1998).

Reducing the high cost of SiC substrates while conserving the high thermal conductivity is possible by inserting a thin film of monocrystalline SiC onto polycrystalline SiC (Pecz et al., 2008) which has the thermal conductivity lower than that of single crystal but close to it (Franko & Shanafield, 2004). Polycrystalline SiC is produced by the low cost process of the SiC powder sintering (Medraj et al., 2005). Bonding the thin sapphire layer to polycrystalline AlN (P-AlN) retains the epitaxial template for the growth while improving the thermo-mechanical properties of the substrate (Pinnington et al., 2008). Another example of composite substrates is SiC semiconductor-on-insulator (SOI) structures that are the low cost 3C-SiC (111) substrates fabricated by the conversion of silicon SOI by carbonization of the surface —a reaction with propane and hydrogen at high temperatures (Steckl et al., 1997).

Epitaxy of InGaN/GaN layers on foreign substrates with the large lattice mismatch requires a preliminary growth of a buffer layer or multiple buffer layers (Miskys et al., 2003; Xi et al., 2006). The common example of the use of a double buffer layer is the growth of AlGaN with a starting low-temperature AlN layer on the GaN templates on sapphire which themselves contain the low-temperature GaN buffer layer (Kuwano et al., 2004) or an introduction of an insulating AlN sub-buffer layer on the semi-insulating SiC substrate under the GaN buffer layer (Shealy et al., 2002).

The buffer layer frequently has an amorphous-like structure with small crystallites providing the very smooth morphology (Matsuoka, 2005). AlN is frequently used as the buffer layer material for sapphire and SiC substrates; buffer layers from GaN, AlGaN (including graded AlGaN conducting buffer layers (Moran et al., 2000)), ZnO (Jain et al., 2000; Ougazzaden et al., 2008) or platinum nanocluster (Oh et al., 2009) are used for sapphire while HfN buffer layers for Mo substrates (Okamoto et al., 2009).

The threading dislocation density could be greatly reduced using the lateral epitaxial overgrowth (LEO, the term ELOG is also used) developed in 1990s (Beaumont et al., 1998; Davis et al., 2002; Nam et al., 1997; 1998). This method involves patterning of the substrate

and the initial vertical growth in the "windows" etched in the mask with subsequent growth of material of higher quality laterally over the mask patches. The modification of LEO - the so called pendeo-epitaxy (Davis et al., 2001; Zheleva et al., 1999) - is free of the two major LEO drawbacks (cracking of thick layers and void formation on the top of mask stripes (Wang et al. , 2001)). The crystallographic tilt in the overgrown material is also significantly reduced and the diffusion of impurities from the mask is avoided (Davis et al., 2002).
The extremely low threading dislocation density could be obtained via two sussessive ELO steps with the mask of the second step positioned over the windows etched in the mask during the first step (Pearton et al., 2000).
Growth of GaN or AlN directly on Si surface usually results in the formation polycrystalline films, probably due to the prior formation of amorphous SiN_x layer (Davis et al., 2001a). The large lattice (17%) and TEC (33%) mismatch between Si and GaN cannot be fully accommodated by an AlN buffer layer (Lin et al., 2008), still, involving additionally nitridation (Yamabe et al., 2009) or carbonization of the surface or SiC coating (Kukushkin et al., 2008), some GaN device grown on Si substrates such as HEMTs show an acceptable performance.
In MBE the best results (an order of magnitude decrease of the threading dislocation density in GaN epilayers) are obtained when the growth is initiated by exposing the Si substrate to NH_3 first (Louarn et al., 2009). Usually Si (111) surface (Davis et al., 2001a) is used while an integration with Si microelectronics requires growth on Si (100). In the GaN layers grown on (111) Si substrate inclusions of the cubic nitride phase are frequently observed (Jain et al., 2000).

1.3 Templates & *pseudo bulk* substrates

Considerable efforts were directed to the development of *templates* (the foreign substrates with the deposited thin nitrides layers) (Gautier et al., 2008; Miskys et al., 2003) sometimes referred to as *MOVPE-derived GaN substrates* (Davis et al., 2002) and *pseudo bulk* or *freestanding* nitride substrates (Lee et al., 2004; Nikolaev et al., 2000; Weyers et al., 2008) obtained by separation of the thick nitride layers from the sacrificial substrate after the growth (by laser-assisted lift-off (LLO) (Lee et al., 2004; Paskova et al., 2004) or by etching, e.g, in aqua regia for GaAs substrates) or during the growth on sapphire substrates with patterned GaN seeds by spontaneous self-separation (Tomita et al., 2004).
The quality of the epitaxial layers grown on the templates in comparison to those grown on the basis substrate is under discussion. Recently Ashraf et al. (Ashraf et al., 2008) have used a number of characterization techniques (diffraction interference contrast optical microscopy, scanning electron microscopy, micro-Raman scattering, X-ray diffraction) to asses the quality of the thick GaN layers grown by HVPE directly on the sapphire substrate (with the optimal nucleation layer deposition at low temperatures and low pressures on the nitridated surface) and on the GaN/Al_2O_3 templates prepared by GTS-metalorganic chemical vapour deposition (MOCVD) process (Grzegorczyk et al., 2005) (GTS stands for the gallium treatment step which is a long exposition of the substrate surface to TMGa at high temperature). The authors found that the layers did not significantly differ in the surface morphology and the structural quality, but the layers grown on the MOCVD templates suffered from cracking in few cases while no cracking occurred in the HVPE layers directly grown on sapphire.

1.4 Epitaxial layers and devices on single-crystal native III-nitride substrates

Characterization of AlGaN epilayers with 40 and 50 % concentration of aluminum grown on the single-crystal AlN substrates by Migration Enhanced Metal Organic Chemical Vapour Deposition (MEMOCVD) using the observation of atomic steps in atomic force microscope scans of epilayers and the measurements of FWHM of X-ray diffraction curves demonstrated an excellent crystallographic quality of the epilayers, the dislocation density of AlGaN layers was estimated to be in mid 10^6cm^{-2} range (Schowalter et al., 2006). The comparison of photoluminescence of the GaN layers deposited on the Al face and on the N face of the single-crystal AlN substrate showed that in the former case photoluminescence is consistent with that of the homoepitaxial Ga-face GaN while in the latter an existance of the tail localized states was found (Tamulatis et al., 2003).

The studies of deep-UV emission of AlGaN quantum wells (Gaska et al., 2002) as well as AlGaN UV (Nishida et al., 2004a; Ren et al., 2007; Xi et al. , 2006a) and InGaN MQW green (Cartwright et al., 2006) LEDs grown on the bulk AlN substrates and blue and UV LED on the bulk GaN substrates (Cao et al., 2004; Cao et al. , 2004a; Du, Lu, Chen, Xiu, Zhang & Zheng, 2010) as well as cyan and green LEDs grown on *a*-plane (Detchprohm et al., 2008) and *m*-plane (Detchprohm et al., 2010) GaN bulk substrates prove the superiority of the native substrates, e.g., the luminescence intensity of the quantum well grown on bulk AlN was higher that that of the quantum well grown on SiC by a factor of 28, the noticeable improvement over LEDs grown on sapphire in device impedance and thermal characteristics (Ren et al. , 2007a), the reduction in current-voltage differential resistance and in turn-on voltage (Paskova et al., 2010). The emission spectrum of AlGaN-based UV-LEDs on a bulk AlN substrate under the high current injection is much more stable than that of LEDs fabricated on the conventional substrate (Nishida et al., 2004a;b). Recently J. J. Grandusky et al. have demonstrated mid-UV LED fabricated from pseudomorphic layers on the bulk AlN substrates (Grandusky et al., 2010).

High-quality green (Miyoshi et al., 2009) and violet and near-UV (Perlin et al., 2005) laser diodes have been fabricated on the bulk GaN substrates, in the latter case substrate were grown by the HNPSG.

Studies of high-electron mobility transistors (HEMTs) with the ALGaN channel[7] grown on different substrates also demonstrate the superiority of the single crystal AlN substrates (Hu et al., 2003), in particularity, the use of AlN substrates improved the crystalline quality of the AlGaN layer and lowered the sheet resistance of the 2-dimensional electron gas (Hashimoto et al., 2010).

The substrates cut from the bulk crystals along the specific crystallographic plane can have different orientation (polar, semipolar or nonpolar) (Lu et al., 2008)[8], enhancing the freedom of the device design (Mymrin et al., 2005; Schowalter et al., 2006; Stutzmann et al., 2001): an engineer, using the spontaneous and piezoelectric polarization that is a nonlinear function of the strain and the composition of nitride materials, can tailor the surface and interface charges

[7] The channel layer substitution of a wider bandgap AlGaN for a conventional GaN in high electron mobility transistors is an effective method of drastically enhancing the breakdown voltage (Nanjo et al., 2008).

[8] For a long time attempts to grow the nitride epitaxial layers on the nonpolar planes were unsuccessful producing the low quality films (Karpov, 2009). However, recently I. Satoh et al. (Satoh et al., 2010) demonstrated the possibility to fabricate the non-polar AlN substrate by heteroepitaxial growth on *m*-plane SiC substrates.

Although the dislocation density of the epitaxial layers grown on *m*-plane was an order of magnitude higher than that on *c*-plane substrate, the emission intensity was 25 times greater, evidently due to the elimination of the spatial separation of the electron and holes wavefunctions in the quantum wells induced by the built-in polarization field that leads to the decrease of the radiation recombination rate - see, for example, (Chakraborty et al., 2005; McAleese et al., 2006; Ram-Mohana et al., 2006).

High efficiency non-polar *m*-plane InGaN light emitting diodes and laser diodes have been demonstrated also by M. S. Schmidt et. al. (Schmidt et al. , 2007;a).

to get the desired properties of the heterostructure, for example, to achieve two-dimensional electron gas without modulation doping (Ambacher et al., 2003).
Note that an "effective" cost of single crystal bulk AlN substrates could be rather low if repeated use (removal of the AlN substrate by laser-lift off and recycling[9]) is realized.
The most mature growth technology for bulk group III - nitride crystals is the sublimation growth. Currently, the reproducible growth of large AlN bulk single crystals has been achieved (Bondokov et al., 2008; Helava et al., 2007; Mokhov et al., 2005; Raghothamachar et al., 2006; 2003). Still, a number of problems such as improvement of the crystal seeding and stoichiometry, the reduction of impurities concentration, the maintenance of stable conditions during the long growth remains.
The present chapter reports advances in AlN sublimation growth and its modelling. Both numerical and experimental approaches to understanding and optimization of the sublimation growth of AlN bulk single crystals are reviewed. A developed two-stage technology for the growth of large AlN crystals on SiC seeds that allows to exploit the best features of crucibles made from different materials is described. The superiority of single crystal AlN substrates for the growth of the group-III nitride epitaxial layers and the device performance is demonstrated. The chapter is organized as follows. The next section contains a review of approaches to growth of bulk group III - nitride crystals from liquid - melt or solution - and vapour phases. The sublimation growth of bulk AlN crystals and modelling of this process are considered in the sections 3 & 4, respectively. The developed by the authors two-stage growth procedure based on using SiC seeds and two crucibles made from different materials is described in the section 5. Results of the assessment of the quality of the grown crystals and fabricated single-crystal 2in AlN substrates are presented in the section 6. The section 7 summarizes the results.

2. Growth of bulk group III nitride crystals from liquid and vapour phases

Although the first AlN crystals were synthesized by F. Briegler and A. Gúther using the reaction between molten aluminum and nitrogen about a century and a half ago (see, for example, (Dalmau, 2005; Dalmau & Sitar, 2005) and the small needles of GaN were synthesized by R. Juza and H. Hahn in 1938 by passing ammonia over hot gallium (Jain et al., 2000) (and, similar, the AlN needle crystals were obtained by flowing nitrogen over the compacted AlN powder), growth of bulk GaN, AlN and AlGaN crystals is difficult (Denis et al., 2006); InN bulk crystals have not been demonstrated so far - the thermal instability of the group III - nitride compound increases as one goes down the group III column of the Periodic Mendeleev system (Schowalter et al., 2004).
Bulk group-III nitride crystals could not be congruently grown from the stoichiometric melt under practically acceptable environment conditions (temperature and pressure) as most semiconductors not due to the high melting temperature itself, as sometimes stated, but due to the decomposition of the crystals occurring at much lower temperature resulting from the strong bonding of nitrogen molecule and the low free energy of the crystal (Krukowski,

[9] LLO is usually performed to separate group III-nitride structure from the sapphire substrate by a short pulse of UV laser — either the eximer KrF laser at 248 nm or the third harmonic (255 nm) of the Nd:YAG laser — that locally heats the nitride layer causing its decomposition into metal and nitrogen. In case of AlN substrate having a relatively low transparency at short wavelengths an additional operation could be needed — a preliminary thinning of the substrate by chemical etching (Schujman & Schowalter, 2010).

1997)[10]. The congruent melting of gallium nitride has been achieved recently under the severe experimental conditions (the pressure of 6 Gpa and temperature about 2200° C (Utsumi et al., 2003) (so far grown crystals are smaller than 100 μm).
Bulk single group III- nitride crystals could be grown either from solution or from vapour phase. The former is known in three variants: High Nitrogen Pressure Solution Growth (HNPSG), and two "low" pressure techniques — ammonothermal growth and flux growth. Two vapour phase growth methods are halide (hydride) vapour phase epitaxy (HVPE) and sublimation growth. These growth techniques are briefly considered in this section, except sublimation growth that is treated in the next section.

2.1 High Nitrogen Pressure Solution Growth

The solubility of nitrogen in the Ga melt is very low (Nord et al., 2003) and the formation of the N_2 bubbles is possible in the supersaturated GaN liquid (Krukowski, 1997). The nitrogen dissolution in metal melt could be increased by two orders of magnitude —as well known in iron- and steelmaking — by dissolving the nitrogen radicals instead of nitrogen itself, thus ammonia is preferable as an ambient gas (Kawahara et al., 2005).
The growth of the centimeter-size GaN crystals was achieved at high temperatures and ultra-high N_2 pressures that provide a sufficient concentration of nitrogen in the Ga melt (High Nitrogen Pressure Solution Growth - HNPSG) (Grzegory, 2001; Porowski & Grzegory, 1997). The GaN crystals grown by this method have a very low threading dislocation density of $10^2 cm^{-2}$. Similar temperature and pressure are used in the Pressure-Controlled Solution Growth - PCSG (Denis et al., 2006). The HNPSG without an intentional seeding produces the needle-like AlN crystals (Bockowski, 2001). Both needle-like and bulk form of AlN single crystals up to 1 cm and 1 mm, respectively, were grown at high nitrogen pressure of the order of 1 GPa and temperatures up to 2000 K (Bockowski et al., 2001).
Recently (Al,Ga)N bulk single crystals have been grown from the Al/Ga melt under high nitrogen pressure (up to 10 kbar) at high temperature (1425 - 1780 °C) with an aluminum content from 0.22 to 0.91 (Belousov, 2010; Belousov et al., 2010; 2009). The largest crystal was $0.8 \times 0.8 \times 0.8$ mm^3. The distinct feature of this study is the use of pre-reacted polycrystalline (Al,Ga)N or AlN pellets. The composition of the growing crystal was controlled by the proper choice of the pressure and temperature.
Note, however, that the high-pressure requirement limits the scalability of the HNPSG to the growth of small area crystals.

2.2 Ammonothermal growth and flux growth

The extreme parameters needed for the HNPSG and the PCSG are reduced in the ammonothermal (Purdy et al., 2003; Yoshikawa et al., 2004) and the alkali metal flux methods (Aoki et al., 2002; Onda et al., 2002; Song et al., 2003; Yano et al., 2000). The former is similar to the well-known hydrothermal growth of quartz crystals (Iwasaki & Iwasaki, 2002) (that closely reproduces the growth of amethyst in nature (Carter & Norton, 2007)) with supercritical ammonia instead of water. This method belongs to a wide class of methods called *solvothermal*, another member of this class of growth techniques is *glycothermal* growth from glycerinated solutions (Adekore et al., 2006; Callahan & Chen, 2010). Being a low-temperature

[10] For example, the melting point of GaN is 2300° C at the pressure of 6GPa; at lower pressure GaN dissociates into metallic gallium and nitrogen gas or a state where nitrogen is dissolved in liquid gallium (Ohtani et al., 2007); at atmospheric pressure GaN decomposes at 1150 K (Ehrentraut & Fukuda, 2008).

process, solvothermal growth minimize the incidence of the temperature-induced point defects.

The relatively low growth rates of the solvothermal methods are partially compensated by the ability to grow multiple crystals (for example, over a hundred in the case of ZnO and over a two thousands in the case of quartz) in a single run.

Hundreds of different compounds are grown by the hydrothermal method, some of them at ambient conditions such as aluminum potassium sulfate and potassium dihydrogen sulfate (KDP). Hydrothermal growth of the low-defect ZnO crystals requires a high oxygen overpressure (about 50 atm) (Nause & Nemeth, 2005), but thus far remains a unique example of the industrial growth of widebandgap semiconductors by the solvothermal method (Ehrentraut et al., 2006).

In solvothermal method a liquid polar solvent (water in hydrothermal and ammonia in ammonothermal growth) forms metastable intermediate products with the solute (nutrient). Mineralizers are needed to increase the solubility of the nutrient.

No growth of GaN crystals is observed in the pure Ga solution. It is necessary, similar to other hydrothermal-type processes, to add either lithium as a transporting agent or either acidic or basic *complexing agents - mineralizers* (Callahan & Chen, 2010) such NH_4X (where X= Cl, Br, I) (Purdy et al., 2003; Yoshikawa et al., 2004) or gallium triiodide with CuI or LiI (Purdy et al., 2003). In the first approach chemical reactions occur in the solution involving such compounds as $LiGa(NH_2)_2$ and $LiNH_2$. Acidic mineralizers effectively increase the reaction rate by increasing the amount of anions in the solution. A mixture of alkali metal amide and iodide has been successfully used in (Ketchum & Kolis, 2001) while neither of these mineralizers alone could provide GaN growth. The growth mechanism involves formation of the intermediate soluble Ga-imide complex. For the growth of AlN crystals a Ca_3N_2 or Na flux has been used (Yano et al., 2000).

In the sodium flux method Na acts as a catalyst that releases electrons easily. It is speculated that nitrogen in N_2 molecule absorbed onto the Ga-Na melt surface receives electrons from Na, that weakens the N_2 bonds and causes the dissociation of N_2 into two negatively charged radicals at much lower temperature and pressure (Aoki et al., 2002). The use of lithium instead of sodium is more promising since the ability of the former to fix nitrogen is higher and that allows one to achieve the growth of GaN single crystals under the pressure of 1-2 atm (Song et al., 2003).

The sodium flux growth is performed in either a closed or in an open tube (Aoki et al., 2002; Onda et al., 2002). In the former the only source of nitrogen is a solid precursor such as NaN_3 powder, thus the pressure decreases during the process as nitrogen is being consumed for the GaN growth. In the latter N_2 or its mixture with NH_3 serves as a nitrogen source and the pressure can be either kept constant or varied with the time by a prescribed law (Onda et al., 2002). NH_3 is superior over N_2, however, the size of GaN crystals grown in $NH_3 - N_2$ mixture is smaller then in pure N_2.

Cathodoluminescence spectra show that the GaN crystals grown in the open system are of higher quality (Aoki et al., 2000). Sometimes the formation of the intermetallic compound $Ga_{39}Na_{22}$ is observed (Aoki et al., 2000). Black color of grown crystals is explained either by the nitrogen deficiency or by the oxygen impurity (Aoki et al. , 2002a; Aoki et al., 2000).

Change of the crystal shape from prismatic to platelet with increasing the Na/(Na+Ga) ratio has been studied in (Aoki et al., 2000; Yamane et al., 1998). An agglomeration of small crystals at high values of this ratio is explained by the drastic increase of the supersaturation. It can lead to the growth instability due to the constitutional supercooling similar to processes in pure Ga melt HNPSG (Grzegory, 2001; Grzegory et al., 2002). Polycrystallization occurs in the

seeded growth when the pressure exceeds the threshold value for the unseeded nucleation (Iwahashi et al., 2003). The growth rate of GaN crystals is anisotropic being higher in c direction. Usually N-polar face is smooth while Ga polar face is rough, corrugated with macrosteps (Frayssinet et al., 2001; Skromme et al., 2002). However, the reverse pattern could also be observed (Yamane et al., 1998) probably due to the impurity incorporation: in (Grzegory, 2001) the growth instability has been observed on the Ga-polar face without doping and on the N-polar face in the presence of Mg.
It is speculated (Grzegory et al., 2002) that the nucleation rates at the different faces can vary greatly due to the different geometry of the 2D nuclei (hexagonal or square, differing in the number of atoms in the nucleus as well as the number of the broken bonds). In the Li flux method (Song et al., 2003) liquid gallium infiltrates into porous Li_3N and reacts to produce $GaLi_3N_2$ and metal Li. In the case of the excess of Li_3N no growth of GaN has been observed and only $GaLi_3N_2$ has been formed. The authors consider two possible mechanisms of GaN growth - direct reaction of $GaLi_3N_2$ with gallium and dissolution of $GaLi_3N_2$ in Li-Ga melt to form ternary system - and conclude that the latter is the one that is most probable.
The growth rate of GaN in the ammonothermal technique is rather small (not greater than 2μ/hour (Fukuda & Ehrebtraut, 2007)). Frequently a columnar growth occurs yielding the crystals of poor quality (Waldrip, 2007; Wang et al. , 2001a). Still, "tremendous progress over the last decade" has been recently reviewed in (Avrutin et al., 2010; Ehrentraut & Fukuda, 2010). It is claimed that since the ammothermal growth occurs at near thermodynamic equilibrium with the extremely low supersaturation, the high crystalline quality can be expected (Ehrentraut & Kagamitanii, 2010). However, the growth of the large GaN crystals by the ammonothermal method using the HVPE-grown free-standing substrates gives disappointing results: the density of dislocations in the grown crystals are two order of magnitude larger than in HVPE seed (Callahan & Chen, 2010; Ohtani et al., 2007).
A new method called Electrochemical Solution Growth (ESG) based on the transport of the nitrogen ion N^{3-} in the molten chloride salt is being developed now (Waldrip, 2007); so far only millimeter-size GaN crystals were produced. A reaction between Ga and Li_3N under NH_3 atmosphere via the formation of $LiNH_2$ is used to grow GaN crystals by T. Hirano et al. (Hirano et al., 2009).
The ammonothermal growth of polycrystalline Aln at temperatures between 525° and 550° in alkaline conditions (using potassium azide KN_3 as the mineralizer) was reported by B. T. Adekore et al. (Adekore et al., 2006). The thickness of the layers grown on the GaN seed in 21 days varied from 100 to 1500 μm. For the growth of AlN crystals a Ca_3N_2 or Na flux has been used (Yano et al., 2000). The growth using AlN wires as a starting material shows the very high contamination of oxygen, probably due to the intrinsic oxidized Al surface. Precipitation of single crystalline AlN from Cu-Al-Ti solution was studied in (Yonemura et al., 2005). The larges pencil type crystal has 3mm in length and 0.2 mm in its diameter. An AlN platelet (1.5 mm diameter) was also obtained by the regrowth technique. Evidently, the solution growth of AlN cannot be developed in a near future to the production scale due to the difficult control of process and low growth rate.
The alkali metal flux growth has been used in the liquid phase epitaxy (LPE) (Kawahara et al., 2005). 3 μm-thick MOCVD-GaN layers with the threading dislocation density $1.3 \cdot 10^6$ were used to grow the 500 μm crystals that were almost dislocation free. PL intensity of the LPE-GaN with Na flux and Ca-Na mixed flux was 47 and 86 times, respectively, greater than that of the seed crystal. LPE was also used to grow the hexagonal or prismatic platelets at ambient pressure with NH_3 as a nitrogen source; the growth anisotropy was found to be

comparable to that in bulk Na flux growth and much smaller than in HNPSG (Meissner et al., 2004).

2.3 Halide (hydride) vapour phase epitaxy - HVPE

A HVPE reactor consists of the two main zones: the source zone where chloride gas of group III metal is formed and the growth zone where it is mixed with NH_3 to grow the nitride crystal. This method, including its variant *iodine* vapour phase epitaxy (IVPE) (Cai et al., 2010), and corresponding mathematical models are well documented (see, in addition to the just cited chapter, for example (Dmitriev & Usikov, 2006; Hemmingsson et al., 2010; Segal et al., 2009) and the references therein), thus only a few comments are in order here.

The uniqueness of HVPE is the applicability of this method to both growth of thick substrates and epitaxial heterostructures due to an extremely wide range of growth rates (1 - 150 μm/hour), the low cost compared to the MOCVD, an ability to grow the heavily doped p-layers, an absence of the carbon contamination.

At present, however, the freestanding AlN films grown by HVPE are of inferior crystalline quality: the typical value of the x-ray rocking curve Full Width at Half Maximum (FWHM) is at least an order of magnitude larger than that of AlN substrate cut out from bulk AlN boule (Cai et al., 2010; Freitas, 2010); the self-separated thick (85 μm) AlN films grown recently by the three-step modification of HVPE that include the formation of numerous voids at the interface between an AlN layer and the sapphire substrate has the dislocation density on order of 10^9cm^{-2} (Kumagai et al., 2008); the sublimation-grown bulk AlN crystals usually are transparent while the HVPE-grown ones are opaque (Cai et al., 2010), Tabl. 37.16.

The reverse breakdown voltage of the m-i-m structure on bulk AlN was an order of magnitude greater than that on free-standing GaN (Luo et al., 2002) proving high potential of Al-Ga-N system for high power rectifiers.

3. Sublimation growth of AlN crystals

Sublimation[11] (also *sublimation - recondensation*) growth (or physical vapour transport - PVT) of AlN is the most mature technology of the bulk nitride crystal growth (Dalmau & Sitar, 2005; 2010) (sublimation is also used to grow AlN fibers (Bao et al., 2009) and other nitride compounds, for example, the titanium nitride crystals (Du, Edgar, Kenik & Meyer, 2010); sublimation growth of GaN crystals is less successful (Kallinger et al., 2008)[12]). Probably the first application of PVT is the growth of the single cadmium sulfide crystals more than a half of a century ago.

The growth of other wide bandgap materials such as ZnO using physical vapour transport also was reported (Rojo et al., 2006). Note that the earlier attempts to grow ZnO by sublimation were performed using much lower temperatures and the sublimation activators such as H_2O,

---

[11] "Sublimation" refers to the direct formation of the vapour from the solid phase; however, usually it is implicitly assumed that solid and vapour are the same substance. Thus the use of this term for the process in question is not strictly correct: AlN does not *sublime* but rather *decomposes*.

[12] Evidently, the source of the problem is the nitrogen pressure over GaN surface that is six orders of magnitude higher than that over AlN (Freitas, 2010).
The gallium vapour is generated either from the molten gallium or by the thermal decomposition of the GaN powder (Waldrip, 2007). In order to suppress the dissociation of GaN, NH_3 gas is used in addition to nitrogen. It is possible to grow only pellets of GaN with the size up to several square millimeters due to the depletion of the source (Ohtani et al., 2007). There was a great interest in this method in 1960s and 1970s that has been lost due to the great progress in producing pseudo-bulk GaN substrates by HVPE.

$HgCl_2$ or ZnX_2 (where X = Cl, Br or I) gases so that both sublimation and growth involve reversible chemical reaction (Rojo et al., 2006).
The simplest case for the analysis is the so called congruent (diffusionless) vapor transport (Abernathy et al., 1979): sublimation at the source and condensation at the growing crystal surface are congruent (i.e. there is no change in the composition), no foreign gases are involved and the vapour stoichiometry is preserved across the growth facility. Thus there is no relative motion of the vapour components - the diffusion plays no role, the transport is provided by the "Stefan wind" ("drift transport" (Karpov et al., 1999)) and the growth rate is maximal (Brinkman & Carles, 1998). An experimental path to the "Dryburgh" regime is the decreasing the pressure in the reactor (Wolfson & Mokhov, 2010). The growth rate under the "vacuum" conditions (the growth cell was placed in a special container with a background pressure maintained at the level about 10^{-4} Torr) corresponds to the growth rate in nitrogen atmosphere at temperatures about 350–400 K higher (Karpov et al., 1999).
In the other extreme case where inert gas is the predominant component in the vapour, the growth rate is directly proportional to the partial pressure difference at the source and at the crystal. Polycrystalline AlN frequently is grown by the sublimation method with the grain size increasing and the number of grain per unit area decreasing in the first few mm of growth (Noveski et al., 2004b).
The overall reaction of AlN sublimation growth can be written as

$$(AlN)_{\text{solid}} \xrightarrow[T_{growth}+\Delta T]{Sublimation} Al_{vapour} + \frac{1}{2}N_2 \xrightarrow[T_{growth}]{Deposition} (AlN)_{\text{solid}}$$

This method developed by G.A. Slack and T.F. McNelly in 1976 (Slack & McNelly, 1977) (whose largest crystal was 10 mm long by 3mm diameter) now provides the growth rates up to 1 mm/hr (Rojo et al., 2002)) and the high crystal quality (the threading dislocation density is lower than 1000 cm^2 and FWHM of the rocking curve is less than 10 arcsec in the best samples (Raghothamachar et al., 2003)).
Either a spontaneous nucleation (a self-seeding growth) without any attempt to control the crystal orientation or an intentional seeding (homoepitaxial (Hartmann et al., 2008) or heteroepitaxial (Lu, 2006; Miyanaga et al., 2006)) can be exploited. The SiC substrates are often used (Lu et al., 2008; Mokhov et al., 2002), other substrates —sapphire, tantalum carbide (TaC) and niobium carbide (NbC) —also have been tried (Lee, 2007).
The decomposition of SiC at high temperature affects the growth morphology and could provoke the growth of polycrystalline AlN (Noveski et al., 2004a). Fig. 1 clearly shows the graphitization of the SiC substrate (silicon evaporation) propagating from its lower side.
Both the crucible and the source usually are cylindrical, however, the conical crucibles are used sometimes (Slack & McNelly, 1977) as well a central hole in the powder source to increase the source surface area (Wang et al., 2006).
In contrast to the bulk SiC sublimation growth, there is no evidence of the polytypism in the bulk AlN wurtzite 2H polytype structure that has the lowest formation energy (Bondokov et al., 2007).
The high growth temperature and the highly reactive Al vapour create a problem in selection of crucible material that should have melting point well above 2300 C, a reasonable degree of chemical compatibility with AlN, relatively low vapor pressures, and the relatively small thermal expansion coefficient (Slack et al., 2004).
Different crucibles have been tried including ones made from refractory transition metals (W, Ta, Nb, Zr) and graphite coated with SiC, NbC or TaC (Dalmau & Sitar, 2005; Lu, 2006) revealing their weak points: e.g., the boron nitride growth environment results in the highly

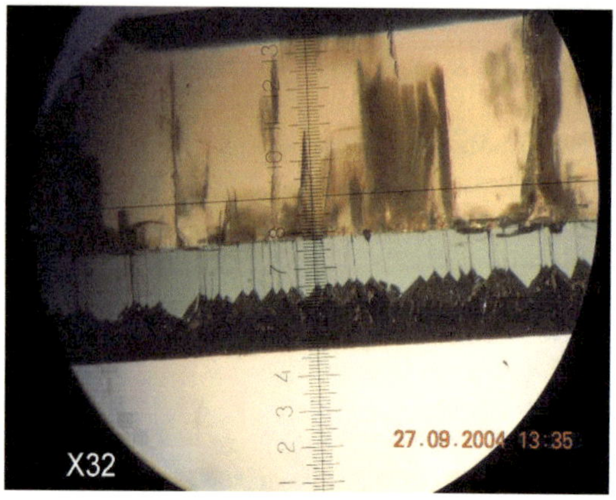

Fig. 1. Graphitization of the SiC substrate during AlN growth.

anisotropic growth (Dalmau & Sitar, 2005), the coated graphite crucibles deteriorate at high temperatures and carbon detrimentally influence the growth morphology as in the pure C crucibles; the crucibles made from nitrides or carbides suffer severe cracking. Among the most successful ones are the crucibles made from transition metals and their carbides (Lu, 2006).

The nature of other components of the growth facility such as heating elements along with the crucible material defines an environment that is responsible for the crystal contamination with different impurities (compatibility of the reactor materials has been considered, among others, by B. Epelbaum et al. (Epelbaum et al., 2002) and C.M. Balkas et al. (Balkas et al., 1997)).

The duration of growth process is limited by the degradation of the crucible (de Almeida & Rojo, 2002; Wang et al., 2006) and of the graphite insulation (Cai et al., 2006) as well as by continuous operation of the source that requires the temperature gradients within the source to be as low as possible (Bogdanov et al., 2003; 2004).

The growth temperature influence the size of AlN nuclei (Yazdi et al., 2006) and thus the crystal morphology (Sitar et al., 2004). The dislocations in the crystal can arise both during growth or after the growth in the course of the thermomechanical stress relaxation (Bogdanov et al., 2003; Klapper, 2010; Kochuguev et al., 2001; Zhmakin et al., 2000). The effect of substrate misorientation and buffer layers on growth modes and defects in AlN sublimed onto 6H-SiC substrates were studied in Refs. (Shi et al., 2001), (Yakimova et al., 2005); different growth modes were related to the low mobility of AlN adatoms on the crystal surface.

The grown bulk AlN crystals (the typical growth rate is about 100 μ/hour) usually have the rough side surface while the top surface could be facetted. The crystals are transparent with colour from yellow or amber to glass-clear (Fig. 2) having the Bragg FWHM 60 - 150 arc sec (Helava et al., 2007).

The reddish samples turned out to contain Fe^{2+} impurity (Ilyin et al., 2010). The below band-gap absorption bands limiting UV transparency are attributed to the point defects (Bickermann et al., 2010), for example, the band-to-impurity absorption manifesting itself as yellow coloration is thought to be related to either the doubly negative charged state V_{Al}^{2-}, the isolated aluminum vacancies $(V_{Al})^{3-/2-}$ (Hung et al., 2004; Sedhain et al., 2009) or the Al vacancy-impurity complexes (Lu et al., 2008).

(a) (b)

(c)

Fig. 2. Examples of grown bulk AlN crystals.

4. Modelling of bulk AlN growth

Since the AlN sublimation growth is implemented in a tightly closed crucible under high-temperature, it is difficult to study and control the growth process *in situ*, thus the importance of the mathematical modelling is evident (Bogdanov et al., 2003; 2004; Chen et al., 2008; de Almeida & Rojo, 2002; Ern & Guermond, 2004; Wellmann et al., 2006; Wu et al., 2005; Zhmakin, 2004).

The aims of numerical simulation are to explain and to predict the growth behaviour. Numerical simulation is not a substitute for experiment, but a complement to it. Numerical models can provide detailed information on the flow, temperature and concentration fields, strain in the crystals etc. which can be measured experimentally only partly or not at all. On the other hand, numerical models depend on experimental data (materials properties, boundary conditions and so on). Moreover, numerical predictions are unreliable unless models are validated using experimental data in the widest possible range of macroscopic parameters.

"Modelling" and "Simulation" are frequently used as synonyms. In Computational Fluid Dynamics (CFD) community, however, the former usually refers to the development or modification of a model while the latter is reserved for the application of the model (AIAA, 1998).

A model should relate the process specification (equipment geometry, materials properties and transport coefficients, technological parameters such as the heating power and heater position, external electromagnetic fields, orientation of the growth facility etc.) to its outputs: crystals yield, crystal quality, process duration and production costs.
If one needs a single criteria (ideally, quantifiable) to estimate the practical usefulness of the simulation, the best choice is probably the reliability of a computer prediction (Oden, 2002). It should be stressed, however, that this parameter characterizes not the model itself, but the simulation, being depending on the adequacy of the model and the accuracy of the computations as well as on the particular aim of the simulations. Evidently, the same results could be considered successful if one is interested in unveiling some trend - and unsatisfactory if the goal is to find, for example, the exact position of the inductor coils in the growth furnace. The straightforward use of the model is referred to as a direct problem (see Table 1).

Process Specification		Process Output
Equipment geometry	$\Rightarrow$ Direct (Insight)	Growth rate
Materials properties		Crystal quality
Process parameters		Process duration
	$\Leftarrow$ Inverse (Optimization)	Production costs

Table 1. Simulation scheme

From a practical point of view the reversed formulation is more useful: how one should change the equipment design or the process parameters to improve the crystal quality or to reduce production costs, for example. The simplest way is a "try-and-error" approach: to use one's intuition to introduce changes into the process specification or into the growth facility design/size, perform simulation and evaluate results. Sometimes, especially at the early development stages, an even simpler "blind search" (Luft et al., 1999) approach is exploited which essentially is a screening of a range of parameters.
A more systematic way is to state an inverse problem by indicating

1. which geometry characteristics of the reactor or operating conditions (control parameters) could be varied and
2. what criteria should be used to measure the success of the optimization.

The *inverse* problems are, unfortunately, ill-conditioned (*not-well-posed*) and their solution requires some kind of a regularization (Tihonov & Arsenin, 1977) that frequently is just the restriction on the space of possible solutions.
Mathematical models, as well as numerical methods, used for simulation of crystal growth are essentially the same as in other Computational Continuum Mechanics (CCM) applications (heat transfer, fluid dynamics, electromagnetics, elasticity). Both the block-structured and the unstructured grids are used to solve practical problems.
The advantage of the unstructured grids is the relative ease with which the complex geometry can be treated. This approach needs the minimum input description of the domain to be discretized and is not tied closely to its topology in contrast to the block-structured grid. The required CPU time to attain the prescribed accuracy may be less than for the block-structured approach due to the much lesser total number of the grid cells as a direct sequence of the second advantage of the unstructured grids — the easiness of an adaptive mesh refinement, which allows one to place the cells exactly where needed.
The main difference between simulation of the thin film growth and the bulk crystal growth is that in the former case the computational domain can be considered fixed due to the small

thickness of the epitaxial layer. Numerical study of the bulk crystal growth requires the use of either the moving grids or a regeneration of the grid. The latter approach is attractive when one can exploit a quasi-stationary approximation of the growth processes (the characteristic time of the crystal shape changes is large compared to the hydrodynamic/thermal time).
A simulation of the crystal growth requires solution of the conjugated multidisciplinary problem. The key sub-problem is the computation of the fluid flow coupled to the global heat transfer in the growth facility. Frequently, the global solution is used to the specify boundary conditions for a smaller imbedded computational domain where a more elaborate physical model is considered.

4.1 Low-Mach number (hyposonic) equations

The low-Mach number Navier–Stokes equations seem to be the most adequate model for gas flows with essentially subsonic velocities and large temperature variations (Makarov & Zhmakin, 1989). These equations provide the results identical to the full compressible Navier–Stokes computations while reducing greatly CPU time. Often (when gas mixtures used are not diluted) a CFD problem can not be decoupled from the mass transfer one (Egorov & Zhmakin, 1998).
The hyposonic flow equations follow from the full compressible Navier–Stokes equations under the following assumptions (Makarov & Zhmakin, 1989):
1) the Mach number is small $M^2 \ll 1$;
2) the hydrostatic compressibility parameter $\varepsilon = gL/R_g T_0$ is small $e \ll 1$;
3) the characteristic time τ is large compared to an acoustic time scale $\tau \gg L/a$
and for N_s-species mixture flows may be written in the following vector form:

$$\nabla \cdot \rho \mathbf{V} = 0$$
$$\nabla \cdot (\rho \mathbf{V}\mathbf{V} + p\hat{I} - \hat{\tau}) - (\rho - \rho_0)\,\mathbf{g} = 0$$
$$\nabla \cdot (\rho \mathbf{V} h + \mathbf{q}) = 0$$
$$\nabla \cdot (\rho \mathbf{V} c_s + \mathbf{J}_s) = W_s\,,\ s = 1,2,\ldots,N_s$$

where ρ is the mixture density, $\mathbf{V}$ is the mixture mass-averaged velocity, h is the mixture specific enthalpy, c_s is the mass-fraction of s-th species, p is the dynamic pressure, τ is the viscous stress tensor, $\mathbf{q}$ is the heat flux, $\mathbf{J_s}$ is the mass diffusion flux of the s-th species.
The constitutive relations required to close the system are the state equations for the perfect gas mixture with the variable specific heat:

$$\rho \frac{R}{m} T = p_0 = const\,,\ 1/m = \sum_{s=1}^{N_s} (c_s/m_s)$$

$$h = \sum_{s=1}^{N_s} c_s h_s(T)\,,\ h_s(T) = h_s^0 + \int_{T^0}^{T} Cp_s(T)dT$$

and the relations for the molecular transfer fluxes: the viscous stress tensor

$$\hat{\tau} = -\frac{2}{3}\mu\,(\nabla \cdot \mathbf{V})\,\hat{I} + \mu\,(\nabla \mathbf{V} + \mathbf{V}\nabla)$$

the heat flux

$$\mathbf{q} = -\lambda \nabla T + \sum_{s=1}^{N_s} h_s \mathbf{J}_s + p_0 \sum_{s=1}^{N_s} k_s^T \mathbf{J}_s / (\rho c_s)$$

and the diffusion flux of sth component

$$\mathbf{J}_s = -\rho D_s \left(\nabla c_s + \frac{m_s}{m} k_s^T \frac{\nabla T}{T} \right)$$

Generally both homogeneous and heterogeneous chemical reactions are to be taken into account, the latter resulting in the highly nonlinear boundary conditions.

4.2 Conjugate heat transfer

The steady-state temperature distribution inside the solid block without heat sources satisfies the scalar equation of thermal conductivity:

$$\nabla \cdot (-\lambda_{solid} \nabla T) = 0$$

The heat conduction is the simplest heat transfer mechanism. Still, two aspects of heat conduction in crystal growth problems should be mentioned.

Firstly, one needs to account for the anisotropic thermal conductivity for certain crystals and for the solid blocks manufactured from the pyrolytic graphite. In the first case the degree of the anisotropy, being determined by the crystal composition and the crystallographic symmetry, is usually not large. The pyrolytic graphite is obtained by the pyrolysis of hydrocarbon gas at high temperature in the vacuum furnace and has a layered structure with the highly ordered hexagonally arranged carbon atoms in planes and the randomly oriented atoms in the perpendicular direction. The ratio of the values of the thermal conductivity in the different directions is 100-400, depending on the material quality.

Whether anisotropy forces one to consider a three-dimensional problem for the geometrically two-dimensional configuration depends on the crystal symmetry and the orientation of its principal axes. For example, if the symmetry axis coincides with the [0001] axis of a hexagonal crystal such as SiC, the solution should be isotropic with respect to rotations around the axis and the two-dimensional formulation is valid.

Secondly, some parts of the facility could be modelled as the porous medium. Powder source is used in a number of techniques such as the metal flux method for growth of bulk GaN crystals from Li-Ga-N liquid phase (Song et al., 2004), the ammonothermal method for GaN (Yoshikawa et al., 2004), the sublimation growth of single crystals of wide bandgap semiconductors (SiC, AlN) (Bogdanov et al., 2003; Dhanaraj et al., 2003). The granular or fibrous medium is often used for the insulation. The usual approach in the computation of the global heat transfer in the facility is to treat the porous media using the effective thermal conductivity. For a given porous structure this quantity is a function of the pressure and the temperature that determine the relative contribution of the solid matrix conduction, heat conduction through the medium (gas) filling the pores and radiation to the total heat transfer. Experimental data being rather scarce, especially at high temperatures, the main problem is to formulate a model that could adequately extrapolate the effective thermal conductivity beyond the measured range of the pressures and temperatures (Daryabeigi, 1999; Kitanin et al., 1998). The effective thermal conductivity could be two orders of magnitude smaller than that of bulk material that is evidently favorable for the use of the porous media as insulation. However, it has a detrimental effect on the optimal heating of the SiC powder source in the

sublimation method (Kitanin et al., 1998). The composition, the porosity and the thermal conductivity of the SiC powder vary during the growth process (Karpov et al., 2001a).
Often radiative heat exchange through a non-participating fluid between solid surfaces can be accounted for under the assumption of the gray-diffusive surface radiation. All solid blocks are assumed to be opaque, while the external boundaries of the gaseous domain may be semi-transparent. Computation of the total radiative flux incoming to the given small surface element requires knowledge of the configuration factors (view factors). Calculation of these view factors via an integration over the complex geometry of the emitting area with account for the shadowing effect is described in details in (Dupre et al., 1990). If the view factors are known, the total radiative flux incoming to the surface element $i(i = \overline{1, N_e}$, where N_e is the total number of surface elements on the boundary) can be calculated as a sum

$$q_i^{in} = \sum_{j=1}^{N_e} q_j^{out} F_{ij},$$

where q^{in} and q^{out} are the radiation fluxes to the wall and from the wall, F_{ij} are the view factors. For the semitransparent external boundary the radiative flux out from the objects inside the region could be calculated by the Stefan-Boltzmann law and definitions of emissivity (E), reflectivity (R) and transmissivity (T) as

$$q_I^{out} = \sigma E_i T_I^4 + R_i q_I^{in} + T_I \sigma T_a^4,$$

where σ is the Stefan-Boltzmann constant and T_a is the ambient temperature.

4.3 Boundary conditions

Boundary conditions at the surfaces where heterogeneous reactions occur are formulated under assumptions that growth is limited by mass transport to the surface, the properties of the adsobtion layer are identical to those of the solid phase and are described by the following equations:
total zero flux of the inert gas

$$\rho u c_0 + J_0 = 0$$

equations relating the total species fluxes and the rates of the heterogeneous reactions

$$\rho u c_i + J_i = M_i \sum_{r=1}^{N_r} \nu_{ir} \dot{w}_r \, , i = 1, \ldots, N_k^r$$

the mass action law

$$\prod_{i=1}^{N_k^r} X_i^{\nu_{ir}} = K_r \, , r = 1, \ldots, N_r$$

the normalization condition

$$\sum_{i=0}^{N_k^r} X_i = 1,$$

where N_k^r is the number of the gas phase species participating in the surface reactions, subscript "0" refers to the inert ('carrier') gas, N_E is the total number of elements, N_S is the number of solid state phases, $N_r = N_k^r - N_E + N_S$ is the number of reactions, ν_{ir} is the stoichiometric coefficient of ith component in rth reaction, $\dot{w}_r$ is the rate of rth heterogeneous reaction, u is Stephan velocity, c_i and M_i are mass concentration and molar mass of ith component, J_i is the normal diffusion flux of ith component, K_r is the equilibrium constant of rth heterogeneous chemical reaction.

When experimental data on the reaction constants are absent, their values could be estimated using thermodynamic properties of individual materials as

$$\ln K_r = \frac{1}{RT}\sum_{i=1}^{N_k^r+N_s} \nu_{ir} G_i - \ln\left(\frac{p_s}{p}\right)\sum_{i=1}^{N_k^r} \nu_{ir},$$

where G_i is the Gibb's potential of the Ith component at normal pressure.

Theoretical analysis of AlN PVT growth was probably first performed in (Dryburgh, 1992). In this paper, the surface decomposition of low-reactive nitrogen was noticed as the rate-limiting stage of the AlN evaporation/ growth. kinetic mechanism. An one-dimensional model of the process was developed, no transport effects being taken into account. In contrast, most of the later studies assumed the AlN growth rate to be limited by the species transport (Noveski et al., 2004b; Wu et al., 2004) and the AlN growth rate was found using the Hertz-Knudsen equation for the interface Al fluxes at the AlN surfaces. Under the additional assumption of a low Al content in the vapor, an approximate explicit relationship for the AlN growth rate was derived.

These studies, unfortunately, neither clarified the boundaries of the kinetically- and transport-limited approximations nor accounted for the mass exchange between the crucible and the ambient through the small gaps and/or through the porous crucible walls that may essentially affect the process. Besides, they do not consider the evolution of the AlN crystal and of the source and the corresponding gradual change of the growth conditions. The evolution effects are also important as they determine the crystallization front shape that, in turn, affects distributions of dislocations and other defects in the crystal (a slightly convex crystallization front is preferable). A model of AlN sublimation growth that does not rely on the kinetically or transport limited approximations and describes all the above effects within a single approach was developed in (Karpov et al., 2001; Karpov et al., 1999; Segal et al., 2000) where an one-dimensional stationary model was considered.

The developed evolutionary model for AlN growth relies on the following assumptions:

- there are only Al and N_2 species in the gas phase (volatile impurities are negligible);
- the growth rate of the AlN crystal is determined by the local vapor composition and temperature but independent of the surface orientation (the isotropic growth);
- the evaporation of the AlN source occurs from the surface only (dense polycrystalline sources is used rather than porous sources);
- the evolution of the AlN source and the crystal occurs much slower than the transfer processes (the quasi-stationary transfer).

The model of AlN sublimation growth is based on the conventional description of heat and radiation transfer, gas flow dynamics, and species diffusion in the growth system coupled with the reduced quasi-thermodynamic description of the surface kinetics at the AlN surfaces. The latter was earlier applied to the modelling of other growth techniques (see, for example, (Segal et al., 2004) and references therein). As applied to AlN sublimation growth, it utilizes the extended Hertz-Knudsen relationships (Segal et al., 1999) for two reactive gaseous species, Al and N_2

$$J_i = \alpha_i(T)\beta_i(T)(P_i^w - P_i^e)$$

Here, J_i are the interface molar fluxes, $\alpha_i(T)$ are the temperature-dependent sticking probabilities,

$$\beta_i(T) = (2\pi\mu_i RT)^{-1/2}$$

are the Hertz-Knudsen collision factors, μ_i are the molar masses, R is the gas constant, P_i^w are the species partial pressures at the interface, P_i^e are the quasi-equilibrium (thermodynamic) species pressures, and subscript i indicates a particular species (Al or N_2).
The Al sticking probability is assumed unity due to its high reactivity. In contrast, the N_2 sticking probability is very low. In (Karpov et al., 2001), it was fitted as a function of temperature using data of (Dreger et al., 1962) on the AlN evaporation in vacuum (more recent data of (Fan & Newman, 2001) confirmed the derived approximation)

$$\alpha_{N_2}(T) = \frac{3.5 \exp -30000/T}{1 + 8 \cdot 10^{15} \exp 55000/T}.$$

The pressures P_i^e satisfy the mass-action law equation

$$\left(P_{Al}^e\right)^2 P_{N_2}^e = K(T),$$

where $K(T)$ is the equilibrium constant for the surface reaction

$$2Al + N_2 \Leftrightarrow 2AlN(s)$$

The model was validated in (Segal et al., 2000) by the comparison with the experimental data on the AlN growth rate as a function of the temperature and the pressure; it is implemented as software package Virtual Reactor^{TM} (Bogdanov et al., 2001; STR-soft, 2000). Virtual Reactor^{TM} provides an accurate simulation of all major physical-chemical phenomena relevant to this method such as resistive or RF heating ; conductive, convective and radiative heat transfer; mass transfer in gas and porous media; heterogeneous chemical reactions at the catalytic walls and on the surface of powder granules; deposits formation; formation of elastic strain and dislocations in the growing crystal; the evolution of the crystal and of the deposit shape, including partial facetting of the crystal surface. The problem is solved using a quasi-stationary formulation.
Temperature distributions in the tungsten and in the graphite furnaces for PVT growth of 2 inch diameter AlN boules are shown on Figs. 3, 4, respectively.
Computations revealed, in particularity, that the growth characteristics are extremely sensitive to the temperature distribution in the crucible, for which reason an accurate prediction of this distribution is of primary importance for successful modeling. Figure 5 illustrates the high accuracy of temperature prediction.
Good agreement with experimental data proves the adequacy of the model. The small deviation of the points and the curve is probably due to some uncertainty in the thermal and optical properties of the materials involved at high temperatures.
The distribution of the Al vapor molar fraction in the gaps between the AlN source (bottom), the seed (top), and the walls of the carbonized tantalum crucible in the graphite furnace is shown on Fig. 6.
If the growth occurs in a hermetically closed and chemically inert crucible, the inside static pressure is spontaneously established to provide the conservation of the initial difference between the total numbers of aluminum and nitrogen atoms in the crucible. This quantity is constant since the moment of the sealing of the crucible because the vapour-solid mass

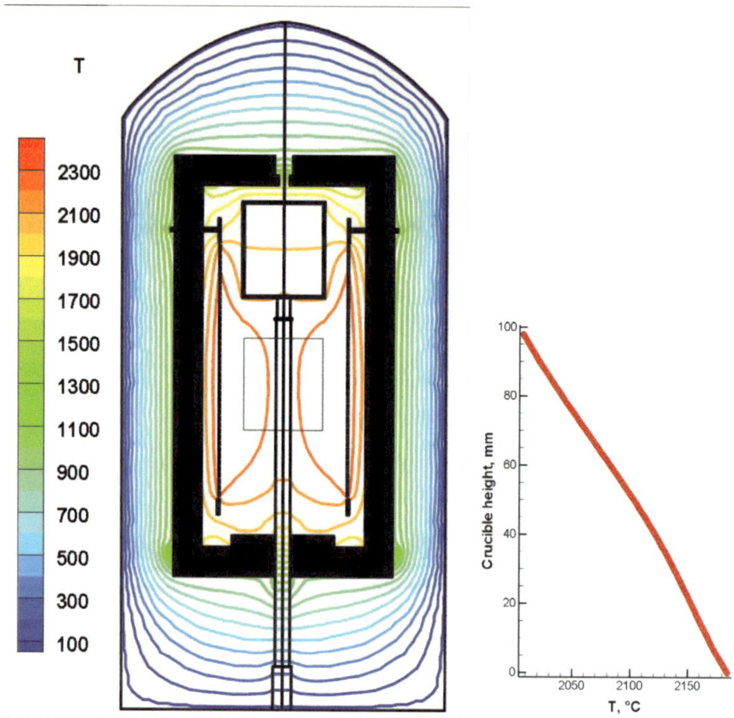

(a) Temperature distribution in the tungsten furnace

(b) Temperature profile along the axis of the tungsten crucible

Fig. 3. Temperature distribution in the tungsten furnace.

exchange occurs stoichiometrically (at the AlN surfaces) or does not occur at all (at the inert crucible' walls). 1
Due to the existence of the small gap between the crucible body and lid, the ambient pressure strongly affects the growth process, which is largely related to the notion of critical pressure. In equilibrium

$$
\begin{aligned}
J_i &= 0 \\
P_i^e &= P_i^w
\end{aligned}
$$

and the partial pressures of the two species can be found from the system of two equations:

$$
\begin{aligned}
(P_{Al}^w)^2 P_{N_2}^w &= K(T) \\
P_{Al}^w + P_{N_2}^w &= P,
\end{aligned}
$$

where P is the total pressure in the crucible. Analysis shows that if

$$P > P^{\star}(T) = 3/2[2K(T)]^{1/3}$$

with $P^{\star}$ denoting the critical pressure, then the system has two solutions corresponding to the Al-rich and N-rich vapor. Since the vapor composition in the crucible is established due to

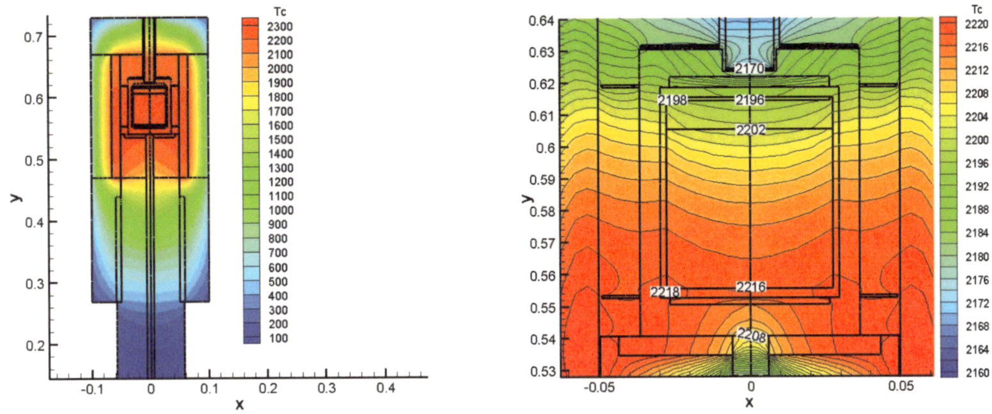

(a) Temperature distribution in the graphite furnace (b) Temperature distribution in the carbonized tantalum crucible

Fig. 4. Temperature distribution in the graphite furnace.

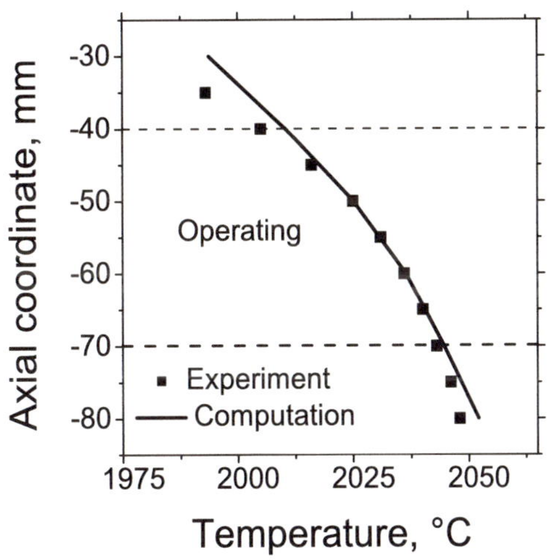

Fig. 5. Temperature at the center of the crucible lid vs. crucible coordinate at the vertical axis, computed (solid line) and pyrometrically measured (points).

mass exchange with the nitrogen ambient, the N-rich branch should be selected. If $P = P^{\star}(T)$, then there is the unique solution corresponding to the stoichiometric vapor, and if $P < P^{\star}(T)$, there is no solution. The latter means that in a hermetically closed crucible the equilibrium pressure is always higher than the critical pressure.

In a non-hermetic crucible, the equilibrium is impossible if the ambient pressure is so low that the related pressure inside the crucible is lower than the critical pressure. In this case, both the AlN source and seed evaporate, with the Al/N_2 vapor coming out from the crucible

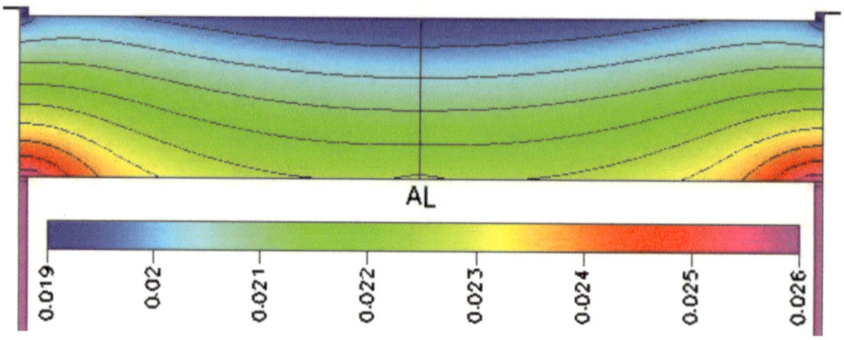

Fig. 6. Al vapor molar fraction.

to the ambient (see Fig. 7 where the pressure inside the crucible and AlN crystal growth rate vs. ambient pressure are shown for different gaps between the side crucible wall and the lid). Mass exchange between the crucible and the ambient occurs through a narrow

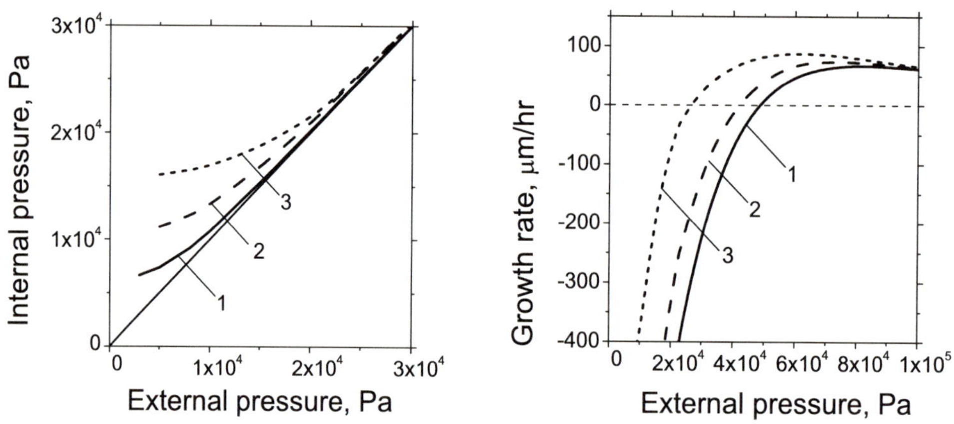

(a) Pressure inside the crucible

(b) AlN crystal growth rate

Fig. 7. Pressure inside the crucible (a) and AlN crystal growth rate (b) vs. ambient pressure at different gaps between the side crucible wall and the lid: 1 - gap length λ is 1 mm, gap thickness δ is 100 μm, 2 - $\lambda = 1$ mm, $\delta = 50\mu m$, 3 - $\lambda = 10$ mm, $\delta = 100\mu m$.

ring gap between the crucible side wall and the lid. The internal and external pressures are close to each other at a sufficiently high external pressure but considerably deviate as it decreases, depending on the gap hydraulic resistance ζ that is proportional to the gap length and inversely proportional to the third power of the gap thickness. Al and N_2 evaporate from the AlN source and then either deposit on the AlN crystal or escape from the crucible. The ratio of the deposited and escaped material depends on the external pressure and on ζ. Figure 7 shows the computed dependencies of the AlN growth rate on the ambient pressure at the crystal center.

As the external pressure decreases, a higher fraction of the material escapes from the crucible and the growth rate decreases. At a sufficiently low ambient pressure P_a^0, the crystal begins evaporating, with all vapors coming out from the crucible through the gap (negative growth rates). The value P_a^0 decreases with the gap hydraulic resistance (at a sufficiently high ζ, the crystal does not evaporate at an arbitrary small ambient pressure). The local AlN evaporation/growth rate is determined by the local supersaturation, both for the source and seed. The local interface flux on an AlN surface can be approximated as

$$J \approx \frac{\left(P_{Al}^{w}\right)^2 P_{N_2}^{w}/K(T) - 1}{4/[3\beta_{Al}(T)P_{Al}^{w}] + 1/[3\alpha_{N_2}\beta_{N_2}(T)P_{N_2}^{w}]}$$

Here, the quantity in the numerator is the local supersaturation that represents the driving force for the local AlN evaporation/growth while the denominator corresponds to the local kinetic resistance. The distribution of the supersaturation over the crystal surface determines its evolution (Fig. 8). Black lines with arrows are the streamlines, gray lines are the

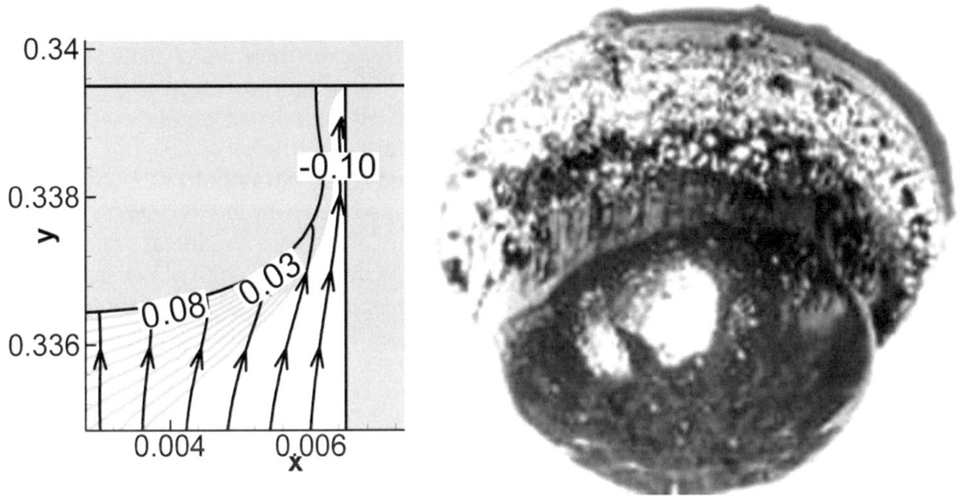

(a) Computations (b) Experiment

Fig. 8. Shape of AlN boule after 20 hrs of growth.

supersaturation isolines, numbers at the crystal surface are the local supersaturation values. Note that the absolute values of supersaturation are rather small due to the smallness of the relative temperature difference between the source and the seed.

The higher the local supersaturation is, the faster the crystal grows there. At the very periphery, supersaturation is negative due to the species exchange between the crucible and the ambient through the ring gap (here, the vapor is enriched by N_2 and depleted in Al) and the crystal evaporates, taking a mushroom shape; the vapor escapes from the crucible through the gap (the last right streamline is directed to the gap). Computational results are in good agreement with the observed crystal shapes. Using this model, we have optimized the growth conditions and crucible design, which eventually favored the growth of 2″ diameter and up to 10 mm long AlN boules with a slightly convex shape providing a low defect content in the crystal.

5. Experimental set up for AlN bulk growth

Before AlN seeds become available, SiC (4H and 6H of both (0001)C and (0001)Si orientations (Mokhov et al., 2002)) seeds were used. SiC has a small *a* -lattice mismatch with AlN - 0.96% for 6H-SiC and 1.2% for 4H-SiC. However, SiC substrates are known to degrade at high temperatures required for the AlN growth.
A two-stage technology to exploit the best features of different crucibles by avoiding interaction of W with Si and C that form easily melted eutectics and by limiting the incorporation of Si and C in C-rich environment was applied (the AlN crystals grown on SiC seeds in C-containing ambient contain a lot of Si and C impurities that, in particularity, determine the color of the crystal):

1. seeding and initial growth of the 2-3 mm long AlN crystals on the SiC seeds in the TaC crucibles in graphite equipment and
2. growth of bulk AlN crystals on the AlN seeds in tungsten crucibles and equipment.

High-quality AlN seeds of large diameter are currently unavailable while use of AlN seeds of a smaller diameter requires long multi-time lateral overgrowth of the crystals to reach the desired diameters, as the lateral overgrowth angle was found not to exceed 10-15 degrees — the diameter enlargement of AlN boules is often associated with defect generation (Bondokov et al., 2006) or the crack formation (Schujman & Schowalter, 2010).
Another kind of the two-stage procedure was reported by M. Strassburg et al. (Strassburg et al., 2004) where the temperature was gradually ramped between the two stages and by Z. Sitar et al. (Sitar et al., 2004) who used vaporization of Al (use of the metal vapour source is called *direct synthesis method* by K. Nishino et al. (Nishino et al., 2002))[13] at the first and the AlN powder source at higher temperature at the second stage; two-stage growth was also used by R. Dalmau et al. (Dalmau et al., 2005) with stages differing in the growth temperature and, thus, growth rate.

5.1 Pre-growth processing

Preparation of W crucibles includes annealing of W ones to remove the adsorbed impurities, while for Ta crucibles a pre-carbonization is necessary (Fig. 9). These crucibles are remarkably thermally and chemically stable and can endure over 3000 hours of the cumulative AlN growth in graphite (tungsten) equipment.
The presence of oxygen is detrimental in AlN growth due to the formation of oxynitrides and enhanced formation of stacking faults (de Almeida & Rojo, 2002; Majewski & Vogl, 1998) that can induce shallow electronic states (Northrup, 2005) and decrease the thermal conductivity. Oxygen has a negligible effect on the growth rate itself, but it can, at low temperature (e.g., during the heating of the system), provoke generation of Al_2O_3 inclusions (Karpov et al., 2003) that, in turn, causes surface roughness (Kazan et al., 2006). The addition of the hydrogen to the nitrogen during growth is beneficial (Karpov et al., 2001); alternatively, the sources could be processed to reduce the oxygen content. The source of oxygen in the sublimation AlN growth is the hydroxides and oxides on the surface of AlN particles (Edgar et al., 2008).
High-purity AlN sources were prepared from the commercially available AlN powders either by the annealing in the N_2 atmosphere or by the sublimation-recrystallization (Epelbaum

[13] Such method provides a high growth rate (5 mm/hour) but the long-term growth is impossible due to the formation of a nitride layer over the metallic Al source and resulting drop of the Al vapour pressure (Lu, 2006).

Fig. 9. Ta (dark gray) crucible (diameter 63 mm) completely covered with a layer of TaC-Ta_2C ceramics (gold) that is converted into TaCN during the growth to prevent the formation of easily melted eutectics (Ta-Si and Ta-Al) (Vodakov et al., 2003).

et al., 2004; Helava et al., 2007), producing, respectively, dense porous or polycrystalline AlN source[14].

The content of impurities in the AlN samples was studied using glow discharge mass spectrometry (GDMS). Sublimation-recrystallization was found to be superior and accepted as a standard technique.

SiC seeds were cut from 6H SiC bulk crystals as 0.5 mm thick round plates of different diameters (15-50 mm). The SiC seeds are mechanically lapped and mounted onto the crucible lid using the C-based glue. The technique of the lateral overgrowth of AlN crystals from small-diameter AlN seeds can also be attempted starting from 0.5 mm thick and 15-18 mm diameter AlN seeds from previously grown AlN bulk single crystals. These seeds are similarly mechanically processed and mounted onto the crucible lid with the AlN-based glue.

[14] In a two-step procedure developed by L. Du & J.H. Edgar (Du & Edgar, 2010) the low temperature ($<$ 1000 ° C) annealing was followed by a high temperature ($>$ 1900 ° C) sintering aimed at the reduction of the specific surface area of the AlN powder through the particle agglomeration. Sintering could be applied not only to raw AlN powder, but also to the flakes obtained by pressing the AlN powder (Han et al., 2008).
Evidently, the initial contamination of the AlN powder depends on the process of its synthesis. AlN powder is usually obtained either by the carbothermal reduction of Al_2O_3 or by direct nitridation of metals, both methods having very long reaction times (hours or even days) - see (Radwan & Miyamoto, 2006) and references therein. The recently proposed microwave-assisted synthesis with addition of ammonium chloride to produce HCl as an intermediate product requires much shorter time - tens of minutes (Angappan et al., 2010). Another fast method is the combustion of ultrafine aluminum powder in air; in this case additives could increase the yield of AlN (Gromov et al., 2005).

5.2 Seeding and initial growth

Use of SiC seeds in pre-carbonized Ta crucibles in graphite RF-heated furnaces is necessarily accompanied by the diffusion of Si and, to a lesser degree, of C into AlN. This process along with the lattice and TEC mismatch results in generation of defects at the SiC-AlN interface, such as micropipes inherited from SiC, dislocations, cracks, and others. Thus in-house sublimation grown *thick* low dislocation density (dislocation etch pits density $1\text{-}4{\cdot}10^3 cm^{-2}$ after 30 min etching in molten KOH) micropipe-free SiC substrates (60 mm in diameter) have been used (Fig. 10). X-ray diffractometry and topography of the grown AlN layers show that

Fig. 10. Thick micropipe-free 2in SiC substrates.

FWHMs of the rocking curves in ω-scan to lie in the range of 60-120 arcsec. At the same time, the c-parameter of the crystal lattice is found to vary as 4.984-4.988 A, which suggests that the AlN layers contain much impurity (the reference value of the c-parameter for pure AlN is 4.982). X-ray microanalysis with SEM shows 5-6% wt of Si and 1% wt of C in the AlN layers. Concentration of other impurities is less than 100 ppm.

It is known that the typical temperature of AlN growth on SiC seeds is lower by 200-300 °C than that on AlN seeds for the comparable growth rates (Epelbaum et al., 2001). Analysis of the three-phase thermodynamic equilibrium in the system Al/N/Si/C(vapor)-AlN(solid)-SiC(solid) allows finding gaseous species that may be responsible for the more intensive AlN growth on SiC. Calculation of the equilibrium partial pressures of the most volatile species in this system (more than 30 gaseous species were considered) showed that there are two "cross" volatile species (AlNC and Si_2N) that may intensify AlN growth on SiC.

5.3 Growth of bulk AlN crystals starting from AlN seeds

The 2-3 mm thick AlN layers (Fig. 11) are separated from the SiC substrate and mounted on the tungsten crucible lid with the AlN-based glue. AlN bulk crystals 10-20 mm long are

Fig. 11. 2in AlN seed grown on 60 mm micropipe-free SiC substrate.

grown in the W crucibles in the W resistively heated furnace (Mokhov et al., 2005), the growth temperature was varied in the range of 2050-2250 °C, the N_2 pressure was varied in the range of 0.5-1 atm, and the AlN growth rate was of 50-150 $\mu m/\text{hr}$. Long AlN crystals grow for several dozens of hours and often in several runs.

The evolution of the crystal quality via the improvement of the growth regime from the 2-inch diameter 10 mm long bulk AlN crystal having a single-crystal core of about 40 mm diameter and a polycrystalline rim (Avdeev et al., 2010; 2011) to the "good enough" single crystal (and, hence, the substrate) is illustrated by Fig. 12.

6. Properties of sublimation-grown AlN

Currently the technology provides stable reproducible growth of up to 2" diameter and 10-15 mm long AlN single crystals. Post-growth processing includes calibration, slicing into wafers, mechanical lapping-polishing, finishing chemical-mechanical polishing (CMP) to remove the subsurface damage due to the mechanical polishing that extends up to 4000 Å below the surface (Chen et al., 2008; Freitas, 2005), and characterization.

Impurities such as oxygen, silicon, carbon, boron contribute to the absorption and emission bands below the bandgap (Senawiratne et al., 2005) and thus reduce the AlN transparency in the UV spectral range. The content of impurities in the AlN seeds grown on SiC substrates

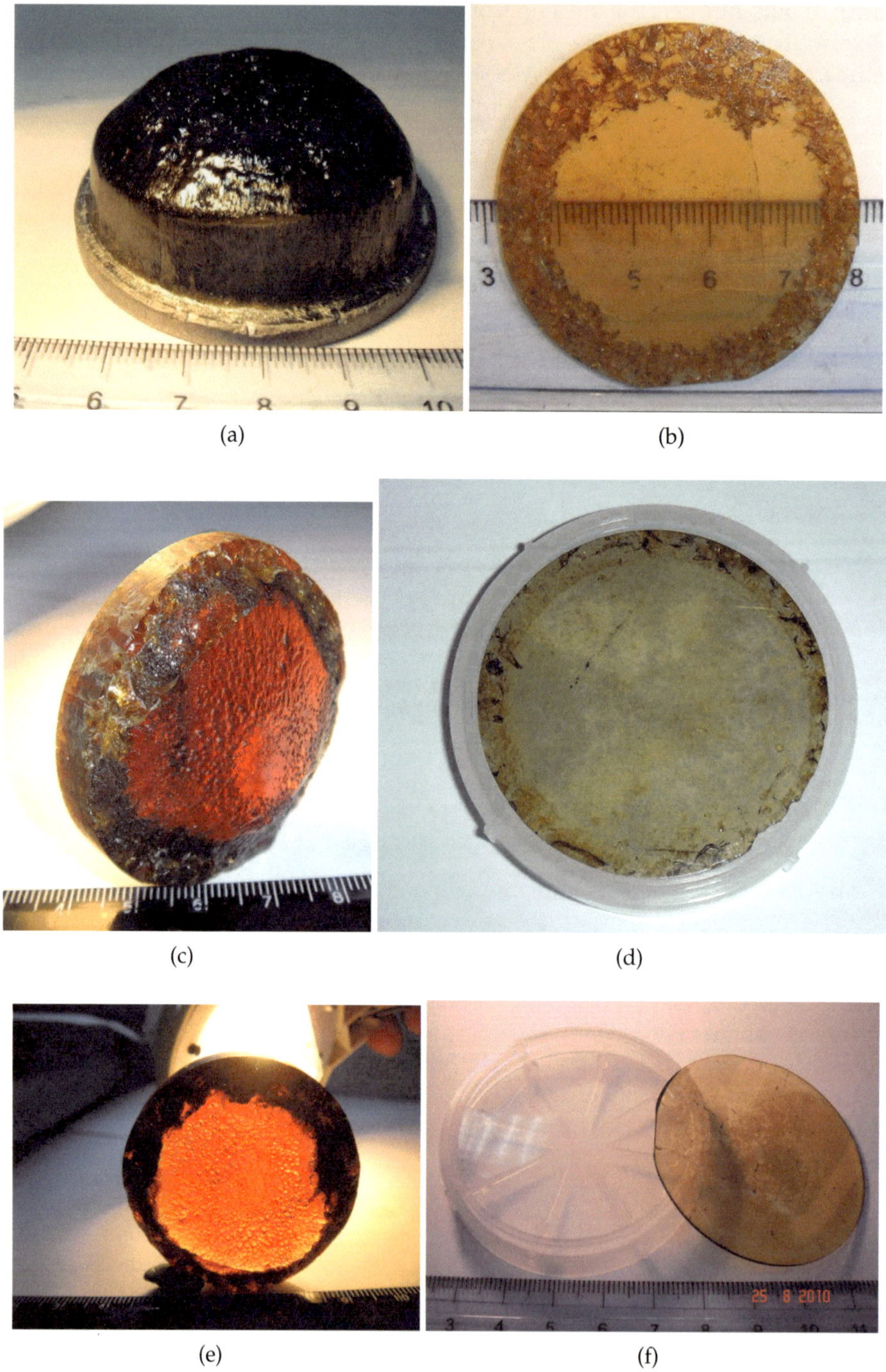

Fig. 12. Evolution of the quality of 2″ bulk AlN crystal (a, c, e) & corresponding substrate (b, d, f).

are rather high: about 5% Si and about 0.6 % C. However, the impurity concentration rapidly drops during the subsequent growth of the bulk AlN as could be seen from Fig. 13 where the variation of the concentration of aluminum, nitrogen, carbon, silicon and oxygen along the crystal thickness is shown.

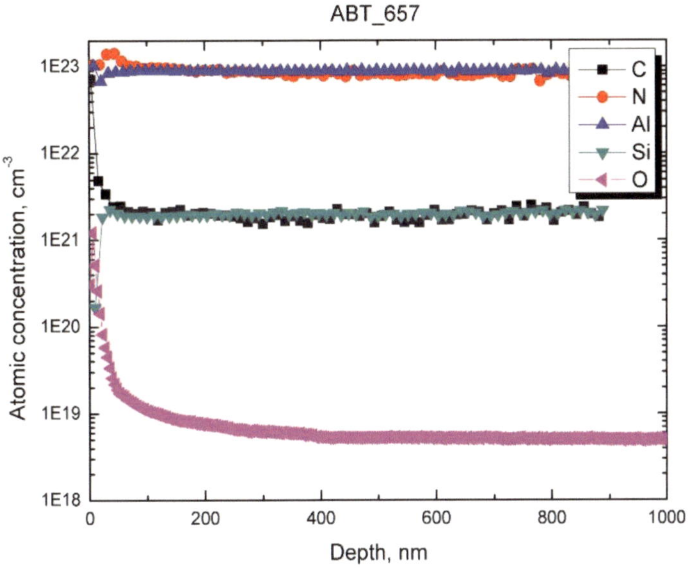

Fig. 13. Concentration of impurities along the crystal (Courtesy of V.V. Ber, Ioffe Physical Technical Institute, Russian Academy of Sciences).

X-ray diffractometry of the AlN substrates gives FWHMs of the rocking curves in ω-scan of about 2-5 arcmin. No impurities in concentrations higher than 100 ppm were found in the substrates. Typical data of X-ray analysis are presented in (Mokhov et al., 2005).
The resistivity of Si-doped AlN is generally lower than 10^5 Ohm·cm. These values are much lower than the reference resistivity of undoped semi-insulating AlN (10^{11}-10^{13} Ohm·cm). This result is attributed to the effect of the residual impurities (primarily Si) that still remain in the AlN bulk crystals grown on the SiC seeds. Repeated use of the initial AlN layer separated from the "primary" crystal (Chemekova et al., 2008) results in high purity material with the resistivity in the range of $3 \cdot 10^9$ - $3 \cdot 10^{11}$ Ohm · cm, which is rather close to the reference values.
Measurements of the transmittance spectrum of the substrates with thickness of 400 μm in the UV range have shown that most of the crystals have the average transmittance of 50-60% and demonstrate the sharp cut-off between 250nm and 320nm.
Selective etching in KOH/NaOH eutectic solution at 450°C reveals the presence of large grains with dislocation grain boundaries and individual dislocations with a low density (Fig. 14). The grain dimensions and the local dislocation densities can vastly differ.
The high crystallographic quality of the single crystal AlN substrate is confirmed by the Laue photo (Fig. 15, a) and the photo of the substrate in polarized light (Fig. 15, b).
The typical X-ray rocking curves of the 2in substrate at two different points are presented in Fig. 16.

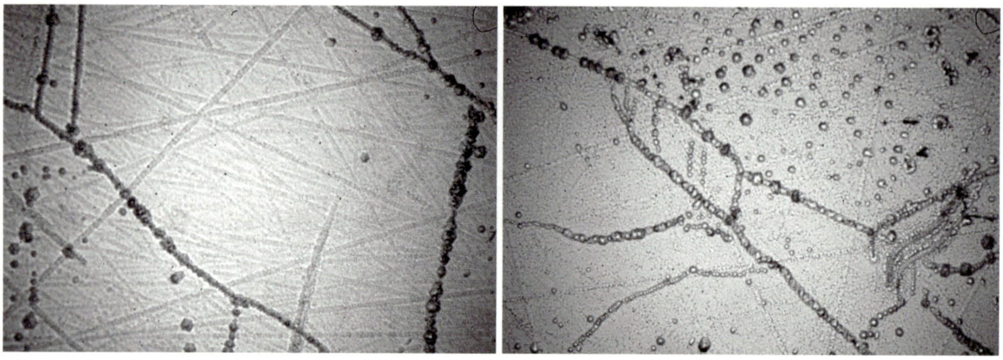

(a) (b)

Fig. 14. Etched AlN surface (Courtesy of A. Polyakov, GIREDMET, Moscow).

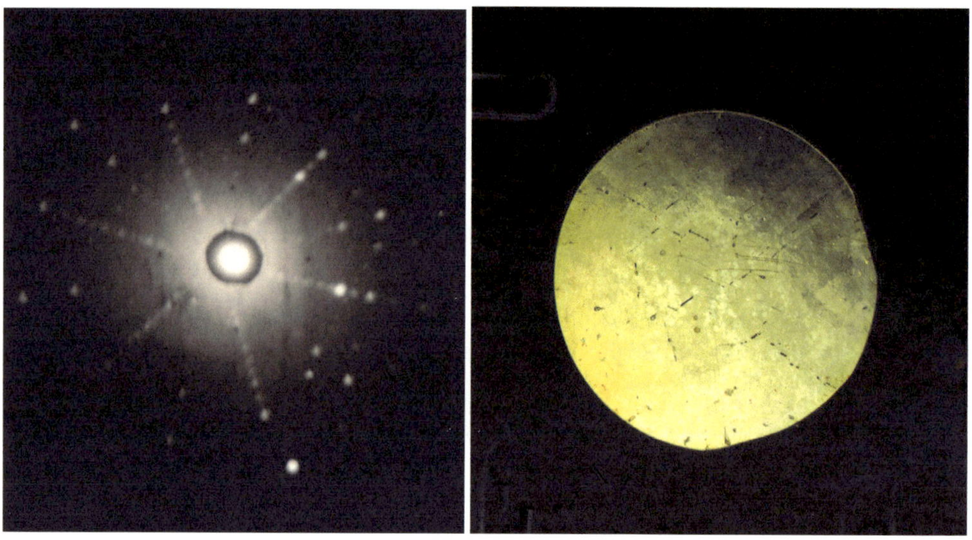

(a) (b)

Fig. 15. Laue photo (a) and photo of 2in AlN substrate in polarized light (b).

6.1 Epitaxial layers

The best characteristic of the substrate is the quality of the grown epitaxial layers and the performance of the fabricated devices. The epitaxial structures deposited at bulk AlN have been studied using transmittance and reflectance optical microscopy, high-resolution XRD diffraction, cross-polarization and cathodoluminescence. The CL spectrum of the MOCVD grown 0.3 μm AlGaN layer (Fig. 17) has FWHM of about 10 nm that indicates layer-by-layer growth of the epitaxial film.

The morphology of 0.5μm AlN epitaxial layer grown on the 2in wafer could be assessed by the AFM images shown on Fig. 18. One could see large step bunches associated with the miscut of the wafer (1° for this wafer) off the (0001) axis. On the terraces between step bunches there are nice atomic steps.

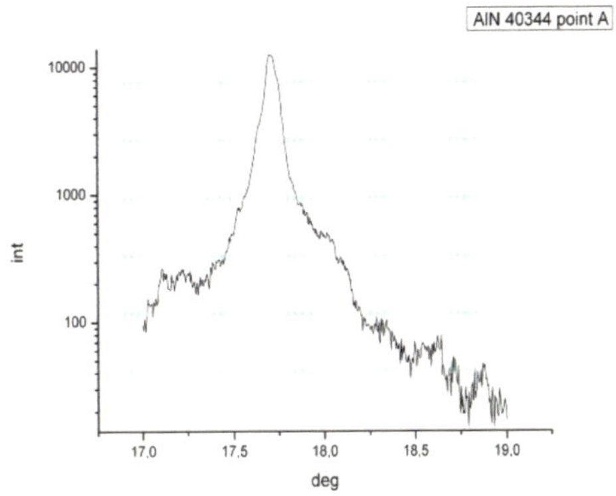

(a)

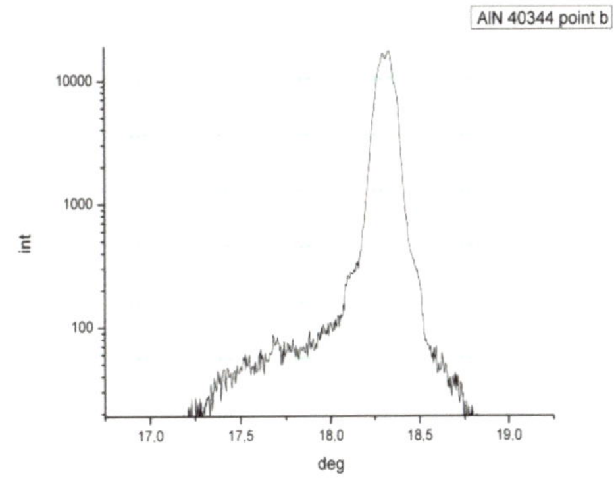

(b)

Fig. 16. X-ray rocking curves at different points of 2in substrate.

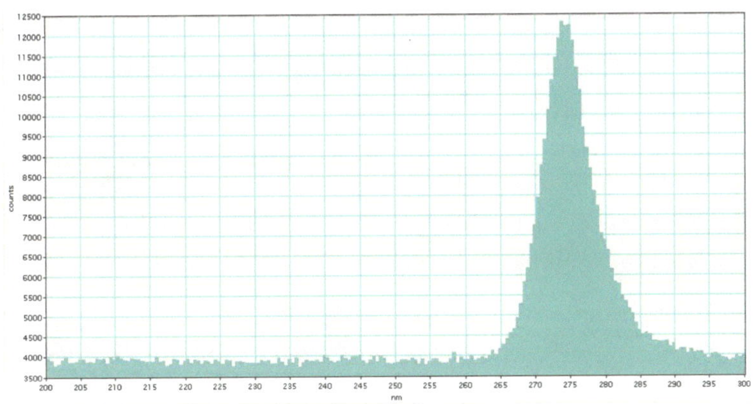

Fig. 17. CL spectrum of LED structure AlGaN/AlN (Courtesy of Prof. Asif Khan , University of South Carolina).

6.2 Light-emitting diodes

The emission of 5QW UV LED grown on bulk AlN substrate has a peak about 352 nm with FWHM 8 nm. The emission intensity was 4 times greater than that of the identical structure grown on sapphire (Fig. 19).

It should be stressed that the MOCVD growth procedure on the single crystal AlN substrate was not specially developed - the one optimized for the growth group III nitride epitaxial layers on sapphire was used.

UV LED emitting at 360 nm was grown by chloride VPE is shown in Fig. 20. This LED has a rather long lifetime: only a slight degradation is observed during long time operation (Fig. 21).

6.3 SAW applications

The surface acoustic wave velocity of the grown bulk AlN crystals measured at different frequencies and extrapolated to the zero frequency yields a value of about 5750 m/s.

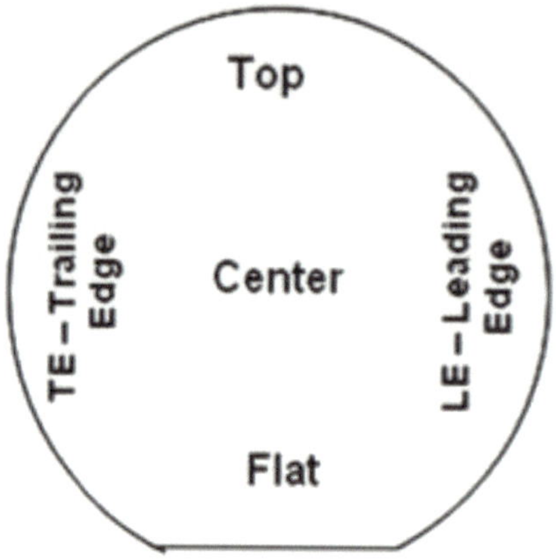

(a) Image location on the substrate

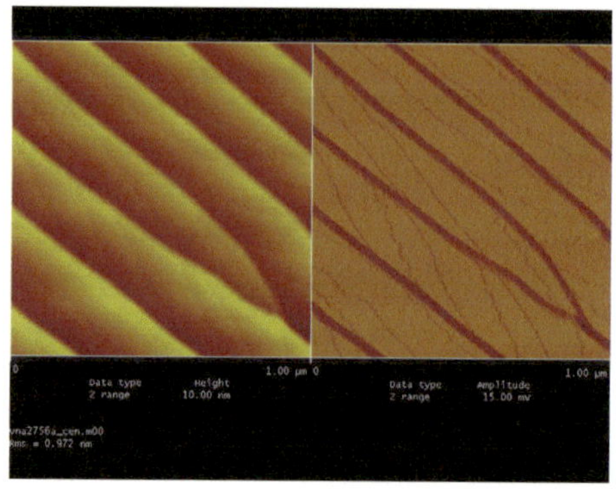

(b) Center: 1 x 1 μm, Rms 0.97 nm

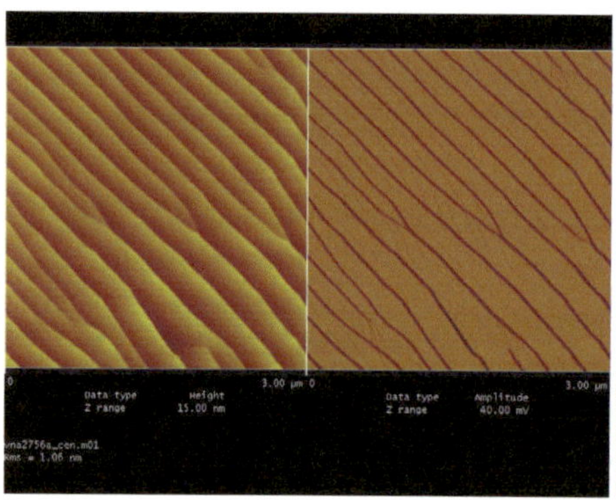

(c) Center: 3 x 3 μm, Rms 1.06 nm

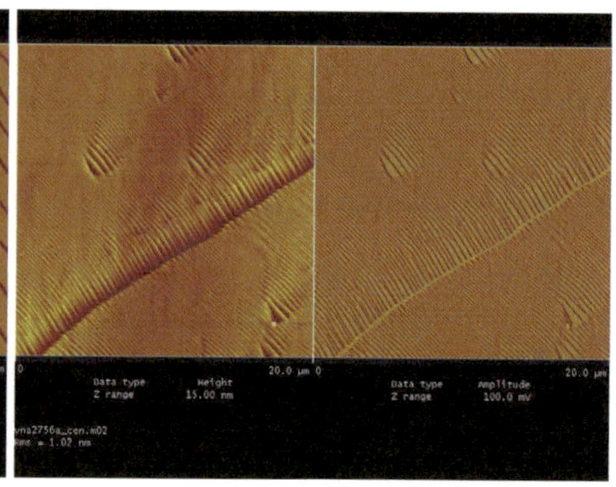

(d) Center: 20 x 20 μm, Rms 1.02 nm

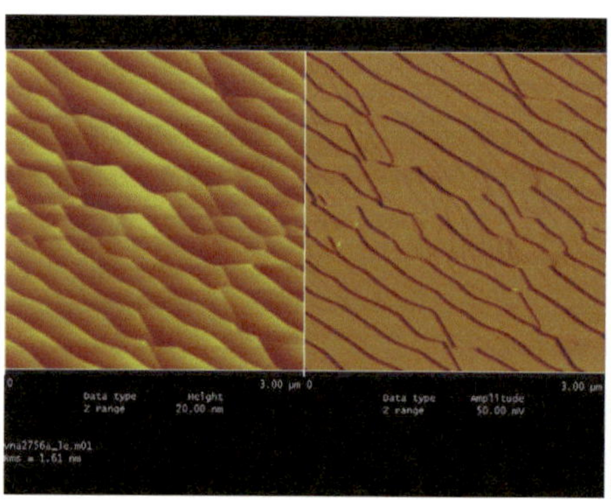

(e) Leading edge: 3 x 3 μm, Rms 1.61 nm

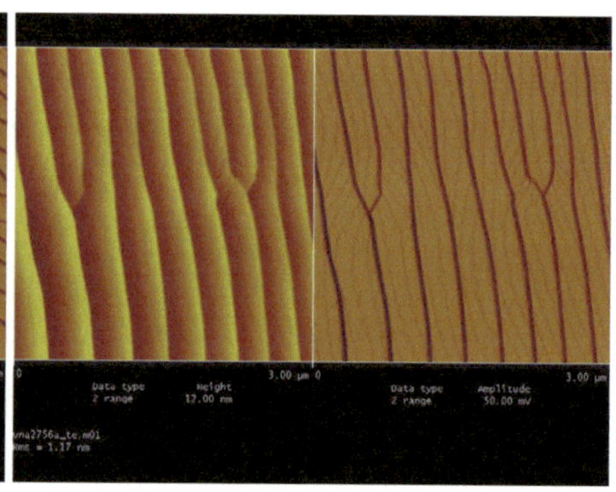

(f) Trailing edge: 3 x 3 μm, Rms 1.17 nm

Fig. 18. AFM of AlN epitaxial layer on 2 inch bulk AlN substrate (Courtesy of A. Allerman, Sandia National Laboratory)

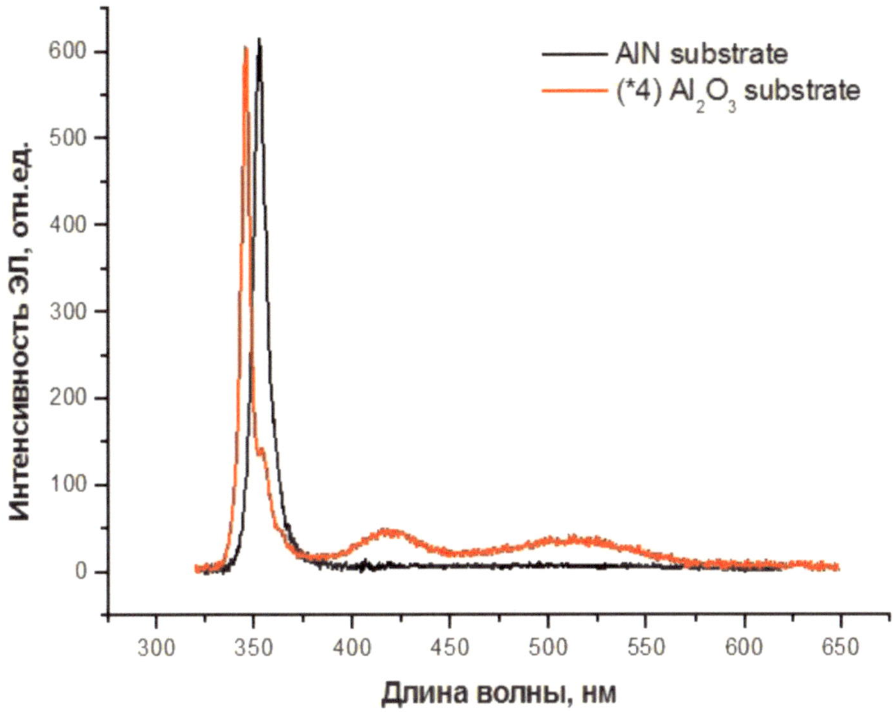

Fig. 19. EL intensity, that of UV LED on the sapphire substrate is increased fourfold (Courtesy of V.V. Lundin, Ioffe Physical Technical Institute, Russian Academy of Sciences).

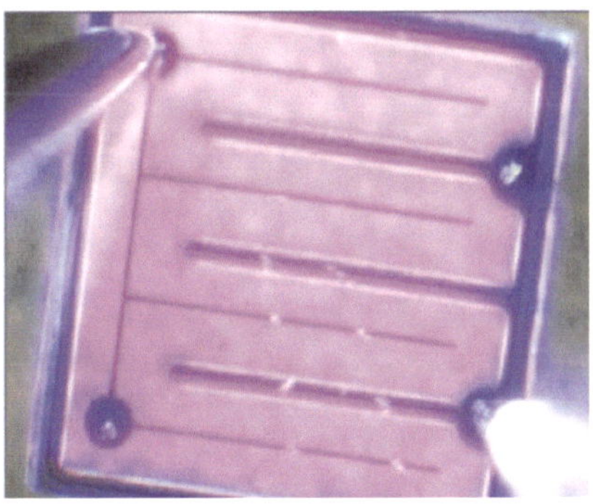

(a) UV LED chip

(b) Packaged LED

Fig. 20. 360 nm UV LED (Courtesy of The Fox Group, Inc.).

A simple regular electrode structure for SAW devices has been proposed in ref. (Biryukov et al., 2007). The structure consists of an interdigital transducer in the form of a ring placed on the Z cut of a hexagonal piezoelectric crystal (Fig. 22). Finite thickness electrodes produce the known slowing effect for a SAW in comparison with this SAW on a free surface. The closed

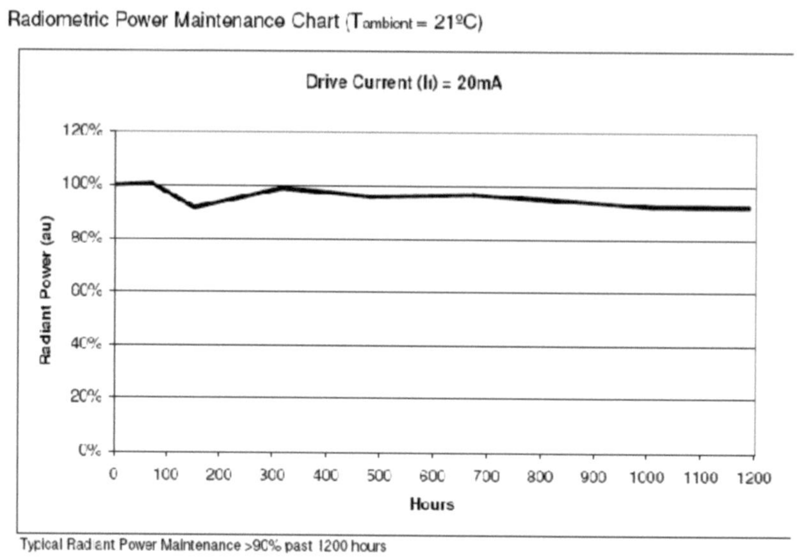

Fig. 21. Radiant power degradation (Courtesy of The Fox Group, Inc.).

slow electrode region with the *fast* surrounding region forms an open waveguide resonator structure with the acoustic field concentrated in the electrode region. If the radius of the structure - ring waveguide resonator (RWR) - is large enough for a given wavelength, an acceptable level of radiation losses can be reached. The electrical admittance of such resonator does not have sidelobes. Such device has been manufactored using a transparent pale brown

Fig. 22. Ring waveguide resonator on surface acoustic waves.

colored AlN single crystal wafer of 21 mm diameter and a thickness of 850 μm (Biryukov et al., 2009). Excitation of radial modes has been investigated. Experiments demonstrated the excellent device performance (high selectivity and a large Q-factor, estimated to be about 2700). The electrical admittance frequency dependence did not have sidelobes.

7. Conclusions

The technology of sublimation growth of AlN bulk crystals on the SiC seeds based on the successive use of crucibles made from different materials is described. Currently the technology provides stable growth of crystals up to 2" diameter and producing of single crystal 2in AlN substrates. The superiority of such "good enough" AlN substrates for crystalline quality of the grown epitaxial layers and the device performance is demonstrated.

8. Acknowledgments

The authors thank J.H. Edgar (Kansas State University), T. Bogart (Penn State University), D. Yoo (Georgia Institute of Technology), A. Allerman (Sandia National Laboratory), Asif Khan (University of South Carolina), V.V. Ber and V.V. Lundin (Ioffe Physical Technical Institute, Russian Academy of Sciences), A. Polyakov (GIREDMET, Moscow), The Fox Group, Inc. (Warrenton, VA) for experimental data.
The authors are grateful to M.V. Bogdanov, S.Yu. Karpov, A.V. Kulik, and M.S. Ramm (Soft-Impact, Ltd.) for useful discussions.

9. References

Abernathy, J. R. et al. (1979). Congruent (diffusionless) vapour transport, *J. Crystal Growth* 47: 145–154.

Adekore, B. T. et al. (2006). Ammonothermal synthesis of aluminum nitride crystals on group III-nitride templates, *J. Electron. Mater.* 35: 1104–1111.

AIAA (1998). Guide for the verification and validation of computational fluid dynamics simulations, AIAA G-077-1998.

Akasaki, I. & Amano, H. (2006). Breakthroughs in improving crystal quality of GaN and invention of the p-n junction blue-light-emitting diode, *Jap. J. Appl. Phys.* 454: 9001–9010.

Amano, H. et al. (2003). Group III nitride-based UV light emitting devices, *phys. stat. sol. (a)* 195: 491–495.

Ambacher, O. et al. (2003). Electronics and sensors based on pyroelectric AlGaN/GaN heterostructures. Part A: Polarization and pyroelectronics, *phys. stat. sol. (c)* 0: 1878–1907.

Angappan, S. et al. (2010). Aluminum nitride by microwave-assisted synthesis: Effect of added ammonia chloride, *Int. J. Self-Propag. High-Temp. Synthesis* 19: 214–220.

Aoki, M. et al. (2002). CrN single-crystal growth using Cr-Ga-Na ternary melt, *J. Crystal Growth* 246: 133–138.

Aoki, M. et al. (2002a). GaN single crystal growth using high-purity Na as a flux, *J. Crystal Growth* 242: 70–76.

Aoki, M. et al. (2000). Growth of GaN single crystals from a Na-Ga melt at 750° C and 5 MPa of N_2, *J. Crystal Growth* 218: 7–12.

Ashraf, H. et al. (2008). Thick GaN layers grown by HVPE: Influence of the templates, *J. Crystal Growth* 310: 3957–3963.

Avdeev, O. V. et al. (2010). Manufacturing of bulk AlN substrates, *in* P. Capper & P. Rudolph (eds), *Crystal Growth Technology: Semiconductors and Dielectrics*, Wiley-VCH Verlag GmbH & Co. KGaA, ISBN: 978-3-527-32593-1, pp. 121–136.

Avdeev, O. V. et al. (2011). Growth of bulk AlN crystals, *in* P. K. Bhattacharya et al. (eds), *Comprehensive Semiconductor Science and Technology*, Vol. 3, Elsevier Science Ltd, ISBN-10: 0444531432, ISBN-13: 9780444531438, pp. 282–301.

Avrutin, V. et al. (2010). Growth of bulk GaN and AlN: Progress and challenges, *Proceedings of the IEEE* 98: 1302 – 1315.

Bakhtizin, R. Z. et al. (2004). Scanning tunneling microscopy studies of III-nitride thin film heteroepitaxial growth, *Phys. Usp.* 47: 371–424.

Balkas, C. M. et al. (1997). Sublimation growth and characterization of bulk aluminum nitride single crystals, *J. Crystal. Growth* 179: 363–370.

Bao, H. Q. et al. (2009). The sublimation growth of AlN fibers: transformations in morphology & fiber direction, *Appl. Phys. A* 94: 173–177.

Beaumont, B. et al. (1998). Mg-enhanced lateral overgrowth of GaN on patterned GaN/sapphire substrate by selective metal organic vapor phase epitaxy, *MRS Internet J. Nitride Semicond. Res.* 3: 20.

Belousov, A. (2010). High pressure crystal growth, thermodynamics and physical properties of $Al_xGa_{1-x}N$ semiconductors, Dr. Sci. thesis, Eidgenössische Technische Hochschule, Zürich.

Belousov, A. et al. (2010). $Al_xGa_{1-x}N$ bulk crystal growth: Crystallographic properties and p-T phase diagram, *J. Crystal Growth* 312: 2585–2592.

Belousov, A. et al. (2009). Bulk single-crystal growth of ternary $Al_xGa_{1-x}N$ from solution in gallium under high pressure, *J. Crystal Growth* 311: 3971–3974.

Bennett, H. S. et al. (2004). Radio-frequency and analog/mixed-signal circuits and devices for wireless communications, *IEEE Circuits & Devices Magazine* pp. 38–51.

Bickermann, M. et al. (2010). UV transparent single-crystalline bulk AlN substrates, *phys. stat. sol. (c)* 7: 21–24.

Biryukov, S. V. et al. (2007). Ring waveguide resonator on surface acoustic waves, *Appl. Phys. Lett.* 90: 173503.

Biryukov, S. V. et al. (2009). Ring waveguide resonator on surface acoustic waves: First experiments, *J. Appl. Phys.* 106: 126103.

Bockowski, M. (2001). Growth and doping of GaN and AlN single crystals under high nitrogen pressure, *Cryst. Res. Technol.* 36: 771–787.

Bockowski, M. et al. (2001). Crystal growth of aluminum nitride under high pressure of nitrogen, *Mater. Sci. Semicond. Processing* 4: 543–548.

Bogdanov, M. V. et al. (2008). Coupled modeling of current spreading, thermal effects and light extraction in III-nitride light-emitting diodes, *Semicond. Sci. Technol* 23: 125023.

Bogdanov, M. V. et al. (2003). Advances in modeling of wide-bandgap bulk crystal growth, *Cryst. Res. Techn.* 38: 237–249.

Bogdanov, M. V. et al. (2001). Virtual reactor: a new tool for SiC bulk crystal growth study and optimization, *Mat. Sci. Forum* 353-356: 57–61.

Bogdanov, M. V. et al. (2004). Industrial challenges for numerical simulation of crystal growth, *Centr. Europ. J. Phys.* 2: 183.

Bondokov, R. T. et al. (2006). Defect content evaluation in single-crystal AlN wafers, *Mater. Res. Soc. Symp. Proc.* 892: FF30.

Bondokov, R. T. et al. (2007). Fabrication and characterization of 2-inch diameter AlN single-crystal wafers cut from bulk crystals, *Mater. Res. Soc. Symp. Proc.* 955: I03.

Bondokov, R. T. et al. (2008). Large-area AlN substrates for electronic applications: An industrial perspective, *J. Crystal Growth* 310: 4020–4026.

Brinkman, A. W. & Carles, J. (1998). The growth of crystals from the vapour, *Progr. Crystal Growth Character. Mater.* pp. 169–209.

Bulashevich, K. A. et al. (2007). Current spreading and thermal effects in blue LED dice, *Phys. Stat Solidi (c)* 4: 45–48.

Cai, D. et al. (2006). Thermal environment evolution and its impact on vapor deposition in large diameter AlN bulk growth. AIAA 2006-3825.

Cai, D. et al. (2010). Vapor growth of III nitrides, In: Springer Handbook of Crystal Growth, ISBN 978-3-540-74182-4, Eds. Dhanaraj et al., part F, 1243–1280.

Callahan, M. J. & Chen, Q. S. (2010). Hydrothermal and ammonothermal growth of ZnO and GaN, Ibid., part C, 655–689.

Cao, X. A. et al. (2004). Growth and characterization of blue and near-ultraviolet light-emitting diodes on bulk GaN, *Proc. SPIE* 5530: 48–53.

Cao, X. A. et al. (2007). Homoepitaxial growth and electrical characterization of GaN-based Schottky and light-emitting diodes, *J. Crystal Growth* 300: 382–386.

Cao, X. A. et al. (2004a). Electrical characteristics of InGaN/GaN light-emitting diodes grown on GaN and sapphire substrates, *Appl. Phys. Lett.* 85: 7–9.

Carter, C. B. & Norton, M. G. (2007). *Ceramic Materials, ISBN 978-0-387-46271-4*, Springer New York.

Cartwright, A. N. et al. (2006). Ultrafast carrier dynamics and recombination in green emitting InGaN MQW LED, *Mater. Res. Soc. Symp. Proc.* 916: DD04.

Chakraborty, A. et al. (2005). Demonstration of nonpolar *m*-plane InGaN/GaN light-emitting diodes on free-standing *m*-plane GaN substrates, *Jap. J. Appl. Phys.* 44: L173–L175.

Chemekova, T. Y. et al. (2008). Sublimation growth of 2 inch diameter bulk AlN crystals, *phys. stat. sol. (c)* 5: 1612–1614.

Chen, J. R. et al. (2007). Theoretical analysis of current crowding effect in Metal/AlGaN/GaN Schottky diodes and ilts reduction by using polysilicon in anode, *Chin. Phys. Lett.* 24: 2112–2114.

Chen, X. F. et al. (2008). Surface preparation of AlN substrates, *Cryst. Res. Technol.* 43: 651–655.

Cherns, D. (2000). The structure and optoelectronic properties of dislocations in GaN, *J. Phys.: Condens. Matter* 12: 10205–10212.

Cho, H. K. et al. (2000). Phase separation and stacking faults of $In_xGa_{1-x}N$ layers grown on thick GaN and sapphire substrate by metalorganic chemical vapor deposition, *J. Crystal Growth* 220: 197–203.

Choi, Y. S. et al. (2004). Effect of dislocations on the luminescence of GaN/InGaN multi-quantum-well light-emitting-diode layers, *Mater. Lett.* 58: 2614–2617.

Christen, J. et al. (2003). Optical micro-characterisation of group-III-nitrides: correleation of structural, electronic and optical properties, *phys. stat. sol. (c)* 0: 1795–1815.

Cleland, A. N. et al. (2001). Single-crystal aluminum nitride nanomechanical resonators, *Appl. Phys. Lett.* 79: 2070–2072.

Craford, M. G. (2005). LEDs for solid state lighting and other emerging applications: status, trends and challenges, *Proc. of SPIE* 5941: 594101.

Dalmau, R. F. (2005). Aluminum nitride biulk crystal growth in a resistively heated reactor, Ph. D. Thesis, North Carolina State University.

Dalmau, R. et al. (2005). AlN bulk crystals grown on SiC seeds, *J. Crystal Growth* 281: 68–74.

Dalmau, R. & Sitar, Z. (2005). Sublimation growth of AlN crystals, In: Encyclopedia of Materials: Science and Technology, Eds. K. H. J. Buschow et. al., pp. 1-9, Elsevier.

Dalmau, R. & Sitar, Z. (2010). AlN bulk crystal growth by physical vapor transport, In: Springer Handbook of Crystal Growth, ISBN 978-3-540-74182-4, Eds. Dhanaraj, G. et al., part D, 821–843.

Daryabeigi, K. (1999). Analysis and testing of high temperature fibrous insulation for reusable launch vehicles, AIAA 99-1044.

Davis, R. F. et al. (2001). Pendeo-epitaxial growth of thin films of gallium nitride and related materials and their characterization, *J. Crystal Growth* 225: 134–140.

Davis, R. F. et al. (2001a). Conventional and pendeo-epitaxial growth of GaN (0001) thin films on Si (111) substrates, *J. Crystal Growth* 231: 335–341.

Davis, R. F. et al. (2002). Gallium nitride materials - progress, status, and potential roadblocks, *Proc. IEEE* 90: 993–1005.

de Almeida, V. F. & Rojo, J. C. (2002). Simulation of transport phenomena in aluminum nitride single-crystal growth, Oak Ridge Nat. Lab., Tech. Rep. ORNL/TM-2002/64, 1–32.

Denis, A. et al. (2006). Gallium nitride bulk crystal growth processes; a review, *Mater. Sci. Eng.* R50: 167–194.

Detchprohm, T. et al. (2008). Green light emitting diodes on *a*-plane GaN bulk substrates, *Appl. Phys. Lett.* 92: 241109.

Detchprohm, T. et al. (2010). Wavelength-stable cyan and green light emitting diodes on nonpolar *m*-plane GaN bulk substrates, *Appl. Phys. Lett.* 96: 051101.

Dhanaraj, G. et al. (2003). Silicon carbide crystals - part I: Growth and characterization, *in* K. Byrappa & T. Ohashi (eds), *Crystal Growth Technology*, William Andrew Publishing, pp. 181–232.

Dinwiddie, R. B. et al. (1989). Thermal conductivity, heat capacity, and thermal diffusivity of selected commercial AlN substrates, *Int. J. Thermophys.* 10: 1075.

Dmitriev, V. & Usikov, A. (2006). Hydride vapor phase epitaxy of group III nitride materials, *in* Z. C. Feng (ed.), *III-nitride semiconductor materials*, ISBN-10: 1860946364, ISBN-13: 978-1860946363, World Scientific Publishing Company, pp. 1–40.

Dreger, L. H. et al. (1962). Sublimation and decomposition studies on boron nitride and aluminum nitride, *J. Phys. Chem.* 66: 1556–1559.

Dryburgh, P. M. (1992). The estimation of maximum growth rate for aluminium nitride crystals grown by direct sublimation, *J. Crystal Growth* 125: 65–68.

Du, L. & Edgar, J. H. (2010). Thermodynamic analysis and purification for the source materials in sublimation crystal growth of aluminum nitride, *Mater. Res. Soc. Symp.* 1202: I05–08.

Du, L. et al. (2010). Sublimation growth of titanium nitride crystals, *J. Mater. Sci.: Mater. Electron.* 21: 78–87.

Du, X. Z. et al. (2010). UV light-emitting diodes at 340 nm fabricated on a bulk GaN substrate, *Chin. Phys. Lett.* 27: 088105.

Dupret, F. et al. (1990). Global modeling of heat transfer in crystal growth furnaces, *Int. J. Heat Mass Transfer* 33: 1849–1871.

Edgar, J. H. et al. (2008). Native oxides and hydroxides and their implications for bulk AlN crystal growth, *J. Crystal Growth* 310: 4002–4006.

Edgar, J. H. et al. (eds) (1999). *Properties, processing and applications of GaN and related semiconductors*, INSPEC, the IEE, London.

Egorov, Y. & Zhmakin, A. (1998). Numerical simulation of low-Mach number gas mixture flows with heat and mass transfer using unstructured grid, *Comput. Mat. Sci.* 11: 204–220.

Ehrentraut, D. & Fukuda, T. (2008). Bulk zinc oxide and gallium nitride crystals by solvothermal techniques, *in* Y. Fujikawa et al. (eds), *Frontiers in Materials Research*, Springer, pp. 111–120.

Ehrentraut, D. & Fukuda, T. (2010). The ammonothermal crystal growth of gallium nitride - a technique on the up rise, *Proceedings of the IEEE* 98: 1316 – 1323.

Ehrentraut, D. & Kagamitanii, Y. (2010). Acidic ammonothermal growth technology for GaN, *in* Hull, R. et al. (eds), *Technology of Gallium Nitride Crystal Growth, , ISBN 978-3-642-04830-2*, Vol. 133 of *Springer Series in Materials Science*, Springer Berlin Heidelberg, pp. 183–203.

Ehrentraut, D. et al. (2006). Solvothermal growth of ZnO, *Prog. Cryst. Growth Charact. Mater.* 52: 280–335.

Eisenberg, H. R. & Kandel, D. (2002). Origin and properties of the wetting layer and early evolution of epitaxially strained thin films, *Phys. Rev. B* 66: 155429.

Epelbaum, B. et al. (2002). Sublimation growth of bulk AlN crystals: materials compatibility and crystal quality, *Mater. Sci. Forum* 433–436: 983–986.

Epelbaum, B. et al. (2001). Seeded PVT Growth of Aluminum Nitride on Silicon Carbide, *Mat. Sci. Form* 389-393: 1445–1448.

Epelbaum, B. M. et al. (2004). Growth of bulk AlN crystals for SAW devices, Proc. Second Int. Symp.on Acoustic Wave Devices for Future Mobile Comm. Systems, Chiba, March 2004, 157–162.

Ern, A. & Guermond, J. L. (2004). Accurate numerical simulation of radiative heat transfer with application to crystal growth, *Int. J. Numer. Meth. Engineer.* 61: 559–583.

Evstratov, I. Y. et al. (2006). Current crowding effects on blue LED operation, *Phys. stat. sol.(c)* 3: 1645 – 1648.

Fan, Z. Y. et al. (2006). AlGaN/GaN/AlN quantum-well field-effect transistors with highly resistive AlN epilayers, *Appl. Phys. Lett.* 88: 073513.

Fan, Z. Y. & Newman, N. (2001). Experimental determination of the rates of decomposition and cation desorption from AlN surfaces, *Mat. Sci. Eng. B* 87: 244–248.

Fara, A. et al. (1999). Theoretical evidence for the semi-insulating character of AlN, *Appl. Phys. Lett.* 85: 2001–2003.

Fini, P. et al. (1998). The effect of growth environment on the morphological and extended defect evolution in GaN grown by metalorganic chemical vapor deposition, *Jap. J. Appl. Phys.* 37: 4460–4466.

Frank, F. C. & van der Merwe, J. (1949). One dimensional dislocations. II. Misfit monolayers and oriented overgrowth, *Proc. Roy. Soc.(London) A* 198: 216–225.

Franko, A. & Shanafield, D. J. (2004). The thermal conductivity of polycrystalline aluminum nitride (AlN) ceramics, *Ceramica* 50: 247–253.

Frayssinet, E. et al. (2001). Evidence of free carrier concentration gradient along the c-axis for undoped GaN single crystals, *J. Crystal Growth* 230: 442–447.

Freitas, J. A. (2005). Bulk and homoepitaxial films of III-V nitride semiconductors: Optical studies, *J. Ceramic Process. Res.* 6: 209–217.

Freitas, J. A. (2010). Properties of the state of the art of bulk III-V nitride substrates and homoepitaxial layers, *J. Phys. D: Appl. Phys.* 43: 073301.

Freund, L. (1995). Evolution of waviness of the surface of a strained elastic solid due to stress-driven diffusion, *Int. J. Solid Structures* 32: 911–923.

Fritze, H. (2011). High-temperature piezoelectric crystals and devices, *J. Electroceram.* 26: 122–161.

Fukuda, T. & Ehrebtraut, D. (2007). Prospects for the ammonothermal growth of large GaN crystals, *J. Crystal Growth* 305: 304–310.

Fukuyama, H. et al. (2006). Single crystalline aluminum nitride films fabricated by nitriding α-SiO_2, *Appl. Phys. Lett.* 100: 024905.

Gaska, R. et al. (2002). Deep-ultraviolet emission of AlGaN/AlN quantum wells on bulk AlN, *Appl. Phys. Lett.* 81: 4658–4660.

Gautier, S. et al. (2008). AlGaN/AlN multiple quantum wells grown by MOVPE on AlN templates using nitrogen as a carrier gas, *J. Crystal Growth* 310: 4927–4931.

Gilmer, G. H. et al. (1998). Thin film deposition: fundamentals and modeling, *Comput. Mater. Sci.* 12: 354–380.

Golan, Y. et al. (1998). Morphology and microstructural evolution in the early stages of hydride vapor phase epitaxy of GaN on sapphire, *Appl. Phys. Lett.* 73: 3090–3092.

Grandusky, J. J. et al. (2010). Properties of mid-ultraviolet light emitting diodes fabricated from pseudomorphic layers on bulk aluminum nitride substrates, *Appl. Phys. Express* 3: 072103.

Grandusky, J. J. et al. (2009). Pseudomorphic growth of thick n-type $Al_xGa_{1x}N$ layers on low-defect-density bulk AlN substrates for UV LED applications, *J. Crystal Growth* 311: 2864–2866.

Gromov, A. A. et al. (2005). Nitride formation during combustion of ultrafine aluminum powder in air. I. Effect of additives, *Combustion, Explosion and Shock Waves* 41: 303–314.

Grzegorczyk, A. P. et al. (2005). Influence of sapphire annealing in trimethylgallium atmosphere on GaN epitaxy by MOCVD, *J. Crystal Growth* 283: 72–80.

Grzegory, I. (2001). High pressure growth of bulk GaN from solutions in gallium, *J. Phys.: Condens Matter* 13: 6875–6892.

Grzegory, I. et al. (2002). Mechanisms of crystallization of bulk GaN from the solution under high N_2 pressure, *J. Crystal Growth* 246: 177–186.

Han, Q. et al., Q. H. (2008). Polarity analysis of self-seeded aluminum nitride crystals grown by sublimation, *J. Electron. Mater.* 37: 1058–1063.

Hartmann, C. et al. (2008). Homoepitaxial seeding and growth of bulk AlN by sublimation, *J. Crystal Growth* 310: 930–934.

Hashimoto, S. et al. (2010). Epitaxial layers of AlGaN channel HEMTs on AlN substrates, *SEI Techn. Rev.* 71: 83–87.

Helava, H. et al. (2007). Growth of bulk aluminum nitride crystals, *phys. stat. sol. (c)* 4: 2281–2284.

Hemmingsson, C. et al. (2010). Growth of III-nitrides with halide vapor phase epitaxy (HVPE), In: Springer Handbook of Crystal Growth, ISBN 978-3-540-74182-4, Eds. Dhanaraj, G. et al., part D, 869–896.

Herro, Z. D. et al. (2006). Growth of large AlN single crystals along the [0001] direction, *Mater. Res. Soc. Symp. Proc.* 892: FF21.

Heying, B. et al. (1999). Dislocation mediated surface morphology of GaN, *J. Appl. Phys.* 85: 6470–6476.

Hirano, T. et al. (2009). Synthesis of GaN crystal by the reaction of Ga with Li_3N in NH_3 atmosphere, *J. Crystal growth* 311: 3040-3043.

Hovakimian, L. B. (2009). Strong-coupling theory of mobility collapse in GaN layers, *Appl. Phys. A* 96: 255–257.

Hu, X. et al. (2003). AlGaN/GaN heterostructure field-effect transistors on single-crystal bulk AlN, *Appl. Phys. Lett.* 82: 1299–1301.

Hull, R. et al. (2002). Interaction between surface morphology and misfit dislocations as strain relaxation modes in lattice-mismatched heteroepitaxy, *J. Phys.: Condens. Matter* 14: 12829–12841.

Hull, R. & Stach, E. A. (1996). Equilibrium and metastable strained layer semiconductor structures, *Current Opinion in Solid State & Mater. Sci.* 1: 21–28.

Hung, A. et al. (2004). An *ab initio* study of structural properties and single vacancy defects in wurtzite AlN, *J. Chem. Phys.* 120: 4890–4896.

I-hsiu, H. & Stringfellow, G. B. (1996). Solid phase immiscibility in GaInN, *Appl. Phys. Lett.* 69: 2701–2703.

Ilyin, I. V. et al. (2010). Deep-level defects in AlN single crystals: EPR studies, *Mater. Sci. Forum* 645–648: 1195–1198.

Iwahashi, T. et al. (2003). Effects of ammonia gas on threshold pressure and seed growth for bulk GaN single crystals by Na flux method, *J. Crystal Growth* 253: 1–5.

Iwasaki, F. & Iwasaki, H. (2002). Historical review of quartz crystal growth, *J. Crystal Growth* 237-239: 820.

Jain, S. C. et al. (1997). Misfit strain and misfit dislocations in lattice mismatched epitaxial layers and other systems, *Phil. Mag. A* 75: 1461–1515.

Jain, S. C. & Hayes, W. (1991). Structure, properties and applications of Ge_xSi_{1x} strained layers and superlattices, *Semicond. Sci. Technol.* 6: 547–576.

Jain, S. C. et al. (2000). III-nitides: growth, characterization, and properties, *J. Appl. Phys.* 87: 965–1006.

Kallinger, B. et al. (2008). Vapor growth of GaN using GaN powder sources and thermographic investigations of the evepoarting behaviour of the source material, *Cryst. Res. Techn.* 43: 14–21.

Karpov, D. et al. (2001a). Mass transport and powder source evolution in sublimation growth of SiC bulk crystals, *Mat. Sci. Forum.* 353-356: 37–40.

Karpov, S. Y. (1998). Suppression of phase separation in InGaN due to elastic strain, *MRS Internet J. Nitride Semicond. Res.* 3: 16.

Karpov, S. Y. (2009). Bandgap engineering of III-nitride devices on low-defect substrates, *in* Z. C. Feng (ed.), *III-Nitride Devices and Nanoengineering*, Imperial College Press, London, pp. 367–398.

Karpov, S. Y. et al. (2001). Effect of reactive ambient on AlN sublimation growth, *phys. stat. sol. (a)* 188: 763–767.

Karpov, S. Y. et al. (2003). Role of oxygen in AlN sublimation growth, *phys. stat. sol. (c)* 0: 1989–1992.

Karpov, S. Y. et al. (1999). Sublimation growth of AlN in vacuum and in a gas atmosphere, *phys. stat. sol. (a)* 176: 435–438.

Kasap, S. & Capper, P. (eds) (2006). *Springer Handbook of Electronic and Photonic Materials, ISBN 978-0-387-29185-7*, Springer.

Kawahara, M. et al. (2005). A systematic study on the growth of GaN single crystals using Na-based flux method, *J. Ceramic Process. Res.* 6: 146–152.

Kazan, M. et al. (2006). What causes rough surface in AlN crystals?, *J. Crystal Growth* 290: 44–49.

Kazan, M. et al. (2006a). Phonon dynamics in AlN lattice contaminated by oxygen, *Diamond Rel. Mater.* 15: 1525–1534.

Ketchum, D. R. & Kolis, J. W. (2001). Crystal growth of gallium nitride in supercritical ammonia, *J. Crystal Growth* 222: 431–434.

Kirchner, C. et al. (1999). Homoepitaxial growth of GaN by metal-organic vapor phase epitaxy: A benchmark for GaN technology, *Appl. Phys. Lett.* 75: 1098–1100.

Kitanin, E. et al. (1998). Heat transfer through source powder in sublimation growth of SiC crystal, *Mater. Sci. Eng.* B55: 174–183.

Klapper, H. (2010). Generation and propagation of defects during crystal growth, *in* G. Dhanaraj et al.(eds), *Springer Handbook of Crystal Growth, ISBN 978-3-540-74182-4*: 93–116.

Kochuguev, S. K. et al. (2001). Solution of crystal growth problems using adaptive unstrucured grids, *in* M. Baines (ed.), *Numerical Methods for Fluid Dynamics VII*, ISBN-10: 095249292X, ISBN-13: 978-0952492924, Oxford University Press, pp. 363–369.

Korcak, S. et al. (2007). Structural and optical properties of an $In_xGa_{1-x}N/GaN$ nanostructure, *Surf. Sci.* 601: 3892–3897.

Krukowski, S. (1997). Thermodynamics and high-pressure growth of (Al, Ga, In)N single crystals, *Diamond and Related Materials* 6: 1515–1523.

Kukushkin, S. A. et al. (2008). Substrates for epitaxy of gallium nitride; new materials and techniques, *Rev. Adv. Mater. Sci* 17: 1–32.

Kumagai, Y. et al. (2008). Self-separation of a thick AlN layer from a sapphire substrate via interfacial voids formed by the decomposition of sapphire, *Appl. Phys. Express* 1: 045003.

Kuwano, N. et al. (2004). Formation and annihilation of threading dislocations associated with stress in hetero-structure of GaN and AlGaN, *IPAP Conf. Series* 4: 21–24.

Lee, H. et al. (2004). Preparation of freestanding GaN and GaN template by hybride vapor phase epitaxy, *IPAP Conf. Series* 4: 25–27.

Lee, R. G. (2007). Transmission electron microscopy and thermal residual stress analysis of AlN crystal. MS thesis, Texas Tech UniverSity.

Lim, Y. S. et al. (2000). Strain-induced diffusion in a strained $Si_{1-x}Ge_x/Si$ heterostructure, *Appl. Phys. Lett.* 77: 4157–4159.

Limb, J. B. (2007). Design, fabrication and characterization of III-nitride pn junction devices, PhD thesis, Georgia Institute of Technology.

Lin, G. Q. et al. (2008). Influence of AlN buffer thickness on GaN grown on Si(111) by gas source molecular beam epitaxy with ammonia, *Chin. Phys. Lett.* 25: 4097–4100.

Liu, W. & Balandin, A. A. (2005). Thermal conduction in $Al_xGa_{1-x}N$ alloys and thin films, *J. Appl. Phys.* 97: 073710.

Liu, Y. S. (2009). An overview of the development of major light sources; from light bulbs to solid state lighting, *in* Z. C. Feng (ed.), *III-Nitride Devices and Nanoengineering*, Imperial College Press, London, pp. 1–20.

Louarn, A. L. et al., Vézian, S., Semond, F. & Massies, J. (2009). AlN buffer layer growth for GaN epitaxy on (111) Si: Al or N first?, *J. Crystal Growth* 311: 3278–3284.

Lu, P. (2006). Sublimation growth of AlN bulk crystals and high-speed CVD growth of SiC epilayers, and their characterization, Ph. D. thesis, Kansas State Univ.

Lu, P. et al. (2008). Seeded growth of AlN on SiC substrates and defect characterization, *J. Crystal Growth* 310: 2464–2470.

Luft, J. et al. (1999). Microbatch macromolecular crystallization on a thermal gradient, *J. Crystal Growth* 196: 447–449.

Luo, B. et al. (2002). High breakdown M-I-M structures on bulk AlN, *Solid-State Electron.* 46: 573–576.

Majewski, J. A. & Vogl, P. (1998). Polarization and band offsets of stacking faults in AlN and GaN, *MRS Internet J. Nitride Semicond. Res.* 3: 21.

Makarov, Y. & Zhmakin, A. (1989). On flow regimes in VPE reactors, *J. Crystal Growth* 94: 537–551.

Matsuoka, T. (2005). Progress in nitride semiconductors from GaN to InN - MOVPE growth and characteristics, *Superlatt. and Micristruct.* 37: 19–32.

Matthews, J. W. & Blakeslee, A. E. (1974). Defects in epitaxial multilayers.I. Misfit dislocations, *J. Crystal Growth* 27: 118–125.

McAleese, C. et al. (2006). Electric fields in AlGaN/GaN quantum well structures, *phys. stat. sol. (b)* 243: 1551–1559.

Medraj, M. et al. (2005). Understanding of AlN sintering through computational thermodynamics combined with experimental investigation, *J. Mat. Process. Technol.* 161: 415–422.

Meissner, E. et al. (2004). Growth of GaN crystals and epilayers from solutions at ambient pressure, *IPAP Conf. Series* 4: 46–49.

Miskys, C. R. et al. (2003). Freestanding GaN-substrates and devices, *phys. stat. sol. (c)* 0: 1627–1650.

Miyajima, T. et al. (2001). GaN-based blue laser diodes, *J. Phys.: Condens. Matter* 13: 7099–7114.

Miyanaga, M. et al. (2006). Single crystal growth of AlN by sublimation method, *SEI Techn. Rev.* 63: 22–25.

Miyoshi, T. et al. (2009). 510–515 nm InGaN-based green laser diodes on *c*-plane GaN substrates, *Appl. Phys. Express.* 2: 062201.

Mokhov, E. N. et al. (2005). Sublimation growth of AlN bulk crystals in Ta crucibles, *J. Crystal Growth* 281: 93–100.

Mokhov, E. N. et al. (2002). Growth of AlN bulk crystals by sublimation sandwich method, *Mat. Sci. Forum* 433-436: 979–982.

Moran, B. et al. (2000). Growth and characterization of graded AlGaN conducting buffer layers on n^+ SiC substrates, *J. Crystal Growth* pp. 301–304.

Morkoc, H. (1998). *Wide Band Gap Nitrides and Devices*, Springer, Berlin.

Mukai, T. et al. (2001). Nitride light-emitting diodes, *J. Phys.: Condens. Matter* 13: 7089–7098.

Müller, E. et al. (2006). Electrical activity of dislocations in epitaxial ZnO- and GaN-layers analyzed by holography in a transmission electron microscope, *Mat. Sci. in Semicond. Processing* 9: 127–131.

Mymrin, V. F. et al. (2005). Bandgap engineering of electronic and optoelectronic devices on native AlN and GaN substrates: a modelling insight, *J. Crystal Growth* 281: 115–124.

Nakamura, S. et al. (2000). *The Blue Laser Diode (2nd ed.)*, ISBN-10: 3642085792, ISBN-13: 978-3642085796, Springer, N. Y.

Nam, O. H. et aL. (1997). Lateral epitaxy of low defect density GaN layers via organometallic vapor phase epitaxy, *Appl. Phys. Lett.* 71: 2638–2640.

Nam, O. H. et al. (1998). Lateral epitaxial overgrowth of GaN films on SiO_2 areas via metal-organic vapor phase epitaxy, *J. Electron. Mater.* 27: 233–237.

Nanjo, T. et al. (2008). Remarkable breakdown voltage enhancement in AlGaN channel high electron mobility transistors, *Appl. Phys. Lett.* 92: 263502.

Nause, J. & Nemeth, B. (2005). Pressurized melt growth of ZnO boules, *Semicond. Sci. Technol.* 20: S45–S48.

Nikolaev, A. et al. (2000). AlN wafers fabricated by hydride vapor phase epitaxy, *MRS Internet J. Nitride Semicond. Res.* 5S1: W6.5.

Nishida, T. et al. (2004a). AlGaN-based ultraviolet light-emitting diodes grown on bulk AlN substrates, *Appl. Phys. Lett.* 84: 1002–1003.

Nishida, T. et al. (2004b). High current injection to a UV-LED grown on bulk AlN substrates, *MRS Proc.* 798: Y1.3.

Nishino, K. et al. (2002). Bulk GaN by direct synthesis method, *J. Crystal Growth* 237-239: 922–925.

Nord, J. et al. (2003). Modelling of compound semiconductors: analytical bond-order potential for gallium, nitrogen and gallium nitride, *J. Phys.: Condens Matter* 15: 5649.

Northrup, J. E. (2005). Shallow electronic states induced by prismatic stacking faults in AlN and GaN, *Appl. Phys. Lett.* 86: 071901.

Noveski, V. et al. (2004a). Growth of AlN crystals on AlN/SiC seeds by AlN powder sublimation in nitrogen atmosphere, *MRS Internet J. Nitride Semicond. Res.* 9: 2.

Noveski, V. et al. (2004b). Mass transfer in AlN crystal growth at high temperatures, *J. Crystal Growth* 266: 369–378.

Oden, J. T. (2002). The promise of Computational Engineering and Science: will it be kept?, *IACM Express* (12): 12–15.

Oh, T. S. et al. (2009). Epitaxial growth of improved GaN epilayer on sapphire substrate with platinum nanocluster, *J. Crystal Growth* 311: 2655–2658.

Ohtani, N. et al. (2007). Crystal growth, *in* K. Nakahashi et al., *Wide Bandgap Semiconductors. Fundamental Properties and Modern Photonic and Electronic Devices. ISBN 10-3-540-47234-7*, Springer Berlin Heidelberg New York, pp. 329–446.

Okamoto, K. et al. (2009). Epitaxial growth of GaN on single-crystal Mo substrates using HfN buffer layers, *J. Crystal Growth* 311: 1311–1315.

Onda, M. et al. & Sasaki, T. (2002). Influence of pressure control on the growth of bulk GaN single crystal using a Na flux, *J. Crystal Growth* 237-239: 2112–2115.

Orton, J. W. & Foxon, C. T. (1998). Group III nitride semiconductors for short wavelength light-emitting devices, *Rep. Prog. Phys.* 61: 1–75.

Ougazzaden, A. et al. (2008). Growth of GaN by metalorganic vapor phase epitaxy on ZnO-buffered c-sapphire substrates, *J. Crystal Growth* 310: 944–947.

Paskova, T. et al. (2010). GaN substrates for III-nitride devices, *Proc. IEEE* 98: 1324–1338.

Paskova, T. et al. (2009). Polar and non-polar HVPE GaN substrates: Impact of doping on the structural, electrical and optical characteristics, *phys. stat. sol.* pp. S344–S347.

Paskova, T. et al. (2004). Growth, separation and properties of HVPE grown GaN films using different nucleation schemes, *IPAP Conf. Series* 4: 14–20.

Pearton, S. J. et al. (2000). Fabrication and performance of GaN electronic devices, *Mater. Sci. Eng.* 30: 55–212.

Pecz, B. et al. (2008). Composite substrates for GaN growth, *in* A. G. Cullis & P. A. Midgley (eds), *Microscopy of Semiconducting Materials, ISBN 978-1-4020-8614-4, 2007*, Springer Netherlands, pp. 53–56.

Perlin, P. et al. (2005). Properties of InGaN blue laser diodes grown on bulk GaN substrates, *J. Crystal Growth* 281: 107–114.

Pinnington, T. et al. (2008). InGaN/GaN multi-quantum well and LED growth on wafer-bonded sapphire-on-polycrystalline AlN substrates by metalloorganic chemical vapor deposition, *J. Crystal Growth* 310: 2514–2519.

Porowski, S. & Grzegory, I. (1997). Thermodynamical properties of III-V nitrides and crystal growth of GaN at high N_2 pressure, *J. Crystal Growth* 178: 174–188.

Potin, V. et al. (1998). Extended defects in nitride semiconductors, *J. Electron. Mater.* 27: 266–275.

Purdy, A. P. et al. (2003). Synthesis of GaN by high-pressure ammonolysis of gallium triiodide, *J. Crystal Growth* 252: 136–143.

Radwan, M. & Miyamoto, Y. (2006). Self-propagating high-temperature synthesis of AlN nanostructures and their sintering properties, *Trans. of JWRI* 35: 43–46.

Raghothamachar, B. et al. (2006). Characterization of bulk grown GaN and AlN single crystal materials, *J. Crystal Growth* 287: 349–353.

Raghothamachar, B. et al. (2003). X-ray characterization of bulk AlN single crystals grown by the sublimation technique, *J. Crystal Growth* 250: 244–250.

Ram-Mohana, L. R. et al. (2006). Wavefunction engineering of layered wurtzite semiconductors grown along arbitrary crystallographic directions, *Superlattices and Microstructures* 39: 455–477.

Ren, Z. et al. (2007). Heteroepitaxy of AlGaN on bulk AlN substrates for deep ultraviolet light emitting diodes, *Appl. Phys. Lett.* 91: 051116.

Ren, Z. et al. (2007a). AlGaN deep ultraviolet LEDs on bulk AlN substrates, *phys. stat. sol. (c)* 4: 2482–2485.

Rojas, T. C. et al. & García, R. (1998). Relaxation mechanism of InGaAs single and graded layers grown on (111)B GaAs, *Thin Solid Films* 317: 270–273.

Rojo, C. J. et al. (2006). Physical vapor transport crystal growth of ZnO, *Proc. of SPIE* 6122: 61220Q1.

Rojo, J. C. et al. (2001). Single-crystal aluminum nitride substrate preparation from bulk crystals, *Mat. Res. Soc. Symp. Proc.* 680: E2.1.

Rojo, J. C. et al. (2002). Progress in the preparation of aluminum nitride substrates from bulk crystals *Mat. Res. Soc. Symp. Proc.* 722: K1.1.1.

Roycroft, B. et al. (2004). Origin of power fluctuations in GaN resonant-cavity light-emitting diodes, *Opt. Express* 12: 736–741.

Russel, P. (2006). SiC, sapphire and GaN materials status in Opto and RF business, Proc. CS MANTECH Conf., Vancouver, 231-232.

Satoh, I. et al. (2010). Development of aluminum nitride single-crystal substrates, *SEI Techn. Rev.* (71): 78–82.

Schmidt, M. C. et al. (2007). Demonstration of *m*-plane InGaN/GaN laser diodes, *Jap. J. Appl. Phys.* 46: L190–L191.

Schmidt, M. C. et al. (2007a). High power and high external efficiency *m*-plane InGaN light emitting diodes, *Jap. J. Appl. Phys.* 46: L126–L128.

Schowalter, L. J. et al. (2006). Development of native, single crystal AlN substrates for device applications, *phys. stat. sol. (a)* 203: 1667âĂŞ1671.

Schowalter, L. J. et al. (2004). Fabrication of native, single-crystal AlN substrates, *IPAP Conf. Series* 4: 38–40.

Schubert, E. F. & Kim, J. K. (2005). Solid-state light sources getting smart, *Science* 308: 1274–1278.

Schubert, E. F. (2006). *Light-Emitting Diodes (2nd ed.)*, Cambridge University Press.

Schujman, B. S. & Schowalter, L. J. (2010). GaN-ready aluminum nitride substrates for cost-effective, very low dislocation density III-nitride LEDs, Crystal IS, Inc., Final Scientific/Technical Report DE-FC26-08-NT01578.

Schultz, T. (2010). Defect analysis of aluminum nitride, Dr. rer. nat. thesis, Technishen Universität, Berlin.

Sedhain, A. et al. (2009). The origin of 2.78 eV emission and yellow coloration in bulk AlN substrates, *Appl. Phys. Lett.* 95: 262104.

Segal, A. S. et al. (2009). Modeling analysis of AlN and AlGaN HVPE, *phys. stat. sol. (c)* 6: S329–332.

Segal, A. S. et al. (2000). On mechanisms of sublimation growth of AlN bulk crystals, *J. Crystal Growth* 211: 68.

Segal, A. S. et al. (2004). Surface chemistry and transport effects in GaN hydride vapor phase epitaxy, *J. Crystal Growth* 270: 384–395.

Segal, A. S. et al. 1999). Transport phenomena in sublimation growth of SiC bulk crystals, *Mater. Sci. Engineer.* B61–62: 40–43.

Senawiratne, J. et al. (2005). Raman, photoluminescence and absorption studies on high quality AlN single crystals, *phys. stat. sol. (c)* 2: 2774–2778.
Seol, D. J. et al. (2003). Computer simulation of spinodal decomposition in constrained films, *Acta Mater.* 51: 5173–5185.
Shealy, J. R. et al. (2002). An AlGaN/GaN high-electron mobility transistor with an AlN sub-bufer layer, *J. Phys.: Condens. Matter* 14: 3499–3509.
Shi, Y. et al. & Kuball, M. (2001). New technique for sublimation growth of AlN single crystals, *MRS Internet J. Nitride semicond. Res.* 6: 7.
Shur, M. S. (1998). GaN based transistors for high power applications, *Solid-State Electron.* 42: 2131–2138.
Sitar, Z. et al. (2004). Growth of AlN crystals by vaporization of Al and sublimation of AlN powder, *IPAP Conf. Series* 4: 41–45.
Skierbiszewski, C. (2005). From high electron mobility GaN/AlGaN heterostructures to blue-violet InGaN laser diodes. perspectives of MBE for nitride optoelectronics, *Acta Physica Polonia* 108: 635–651.
Skromme, B. et al. (2002). Optical characterization of bulk GaN grown by a Na-Ga melt technique, *J. Crystal Growth* 246: 299–306.
Slack, G. A. & McNelly, T. F. (1977). Growth of high purity AlN crystals, *J. Crystal Growth* 34: 263–279.
Slack, G. A. et al. (2002). Some effects of oxygen impurities on AlN and GaN, *J. Crystal Growth* 246: 287–298.
Slack, G. A. et al. (2004). Properties of crucible materials for bulk growth of AlN, *Mat. Res. Soc. Symp. Proc.* 798: Y10.74.
Song, Y. et al. (2004). Preparation and characterization of bulk GaN crystals, *J. Crystal Growth* 260: 327–330.
Song, Y. et al. (2003). Bulk GaN single crystals: growth conditions by flux method, *J. Crystal Growth* 247: 275–278.
Speck, J. S. (2001). The role of threading dislocations in the physical properties of GaN and its alloys, *Mater. Sci. Forum* 353-356: 769–778.
Speck, J. S. & Rosner, S. J. (1999). The role of threading dislocations in the physical properties of GaN and its alloys, *Physica B: Condensed Matter* 273-274: 24–32.
Srikant, V. et al. (1997). Mosaic structure in epitaxial thin films having large lattice mismatch, *J. Appl. Phys.* 82: 4286–4295.
Steckl, A. J. et al. (1997). Growth and characterization of GaN thin films on SiC SOI substrates, *J. Electron. Mater.* 26: 217 – 223.
STR-soft (2000). http://www.str-soft.com/products/Virtual_Reactor/.
Strassburg, M. et al. (2004). The growth and optical properties of large, high-quality AlN single crystals, *J. Appl. Phys.* 96: 5870–5876.
Stutzmann, M. et al. (2001). Playing with polarity, *phys. stat. sol. (b)* 228: 505–512.
Taguchi, T. (2003). Present status of white LED lighting technology in Japan, *J. Light & Vis. Env.* 27: 131–138.
Tamulatis, G. et al. (2003). Photoluminescence of GaN deposited on single-crystal bulk AlN with different polarities, *Appl. Phys. Lett.* 83: 3507–3509.
Tihonov, A. & Arsenin, V. (1977). *Solution of Ill-posed problems*, ISBN-10: 0470991240, ISBN-13: 978-0470991244, Wiley, New York.
Tomita, K. et al. (2004). Self-separation of freestanding GaN from sapphire substrates with stripe-shaped GaN seeds by HVPE, *IPAP Conf. Series* 4: 28–31.

Usikov, A. S. et al. (2003). Indium-free violet LEDs grown by HVPE, *phys. stat. solidi (c)* 0: 2265–2269.

Utsumi, W. et al. (2003). Congruent melting of gallium nitride at 6 GPa and its application to single crystal growth, *Nat. Mater.* 2: 735–738.

Vaudo, R. P. et al. (2003). Characteristics of semi-insulating Fe doped GaN substrates, *phys. stat. sol. (a)* 200: 18–21.

Vodakov, Y. A. et al. (2003). Tantalum crucible fabrication and treatment, US Patent 6,547,87.

Wadley, H. G. et al. (2001). Mechanisms, model and methods of vapor deposition, *Progr. Mater. Sci.* 46: 329–377.

Waldrip, K. E. (2007). Initial Exploration of Growth of InN by Electrochemical Solution Growth, SAND2010-0952, 20 pp.

Waldrip, K. E. (2007). Molten salt-based growth of bulk GaN and InN for substrates, SAND2007-5210, 27 pp.

Waltereit, P. et al. (2002). Heterogeneous growth of GaN on 6H-SiC (0001) by plasma-assisted molecular beam epitaxy, *phys. stat. sol. (a)* 194: 524–527.

Wang, C. et al. (2001). Influence of growth parameters on crack density in thick epitaxial lateral overgrown GaN layers by hydride vapor phase epitaxy, *J. Crysytal Growth* 230: 377–380.

Wang, T. C. et al. (2001a). Dislocation evolution in epitaxial multilayers and graded composition buffers, *Acta. Mater.* 49: 1599–1605.

Wang, X. et al. (2006). Powder sublimation and porosity evolution in sublimation crystal growth, AIAA 2006-3270.

Wellmann, P. et al. (2006). Modeling and experimental verification of SiC M-PVT bulk crystal growth, *Mater. Sci. Forum* 527–529: 75–78.

Weyers, M. et al. (2008). GaN substrates by HVPE, *Proc. of SPIE* 6910: 691001.

Wolfson, A. A. & Mokhov, E. N. (2010). Dependence of the growth rate of an AlN layer on nitrogen pressure in a reactor for sublimation growth of AlN crystals, *Semicond.* 44: 1383–1385.

Wu, B. et al. (2005). Design of an RF-heated bulk AlN growth reactor: Induction heating and heat transfer modeling, *Crystal Growth & Design* 5: 1491–1495.

Wu, B. et al. (2004). Application of flow-kinetics model to the PVT growth of SiC crystals, *J. Crystal Growth* 266: 303–312.

Wu, X. H. et al. (1996). Defect structure of metal-organic chemical vapor deposition-grown epitaxial (0001) GaN/Al_2O_3, *J. Appl. Phys.* 80: 3228 - 3237.

Wu, X. H. et al. (1998). Dislocation generation in GaN heteroepitaxy, *J. Crystal Growth* pp. 231–243.

Xi, Y. A. et al. (2006). Very high quality AlN grown on (0001) sapphire by metal-organic vapor phase epitaxy, *Appl. Phys. Lett* 89: 103106.

Xi, Y. A. et al. (2006a). AlGaN UV light-emitting diodes emitting at 340 nm grown on AlN bulk substrates, Proc. L. Eastman Conf. High Performance Devices, Cornell Univ., 54–55.

Xing, H. et al. (2001). Gallium nitride based transistors, *J. Phys.: Condens. Matter* 13: 7139–7157.

Yakimova, R. et al. (2005). Sublimation growth of AlN crystals: Growth mode and structure evolution, *J. Crystal Growth.* 281: 81–86.

Yamabe, N. et al. (2009). Nitridation of Si(1 1 1) for growth of 2H-AlN(0 0 0 1)/Si_3N_4 /Si(1 1 1) structure, *J. Crystal Growth*: 3049-3053.

Yamane, H. et al. (1998). Morphology and characterization of GaN single crystals grown in a Na flux, *J. Crystal Growth* 186: 8–12.

Yan, F. et al. (2007). High-efficiency GaN-based blue LEDs grown on nano-patterned sapphire substrates for solid-state lighting, *Proc. of SPIE* 6841: 684103.

Yano, M. et al. (2000). Growth of nitride crystals, BN, AlN and GaN by using a Na flux, *Diamond and Related Materials* 9: 512–515.

Yazdi, G. R. et al. (2006). Growth and morhology of AlN crystals, *Phys. Scr.* T126: 127–130.

Yonemura et al. (2005). Precipitation of single crystalline AlN from Cu-Al-Ti solution under nitrogen atmosphere, *J. Mater. Sci.: Mater. In Electron.* 16: 197–201.

Yoshikawa, A. et al. (2004). Crystal growth of GaN by ammonothermal method, *J. Crystal Growth* 260: 67–72.

Zheleva, T. et al. (1999). Pendeo-epitaxy - a new approach for lateral growth of gallium nitride films, *J. Electron. Mater.* 28: L5–L8.

Zhmakin, A. I. (2004). Heat transfer problems in crystal growth, Keynote lecture. 3rd Int. Symp. Adv. in Comp. Heat Transfer, 2004, Norway. CD Proc., Begell House, 24 pp.

Zhmakin, A. I. (2011a). Enhancement of light extraction from light-emitting diodes, *Phys. Reports* 498: 189–241.

Zhmakin, A. I. (2011b). Strain relaxation models, arXiv:1102.5000v1 [cond-mat.mtrl-sci].

Zhmakin, I. A. et al. (2000). Evolution of thermoelastic strain and dislocation density during sublimation growth of silicon carbide, *Diamond and Related Materials* 9: 446–451.

Zukauskas, A. et al. (2002). *Introduction to Solid State Lighting*, ISBN-10: 0471215740, ISBN-13: 978-0471215745, Wiley, New York.

11

Crystal Growth and Stoichiometry of Strongly Correlated Intermetallic Cerium Compounds

Andrey Prokofiev and Silke Paschen
Institute of Solid State Physics, Vienna University of Technology,
Wiedner Hauptstr. 8-10, 1040 Vienna
Austria

1. Introduction

Strongly correlated electron systems are among the most active research topics in modern condensed matter physics. In strongly correlated materials the electron interaction energies dominate the electron kinetic energy which leads to unconventional properties. Heavy fermion compounds form one of the classes of such materials. In heavy fermion compounds the interaction of itinerant electrons with local magnetic moments generates quasiparticles with masses up to several 1000 electron masses. This may be accompanied by exciting properties, such as unconventional superconductivity in a magnetic environment, non-Fermi liquid behavior and quantum criticality. Strong electronic correlations are responsible for physical phenomena on a low energy scale. Consequently, these phenomena have to be studied at low temperatures. This, in turn, requires ultimate quality of single crystals to avoid that the low temperature intrinsic properties are covered by extrinsic effects due to off-stoichiometry, impurities or other crystal imperfections.

The overwhelming majority of heavy fermion systems are cerium and ytterbium intermetallic compounds. In the present paper we discuss the crystal growth of three cerium compounds, $Ce_3Pd_{20}Si_6$, $CeRu_4Sn_6$ and CeAuGe. $Ce_3Pd_{20}Si_6$ undergoes an antiferromagnetic phase transition at low temperatures and shows a magnetic field induced quantum critical point [Takeda et al (1995), Strydom et al (2006)]. $CeRu_4Sn_6$ [Das & Sampathkumaran (1992)] appears to be a Kondo insulator [Aepli & Fisk (1992)] with a highly anisotropic energy gap. CeAuGe is one of a the few cerium compounds showing a ferromagnetic phase transition at low temperatures [Pöttgen et al (1998), Mhlungu & Strydom (2008)]. This is of special interest in the context of quantum criticality, since the occurrence of quantum criticality on the verge of a ferromagnetic ground state is a subject of current debate.

Much attention in this paper is paid to the problem of stoichiometry. Single crystals of intermetallic compounds are grown at high temperatures, which facilitates the formation of thermal defects realized often as deviation from the stoichiometric composition. Thermal instabilities of some intermetallic phases require the use of flux techniques, i.e., growth from off-stoichiometric melts, which is another source of non-stoichiometry.

Sizeable non-stoichiometries can be detected by measuring the composition by chemical and physical analytical techniques, e.g. energy dispersive X-ray spectroscopy analysis (EDX). Tiny deviations from the stoichiometry, on the other hand, can be found only from an analysis of the physical properties of single crystals. Thus physical property measurements are not only the

final purpose of a crystal growth but also a valuable diagnostic tool for further improvement of their quality. Therefore, in the paper the consideration of crystal growth is accompanied by the discussion of their physical properties.
Physical property measurements on $Ce_3Pd_{20}Si_6$ single crystals grown by the floating zone technique have been reported earlier [Goto et al (2009), Prokofiev et al (2009), Mitamura et al (2010)]. In Ref. Prokofiev et al (2009) a systematic study of the relationship between the growth technique, stoichiometry and physical properties of single crystals has been done. The crystal growth and stoichiometry of $CeRu_4Sn_6$ and CeAuGe are reported for the first time. The physical properties of the $CeRu_4Sn_6$ single crystals were published partially in an author's earlier paper [Paschen et al (2010)].

2. $Ce_3Pd_{20}Si_6$

The recent observation of magnetic-field induced quantum criticality [Paschen et al (2007); Strydom et al (2006)] in the cubic heavy fermion compound $Ce_3Pd_{20}Si_6$ [Takeda et al (1995)] has attracted considerable attention. $Ce_3Pd_{20}Si_6$ crystallizes in a cubic $Cr_{23}C_6$-type structure with the space group $Fm\bar{3}m$ [Gribanov et al (1994)]. The cubic cell with a= 12.161 Å [Gribanov et al (1994)]; 12.280 Å [Takeda et al (1995)] contains 116 atoms. The Ce atoms occupy two distinct sites of cubic point symmetry. At the octahedral *4a* (tetrahedral *8c*) site Ce atoms are situated in cages formed by 12 Pd and 6 Si atoms (16 Pd atoms). These high coordination numbers allow to classify $Ce_3Pd_{20}Si_6$ as a cage compound. Strongly correlated cage compounds are of much interest as potential candidates for thermoelectric applications [Paschen (2006)].
Similar to the isostructural germanide compound $Ce_3Pd_{20}Ge_6$, two phase transitions - a presumably antiferromagnetic one at T_L of 0.15 K [Takeda et al (1995)], 0.17 K [Goto et al (2009)], or 0.31 K [Strydom et al (2006)] and a possibly quadrupolar one at T_U of 0.5 K [Strydom et al (2006)] - have been found in the silicide compound $Ce_3Pd_{20}Si_6$. Similar to the effect of magnetic field applied along [100] or [110] in $Ce_3Pd_{20}Ge_6$ [Kitagawa (1998)], a magnetic field shifts the two transitions of polycrystalline $Ce_3Pd_{20}Si_6$ in opposite directions: At the critical field of about 1 T the transition at T_L is suppressed to zero whereas the transition at T_U shifts to 0.67 K [Strydom et al (2006)]. The non-Fermi liquid behavior of the electrical resistivity observed at the critical field in the temperature range 0.1-0.6 K is an indication for the existence of a field-induced quantum critical point [Paschen et al (2007)].
Neutron scattering experiments on polycrystalline $Ce_3Pd_{20}Si_6$ have to date failed to detect any kind of magnetic order [Paschen et al (2008)]. Thus, to clarify the nature of the phases below T_L and T_U large single crystals of high quality are needed. In fact, in the first neutron scattering study on $Ce_3Pd_{20}Si_6$ single crystals [Paschen et al (2008)] the absence of signatures of magnetic order was attributed to a too low T_L value of the investigated specimen.
Since both phase transitions take place at rather low temperatures, disorder may influence them significantly. The discrepancy in the reported ordering temperatures (e.g. [Goto et al (2009); Strydom et al (2006); Takeda et al (1995)]) demonstrates this delicate dependence. Also the non-negligible difference in the reported unit cell constants [Gribanov et al (1994), Takeda et al (1995)] needs a clarification. This has motivated us to undertake a systematic investigation of the relation between crystal growth techniques/conditions, sample quality, and the resulting physical properties down to dilution refrigerator temperatures.
We show here that both phase transitions are extremely sensitive to small stoichiometry variations that result from different growth procedures.

To elucidate the discrepancies in the low-temperature data reported on the quantum critical heavy fermion compound $Ce_3Pd_{20}Si_6$ and reveal the compound's intrinsic properties, single crystals of varying stoichiometry were grown by various techniques – from the stoichiometric and slightly off-stoichiometric melts as well as from high temperature solutions using fluxes of various compositions. The results of this work on $Ce_3Pd_{20}Si_6$ have been partially reported earlier [Prokofiev et al (2009)]. Here a more detailed analysis including also information on new crystal growth experiments as well as the physical property measurements on new single crystals are reported.

2.1 Growth from the stoichiometric melt

To investigate the melting character of $Ce_3Pd_{20}Si_6$ a differential thermal analysis (DTA) run up to 1350°C was carried out (Fig. 1). There is only a single peak both on the heating and on the cooling curve, indicating congruent melting as previously reported [Gribanov et al (1994)]. A closer inspection of the shape of the melting peak (Fig. 1, inset) might, however, suggest merely a quasi-congruent melting character. The onset of melting occurs at $T_M \approx 1250$°C. Because of undercooling the crystallization begins about 100°C lower than the melting.

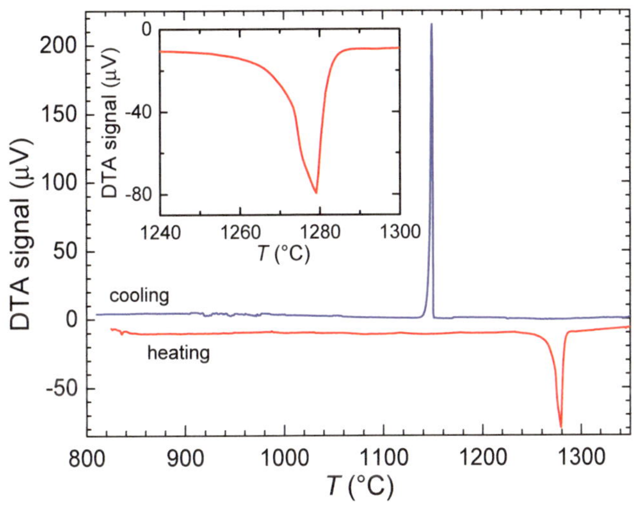

Fig. 1. DTA heating and cooling curves of $Ce_3Pd_{20}Si_6$. The inset shows a magnification of the melting peak. From Ref. Prokofiev et al (2009).

The floating zone technique with optical heating was used for crystal growth from the melt, with a pulling rate of 10 mm per hour and an upper rod rotation speed of 5 rpm. Due to the high density and the low surface tension of the $Ce_3Pd_{20}Si_6$ melt the floating zone was rather unstable, and the melt dropped down repeatedly (Fig. 2, right). However, one growth could be kept running long enough to grow an ingot of 25 mm length (Fig. 2, left).

Over the growth run the originally shiny and clear surface of the melt became more and more opaque, and a crust on the surface could be seen after some time. This is a sign of incongruent melting supposedly due to a peritectic reaction. The crust is the higher temperature phase, therefore it forms on the optically heated surface which is the hottest place of the melt. This observation can be explained by close proximity of the peritectic point to the temperature of the complete melting of the system with the total composition $Ce_3Pd_{20}Si_6$, as indicated already by the peculiar shape of the DTA curve (Fig. 1 inset). Therefore the formation of

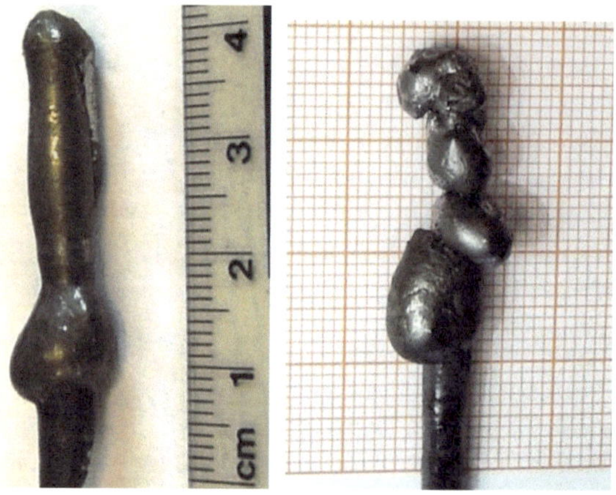

Fig. 2. Samples resulting from crystal growth experiments from the stoichiometric melt with upper and lower rod rotation. Left - Successfully grown single crystal (*sc*1); Right - The crystallized ingot experienced repeated dropping down of the melt because of the unfavorable combination of the high density and of the low surface tension of the melt.

a small amount of foreign phase on the hot surface results in a slight shift of the melt composition, which returns the crystallization process in the melt deep into the primary crystallization field of the $Ce_3Pd_{20}Si_6$ phase.

Due to the thin foreign phase film the surface of the recrystallized (lower) rod was yellowish. The EDX measurement detected Ce ($\sim$ 70 at.%), Pd ($\sim$ 22 at.%), and Si ($\sim$ 8 at.%) in this film. However no inclusions of foreign phases were found by scanning electron microscopy (SEM) inside the crystal. A single crystal grown in this way is specified here as *sc*1 (see Table 1 for the nomenclature of all crystals). In order to trace segregation effects we differentiate additionally between the part of this crystal grown at the beginning of the growth process (bottom part of the ingot, *sc*1*b*) and that grown at the end (top part, *sc*1*t*). For the growth of *sc*1 the lower purity starting materials (Ce 99.99 at.%, Pd 99.95 at.%) were used.

Further growth runs with rotation of the upper rod led to a permanent breaking of the surface crust. The melt leaked out through the cracks of the crust. Hence, the crust may serve as a quasi-crucible if it remains intact during the whole growth time. Based on this observation a growth run without rotation of the upper rod was carried out with the same growth parameters. The melt zone was quite stable in this experiment. A crystal grown by this procedure will be specified as *sc*2. For the growth of this crystal higher purity starting materials (Ce 99.99 at.%, Pd 99.998 at.%) were used.

Laue investigations show very good crystallographic perfection of both types of crystals grown from the melt. Before annealing an SEM/EDX investigation of both crystals was carried out. The polished cross-section of the *sc*1 rod is more homogeneous in composition than that of *sc*2. The surface of the latter had a well distinguishable 300 μm thick outer shell where the concentration of Ce was about 5% higher and that of Si somewhat lower than in the core region (Fig. 3, left). This shell may result from a partial dissolution and diffusion of the crust into the core of the rod. In the core region of *sc*2 diffuse 10 μm inclusions of supposedly the same shell phase, however with lower Ce concentration, were found (Fig. 3, right). These inclusions were not detected by X-ray powder diffraction and they disappeared after annealing, according to SEM analysis.

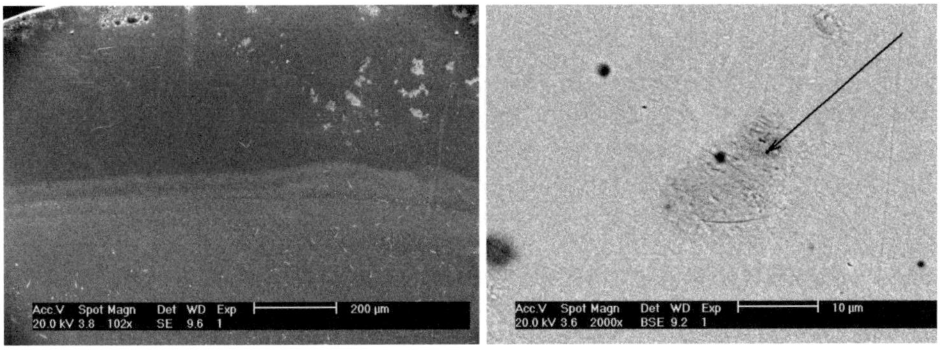

Fig. 3. Microstructure of *sc*2 before annealing.
Left - Surface shell (darker top part) with a 5% higher Ce concentration; Right - An inclusion with a lower Ce concentration in the core (shown by arrow).

After annealing of both crystals for 3 weeks at 900°C a second EDX investigation was carried out. At first we did the analysis *without* any reference sample. This yielded a stoichiometric Ce content (10.3 at.%) but an over-stoichiometric Si content (22.0 instead of 20.7 at.%) and an under-stoichiometric Pd content (67.7 instead of 69.0 at.%) for both *sc*1 and *sc*2, corresponding to a partial substitution of Pd by Si on their sites. These results motivated our efforts to grow crystals using the flux technique, where the stoichiometry of the crystals can be tuned through the variation of the flux composition. The results of further EDX investigations *with* a reference sample will be discussed in Sect. 2.4.

Sample name	Growth technique details
*sc*1*t*	from melt with rod rotation, top part
*sc*1*b*	from melt with rod rotation, bottom part
*sc*2	from melt without rod rotation
*sc*3	from Pd_5Si flux, molar $Ce_3Pd_{20}Si_6$ to Pd_5Si ratio 2:1
*sc*4	from Pd_5Si flux, molar $Ce_3Pd_{20}Si_6$ to Pd_5Si ratio 1:2
*sc*5	from slightly off-stoichiometric melt, with a 0.3 at.% Ce-excess and a 0.2 at.% Pd-deficiency
Sn flux crystals	using Sn or Sn-Pd fluxes

Table 1. Nomenclature of the $Ce_3Pd_{20}Si_6$ crystals grown by various techniques.

2.2 Flux growth

2.2.1 Tin flux

We tried at first a crystal growth with standard fluxes. The related compound $Ce_2Pd_3Si_5$ can be grown with Sn as flux at below 1100°C [Dung et al (2007)]. To check for the compatibility of Sn flux with $Ce_3Pd_{20}Si_6$, a mixture of $Ce_3Pd_{20}Si_6$ and Sn was heated up to 1100°C in a boron nitride crucible and then cooled slowly (1°C/h) down to 700°C. The solute-to-solvent ratio was 2:1. After crystallization the ingot was cut, polished, and investigated by SEM/EDX and X-ray powder diffraction (XRD).
The crystallization yielded relatively large single crystals of the non-stoichiometric phase $CePd_{2-x}Si_{2+x}$ with $x \approx 0.25$, incorporated in a matrix of Sn_4Pd. In addition, small inclusions

of other phases were found. The experiment thus shows the inapplicability of Sn as a flux because its affinity to Pd leads to a destruction of the $Ce_3Pd_{20}Si_6$ phase.
However, other Sn-based flux compositions with a lower affinity to Pd can be found in the Sn-Pd binary phase diagram [Chandrasekharaiah (1990)]. A series of Sn-Pd compounds - Sn_4Pd, Sn_3Pd, Sn_2Pd - with low melting points (below 600°C) exists.
Similar experiments as the one with pure Sn flux were carried out using the above Sn-Pd compositions. In all cases the primarily crystallized phase was $CePd_{2-x}Si_{2+x}$ according to SEM/EDX. The x-value diminished with increasing Pd content in the flux, reaching about 0.05 for Sn_2Pd flux. The stoichiometric $CePd_2Si_2$ phase has the same Ce/Si ratio as the $Ce_3Pd_{20}Si_6$ phase but a strongly reduced Pd content. However, further increasing of the Pd concentration in the flux for tuning of Pd content in crystals was impossible: with the higher melting compound SnPd, only partial melting of the crucible content was observed at 1100°C. Thus, foreign flux growth was not successful.

2.2.2 Self-flux Pd_5Si

The reason for using the flux method was, on one hand, to obtain single crystals with exact stoichiometric composition and, on the other hand, the expectation that the off-stoichiometric melt would have a higher surface tension and hence the floating zone would be more stable than without flux. To avoid a contamination by foreign atoms we first opted for self flux growth. Since the concentration of thermal defects (e.g., Si - Pd substitutions) in the crystal is expected to decrease with decreasing growth temperature, we searched in the Ce-Pd-Si phase diagram (Fig. 4a) for low-melting (at first binary) compositions with an over-stoichiometric Pd content.
The phase Pd_5Si [Seropegin (2001)] which, according to a later study [Gribanov et al (2006)], appears to consists or two scarcely distinguished phases $Pd_{14}Si_3$ and $Pd_{84}Si_{16}$, fulfills all requirements: it melts at a relatively low temperature of 835°C [Chandrasekharaiah (1990)], has an over-stoichiometric (> 20:6) Pd/Si ratio, and there are no stable Ce-containing intermediate phases between Pd_5Si and $Ce_3Pd_{20}Si_6$ in the Ce-Pd-Si ternary phase diagram (cross-section at 600°C [Gribanov et al (2006); Seropegin (2001)]).
As above, the floating zone technique with optical heating was used. The feed and the seed rods had the stoichiometric starting composition $Ce_3Pd_{20}Si_6$, and the zone was a molten mixture of $Ce_3Pd_{20}Si_6$ and Pd_5Si. Contrary to the melt growth, the floating zone was very stable, and its surface remained clear during the entire growth run. The latter means that the growth occurred within the $Ce_3Pd_{20}Si_6$ primary crystallization field. The stability of the melt zone allowed the rotation of the upper rod. The pulling rate was 0.6 mm/h. Two growth runs with different flux compositions (molar ratios $Ce_3Pd_{20}Si_6$ to Pd_5Si of 2:1 and 1:2) were carried out (Fig. 4b). The crystals were annealed for 3 weeks at 900°C. The corresponding samples are referred to as *sc*3 and *sc*4, respectively.

2.3 Growth from slightly off-stoichiometric melt

As the analysis of the composition and the properties of the crystals *sc*1 - *sc*4 grown from the melt and from flux has shown (see Sections 2.4, 2.5) the stoichiometry was strongly sensitive to the starting composition of the melt. For a fine correction of non-stoichiometry a growth from a slightly off-stoichiometric melt was carried out, too (*sc*5). For reasons discussed later, the feed rod for *sc*5 contained 0.3 at.% excess of Ce and 0.2 at.% deficiency of Pd. To avoid the floating zone instability the growth was carried out without rod rotation, as in case of *sc*2.

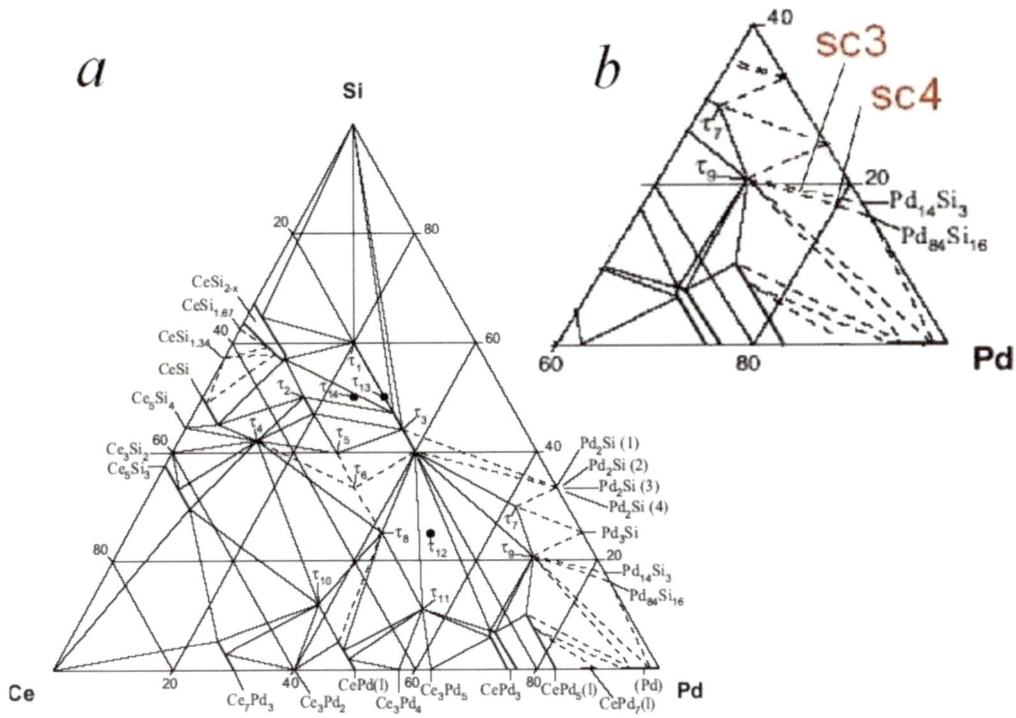

Fig. 4. a) Phase diagram of the Ce-Pd-Sn ternary system (from Gribanov et al (2006)). Here $Ce_3Pd_{20}Si_6$ is denoted as τ_9-phase. b) Magnified part of the phase diagram with the compositions of the solutions from that *sc*3 and *sc*4 were grown.

The pulling rate was 4 mm/h. Similar to *sc*1 we differentiate additionally between the bottom and top parts of the ingot, *sc*5*b* and *sc*5*t*.

2.4 Composition of the grown crystals

The usual EDX technique without standards cannot determine absolute atomic concentrations with sufficient accuracy. This problem can, in principle, be solved by using a standard of exactly known composition. However, as such a sample is not readily available, we used, instead, the polycrystalline sample (*pc*) which will be shown below to be of best quality, according to the physical property measurements. Irrespective of whether its composition may be identified with the *exact* stoichiometry 3:20:6 or not it served as a practical guideline to establish the crystal composition–property relationship.

With additional improvements of our EDX setup (improved measurement statistics, counting time, beam current control) we can measure, with a high accuracy, *deviations* from the reference sample stoichiometry. The results are summarized in Table 2 where the measured lattice parameters are given, too. Because crystals grown from the Sn-containing fluxes are not the title phase and their compositions vary sizably they are simply represented by $CePd_{2-x}Si_{2+x}$ in Table 2.

Even though, in the absence of a real standard, there remains some uncertainty about the *absolute* values measured by EDX (which even for the reference sample *pc* differ from the calculated stoichiometry 3:20:6), *trends* in the composition of the investigated series of crystals

Sample name	Composition (at.%)			a (Å)
	Ce	Pd	Si	
sc1t	10.5	67.6	21.9	12.276
sc1b	9.9	67.7	22.4	12.272
sc2	9.9	68.0	22.1	12.272
sc3	9.2	68.7	22.1	12.233
sc4	7.4	70.1	22.5	12.180
sc5	10.7	68.0	21.3	12.277
pc	10.1	67.7	22.2	12.280
Exact 3:20:6 stoichiometry	10.3	69.0	20.7	
Sn flux crystals	$CePd_{2-x}Si_{2+x}$			
Standard deviation	0.24	0.20	0.20	
Systematic error				<0.002

Table 2. Composition and lattice parameter a of the investigated $Ce_3Pd_{20}Si_6$ samples.

can be discussed. While the Si content varies only weakly a stronger variation of the Ce and Pd content is observed, the Ce concentration nearly anticorrelating with the one of Pd (Fig. 5).

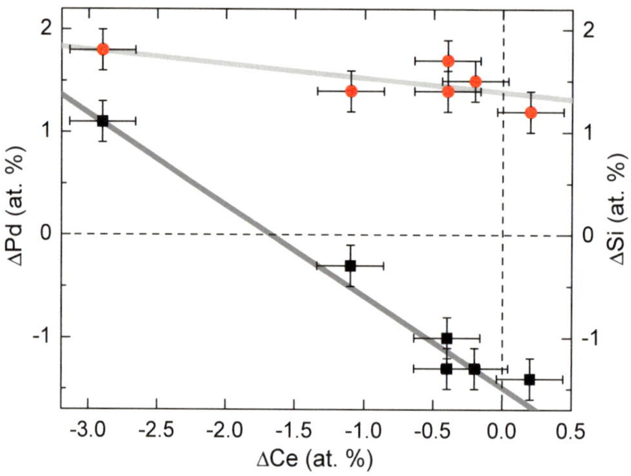

Fig. 5. Deviation of the Pd and Si content, ΔPd and ΔSi, from the exact 3:20:6 stoichiometry as function of the Ce non-stoichiometry ΔCe. The dashed lines indicate the exact stoichiometry. The meaning of the broad grey lines is discussed in the text. Error bars indicate the standard deviations given in Table 2. From Ref. Prokofiev et al (2009).

Ce *over*-stoichiometry can be realized by a substitution of Pd or Si atoms on their crystallographic sites by excess Ce-atoms or, alternatively, by vacancies on the Pd and/or Si sites. Since the lattice parameter of *sc1t* and *sc5* is smaller than that of the *pc*, the latter option is more probable, but a combination of both options cannot be excluded. If only Pd/Si vacancies were present, the Ce sublattice would remain occupied and fully ordered. Ce *under*-stoichiometry, on the other hand, can be associated either with Ce vacancies or, which is more probable taking into account the approximate anti-correlation between the Ce and Pd contents, with a partial Pd substitution on Ce sites. The broad dark grey line in Fig. 5 represents the Ce-Pd concentration relation for a crystallochemical model which, as an example, assumes a half-filling of the Ce vacancies V by excess Pd atoms:

$[Ce_{1-x}V_{0.5x}Pd_{0.5x}]_3Pd_{20}Si_6$. The same model was used to describe the ΔSi vs ΔCe relation (light grey line). The agreement with the data is excellent.

2.5 Influence of the growth technique on the physical properties

The resulting stoichiometry of the crystals as well as their physical properties show sizable dependence on the different growth techniques.

Neutron diffraction experiments were carried out on an oriented sample cut from *sc1b*. The crystal was confirmed to have excellent crystallinity but, contrary to our expectations from the investigations on polycrystalline samples, showed no phase transition down to the lowest accessed temperature of about 0.15 K [Paschen et al (2008)]. The present investigation shows that, for this very single crystal, this temperature was still too high.

The specific heat C_p was measured for *sc*1 - *sc*3 and *sc*5. Figure 6a shows the temperature dependence of C_p/T of these crystals, together with published data of a polycrystalline sample [Strydom et al (2006)]. $C_p/T(T)$ of the polycrystalline sample *pc* has a sharp peak with a maximum at $T_L = 0.31$ K and a shoulder at $T_U \sim 0.5$ K [Strydom et al (2006)]. These features get successively broadened and suppressed to lower temperatures in the crystals *sc*1*t*, *sc*5*b*, *sc*1*b*, and *sc*2, respectively. In *sc*3 no clear signature of T_L can be identified, suggesting that it has shifted further to lower temperatures and was further broadened or, alternatively, has vanished altogether. Due to the suppression of this lower temperature feature the anomaly at T_U, identified in all other samples as shoulder, now appears as a maximum.

From the specific heat measurements the best single crystals thus appear to be *sc*1*t* and *sc*5. With their rather sharp peaks at 0.22 K and 0.20 K, respectively, and a shoulder at about 0.4 K they demonstrate all features associated with the intrinsic behavior of $Ce_3Pd_{20}Si_6$.

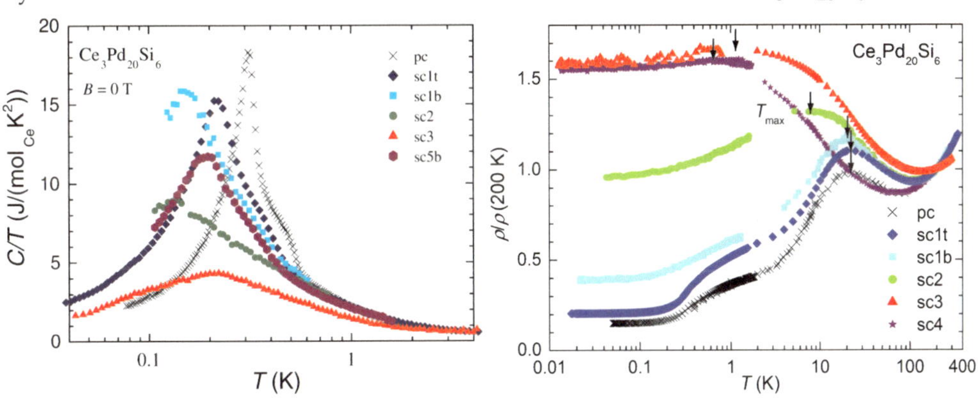

(a) (b)

Fig. 6. a) Specific heat divided by temperature C_p/T plotted for all single crystals prepared here, and for a polycrystal [Strydom et al (2006)] for comparison, as function of temperature T on a logarithmic scale. The lower transition temperatures T_L are taken here as position of the maxima. The maximum in the data for *sc*3 can be identified with T_U (see text).
b) Electrical resistivity of the $Ce_3Pd_{20}Si_6$ single crystals grown here normalized to the respective values $\rho_{200\,K}$ vs temperature T. The arrows indicate the positions T_{max} of the (local) maxima. Data of a polycrystal [Strydom et al (2006)] are shown for comparison. From Ref. Prokofiev et al (2009).

The electrical resistivities of all single crystalline samples and of one polycrystalline sample are shown in Figs. 6b and 7 as function of temperature. While the resistivity of the polycrystalline sample drops to very low values at low temperatures, the residual resistivities of the single crystals are considerably higher. They increase in the sequence *pc* $\rightarrow$ *sc5t* $\rightarrow$ *sc1t* $\rightarrow$ *sc5b*$\rightarrow$ *sc1b* $\rightarrow$ *sc2* $\rightarrow$ *sc3*. This is about the same sequence in which the temperature and the sharpness of the low-temperature phase transition of the C_p/T curves decreases (Fig. 6a). Thus, it is natural to assume that in the sequence *pc* $\rightarrow$ (*sc1t* and *sc5*)$\rightarrow$ *sc1b*$\rightarrow$ *sc2* $\rightarrow$ *sc3* the lattice disorder increases. Since the starting material purity of *sc1t* and *sc1b* was lower than that of *sc2*, we conclude that the main reason of the disorder is a deviation of the sample composition from the exact stoichiometry 3:20:6 and not the concentration of foreign atoms. For single crystal *sc4* the high temperature minimum occurs at lower temperatures than expected, leading to a slightly lower residual resistivity than for *sc3*.

In *sc1* a remarkable increasing of the quality from the bottom to the top part of the crystal is observed. Only a small (top) part of the total crystal has an excellent quality and can be used for physical property investigations. *sc5* is indeed ranked after *sc1t* according to the C_p data but it is more homogeneous throughout the whole volume of the batch. A scenario which can explain the difference between the bottom (*sc1b*) and the top (*sc1t*) parts of the crystal *sc1* is as follows. At first (bottom part of the ingot) the crystal phase captures less Ce and more Pd from the stoichiometric melt (Table 1). While the crystal grows the melt gets enriched by Ce and depleted by Pd. At the end of the crystallisation (top part) this change in the melt composition results in a shift of the crystal composition to a more stoichiometric one, in accordance with the law of mass action. It was this observation that motivated us to perform the off-stoichiometric growth with a little Ce excess and Pd deficiency that resulted in *sc5*. As Fig. 7 shows the low temperature (down to 2 K) relative resistivity of *sc5* is comparable with the best sample *sc1t* and the spatial (top-bottom) difference in the resistivity is much smaller for *sc5* than for *sc1*.

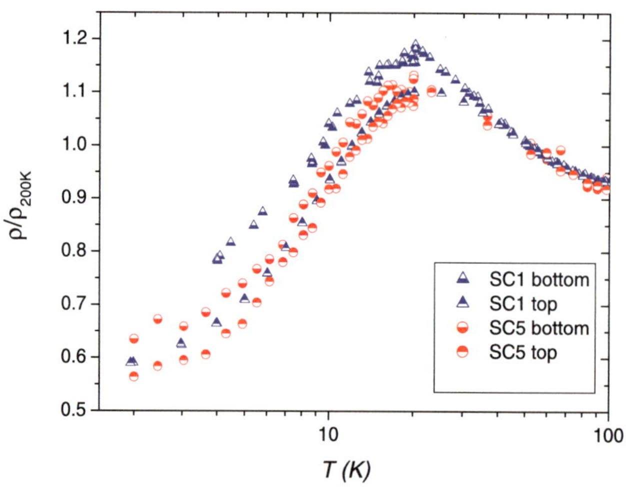

Fig. 7. Relative resistivities of the bottom and top parts of *sc1* and *sc5*.

The highest residual resistance ratio (*RRR*, defined here as $\rho(200\ \text{K})/\rho(50\ \text{mK})$) and the sharpest and most pronounced phase transition features in C_p/T are found for the polycrystalline sample (*pc*) which therefore appears to be the most stoichiometric one. This can be easily understood by the specifics of preparation. The accuracy of the stoichiometric *total* composition of a polycrystalline sample is limited only by the accuracy of the weighing process of starting materials and by their purity. A possible high temperature non-stoichiometry of the main phase of an as-cast polycrystal is compensated by the presence of minor impurity phases, the phase separation being heterogeneous on a microscopic scale. This heterogeneity can be lifted by annealing at lower temperatures due to the short diffusion path. During single crystal growth, however, a macroscopic phase separation can occur, making annealing very inefficient. Actually the resistivity curves of *sc1b* before and after annealing were practically identical.

The Ce content ΔCe varies by more than 3 at.% among all grown single crystals. We have revealed a systematic dependence of the residual resistance ratio, the lattice parameter, the (lower) phase transition temperature T_L, and the maximum in the temperature dependent electrical resistivity T_{max} with ΔCe. This clarifies the sizable variation in the values of T_L reported in the literature. We discuss the physical origin of the observed composition–property relationship in terms of a Kondo lattice picture. We predict that a modest pressure can suppress T_L to zero and thus induce a quantum critical point.

While no clear correlation between the physical properties and the Si content can be demonstrated, a pronounced dependence on the Ce content (or anti-correlated Pd content) is observed. Figure 8a shows the dependencies of the lattice parameter a and the residual resistance ratio RRR as function of the deviation from stoichiometry ΔCe. The largest deviations from the "intrinsic behavior" (which is most closely met by the polycrystal) are seen for *sc*4. It has the lowest Ce content of 7.4 at. % (ΔCe = 2.9 at. %). The polycrystalline sample (*pc*) which demonstrates the most pronounced phase transition features and *sc1t* with the second-sharpest features have the highest Ce contents, – and the lowest deviation from the exact Ce stoichiometry. The deviations ΔCe of *sc1t* and *sc*5 lie on the other side of the stoichiometry line. One may argue that only the absolute value of ΔCe is relevant with respect to the composition–disorder–property relationship since *any* non-stoichiometry is usually associated with lattice imperfection. To test this conjecture we plot, in Fig. 8a, the data point for *sc1t* also mirrored through the ΔCe = 0 axis (open symbol). Indeed this point fits nicely into the overall $RRR(\Delta\text{Ce})$ behavior.

As it was pointed out in Sec. 2.4, the Ce-understoichiometry may be realized by two ways: either by Ce-vacancies or by Ce-vacancies partially (to 50%) filled by Pd atoms. Both cases of Ce under-stoichiometry correspond, in a Kondo lattice description, to Kondo holes. In the Kondo coherent state at low temperatures Kondo holes act as strong scattering centers, decreasing the RRR. This effect is seen in Fig. 8a. Since also *sc1t* has a reduced RRR, over-stoichiometry appears to be indeed realized by the combination of excess Ce and Pd/Si vacancies. Excess Ce, just as Ce holes, creates Kondo disorder. In addition to reducing RRR, Kondo disorder is expected to successively suppress the temperature T_{max} where the high-temperature incoherent Kondo scattering with an approximate $\rho \propto -\ln T$ behavior crosses over to coherent Kondo scattering at low temperatures. In samples *sc*3 and *sc*4 Kondo disorder is so strong that $\rho \propto -\ln T$ is observed in a wide temperature range, and $\rho(T)$ shows only a tiny drop below 2 K. Figures 6b and 7 show that a sizable suppression of T_{max} occurs in our sample series *pc* – (*sc1t* and *sc*5) – *sc1b* – *sc*2 – *sc*3 – *sc*4 only from *sc*2 on. Thus, in addition to Kondo disorder, there appears to be a second effect influencing T_{max} in the opposite direction. This can be identified as a volume effect: the Ce vacancies or (smaller) Pd atoms on Ce sites in Ce under-stoichiometric samples as well as the Pd or Si vacancies in the over-stoichiometric sample *sc1t* lead to a shrinkage of the crystal lattice, which is seen in Table 2 and in Fig. 8a. The decrease in volume with decreasing Ce content corresponds to positive (chemical) pressure. Using the bulk modulus of the closely related compound $Ce_3Pd_{20}Ge_6$ [$\sim$ 137.5 GPa at 150 K; Nemoto et al (2003)] we can convert the lattice shrinkage in our off-stoichiometric single crystals into a hypothetical external pressure p.

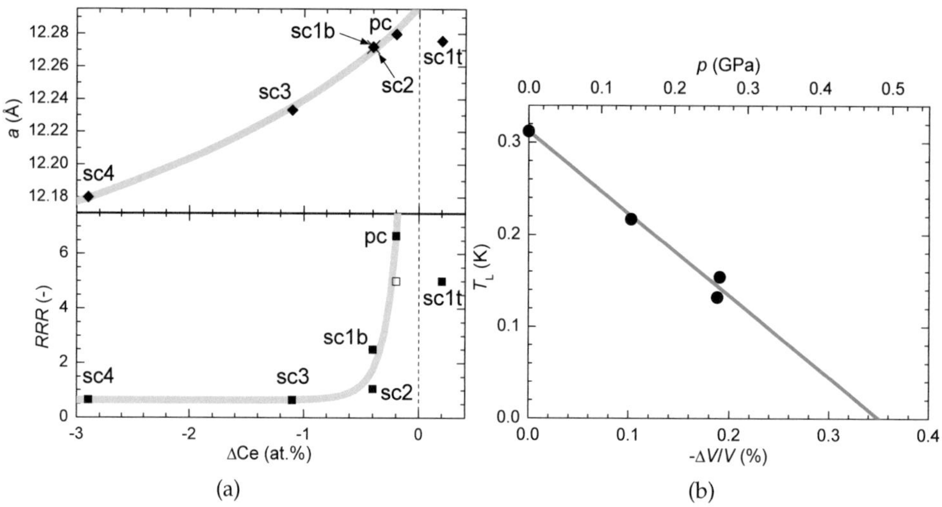

Fig. 8. a) Dependence of the lattice parameter a and the residual resistance ratio RRR on the deviation ΔCe from the stoichiometric Ce content. The open symbol in the lower panel represents the data point of $sc1t$, mirrored through the $\Delta\text{Ce} = 0$ line. The thick grey lines are guides to the eye. b) Lower transition temperature T_L vs relative volume shrinkage $-\Delta V/V$ of the different $Ce_3Pd_{20}Si_6$ single crystals with respect to the volume of the polycrystal pc. On the upper axis the corresponding pressure as estimated via the bulk modulus of $Ce_3Pd_{20}Ge_6$ [Nemoto et al (2003)] is given. The line represents a linear fit to the data and its extrapolation to $T_L = 0$. From Ref. Prokofiev et al (2009).

Figure 9 shows T_{max} (full dots) vs the relative volume shrinkage $-\Delta V/V$ of our single crystals with respect to our polycrystal (pc) (bottom axis) and vs a hypothetical external pressure calculated via the bulk modulus [Nemoto et al (2003)] (top axis). The step-like, as opposed to continuous, decrease of T_{max} with $-\Delta V/V$, addressed already above, is clearly seen. In order to understand this behavior we compare our results to a pressure study [Hashiguchi et al (2000)]. We include the T_{max} vs p data from this investigation on a polycrystalline sample under quasi-hydrostatic pressure in Fig. 9 (crosses and grey fit to these). From a value $T_{max} \approx 20$ K at $p = 0$, similar to the value for our polycrystal, T_{max} *in*creases continuously with increasing pressure. This behavior is typical for Ce-based heavy fermion compounds: with increasing pressure the hybridization between the $4f$ electron of Ce and the conduction electrons increases and thus the Kondo temperature which is proportional to T_{max} increases. Our data follow this trend only at low non-stoichiometry (low $-\Delta V/V$ values) while at higher non-stoichiometry T_{max} decreases quickly. Thus we conclude that the volume effect on T_{max} dominates in our sample series pc – $sc1t$ – $sc1b$ – $sc2$ – $sc3$ – $sc4$ up to $sc1b$ while the Kondo disorder effect dominates from $sc2$ on.

Next we analyze the evolution of the ordering temperature T_L along our sample series. For this purpose T_L, extracted from the specific heat data in Fig. 6a, is plotted vs RRR in Fig. 9b and vs $-\Delta V/V$ (bottom axis) or p (top axis) in Fig. 8b. T_L decreases continuously with decreasing RRR (Fig. 9b). This might be taken as indication for the strong influence of disorder on T_L. However, in our sample series, the variation of RRR is intimately linked to a variation of ΔCe (Fig. 8a) and thus to a variation of the lattice parameter a, the relative volume change $-\Delta V/V$, and the corresponding pressure p. Thus, an alternative

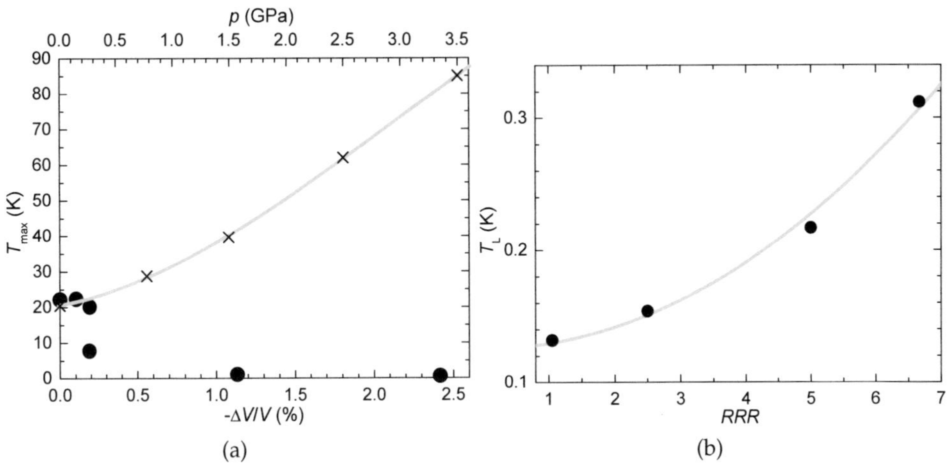

Fig. 9. a) Temperature T_{max} of the maximum in the electrical resistivity *vs* relative volume shrinkage $-\Delta V/V$ of our $Ce_3Pd_{20}Si_6$ single crystals with respect to the polycrystal (dots) and, for comparison, T_{max} *vs* pressure p for the polycrystal of Hashiguchi et al (2000) (crosses and grey line, which is fit to the data). $\Delta V/V$ and p are related to each other via the bulk modulus of $Ce_3Pd_{20}Ge_6$. b) Lower transition temperature T_L vs residual resistance ratio RRR for the different $Ce_3Pd_{20}Si_6$ samples. The line represents a quadratic fit to the data. From Ref. Prokofiev et al (2009).

view of the situation is that not increasing disorder but decreasing volume is responsible for suppressing T_L (Fig. 8b). To decide which effect is dominant we come back to our above discussion on Fig. 9a which revealed that the volume effect on T_{max} dominates in the series $pc - sc1t - sc1b - sc2 - sc3 - sc4$ up to $sc1b$ while the disorder effect dominates from $sc2$ on. If this holds true also for the effect on T_L it is Fig. 8b (without $sc2$) that captures the essential physics while Fig. 9b only displays implicit dependencies (except for $sc2$).
In a Kondo lattice picture the physical origin of the suppression of T_L with p is that pressure increases the hybridization between conduction electrons and $4f$ states, thus strengthening Kondo compensation, diminishing Ce magnetic moments, and suppressing the (most probably magnetic) ordering temperature T_L. Extrapolating the T_L vs p dependence of Fig. 8b to higher p suggests that a pressure induced quantum critical point is at reach for $Ce_3Pd_{20}Si_6$. A linear extrapolation of the fit shown in Fig. 8b yields $p_c \approx 0.5$ GPa. Of course, actual low-temperature pressure measurements are needed to verify this possibility.
Finally, we comment on a related study on the germanide compound $Ce_3Pd_{20}Ge_6$ where the influence of the starting composition on the physical properties was investigated [Kitagawa et al (1999)]. A strong composition effect on C_p and ρ was found to occur inside a narrow homogeneity range where a volume contraction of up to 0.7% takes place with increasing Pd content. In that work the concentration of Pd and not of Ce was concluded to be variable and responsible for the changing physical properties. It should be noted that the real composition of the resulting phases was not investigated. Thus, it is plausible that the increasing Pd content is accompanied by a decreasing Ce content. Under this assumption the composition effect on the properties of $Ce_3Pd_{20}Ge_6$ and $Ce_3Pd_{20}Si_6$ are indeed comparable.

3. $CeRu_4Sn_6$

In Kondo insulators (frequently also referred to as heavy fermion semiconductors) a narrow gap develops at low temperatures in the electronic density of states at the Fermi level [Aepli & Fisk (1992)]. While most Kondo insulators known to date adopt a cubic crystal structure (e.g., YbB_{12}, SmB_6, FeSi, $Ce_3Bi_4Pt_3$) a few compounds (e.g., CeNiSn, CeRhSb) are orthorhombic. These latter show anisotropic properties which suggest that the energy gap vanishes along certain directions in *k*-space. $CeRu_4Sn_6$, first synthesized by Das and Sampathkumaran [Das & Sampathkumaran (1992)], crystallizes in a tetragonal structure of space group $I\bar{4}2m$ (a = 6.8810 Å, c = 9.7520 Å, c/a = 1.4172) [Venturini et al (1990)]. A peculiarity of this compound is that, in addition to the tetragonal (body-centered) cell with lattice parameters a and c there exists a quasi cubic (face-centered) cell with lattice parameters c' and c, where c' is the diagonal of the tetragonal plane which differs by only 0.2% from c. This makes it very difficult to orient single crystals in an unambiguous way. On the other hand it allows us to study a "tetragonal" and a "quasi-cubic" Kondo insulator within the same material which is very appealing.

3.1 Choice of the growth method and the growth procedure

The compound $CeRu_4Sn_6$ melts incongruently. Melting followed by rapid quenching yields $CeRuSn_3$, Ru_3Sn_7 and a tiny amount of the title phase in the solidified material. The incongruent melting was confirmed by a DTA experiment which showed multiple peaks on the heating and on the cooling curves (Fig. 10). Since crystal growth from the melt is impossible for an incongruently melting compound we searched for an appropriate flux. Usually low melting metals are used as high temperature solvents for intermetallic compounds. For $CeRu_4Sn_6$ this might be tin. However $CeRu_4Sn_6$ was shown to react with tin with the formation of other phases. To avoid possible contaminations by foreign elements we searched thereafter for a flux composition in the ternary Ce-Ru-Sn system (self-flux). Unfortunately, the Ce-Ru-Sn phase diagram which would be helpful for the choice of the best flux composition has not been reported. A Ce-rich flux for the Ce-poor $CeRu_4Sn_6$ is expected to lead to the formation of the highly stable phase $CeRuSn_3$. Therefore, a Ce-free binary Ru-Sn mixture is more appropriate. Taking into account the Ru:Sn ratio of 2:3 in

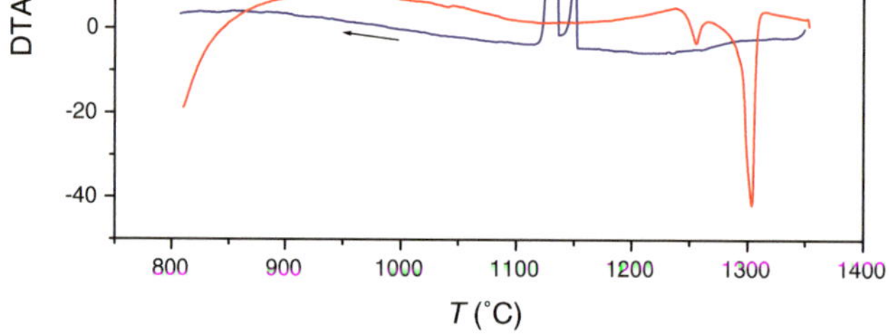

Fig. 10. DTA heating (red) and cooling (blue) curves of $CeRu_4Sn_6$.

$CeRu_4Sn_6$ the flux composition "Ru_2Sn_3" is optimal to maintain the Ru:Sn stoichiometry in the crystals. Moreover, the lowest melting (1160°C) composition in the Ru-Sn binary diagram is the eutectic with the composition 42.5 at.% Ru and 57.5 at.% Sn [Massalski (1990)] which is close to the element ratio 2:3. According to Ananthasivan (2002) the system is a (partially immiscible) liquid above 1200°C in the composition range of 37–75 at.% Sn. Finally, the large excess of Ru and Sn in the melt is expected to suppress the formation of the stable Ce-rich $CeRuSn_3$ phase.

The crystals were grown by the floating-solution-zone traveling heater method (THM) in a mirror furnace. The seed and the feed rods had the stoichiometric composition whereas the melt zone was a solution of $CeRu_4Sn_6$ in Ru_2Sn_3. The growth rate was 0.3–1.0 mm/h. The composition of the crystals along their length in the growth direction was investigated by SEM/EDX analysis. No noticeable deviation from the stoichiometric ratio 1:4:6 can be observed (Fig. 11).

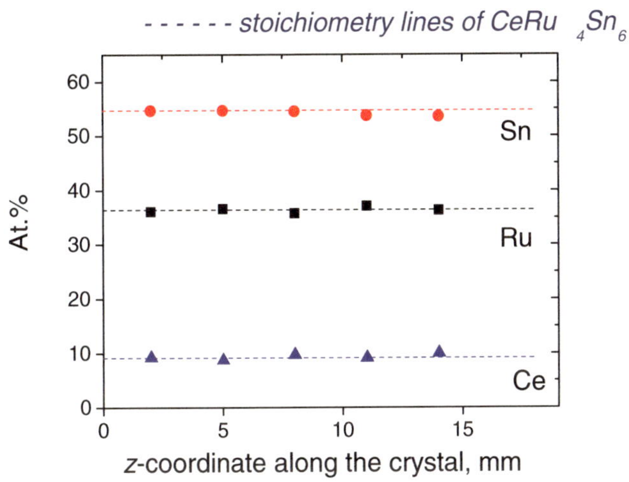

Fig. 11. Element concentration profiles of the $CeRu_4Sn_6$ single crystal in the growth direction.

3.2 Physical properties of $CeRu_4Sn_6$ single crystals

Figure 12 shows the temperature dependence of the electrical resistivity, $\rho(T)$, of $CeRu_4Sn_6$ on a semi-logarithmic scale. With decreasing temperature ρ first increases steeply (range 1), then passes over a maximum at about 10 K, increases again, albeit less steeply (range 2), and finally tends to saturate at the lowest temperatures. A possible explanation of this behaviour is a double-gap structure frequently encountered in simple semiconductors: a larger intrinsic gap visible at high temperatures (range 1) and a smaller extrinsic gap between impurity states and the band edge that dominates the low-temperature behaviour (range 2). Approximating $\rho(T)$ between 120 and 300 K with exponential behaviour ($\rho = \rho_0 \exp(\Delta_1/2k_BT)$, Arrhenius law) yields an energy gap Δ_1/k_B = 125 K, sizeably larger than previously reported for polycrystalline samples [Brünig et al (2006); Das & Sampathkumaran (1992)]. Fitting the data between 0.8 and 1.8 K with the same function yields $\Delta_2/k_B = 0.1$ K (see full red lines in Fig. 12 for both fits). While this gap value may seem incongruous with respect to the fit range, it has to be bourn in mind that the influence of the low-T gap on $\rho(T)$ is expected to diminish by eventual thermal depopulation of the upper states towards $T = 0$ and hence the observed saturation in $\rho(T \to 0)$. Thus, a more complete fitting function is needed to account for these effects.

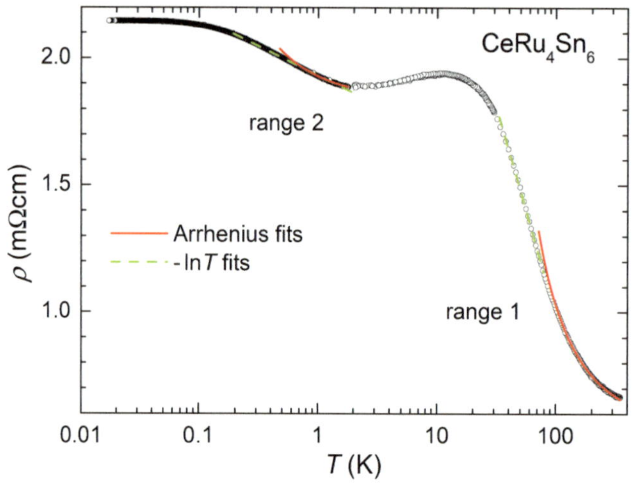

Fig. 12. Temperature dependence of the electrical resistivity, $\rho(T)$, of the $CeRu_4Sn_6$ single crystal on a semi-logarithmic scale. From Ref. Paschen et al (2010).

Now, having single crystals an investigation of the anisotropy of $CeRu_4Sn_6$ became feasible. The magnetic susceptibility was measured on an oriented single crystal in two mutually perpendicular directions (Fig. 13a). One of these directions is the crystallographic c axis, the other one is situated within the tetragonal plane. A pronounced difference is seen. For both directions Curie Weiss-type behaviour is observed at high temperatures, with an effective magnetic moment that is roughly consistent with the full effective moment of Ce^{3+}, and with the paramagnetic Weiss temperatures $\Theta_{\|c} \approx 395$ K and $\Theta_{\perp c} \approx 155$ K for H$\|c$ and H$\perp c$, respectively.

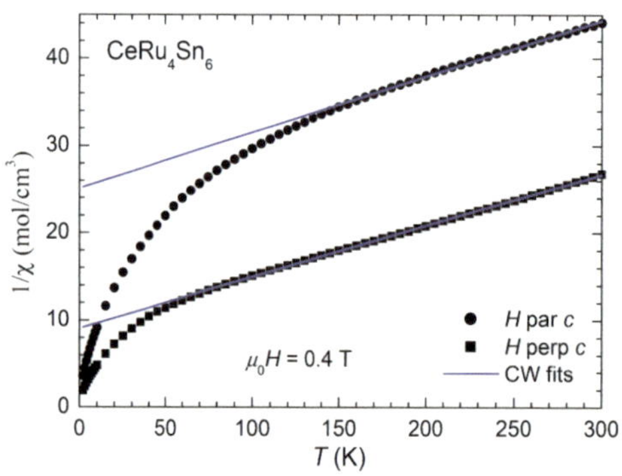

(a)

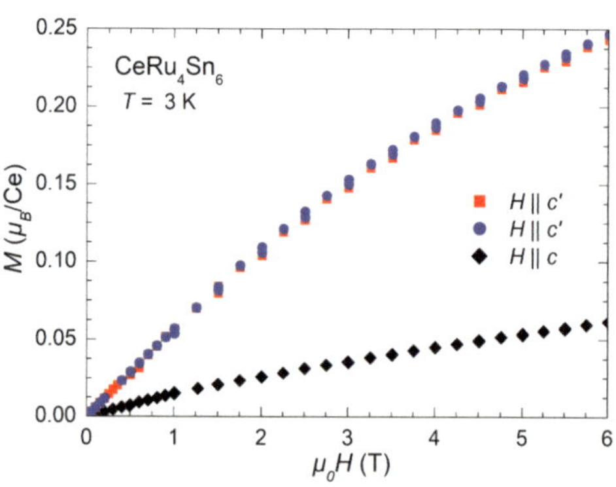

(b)

Fig. 13. a) Temperature dependence of the inverse magnetic susceptibility, $\chi^{-1}(T)$, of $CeRu_4Sn_6$ for the magnetic field $\mu_0 H = 0.4$ T applied along the crystallographic c axis and within the tetragonal plane. b) Magnetic field dependence of the magnetization, $M(\mu_0 H)$, taken at 3 K for fields applied along the crystallographic c and c' axes. Fig. 13b from Ref. Paschen et al (2010).

In order to test whether anisotropy also exists within the quasi-cubic cell we prepared small single crystalline platelets (with geometries which allowed for specific heat measurements only) cut from one piece in such a way that three mutually perpendicular directions were obtained: for two samples a c' axis is perpendicular to the platelet plane, for one sample it is c. As explained above, our Laue diffractograms cannot identify which sample is which. Our specific heat measurements, however, allow to clearly identify the c- and c'-oriented samples: while the zero field data are very similar for all three samples a magnetic field applied perpendicular to the platelet planes induces sizable anisotropy. The difference is best seen by plotting the relative difference in specific heat induced by a magnetic field, $(c_p(B) - c_p(0))/c_p(0)$, which reaches a maximum of more than 70% at 3.5 K for $B||c'$ but is below 15% for $B||c$ at this temperature (Fig. 14). Two samples (*sc*1 and *sc*2) show very similar behavior and must thus be c'-oriented while sc3 shows distinctly different behavior and is thus identified as the c-oriented sample.

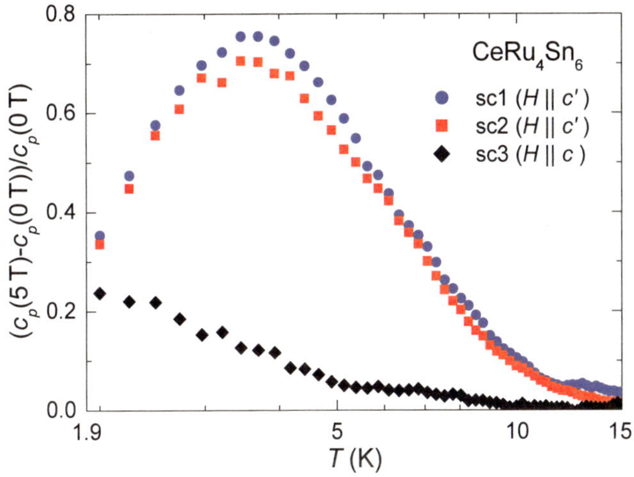

Fig. 14. Temperature dependence of the surplus specific heat induced by a magnetic field of 5 T, $(c_p(5T) - c_p(0))/c_p(0)$, of the three $CeRu_4Sn_6$ single crystals *sc*1, *sc*2, and *sc*3. From Ref. Paschen et al (2010).

A possible interpretation of these data is that a narrow energy gap which is present along c' in zero field is suppressed/diminished by a field of 5 T in this direction. Since the same field applied along c leads to a much weaker increase of c_p we believe that no or a much less field sensitive gap is present along c [Paschen et al (2010)].

4. CeAuGe

A sizable number of Ce- and Yb-based intermetallic compounds demonstrate quantum critical behaviour. Most of the compounds are antiferromagnets, whereas only a handful of Ce and Yb compounds with a ferromagnetic phase transition at low temperatures is known. The occurrence of quantum criticality in a ferromagnetic ground state is a subject of current debate. CeAuGe orders ferromagnetically at a relatively low Curie temperature of $T_C = 10.9$ K [Sondezi-Mhlungu et al (2009), Mhlungu & Strydom (2008)]. Thus it can be expected that the magnetic order can be tuned or fully suppressed by modest variations in magnetic field or pressure.

CeAuGe is a phase with a homogeneity range. Stoichiometric and nearly stoichiometric CeAuGe adopts the NdPdSb structure, an ordered variant of the AlB_2 structure type. The unit cell is hexagonal (space group $P6_3mc$) with the lattice parameters a= 4.4569 Å and c= 7.9105 Å [Pöttgen et al (1998)]. At larger deviations from the elemental 1:1:1 ratio a phase with a slightly different structure forms. The non-stoichiometric $CeAu_{1-x}Ge_{1+x}$ crystallizes in the true AlB_2 structure type and has about twice smaller unit cell. Unlike the ordered 1:1:1 phase, Au and Ge atoms are distributed statistically and the Au/Ge layers are not puckered but planar [Jones et al (1997)].
Since CeAuGe is not cubic it is especially important to investigate its physical properties on oriented single crystalline samples. We have investigated the crystal growth of this phase of various stoichiometries by the floating zone technique. In the course of the growth experiments we encountered a severe non-stoichiometry problem. We report here on our efforts to diminish the deviation from the 1:1:1 stoichiometry and the segregation effects resulting from it.

4.1 Crystal growth using stoichiometric feed rods

As a starting point, growth from the stoichiometric melt was tried. We studied the evolution of the crystal composition during the crystallization by measuring the composition at the starting, middle and final part of the crystallized ingot by the EDX technique. The concentration profiles are represented in Figs. 15 and 16. At the left border of each panel the initial compositions of the respective polycrystalline feed rod is shown. The beginning of the crystallization corresponds thus to the length coordinate $z = 0$. As Fig. 15a shows, the crystals primarily crystallized from the *stoichiometric melt* (i.e. the first portion of the crystalline phase) have a non-stoichiometric composition with a reduced Au content and increased Ce and Ge contents. This leads to a change of the melt composition with an accumulation of Au and a depletion of Ce and Ge. As a consequence, the Au content increases in the crystals and the Ce and Ge contents decrease in the course of further crystallization (Fig. 15a).
Due to the composition change the lattice parameters change too (Fig. 15b). For the first solidified crystals the a-parameter is lower and the c-parameter is higher than the stoichiometric values (Fig. 15b). Further crystallization leads to a decrease of c and to an increase of a.
Thus, the growth using a stoichiometric feed rod results in a non-stoichiometric single crystalline ingot which, along its length, is macro-inhomogeneous with respect to all three constituting elements.

4.2 Crystal growth using off-stoichiometric feed rods

In order to suppress the above discussed starting deviation in the element concentration in the crystals with respect to the melt we used a feed rod enriched in Au and depleted in Ge content, with the off-stoichiometric composition $CeAu_{0.96}Ge_{1.04}$. The *primary* crystals obtained from this run appear to be nearly stoichiometric (Fig. 16a, z=0 mm), but in the course of further crystallization the composition again drifts away from the 1:1:1 stoichiometry: the Au content increases and the Ge content decreases, both summing up to a constant value. However, the Ce content remains constant along the whole solidified ingot (Fig. 16a). This fact is favorable for physical investigations on large crystals because non-stoichiometry of Ce is usually the most disturbing factor in heavy fermion systems. The crucial role of the Ce stoichiometry was demonstrated in Section 2 on $Ce_3Pd_{20}Si_6$. Crystallization from the off-stoichiometric melt $CeAu_{0.96}Ge_{1.04}$ seems to be most promising for the growth of stoichiometric homogeneous

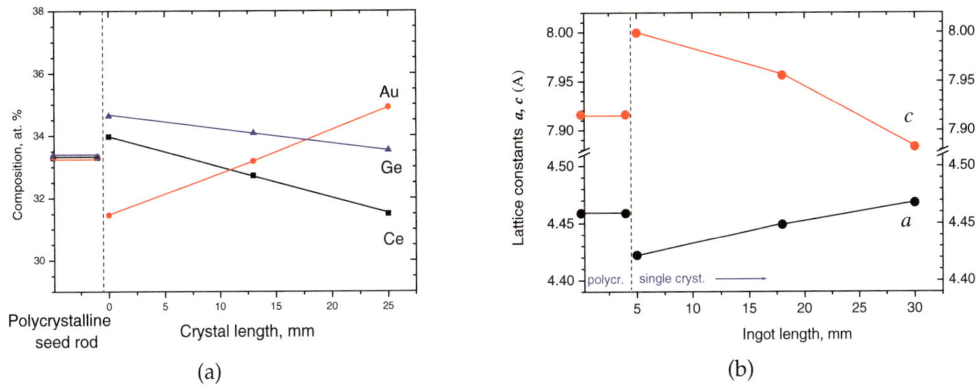

Fig. 15. a) Element concentration profiles along the growth direction at the crystallization from a stoichiometric feed rod. b) Profiles of the unit cell parameters a and c.

CeAuGe crystals, provided that the growth technique should be modified. In the floating zone method the melt-to-crystal volume ratio is very small. Therefore the segregation phenomena have very strong impact on the composition of the melt zone, and the crystal composition varies strongly during the growth. In other techniques, e.g. the Czochralski method, the melt-to-crystal volume ratio can be rather large, and the growth of homogeneous (also with respect to the Au/Ge ratio) crystals from the melt with a practically constant composition appears feasible.

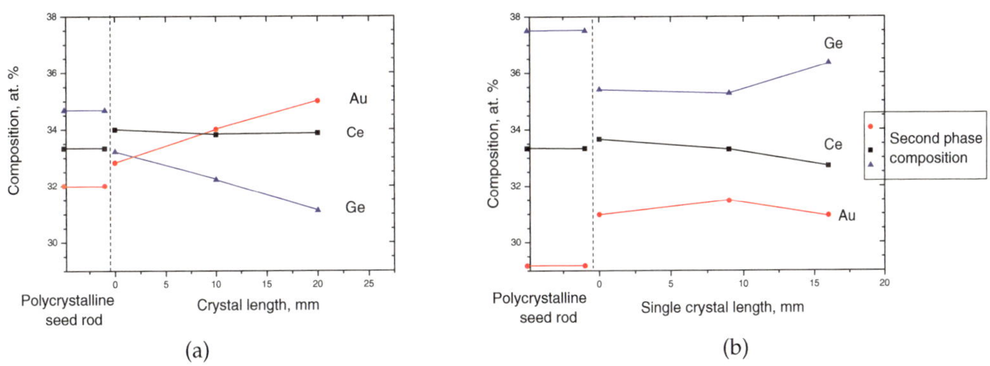

Fig. 16. Element concentration profiles along the growth direction for crystal growth from off-stoichiometric feed rods with the compositions (a) $CeAu_{0.96}Ge_{1.04}$ and (b) $CeAu_{0.88}Ge_{1.12}$.

Crystallization from a more strongly off-stoichiometric melt $CeAu_{0.88}Ge_{1.12}$ yielded strongly non-stoichiometric crystals with a large excess of Ge and a lack of Au, the Ce content remaining at nearly stoichiometric level (Fig. 16b.)

While the primarily crystallized material is, according to SEM, single phased, the finally solidified ingot consists of two phases (Fig. 17, right). These two phases are the ordered and the disordered variants of the AlB_2 structure (the NdPdSb and the true AlB_2 types). The material from the middle part of the ingot seems to be single-phased at first glance (Fig. 17, upper left), but a closer inspection under higher magnification reveals a two-phase pattern (Fig. 17, lower left). The fine grained microstructure of the material (especially apparent in comparison with that of the final part (Fig. 17, right)) evidences for crystallisation of a single phase at first which was decaying into two phases later in the course of cooling. This behavior is the sign of a strong temperature dependence of the homogeneity range of the NdPdSb-type phase.

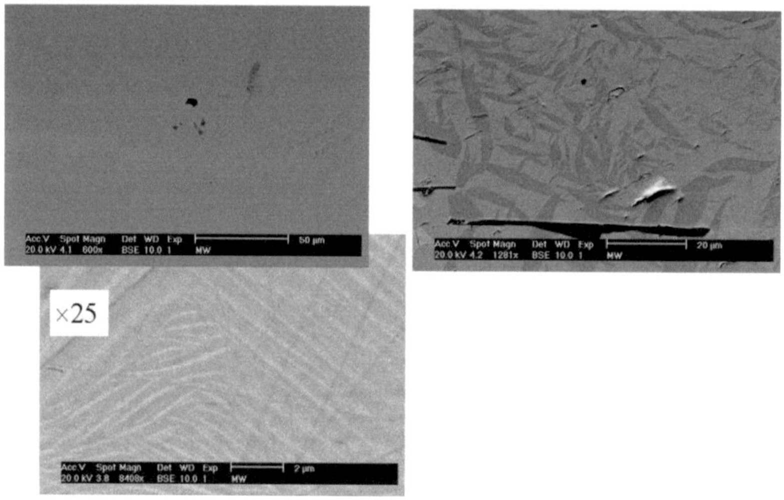

Fig. 17. Microstructure of the ingot crystallized from the feed rod with the composition $CeAu_{0.88}Ge_{1.12}$. Left - middle part, right - upper part. The magnified (×25) image of the middle part shows a fine inhomogeneity of the material.

5. Conclusions

Single crystals of the $Ce_3Pd_{20}Si_6$ phase were grown from the melt (stoichiometric and slightly off-stoichiometric) under various growth conditions and from high-temperature solutions using Pd_5Si as a flux. $Ce_3Pd_{20}Si_6$ melts quasi-congruently, i.e., the peritectic temperature is very close to the temperature of the complete melting. This fact follows from our DTA experiments and the observation of the melting zone during the growth process. The floating zone with stoichiometric composition was very unstable because of the low surface tension, which made the melt growth problematic. In contrast, the off-stoichiometric flux growth ran stably but resulted in non-stoichiometric single crystals.

While the Si content varies only slightly for different crystals, the Ce and Pd contents do so sizably, the decrease of Ce being partially compensated by an increase in Pd. The existence of a homogeneity range in $Ce_3Pd_{20}Si_6$ is the reason for the strong variation of the properties of single crystals grown by different techniques. The sharpness of the lower (presumably antiferromagnetic) phase transition, its transition temperature T_L, the residual resistivity, and the temperature T_{max} of the (local) maximum in $\rho(T)$ were shown to be measures of the crystal quality. Based on all these properties the upper part of the crystal grown from the stoichiometric melt (*sc*1*t*) and the whole volume of the crystal *sc*5 grown from a slightly off-stoichiometric melt can be ranked as having the highest perfection among all the grown single crystals. Their lattice parameters together with their compositions indicate that the slight off-stoichiometry is not dominated by Ce on the Pd sites but by Pd vacancies, which do not directly disturb the $4f$ lattice.

$CeRu_4Sn_6$ melts incongruently. Single crystals of $CeRu_4Sn_6$ can be grown from Ru_2Sn_3 flux. The grown single crystals show no marked deviation from the ideal stoichiometry, which indicates a very narrow homogeneity range of the phase. On the grown crystals anisotropies of the magnetic properties of two types were demonstrated: along the tetragonal unit cell axes and along the axes of a quasi-cubic unit cell.

Single crystal growth of CeAuGe is complicated by a wide homogeneity range of the phase. Growth from the stoichiometric melt yields single crystals with an essential deviation from the stoichiometry. The composition of single crystals varies strongly along the growth direction. The crystal composition depends complexly on the melt composition and on the crystallisation temperature (the latter is a function of the former). Growth from the off-stoichiometric melt $CeAu_{0.96}Ge_{1.04}$ is optimal for nearly stoichiometric crystals provided that the melt-to-crystal volume ratio is large enough for keeping the melt composition quasi-constant during the entire growth run.

6. Acknowledgement

We are grateful to A. Strydom and co-workers for feed rod preparation of the CeAuGe phases, to K. Neumaier for C_p measurements on *sc*5, and to H. Ossmer for resistivity measurements on *sc*5. We thank M. Waas and V. Peter for the SEM/EDX measurements and for metallographic sample preparation. The work was financially supported by the Austrian Science Foundation (project P19458-N16) and the European Research Council (Advanced Grant n°227378).

7. References

Aeppli G. and Fisk Z., Comments Condens. Matter Phys. 16, 155 (1992).

Ananthasivan K., Kaliappan I., Vasudeva Rao P.R., Sudha C., Terrance A.L.E.. J. Nucl. Mater., 305, 97-105 (2002)

Brünig E. M., Baenitz M., Gippius A. A., Paschen S., Strydom A. M., and Steglich F., Physica B 378Ű380, 839 (2006).

Chandrasekharaiah M., in *Binary Alloy Phase Diagrams*, 2nd ed., edited by T. Massalski (William W.Scott, Jr., USA, 1990).

Das I. and Sampathkumaran E. V., Phys. Rev. B 46, 4250 (1992).

Dung N., Haga Y., Matsuda T., Yamada T., Thamizhavel A., Okuda Y., Takeuchi T., Sugiyama K., Hagiwara M., Kindo K., Settai R., and Onuki Y., J. Phys. Soc. Jpn. 76, 024702 (2007).

Goto T., Watanabe T., Tsuduku S., Kobayashi H., Nemoto Y., Yanagisawa T., M.Akatsu, Ano G., Suzuki O., Takeda N., Dönni A., and Kitazawa H., J. Phys. Soc. Jpn. 78, 024716 (2009).

Gribanov A. V., Seropegin Y. D., and Bodak O. J., J. Alloys Compd. 204, L9 (1994).

Gribanov A. V., Rogl P., and Seropegin Y. D., in Noble metal systems, in *Landolt-Börnstein, New Series* (Springer, Berlin, 2006), Vol. IV/11B, p. 340.

Hashiguchi T., Takeda N., Kitagawa J., Wada N., Takayanagi S., Ishikawa M., and Mori N., J. Phys. Soc. Jpn. 69, 667 (2000).

Jones C.D.W., Gordon R.A., DiSalvo F.J., Pöttgen R., Kremer R.K. J. Alloys & Comp. 260, 50 (1997)

Kitagawa J., Takeda N., and Ishikawa M., Phys. Rev. B 53, 5101 (1996).

Kitagawa J., Takeda N., Ishikawa M., Yoshida T., Ishiguro A., Kimura N., and T.Komatsubara, Phys. Rev. B 57, 7450 (1998).

Kitagawa J., Takeda N., Sakai F., and Ishikawa M., J. Phys. Soc. Jpn. 68, 3413 (1999).
In: *Binary alloy phase diagrams*, T.B. Massalski, (Ed.), V. 3, 3255, William W. Scott, Jr., ISBN 0-87170-406-4.
Mhlungu B.M., Strydom A.M.. Physica B 403, 862 (2008)
Mitamura H., Tayama T., Sakakibara T., Tsuduku S., Ano G., Ishii I., Akatsu M., Nemoto Y., Goto T., Kikkawa A., Kitazawa H., J. Phys. Soc. Jpn. 79, 074712 (2010).
Nemoto Y., Yamaguchi T., Horino T., Akatsu M., Yanagisawa T., Goto T., O.Suzuki, Dönni A., and Komatsubara T., Phys. Rev. B 68, 184109 (2003).
Paschen S., in *Thermoelectrics Handbook* (ed. D. M. Rowe, CRC Press, Boca Raton, 2006), Chap. 15 (Thermoelectric aspects of strongly correlated electron systems).
Paschen S., Müller M., Custers J., Kriegisch M., Prokofiev A., Hilscher G., Steiner W., Pikul A., Steglich F., and Strydom A. M., J. Magn. Magn. Mater. 316, 90 (2007).
Paschen S., Laumann S., Prokofiev A., Strydom A. M., Deen P. P., Stewart J. R., Neumaier K., Goukassov A., and Mignot J.-M., Physica B 403, 1306 (2008).
Paschen S., Winkler H., Nezu T., Kriegisch M., Hilscher G., Custers J., Prokofiev A., Strydom A. J. Phys.: Conf. Ser. 200, 012156 (2010)
Pöttgen R., Bormann H., Kremer R.K., J. Magn. Magn. Mater. 152, 196 (1996)
Prokofiev A., Custers J., Kriegisch M., Laumann S., Müller M., Sassik H., Svagera R., Waas M., Neumaier K., Strydom A. M., Paschen S., Phys. Rev. B 80, 235107 (2009).
Seropegin Y. D., Gribanov A. V., Kubarev O. L., Tursina A. I., and Bodak O. I., J. Alloys Comp. 317-318, 320 (2001).
Sondezi-Mhlungu B.M., Adroja D.T., Strydom A.M., Paschen S., Goremychkin E.A.. Physica B 404, 3032 (2009)
Strydom A. M., Pikul A., Steglich F., and Paschen S., J. Phys.: Conf. Series 51, 239 (2006).
Takeda N., Kitagawa J., and Ishikawa M., J. Phys. Soc. Jpn. 64, 387 (1995).
Venturini G., Chafik El Idrissi B., Marêché J., and Malaman B., Mater. Res. Bull. 25, 1541 (1990).

Part 2

Growth of Thin Films and Low-Dimensional Structures

12

Controlled Growth of C-Oriented AlN Thin Films: Experimental Deposition and Characterization

Manuel García-Méndez
Centro de Investigación en Ciencias Físico-Matemáticas, FCFM de la UANL Manuel L. Barragán S/N, Cd. Universitaria, México

1. Introduction

Nowadays, the science of thin films has experienced an important development and specialization. Basic research in this field involves a controlled film deposition followed by characterization at atomic level. Experimental and theoretical understanding of thin film processes have contributed to the development of relevant technological fields such as microelectronics, catalysis and corrosion.

The combination of materials properties has made it possible to process thin films for a variety of applications in the field of semiconductors. Inside that field, the nitrides *III-IV* semiconductor family has gained a great deal of interest because of their promising applications in several technology-related issues such as photonics, wear-resistant coatings, thin-film resistors and other functional applications (Moreira et al., 2011; Morkoç, 2008).

Aluminium nitride (*AlN*) is an *III-V* compound. Its more stable crystalline structure is the hexagonal würzite lattice (see figure 1). Hexagonal *AlN* has a high thermal conductivity (260 $Wm^{-1}K^{-1}$), a direct band gap (E_g=5.9-6.2 eV), high hardness (2 x 10^3 kgf mm^{-2}), high fusion temperature (2400°C) and a high acoustic velocity. *AlN* thin films can be used as gate dielectric for ultra large integrated devices (ULSI), or in GHz-band surface acoustic wave devices due to its strong piezoelectricity (Chaudhuri et al., 2007; Chiu et al., 2007; Jang et al., 2006; Kar et al., 2006; Olivares et al., 2007; Prinz et al., 2006). The performance of the *AlN* films as dielectric or acoustical/electronic material directly depends on their properties at microstructure (grain size, interface) and surface morphology (roughness). Thin films of *AlN* grown at a *c-axis* orientation (preferential growth perpendicular to the substrate) are the most interesting ones for applications, since they exhibit properties similar to monocrystalline *AlN*. A high degree of *c-axis* orientation together with surface smoothness are essential requierements for *AlN* films to be used for applications in surface acoustic wave devices (Jose et al., 2010; Moreira et al., 2011).

On the other hand, the oxynitrides MeN_xO_y (Me=metal) have become very important materials for several technological applications. Among them, aluminium oxynitrides may have promising applications in diferent technological fields. The addition of oxygen into a growing *AlN* thin film induces the production of ionic metal-oxygen bonds inside a matrix of covalent metal-nitrogen bond. Placing oxygen atoms inside the würzite structure of *AlN* can produce important modifications in their electrical and optical properties of the films, and thereby changes in their thermal conductivity and piezoelectricity features are produced too (Brien & Pigeat, 2008; Jang et al., 2008). Thus, the addition of oxygen would allow to tailor the properties of the AlN_xO_y films between those of pure aluminium oxide (Al_2O_3) and nitride (AlN), where the concentration of Al, N and O can be varied depending on the specific application being pursued (Borges et al., 2010; Brien & Pigeat, 2008; Ianno et al., 2002; Jang et al., 2008). Combining some of their advantages by varying the concentration of *Al*, *N* and *O*, aluminium oxynitride films (*AlNO*) can produce applications in corrosion protective coatings, optical coatings, microelectronics and other technological fields (Borges et al., 2010; Erlat et al., 2001; Xiao & Jiang, 2004). Thus, the study of deposition and growth of *AlN* films with the addition of oxygen is a relevant subject of scientific and technological current interest.

Thin films of *AlN* (pure and oxidized) can be prepared by several techniques: chemical vapor deposition (*CVD*) (Uchida et al., 2006; Sato el at., 2007; Takahashi et al., 2006), molecular beam epitaxy (*MBE*) (Brown et al., 2002; Iwata et al., 2007), ion beam assisted deposition (Lal et al., 2003; Matsumoto & Kiuchi, 2006) or direct current (*DC*) reactive magnetron sputtering.

Among them, reactive magnetron sputtering is a technique that enables the growth of *c-axis AlN* films on large area substrates at a low temperature (as low as 200°C or even at room temperature). Deposition of *AlN* films at low temperature is a "must", since a high-substrate temperature during film growth is not compatible with the processing steps of device fabrication. Thus, reactive sputtering is an inexpensive technique with simple instrumentation that requires low processing temperature and allows fine tuning on film properties (Moreira et al., 2011).

In a reactive *DC* magnetron process, molecules of a reactive gas combine with the sputtered atoms from a metal target to form a compound thin film on a substrate. Reactive magnetron sputtering is an important method used to prepare ceramic semiconducting thin films. The final properties of the films depend on the deposition conditions (experimental parameters) such as substrate temperature, working pressure, flow rate of each reactive gas (Ar, O_2, N_2), power source delivery (voltage input), substrate-target distance and incidence angle of sputtered particles (Ohring, 2002). Reactive sputtering can successfully be employed to produce *AlN* thin films of good quality, but to achieve this goal requires controlling the experimental parameters while the deposition process takes place.

In this chapter, we present the procedure employed to grow *AlN* and *AlNO* thin-films by *DC* reactive magnetron sputtering. Experimental conditions were controlled to get the growth of *c-axis* oriented films.

The growth and characterization of the films was mainly explored by way of a series of examples collected from the author´s laboratory, together with a general reviewing of what already has been done. For a more detailed treatment of several aspects, references to highly-respected textbooks and subject-specific articles are included.

One of the most important properties of any given thin film system relies on its crystalline structure. The structural features of a film are used to explain the overall film properties, which ultimately leads to the development of a specific coating system with a set of required properties. Therefore, analysis of films will be concerned mainly with structural characterization.

Crystallographic orientation, lattice parameters, thickness and film quality were characterized through X-ray Diffraction (XRD) and UV-Visible spectroscopy (UV-Vis). Chemical indentification of phases and elemental concentration were characterized through X-ray photoelectron spectroscopy (*XPS*). From these results, an analysis of the interaction of oxygen into the *AlN* film is described. For a better understanding of this process, theoretical calculations of Density of States (DOS) are included too.

The aim of this chapter is to provide from our experience a step wise scientific/technical guide to the reader interested in delving into the fascinating subject of thin film processing.

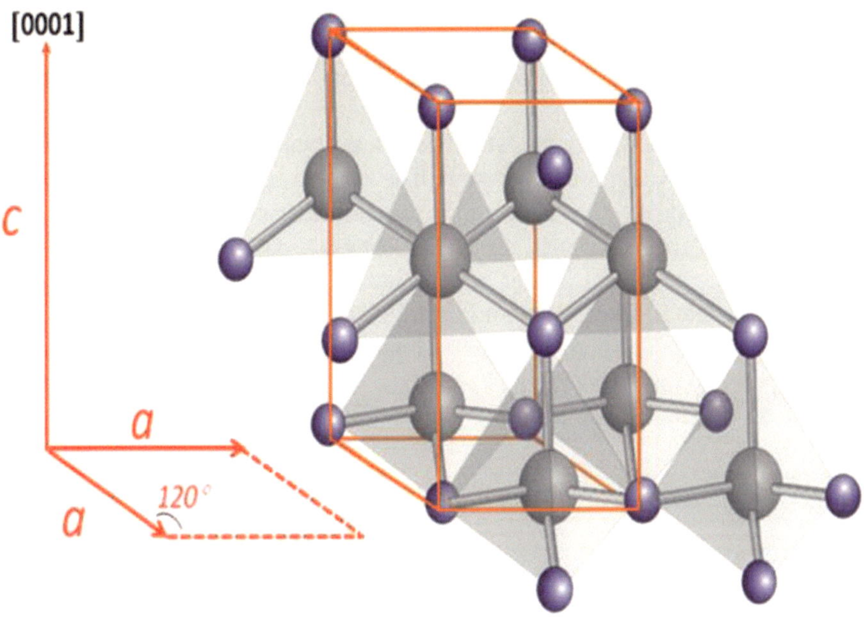

Fig. 1. Würzite structure of *AlN*. Hexagonal *AlN* belongs to the space group 6mm with lattice parameters c=4.97 Å and a=3.11 Å.

2. Deposition and growth of AlN films

The sputtering process consists in the production of ions within generated plasma, on which the ions are accelerated and directed to a target. Then, ions strike the target and material is ejected or sputtered to be deposited in the vicinity of a substrate. The plasma generation and sputtering process must be performed in a closed chamber environment, which must be maintained in vacuum. To generate the plasma gas particles (usually argon) are fed into the chamber. In *DC* sputtering, a negative potential U is applied to the target (cathode). At critical applied voltage, the initially insulating gas turns to electrical conducting medium. Then, the positively charged Ar^{+} ions are accelerated toward the cathode. During ionization, the cascade reaction goes as follows:

$$e^- + Ar \rightarrow 2e^- + Ar^+$$

where the two additional (secondary) electrons strike two more neutral ions that cause the further gas ionization. The gas pressure "P" and the electrode distance "d" determine the breakdown voltage "V_B" to set the cascade reaction, which is expressed in terms of a product of pressure and inter electrode spacing:

$$V_B = \frac{APd}{\ln(Pd) + B} \tag{1}$$

where A and B are constants. This result is known as Paschen´s Law (Ohring, 2002).
In order to increase the ionization rate by emitted secondary electrons, a ring magnet (magnetron) below the target can be used. Hence, the electrons are trapped and circulate over the surface target, depicting a cycloid. Thus, the higher sputter yield takes place on the target area below this region. An erosion zone (trace) is "carved" on the target surface with the shape of the magnetic field.
Equipment description: Films under investigation were obtained by *DC* reactive magnetron sputtering in a laboratory deposition system. The high vacuum system is composed of a pirex chamber connected to a mechanic and turbomolecular pump. Inside the chamber the magnetron is placed and connected to a DC external power supply. In front of the magnetron stands the substrate holder with a heater and thermocouple integrated. The distance target-substrate is about 5 cm and target diameter 1". The power supply allows to control the voltage input (*Volts*) and an external panel display readings of current (*Amperes*) and sputtering power (*Watts*) (see Figure 2).

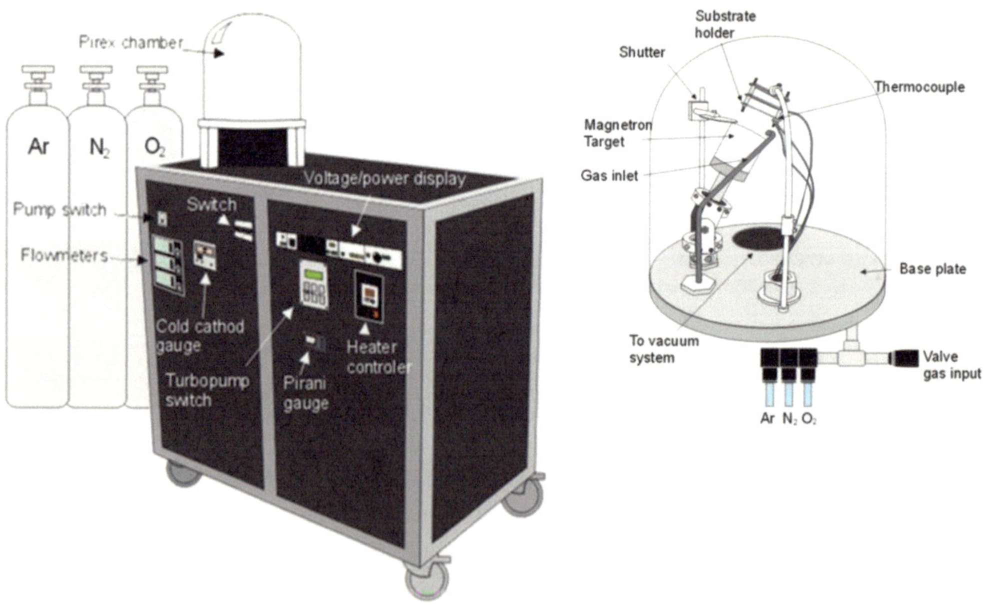

Fig. 2. Schematic diagram of the equipment utilized for film fabrication.

Deposition procedure: A disc of *Al* (2.54 cm diameter, 0.317 cm thick, 99.99% purity) was used as a target. Films were deposited on silica and glass substrates that were ultrasonically cleaned in an acetone bath. For deposition, the sputtering chamber was pumped down to a base pressure below $1x10^{-5}$ Torr. When the chamber reached the operative base pressure, the *Al* target was cleaned *in situ* with Ar^{+} ion bombardment for 20 minutes at a working pressure of 10 mTorr (20 *sccm* gas flow). A shutter is placed between the target and the substrate throughout the cleaning process. The Target was systematically cleaned to remove any contamination before each deposition.

Sputtering discharge gases of *Ar*, N_2 and O_2 (99.99 % purity) were admitted separately and regulated by individual mass flow controllers. A constant gas mixture of *Ar* and N_2 was used in the sputtering discharge to grow *AlN* films; a gas mixture of *Ar*, N_2 and O_2 was used to grow *AlNO* films.

A set of eight films were prepared: four samples on glass substrates (*set 1*) and four samples on silica substrates (*set 2*). From *set 1*, two samples correspond to *AlN* (15 min of deposition time, labeled *S1* and *S2*) and two to *AlNO* (10 min of deposition time, labeled *S3* and *S4*). From *set 2*, three samples correspond to *AlN* (10 min of deposition time, labeled *S5*, *S6* and *S7*) and one to *AlNO* (10 min of deposition time, labeled *S8*). All samples were deposited using an *Ar* flow of 20 *sccm*, an N_2 flow of 1 *sccm* and an O_2 flow of 1 *sccm*. In all samples (excluding the ones grown at room temperature.), the temperature was supplied during film deposition.

Tables 1 (a) (*set 1*) and 1 (b) (set 2) summarize the experimental conditions of deposition. Calculated optical thickness by formula 4 is included in the far right column.

Sample	*(Film) °C - time*	*V (Volts)*	*P (Watts)*	*Thickness (nm)*
S1	(*AlN*) RT-15 min	360	120	980
S2	(*AlN*) 100°C-15 min	360	130	970
S3	(*AlNO*) RT-10 min	360	190	820
S4	(*AlNO*) 120°C-10 min	360	185	940

Table 1a. Deposition parameters for *DC* sputtered films grown on glass substrates (*set 1*)

Sample	*(Film) °C - time*	*V (Volts)*	*P (Watts)*	*Thickness (nm)*
S5	(*AlN*)RT-10 min	340	100	630
S6	(*AlN*)100°C-10 min	330	110	630
S7	(*AlN*)200°C-10 min	340	120	730
S8	(*AlNO*)RT-10 min	380	140	490

Table 1b. Deposition parameters for *DC* sputtered films grown on silica substrates (*set 2*).

3. Structural characterization

XRD measurements were obtained using a Philips X'Pert diffractometter equipped with a copper anode *Kα* radiation, λ =1.54 Å. High resolution *theta/2Theta* scans (Bragg-Brentano geometry) were taken at a step size of 0.005°. Transmission spectra were obtained with a UV- Visible double beam Perkin Elmer 350 spectrophotometer.
Figure 3 (a) and (b) display the *XRD* patterns of the films deposited on glass (*set 1*) and silica (*set 2*) substrates, respectively.
The diffraction pattern of films displayed in figure 3 match with the standard *AlN* würzite spectrum (JCPDS card 00-025-1133, a=3.11 Å, c=4.97 Å) (Powder Diffraction file, 1998). The highest intensity of the *(002)* reflection at $2\theta \approx 35.9^0$ indicates an oriented growth along the *c-axis* perpendicular to substrate.
From *set 1*, it can be observed that the intensity of *(002)* diffraction peak is the highest in *S2*. In this case, the temperature of 100ºC increased the crystalline ordering of film. In *S3* and *S4* the intensity of *(002)* diffraction and grain size are very similar for both samples, which shows that applied temperature on *S4* had not effect in improving its crystal ordering.
From *set 2*, it can be observed that the intensity of *(002)* diffraction peak is the highest in *S5*. Generally, temperature gives atoms an extra mobility, allowing them to reach the lowest thermodynamically favored lattice positions hence, the crystal size becomes larger and the crystallinity of the film improves. However, the temperature applied to *S6* and *Ss* makes no effect to improve their crystallinity. In this case, a substrate temperature higher than 100°C can trigger a re-sputtering of the atoms that arrive at the substrate´s surface level and crystallinity of films experiences a downturn.
From *set 1* and *set 2*, *S2* and *S5*, respectively, were the ones that presented the best crystalline properties. A temperature ranging from RT to 100°C turned out to be the critical experimental factor to get a highly oriented crystalline growth.

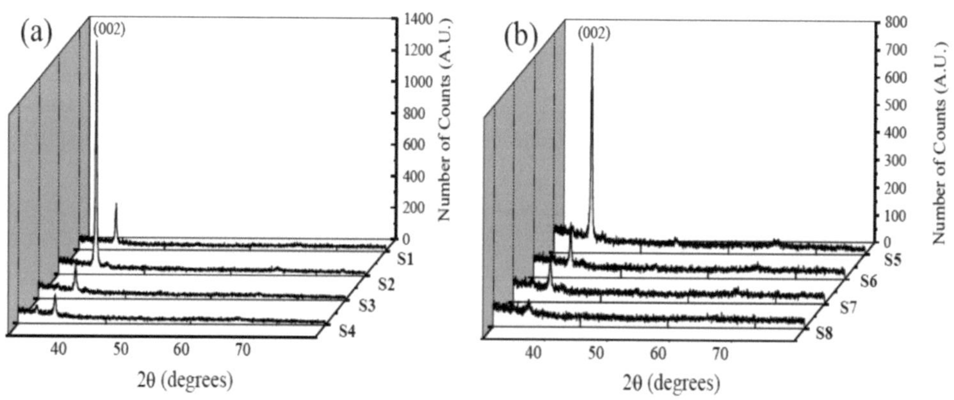

Fig. 3. *XRD* patterns of films deposited on (a) glass and (b) silica substrates.

In terms of the role of oxygen, for *S3*, *S4* and *S8*, the presence of alumina (γ-Al_2O_3: *JCPDS file 29-63*) or spinel (γ-*AlON*: *JCPDS files 10-425 and 18-52*) compounds in the diffraction patterns was not detected. However, it is known from thermodynamic that elemental aluminium reacts more favorably with oxygen than nitrogen: it is more possible to form Al_2O_3 by gaseous phase reaction of $Al+(3/2)O_2$ than AlN of $Al+(1/2)N$ since $\Delta G(Al_2O_3)=-1480\ KJ/mol$ and $\Delta G(AlN)=-253\ KJ/mol$ (Borges et al., 2010; Brien & Pigeat, 2007). Therefore, the existence of Al_2O_3 or even spinel *AlNO* phases in samples cannot be discarded, but maybe in such a small proportions as to be detected by *XRD*.

S1, *S2* and *S5* show a higher crystalline quality than *S3*, *S4* and *S8*. For these last samples, the extra O_2 introduced to the chamber promotes the oxidation of the target-surface (target poisoning). In extreme cases when the target is heavily poisoned, oxidation can cause an arcing of the magnetron system. Formation of aluminium oxide on the target can act as an electrostatic shell, which in turn can affect the sputtering yield and the kinetic energy of species which impinge on substrate with a reduction of the sputtering rate: The lesser energy of species reacting on substrate, the lesser crystallinity of films.

Also, the oxygen can enter in to the *AlN* lattice through a mechanism involving a vacancy creation process by substituting a nitrogen atom in the weakest *Al-N bond* aligned parallel to [0001] direction. During the process, the mechanism of ingress of oxygen into the lattice is by diffussion (Brien & Pigeat, 2007; Brien & Pigeat, 2008; Jose et al., 2010). On the other hand, the ionic radius of oxygen ($r_O=0.140\ nm$) is almost ten times higher than that of nitrogen ($r_N=0.01\text{-}0.02\ nm$) (Callister, 2006). Thus, the oxygen causes an expansion of the crystal lattice through point defects. As the oxygen content increases, the density of point defects increases and the stacking of hexagonal *AlN* arrangement is disturbed . It has been reported that the *Al* and *O* atoms form octahedral atomic configurations that eventually become planar defects. These defects usually lie in the basal *{001}* planes (Brien & Pigeat, 2008; Jose et al., 2010).

As was mentioned, during the deposition of thin films, the oxygen competes with the nitrogen to form an oxidized *Al-compound*. The resulting films are then composed of separated phases of *AlN* and Al_xO_y domains. The presence of Al_xO_y domains provokes a disruption in the preferential growth of the film.

For example, in *S4*, the applied temperature of 120ºC can promote an even more efficient diffusive ingress of oxygen into the *AlN* lattice and such temperature was not a factor contributing to improve crystallinity. In *S3* and *S8*, oxygen by itself was the factor that provoked a film´s low crystalline growth.

By using the Bragg angle (θ_b) as variable that satisfies the Bragg equation:

$$2d_{hkl}Sen\,\theta_b=n\lambda \tag{2}$$

and the formula applied for hexagonal systems:

$$\frac{1}{d_{hkl}^{\ 2}}=\frac{4}{3}\left(\frac{h^2+hk+k^2}{a^2}\right)+\frac{l^2}{c^2} \tag{3}$$

the length of the lattice parameters "a" and "c" can then be obtained from the experimental data.

As films crystallized in a hexagonal würzite structure, *XRD* patterns were processed with a software program in order to obtain the lattice parameters "*a*" and "*c*". The *AlN* würzite structure from the *JCPDS* database (*PDF file 00-025-1133, c= 4.97 Å, a=3.11 Å*) was taken as a reference (Powder Diffraction File, 1998). For the fitting, input parameters of *(h k l)* planes with their corresponding *theta-angle* are given. By using the Bragg formula and the equation of distance between planes (for a hexagonal lattice), the lattice parameters are then calculated by using a multiple correlation analysis with a least squares minimization. The 2θ angles were set fixed while lattice parameters were allowed to fit. Calculated lattice parameters "*a*" and "*c*" and grain size "*L*" by formula (4) are included in Table 2.

	a	*c*	*c/a*	*L*
S1	3.11	4.99	1.60	21
S2	3.11	4.98	1.60	23
S3	3.13	5.0	1.59	21
S4	3.14	5.0	1.59	20
S5	3.13	4.98	1.59	24
S6	3.13	4.99	1.59	20
S7	3.11	4.99	1.60	17
S8	3.11	5.0	1.60	17

Table 2. Lattice parameters "*a*" (nm) and "*c*" (nm) obtained from *XRD* measurements.

The average grain size "*L*" is obtained through the Debye-Scherrer formula (Patterson, 1939):

$$L = \frac{K\lambda}{B\cos\theta_b} \tag{4}$$

where K is a dimensionless constant that may range from 0.89 to 1.30 depending on the specific geometry of the scattering object.

For a perfect two dimenssional lattice, when every point on the lattice produces a spherical wave, the numerical calculations give a value of K=0.89. A cubic three dimensional crystal is best described by K=0.94 (Patterson, 1939).

The measure of the peak width, the full width at half maximum (FWHM) for a given θ_b is denoted by B (for a gaussian type curve).

From table 2, it can be observed that the calculated lattice parameters differ slightly from the ones reported from the *JCPDS* database, mainly the "c" value, particularly for *S3*, *S4* and *S8*. Introduction of oxygen into the *AlN* matrix along the {001} planes also modifies the lattice parameters. As expected, the "c" value is the most affected.

The quality of samples can also be evaluated from UV-Visible spectroscopy (Guo et al., 2006). By analysing the measured T *vs* λ spectra at normal incidence, the absorption coefficient (α) and the film thickness can be obtained.

If the thickness of the film is uniform, interference effects between substrate and film (because of multiple reflexions from the substrate/film interface) give rise to oscillations. The number of oscillations is related to the film thickness. The appearence of these oscillations on analized films indicates uniform thickness. If the thickness "t" were not uniform or slightly tappered, all interference effects would be destroyed and the T *vs* λ spectrum would look like a smooth curve (Swanepoel, 1983).

Oscillations are useful to calculate the thickness of films using the formula (Swanepoel, 1983; Zong et al., 2006):

$$t = \frac{1}{2n\left(\frac{1}{\lambda_2} - \frac{1}{\lambda_1}\right)} \tag{5}$$

Where t is the thickness of film, n the refractive index, λ_1 and λ_2 are the wavelength of two adjacent peaks. Calculated optical thickness of samples using the above mentioned formula, are included in Tables 1(a) and (b).

Regarding the absorbance (α), a T *vs* λ curve can be divided (grossly) into four regions. In the transparent region $\alpha=0$ and the transmitance is a function of n and t through multiple reflexions. In the region of weak absorption α is small and the transmission starts to reduce. In the region of medium absorption the transmission experiences the effect of absoption even more. In the region of strong absorption the transmission decreases abruptly. This last region is also named the absorption edge.

Near the absorption edge, the absorption coefficient can expressed as:

$$h\nu\alpha=\beta(h\nu-E_g)^{\gamma} \tag{6}$$

where $h\nu$ is the photon energy, E_g the optical band gap and γ is the parameter measuring the type of band gap (direct or indirect) (Guerra et al., 2011; Zong et al., 2006).

Thus, the optical band gap is determined by applying the Tauc model and the Davis and Mott model in the high absorbance region. For *AlN* films, the transmittance data provide the best linear curve in the band edge region, taking n=1/2, implying that the transition is direct in nature (for indirect transition n=2). Band gap is obtained by plotting $(\alpha h\nu)^2$ vs $h\nu$ by extrapolating the linear part of the absorption edge to find the intercept with the energy axis. By using UV-Vis measurements for *AlNO* films on glass sustrates, authors of ref. (Jang et al., 2008) found band gap values between 6.63 to 6.95 eV, depending the Ar:O ratio.

From our measurements, figure 4 displays the optical spectra (T *vs* λ curve) graphs. The oscillations detected in the curves attest the high quality in homogeneity of deposited films. All the samples have oscillation regardless their degree of crystallinity. An important feature to note is that curves present differences in the "sharpness", at the onset of the strong absorption zone. These differences are attributed to deposition conditions, where final density of films, presence of deffects and thickness, modify the shape of the curve at the band edge.

A FESEM micrograph cross-section of *S2* is displayed on figure 5. From figure, it is possible to identify a well defined substrate/film interface and a section of film with homogeneous thickness. Together with micrographs, *in-situ* EDAX analyses were conducted in two specific regions of the film. An elemental analysis by EDAX allows to distinguish the differences in elemental concentration depending on the analized zone. In the film zone , an elemental concentration of *Al* (54.7 %) and *N* (45.2 %) was detected, as expected for *AlN* film. Conversely, in the substrate zone, elemental concentration of Si and O with traces of Ca, Na, Mg was detected, as expected for glass.

At this stage, we can establish that during the sputtering process, the oxygen diffuses in to the growing *AlN* films. Then, the oxygen attaches to available Al, forming Al_xO_y phases. Dominions of these phases, contained in the whole film, can induce defects. These defects are piled up along the c-axis. From X-ray diffractograms, a low and narrow intensity at the (0002) reflection indicates low crystallographic ordering. By calculating lattice parameters "a" and "c" and evaluating how far their obtained values deviate from the JCPDF standard (mainly the "c" distance), also provides evidence about the degree of crystalline disorder. In films, a low crystallographic ordering does not imply a disruption in the homogeneity, as was already detected by UV-Visible measurements. A more detailed analysis concerning the identification and nature of the phases contained in films were performed with a spectroscopic technique.

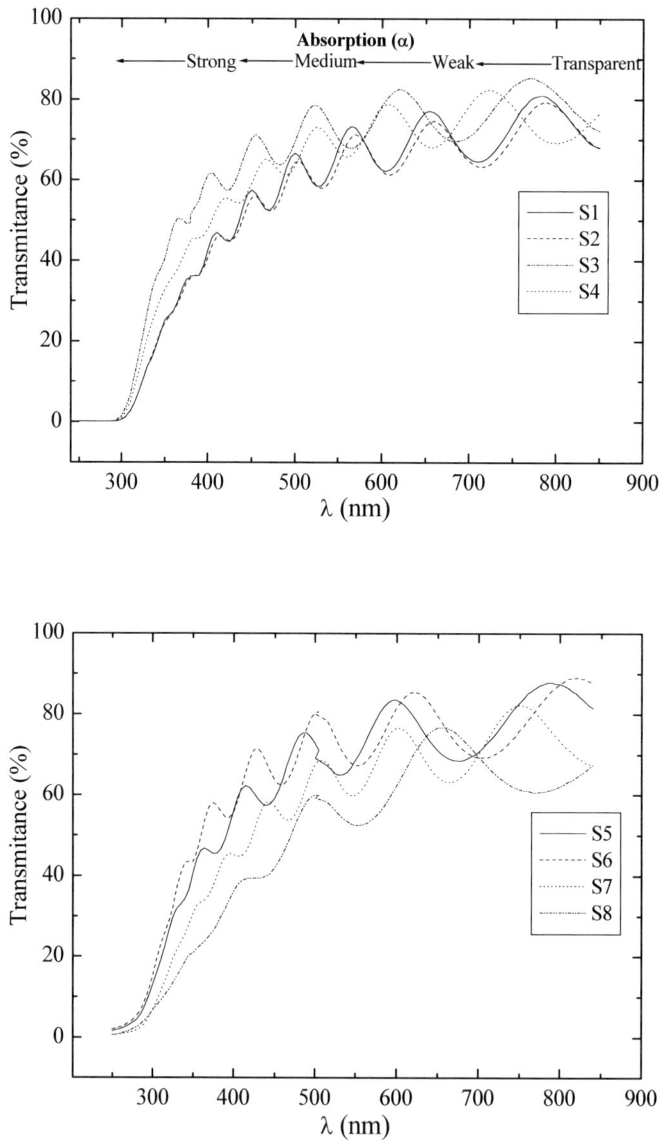

Fig. 4. Optical transmission spectra of deposited films.

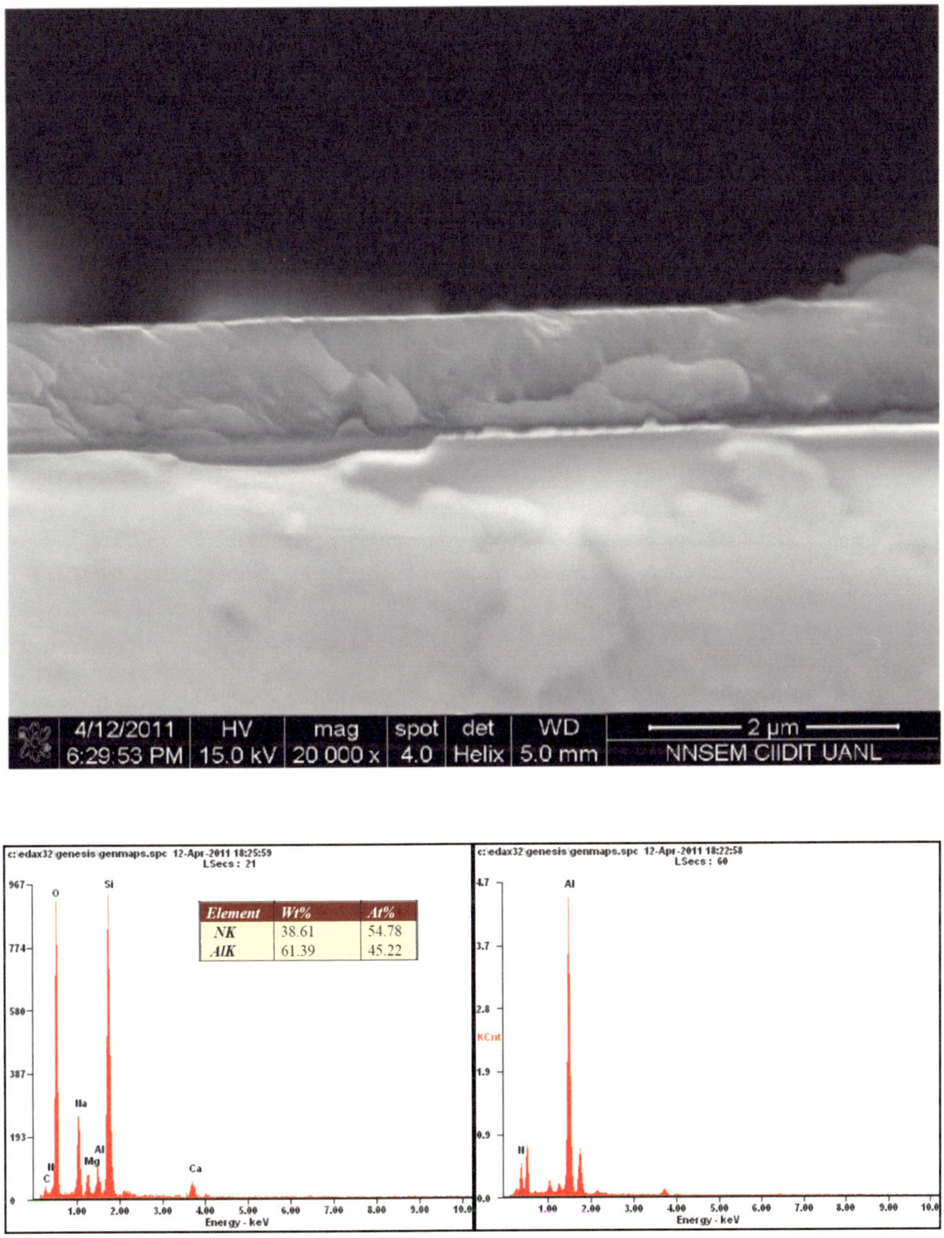

Fig. 5. Cross section FESEM micrograph of AlN film (*S2*). An homogeneous film deposition can be observed. In the right column an EDAX analysis of (a) film zone and (b) substrate zone is included.

4. Chemical characterization

The process of oxidation is a micro chemical event that was not completely detected by *XRD*. Because of that, *XPS* analyses were performed in order to detect and identify oxidized phases.

XPS measurements were obtained with a Perkin-Elmer PHI 560/ESCA-SAM system, equipped with a double-pass cylindrical mirror analyzer, and a base pressure of 1×10^{-9} Torr. To clean the surface, Ar^+ sputtering was performed with 4 keV energy ions and 0.36 $\mu A/cm^2$ current beam, yielding to about 3 nm/min sputtering rate. All XPS spectra were obtained after Ar^+ sputtering for 15 min. The use of relatively low current density in the ion beam and low sputtering rate reduces modifications in the stoichiometry of the *AlN* surface. For the XPS analyses, samples were excited with 1486.6 eV energy $Al_{K\alpha}$ X-rays. XPS spectra were obtained under two different conditions: (*i*) a survey spectrum mode of 0-600 eV, and (*ii*) a multiplex repetitive scan mode. No signal smoothing was attempted and a scanning step of 1 eV/step and 0.2 eV/step with an interval of 50 ms was utilized for survey and multiplex modes, respectively. The spectrometer was calibrated using the $Cu\ 2p_{3/2}$ (932.4 eV) and $Cu\ 3p_{3/2}$ (74.9 eV) lines. Al films deposited on the glass and silica substrates were used as additional references for Binding energy. In both kind of films, the BE of metallic (Al^0) *Al2p*-transition gave a value of 72.4 eV respectively. On these films, the *C1s*-transition gave values of 285.6 eV and 285.8 eV for glass and silica substrates, respectively. These values were set for *BE* of *C1s*. The relative atomic concentration of samples was calculated from the peak area of each element (*Al2p, O1s, N1s*) and their corresponding relative sensitivity factor values (*RSF*). These *RSF* were obtained from software system analysis (Moulder, 1992). Gaussian curve types were used for data fitting.

Figure 6 displays the *XPS* spectra of films. The elemental attomic concentration (atomic percent) calculated from the *O1s*, *N1s* and *Al2p* transitions is also included in the figure. Figure 6a shows the *Al2p* high-resolution photoelectron spectrum of *S1*. The binding energies (BE) from the acquired *Al2p* photoelectron transition are presented in table 3.

The survey spectra show the presence of oxygen in all films, regardless of the fact that some samples were grown without oxygen during deposition. From the *XPS* analysis, *S2* and *S5*, our films with the best crystalline properties, a concentration of oxygen of 26.3% and 21.6% atomic percent respectively, was measured. The highest measured concentration of oxygen was of about 36.6%, corresponding to *S8*. This occurrence of oxidation was not directly detected by the *XRD* analysis, since these oxidized phases can be spread in a low amount throughout the film.

The nature of these phases can be inferred from the deconvoluted components of the *Al2p* transition. In Figure 6a, the *Al2p* core level spectrum is presented. This spectrum is composed of contributions of metallic *Al* (*BE=72.4 eV*), nitridic *Al* in *AlN* (*BE=74.7 eV*) and oxidic *Al* in Al_2O_3 (*BE=75.6 eV*).

Despite the differences in experimental conditions, aluminium reacted with the nitrogen and the oxygen in different proportions. Even in *S2*, the thin film with the best crystalline properties, a proportion of about 30.6 % of aluminum reacted with oxygen to form an aluminium oxide compound. In *S7*, the relative contribution of *Al* in nitridic and oxidic state is almost similar, of *42.2%* and *49.5%*, respectively. A tendency, not absolute but in general, indicates that the higher the proportion of *Al* in oxidic state, the more amorphous the film.

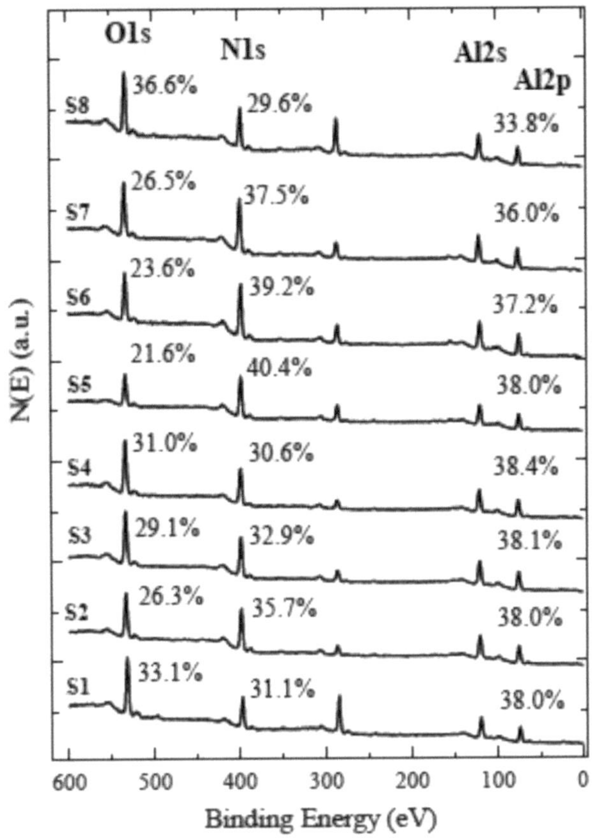

Fig. 6. XPS survey spectra of *dc* sputtered films. In this figure, the *O1s*, *N1s* and *Al2p* core-level principal peaks can be observed.

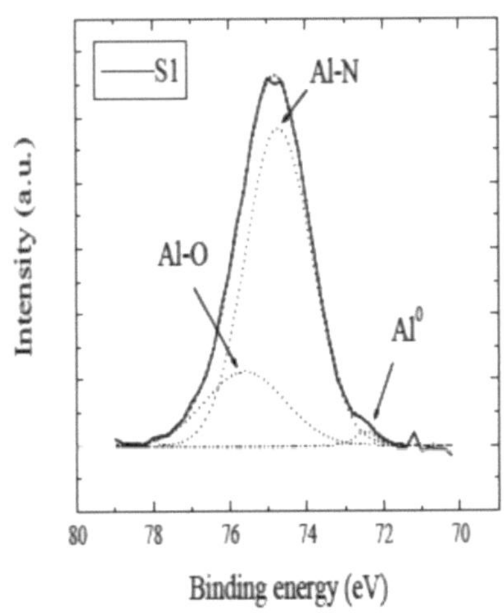

Fig. 6a. *Al2p* XPS spectrum of *S1*. The *Al2p* peak is composed of contributions of metallic aluminium (Al^O), aluminium in nitride (*Al-N*) and oxidic (*Al-O*) state.

	Binding energy (eV)			*Contribution (%)*		
	Al^0	*Al-N*	*Al-O*	*Al-N*	*Al-O*	*ΔE (eV)*
S1	72.4	74.7	75.6	77.3	21.6	0.9
S2		74.7	75.7	69.3	30.6	1.0
S3		74.7	75.2	74.7	22.9	0.5
S4	72.1	74.4	75.1	58.4	39.9	0.7
S5		74.7	75.5	64.4	35.1	0.8
S6	72.6	74.7	75.5	60.2	37.3	0.8
S7		74.7	75.4	42.2	49.5	0.7
S8		74.4	75.5	58.7	36.3	1.1

Table 3. Binding energy (eV) of metallic aluminium (*Al^O*), aluminium in nitridic (*Al-N*) and oxidic (*Al-O*) state obtained from deconvoluted components of *Al2p* transition. Percentage (relative %) of *Al* bond to *N* and *O* is also displayed.

For comparison purposes, some relevant literature concerning the binding energies of *metallic-Al*, *AlN* and *Al_2O_3* has been reviewed and included in table 4. Aluminium in metallic state lies in the range of *72.5-72.8 eV*. Aluminium in nitridic state lies in the range of *73.1-74.6 eV*, while aluminium in oxidic state lies in the range of *74.0-75.5 eV*. Also, there is an *Al-N-O* spinel-like bonding, very similar in nature to oxidic aluminium with a BE value of *75.4 eV*. Another criteria used by various authors for phase identification, is to take the difference (*ΔE*) in *BE* of the *Al2p* transition corresponding to *Al-N* and *Al-O bonds*. This difference can take values of about *0.6 eV* up to *1.1 eV* (see Table 4).

Binding energy (eV)				
Al^0	*Al-N*	*Al-O*	*ΔE (eV)*	References
72.8	74.1 74.3	74.7 75.2	0.6 0.9	(Stanca, 2004)
	74.6	75.4 (*spinel*)	0.8	(Sohal et al., 2006)
72.8	74.1 73.6	74.7 74.6	0.7 1.0	(Jose et al., 2010)
72.8 72.7 72.8 72.5	74.4 74.5 74.6	75.2 75.5 75.6	0.8 1.0 1.0	(Wang et al., 1997) (Gredelj et al., 2001)
	73.1	75.1 74.2 (*spinel*)	2.0 1.1	(Richthofen et al., 1996)

Table 4. Binding energy of (eV) of metallic aluminium (*Al^O*), aluminium in nitridic (*Al-N*) and oxidic (*Al-O*) state obtained from literature

In films, only small traces of metallic aluminium were detected in *S1* at *72.4 eV*. For *S4* and *S8*, *BE* of *Al* in nitride gave a value of *74.4 eV*, just below the *BE* of *74.7 eV*, detected for the rest of the samples. This value of *74.4 eV* can be attributed to a substoichiometric AlN_x phase (Robinson et al., 1984; Stanca, 2004). On the other hand, the *BE* for aluminium in oxydic state varies from *75.1 eV* to *75.7 eV*. The lowest values of *BE* of about *75.1 eV* and *75.2 eV*, corresponding to *S3* and *S4*, respectively, could be attributed to a substoichiometric Al_xO_y phase, although in our own experience, the reaction of aluminium with oxygen tends to form the stable γ-Al_20_3 phase, which possesses somewhat higher value in *BE*. These finding agree with those reported in other works, where low oxidation states such as Al^{+1}, Al^{+2} can be found at a *BE* lower than the one of Al^{+3} (Huttel et al., 1993; Stanca, 2004). Oxidation states lower than +3 confer an amorphous character to the aluminium oxide (Gutierrez et al., 1997).

5. Theoretical calculations

Experimental results provided evidence that oxygen can induce important modifications in the structural properties of sputtered-deposited *AlN* films. In this way, theoretical calculations were performed to get a better understanding of how the position of the oxygen into the *AlN* matrix can modify the electronic properties of the film system.

The bulk structure of hexagonal AlN was illustrated in Figure 1. Additionally, hexagonal AlN can be visualized as a matrix of distorted tetrahedrons. In a tetrahedron, each Al atom is surrounded by four N atoms. The four bonds can be categorized into two types. The first type is formed by three equivalent Al-Nx, (x=1,2,3) bonds, on which the N atoms are located in the same plane normal to the [0001] direction. The second type is the Al-N_0 bond, on which the Al and N atoms are aligned parallel to the [0001] direction (see figure 7). This last bond is the most ionic and has a lower binding energy than the other three (Chaudhuri et al., 2007; Chiu et al., 2007; Zhang et al, 2005). When an AlN film is oxidized, the oxygen atom can substitute the nitrogen atom in the weakest Al-N0 bond while the displaced nitrogen atom can occupy an interstitial site in the lattice (Chaudhuri et al., 2007). For würzite AlN, there are four atoms per hexagonal unit cell where the positions of the atoms for Al and N are: Al(0,0,0), (2/3,1/3,1/2); N(0,0,u), (2/3,1/3, u+1/2), where "u" is a dimensionless internal parameter that represents the distance between the Al-plane and its nearest neighbor N-plane, in the unit of "c", according to the JCPDS database (Powder diffraction file, 1998).

The calculations were perfomed using the tight-binding method (Whangbo & Hoffmann, 1978) within the extended Hückel framework (Hoffmann, 1963) using the computer package YAeHMOP (Landrum, 1900). The extended Hückel method is a semiempirical approach that solves the Schrödinger equation for a system of electrons based on the variational theorem (Galván, 1998). In this approach, explicit correlation is not considered except for the intrinsic contributions included in the parameter set. For a best match with the available experimental information, experimental lattice parameters were used instead of optimized values. Calculations considered a total of 16 valence electrons corresponding to 4 atoms within the unit cell for *AlN*.

Band structures were calculated using 51 k-points sampling the first Brillouin zone (FBZ). Reciprocal space integration was performed by *k-point* sampling (see figure 8). From band structure, the electronic band gap (E_g) was obtained.

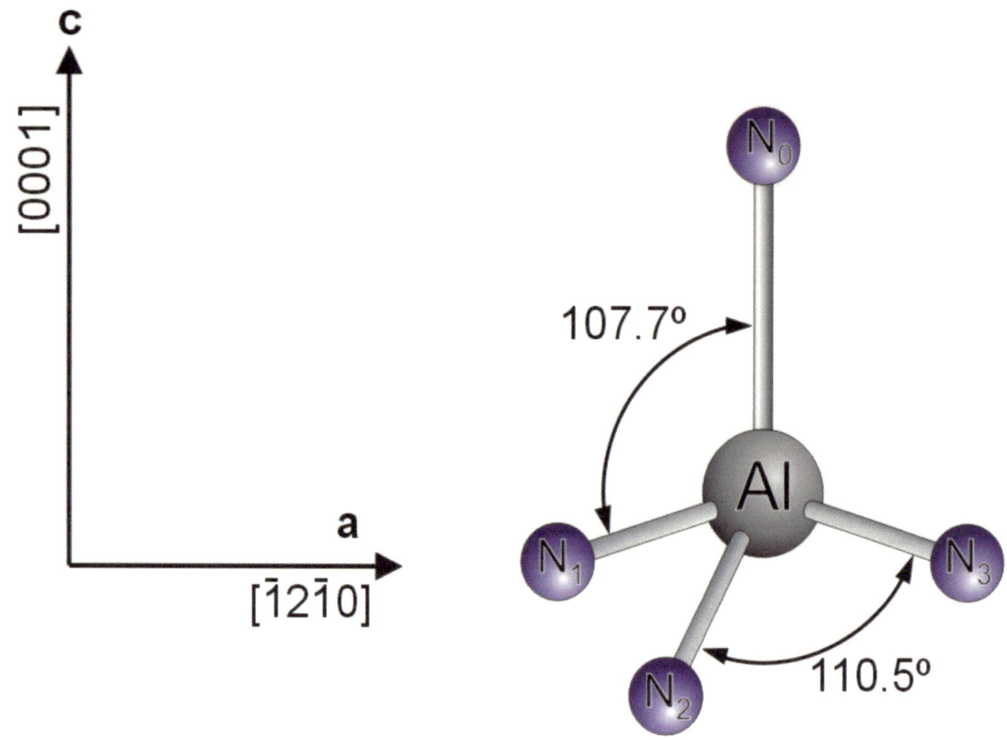

Fig. 7. Individual tetrahedral arrangement of hexagonal *AlN*.

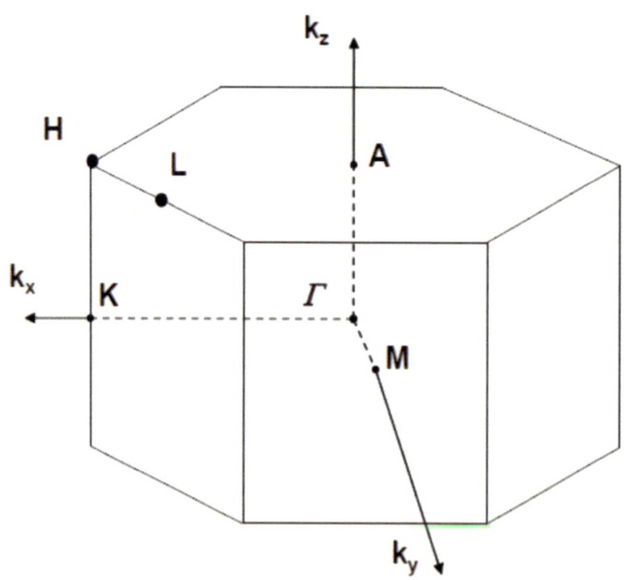

Fig. 8. Hexagonal lattice in k-space.

Calculations were performed considering four scenarios:

1. A wurzite-like *AlN* structure with no oxygen in the lattice
2. An oxygen atom inside the interstitial site of the tetrahedral arrangement (interstitial)
3. An oxygen atom in place of the *N* atom in the weakest $Al\text{-}N_0$ bond (substitution)
4. An oxygen atom on top of the *AlN* surface (at the surface).

Theoretical band-gap calculations are summarized in Table 5. Values are given in electron volts (eV).

	Energy gap (eV)
(1) *AlN* Hexagonal	7.2 – Direct *M*
(2) Interstitial *O*	1.3 – Direct *M*
(3) Substitution of *N* by *O*	0.82 – Direct *H*
(4) Surface *O*	6.2 – Indirect *Γ-M*

Table 5. Calculated energy gaps for pure *AlN* (würzite) and with oxygen in different atomic site positions.

For *AlN* hexagonal, a direct band gap of ≈ 7.2 eV at *M* was calculated (see Figure 9). When oxygen was taken into account in the calculations, the band gap value undergoes a remarkable change: *1.3 eV* for *AlN* with intercalated oxygen (2) and *0.8 eV* for *AlN* with oxygen substitution (3). In terms of electronic behavior, the system transformed from insulating (7.2 eV) to semiconductor (1.3 eV), and then from semiconductor (1.3 eV) to semimetal (0.82 eV).

This change in the electronic properties is explained by the differences between the ionic radius of Nitrogen (r_N) and Oxygen (r_O). The ionic radius of the materials involved was: r_N=*0.01-0.02 nm*, r_O=*0.140 nm* (Callister, 2006). Comparing these values, it can be noted that r_O is almost ten times higher than r_N. This fact would imply that when the oxygen atom takes the place of the nitrogen atom (by substitution o intercalation of *O*), the crystalline lattice expands because of the larger size of oxygen. Any change in the distance among atoms and the extra valence electron of the oxygen will alter the electronic interaction and in consequence, the band gap value

In calculation (4), the atoms of *Al* and *N* are kept in their würzite atomic positions while the oxygen atom is placed on top of the *AlN* lattice. In this case, the calculated band gap (6.31 eV) is closer in value to pure *AlN* (7.2 eV) than the calculated ones for interstitial (1.3 eV) and substitution (0.82 eV). In this case, theoretical results predicts that when the oxygen is not inside the Bravais lattice, the band gap will be close in value to the one of hexagonal *AlN*; conversely, the more the oxygen interacts with the *AlN* lattice, the more changes in electronical properties are expected; However, in energetic terms, competition between *N* and *O* atoms to get attached to the *Al* to form separated phases of *AlN* and Al_xO_y is the most probable configuration, as far as experimental results suggests.

Theoretical calculations of band structure for würzite *AlN* have been performed using several approaches; For comparison purposes, some of them are briefly described in Table 6.

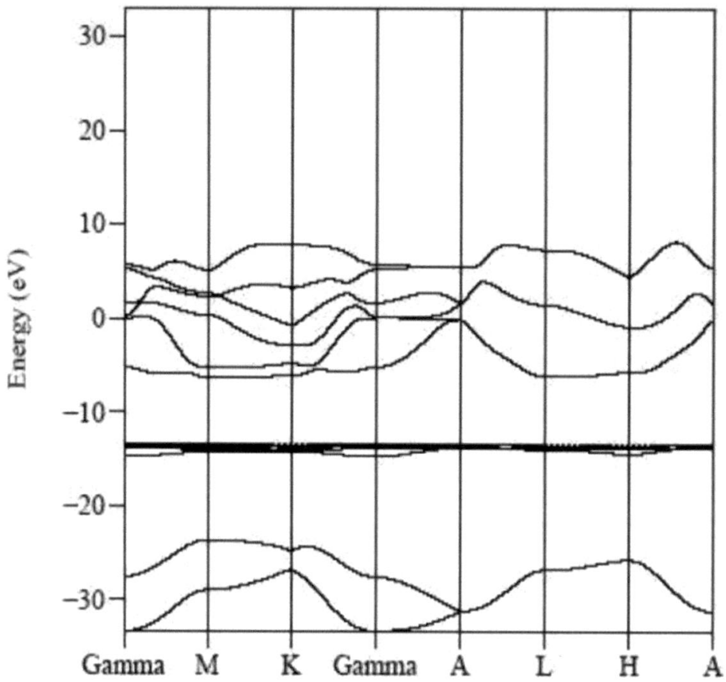

Fig. 9. Band structure for 2H-AlN hexagonal, sampling the first Brillouin zone (FBZ).

Energy band gap (eV)	Method/Procedure	Reference
6.05	Local density approximation (LDA) within the density functional theory (DFT) with a correction Δg, using a quasi-particle method: LDA+Δg	(Ferreira et al., 2005)
6.2	Empirical pseudopotential method (EPM). An analytical function using a fitting procedure for both symmetric and antisymmetric parts, and a potential is constructed	(Rezaei et al., 2006)
4.24	Full potential linear muffin-tin orbital (FPLMTO)	(Persson et al., 2005)
6.15	FPLMTO with a corrected band gap Δg	(Persson et al., 2005)

Table 6. AlN energy band gap values obtained from theoretical calculations.

From our results, the calculated band gap for *AlN* was 7.2 eV: slightly different to the reported experimental-value of ≈ 6.2 eV. About this issue, is important to take into account that in our calculations spin-orbit effects were not considered. Therefore, some differences arise, especially when an energy-band analysis is performed. Some bands could be shifted up or down in energy due to these contributions. However, it must be stressed out that our proposed method is simple, computationally efficient and the electronic structures obtained can be optimized to closely match the experimental and/or *ab-initio* results. More specific details about DOS graphs and PDOS calculations can be found in reference (García-Méndez et al., 2009), of our authorship.

6. Conclusions

In this chapter, the basis of *DC* reactive magnetron sputtering as well as experimental results concerning the growth of *AlN* thin films has been reviewed.

For instance, films under investigation were polycrystalline and exhibit an oriented growth along the [0002] direction. XRD measurements showed that films are composed mainly by crystals of hexagonal *AlN*. From XPS measurements, traces of aluminium oxides phases were detected. Films deposited without flux of oxygen presented the best crystalline properties, although phases of aluminum oxide were detected on them too. In this case, even in high vacuum, ppm levels of residual oxygen can subside and react with the growing film. Oxygen induces on films structural disorder that tends to disturb the preferential growth at the *c-axis*.

In other works of reactive magnetron sputtering, authors of ref. (Brien & Pigeat, 2008) found that for contamination of oxygen atoms (from 5% to 30 % atomic), *AlN* films can still grow in würzite structure at room temperature, with no formation of crystalline *AlNO* or Al_2O_3 phases, just only traces of amorphous AlO_x phases, that leave no signature in diffraction recordings, which is consistent with our results, where a dominant *AlN* phase in the whole film was detected. On the other hand, authors of ref. (Jose et al., 2010) reports that even in high vacuum, ppm levels of oxygen can stand and promote formation of alumina-like phases at the surface of AlN films, where these phases of alumina could only be detected and quantified by XPS and conversely, X-ray technique was unable to detect. In other report, authors of ref. (Borges et al., 2010), stablished three regions: Metallic (zone M), transition (zone T) and compound (zone C), where chemical composition of *AlNO* films varies depending the reactive gas mixture in partial pressure of N_2+O_2 at a fixed *Ar* gas partial pressure. Then, they found that when film pass from zone M to zone C, films grow from crystalline-like to amorphous-type ones, and the lattice parameters increase as more oxygen and nitrogen is incorporated into the films, which also represents the tendency we report in our results.

Thus, the versatility of the reactive *DC* magnetron sputtering that enables the growth of functional and homogeneous coatings in this case *AlN* films has been highlighted. To produce suitable films, however, it is necessary to identify the most favourable deposition parameters that maximize the sputtering yield, in order to get the optimal deposition rate: the sputter current that determines the rate of deposition process, the applied voltage that determines the maximum energy at which sputtered particles escape from target, the pressure into the chamber that determines the mean free path for the sputtered material, together with the target-substrate distance that both determines the number of collisions of particles on their way to the substrate, the gas mixture that determines the stoichiometry, the substrate temperature, all together influence the crystallinity, homogeneity and porosity of deposited films. As the physics behind the sputtering process and plasma formation is not simple, and many basic and technological aspects of the sputtering process and *AlN* film growth must be explored (anisotropic films, preferential growth, band gap changes), further investigation in this area is being conducted.

7. Acknowledgment

This work was sponsored by PAICyT-UANL, 2010.

8. References

Moreira, M.A.; Doi, I; Souza, J.F.; Diniz, J.A. (2011). Electrical characterization and morphological properties of AlN films prepared by dc reactive magnetron sputtering. Microelectronic Engineering, Vol. 88, No. 5, (May 2011), pp. 802-806, ISSN 0167-9317

Morkoç, H. (2008) *Handbook of Nitride Semiconductors and Devices,* Vol. 1: Materials Properties, Physics and Growth (2008) Weinheim Wiley-VCH. ISBN 978-3-527-40837-5, Weinheim, Germany.

Chaudhuri, J.; Nyakiti, L.; Lee, R.G.; Gu, Z.; Edgar, J.H.; Wen, J.G. (2007). Thermal oxidation of single crystalline aluminium nitride. *Materials Characterization*, Vol. 58, No. 8-9, (August 2007), pp. 672-679, ISSN 1044-5803

Chiu, K.H.; Chen, J.H.; Chen, H.R.; Huang, R.S. (2007). Deposition and characterization of reactive magnetron sputtered aluminum nitride thin films for film bulk acoustic wave resonator. *Thin Solid Films*, Vol. 515, No. 11, (April 2007), pp. 4819-4825, ISSN 0040-6090

Jang, K.; Lee, K.; Kim, J.; Hwang, S.; Lee, J.; Dhungel, S.K.; Jung, S.; Yi, J. (2006). Effect of rapid thermal annealing of sputtered aluminium nitride film in an oxygen ambient. *Mat Sci Semicon Proc*, Vol. 9, No. 6, (December 2006), pp. 1137-1141, ISSN 1369-8001

Kar, J.P.; Bose, G.; Tuli, S. (2006). A study on the interface and bulk charge density of AlN films with sputtering pressure. *Vacuum*, Vol. 81, No. 4, (November 2006), pp. 494-498, ISSN 0042-207X

Olivares, J.; González-Castilla, S.; Clement, M.; Sanz-Hervás, A.; Vergara, L.; Sangrador, J.; Iborra, E. (2007). Combined assessment of piezoelectric AlN films using X-ray diffraction, infrared absorption and atomic force microscopy. *Diamond & Related Materials*, Vol. 16, No. 4-7, (April 2007), pp. 1421-1424, ISSN 0925-9635

Prinz, G.M.; Ladenburger, A; Feneberg, M.; Schirra, M.; Thapa, S.B.; Bickermann, M.; Epelbaum, B.M.; Scholz, F.; Thonke, K.; Sauer, R. (2006). Photoluminescence, cathodoluminescence, and reflectance study of AlN layers and AlN single crystals. Superlattices *& Microstructures*, Vol. 40, No. 4-6, (December 2006), pp. 513-518, ISSN 0749-6036

Jose, F.; Ramaseshan, R.; Dash, S.; Tyagi, A.K.; Raj, B. (2010). Response of magnetron sputtered AlN films to controlled atmosphere annealing. Journal of Physics D: Applied Physics, Vol. 43, No. 7, (February 2010), pp. 075304-10, ISSN 0022-3727

Brien, V.; Pigeat, P. (2008). Correlation between the oxygen content and the morphology of AlN films grown by r.f. magnetron sputtering. *Journal of Crystal Growth*, Vol. 310, No. 16, (August 2008), pp. 3890-3895, ISSN 0022-0248

Jang, K.; Jung, S.; Lee, J.; Lee, K.; Kim, J.; Son, H.; Yi, J. (2008). Optical and electrical properties of negatively charged aluminium oxynitride films. *Thin Solid Films*, Vol. 517, No. 1, (November 2008), pp. 444-446, ISSN 0040-6090

Borges, J.; Maz, F.; Marques, L. (2010). AlNxOy thin films deposited by DC reactive magnetron sputtering. *Applied Surface Science*, Vol. 252, No. 257, No. 5, (December 2010), pp. 1478-1483, ISSN 0169-4332

Ianno, N.J.; Enshashy, H.; Dillon, R.O. (2002). Aluminum oxynitride coatings for oxidation resistance of epoxy films. *Surface and Coatings Technology*, Vol. 155, No. 2-3, (June 2002), pp. 130-135, ISSN 0257-8972

Erlat, A.G.; Henry, B.M.; Ingram, J.J.; Mountain, D.B.; McGuigan, A.; Howson, R.P.; Grovenor, C.R.M.; Briggs, G.A.D.; Tsukahara, T. (2001). Characterisation of aluminium oxynitride gas barrier films. *Thin Solid Films*, Vol. 388, No. 1-2, (June 2001), pp. 78-86, ISSN 0040-6090

Xiao, W.; Jiang X. (2004). Optical and mechanical properties of nanocrystalline aluminum oxynitride films prepared by electron cyclotron resonance plasma enhanced chemical vapor deposition. *Journal of Crystal Growth*, Vol. 263, No. 1-3, (March 2004), pp. 165-171, ISSN 0022-0248

Uchida, H.; Yamashita, M.; Hanaki, S.; Fujimoto, T. (2006). Structural and chemical characteristics of (Ti,Al)N films prepared by ion mixing and vapor deposition. *Vacuum*, Vol. 80, No. 11-12, (September 2006), pp. 1356-1361, ISSN 0042-207X

Sato, A.; Azumada, K.; Atsumori, T.; Hara, K. (2007). Characterization of AlN:Mn thin film phosphors prepared by metalorganic chemical vapor deposition. *Journal of Crystal Growth*, Vol. 298, (January 2007), pp. 379-382, ISSN 0022-0248

Takahashi, N.; Matsumoto, Y.; Nakamura, T. (2006). Investigations of structure and morphology of the AlN nano-pillar crystal films prepared by halide chemical vapor deposition under atmospheric pressure. *Journal of Physical and Chemistry of Solids*, Vol. 67, No. 4, (April 2006), pp. 665-668, ISSN 0022-3697

Brown, P.D.; Fay, M.; Bock, N., Marlafeka, M.; Cheng, T.S.; Novikov, S.V.; Davis, C.S.; Campion, R.P., Foxon, C.T. (2002). Structural characterisation of Al grown on group III-nitride layers and sapphire by molecular beam epitaxy. *Journal of Crystal Growth*, Vol. 234, No. 2-3, (January 2002), pp. 384-390, ISSN 0022-0248

Iwata, S.; Nanjo, Y.; Okuno, T.; Kurai, S.; Taguchi, T. (2007). Growth and optical properties of AlN homoepitaxial layers grown by ammonia-source molecular beam epitaxy. *Journal of Crystal Growth*, Vol. 301-302, (April 2007), pp. 461-464, ISSN 0022-0248

Lal, K.; Meikap, A.K. ; Chattopadhyay, S.K.; Chatterjee, S.K.; Ghosh, P.; Ghosh, M.; Baba, K.; Hatada, R. (2003). Frequency dependent conductivity of aluminium nitride films prepared by ion beam-assisted deposition. *Thin Solid Films*, Vol. 434, No. 1-2, (June 2003), pp. 264-270, ISSN 0040-6090

Matsumoto, T.; Kiuchi, M. (2006). Zinc-blende aluminum nitride formation using low-energy ion beam assisted deposition. *Nuclear Instruments and Methods in Physics Research*, Vol. 242, No. 1-2, (January 2006), pp. 424-426, ISSN 0168-583X

Ohring, M. (2002). *Materials Science of Thin Films: Deposition and Structure*, London Academic Press. ISBN 0-12-524975-6

Powder Diffraction File, JCPDS International Centre for Diffraction Data, PA, 1998 (www.icdd.com)

Brien, V.; Pigeat, P. (2007). Microstructures diagram of magnetron sputtered AlN deposits: Amorphous and nanostructured films. *Journal of Crystal Growth*, Vol. 299, No. 1, (February 2007), pp. 189-194, ISSN 0022-0248

Callister Jr, W.D. (2006). Materials Science & Engineering: an introduction. 6th edition (2006) Wiley & Sons. New York. ISBN 0471135763

Patterson, A.L. (1939). The Scherrer Formula for X-Ray Particle Size Determination. *Physical Review*, Vol. 56, No. 1, (July 1939), pp. 978-982, ISSN 1098-0121

Guo, Q.X.; Tanaka, T.; Nishio, M.; Ogawa, H. (2006). Growth properties of AlN films on sapphire substrates by reactive sputtering. *Vacuum*, Vol. 80, No. 7, (May 2006), pp. 716-718. ISSN 0042-207X

Swanepoel, R. (1983). Determination of the thickness and optical constants of amorphous silicon. *Journal of Physics E: Scientific Instruments*, Vol. 16, No. 12, (May 1983), pp. 1214-1222. ISSN 0022-3735

Zong, F.; Ma, H.; Du, W.; Ma, J.; Zhang, X.; Xiao, H.; Ji, F.; Xue, Ch. (2006). Optical band gap of zinc nitride films prepared on quartz substrates from a zinc nitride target by reactive rf magnetron sputtering. *Applied Surface Science*, Vol. 252, No. 22, (September 2006), pp. 7983-7986.ISSN 0169-4332

Guerra, J.A.; Montañez, L.; Erlenbach, O.; Galvez, G.; de Zela, F.; Winnacker, A.; Weingärtner, R. (2011). Determination of optical band gap of thin amorphous SiC and AlN films produced by RF magnetron sputtering. *Journal of Physics: Conference series*, Vol. 274, (September 2010), pp. 012113-012118, ISSN 1742-6588

Moulder, J.F.; Sticke, W.F.; Sobol, P.E.; Bomben, K.D. (1992) "Handbook of X-ray Photoelectron Spectroscopy". 2nd edition. Perkin-Elmer, Physical Electronics Division. Eden Prairie. ISBN 0962702625

Stanca, I. (2004). Chemical structure of films grown by AlN laser ablation: an X-ray photoelectron spectroscopy stydy. *Romanian Journal of Physics*, Vol. 49, No. 9-10, (May 2004), pp. 807-816, ISSN 1221-146X

Sohal, R.; Torche, M.; Henkel, K.; Hoffmann, P.; Tallarida, M.; Schmeiber, D. (2006). Al-oxynitrides as a buffer layer for Pr2O3/SiC interfaces. *Materials Science in Semiconductor Processing*, Vol. 9, No. 6, (December 2006), pp. 945-948, ISSN 1369-8001

Wang, P.W.; Sui, S.; Wang, W.; Durrer, W. (1997). Aluminum nitride and alumina composite film fabricated by DC plasma processes. *Thin Solid Films*, Vol. 295, No. 1-2, (February 1997), pp. 142-146, ISSN 0040-6090

Gredelj, S.; Gerson, A.R.; Kumar, S.; Cavallaro, G.P. (2001). Characterization of aluminium surfaces with and without plasma nitriding by X-ray photoelectron spectroscopy. *Applied Surface Science*, Vol. 174, No. 3-4, (April 2001), pp. 240-250, ISSN 0169-4332

Richthofen, A. von; Domnick, R. (1996). Metastable single-phase polycrystalline aluminium oxynitride films grown by MSIP: constitution and structure. *Thin Solid Films*, Vol. 283, No. 1-2, (September 1996), pp. 37-44, ISSN 0040-6090

Robinson, K.S.; Sherwood, P.M.A. (1984). X-Ray photoelectron spectroscopic studies of the surface of sputter ion plated films. *Surface and Interface Analysis*, Vol. 6, No. 6, (December 1984), pp. 261-266, ISSN: 0142-2421

Huttel, Y.; Bourdie, E.; Soukiassian, P.; Mangat, P.S.; Hurych, Z. (1993). Promoted oxidation of aluminum thin films using an alkali metal catalyst. *Journal of Vacuum Science and Technol A*, Vol. 11, No. 4, (July 1993), pp. 2186-2192, ISSN 0734-2101

Gutierrez, A.; Lopez, M.F.; Garcia-Alonso, C.; Escudero, M. In: I. Olefjord, L. Nyborg, D. Briggs, *7th European Conf. Applications on Surface & Interface Analysis*, Göteborg (1997) 1035-1038, John Wiley & Sons, Canada. ISBN 0471978272

Zhang, J.X.; Cheng, H.; Chen, Y.Z.; Uddin, A.; Yuan, S.; Geng, S.J.; Zhang, S. (2005). Growth of AlN films on Si (100) and Si (111) substrates by reactive magnetron sputtering. *Surface and Coatings Technology*, Vol. 198, No. 1-3, (August 2005), pp. 68-73, ISSN 0257-8972

Whangbo M.H.; Hoffmann, R. (1978). The band structure of the tetracyanoplatinate chain. *J. Am. Chem. Soc.*, Vol. 100, No. 19, (September 1978), pp. 6093-6098, ISSN 0002-7863

Hoffmann, R. (1963). An Extended Hückel Theory I Hydrocarbons. *J. Chem. Phys.*, Vol. 39, No. 6, (September 1963), pp. 1397-1413, ISSN 0021-9606

Landrum, G. A. YAeHMOP package:
http://overlap.chem.Cornell.edu:8080/yaehmop.html

Galván, D.H. (1998). Extended Hückel Calculations on Cubic Boron Nitride and Diamond. *Journal of Materials Science Letters*, Vol 17, No. 10 (May 2008), pp. 805-810, ISSN 1573-4811

Ferreira da Silva, A.; Souza Dantas, N.; de Almeida, J.S.; Ahuja, R.; Person C. (2005). Electronic and optical properties of würtzite and zinc-blende TlN and AlN. *Journal of Crystal Growth*, Vol. 281, No. 1, (July 2005), pp. 151-160, ISSN 0022-0248

Rezaei, B.; Asgari, A.; Kalafi, M. (2006). Electronic band structure pseudopotential calculation of wurtzite III-nitride materials. *Physica B*, Vol. 371, No. 1, (January 2006), pp. 107-111, ISSN 0921-4526

Persson, C.; Ahuja, R.; Ferreira da Silva, A.; Johansson, B. (2005). First-principle calculations of optical properties of wurtzite AlN and GaN. *Journal of Crystal Growth*, Vol. 231, No. 3, (October 2001), pp. 407-414, ISSN 0022-0248

García-Méndez, M.; Morales-Rodríguez, S.; Galván, D.H.; Machorro, R. (2009). Characterization of AlN thin films fabricated by reactive DC sputtering: experimental measurements and Hückel calculations. *International Journal of Modern Phisics B*, Vol. 33, No. 9, (April 2009), pp. 2233-2251, ISSN: 0217-9792

13

Three-Scale Structure Analysis Code and Thin Film Generation of a New Biocompatible Piezoelectric Material $MgSiO_3$

Hwisim Hwang, Yasutomo Uetsuji and Eiji Nakamachi
Doshisha University
Japan

1. Introduction

In this study, three subjects were investigated for a new biocompatible piezoelectric material generation:

1. Development of a numerical analysis scheme of a three-scale structure analysis and a process crystallographic simulation.
2. Design of new biocompatible piezoelectric materials.
3. Generation of $MgSiO_3$ thin film by using radio-frequency (RF) magnetron sputtering system.

Until now, lead zirconate titanate ($Pb(Zr,Ti)O_3$: PZT) has been used widely for sensors (Hindrichsena et al., 2010), actuators (Koh et al., 2010), memory devices (Zhang et al., 2009) and micro electro mechanical systems (MEMS) (Ma et al., 2010), because of its high piezoelectric and dielectric properties. The piezoelectric thin film with aligned crystallographic orientation shows the highest piezoelectric property than any polycrystalline materials with random orientations. Sputtering (Bose et al., 2010), chemical or physical vapor deposition (CVD or PVD) (Tohma et al., 2002), pulsed laser deposition (PLD) (Kim et al., 2006) and molecular beam epitaxy (MBE) (Avrutin et al., 2009) are commonly used to generate high performance piezoelectric thin films. Lattice parameters and crystallographic orientations of epitaxially grown thin films on various substrates can be controlled by these procedures. K. Nishida et al. (Nishida et al., 2005) generated [001] and [100]-orientated PZT thin films on MgO(001) substrate by using CVD method. They succeeded to obtain a huge strain caused by the two effect: the synergetic effect of [001] orientation with the piezoelectric strain; and the strain effect of [100] orientation caused by switching under conditions of the external electric field. Additionally, PZT-based piezoelectric materials, such as $Pb(Zn_{1/3}Nb_{2/3})O_3$-$PbTiO_3$ (Geetika & Umarji, 2010) and $PbMg_{1/3}Nb_{2/3}O_3$-$PbTiO_3$ (Kim et al., 2010), have also been developed.

However, lead, which is a component of PZT-based piezoelectric material, is the toxic material. The usage of lead and toxic materials is prohibited by the waste electrical and electronic equipment (WEEE) and the restriction on hazardous substances (RoHS).

For alternative piezoelectric materials of the PZT, lead-free piezoelectric materials have been studied. J. Zhu et al. (Zhu et al., 2006) generated [111]-orientated $BaTiO_3$ on $LaNiO_3(111)$ substrate, which had a crystallographic orientation with maximum piezoelectric strain constants. S. Zhang et al. (Zhang et al., 2009) doped Ca and Zr in $BaTiO_3$ and succeeded in generating the piezoelectric material with high piezoelectric properties. Further, P. Fu et al. (Fu et al., 2010) doped La_2O_3 in Bi-based $(Bi_{0.5}Na_{0.5})_{0.94}Ba_{0.06}TiO_3$ and succeeded in generating a high performance piezoelectric material. However, their goals were to develop an environmentally compatible piezoelectric material, and the biocompatibility of their piezoelectric materials has not been investigated. Therefore, their piezoelectric materials could not be applied for Bio-MEMS devices.

Recently, the Bio-MEMS, which can be applied to the health monitoring system and the drug delivery system, is one of most attractive research subject in the development of the nano- and bio-technology. Therefore, the biocompatible actuator for the micro fluidic pump in Bio-MEMS is strongly required. However, they remain many difficulties to design new biocompatible materials and find an optimum generation process. Especially, it is difficult to optimize the thin film generation process because there are so many process factors, such as the substrate material, the substrate temperature during the sputtering, the target material and the pressure in a chamber. Therefore, the numerical analysis scheme is necessary to design new materials and optimize the generation process.

The analysis scheme based on continuum theory is strongly required, due to time consuming experimental approach such as finding an optimum sputtering process and a substrate crystal structure through enormous experimental trials. The analysis scheme should predict the thin film deformation, strain and stress, which are affected by the imposed electric field and are constrained by the substrate.

Until now, the conventional analysis schemes, such as the molecular dynamics (MD) method (Rubio et al, 2003) and the first-principles calculation based on the density functional theory (DFT) (Lee & Chung, 2006), have been applied to the crystal growth process simulations. The MD method has been used mainly to analyze the crystal growth process of pure atoms. J. Xu et al. (Xu & Feng, 2002) calculated the Ge growth on Si(111). In the cases of the perovskite compounds, the MD method has been applied to analyze the phase transition, the polarization switching and properties of crystal depending on temperature and pressure. J. Paul et al. (Paul et al., 2007) analyzed the phase transition of $BaTiO_3$ caused by rising temperature and S. Costa et al. (Costa et al., 2006) analyzed the one of $PbTiO_3$ caused by rising temperature and pressure. However, the reliability of its numerical results is poor due to its uncertain inter-atomic potentials for the various combinations of atoms. The MD method could not predict the differences of poly-crystal structures and material properties caused by changing combinations of the crystals and the substrates. It can be concluded that the conventional MD method has many problems for the crystal growth prediction of perovskite compounds grown on the arbitrarily selected substrates.

On the other hand, the DFT can treat interactions between electrons and protons, therefore the reliable inter-atomic potentials can be obtained. The first-principles calculations based on the DFT were applied to the epitaxial growth of the ferroelectric material by O. Diegueaz et al. and I. Yakovkin et al.. O. Diegueaz et al. (Diegueaz et al., 2005) evaluated the stress increase and the polarization change caused by the lattice mismatch between a substrate and a thin film crystal, such as $BaTiO_3$ and $PbTiO_3$. Similarly, I. Yakovkin et al. (Yakovkin & Gutowski, 2004) has investigated in the case of $SrTiO_3$ thin film growth on Si substrate. However, these analyses adopted limited assumptions, such as fixing the conformations of thin film crystals and the growth orientations on the substrates. In this conventional algorithm, the grown orientation is determined by the purely geometrical lattice mismatch between thin films and substrates. This algorithm is not sufficient to predict accurately the preferred orientation of the thin film.

In order to generate the new piezoelectric thin film, a crystal growth process of the thin film should be predicted accurately. The stable crystal cluster of the thin film, which consists geometrically with substrate crystal, is grown on the substrate. Generally, the crystal cluster is an aggregate of thin film crystals. Their morphology and orientations were varied according to the combination of the thin film and the substrate crystals. Therefore, the numerical analysis scheme of the crystal growth process, which can find the best combination of the thin film and the substrate crystal, is strongly required, to optimize the new piezoelectric thin film.

In this chapter, following contents are discussed to develop the new biocompatible $MgSiO_3$ piezoelectric thin film.

1. The three-scale structure analysis algorithm, which can design new piezoelectric materials, is developed.
2. The best substrate of the $MgSiO_3$ piezoelectric thin film is found by using the three-scale structure analysis code.
3. The $MgSiO_3$ thin film is grown on the best substrate by using the RF magnetron sputtering system, and piezoelectric properties are measured.
4. An optimum generating condition of the $MgSiO_3$ piezoelectric thin film is found by using the response surface method.

Section 2 provides the description to the algorithm of the three-scale structure analysis code on basis of the first-principles calculation, the process crystallographic simulation and the crystallographic homogenization theory. Section 3 provides the best substrate of the new biocompatible $MgSiO_3$ piezoelectric thin film calculated by the three-scale structure analysis code. In section 4, the optimum generating condition of $MgSiO_3$ piezoelectric thin film is found. Finally, conclusions are given in section 5.

2. A three-scale structure analysis code

This section describes the physical and mathematical modelling of the three-scale structure and the numerical analysis scheme of three-scale structure analysis to characterize and design epitaxially grown piezoelectric thin films. The existing two-scale finite element analysis is the effective analysis tool for characterization of existing piezoelectric materials. This is because virtually or experimentally determined crystal orientations can be employed for calculation of piezoelectric properties of the macro continuum structure (Jayachandran et al., 2009). However, it can not be applied to a new piezoelectric material, due to unknown crystal structure and material properties.

Figure 1 shows the schematic description of the three-scale modelling of a new piezoelectric thin film, which is grown on a substrate. It shows the three-scale structures, such as a "crystal structure", a "micro polycrystalline structure" and a "macro continuum structure". In the crystal structure analysis, stable structures and crystal properties are evaluated by using the first-principles calculation. Preferred orientations and their fraction are calculated by using the process crystallographic simulation in the micro polycrystalline structure analysis. The macro continuum structure analysis provides the piezoelectric properties of the thin film by using the finite element analysis on basis of the crystallographic homogenization theory. Therefore, the three-scale structure analysis can predict the epitaxial growth process of not only the existent piezoelectric materials but also the new ones.

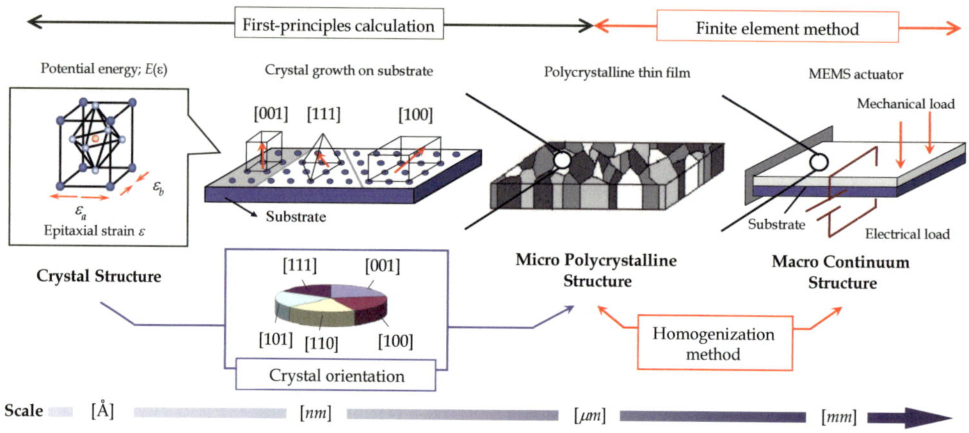

Fig. 1. Three-scale modelling of piezoelectric thin film as a process crystallography.

2.1 Crystal structure analysis by using the first-principles calculation

2.1.1 Stable crystal structure analysis

The stable structures of the perovskite cubic are calculated by the first-principles calculation based on the density functional theory (DFT) by using the CASTEP code (Segal et al., 2002). The stable structures are, then, computed using an ultra-soft pseudo potential method under the local density approximation (LDA) for exchange and correlation terms. A plane-wave basis set with 500eV cutoff energy is used and special k-points are generated by a 8x8x8 Monkhorst-pack mesh (Monkhorst et al., 1976).

The perovskite-type compounds ABX_3 provide well-known examples of displacive phase transitions. They are in a paraelectric non-polar phase at high temperature and have a cubic crystal structure (lattice constant $a = c$). The cubic crystal structure consists of A cations in the large eightfold coordinated site, B cations in the octahedrally coordinated site, and X anions at equipoint. The stability of cubic crystal structure can be estimated by an essential geometric condition, tolerance factor t. If ion radiuses of A, B and X are indicated with r_A, r_B, r_X, tolerance factor t can be described as

$$t = \frac{r_A + r_X}{\sqrt{2}\,(r_B + r_X)} \tag{1}$$

When tolerance factor t consists in the range from 0.75 to 1.10, the perovskite-type crystal structure has high stability. The cubic crystal structure often distorts to ferroelectric phase of lower symmetry at decreased temperature, which is a tetragonal crystal structure ($a > c$) with spontaneous polarization. These ferroelectric distortions are caused by a soft-mode of phonon vibration in cubic crystal structure, and it brings to good piezoelectricity. The soft-mode can be distinguished from other phonon vibration modes with negative eigenfrequency, and the transitional crystal structure depends strongly on the eigenvectors. Consequently, new biocompatible piezoelectric materials are searched according to the flowchart in Fig. 2. Firstly, biocompatible elements are inputted to A and B cations while halogens and chalcogens are set to X anion for the perovskite-type compounds. The combination of three elements is determined to satisfy the stable condition of the tolerance factor. The stable cubic structure of perovskite-type oxides is calculated to minimize the total energy. Next, the phonon vibration in the stable cubic structure is analyzed to catch the soft-mode which causes a phase transition from the paraelectric non-polar phase (cubic structure) to the ferroelectric phase (tetragonal structure). When the eigenfrequency of phonon vibration is positive, it is considered that the cubic structure is the most stable phase and does not change to other phase. On the other hand, in case that the eigenfrequency is negative, the cubic structure is guessed to be an unstable phase and change to other phase corresponding to the soft-mode. Additionally, if all eigenvectors of constituent atoms are parallel to c direction in crystallographic coordinate system, it is supposed to change from cubic to tetragonal structure. If not, it is supposed to transit to other structures except tetragonal one. On the base of phonon properties, the stable tetragonal structure with minimum total energy is searched using the eigenvector components for the initial atomic coordinates.

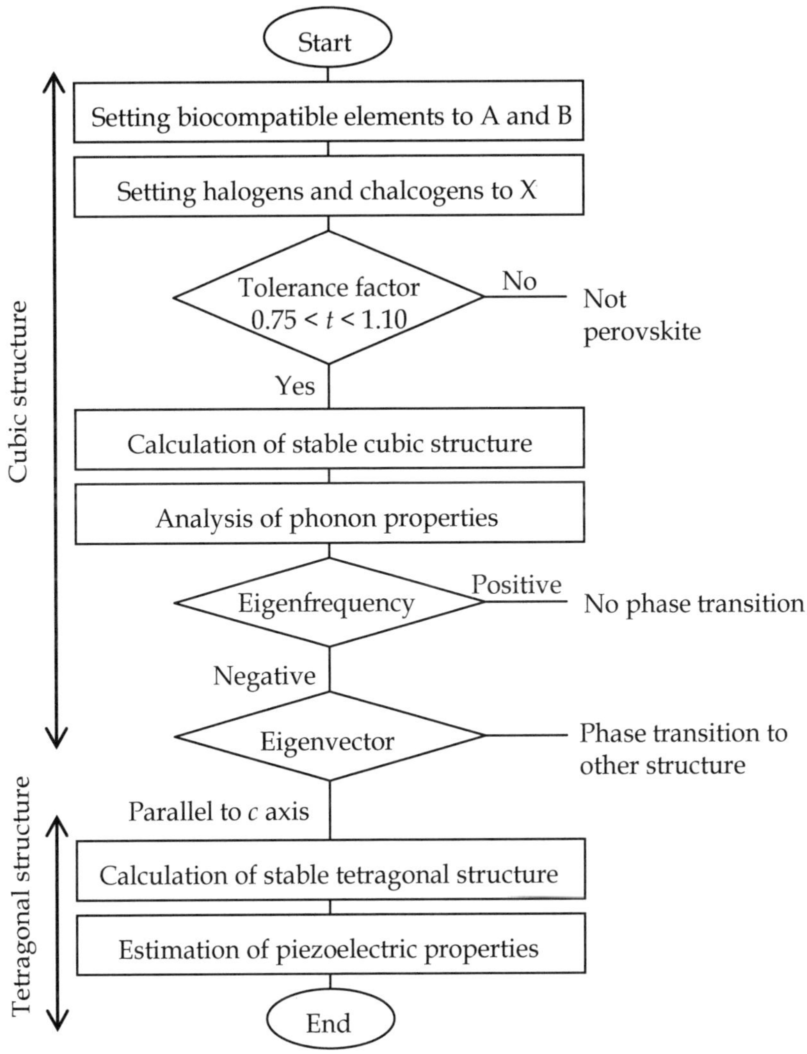

Fig. 2. The flowchart of searching new piezoelectric materials by the first-principles DFT.

Recently, many perovskite cubic crystals such as $SrTiO_3$ and $LaNiO_3$ have been reported. However, most of these materials could not be transformed into a tetragonal structure below Curie temperature, because most of perovskite cubic crystals are more stable than tetragonal crystals. Therefore, the tetragonal structure indicates a soft-mode of the phonon oscillation in cubic structure. Lattice parameters and piezoelectric constants of the tetragonal structure are calculated using the DFT.

2.1.2 Characterization of piezoelectric constants

The total closed circuit (zero field) macroscopic polarization of a strained crystal P_i^T can be described as,

$$P_i^T = P_i^S + e_{i\nu}\varepsilon_\nu \tag{2}$$

where P_i^S is the spontaneous polarization of the unstrained crystal (Szabo et al., 1998, 1999). Under Curie temperature, ferroelectric crystal with tetragonal structure has a polarization along the *c* axis. The three independent piezoelectric stress tensor components are $e_{31} = e_{32}$, e_{33} and $e_{15} = e_{24}$. $e_{31} = e_{32}$ and e_{33} describe the zero field polarization induced along the *c* axis, when the crystal is uniformly strained in the basal *a-b* plane or along the *c* axis, respectively. $e_{15} = e_{24}$ measures the change of polarization perpendicular to the *c* axis induced by the shear strain. This latter component is related to induced polarization by $P_1 = e_{15}\varepsilon_5$ and $P_2 = e_{15}\varepsilon_4$. The total induced polarization along *c* axis can be described by a sum of two contributions.

$$P_3 = e_{33}\varepsilon_3 + e_{31}(\varepsilon_1 + \varepsilon_2) \tag{3}$$

where $\varepsilon_1 = (a - a_0)/a_0$, $\varepsilon_2 = (b - b_0)/b_0$ and $\varepsilon_3 = (c - c_0)/c_0$ are strains along the *a*, *b* and *c* axes, respectively, and a_0, b_0 and c_0 are lattice parameters of the unstrained structure.

The electronic part of the polarization is determined using the Berry's phase approach (Smith & Vandelbilt, 1993), a quantum mechanical theorem dealing with a system coupled under the condition of slowly changing environment. One can calculate the polarization difference between two states of the same solid, under the necessary condition that the crystal remains an insulator along the path, which transforms the two states into each other through an adiabatic variation of a crystal Hamiltonian parameter λ_H. The magnitude of the electronic polarization of a system in state λ_H is defined only modulo $e\mathbf{R}/\Omega$, where $\mathbf{R}$ is a real-space lattice vector, Ω the volume of the unit cell, and *e* the charge of electron. In practice, the $e\mathbf{R}/\Omega$ factor can be eliminated by careful inspection, in the condition where the changes in polarization are described as $|\Delta P| << |e\mathbf{R}/\Omega|$. The electronic polarization can be described as,

$$\mathbf{P}^{\mathrm{el}}(\lambda_H) = -\frac{2e}{(2\pi)^3}\int_{BZ} d\mathbf{k} \frac{\partial}{\partial \mathbf{k}'} \phi^{(\lambda_H)}(\mathbf{k},\mathbf{k}')\bigg|_{\mathbf{k}'=\mathbf{k}} \tag{4}$$

where the integration domain is the reciprocal unit cell of the solid in state λ_H and $\phi^{(\lambda_H)}$ is quantum phase defined as phases of overlap-matrix determinants constructed from periodic parts of occupied valence Bloch states $v_n^{(\lambda_H)}(\mathbf{k})$ evaluated on a dense mesh of $\mathbf{k}$ points from $\mathbf{k}_0$ to $\mathbf{k}_0+\mathbf{b}$, where $\mathbf{b}$ is the reciprocal lattice vector.

$$\phi^{(\lambda_H)}(\mathbf{k},\mathbf{k}') = \mathrm{Im}\left\{\mathrm{In}[\det\left\langle v_m^{(\lambda_H)}(\mathbf{k})\middle| v_m^{(\lambda_H)}(\mathbf{k}')\right\rangle]\right\} \tag{5}$$

The electronic polarization difference between two crystal states can be described as,

$$\Delta \boldsymbol{P}^{\mathrm{el}} = \boldsymbol{P}^{\mathrm{el}}(\lambda_{H2}) - \boldsymbol{P}^{\mathrm{el}}(\lambda_{H1}) \tag{6}$$

Common origins to determine electronic and core parts are arbitrarily assigned along the crystallographic axes. The individual terms in the sum depend on the choice, however, the final results are independent of the origins.
The elements of the piezoelectric stress tensor can be separated into two parts, which are a clamped-ion or homogeneous strain u , and a term that is due to an internal strain such as relative displacements of differently charged sublattices

$$\boldsymbol{e}_{i\nu} = \left.\frac{\partial \boldsymbol{P}_i^T}{\partial \varepsilon_\nu}\right|_u + \sum_k \left.\frac{\partial \boldsymbol{P}_i^T}{\partial \boldsymbol{u}_{k,i}}\right|_\varepsilon \frac{\partial \boldsymbol{u}_{k,i}}{\partial \varepsilon_\nu} \tag{7}$$

where $\boldsymbol{P}_i^T$ is the total induced polarization along the ith axis of the unit cell.
Equation (7) can be rewritten in terms of the clamped-ion part and the diagonal elements of Born effective charge tensor.

$$\boldsymbol{e}_{i\nu} = \boldsymbol{e}_{i\nu}^{(0)} + \sum_k \frac{\boldsymbol{e}\boldsymbol{a}_i}{\Omega} \boldsymbol{Z}_{k,ii}^* \frac{\partial \boldsymbol{u}_{k,i}}{\partial \varepsilon_\nu} \tag{8}$$

where a_i is the lattice parameter, the clamped-ion term $e^{(0)}$ is the first term of Eq. (8). $e^{(0)}$ is equal to the sum of rigid core $e^{(0)}$, core and valence electronic $e^{(0),\mathrm{el}}$ contributions. Subscript k corresponds to the atomic sublattices. Z^* is the Born effective charge described as,

$$\boldsymbol{Z}_{k,i\nu}^* = \boldsymbol{Z}_k^{core} + \boldsymbol{Z}_{k,i\nu}^{*,\mathrm{el}} = \left.\frac{\Omega}{\boldsymbol{e}\boldsymbol{a}_i}\frac{\partial \boldsymbol{P}_i}{\partial \boldsymbol{u}_{k,\nu}}\right|_\varepsilon \tag{9}$$

Piezoelectric response includes two contributions, that appear in linear response for finite distortional wave vectors **q**, and contributions which appear at **q**= 0. Improper polarization changes arise from the rotation or dilation of the spontaneous polarization $\boldsymbol{P}_i^s$. The proper polarization of a ferroelectric or pyroelectric material is given by

$$\boldsymbol{P}_i^P = \boldsymbol{P}_i^T - \sum_j \left(\varepsilon_{ij}\boldsymbol{P}_j^s - \varepsilon_{jj}\boldsymbol{P}_i^s\right) \tag{10}$$

Proper piezoelectric constants $\boldsymbol{e}_{i\nu}^P$ can be described as,

$$\boldsymbol{e}_{31}^P = \frac{\partial \boldsymbol{P}_3^T}{\partial \varepsilon_1} + \boldsymbol{P}_3^s \tag{11}$$

$$\boldsymbol{e}_{15}^P = \frac{\partial \boldsymbol{P}_1^T}{\partial \varepsilon_5} - \boldsymbol{P}_3^s \tag{12}$$

and $e_{33}^{P} = e_{33}^{T}$, because the improper part of e_{33}^{T} is zero. The difference between proper polarization and total one is due to only homogeneous part, which can be described in the following equation for e_{31} ($e_{31}^{P,\text{hom}}$).

$$e_{31}^{P,\text{hom}} = e_{31}^{\text{hom}} + P_3^s = \frac{\partial P_3^{\text{el},T}}{\partial \varepsilon_1} + P_3^{\text{el},s} \tag{13}$$

This equation can use the similar expression for $e_{15}^{P,\text{hom}}$. The homogeneous part appears as a pure electronic term in the expression for the proper piezoelectric constants, which differ in crystal with nonzero polarization in the unstrained state.
The first term in Eq. (8) can be evaluated by polarization differences as a function of strain, with the internal parameters kept fixed at their values corresponding to zero strain. The second term, which arises from internal microscopic relaxation, can be calculated after determining the elements of the dynamical transverse charge tensors and variations of internal coordinate u_i as a function of strain. Generally, transverse charges are mixed second derivatives of a suitable thermodynamic potential with respect to atomic displacements and electric field. They evaluate the change in polarization induced by unit displacement of a given atom at the zero electric field to linear order. In a polar insulator, transverse charges indicate polarization increase induced by relative sublattice displacement. While many ionic oxides have Born effective charges close to their static value, ferroelectric materials with perovskite structure display anomalously large dynamical charges.

2.2 Micro polycrystalline structure analysis by using the process crystallographic simulation

2.2.1 Evaluation method of the total energy

The tetragonal crystal structure of perovskite compound and its five typical orientations [001], [100], [110], [101] and [111] are shown in Fig. 3. Considering a epitaxial growth of the crystal on a substrate, the lattice constants including a, b, c, θ_{ab}, θ_{bc} and θ_{ca} are changed because of the lattice mismatch with the substrate. These crystal structure changes can be determined by considering six components of mechanical strain in crystallographic coordinate system such as ε_a, ε_b, ε_c, γ_{ab}, γ_{bc} and γ_{ca}. In a general analysis procedure, the lattice mismatch in the specific direction was calculated and the crystal growth potential was derived. However, the epitaxially grown thin film crystal is in a multi-axial state. Therefore, the numerical results of the crystal energy of thin films are not corrected when considering only uni-axis strain.
In this study, the total energy of a crystal thin film with multi-axial crystal strain states is calculated by using the first-principles calculation, and is applied to the case of the epitaxial growth process. An ultra-soft pseudo-potentials method is employed in the DFT with the condition of the LDA for exchange and correlation terms. Total energies of the thin film crystal as the function of six components of crystal strain are calculated to find a minimum value. Total energies are calculated discretely and a continuous function approximation is introduced. A sampling area is selected by considering the symmetry between a and b axes in a tetragonal crystal structure. Sampling points are generated by using a latin hypercube sampling (LHS) method (Olsson & Sandberg, 2002), which is the efficient tool to get nonoverlap sapling points. The following global function model is generated by using a kriging polynomial hybrid approximation (KPHA) method (Sakata et al., 2007).

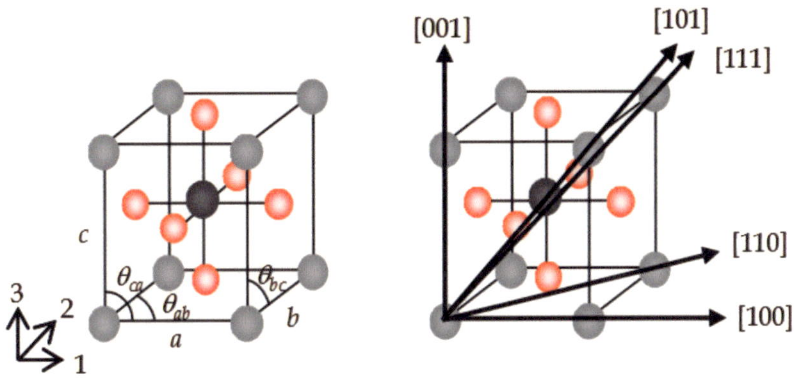

Fig. 3. Crystal structure and orientations of perovskite compounds.

$$E = A_h \varepsilon_h^2 + B_{ij} \varepsilon_i \varepsilon_j + C_l \varepsilon_l + E_{T0} \ (h,i,j = a,b,c,ab,bc,ca) \tag{14}$$

where E_{T0} is the total energy of the stable crystal, ε_h, ε_i and ε_j epitaxial strains and A_h, B_{ij} and C_l coefficients generated by KPHA method. A gradient of total energy at each sampling point is calculated to generate an approximate quadratic function. The minimum point of a total energy can be found by using this function.

2.2.2 Algorithm of the process crystallographic simulation

In the process crystallographic simulation, it is assumed that several crystal unit cells of crystal clusters, which have certain conformations, can grow on a substrate as shown Fig. 4. The left-hand side diagram in Fig. 4 shows an example of conformation in cases of [001], [100], [110] and [101] orientations, and the right-hand side shows [111] orientation. O, A and B are points of substrate atoms corresponding to thin films ones within the allowable range of distance. l_{OA} and l_{OB} indicate distances of A and B from O, respectively. θ_{AOB} indicates the angle between lines OA and OB.

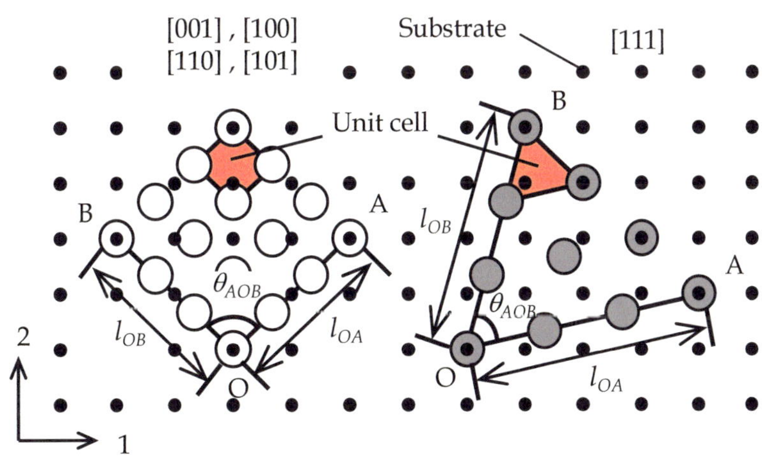

Fig. 4. Schematic of crystal conformations on a substrate.

	[001]	[100]	[110]	[101]	[111]
l_{OA}	a	c	c	$\sqrt{a^2+c^2}$	$\sqrt{a^2+c^2}$
l_{OB}	b	b	$\sqrt{a^2+b^2}$	b	$\sqrt{b^2+c^2}$
Epitaxial strain					
ε_a	$\frac{l_{OA}}{ka}-1$	ε_a^*	ε_a^*	ε_a^*	$\frac{\sqrt{(l_{OA}/k)^2-c^2(1+\varepsilon_c)^2}}{a}-1$
ε_b	$\frac{l_{OB}}{kb}-1$	$\frac{l_{OB}}{kb}-1$	$\frac{\sqrt{(l_{OB}/k)^2-a^2(1+\varepsilon_a^*)^2}}{b}-1$	$\frac{l_{OB}}{kb}-1$	$\frac{\sqrt{(l_{OB}/k)^2-c^2(1+\varepsilon_c)^2}}{b}-1$
ε_c	ε_c^*	$\frac{l_{OA}}{kc}-1$	$\frac{l_{OA}}{kc}-1$	$\frac{\sqrt{(l_{OA}/k)^2-a^2(1+\varepsilon_a^*)^2}}{c}-1$	$\frac{k-1}{kc}\sqrt{l_{OA}l_{OB}\cos\theta_{AOB}}-1$
γ_{ab}		γ_{ab}^*	γ_{ab}^*	$\theta_{AOB}-\theta_{ab}$	γ_{ab}^*
γ_{bc}	γ_{bc}^*	γ_{bc}^*	$\theta_{AOB}-\theta_{bc}$	γ_{bc}^*	$\frac{b}{c}\gamma_{ab}^*$
γ_{ca}	γ_{ca}^*	$\theta_{AOB}-\theta_{ca}$	$\theta_{AOB}-\theta_{ca}$	$\frac{a}{c}\gamma_{ab}$	$\frac{a}{c}\gamma_{ab}^*$

ε_i^* and γ_{ij}^* can be given from first-principles calculation to the minimize total energy

Table 1. Relationship of lattice constants and epitaxial strain with crystal orientations.

Table 1 summarizes the relationship between the lattice constants of the thin film and l_{OA} and l_{OB} according to crystal orientations. Additionally, Table 1 shows crystal strains, which can be determined in the corresponded crystal orientations. However, particular crystal strains, such as ε_i^* and γ_{ij}^*, cannot be determined by employing the lattice constants of the thin film and the geometric constants of the substrate. In this numerical analysis scheme, their unknown components are determined by employing the condition of minimum total energy of the crystal unit cell.

Figure 5 shows the flowchart of the crystal growth prediction algorithm. First, lattice constants of the thin film and the substrate are inputted. The following procedure is demonstrated. Substrate coordinates of A and B points, which are indicated as (m_A, n_A) and (m_B, n_B), are updated according to the numerical result under the condition of fixing O point in order to generate candidate crystal clusters with assumed conformations and orientations. The search range of the crystal cluster is settled as $0 < m_A, m_B < m$ and $0 < n_A, n_B < n$ by considering the grain size of the piezoelectric thin film crystal. $\vec{e}_1$ and $\vec{e}_2$ as shown in Figure 5 are unit vectors of the substrate coordinate system. Lattice constants of the crystal cluster are compared with geometrical parameters of the substrate, and candidate crystal clusters, which have extreme lattice mismatches, are eliminated. Crystal strains caused by the epitaxial growth are calculated for every candidates of the grown crystal cluster as shown in Table 1. The total energy of grown crystal cluster is estimated by using the total energy as a function of crystal strains. Total energies of candidate crystal clusters are compared with one of the free-strained boundary condition, in order to calculate total energy increments of candidate crystal clusters.

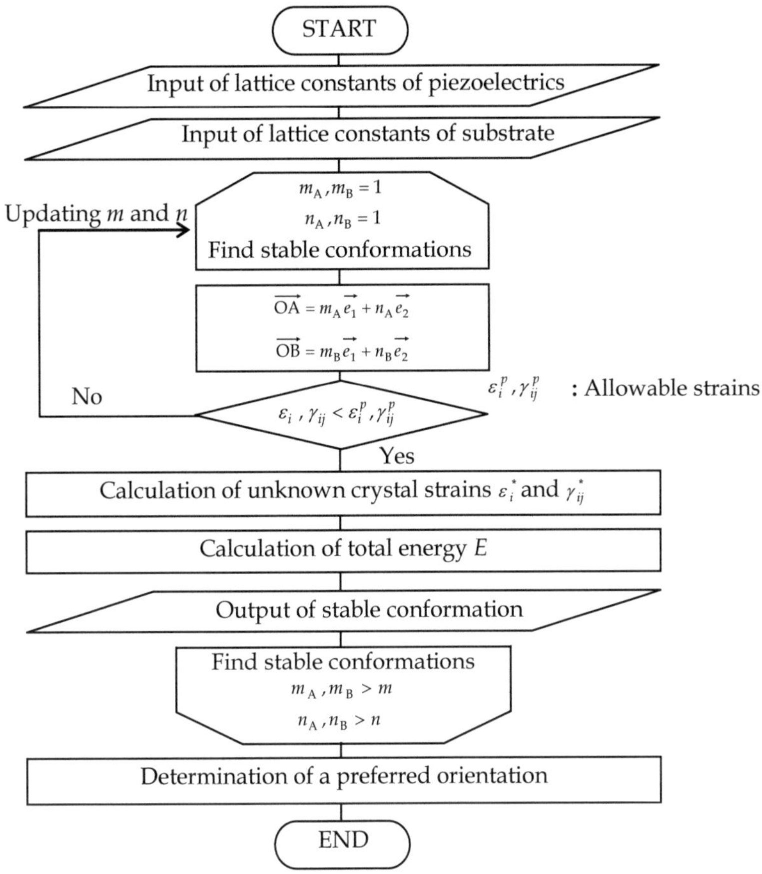

Fig. 5. Flowchart of the process crystallographic simulation.

The fraction of crystal cluster grown on the substrate is calculated by a canonical distribution (Nagaoka et al., 1994).

$$p_i = \frac{\exp\left[-\Delta E_i / k_B T\right]}{\sum_n \exp\left[-\Delta E_n / k_B T\right]} \tag{15}$$

Where ΔE is the total energy increment of the grown cluster, k_B the Boltzmann constant and T the temperature.

2.3 Macro continuum structure analysis by using the crystallographic homogenization method

The crystallographic homogenization method scales up micro heterogeneous structure, such as polycrystalline aggregation, to macro homogeneous structure, such as continuum body. The micro heterogeneous structure has the area Y and microscopic polycrystalline coordinate y, and the macro homogeneous structure has the area X and macroscopic sample coordinate x. Here, it relates to two scales by using the scale ratio λ_h.

$$\lambda_h = \frac{x}{y} \tag{16}$$

where λ_h is an extremely small value. Both coordinates of the micro polycrystalline and the macro continuum structures can be selected independently based on the Eq. (16). Coupling variables are affixed to the superscript λ_h, because the behaviour of the piezo-elastic materials is affected by the polycrystalline structure and λ_h.
The linear piezo-elastic constitutive equation is described as,

$$\sigma_{ij}^{\lambda_h} = C_{ijkl}^{E\lambda_h} \varepsilon_{kl}^{\lambda_h} - e_{kij}^{\lambda_h} E_k^{\lambda_h} \tag{17}$$

$$D_i^{\lambda_h} = e_{ikl}^{\lambda_h} \varepsilon_{kl}^{\lambda_h} + \in_{ik}^{S\lambda_h} E_k^{\lambda_h} \tag{18}$$

The equation of the virtual work of piezoelectric material is written as,

$$\begin{aligned} \int_{\Omega^{\lambda_h}} \left(C_{ijkl}^{E\lambda_h} \varepsilon_{kl}^{\lambda_h} - e_{kij}^{\lambda_h} E_k^{\lambda_h} \right) \frac{\partial \delta u_i^{\lambda_h}}{\partial x_j} \mathrm{d}\Omega + \int_{\Omega^{\lambda}} \left(e_{ikl}^{\lambda_h} \varepsilon_{kl}^{\lambda_h} + \in_{ik}^{S\lambda_h} E_k^{\lambda_h} \right) \frac{\partial \delta \phi^{\lambda_h}}{\partial x_i} \mathrm{d}\Omega \\ = \int_{\Gamma_d^{\lambda}} t_i \, \delta u_i^{\lambda_h} \, \mathrm{d}\Gamma + \int_{\Gamma_e^{\lambda}} \rho \; \delta \phi^{\lambda_h} \, \mathrm{d}\Gamma \end{aligned} \tag{19}$$

where the strain tensor and the electric field tensor are

$$\varepsilon_{ij}^{\lambda_h} = \frac{1}{2}\left(\frac{\partial u_i^0(x)}{\partial x_j} + \frac{\partial u_j^0(x)}{\partial x_i} \right) + \frac{1}{2}\left(\frac{\partial u_i^1(x,y)}{\partial y_j} + \frac{\partial u_j^1(x,y)}{\partial y_i} \right) \tag{20}$$

$$E_i^{\lambda_h} = -\frac{\partial \phi^0(x)}{\partial x_i} - \frac{\partial \phi^1(x,y)}{\partial y_i} \tag{21}$$

It is assumed that the microscopic displacement and the electrostatic potential can be written as the separation of variables:

$$u_i^1(\mathbf{x},\mathbf{y}) = \chi_i^{kl}(\mathbf{x},\mathbf{y}) \frac{\partial u_k^0(\mathbf{x})}{\partial x_l} + \Phi_i^m(\mathbf{x},\mathbf{y}) \frac{\partial \phi^0(\mathbf{x})}{\partial x_m} \tag{22}$$

$$\phi^1(\mathbf{x},\mathbf{y}) = \varphi^{ij}(\mathbf{x},\mathbf{y}) \frac{\partial u_i^0(\mathbf{x})}{\partial x_j} + R^k(\mathbf{x},\mathbf{y}) \frac{\partial \phi^0(\mathbf{x})}{\partial x_k} \tag{23}$$

where $\chi_i^{kl}(x,y)$ is the characteristic displacement of a unit cell, $R^k(x,y)$ the characteristic electrical potential of a unit cell and $\phi^{ij}(x,y)$ and $\Phi_i^m(x,y)$ the characteristic coupling functions of a unit cell. The macroscopic dominant equations are described as,

$$\begin{aligned} \int_Y \left(C_{ijkl}^E \frac{\partial \chi_k^{mn}(x,y)}{\partial y_l} + e_{kij} \frac{\partial \varphi^{mn}(x,y)}{\partial y_k} \right) \frac{\partial \delta u_i^1(x,y)}{\partial y_j} \mathrm{d}Y \\ = -\int_Y C_{ijmn}^E \frac{\partial \delta u_i^1(x,y)}{\partial y_j} \mathrm{d}Y \end{aligned} \tag{24}$$

$$\int_Y \left(e_{ikl} \frac{\partial \chi_k^{mn}(x,y)}{\partial y_l} - \epsilon_{ik}^S \frac{\partial \varphi^{mn}(x,y)}{\partial y_k} \right) \frac{\partial \delta \phi^1(x,y)}{\partial y_i} \mathrm{d}Y = -\int_Y e_{imn} \frac{\partial \delta \phi^1(x,y)}{\partial y_i} \mathrm{d}Y \tag{25}$$

$$\int_Y \left(C_{ijkl}^E \frac{\partial \Phi_k^p(x,y)}{\partial y_l} + e_{kij} \frac{\partial R^P(x,y)}{\partial y_k} \right) \frac{\partial \delta u_i^1(x,y)}{\partial y_j} \mathrm{d}Y = -\int_Y e_{pij} \frac{\partial \delta u_i^1(x,y)}{\partial y_j} \mathrm{d}Y \tag{26}$$

$$\int_Y \left(e_{ikl} \frac{\partial \Phi_k^p(x,y)}{\partial y_l} - \epsilon_{ik}^S \frac{\partial R^P(x,y)}{\partial y_k} \right) \frac{\partial \delta \phi^1(x,y)}{\partial y_i} \mathrm{d}Y = \int_Y \epsilon_{ip}^S \frac{\partial \delta \phi^1(x,y)}{\partial y_i} \mathrm{d}Y \tag{27}$$

where, C_{ijkl}^E is the elastic stiffness tensor at constant electric field, ϵ_{ik}^S the dielectric constant tensor at constant strain and e_{kij} the piezoelectric stress constant tensor. They are calculated by experimentally measured crystal properties. Equations (24) - (27) have the solution under the condition of the periodic boundary. The homogenized elastic stiffness tensor, piezoelectric stress constant tensor and dielectric constant tensor are described by the following characteristic function tensor.

$$C_{ijmn}^{EH} = \frac{1}{|Y|} \int_Y \left(C_{ijmn}^E + C_{ijkl}^E \frac{\partial \chi_k^{mn}(x,y)}{\partial y_l} + e_{kij} \frac{\partial \varphi^{mn}(x,y)}{\partial y_k} \right) \mathrm{d}Y \tag{28}$$

$$\begin{aligned} e_{pij}^H &= \frac{1}{|Y|} \int_Y \left(e_{pij} + e_{kij} \frac{\partial R^P(x,y)}{\partial y_k} + C_{ijkl}^E \frac{\partial \Phi_k^p(x,y)}{\partial y_l} \right) \mathrm{d}Y \\ &= \frac{1}{|Y|} \int_Y \left(e_{pij} + e_{pkl} \frac{\partial \chi_k^{ij}(x,y)}{\partial y_l} - \epsilon_{pk}^S \frac{\partial \varphi^{ij}(x,y)}{\partial y_k} \right) \mathrm{d}Y \end{aligned} \tag{29}$$

$$\epsilon_{ip}^{SH} = \frac{1}{|Y|} \int_Y \left(\epsilon_{ip}^S + \epsilon_{ik}^S \frac{\partial R^P(x,y)}{\partial y_k} - e_{ikl} \frac{\partial \Phi_k^p(x,y)}{\partial y_l} \right) \mathrm{d}Y \tag{30}$$

where superscript *H* means the homogenized value.
The conventional two-scale finite element analysis is based on the crystallographic homogenization method. In this conventional analysis, the virtually determined or experimentally measured orientations are employed for the micro crystalline structure to characterize the macro homogenized piezoelectric properties. However, this conventional analysis can not characterize a new piezoelectric thin film because of unknown crystal structure and material properties.

A newly proposed three-scale structure analysis can scale up and characterize the crystal structure to the micro polycrystalline and macro continuum structures. First, the stable structure and properties of the new piezoelectric crystal are evaluated by using the first-principles DFT. Second, the crystal growth process of the new piezoelectric thin film is analyzed by using the process crystallographic simulation. The preferred orientation and their fractions of the micro polycrystalline structure are predicted by this simulation. Finally, the homogenized piezoelectric properties of the macro continuum structure are characterized by using the crystallographic homogenization theory. Comparing the provability of crystal growth and the homogenized piezoelectric properties of the new piezoelectric thin film on several substrates, the best substrate is found by using the three-scale structure analysis. It is confirmed that the three-scale structure analysis can design not only existing thin films but also new piezoelectric thin films.

3. Three-scale structure analysis of a new biocompatible piezoelectric thin film

3.1 Crystal structure analysis by using the first-principles calculation

The biocompatible elements (Ca, Cr, Cu, Fe, Ge, Mg, Mn, Mo, Na, Ni, Sn, V, Zn, Si, Ta, Ti, Zr Li, Ba, K, Au, Rb, In) were assigned to A cation in the perovskite-type compound ABO_3. Silicon, which was one of well-known biocompatible elements, was employed on B cation. Values of tolerance factor were calculated by using Pauling's ionic radius. Five silicon oxides satisfied the geometrical compatibility condition, where $MgSiO_3$ = 0.88, $MnSiO_3$ = 0.93, $FeSiO_3$ = 0.91, $ZnSiO_3$ = 0.91 and $CaSiO_3$ = 1.01.

The stable cubic structure with minimum total energy was calculated for the five silicon oxides. As the cubic structure has a feature of high symmetry, the stable crystal structure was easy to estimate because of a little dependency on the initial atomic coordinates. Table 2 shows the lattice constants of the silicon oxide obtained by the first-principles DFT.

The phonon properties of cubic structure at paraelectric non-polar phase were calculated to consider phase transition to other crystal structures. Table 3 summarizes the eigenfrequency, the phonon vibration mode and the eigenvector components normalized to unity. $MgSiO_3$, $MnSiO_3$, $FeSiO_3$, $ZnSiO_3$ showed negative values of eigenfrequency. Cubic structures of these four silicon oxides became unstable owing to softening atomic vibration, and they had possibility of the phase transition to other crystal structure. On the other hand, the stable structure of $CaSiO_3$ was the cubic structure due to positive value of eigenfrequency.

The phonon vibration modes are also summarized in Table 3. All eigenvectors of $MgSiO_3$, $MnSiO_3$ and $FeSiO_3$ were almost parallel to c axis in crystallographic coordinate system. These three silicon oxides had a high possibility to change from the cubic structure to the tetragonal structure, which showed superior piezoelectricity. The eigenvector of O_I and O_{II} in $ZnSiO_3$, however, included a component perpendicular to c axis. It was expected that $ZnSiO_3$ changed from cubic structure to other structures consisting of a rotated SiO_6–octahedron, such as the orthorhombic structure with inferior piezoelectricity.

The stable tetragonal structure to minimize the total energy was calculated for the above three silicon oxides, $MgSiO_3$, $MnSiO_3$ and $FeSiO_3$. Total energies of these tetragonal structures were lower than those of the stable cubic structure. Table 4 shows lattice constants and internal coordinates of constituent atoms. In comparison of the aspect ratio among the three silicon oxides, the value of $MgSiO_3$ was larger than 1.0. On the other hand, the aspect ratio of $MnSiO_3$ and $FeSiO_3$ were smaller than 1.0. Generally, the tetragonal structure of typical perovskite-type oxides such as $BaTiO_3$ and $PbTiO_3$ had larger aspect ratio than 1.0. Consequently, the tetragonal structure of $MnSiO_3$ and $FeSiO_3$ could not be existed. The above results have indicated that $MgSiO_3$ was a remarkable candidate for the new biocompatible piezoelectric material.

Material	Lattice constant (nm)
$MgSiO_3$	0.3459
$MnSiO_3$	0.3431
$FeSiO_3$	0.3421
$ZnSiO_3$	0.3454
$CaSiO_3$	0.3520

Table 2. The lattice constants of cubic structure for candidates of the piezoelectric material.

Material	Eigenfrequency (cm^{-1})	Phonon eigenvector				
		Mode	components			
			Atom	ξ_1	ξ_2	ξ_3
$MgSiO_3$	-112	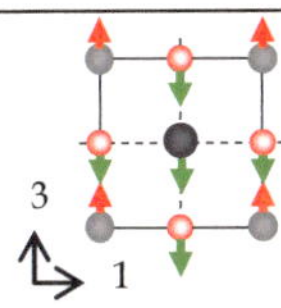	O_I	0.00	0.00	-0.37
			O_{II}	0.00	0.00	-0.37
			O_{III}	0.00	0.00	-0.22
			Si	0.00	0.00	-0.13
			Mg	0.00	0.00	0.88
$MnSiO_3$	-41	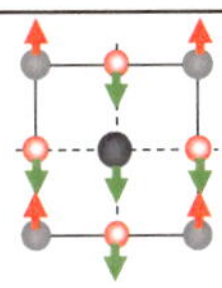	O_I	-0.09	0.00	-0.53
			O_{II}	-0.07	0.00	-0.53
			O_{III}	-0.09	0.00	-0.41
			Si	-0.07	0.00	-0.45
			Mn	0.04	0.00	0.23
$FeSiO_3$	-83	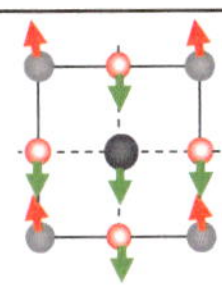	O_I	0.08	0.01	-0.32
			O_{II}	0.04	0.00	-0.32
			O_{III}	0.08	0.00	-0.17
			Si	0.03	0.00	-0.13
			Fe	-0.22	-0.02	0.83
$ZnSiO_3$	-267	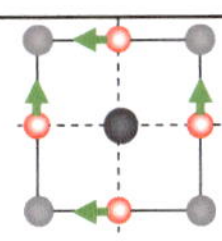	O_I	0.24	0.00	-0.66
			O_{II}	0.00	-0.05	0.66
			O_{III}	-0.24	0.05	0.00
			Si	0.00	0.00	0.00
			Zn	0.00	0.00	0.00
$CaSiO_3$	238	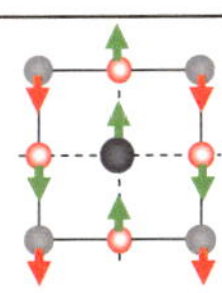	O_I	0.00	0.00	-0.35
			O_{II}	0.01	0.00	-0.35
			O_{III}	0.01	0.00	0.29
			Si	-0.01	0.00	0.79
			Ca	0.00	0.00	-0.23

Table 3. Comparison of phonon properties of cubic structure among $MgSiO_3$, $MnSiO_3$, $FeSiO_3$, $ZnSiO_3$ and $CaSiO_3$.

	$MgSiO_3$	$MnSiO_3$	$FeSiO_3$
Lattice constant (nm)	$a = b = 0.3449$ $c = 0.3538$	$a = b = 0.3547$ $c = 0.3440$	$a = b = 0.3602$ $c = 0.3349$
Aspect ratio	1.026	0.970	0.930
Internal coordinate 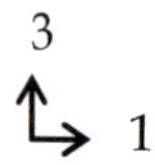	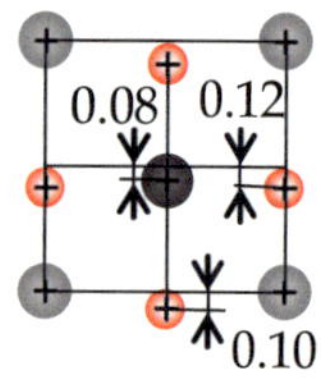	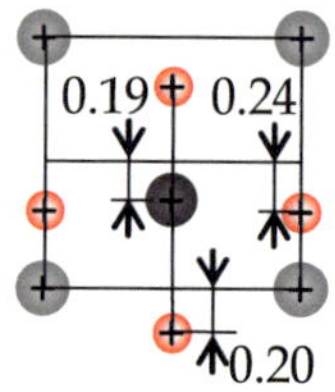	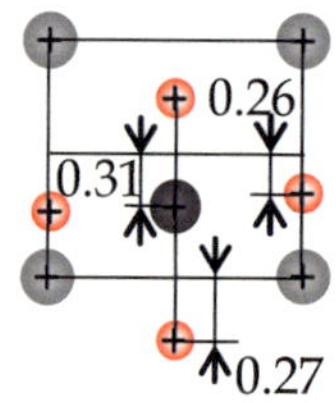

Table 4. Lattice constants and internal coordinates of constituent atoms for tetragonal structure of $MgSiO_3$, $MnSiO_3$ and $FeSiO_3$.

	Mg	Si	O_I	O_{II}	O_{III}
Z^*_{11}	2.331	3.995	-3.024	-1.620	-1.682
Z^*_{22}	2.331	3.995	-1.620	-3.024	-1.682
Z^*_{33}	2.254	4.054	-1.637	-1.637	-3.035

Table 5. Born effective charge in tetragonal structure of $MgSiO_3$ perovskite.

		$MgSiO_3$	$BaTiO_3$	
			DFT	Experiment
Spontaneous polarization (C/m^2)	P_3^S	0.471	0.226	0.260
Piezoelectric stress constant (C/m^2)	e_{33}	4.57	6.11	3.66
	e_{31}	-2.20	-3.49	-2.69
	e_{15}	12.77	21.34	21.30

Table 6. Comparison of the spontaneous polarization and piezoelectric stress constant between $MgSiO_3$ and $BaTiO_3$.

Table 5 shows Born effective charges of each atoms of $MgSiO_3$. Piezoelectric properties, including the spontaneous polarization and piezoelectric stress constants e_{31}, e_{33} and e_{15}, were calculated by these Born effective charges. Table 6 summarizes piezoelectric properties of $MgSiO_3$, those of $BaTiO_3$ calculated by the DFT and observed by the experiment. It could be concluded that $MgSiO_3$ had larger spontaneous polarization than one of $BaTiO_3$. $MgSiO_3$ showed good piezoelectric properties, which were $e_{33} = 4.57$ C/m^2, $e_{31} = -2.20$ C/m^2 and $e_{15} = 12.77$ C/m^2.

3.2 Investigation of the best substrate of the biocompatible piezoelectric material $MgSiO_3$

Three biocompatible atoms, which include Au, Mo and Fe, were selected for the substrate candidate. This is because;

1. these atoms can be used for the under electrode.
2. chemical elements of these atoms have the cubic crystal structure.

Lattice constants of Au with FCC cubic structure are $a = b = c = 0.4080$ nm, and ones of Mo with BCC cubic structure are $a = b = c = 0.3147$ nm. Ones of Fe with BCC cubic structure are $a = b = c = 0.2690$ nm.

Crystal growth process of $MgSiO_3$ thin film on (100) and (111) facets of candidate substrates were demonstrated by using the process crystallographic simulation. Tables 7 - 9 show numerical results of $MgSiO_3$ thin film grown on (100) facets of these four substrates, and Tables 10 - 12 show results of one on (111) facets of the substrates.

Table 13 shows summary of the orientation fractions of $MgSiO_3$ thin film on substrates calculated by canonical distribution. In the case of Mo(100) substrate as shown in Table 8, $MgSiO_3$[100] and [001] were grown, and their orientation fraction were 61.5 % and 38.5 %, respectively. $MgSiO_3$[001] was grown on Au(100) and Fe(100) at 100 % probability. Comparing total energy increments of crystal clusters of $MgSiO_3$ grown on these substrates, $MgSiO_3$[001] on Fe(100) substrate was more stable due to the lowest total energy increment as shown in Table 9. $MgSiO_3$[111] was grown on Au(111) and Mo(111) substrates at 100 % provability.

Orientation	[001]	[001]	[001]
ε_a (%)	0.21	0.48	0.64
ε_b (%)	0.21	0.48	0.64
ε_c (%)	0.00	0.00	0.00
γ_{ab} (%)	0.00	0.00	0.00
γ_{bc} (%)	0.00	0.00	0.00
γ_{ca} (%)	0.00	0.00	0.00
Total energy of the unit cell (eV)	-2398.4025	-2398.3984	-2398.3946
Total energy increment (eV)	0.0749	0.1257	0.3158

Table 7. Analytical results for stable conformations and preferred orientations of $MgSiO_3$ thin film grown on Au(100) substrate.

Orientation	[100]	[001]	[001]
ε_a (%)	0.00	1.13	-0.50
ε_b (%)	1.13	1.13	-0.50
ε_c (%)	-1.41	-0.50	0.00
γ_{ab} (%)	0.00	0.00	0.00
γ_{bc} (%)	0.00	-0.50	0.00
γ_{ca} (%)	0.00	0.00	0.00
Total energy of the unit cell (eV)	-2398.3803	-2398.3772	-2398.3979
Total energy increment (eV)	0.0925	0.1046	0.4447

Table 8. Analytical results for stable conformations and preferred orientations of $MgSiO_3$ thin film grown on Mo(100) substrate.

Comparing total energy increments of crystal clusters of $MgSiO_3$ grown on these two substrates, Au(111) was better substrate than Mo(111) due to low total energy increment. $MgSiO_3$[111] and [001] on Fe(111) were grown at 97.8 % and 2.2 % probability, respectively. Consequently, it could be concluded that four substrates, which included Mo(100), Fe(100) and (111), Au(111), were candidates of the best substrate for $MgSiO_3$ thin film.

Orientation	[001]	[001]	[001]
ε_a (%)	-0.11	0.29	-0.56
ε_b (%)	-0.11	0.29	-0.56
ε_c (%)	0.00	0.00	0.00
γ_{ab} (%)	0.00	0.00	0.00
γ_{bc} (%)	0.00	0.00	0.00
γ_{ca} (%)	0.00	0.00	0.00
Total energy of the unit cell (eV)	-2398.4032	-2398.4016	-2398.3966
Total energy increment (eV)	0.0060	0.0906	0.1092

Table 9. Analytical results for stable conformations and preferred orientations of $MgSiO_3$ thin film grown on Fe(100) substrate.

Orientation	[111]	[001]	[111]
ε_a (%)	2.55	0.06	1.50
ε_b (%)	2.55	0.06	1.50
ε_c (%)	-0.03	0.00	-1.05
γ_{ab} (%)	-0.50	2.20	0.00
γ_{bc} (%)	-0.50	0.00	0.00
γ_{ca} (%)	-0.50	0.00	2.20
Total energy of the unit cell (eV)	-2398.2686	-2398.3450	-2398.3546
Total energy increment (eV)	0.5393	2.1038	2.3891

Table 10. Analytical results for stable conformations and preferred orientations of $MgSiO_3$ thin film grown on Au(111) substrate.

Orientation	[111]	[111]	[001]
ε_a (%)	0.73	0.93	-0.50
ε_b (%)	0.73	0.93	-1.53
ε_c (%)	-1.81	-1.61	0.50
γ_{ab} (%)	0.00	0.00	0.00
γ_{bc} (%)	0.00	0.00	0.50
γ_{ca} (%)	0.00	0.00	0.00
Total energy of the unit cell (eV)	-2398.3724	-2398.3717	-2398.3796
Total energy increment (eV)	0.7738	1.5510	1.9276

Table 11. Analytical results for stable conformations and preferred orientations of $MgSiO_3$ thin film grown on Mo(111) substrate.

Orientation	[111]	[001]	[111]
ε_a (%)	1.33	-0.56	3.19
ε_b (%)	1.33	-0.56	3.19
ε_c (%)	-1.22	0.50	0.60
γ_{ab} (%)	0.00	2.20	-0.50
γ_{bc} (%)	0.00	0.50	-0.50
γ_{ca} (%)	0.00	0.00	-0.50
Total energy of the unit cell (eV)	-2398.3618	-2398.3557	-2398.1793
Total energy increment (eV)	0.6661	0.7639	0.8962

Table 12. Analytical results for stable conformations and preferred orientations of $MgSiO_3$ thin film grown on Fe(111) substrate.

Substrate Atom	Facet	$MgSiO_3$ Orientation	Fraction (%)
Au	(100)	[001]	100.0
	(111)	[111]	100.0
Mo	(100)	[100]	61.5
		[001]	38.5
	(111)	[111]	100.0
Fe	(100)	[001]	100.0
	(111)	[111]	97.8
		[001]	2.2

Table 13. Analytical results of preferred orientations and their fractions for $MgSiO_3$ thin films grown on various substrates.

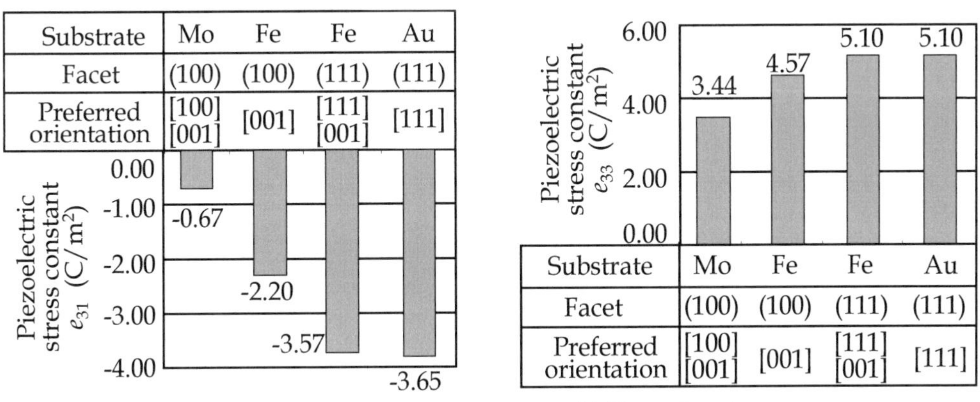

(a) Piezoelectric stress constant e_{31} (b) Piezoelectric stress constant e_{33}

Fig. 6. Homogenized piezoelectric stress constant of $MgSiO_3$ thin film on various substrates.

Analytically determined piezoelectric stress constants and orientation fractions of $MgSiO_3$ were introduced into the macro continuum structure analysis. Homogenized piezoelectric strain constants of the $MgSiO_3$ thin film on four substrate candidates were calculated.

Figure 6(a) shows homogenized e_{31} constants and Fig. 6(b) e_{33}. Substrates, facets of substrates and orientation fractions of $MgSiO_3$ thin film were also shown in figures. $MgSiO_3$[111] on Au(111) substrate indicated the highest piezoelectric stress constants, e_{31} = -3.65 C/m^2 and e_{33} = 5.10 C/m^2. $MgSiO_3$[001] on Fe(100) showed e_{31} = -2.20 C/m^2 and e_{33} = 4.57 C/m^2. e_{31} of $MgSiO_3$[001] on Fe(100) was 39.7 % lower than one on Au(111) and e_{33} of $MgSiO_3$[001] on Fe(100) was 10.4 % lower than one on Au(111). In the case of Fe(111) substrate, e_{33} was equal to one on Au(111) substrate, however e_{31} was smaller than one on Au(111). Furthermore, $MgSiO_3$ on Au(111) was more stable than one on Fe(111) substrate. These results have concluded that Au(111) was the best substrate for $MgSiO_3$ thin film.

4. A new biocompatible piezoelectric $MgSiO_3$ thin film generation

4.1 Experimental method

$MgSiO_3$ tihn film is generated by radio-frequency magnetron sputtering. Three factors are selected for generating perovskite tetragonal structure and high piezoelectric property. These conditions are i) the substrate temperature T_s, ii) the post-annealing temperature T_a and iii) flow rate of oxygen f_{O2}. This is because that i) the substrate temperature contributes configuration and bonding of thin film crystals, and ii) the post-annealing temperature affects crystallization of amorphous crystal. iii) The flow rate of oxygen affects crystal morphology and composition of the $MgSiO_3$ crystal. These generation conditions are set as T_a = 300 °C, 350 °C, 400 °C, T_s = 600 °C, 650 °C, 700 °C, and f_{O2} = 1.0 sccm, 3.0 sccm, 5.0 sccm, respectively. The target material is used the mixed sinter of MgO and SiO_2, the substrate is Au(111)/$SrTiO_3$(110), which is determined by the three-scale structure analysis. The electric power is 100 W, flow rate of Ar gas is 10 sccm and the pressure in chamber during the sputtering is 0.5 Pa. Thin film is sputtered 4 hours and post-annealed an hour after sputtering.

The displacement–voltage curve of $MgSiO_3$ thin film is measured by ferroelectric character evaluation (FCE) system. Generally, displacement-voltage curve of the piezoelectric material shows butterfly-type hysteresis curve. The piezoelectric strain constant d_{33} can be calculated by gradient of the butterfly-type hysteresis curve. The response surface methodology (RSM) (Berger & Maurer, 2002) is employed to find the optimum combination of generation factor levels of the $MgSiO_3$ thin film.

4.2 Generation of the new biocompatible piezoelectric $MgSiO_3$ thin film

Displacement-voltage curves under the conditions of f_{O2} = 1.0, 3.0 and 5.0 sccm are shown in Fig. 7 - 9. All thin films showed the piezoelectric property due to butterfly-type hysteresis curves. The piezoelectric strain constant d_{33} could be calculated by the gradient at cross point of the butterfly-type hysteresis curve, and d_{33} was indicated in all graphs. d_{33} constants of all thin films were larger than the d_{33} constant (= 129.4 pm/V) of $BaTiO_3$, which was commonly used lead-free piezoelectric material generated in our previous study.

The optimum conditions for generating the $MgSiO_3$ thin film were found by using RSM. Figure 10 shows the response surface of d_{33} constant as a function of T_s and T_a under the condition of f_{O2}= 4.0 sccm. Figure 10(a) shows the aerial view and Fig. 10(b) top view. The black point indicates the highest point of d_{33} constant. The optimum condition, for T_s= 300 °C, T_a= 631 °C and f_{O2}= 4.0 sccm, was found.

$MgSiO_3$ thin film was generated at T_s = 250 °C, because the obtained best T_s was lowest temperature in the range of the substrate temperature which was set in this study. However, the displacement-voltage curves were not indicated the butterfly-type hysteresis curve. This is because the thin film was not crystallized to $MgSiO_3$, due to inactive adatoms and low collision rate of adatoms.

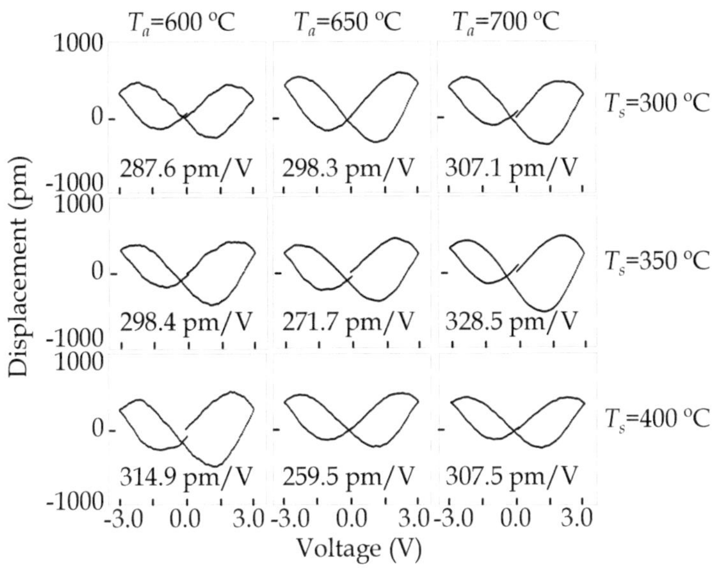

Fig. 7. Displacement-voltage curves of $MgSiO_3$ thin films in the case of f_{O2} = 1.0 sccm.

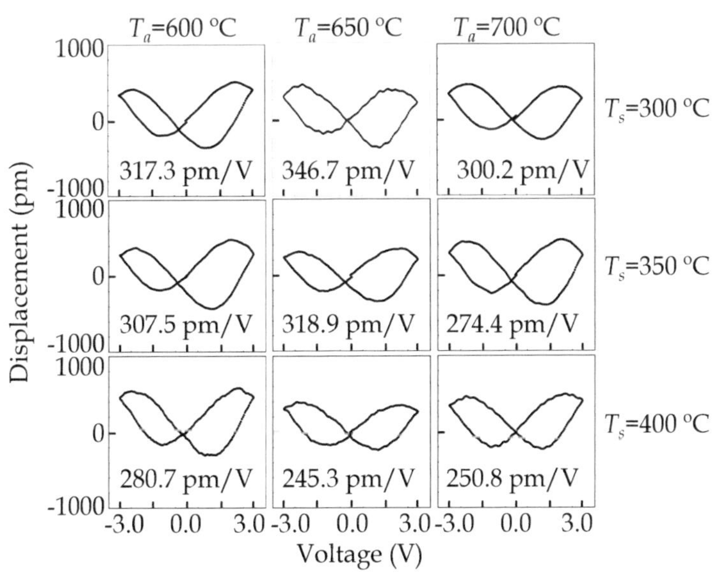

Fig. 8. Displacement-voltage curves of $MgSiO_3$ thin films in the case of f_{O2} = 3.0 sccm.

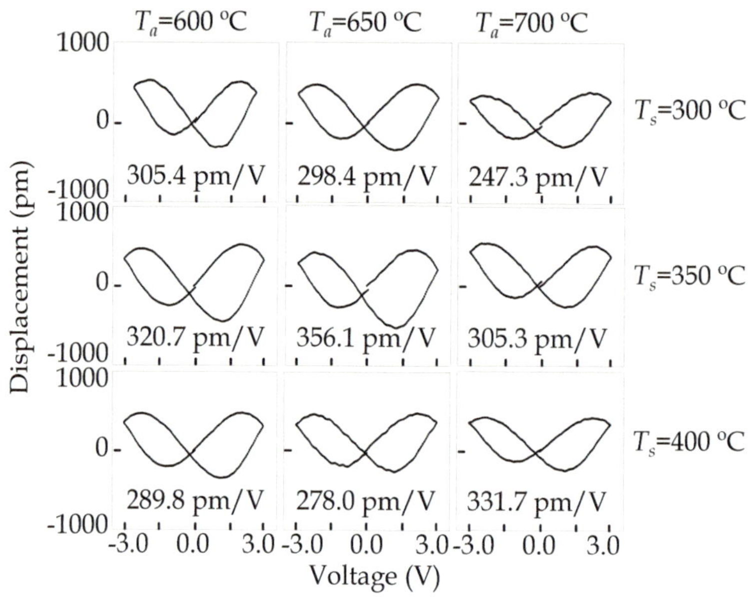

Fig. 9. Displacement-voltage curves of $MgSiO_3$ thin films in the case of f_{O2} = 5.0 sccm.

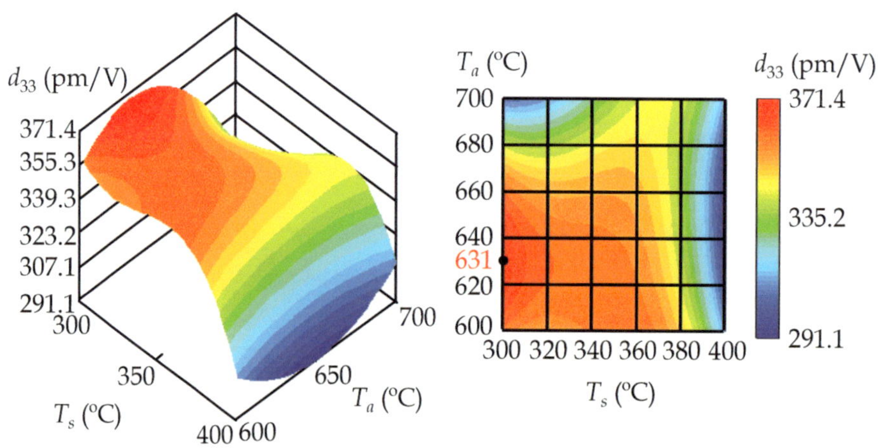

Fig. 10. Piezoelectric strain constant d_{33} as functions of T_s and T_a in the case of f_{O2} = 4.0sccm.

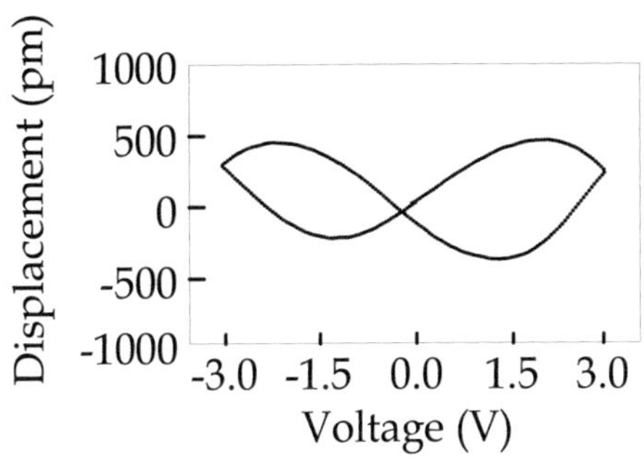

Fig. 11. Displacement-voltage curve of $MgSiO_3$thin film generated at the optimum condition, T_s = 300 °C, T_a = 631 °C and f_{O2} = 4.0 sccm.

Finally, $MgSiO_3$ thin film was generated at the optimum condition. Figure 11 shows its displacement-voltage curve. d_{33} constant was obtained as 359.2 pm/V and this value was higher than one of the pure PZT thin films, d_{33} = 307.0 pm/V, generated by Z. Zhu et al (Zhu et al, 2010).
Consequently, the piezoelectric $MgSiO_3$ thin film was generated successfully and it can be used for sensors and actuators for MEMS or NENS.

5. Conclusion

In this study, the three-scale structure analysis code, which is based on the first-principles density functional theory (DFT), the process crystallographic simulation and the crystallographic homogenization theory, was newly developed. Consequently, a new biocompatible $MgSiO_3$ piezoelectric material was generated by using the radio-frequency (RF) magnetron sputtering system, where its optimum generating condition has been found analytically and experimentally.
Section 2 discussed the algorithm of the three-scale structure analysis, which can design epitaxially grown piezoelectric thin films. This analysis was constructed in three-scale structures, such as a crystal structure, a micro polycrystalline structure and a macro continuum structure. The existing two-scale analysis could evaluate the property of the macro continuum structure by using experimentally observed information of crystal structure, such as crystallographic orientation and properties of the crystal. The three-scale structure analysis can design new biocompatible piezoelectric thin films through three steps, which were to calculate the crystal structure by using the first-principles DFT, to evaluate the epitaxial growth process by using crystallographic simulation, and to calculate the homogenized properties of thin film by using the crystallographic homogenization theory.
In section 3, in order to find a new biocompatible piezoelectric crystal and its best substrate, the three-scale structure analysis was applied to the silicon oxides. Consequently, $MgSiO_3$ had a large spontaneous polarization $P_3^S = 0.471\,C/m^2$ and it could present good piezoelectric stress constants e_{33} = 4.57 C/m^2, e_{31} = -2.20 C/m^2 and e_{15} = 12.77 C/m^2. These results indicated that $MgSiO_3$ was one of the candidates of the new biocompatible piezoelectric thin film. Au(111) was the best substrate of $MgSiO_3$ thin film, because $MgSiO_3$[111] on Au(111) was most stable and showed highest piezoelectric stress constant e_{31}= -3.65 C/m^2 and e_{33}= 5.10 C/m^2.

In section 4, $MgSiO_3$ piezoelectric thin film was generated by using the RF magnetron sputtering system. The optimum condition was found by using the response surface methodology (RSM). Measuring the piezoelectric properties of the thin films by using the ferroelectric character evaluation (FCE) system, all $MgSiO_3$ thin films showed the piezoelectric property due to butterfly-type hysteresis curves. Finally, the optimum condition for T_s = 300 °C, T_a = 631 °C and f_{O2} = 4.0 sccm, was found and the best piezoelectric strain constant d_{33} = 359.2 pm/V was obtained. This value was higher than the one of the pure PZT thin films, d_{33}= 307.0 pm/V, generated by Z. Zhu et al.. Consequently, the piezoelectric $MgSiO_3$ thin film was generated successfully and it can be used for sensors and actuators for MEMS or NENS.

6. References

Hindrichsena, C., Møllerb, R., Hansenc, K. & Thomsena, E. (2010). Advantages of PZT Thick Film for MEMS Sensors, *Sensors and Actuators A: Physical*, (Oct. 2009) pp. 1-6, 0924-4247.

Koh, K., Kobayashi, T., Hsiao, F. & Lee, C. (2010). Characterization of Piezoelectric PZT Beam Actuators for Driving 2D Scanning Micromirrors, *Sensors and Actuators A: Physical*, (Sep, 2009), pp.1-12, 0924-4247.

Zhang, M., Jia, Z. & Ren, T. (2009). Effects of Electrodes on the Properties of Sol-Gel PZT Based Capacitors in FeRAM, *Solid-State Electronics*, Vol.53, (Aug. 2008), pp.473-477, 0038-1101.

Ma, Y., Kong, F., Pan, C., Zhang, Q. & Feng, Z. (2010). Miniature Tubular Centrifugal Piezoelectric Pump Utilizing Wobbling Motion, *Sensors and Actuators A: Physical*, Vol.157, (Jul. 2009), pp.322-327, 0924-4247.

Bose, A., Maity, T., Bysakh, S., Seal, A. and Sen, S. (2010). Influence of Plasma Pressure on the Growth Characteristics and Ferroelectric Properties of Sputter-Deposited PZT Thin Films, *Applied Surface Science*, Vol.256, (Feb. 2010), pp.6205-6212, 0169-4332.

Tohma, T., Masumoto, H. & Goto, T. (2002). Preparation of $BaTiO_3$ Films by Metal-Organic Chemical Vapor Deposition, *Japanese Journal of Applied. Physics Part1*, Vol.41, No.11B, (May 2002), pp.6643-6646, 0021-4922.

Kim, T., Kim, B., Lee, W., Moon, J., Lee, B. & Kim, J. (2006). Integration of Artificial $SrTiO_3/BaTiO_3$ Superlattices on Si Substrates using a TiN Buffer Layer by Pulsed Laser Deposition Method, *Journal of Crystal Growth*, Vol.289, No.2, (Aug. 2006), pp.540-546, 0022-0248.

Avrutin, V., Liu, H., Izyumskaya, N., Xiao, B., Ozgur, U. & Morkoc, H. (2009). Growth of $Pb(Ti,Zr)O_3$ Thin Films by Metal-Organic Molecular Beam Epitaxy, *Journal of Crystal Growth*, Vol.311, (Oct. 2008), pp.1333-1339, 0022-0248.

Nishida, K., Wada, S., Okamoto, S., Ueno, R., Funakubo, H. & Katoda, T. (2005). Domain Distributions in Tetragonal $Pb(Zr,Ti)O_3$ Thin Films Probed by Polarized Raman Spectroscopy, *Applied Physics Letters*, Vol.87, (2005), pp.232902.1-232902.3, 0003-6951.

Geetika & Umarji, A. (2010) The Influence of Zr/Ti Content on the Morphotropic Phase Boundary in the PZT–PZN System, *Materials Science and Engineering B*, Vol.167, (Sep. 2009), pp.171-176, 0921-5107.

Kim, K., Hsu, D., Ahn, B., Kim, Y. & Barnard, D. (2010). Fabrication and Comparison of PMN-PT Single Crystal, PZT and PZT-based 1-3 Composite Ultrasonic Transducers for NDE Applications, *Ultrasonics*, Vol.50, (Feb. 2009), pp.790-797, 0041-624X.

Zhu, J., Zheng, L., Luo, W., Li, Y., & Zhang, Y. (2006). Microstructural and Electrical Properties of $BaTiO_3$ Epitaxial Films on $SrTiO_3$ Substructures with a $LaNiO_3$ Conductive Layer as a Template, *Journal of Physic D*, Vol.39. (Feb. 2006), pp.2438-2443, 1361-6463.

Zhang, S., Zhang, H., Zhang, B. & Zhao, G. (2009). Dielectric and Piezoelectric Properties of $(Ba_{0.95}Ca_{0.05})(Ti_{0.08}Zr_{0.12})O_3$ Ceramics Sintered in A Protective Atmosphere, *Journal of European Ceramics Society*, Vol.29, (Apr. 2009), pp.3235-3242, 0955-2219.

Fu, P., Xu, Z., Chu, R., Li, W., Zhang, G. & Hao, J. (2010). Piezoelectric, Ferroelectric and Dielectric Properties of La_2O_3-doped $(Bi_{0.5}Na_{0.5})_{0.94}Ba_{0.06}TiO_3$ Lead-Free Ceramics, *Materials and Design*, Vol.31, (May 2009), pp.796-801, 0261-3069.

Rubio, J., Jaraiz, M., Bragado, I., Mangas, J., Barblla, J. & Gilmer, G. (2003). Atomistic Monte Carlo Simulations of Three-Dimensional Polycrystalline Thin Films, *Journal of Applied Physics*, Vol.94, (Aug. 2002), pp.163-168, 0021-8979.

Lee, S. & Chung, Y. (2006). Surface Characteristics of Epitaxially Grown Ni Layers on Al Surfaces: Molecular Dynamics Simulation, *Journal of Applied Physics*, Vol.100, No.7, (Feb. 2006), pp.074905.1-074905.4, 0021-8979.

Xu, J. & Feng, J. (2002). Study of Ge Epitaxial Growth on Si Substrates by Cluster Beam Deposition, *Journal of Crystal Growth*, Vol.240, No.3 (Jan. 2002), pp.407-404, 0022-0248.

Paul, J., Nishimatsu, T., Kawazoe, Y. & Waghmare, U. (2007). Ferroelectric Phase Transitions in Ultrathin Films of $BaTiO_3$, *Physical Review Letters*, Vol.99, No.7, (Dec. 2005), pp.077601.1-077601.4, 0031-9007.

Costa, S., Pizani, P., Rino, J. & Borges, D. (2006). Structural Phase Transition and Dynamical Properties of $PbTiO_3$ Simulated by Molecular Dynamics, *Journal of Condensed Matter*, Vol.75, No.6, (Sep. 2005), pp.064602.1-064602.5, 0953-8984.

Dieguez, O., Rabe, K. & Vanderbilt, D. (2005). First-Principles Study of Epitaxial Strain in Perovskites, *Physical Review B*, Vol.72, No.14, (Jun. 2005), pp.144101.1-144101.9, 1098-0121.

Yakovkin, I. & Gutowski, M. (2004). $SrTiO_3$/Si(001) Epitaxial Interface: A Density Functional Theory Study, *Physical Review B*, Vol.70, No.16, (Nov. 2003), pp.165319.1-165319.7, 1098-0121.

Jayachandran, K., Guedes, J. & Rodrigues, H. (2009). Homogenization of Textured as well as Randomly Oriented Ferroelectric Polycrystals, *Computational Materials Science*, Vol.45, (Nov. 2007), pp.816-820, 0927-0256.

Segall, M., Lindan, P., Probert, M., Pickard, C., Hasinp, P., Clark, S. & Payne, M. (2002). First-Principles Simulation: Ideas, Illustrations and the CASTEP Code, *Journal of Physics: Condensed Matter*, Vol.14, (Jan. 2002), pp.2717-2744, 0953-8984.

Monkhorst, H. & Pack, J. (1976). Special Points for Brilloun-Zone Integrations, *Physical Review B*, Vol.13, No.12, (Jan. 1976), pp.5188-5192, 1098-0121.

Szabo, G., Choen, R. & Krakauer, H. (1998). First-Principles Study of Piezoelectricity in $PbTiO_3$, *Physical Review Letters*, Vol.80, No.19, (Oct. 1997), pp.4321-4324, 0031-9007.

Szabo, G., Choen, R. & Krakauer, H. (1999). First-Principles Study of Piezoelectricity in Tetragonal $PbTiO_3$ and $PbZr_{1/2}Ti_{1/2}O_3$, *Physical Review B*, Vol.59, No.20, (Sep. 1998), pp.12771-12776, 1098-0121.

King-Smith, R. & Vanderbilt, D. (1993). Theory of Polarization of Crystalline Solids, *Physical Review B*, Vol.47, No.3, (Jun. 1992), pp.1651-1654, 1098-0121.

Olsson, A. & Sandberg, G. (2002). Latin Hypercube Sampling for Stochastic Finite Element Analysis, *Journal of Engineering Mechanics*, Vol.128, No.1, (Sep. 1999), pp.121-125, 0733-9399.

Sakata, S., Ashida, F. & Zako, M. (2007). Hybrid Approximation Algorithm with Kriging and Quadratic Polynomial-based Approach for Approximate Optimization, *International Journal for Numerical Methods in Engineering*, Vol.70, No.6, (Jul. 2005), pp.631-654, 1097-0207.

Nagaoka, Y. (6th Jul. 1994). *Statistical Mechanics* (in Japanese), Iwanami Shoten, 4-000-07927-1, Japan, Tokyo.

Berger, P. & Maurer, R. (2002). *Experimental Design*, Duxbury Thomson Learning Inc, 0-534-35822-5, USA, Calfornia.

Zhu, Z., Li, J., Liu, Y. & Li, J. (2009). Shifting of the Morphotropic Phase Boundary and Superior Piezoelectric Response in Nb-doped $Pb(Zr, Ti)O_3$ Epitaxial Thin Films, *Acta Materialia*, Vol.57, (Feb. 2009), pp.4288-4295, 1359-6454.

14

The Influence of the Substrate Temperature on the Properties of Solar Cell Related Thin Films

Shadia J. Ikhmayies
Al Isra University, Faculty of Science and Information Technology, Amman, Jordan

1. Introduction

Polycrystalline films are generally considered to consist of crystallites joined together by the grain boundaries. The grain boundary regions are disordered regions, characterized by the presence of a large number of defect states due to incomplete atomic bonding or departure from stoichiometry for compound semiconductors. These states, known as trap states, act as effective carrier traps and become charged after trapping [1]. The density of defects and impurities in the grain boundaries is larger than that within the grains, so as the orientations of the grains change, the density of traps also changes [2]. The density of trap states depends critically on the deposition parameters [1] including the substrate temperature.

By increasing the substrate temperature the grain size increases, grain boundaries become narrower and their number decreases, the height of the potential barrier between grains decreases, and some impurities go out from the grain boundaries and become effectively incorporated in the lattice and other impurities migrate to the grain boundaries. Evaporation of some elements changes stoichiometry and may create other defects. These changes produce changes in the structure and phase of the films. As a result, the electrical, optical and electronic properties will change too. The presence of some of these changes in a film depends on the deposition technique followed in producing the film, raw materials used and deposition conditions.

There are different deposition techniques to prepare thin films in which the deposition temperature is one of the main parameters that should be controlled to get high quality films. These methods include thermal evaporation [3-7], spray pyrolysis (SP) [8-27], chemical bath deposition (CBD) [28-29], dc magnetron sputtering [30] etc.

In the following sections we will discuss the effect of the substrate temperature on the structural, morphological, electrical and optical properties of thin films deposited by different techniques. A review of experimental results obtained by different authors will be performed with discussions and comparisons between different results.

2. Structural properties

There is agreement between authors that the increase in the substrate temperature improves the crystallinity of the films and encourages the change from amorphous to polycrystalline structure and increases the grain size. X-ray diffraction (XRD) is the suitable tool to reveal these changes. For polycrystalline films, the variations of the intensity of Bragg peaks and their width at half maximum (FWHM) with substrate temperature are evidences on the changes in grain size. The narrowing of the lines of crystal growth at the higher substrate temperature (the decrease in FWHM) means that the grain size had increased. The shifts of the positions of the peaks refer to changes in lattice spacing and then lattice parameters. The appearance of some lines and disappearance of others with substrate temperature may mean a phase transition and/or the appearance or disappearance of other phases of the compounds under study or the presence of some elements. In this section different experimental results will be discussed to show the different effects of the substrate temperature on the structure of thin films through XRD diffractogramms.

A lot of experimental results are found about the change from amorphous to polycrystalline structure with substrate temperature. For films prepared by the spray pyrolysis (SP) technique, a lot of workers [8, 12, 14, 18] found that CdS films prepared at substrate temperatures more than or around 200 °C are polycrystalline. Our CdS:In thin films [8] were prepared at T_s = 350-490 °C and they are polycrystalline. Bilgin et al. [12] prepared CdS thin films by the SP technique at substrate temperatures 473-623 K and found them to be polycrystalline. But we reported that [15] SnO_2:F thin films were amorphous at temperatures lower than 360 °C. Gordillo et al. [22] found that SnO_2 films deposited at temperatures lower than 300 °C grow with an amorphous structure, but those deposited at T_s = 430 °C present a polycrystalline structure. Rozati [2] found that increasing the substrate temperature causes the SnO_2 thin films to exhibit a strong orientation along (200).

Films prepared by chemical path deposition (CBD) which is a low temperature technique are in most cases partially or totally amorphous [28]. Liu et al. [28] prepared CdS films by this technique at deposition temperatures in the range 55-85 °C and found that all of the produced films have some amorphous component and an improved crystallinity with the increase of deposition temperature was obtained.

Numerous experimental data showed that the orientations of crystal growth and preferential orientation are sensitive to the substrate temperature. For ZnO spray-deposited thin films of the hexagonal wurtzite-type, Hichou et al. [10] found that the intensity of the diffraction peaks is strongly dependent on the substrate temperature, where they got maximum intensity at T_s = 450 °C. They found that the [002] direction is the main orientation and it is normal to the substrate plane. For these films some orientations of crystal growth appeared and others disappeared with the variation in the substrate temperature. For CdS thin films prepared by SP technique Acosta et al. [14] found that the intensity of the (002) line increases with temperature, while the (101) peak tends to disappear, which is exactly opposite to what we have in our diffractograms for CdS:In thin films [8]. But we also showed that the preferential orientation of the crystal growth is very sensitive to the substrate temperature. At T_s = 350 °C the preferential orientation in our diffractogram [8] is the H(002)/C(111)- The peaks of these two lines are very close to each other, so it is difficult to distinguish them-, and at T_s = 460 °C it is the H(101), but at T_s = 490 °C it is the H(112)/C(311)- also it is difficult to distinguish the peaks of these two lines.

As we see the orientations of crystal growth and the preferential orientations for a certain compound are different for different authors [8, 14]. In some cases [10] the preferential orientation does not change with the substrate temperature. The preferential orientation in Ashour's [12] diffractograms for CdS thin films which showed just the hexagonal phase is the (101) which was not affected by the substrate temperature but all of the other lines are affected by the substrate temperature. Ashour [12], Pence et al. [13] and Bilgin et al. [18] did not find an influence of the substrate temperature on the preferential orientation for CdS films prepared by the SP technique. On the other hand, Abduev et al. [30] found that the position of the preferential orientation (002) of the hexagonal ZnO:Ga films of thickness 300 nm prepared by dc magnetron sputtering was shifted from 34.25° to 34.41° when the substrate temperature was increased from 50 to 300 °C.

A lot of authors observed a phase transition from cubic to hexagonal phase with the increase in substrate temperature [8, 14]. For spray-deposited CdS:In thin films our XRD diffractograms [8] showed a mixed (cubic and hexagonal phase) at T_s = 350 °C which was converted to only hexagonal phase at T_s = 490 °C. Also for CdS:In thin films prepared by the spray pyrolysis technique Acosta et al. [14] found that X-ray diffractograms of the samples prepared with In/Cd = 0.1 in the solution, the intensity of the (002) peak shows a noticeable increase while the (101) tends to disappear for higher T_s. They [14] say that these variations in peak intensity might be related with phase transition from a cubic to a hexagonal structure. For films prepared by CBD the phase change was observed too, where Liu et al. [28] found that the phase of CdS films was ambiguous, at low deposition temperatures. That is, it couldn't be distinguished (cubic or hexagonal) because the positions of the (002) and (110) lines of the hexagonal structure are similar to the (111) and (220) lines of cubic one, making it difficult to conclude whether the film is purely hexagonal or purely cubic or a mixture of the two phases. But the phase was predominantly hexagonal at higher temperatures, where the presence of the lines (102) and (203) of the hexagonal phase are evidences.

On the other hand, Ashour [12] observed spray-deposited CdS thin films with just one phase (wurtzite) in the temperature range 200-400°C. Their [12] XRD diffractograms showed a preferential orientation (002) along the c-axis direction perpendicular to the substrate plane. Also Bilgin et al. [18] observed just the hexagonal phase for CdS thin films prepared by ultrasonic spray pyrolysis (USP) technique onto glass substrates at different temperatures ranging from 473 to 623K in 50K steps.

We conclude that authors who got a preferential orientation that is independent on the substrate temperature, got just one phase (hexagonal), while those who got a change in the preferential orientation with substrate temperature have a phase transition (from cubic to hexagonal). From these results it is confirmed that increasing the deposition temperature promotes phase transformation from cubic to hexagonal and improves the crystallinity in CdS films. Fig.1 displays the XRD diffractograms of spray-deposited SnO_2:F thin films taken at different substrate temperatures by Yadav et al.[31].

The grain size of the polycrystalline films greatly depends on the substrate temperature during deposition [1]. The grain size was found to increase with the substrate temperature for thin films prepared by different deposition techniques [8, 18, 30]. Acosta et al. [14] found that grain size increases with the substrate temperature and presents a smaller dispersion as T_s is increased. This increase is evident from the decrease in FWHM that they observed in their XRD diffractograms. For spray deposited CdS:In thin films, we [8] got an increase in grain size from 10 to 33 nm for the change in the substrate temperatures from 350 to 490 °C, which was calculated by using XRD diffractograms and Sherrer's formula. Bilgin et al. [18] obtained an increase of the grain size of the CdS films from 126 to 336 °A with increasing substrate temperature from 473 to 623K, showing the improvement in the crystallinity of the films. Abduev et al. [30] got an increase from 32 to 36 nm for substrate temperature change from 50 to 300 °C and a decrease in FWHM from 0.27° to 0.24° for the same change in substrate temperature. This change was accompanied by a change in the lattice parameter c which decreased from c = 5.232 Å for the film deposited at 50 °C to c = 5.208 Å for the film deposited at the substrate temperature of 300 °C.

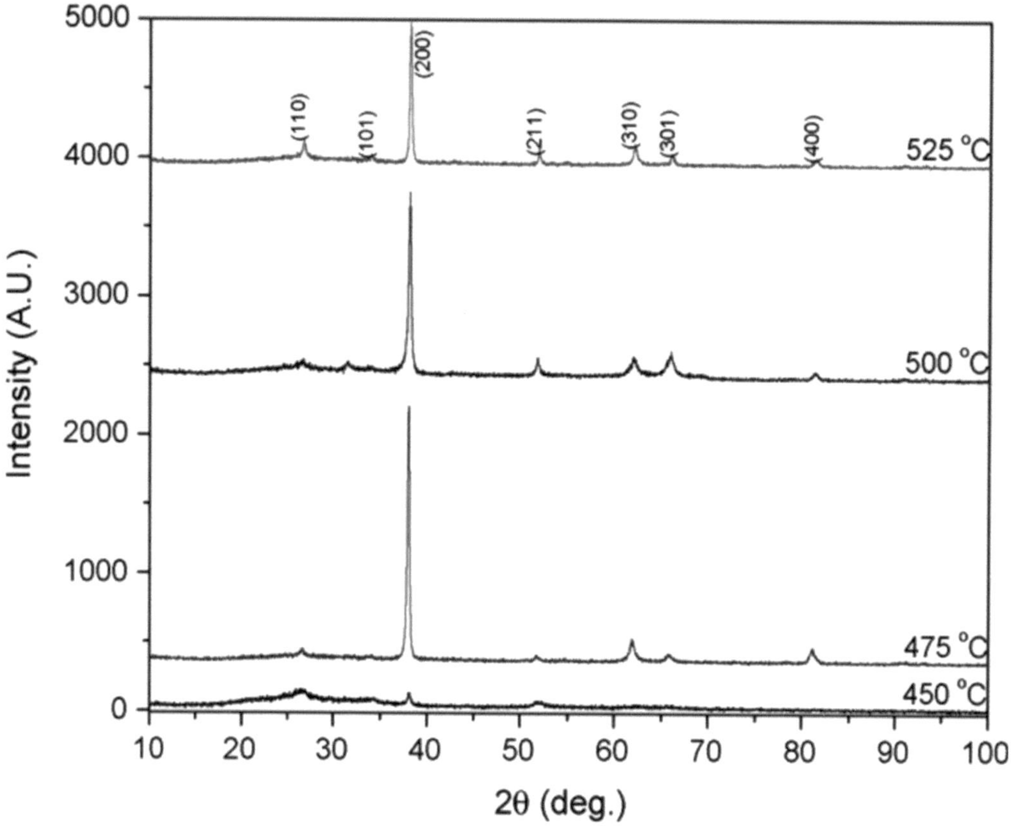

Fig. 1. XRD patterns of spray deposited SnO2:F thin films at various substrate temperatures. Reprinted with permission from Yadav et al. [31]; Copyright © 2009, Elsevier.

Different authors ploted the relation between grain size and the substrate temperature [12, 18]. Bilgin et al. [18] obtained a non-linear relation with the curve concaves down and the grain size increases then becomes constant after a certain value of T_s. Ashour [12] and Patil et al [32] got increasing in grain size with increasing the substrate temperature where the curve concaves up which means that the grain size did not reach a certain size after which there is no increase.

Stress is also varying with substrate temperature as seen by Abduev et al. [30] who found that for gallium doped ZnO films, the film stress had varied with increasing the substrate temperature from –1.915 GPa (the compression condition) at the room substrate temperature to 0.174 GPa (the tension condition) at the substrate temperature T = 300 °C.

In thin film solar cells it is found that the substrate temperature is also an effective parameter on the structure and the grain size. For CdS/CdTe thin film solar cells deposited

on SnO_2-coated Corning 7059 borosilicate glass, or (100) Si wafers, the substrate temperature caused an increase in the grain size of the CdTe layer as found by Al-Jassim et al. [33]. Also for CdS/CdTe thin films Dhere et al. [34] used AFM measurements and showed that there was no CdTe grain growth, for samples deposited at different substrate temperatures after $CdCl_2$ heat-treatment, but samples deposited at lower temperatures have smaller grains and consequently higher grain boundary volume.
Substrate temperature enhances the interdiffusion in the interface region in CdS/CdTe polycrystalline thin films. Al-Jassim et al. [33] found that at deposition temperatures below 450 °C, only small amounts (~1%) of sulfur were detected in the CdTe films in the vicinity of the interface. On the other hand at deposition temperature of 625 °C, sulfur levels exceeding 10% in CdTe films were detected. This clearly indicates that CdTe devices deposited at high temperatures have an alloyed (CdS_xTe_{1-x}) active region. Dhere et al. [34] found that compositional analysis by small-area, energy dispersive X-ray analysis (EDS) revealed significant sulfur diffusion into the CdTe film. The amount of sulfur was below detection limit (<0.1 at.%) at the lowest deposition temperature, and increased with increasing deposition temperature.

3. Film morphology

Substrate temperature is an effective parameter in determining the shape and size of grains, surface roughness, porosity and density of voids as found by different authors.
For CdS:In thin films prepared by the SP technique at different substrate temperatures we [8] observed different surface morphologies. At 350 °C long rods or chains were observed, which are related to complex compounds. At T_s = 460 °C we got open cubes and at T_s = 490 °C spherical grains were observed. For SnO_2:F thin films prepared by the SP technique different morphologies were observed for films prepared at different substrate temperature [15] too. El Hichou et al. [10] observed a change in surface morphology for ZnO spray deposited thin films with the substrate temperature, where they have the larger grains in the film deposited at T_s = 450 °C. The film deposited at the smallest substrate temperature T_s = 350 °C had shown porous structure but films deposited at T_s > 350 °C had a close-packed morphology.
For spray-deposited CdS:In films Acosta et al. [14] got AFM images which are shown in Figs. 2. Besides the grains size and topology details, it can be observed that grains present regular shape and surfaces for T_s values ranging from 300 °C (Fig.2a) to 400 °C (Fig.2b). In samples obtained at T_s = 425 °C (Fig.2c) and 450 °C (Fig.2d) respectively, aggregates of small grains covering grains with bigger sizes are found everywhere. Noting that these results are related with the XRD diffractograms in that reference. Since the substrate temperature is the only parameter that changes, Acosta et al. [14] say that the changes observed in surface morphology might have to do with particular specific thermodynamic parameters during the pyrolysis and nucleation processes.
Besides increasing the grain size, the increase in the substrate temperature decreases the density of voids. Fig.3 illustrates the SEM micrographs of the surfaces of the CdS films deposited by CBD at 55 °C, 65 °C, 75 °C and 85 °C taken by Liu et al. [28]. These micrographs show that increasing the deposition temperature results in an increase in grain size and consequently a decrease in voids. When the deposition temperature is 55 °C, CdS particles of 50 nm dot the surface of the glass substrate attributing to the controlled

nucleation process associated with the low deposition rate. CdS films deposited at 65 °C have spherical particles of about 100 nm in size. The voids with different sizes ranging from 50 nm to 300 nm are still observed, indicating low packing density of the film. The surface of the CdS films deposited at 75 °C is compact and smooth, showing a granular structure with well-defined grain boundaries. It indicates that the increase of the bath temperature is an effective method to diminish voids on the CdS films. But it is noticed that CdS film deposited at higher temperature 85 °C displays a rather rough, inhomogeneous surface with overgrowth grains.

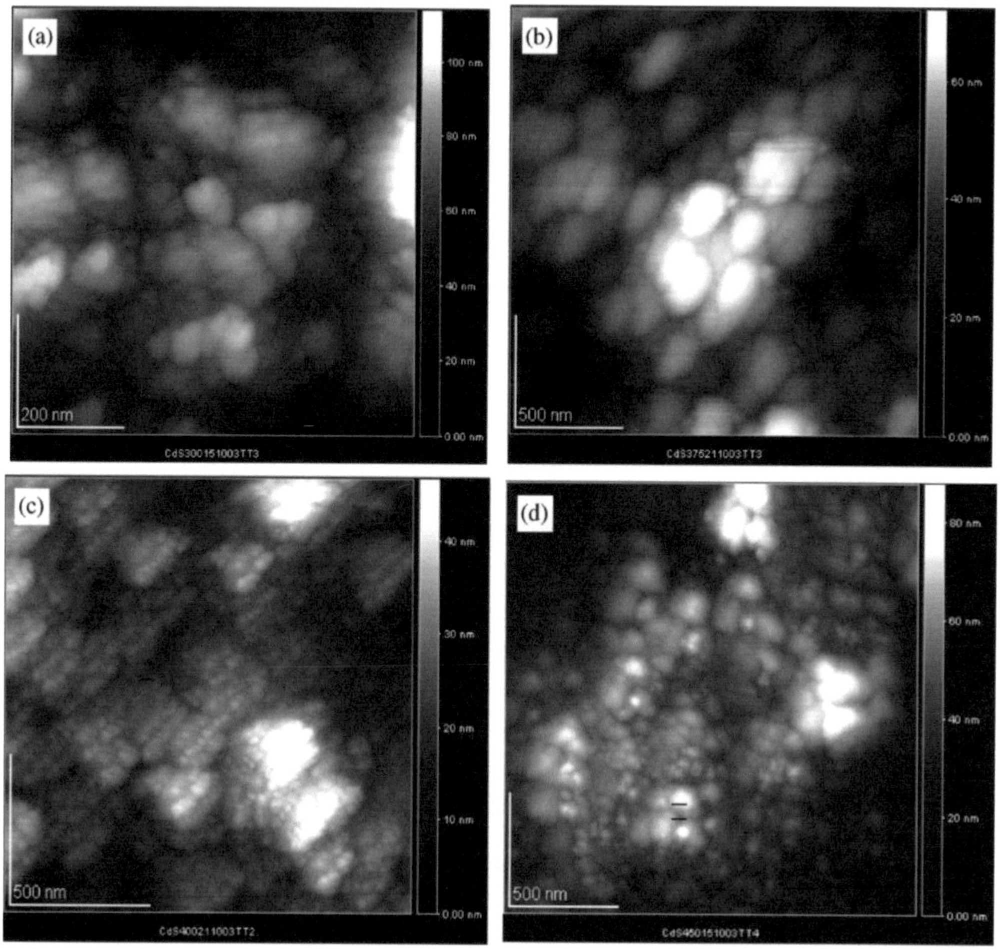

Fig. 2. AFM micrographs of CdS:In deposited by SP technique for different substrate temperatures. a) Ts = 300 °C: The grain size ranges from 50 to 75 nm. b) Ts = 375 °C: The grain size varies between 75 and 225 nm. c) Ts = 400 °C: The grain size between 45 and 60 nm and the size of grain agglomerates is between 170 and 350 nm, d) Ts = 450 °C: The grain size varies between 25 and 65 nm and the size of agglomerates is between 180 and 400 nm, respectively. Reprinted with permission from Acosta et al. [14]; Copyright © 2004, Elsevier.

Other authors showed that roughness increases with the substrate temperature such as Haug et al. [35] who found that the CdTe layers show a higher roughness with increasing substrate temperature, but they are less compact. Atomic Force Microscopy analysis showed that the root mean square (RMS) surface roughness ranges from 100 nm for 500 °C films to 550 nm for 600 °C films. On the other hand some authors found a decrease in roughness with the substrate temperature [14, 36]. For CdS:In thin films prepared by the SP technique Acosta et al. [14] found that as T_s begins to increase, the surface shows a decrease in roughness in zones surrounding pore-like configuration. Also surface roughness was found to decrease with substrate temperature by Abduev et al. [30] for ZnO thin films prepared by magnetron sputtering. Li Zhang et al. [36] also found that surface roughness decreases with the substrate temperature.

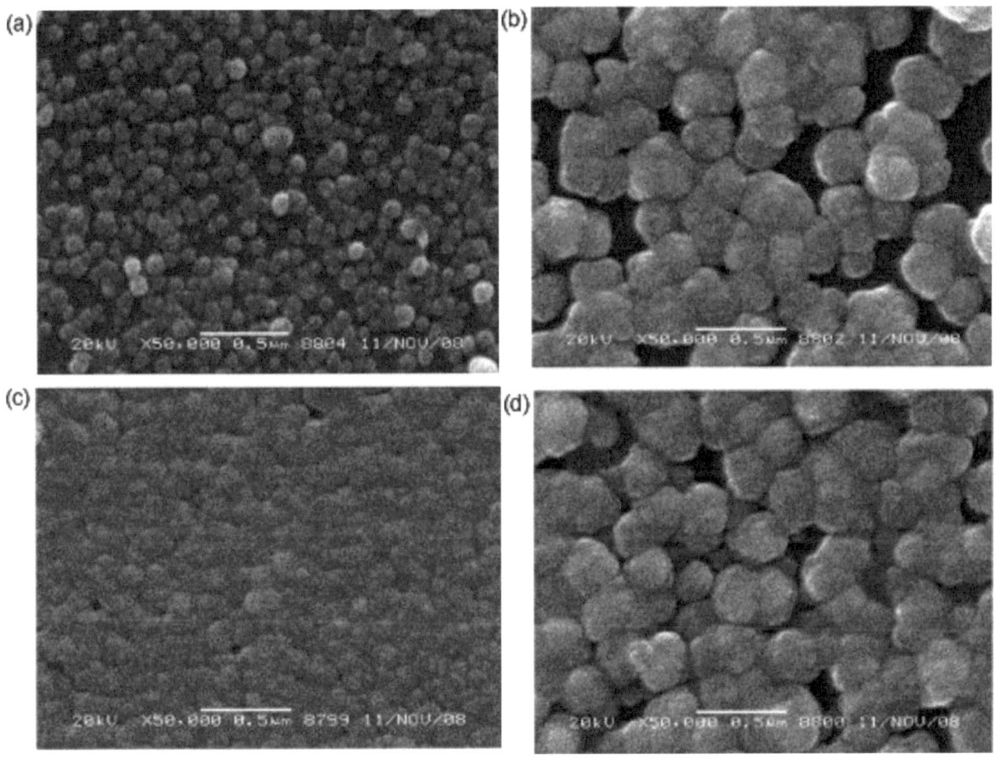

Fig. 3. SEM micrographs of CdS films grown at different temperatures by CBD: a) 55 °C. b) 65 °C. c) 75 °C. d) 85 °C. Reprinted with permission from Liu et al. [28]; Copyright © 2010, Elsevier.

4. Electrical properties

Changes in the structural and morphological properties of the films with the substrate temperature have their effect on the electrical properties of the films. These changes include the phase change, enlargement of grains, diminishing of grain boundaries, motion of impurities from or to the grain boundary region and evaporation of some elements during deposition process. For undoped compound semiconductors, it is found that stoichiometry increases with the substrate temperature due to the reduction in the density of defect states.

These variations will change the number of charge carriers which directly affect the resistivity of the films. They also affect the mechanism of carrier transport and then the linearity of current-voltage characteristics. We will discuss some of the experimental results which include some of these changes. The occurrence of all of these changes or some of them simultaneously has a net effect on the electrical properties as will be seen in the experimental results obtained by different authors.

I-V plots are usually used to investigate the electrical conduction mechanisms and to determine the resistivity of the films. Linear I-V plots mean that the ohmic conduction mechanism is predominant (i.e. electronic conduction through grains). Nonlinear I-V plots mean that other conduction mechanisms are found which are non- electronic and the conduction through the grain boundaries is predominant. It is known that the trap states, which act as effective carrier traps and become charged after trapping result in the appearance of a potential barrier which impedes the flow of majority carriers from one grain to another and affects the electrical conductivity of the films [1]. Three possible mechanisms may govern the grain-to-grain carrier transport through the potential barrier mentioned above:

i. over-the-barrier thermionic emission of carriers having sufficient energy to surmount the barrier;
ii. quantum mechanical tunneling from grain to grain through the barrier by carriers having energy less than the barrier height; and
iii. hopping through the localized states.

The relative magnitudes of the barrier height and the width of the barrier will depend critically on the crystallite size and carrier concentration. Depending on the above, with respect to the energy of carriers, one of the above processes will be operative in charge transport in polycrystalline semiconductor films. The thermionic process is limited by the height of the barrier. The thermionic emission will be temperature dependent, with activation of the order of the barrier height, while the tunneling process would be almost independent of temperature. For films with a barrier height larger than what could be surmounted by the carriers with the energy at lower temperatures, tunneling seems to be the dominant mechanism of charge transform in the films. The films grown at lower temperatures will have a smaller crystallite size, and as such the grain boundary region will be substantially larger than the grains. The grain boundary regions being disordered and highly resistive, the film will look like a conglomeration of crystallites embedded in the amorphous matrix [1].

Linear I-V characteristics were recorded by us at all deposition temperatures under study for CdS:In [8], SnO_2:F thin films prepared by the SP technique on glass substrates [15, 25-26], undoped ZnO thin films [37-38], Al-doped ZnO thin films [39] and CdTe thin films prepared by vacuum evaporation [40]. Bilgin et al. [18] have linear I–V curve for a CdS thin film prepared by ultrasonic spray pyrolysis (USP) technique at 523 K (250 °C) in the voltage range 0-100 V. This means that this film has not got trapped structure and so, the ohmic conduction mechanism is dominant for this film in the whole voltage range. For the sample obtained at 473 K they [18] found four regions with different slopes where the drawing was performed on log-log scale. The ohmic conduction is dominant in the 0.1–8V voltage range where the slope is 1.11. The slope is 2.24 in the second region which is called space charge limited (SCL) region. The existence of this region shows that CdS films have shallow trapped structure. Then, the trap filled limited (TFL) region comes as the third region. This implies that all traps are filled. The last region with a slope of 2.42 shows trap free region. The other two films prepared by Bilgin et al. [18] at 573 and 623K have deep trapped structure. The mechanism in the sample obtained at 623K is more complex. There are three deep trap levels with different energies for this film. The voltage ranges of these three regions are 16–24, 40–56 and 80–100V, respectively.

Numerous experimental results showed that the resistivity of semiconducting thin films decreases with the deposition temperature [8, 12, 15, 23-24]. For spray-deposited CdS:In thin films we [8] got a decrease of the room temperature resistivity in the dark from 1.5×10^8 Ω.cm at T_s = 380 °C to 1.2×10^6 Ω.cm at T_s = 490 °C and we explained this by the encouragement of crystal growth at higher substrate temperature as concluded from the XRD diffractograms. For CdS thin films prepared by SP technique Ashour [12] got a decrease of room temperature resistivity (10^5-10^3) with substrate temperature in the range 200-400 °C and related it to the growth of the grain size and the improvement in film stoichiometry. For SnO_2:F thin films prepared by SP technique [15] we found a rapid decrease of the resistivity with the substrate temperature. The same behavior was also observed by Shanthi et al. [23] for undoped SnO_2 films prepared by the spray pyrolysis, where they recorded a gradual decrease in the resistivity with the deposition temperature in the range 340–540 °C. Also the same behavior was observed by Zaouk et al. [24] for fluorine-doped tin oxide thin films prepared by electrostatic spray pyrolysis at substrate temperatures in the range 400–550 °C.

Other authors found a decrease in resistivity until a certain temperature and then it increases again [15, 18-20, 30]. Abduev et al. [30] found that for ZnO thin films prepared by dc magnetron sputtering, the growth temperature dependence of resistivity is nonmonotonic. They found that the lowest resistivity (3.8×10^{-4} Ω.cm) is attained at the substrate temperature of 250 °C; then, it increases insignificantly. On the other hand some authors found an increase in resistivity followed by a decrease [28, 36]. Liu et al. [28] measured electrical resistivity for CdS thin films prepared by CBD and found that it arises to 5×10^5 Ω.cm level for temperature 55–75 °C, then it decreases to 7.5×10^4 Ω.cm for 80 °C and 8.5×10^3 Ω.cm for 85 °C sharply. They explained this variation as can be due to the cubic-hexagonal transformation in agreement with structural and optical analysis. Li Zhang et al. [36] also observed an increase in the resistivity with substrate temperature followed by a decrease for Cu(In, Ga)Se_2 films prepared by the three-stage co-evaporation process.

From the results of Hall coefficients measurements taken by Liu et al. [28] for CdS thin films prepared by CBD it is found that the CdS films are of n-type conductivity. It is also found that mobility increases from 3.228×10^{-1} $cm^2/(V.s)$ to 6.517 $cm^2/(V.s)$ with the increase of deposition temperature from 55 °C to 75 °C tardily which can be understood by considering the increase of the grain sizes and decrease of the grain boundaries. However, the mobility increases to 6.513×10^1 $cm^2/(V.s)$ at 80 °C and 1.183×10^2 $cm^2/(V.s)$ at 85 °C promptly in contrast with the behavior of resistivity. This behavior can be attributed to the transition from the cubic to the hexagonal phase again, besides the improvement of crystallinity.

The investigation of Hall parameters by Abduev et al. [30] showed that the charge carrier mobility continuously grows with increasing the substrate temperature, and the free carrier concentration has the peak (1.27×10^{21} cm^{-3}) at the substrate temperature of 250 °C. Such a character of the temperature dependence of the free carrier concentration in doped ZnO films is caused by the fact that in zinc oxide there are always intrinsic donor defects in the bulk and at the surface grains in addition to the impurity donors introduced in the ZnO lattice (the substitutional impurity). The multiple experimental and theoretical data indicate that oxygen vacancies play an important role in the conductivity of transparent conducting films. It is observed that the behavior of the resistivity to a large extent is reflected by the carrier density and only little by the mobility; low resistivity corresponds to high carrier density and vice versa.

Abduev et al. [30] explains this behavior by: At low film growth temperatures ($T \leq 150$ °C), the main contribution to the charge carrier concentration is made by intrinsic defects and the efficiency of the Ga incorporation in the ZnO lattice is low, which is confirmed by the small values of Hall mobility in these films. With increasing substrate temperature, the efficiency of impurity atom incorporation into the crystal lattice increases and the concentration of intrinsic defects inside ZnO grains decreases, which is confirmed by the data of the X-ray diffraction analysis and by a substantial increase in the Hall mobility values at a deposition temperature of 200 °C. The increase in Hall mobility at $T \geq 200$ °C is caused also by the lowering of potential barriers for free carriers on the grain boundaries due to the intensification of the process of oxygen thermal desorption from the grain surface during the film growth in vacuum.

For thin film solar cells, the performance is dependent on the substrate temperature. That is, the short-circuit current density J_{sc}, open circuit voltage V_{OC}, Fill factor FF and efficiency η are all dependent on the substrate temperature.

Li Zhang et al. [36] showed that for CIGS solar cells the cell performance increases with the increase in the growth temperature. It is noticed that the cell efficiency increases with increase in the growth temperature. When the substrate temperature is 380 °C the efficiency is very low. The best efficiency at 550 °C is related to the better structural and electrical properties. It is noticed that the effects of the substrate temperature on fill factor (FF) and open circuit voltage (V_{OC}) shows similar trends with cell efficiency. That means the dependence of cell efficiency on the substrate temperature is dominated by the value of FF and V_{OC}. That can be explained by the improvement of carrier concentration and resistivity of CIGS films dominated by Na incorporation diffused from the glass substrate which is expected to be temperature dependent.

Julio et al. [27] investigated the electrical and photovoltaic properties of ZnO/CdTe heterojunctions where ZnO was prepared by the SP technique on CdTe single crystal under the effect of varying the substrate temperature and post deposition temperature for annealing in H_2. For substrate temperatures in the range T_s = 430-490 °C for the spray-pyrolysis deposition the optimum behavior was obtained for T_s = 460 °C. They [27] found that as the substrate temperature is increased from 430 to 460 °C the dark J-V characteristics improved considerably and shifted towards higher bias voltages, remaining almost parallel to one another and exhibiting a strong reduction in J_0 with increasing T_s. The reverse current characteristics show similar improvement. Under simulated illumination, large values of short-circuit current were observed: typically of the order of 20 mA/cm^2 for illumination of 87 mW/cm^2 for T_s less than 470 °C. The solar conversion efficiency increased markedly with increasing T_s up to 460 °C, primarily because of an increase in V_{OC} and a fill factor which can be correlated with the decrease in J_0. The improvement in junction characteristics observed with increasing substrate temperature up to 460 °C according to Jullio et al. [27] may have several explanations: The density of the interface states may depend on the orientation of the film; preferential orientation increases as a characteristic temperature is reached.

5. Optical properties

Since the substrate temperature affects the structural properties of the films including lattice parameters and phase, and the electrical properties including the density of charge carriers and density of traps, the optical properties will change.

The absorption coefficient is dependent on the conductivity which is a function of the density of charge carriers. The change in the absorption coefficient will change the transmittance of the films. Some authors found that the transmittance of thin films increases with the substrate temperature [12, 31]; other workers found a decrease in the transmittance with substrate temperature [14] and others found no change in the transmittance of thin films with the substrate temperature [8].

The increase of transmittance with substrate temperature was recorded by Ashour [12] who found an increase of the transmittance with substrate temperature in the range 200-400 °C for undoped spray pyrolyzed CdS thin films of thickness 500 nm. He attributed this improvement in transmittance with substrate temperature to either the decrease in thickness or the improvement in perfection and stoichiometry of the films. Yadav et al. [31] found an increase in transmission with the increase in the substrate temperature for SnO_2:F thin films prepared by the spray pyrolysis technique on glass substrates at substrate temperatures 450-525 °C (Fig.5a). At lower temperatures, i.e. at 450 °C, relatively lower transmission is due to the formation of whitish milky films due to incomplete decomposition of the sprayed droplets.

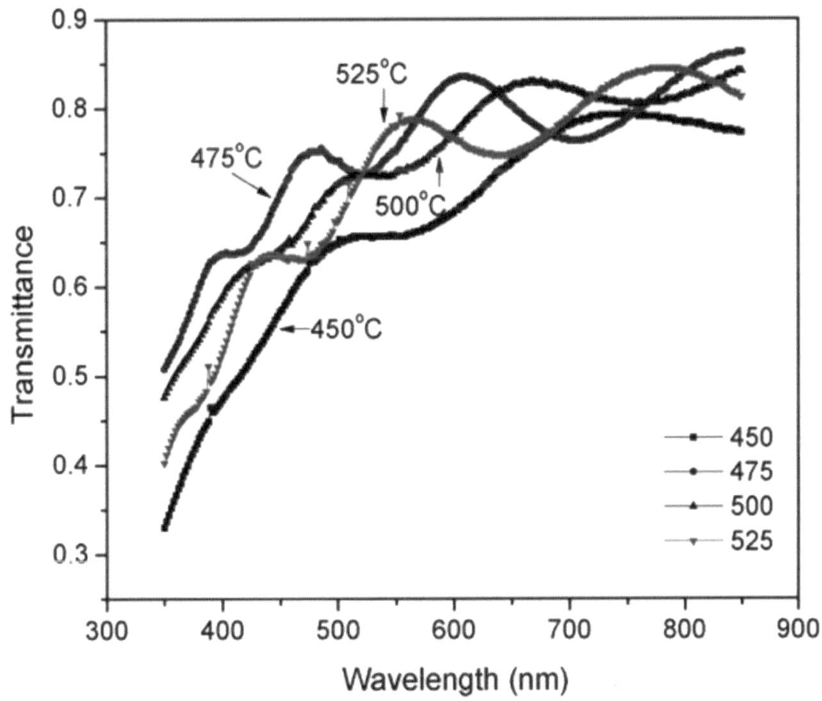

(a)

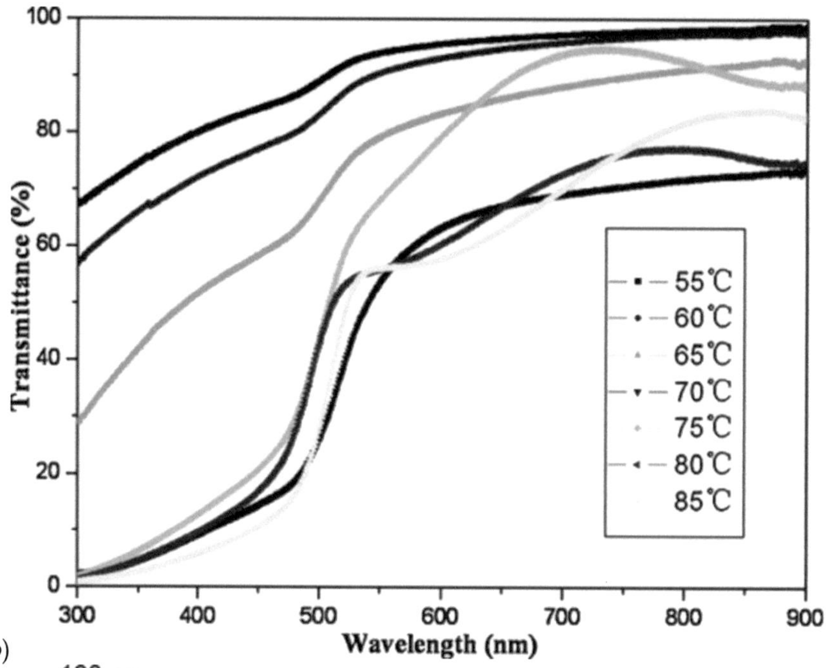

(b)

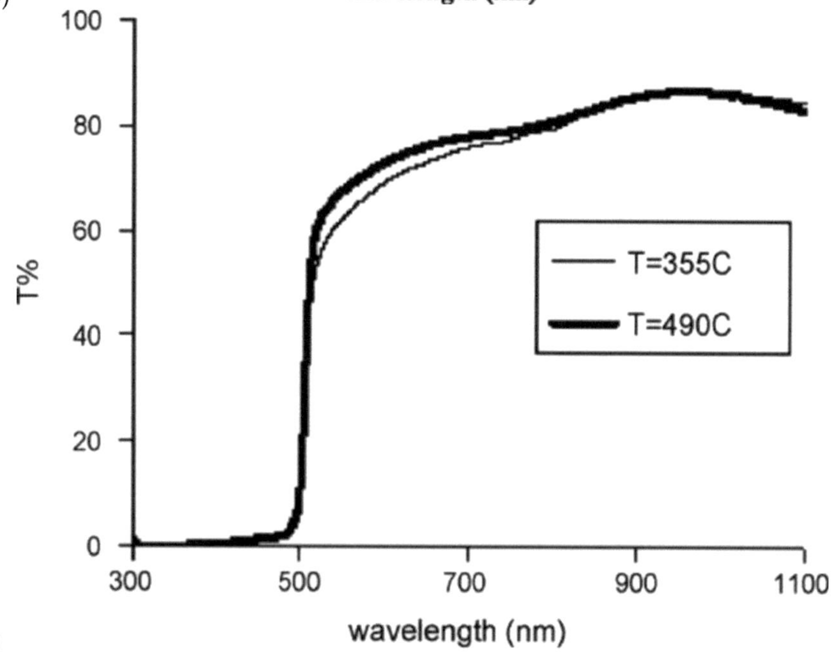

(c)

Fig. 4. The optical transmittance of thin films at different substrate temperatures against the wavelength of incident radiation. a) SnO2:F thin films. Reprinted with permission from Yadav et al. [31]; Copyright © 2009, Elsevier. b) CdS films. Reprinted with permission from Liu et al. [28]; Copyright © 2010, Elsevier. c) CdS:In thin films [8]. permission from [9], S. A. Studenikin et al. Journal of Applied physics, 84 (4) (1998), 2287-2294. Copyright [1998], American Institute of Physics.

The decrease in transmittance with substrate temperature was observed by Acosta et al. [14] for CdS:In thin films prepared by the spray pyrolysis technique, but at the same time they have a variable thickness with substrate temperature (decrease then increase) which may be the main reason of the decrease in transmittance. Also the decrease of transmittance with substrate temperature was recorded by Liu et al. [28] for CdS films prepared by CBD (Fig.5b). It can be observed that the transmittance of the film decreases rapidly with the increase of the deposition temperature from 55 °C up to 70 °C, which is caused by reducing voids and increasing film thickness mainly. For higher deposition temperatures, the transmittance initially increases to 84% for the film deposited at 75 °C due to less light scattering by its smoothest surface. It decreases to about 68% at deposition temperatures above 80 °C, which may be due to either more light scattering on their rough surfaces or the transition of the CdS phase from the cubic to hexagonal structure [28]. Another observation about these transmission spectra is that the absorption edge shifts towards higher wavelength side, suggesting a reduction in the bandgap value, and it becomes steeper with deposition temperature rising.

No dependence of transmittance on the substrate temperature was recorded by us [8] for CdS:In thin films prepared by the spray pyrolysis technique (Fig.5c). We think that the transmittance of our films was independent of the substrate temperature due to the way of spraying that we used. We sprayed for 10 s, waited 1-3 min and then sprayed again. The preparation of a set of films by this method takes a long period of time depending on the required thickness of the films (around 4 h for films of thickness around 1 μm). Ashour [12] did not mention the deposition time that he used or the way he followed in spraying, while Acosta et al. [14] produced their films with a deposition time of 5 min in all cases. From our trials we found that using longer deposition times results in less transparent films, and the short period of spraying results in highly transparent films. Also for transparent conducting gallium_doped ZnO films prepared by magnetron sputtering on glass substrates at T_s = 100-300 °C, Abduev et al. [30] have high transmittance which is approximately independent on the substrate temperature, but they observed a shift of the absorption edge in spectra to shorter wavelengths.

The dependence of the bandgap energy on substrate temperature was recorded by different workers [8, 12, 18, 30]. One reason of this dependence is that stress is greatest in films deposited at low temperatures, which results in wider bandgap than bulk. So the increase in substrate temperature reduces stress and then reduces the bandgap energy. Another reason is the increase in interplanner distances or equivalently the lattice parameter with the substrate temperature which appears as a shift in the XRD diffractogram towards smaller angles. It is well known that the lattice parameter and energy gap have opposite behavior [41]. Other reasons include the change in the density of charge carriers with the substrate temperature and the movement of dopants from grain boundaries to the grains to be effectively incorporated in the crystal lattice.

A slight increase in the optical bandgap energy with substrate temperature was observed by different authors [8, 12, 18, 42]. For spray-deposited CdS:In thin films we [8] found that E_g slightly increases with the substrate temperature. This increase can be related to the phase change from mixed (cubic and hexagonal) to hexagonal phase as seen in XRD diffractograms in reference [8]. We found that the E_g = 2.42 eV for a film deposited at 355 °C and E_g = 2.44 eV for a film deposited at 490 °C. Bilgin et al. [18] observed slight increase of

E_g for CdS films prepared by ultrasonic spray pyrolysis (USP) technique onto glass substrates at different temperatures ranging from 473 to 623K. Ashour [12] got E_g = 2.39 - 2.42 eV for CdS films prepared by chemical spray-pyrolysis technique on glass at substrate temperatures in the range 200-400 °C. Melsheimer and Ziegler [42] observed this increase of E_g with substrate temperature for tin dioxide thin films prepared by the spray pyrolysis technique. Values of E_g = 2.51-3.05 eV where obtained for amorphous and partially polycrystalline thin films prepared at T_s = 340-410 °C, and E_g = 3.35-3.43 eV for polycrystalline tin dioxide thin films produced at T_s = 420-500 °C.

An increase followed by a decrease in bandgap energy with substrate temperature was observed by Abduev et al. [30] for ZnO:Ga thin films prepared by dc magnetron sputtering (from 3.52 to 3.72 eV when T_s increases to 250 °C) then a decrease to 3.65 eV at 300 °C. This result was consistent with their electrical properties. Other authors got a decrease then an increase in the bandgap energy with substrate temperature. For spray-deposited indium doped CdS thin films on glass substrates, Acosta et al. [14] got a decrease in the bandgap energy with substrate temperature from 300-425 °C then it increased at T_s = 450 °C. They interpreted the increase observed in E_g by saying that it might be related with the variations in size and morphology of grains.

The decrease of bandgap energy with substrate temperature was observed by some authors such as Liu et al. [28] who observed this for CdS films prepared by CBD. But it is important to notice that the thickness of their films is not constant, which means that the decrease in bandgap is also related to the increase in film thickness not only to the increase in substrate temperature.

Urbuch tail width E_e which is known to be constant or weakly dependent on temperature and is often interpreted as the width of the tail of localized states in the band gap [43] was also found to be randomly affected by the substrate temperature as shown by Bilgin et al. [18] for CdS thin films prepared by USP technique, where it has values in the range 122-188 meV for substrate temperatures in the range 473-623 K. But Melsheimer and Ziegler [42] observed a decrease of E_e with substrate temperature for spray-deposited tin dioxide thin films. For amorphous and partially polycrystalline films prepared at T_s = 340-410 °C, it decreased from 530 to 350 meV. For polycrystalline films prepared at T_s = 420-500 °C it decreased from 240-200 meV.

Photoluminescence and cathodolumenescence always used to explore defects and traps. But the density of trap states depends critically on the deposition parameters [1] and hence on the substrate temperature. Changes in phase, bandgap and density of traps will be reflected on the photoluminescence and cathodoluminescence spectra. It is found that the luminescence intensity depends strongly on the deposition temperature [10].

Fig.6 displays the photoluminescence (PL) spectra for a set of ZnO films deposited by the SP technique by Studenikin et al. [9] at different substrate temperatures and annealed identically in forming gas at 750 °C for 40 min. As we said before, the photoluminescence intensity depends strongly on the substrate temperature. Fig.7 shows the relation between the PL intensity and the substrate temperature for the green peak in the same reference [9]. As the figure shows, the maximum PL intensity is at T_s = 200 °C. They attributed the green PL to oxygen deficiency. This means that much lower temperatures could be used to produce oxygen-deficient ZnO in a reducing atmosphere. Stoichiometry increases with temperature so the green peak becomes smaller with temperature due to the decrease of oxygen deficiency.

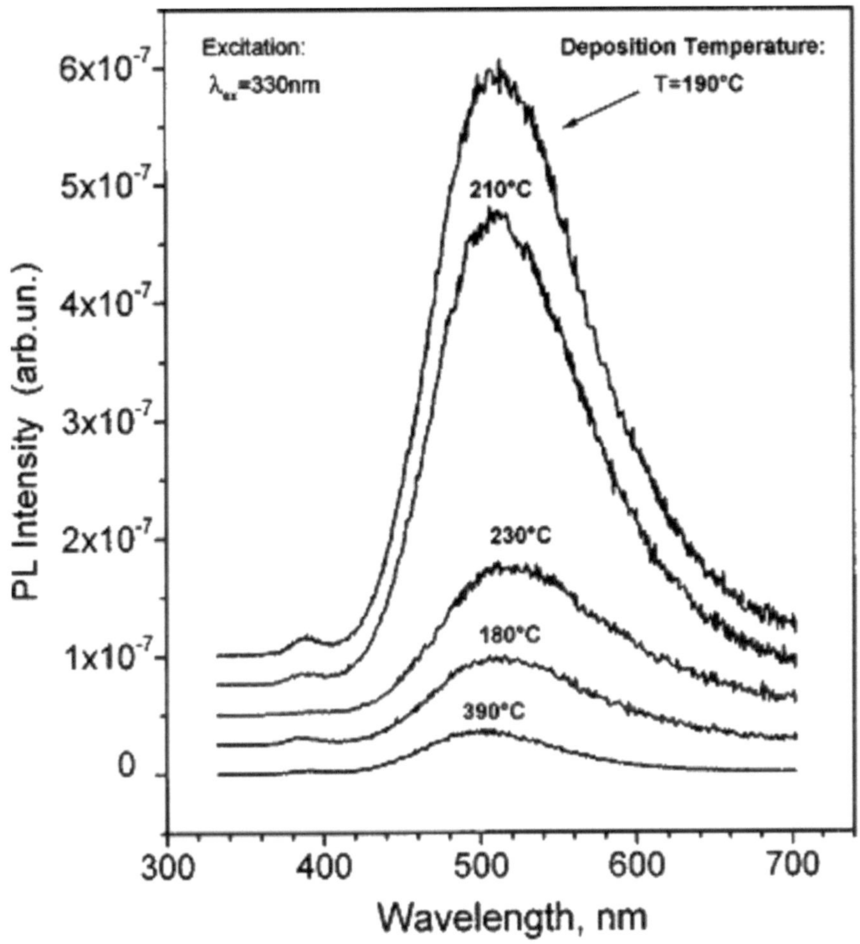

Fig. 5. Photoluminescence spectra of undoped ZnO films grown at different temperatures and annealed in one process in forming gas at 750 °C during 40 minutes. Reprinted with permission from [9], S. A. Studenikin et al. Journal of Applied physics, 84 (4), 2287-2294(1998). Copyright [1998], American Institute of Physics.

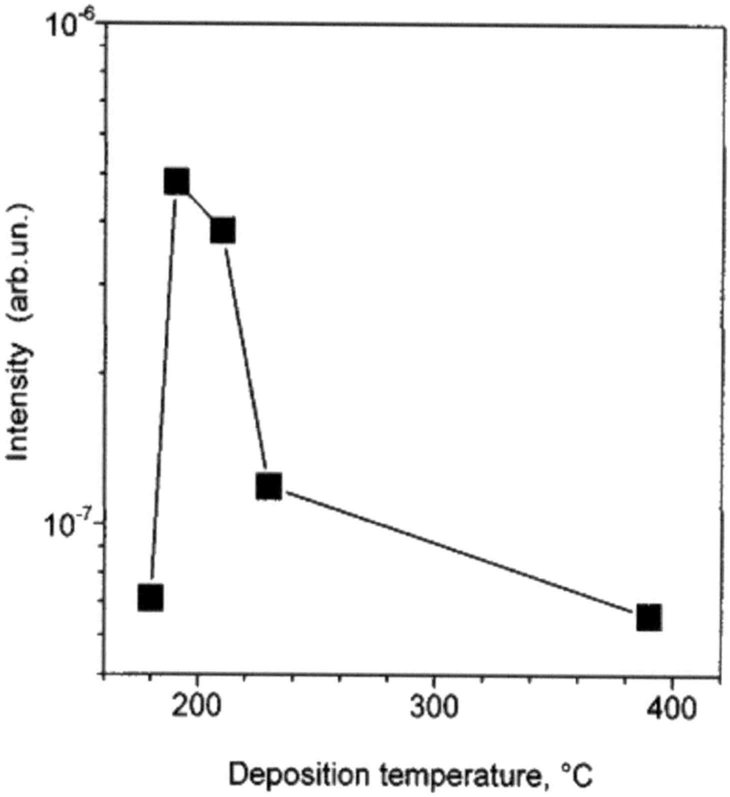

Fig. 6. Intensity of the green photoluminescence of undoped ZnO films as a function of deposition temperature. Reprinted with permission from [9], S. A. Studenikin et al. Journal of Applied physics, 84 (4), 2287-2294(1998). Copyright [1998], American Institute of Physics.

Fig.7 shows the cathodoluminescence spectra for ZnO films prepared by the SP technique at different substrate temperatures taken by El Hichou et. al. [10]. They found that when the substrate temperature increases, the surface of the films is entirely covered by grains and condensed. They observed that extinction of the blue-green emission (centred around 510 nm) is at substrate temperature of 350 and 400 °C, whereas the near UV emission at 382 nm becomes more dominant than other transitions (blue-green and red emissions) at 450 °C. The blue-green emission (510 nm) appears above substrate temperature 450 °C but the red emission (640 nm) appears at different substrate temperature. At T_s = 500 °C, the UV transition shifts to higher wavelength and becomes comparable in cathodoluminescence intensity with blue-green emission. The maximum value of cathodoluminescence intensity for three bands is obtained at T_s = 450 °C [10].

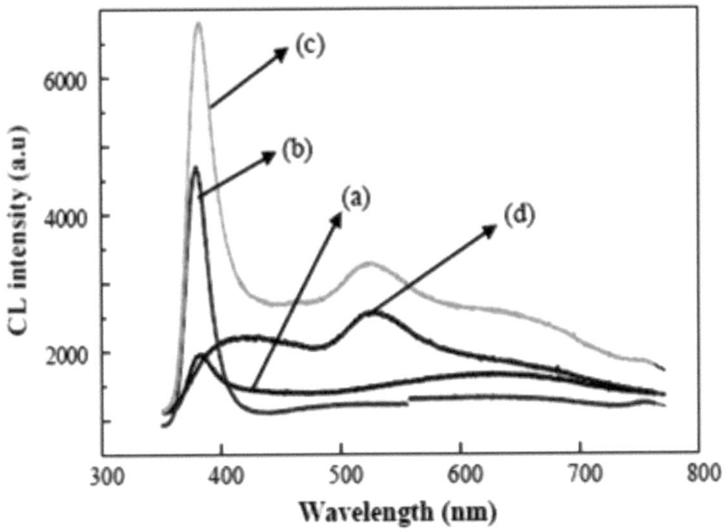

Fig. 7. Cathodoluminescence spectra of ZnO sprayed at flow rate f = 5 ml/min at different substrate temperatures: a) Ts = 350 °C, b) Ts = 400 °C, c) Ts = 450 °C, and d) Ts = 500 °C. Reprinted with permission from El Hichou et al. [10]; Copyright © 2005, Elsevier.

6. Conclusions

Experimental results show that there are influences of the substrate temperature on the properties of semiconducting thin films which are related to solar cells. Change of state from amorphous to polycrystalline, phase change from cubic to hexagonal, increase in grain size and decrease in the number and width of grain boundaries were observed with the increase of the substrate temperature. Morphological changes such as shape of grains, surface roughness, porosity and density of voids were also observed by different authors. The electrical properties were also found to change due to the changes in the density of charge carriers and density of traps with substrate temperature, beside changes in the structural and morphological properties. The optical properties are also sensitive to these changes, and so the transmittance, optical bandgap, width of Urbuch tail, photoluminescence and cathodolumenescence were found to change with changes in the substrate temperatures too.

7. References

[1] D. Bhattacharyya, S. Chaudhuri, and A. K. Pal, Electrical conduction at low temperatures in polycrystalline CdTe and ZnTe films, Materials Chemistry and Physics 40 (1995), 44-49.

[2] S.M. Rozati, The effect of substrate temperature on the structure of tin oxide thin films obtained by spray pyrolysis method, Materials Characterization 57 (2006), 150-153.

[3] O.A. Fouad , A.A. Ismail, Z.I. Zaki and R.M. Mohamed, Zinc oxide thin films prepared by thermal evaporation deposition and its photocatalytic activity, Applied Catalysis B: Environmental 62 (2006), 144–149.

[4] M. Lepek, B. Dogil and R. Ciecholewski, A Study of CdS thin film deposition, Thin Solid Films, 109 (1983), L103-L107.

[5] E. Bertran, J. L. Morenza, J. Esteve and J. M. Codina, Electrical properties of polycrystalline indium-doped cadmium sulfide thin films, J. Physics D, 17(8) (1984), 1679-1685.

[6] D. Patidar, R. Sharma, N. Jain, T. P. Sharma and N. S. Saxena, Optical properties of thermally evaporated CdS thin films, Cryst. Res. Technol., 42(3) (2007), 275-280.

[7] P.P. Sahay, R.K. Nath, S. Tewari, Optical properties of thermally evaporated CdS thin films, Cryst. Res. Technol. 42 (3) (2007), 275-280.

[8] Shadia. J. Ikhmayies, Riyad N. Ahmad-Bitar, The influence of the substrate temperature on the photovoltaic properties of spray-deposited CdS:In thin films, Applied Surface Science 256 (2010), 3541–3545.

[9] S. A. Studenikin, Nickolay Golego and Michael Cocivera, Fabrication of green and orange photoluminescent, undoped ZnO films using spray pyrolysis, Journal of Applied physics, 84 (4) (1998), 2287-2294.

[10] A. El Hichou, M. Addou, J. Ebothé , M. Troyon, Influence of deposition temperature (Ts), air flow rate (f) and precursors on cathodoluminescence properties of ZnO thin films prepared by spray pyrolysis, Journal of Luminescence 113 (2005), 183–190.

[11] Shadia J. Ikhmayies, Production and Characterization of CdS/CdTe Thin Film Photovoltaic Solar Cells of Potential Industrial Use, PhD Thesis, University of Jordan, 2002.

[12] A. Ashour, Physical properties of spray pyrolysed CdS thin films, Turk. J. Phys. 27 (2003), 551-558.

[13] Steve Pence, Elizabeth Varner, Clayton W. Bates Jr., Substrate temperature effects on the electrical properties of CdS films prepared by chemical spray pyrolysis, Mater. Lett. 23 (1995), 13-16.

[14] Dwight R. Acosta, Carlos R. Maganã, Arturo I. Martínez, Arturo Maldonado, Structural evolution and optical characterization of indium doped cadmium sulfide thin films obtained by spray pyrolysis for different substrate temperatures, Solar Energy Mater. Solar Cells 82 (2004), 11-20.

[15] Shadia. J. Ikhmayies, Riyad N. Ahmad-Bitar, Effect of the substrate temperature on the electrical and structural properties of spray-deposited SnO2:F thin films, Materials Science in Semiconductor Processing 12 (2009), 122–125.

[16] M. Krunks, E. Mellikov, E. Sork, Formation of CdS films by spray pyrolysis, Thin Solid Films 145 (1986), 105-109.

[17] L.W. Chow, Y.C. Lee, H.L. Kwok, Structure and electronic properties of chemically sprayed CdS films, Thin Solid Films 81 (1981), 307-318.

[18] V. Bilgin, S. Kose, F. Atay, I. Akyuz, The effect of substrate temperature on the structural and some physical properties of ultrasonically sprayed CdS films, Mater. Chem. Phys. 94 (2005), 103-108.

[19] Chitra Agashe, B. R. Marathe, M. G. Takwale, and V. G. Bhide, Structural properties of SnO2:F films deposited by spray pyrolysis technique. Thin Solid Films 164 (1988), 261- 264.

[20] E. Shanthi, A. Banerjee, V. Dutta, and K. L. Chopra, Electrical and optical properties of tin oxide films doped with F and (Sb+F). J Appl Phys; 53(3) (1982), 1615–1621.

[21] A. E. Rakhshani, Y. Makdisi, and H. A. Ramazaniyan , Electronic and optical properties of fluorine-doped tin oxide films. J Appl Phys 83(2) (1998), 1049–57.
[22] G. Gordillo, L. C. Moreno, W. de la. Cruz., and P. Teheran, Preparation and characterization of SnO2 thin films deposited by spray pyrolysis from SnCl2 and SnCl4 precursors. Thin Solid Films 252 (1994), 61–66.
[23] E. Shanthi, V. Dutta, A. Banerjee, K. L. Chopra, Electrical and optical properties of undoped and antimony-doped tin oxide films. J Appl Phys 51(12) (1980), 6243-6250.
[24] D. Zaouk, Y. Zaatar, A. Khoury, C. Llinares, J.-P. Charles, J. Bechara, Electrical and optical properties of pyrolytically electrostatic sprayed fluorine-doped tin-oxide: dependence on substrate-temperature and substrate-nozzle distance. J Appl Phys 10(87) (2000), 7539-7543.
[25] Shadia J. Ikhmayies, Riyad N. Ahmad-Bitar., Effect of processing on the electrical properties of spray-deposited SnO2:F thin films. Am J. Appl Sci. 5(6) (2008), 672-677.
[26] Shadia J. Ikhmayies, Riyad N. Ahmad-Bitar, The effects of post-treatments on the photovoltaic properties of spray-deposited SnO2:F thin films, Applied Surface Science 255 (2008), 2627–2631.
[27] Julio A. Aranovich, Dolores Golmayo, Alan L. Fahrenbruch, and Richard H. Bube, Photovoltaic properties of ZnO/CdTe heterojunctions prepared by spray pyrolysis. J. Appl. Phys. 51(8) (1980), 4260-4268.
[28] Fangyang Liu, Yanqing Lai , Jun Liu, Bo Wang, Sanshuang Kuang, Zhian Zhang, Jie Li, Yexiang Liu, Characterization of chemical bath deposited CdS thin films at different deposition temperature, Journal of Alloys and Compounds 493 (2010), 305–308.
[29] Hülya Metin, R. Esen., Potoconductivity studies on CdS films grown by chemical bath deposition technique, Erciyes Üniversitesi Fen Bilimleri Enstitüsü Dergisi, 19(1-2) (2003), 96-102.
[30] A. Kh. Abduev, A. K. Akhmedov, A. Sh. Asvarov, A. A. Abdullaev, and S. N. Sulyanov, Effect of growth temperature on properties of transparent conducting gallium_doped ZnO films, Semiconductors 44(1) (2010), 32–36.
[31] A.A. Yadav, E.U. Masumdar, A.V. Moholkar, M. Neumann-Spallart, K.Y. Rajpure, C.H. Bhosale, Electrical, structural and optical properties of SnO2:F thin films: Effect of the substrate temperature, Journal of Alloys and Compounds 488 (2009), 350–355.
[32] P.S. Patil, S.B. Sadale, S.H. Mujawar, P.S. Shinde, P.S. Chigare, Synthesis of electrochromic tin oxide thin films with faster response by spray pyrolysis, Applied Surface Science 253 (2007), 8560–8567.
[33] M.M. Al-Jassim, R.G. Dhere, K.M. Jones, F.S. Hasoon, and P. Sheldon, The morphology, microstructure, and luminescent properties of CdS/CdTe films, Presented at the 2nd World Conference and Exhibition on Photovoltaic Solar Energy Conversion; Vienna, Austria; 6-10 July (1998).
[34] R. Dhere, D. Rose, D. Albin, S. Asher, M. Al-Jassim, H. Cheong, A. Swartzlander, H. Moutinho, T. Coutts, and P. Sheldon, Influence of CdS/CdTe interface properties on the device properties, Presented at the 26th IEEE Photovoltaic Specialists Conference, Anaheim, California, September 29- October 3, (1997).

[35] F.-J. Haug, Zs. Geller, H. Zogg, A. N. Tiwari, and C. Vignali, Influence of deposition conditions on the thermal stability of ZnO:Al films grown by rf magnetron sputtering, J. Vac. Sci. Technol. A 19(1) (2001), 171-174.

[36] Li Zhang, Qing He, Wei-Long Jiang, Fang-Fang Liu, Chang-Jian Li, Yun Sun, Effects of substrate temperature on the structural and electrical properties of Cu(In,Ga)Se2 thin films , Solar Energy Materials & Solar Cells 93 (2009), 114–118.

[37] Shadia J. Ikhmayies, Naseem M. Abu El-Haija and Riyad N. Ahmad-Bitar., Characterization of undoped spray-deposited ZnO thin films of photovoltaic applications, FDMP: Fluid Dynamics & Materials Processing 6(2) (2010), 165-178.

[38] Shadia J. Ikhmayies, Naseem M. Abu El-Haija and Riyad N. Ahmad-Bitar., The influence of annealing in nitrogen atmosphere on the electrical, optical and structural properties of spray- deposited ZnO thin films, FDMP: Fluid Dynamics & Materials Processing 6(2) (2010), 219-232.

[39] Shadia J. Ikhmayies, Naseem M. Abu El-Haija and Riyad N. Ahmad- Bitar, Electrical and optical properties of ZnO:Al thin film prepared by the spray pyrolysis technique, Physica Scripta 81(1) (2010) art. no.015703

[40] Shadia J. Ikhmayies and Riyad N Ahmad-Bitar, Electrical, optical and structural properties of vacuum evaporated CdTe thin films, Collected Proceedings of TMS 2009 138th Annual Meeting & Exhibition, San Francisco, California, USA, February 15-19 (2009), 427-434.

[41] O. De. Melo, L. Hernández., O. Zelaya-Angel, R. Lozada-Morales, M. Becerril, and E. Vasco, Low resistivity cubic phase CdS films by chemical bath deposition technique, Appl. Phys. Lett. 65(10) (1994), 1278-1280.

[42] Melsheimer and D. Ziegler., Band gap energy and Urbach tail studies of amorphous, partially crystalline and polycrystalline tin dioxide. Thin Solid Films, 129 (1985), 35-47.

[43] Y. Natsume, H. Sakata, and T. Hirayama, Low temperature electrical conductivity and optical absorption edge of ZnO films prepared by chemical vapor deposition, phys. stat. Sol. (a) 148 (1995), 485-495.

15

Crystal Growth Study of Nano-Zeolite by Atomic Force Microscopy

H. R. Aghabozorg, S. Sadegh Hassani and F. Salehirad
Research Institute of Petroleum Industry
Iran

1. Introduction

Mesoporous materials possess highly ordered periodic arrays of uniformly sized channels. Therefore, these compounds have attracted considerable attention for many researchers (Luo et al., 2000; Xu & Xue, 2006; Liu et al., 2009 a, Salehirad & Anderson, 1996; 1998a; 1998b; 1998c). Since the shape and texture of the materials strongly affect their properties, these materials due to their large surface areas, controllable pore size and easy functionalization have been used in separation science, drug delivery and various processes such as adsorption and catalysis (Liu et al., 2009a; Duan et al., 2008; Liu et al., 2009 b; Liu & Xue, 2008; Mohanty & Landskron, 2008; Salehirad, 2004; Alibouri, 2009a; 2009b).

Zeolites as crystalline aluminosilicates are microporous materials with well-defined channels and cavities have been used as ion-exchangers, adsorbents, heterogeneous catalysts and catalyst supports (Zhang et al., 2000; Zhang et al., 2000). Many parameters are found to be highly important in the physical appearance of final zeolite products (Aghabozorg et al., 2001). Morphology and crystal size of these compounds have an important role for their specific applications in industries (Zabala Ruiz et al., 2005). The ability to accommodate various organic and inorganic species in zeolites makes them ideal host materials for supramolecular organization (Bruhwiler & Calzaferri, 2004; Schulz-Ekloff et al., 2002; Sadegh Hassani et al., 2010a). In many cases, surface adsorption sites in zeolites lead to interesting photochemical properties for these compounds. (Zabala Ruiz et al., 2005; Hashimoto, 2003). Studying the surface structure of zeolites and dynamic phenomena occurring on the surface of these compounds under various conditions will accelerate the development of these compounds as catalyst or materials with versatile functions (Sugiyama Ono et al., 2001; Anderson, 2001; Sadegh Hassani et al., 2010a). Nucleation mechanisms and growth of zeolites are not well understood for many of the systems involved. In most cases, the problem is compounded with the presence of a gel phase. This gel also undergoes a continuous polymerization type reaction during nucleation and growth. Improved understanding of zeolite growth should enable a more targeted approach to zeolite synthesis in the future and may ultimately lead to the possibility of zeolite design to order. Numerous techniques have been used for study the structure of the zeolites. Among them, scanning probe microscopy (SPM) is an appropriate technique that can be applied in atmospheric condition. However, many SPM techniques can be used in essentially any environment, including ambient, UHV, organic solvent vapour and

biological buffer, making it possible to observe the system in states that are simply inaccessible to other techniques with comparable resolution and they can be also used over a wide range of temperatures (Hobbs, et al., 2009).
Scanning probe microscopy, such as scanning tunneling microscopy (STM) and atomic force microscopy (AFM), has become a standard technique for obtaining topographical images of surface with atomic resolution (Hyon et al., 1999; Klinov & Magonov, 2004; Giessibl, 1995, Sadegh Hassani et al., 2088 a,b; 2010b). In addition, they may be used in many applications such as investigation of mechanical, chemical, electrical, magnetic and optical properties of surfaces, study of friction and adhesion forces, modifying a sample surface, crystal growth study and process controlling (Sundararajan & Bhushan, 2000; Burnham et al., 1991; Aime et al., 1994; Sadegh Hassani & Ebrahimpoor Ziaie, 2006; Ebrahimpoor Ziaie et al., 2008; Sadegh Hassani & Sobat, 2011; Sadegh hassani & Aghabozorg, 2011; Magonov, et al., 1997; Hobbs, et al., 2009; Leggett, et al., 2005; Williams, et al., 1999; Cadby, et al., 2005). There are some disadvantages for these techniques because they are limited to a surface view of a sample. Hence, extreme care has to be taken for interpretation of data for achieving the bulk properties (Franke, 2008; Hobbs, et al., 2009).These nondestructive methods do not require complex preparation of the desired sample and allows processes to be followed in-situ (Hobbs, et al., 2009).
This technique enables very high resolution imaging of nonconducting surfaces (Fonseca Filho et al., 2004) with the ability to measure the height of the surface very accurately and observe zeolite growth features not detectable by conventional methods, i.e. SEM, TEM, etc. (Sadegh Hassani et al., 2010). AFM have been used to image the surface of zeolites such as scolecite, stilbite, faujasite, heulandite and mordenite, since 1990.
In addition, by this method it is possible to follow in situ processes such as crystallization which need non-destructive methods. In addition, the effect of changes in temperature on structures can be monitored at the nanometer scale (Hoobs et al., 2009).
In this chapter, it is focused on using AFM for current progress toward the elucidation of zeolite growth. In this regard, after introducing AFM technique, an overview about controlling of crystal growth of materials, especially zeolites, by AFM is presented.

2. AFM technique

Atomic force microscope is a kind of microscope in which a sharp tip is mounted at the end of a spring cantilever of known spring constant. This microscope employs an optical detection system in which a laser beam is focused onto the backside of a reflective cantilever and is reflected from the cantilever onto a position sensitive photo detector. An image can be obtained based on the interaction between a desired sample and a tip. As the tip scans the surface of the sample, variation in the height of the surface is easily measured as flexing of the cantilever, then variation in the photodiode signal. This gives a 3-D profile map of surface topography.
There are feedback mechanisms that enable the piezo-electric scanners to maintain the tip at a constant force (to obtain height information), or height (to obtain force information) above the sample surface. AFM can be used not only for imaging the surfaces in atomic resolution but also for measuring the forces at nano-newton scale. The force between the tip and the sample surface is very small, usually less than 10^{-9} N. The detection system does not measure force directly. It senses the deflection of the micro cantilever (Ogletree et al., 1996; Sadegh Hassani et al., 2006; 2008c).

Imaging modes of operation for an AFM are dynamic and static modes. In the dynamic mode, including non contact and tapping modes, the cantilever is externally oscillated at or close to its fundamental resonance frequency. The oscillation amplitude, phase and resonance frequency are changed by tip-sample interaction forces. These changes in oscillation with respect to the external reference oscillation provide information about the sample's characteristics. Using dynamic mode, it is possible to monitor both the phase of the drive signal oscillating the cantilever, and the cantilever's response. This phase signal gives access to material properties, a combination of adhesive and viscoelastic properties (Tamayo& Garcia, 1998; Tamayo & Garcia, 1998; Hoobs et al,. 2009) with nanoscale resolution.

In non contact mode, the tip of the cantilever does not contact the sample surface. The cantilever is instead oscillated at a frequency slightly above its resonant frequency where the amplitude of oscillation is typically a few nanometers. The van der Waals forces, or any other long range force which extends above the surface acts to decrease the resonance frequency of the cantilever. This decrease in resonant frequency combined with the feedback loop system maintains a constant oscillation amplitude or frequency by adjusting the average distance between tip and sample surface. By measuring this distance at each point, a topographic image of the sample surface can be obtained.

In tapping mode, the cantilever is driven to oscillate up and down at near its resonance frequency similar to non contact mode. As the tip is approached to the surface the amplitude of oscillation is damped, and it is now the amplitude of the oscillation that is used as a feedback parameter.

In static mode, the cantilever is scanned across the surface of the sample and the topography of the surface are obtained directly using the deflection of the cantilever. In this operation mode, the static tip deflection is used as a feedback signal. Close to the surface of the sample, attractive forces can be quite strong, causing the tip to "snap-in" to the surface. Thus static mode AFM is almost always done in contact where the overall force is repulsive. Consequently, this technique is typically called "contact mode". In contact mode, the force between the tip and the surface is kept constant during scanning by maintaining a constant deflection.

3. Monitoring processes with AFM

The advent of atomic force microscopy (AFM) (Binnig et al. 1986) has provided the new possibilities to investigate the nanometer-sized events occurring at crystal surfaces during crystal growth and recrystallization. This is possible under ex situ and often in situ conditions as it can be suited to observe surfaces under solution. Under ex situ conditions a wide variety of synthetic parameters can be changed but some careful works such as quenching experiments must be performed before transferring the sample to AFM to prevent secondary processes caused by changing growth conditions.

Real-time images of growing crystals have provided the structural details revealing terrace growth, spiral growth, defect and intergrowth structure in a vast variety of growth studies (Mcpherson et al. 2000). The free energy for individual growth processes can be achieved by measuring real-time micrographs at a range of temperatures.

Over the last decade, the application of AFM to follow some processes in soft material has been reported. Hoobe et al. (2009) believed that scanning probe techniques have several capabilities that make them suitable for the investigation of soft materials, organic materials as they pass through a transformation, and directly observe processes at a near molecular scale. For example, in a macromolecular system, such as semicrystalline polymer, phase transitions of large molecule sometimes are in metastable states. To follow these non-equilibrium states which often control the final material properties, is important and can be performed by AFM (Strobl, 2007; Hoobs et al., 2009).

In the early study of soft material using AFM, contact mode was used (Magonov et al., 1997). The development of dynamic modes of operation, such as tapping mode and the subsequent development of phase imaging, allowed a substantial growth in the possibilities of the technique (Garcia & Perez, 2002; Hoobs et al., 2009). For example, atomic force microscope with tapping mode was applied to study the surface morphology of as-grown (111) silicon-face 3C-SiC mesaheterofilms (Neudeck et al., 2004). Their observation showed that wide variations in 3C surface step structure are as a function of film growth conditions and film defect content. The vast majority of as-grown 3C-SiC surfaces consisted of trains of single bilayer height (0.25 nm) steps. They reported that Macrostep formation (i.e., step-bunching) was rarely observed. As supersaturation is lowered by decreasing precursor concentration, terrace nucleation on the top (111) surface becomes suppressed.

AFM technique also has been used for study the structure of the zeolites. In this chapter, it is focused on crystal growth processes in open framework, inorganic materials (i.e. zeolites) studied by AFM. The zeolite crystal growth experiment is important to enhance the understanding of growth of zeolite crystals and nucleation, and controlling the defects in zeolites.

Many studies have been performed to investigate how zeolite crystals grow. Based on the results reported from these studies it is generally found that the growth linearly proceeds during the crystallization of most zeolites and this is applicable for both gel and clear solution syntheses (Subotic et al. 2003; Zhdanov et al. 1980; Bosnar et al. 1999; Iwasaki et al. 1995; Cundy et al. 1995; Bosnar et al. 2004; Cora et al. 1997; Kalipcilar et al. 2000; Schoeman et al. 1997; Caputo et al. 2000). Parameters such as gel composition, aging, stirring and temperature can affect on growth in zeolites (Subotic et al. 2003; Cundy et al. 2005). Imaging of zeolite surface by use of such a very powerful technique, has recently made possible the understanding the exact growth mechanism in the synthesis of these materials. The work performed on cleaved surface (0 0 1) of the natural zeolite scolecite (MacDougall et al. 1991) under ambient conditions has revealed the arrangement of 8-membered ring centers. Parameters such as lattice constants, angles and distances of the zeolite structure have been measured and compared favorably with its crystallographic structure. Feathered terraces with height ~9 Å, half the unit cell dimension of 17.94 Å, have been reported on the cleaved surface (010) of natural zeolite heulandites (Scandella et al. 1993). The outer surface of this zeolite investigated by AFM (Binder et al. 1996) has also revealed growth spiral at screw dislocation with the pitch of ~9 Å (sometimes double), which is consistent with the b-dimension of the zeolite. Further AFM works on heulandite (Yamamoto et al. 1998) crystals has revealed the presence of steps, suggesting a possible birth-and-spread mechanism.

The precipitated sodium aluminosilicate hydrogel has also been analyzed by AFM (Kosanovic et al. 2008). The obtained results have showed that predominantly true amorphous phase of the gel contained small proportions of partially crystalline (quasi-crystalline) or even fully crystalline phase. Some different methods such as FTIR, DTC, electron diffraction have confirmed this finding and also showed partially or even fully crystalline entities of the sample. AFM has also revealed that the particles of the partially or fully crystalline phase are nuclei for further crystallization of zeolite.

Anderson et al. (1996; 1998) have reported the first AFM study of synthetic zeolite Y and observed triangular (1 1 1) facet, revealing approximately triangular terraces. The height of the terraces is 15 Å which is consistent with one faujasite sheet and the step height observed by other techniques. By measuring the area of the terraces they showed that this area was growing at a constant rate which was consistent with a pseudo terrace-ledge-kink mechanism in comparison with the growth in dense phase structures.

Zeolite A due to water softening features is an industrially important zeolite. It has been studied to determine the mechanism of crystal growth, using AFM by some researchers (Agger et al. 1998; 2001; Sugiyama et al. 1999). This zeolite, similar to zeolie Y demonstrates a layer-type growth with terrace heights consistent with simple fractions of unit cell. The authors have reported terrace height of 12 Å, equivalent to half unit cell height consisting of a sodalite cage and double 4-ring (Agger et al. 1998; 2001). Sugiyama et al. (1999) have also reported values corresponding to the individual sodalite cage and double 4-ring with slight differences. Wakihara et al. (2005) have investigated the surface structure of zeolite A by AFM and compared their results with those obtained by the other techniques such as HRTEM and FE-SEM. They have found that the terminal structure of zeolite A is incomplete sodalite cages. These results support one of the terminal structures proposed by Sugiyama et al. (1999), although these findings may not be always applicable to all zeolites A synthesized by various methods.

Irregardless of these differences, the principal conclusions of both studies (Agger et al. 1998; Sugiyama et al. 1999) are the same and a layer-by-layer growth mechanism is operative for zeolite A (Agger et al.1998). The area of the terraces in this zeolite grows at a constant rate confirmed by the parabolic cross-section of the surface. Agger et al. (1998) have also reported a detailed simulation of the growth features in zeolite A and suggested that the rate of growth at kink site is the fastest growth process. They also report that since the terrace edges run parallel to the crystal edges so the rate of growth at edge or ledge sites is considerably less than that at kink sites. These findings of the relative rates of growth processes help to understand and then affect the relative rates to control defects and structure.

Umemura et al. (2008) have presented a computer program that simulates morphology as well as surface topology for zeolite A crystals. They compared favorably the simulation results with those obtained from AFM images on the {1 0 0}, {1 1 0} and {1 1 1} faces of synthetic crystals.

Slilicalite is a siliceous form of zeolite ZSM-5 which is industrially important for catalytic properties, so the control of the synthesis of this inorganic solid is of great interest. An extensive AFM study of crystal growth has been fulfilled for this material (Agger et al. 2001; 2003; Anderson et al. 2000). In the low temperature synthesis of this material (similar to zeolites A and Y) terraced, layer-by-layer growth has been observed on both the (1 0 1) and (1 0 0) facets. The height of the terraces is 10 Å corresponding to half the unit cell dimension in the [0 1 0] direction (or the height of one pentasil chain). A constant-area-deposition mechanism, not dominated by addition at kink site, has resulted in the approximately circular shape of the terraces, indicating no preferential growth direction. Terraces in the high temperature synthesis, which produces large crystal (not similar to the low temperature synthesis, producing small crystals), grow towards the crystal edges and have the height of several hundred angstroms (up to 110 nm high on the (0 1 0) face and up to 20 nm high on the (1 0 0) face). Such terraces have no relation to any structural element of the silicalite. Assuming of a layer growth mechanism the authors have concluded that an obstruction to terrace advance causes a build-up of the layers. They have suggested the defect inclusion mechanism to explain the relative terrace heights on the two faces.

4. Morphology study of Zeolite L by AFM

Zeolite L which was initially determined by Barrer et al. (1969) has hexagonal symmetry (Barrer et al., 1969; Baerlocher et al., 2001) with two-dimensional pores of about 0.71 nm aperture leading to cavities of about $0.48\times1.24\times1.07$ nm^3 and the Si/Al ratio is typically 3.0 (Pichat et al., 1975; Sig Ko & Seung Ahn, 1999). The zeolite crystals consist of cancrinite cages linked by double six-memberd rings, forming columns in the c direction. Connection of these columns gives rise to 12-membered rings with a free diameter varies from 0.71 nm (narrowest part) to 1.26 nm (widest part). The main channels are linked via non-planar eight membered rings forming an additional two-dimensional channel system with openings of about 0.15 nm. Studies have shown that the morphology of the crystals can be approximated by a cylinder, with the entrances of the main channels located at the base. The number of channels is equal to $0.265(dc)^2$, where dc is the diameter of the cylinder in nanometers (Zabala Ruiz et al., 2005; Bruhwiler & Calzaferri, 2005; Breck& Acara, 2005).

Brent and Anderson (2008) studied crystal growth mechanism in zeolite L and control the crystal habit by atomic force microscopy. They claimed that AFM was an excellent tool for determining crystal growth mechanisms in zeolites L and gave a snapshot in time as to how the shape of a crystal had developed, by imaging surface features. In their work, the surface features on both the (1 0 0) side walls, and the (0 0 1) hexagonal faces of zeolite L was investigated.

Sadegh Hassani et al. (2010a) reported synthesis and characterization of nano zeolite L. Nanosized zeolite L was synthesized from a gel mixture at 443 K with different aging times. The molar compositional ratio of the resulting gel was 7.6 Na_2O : 7.2 K_2O : 1.3 Al_2O_3 : 40 SiO_2 : 669 H_2O. This homogeneous gel mixture was transferred to a Teflon-lined autoclave and placed in an air-heated oven at 443K for different synthesis times (24, 45, 110, 160, and 200 h). The autoclave was removed from the oven at the scheduled times and quenched in cold water. The solid product was separated by centrifugation, washed thoroughly a few times with deionized water and oven dried at 353 K for 5 h.

The samples were characterized by XRD, XPS techniques and morphological changes investigated by TEM and AFM.

A commercial atomic force microscope (Solver P47 H, NT-MDT Company) operating in non-contact mode and equipped with a NSG11 cantilever was used to take images in nanometer scale. Samples were dispersed in ethanol by sonication and deposited on a suitable substrate to be applicable for AFM. Atomic force microscope was used to obtain detailed surface images, such as crystal dimensions, by zooming in on a fine particle.

The technique revealed the existence of a multitude of terraces with the height of either ≈ 1 or ≈ 2 nm. Figure 1.a shows a two dimensional image of the zeolite crystals after elapsing 24 h of the synthesis time. This image confirms again the hexagonal geometry of the zeolite crystal. Figure 1.b is a cross sectional profile of the image a. The terraces demonstrate growth direction; consequently, growth fronts develop a hexagonal profile. They are concentric, growing out to the crystal edge from a central nucleation point as shown in figure 1.c. It has been reported that the growth morphology is thermodynamically related to corresponding crystallographic structure, according to the chemical bonding theory of single crystal growth (Xu & Xue, 2006; Yan, 2007; Xue et al., 2009). Work carried out on synthetic zeolite crystals to date suggests crystal growth occurs through deposition and subsequent expansion of layers (Xu & Xue, 2006; Anderson & Agger, 1998). Their findings are in good agreement with this suggestion and show that zeolite L crystals grow via a layer mechanism. Further observation of terraces at atomic level, using a single crystal will help this suggestion.

AFM allowed the detailed observation of nanometer-size events at crystal surfaces. In addition, the images showed layer growth of the zeolite crystal and the height of terraces. Two-dimensional AFM images (Fig. 1.a) showed hexagonal structure which is in good agreement with the TEM results (Fig. 2). Furthermore, Three-dimensional structure of the zeolite crystal (Fig. 1.c) obtained by AFM (not possible by TEM) indicated hexagonal layers. In addition, Figure 3 exhibits the aggregation of zeolite L crystals.

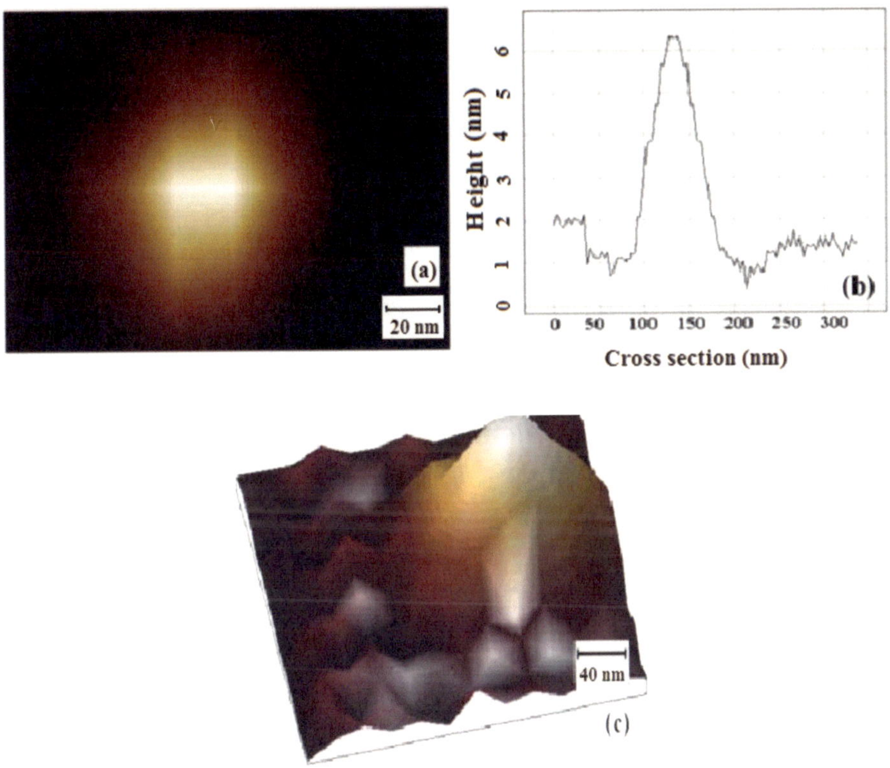

Fig. 1. AFM images of synthesized zeolite L. (a) two dimensional image of the crystal after 24 h and (b) its cross section, (c) three dimensional image of the crystal after 160 h (Sadegh Hassani et al., 2010a).

TEM images of the samples were recorded on CM260-FEG-Philips microscope and samples dispersed in acetone by sonication and deposited on a microgride. Figure 2 shows TEM images of the synthesized samples with various magnifications. These images indicate the average size of the sample crystallites (about 50 nm) possessing hexagonal geometry. A detailed surface image of the zeolite particles (Fig. 2c-d) indicated the parallel one dimensional channels arranged in a uniform pattern with hexagonal symmetry.

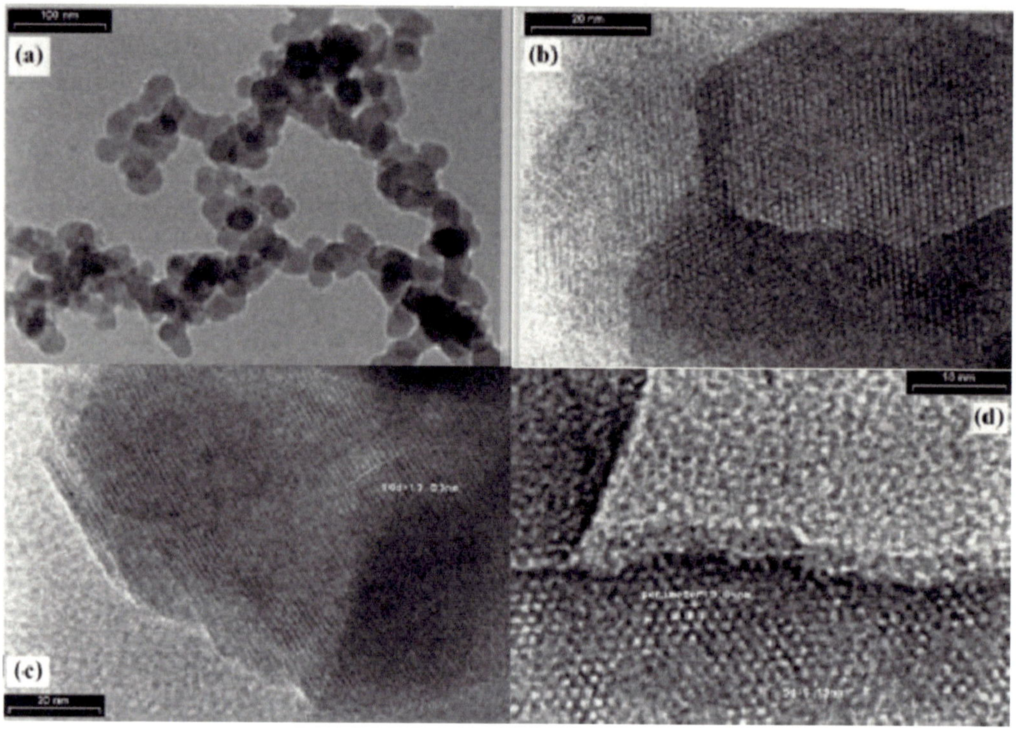

Fig. 2. TEM images of as-synthesized zeolite L with different resolutions for (a) the particle size and (b) morphology of the sample and (c) and (d) detailed surface images of the zeolite (Sadegh Hassani et al., 2010a).

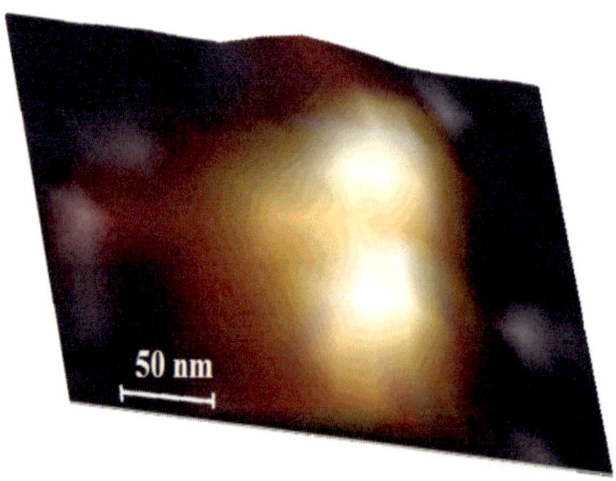

Fig. 3. Three dimensional AFM image of as synthesized zeolite L prepared in 160 h exhibiting aggregation of crystals.

In addition, AFM results exhibited bigger size of the crystal by increasing the synthesis time up to 160 h. Beyond this synthesis time, the size of the crystal decreased (Fig. 4). However, X-Ray diffraction patterns of the samples indicated that the synthesis times up to 110 h maintain almost the same crystallinity, whereas the synthesis times longer than that cause to decrease the crytallinity (89% for 160 h and 63% for 200 h). The apparatus used for XRD study was the powder X-ray diffractometer, Philips PW-1840, with a semi conductor detector and a Ni-filtered Kα (Cu) radiation source attachment.

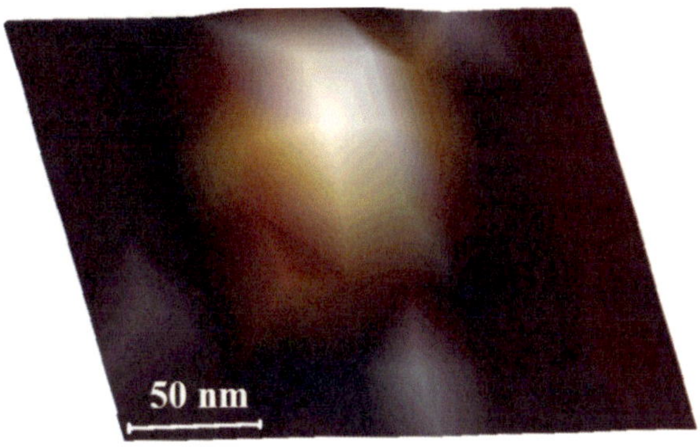

Fig. 4. Three dimensional AFM image of as synthesized zeolite L prepared in 200 h.

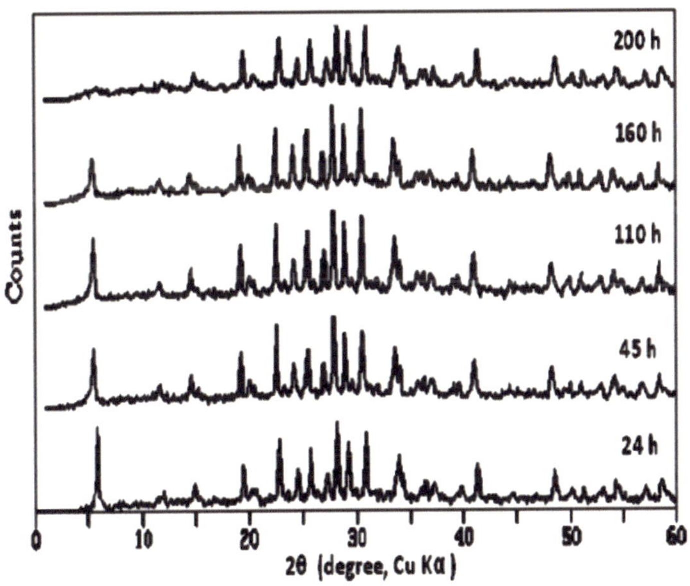

Fig. 5. X-ray diffraction patterns of the synthesized zeolite L obtained at different crystallization time at 443 K (Sadegh Hassani et al., 2010a).

Figure 5 shows the X-ray diffraction patterns of the as synthesized zeolite L samples obtained with different crystallization times. Characteristic XRD peaks showed that the fully crystalline phase (97% crystallinity) was obtained after 24 h. Reflections located at $2\theta \approx 5.5$,

19.4, 22.7, 28.0, 29.1 and 30.7 were used to calculate crystallinity. Results show that the synthesis times up to 110h maintain almost the same crystallinity, whereas the synthesis times longer than that cause to decrease the crytallinity (89% for 160 h and 63% for 200 h).
It was reported that Si/Al molar ratio in zeolites structure could affect on morphology and crystal size of these compounds (Shirazi et al., 2008; Mintova et al., 2006; Celik et al., 2010). Therefore, Sadegh hassani et al. (2010a) using AAS and XPS performed the elemental analyses of the bulk and surfaces of the as synthesized nanozeolite L, respectivly. Elemental surface analysis of the zeolite sample was carried out on the X-ray Photoelectron Spectroscopy (XR3E2 Model-VG Microtech; Concentric Hemispherical Analyzer, EA 10 plus Model-Specsis).
Figure 6 shows the XPS spectrum of the sample. The spectrum depicts the surface analysis of the as synthesized nanozeolite L. The Si/Al ratios of the gel mixture, bulk and surface of the sample are shown in table 1.
The slight difference in Si/Al ratios (0.5) was observed between bulk and surface of the zeolite sample.

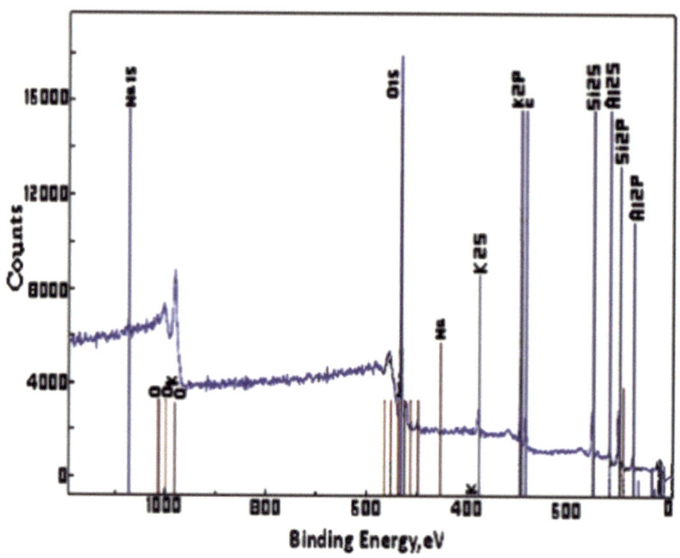

Fig. 6. Surface analysis results of synthesized zeolite L crystal by X-ray photoemission spectroscopy. (Sadegh Hassani et al., 2010a).

Sample	Si/Al
Gel mixture	15.4
Zeolite L (bulk)	3.5
Zeolite L (surface)	3.0

Table 1. Si/Al molar ratios of samples (Sadegh Hassani et al., 2010a).

5. Conclusion

This chapter is focused on the crystal growth study of various samples especially zeolites using AFM. Studies are revealed that atomic force microscopy is a powerful technique to follow in situ processes such as zeolite crystallization. In this regard, a study of crystal growth of nano-zeolite L is focused using atomic force microscopy (AFM). The results are compared with those of obtained from X-ray diffraction (XRD), X-ray photoelectron spectroscopy (XPS) and transmission electron microscopy (TEM) techniques.

TEM and two-dimensional AFM images indicate that the zeolite particles are in a nano-range and they have hexagonal structure. In addition, the AFM images show layer growth of the zeolite crystal and reveal the existence of a multitude of terraces with the height of either ≈1 or ≈2 nm.

The terraces demonstrate growth direction; consequently, growth fronts develop a hexagonal profile. They are concentric, growing out to the crystal edge from a central nucleation point. In addition, the AFM images exhibit layer growth, the height of terraces and the aggregation of zeolite L crystals. A detailed surface image of the zeolite particles indicate the parallel one dimensional channels arranged in a uniform pattern with hexagonal symmetry. AFM results also show bigger size of the crystal by increasing the synthesis time up to 160 h, beyond this synthesis time, the size of the crystal decrease.

6. References

Agger, J. R.; Hanif, N.; Cundy, C. S.; Wade, A. P.; Dennison, S.; Rawlinson, P. A. & Anderson, M. W. (2003). Silicalite crystal growth investigated by atomic force microscopy. *J. Am. Chem. Soc.* 125(3):830-9.

Agger, JR; Hanif, N; Anderson, MW. (2001). Fundamental zeolite crystal growth rate from simulation of atomic force microscopy. *Angew Chem Int* Ed, 40, 4065-4067.

Agger, J. R.; Pervaiz, N.; Cheetham, A. K. & Anderson, M. W. (1998). Crystallization in zeolite A studied by atomic force microscopy. *J. Am. Chem. Soc.* 120, 10754-410759.

Agger, JR; Hanif, N; Anderson, MW.; Terasaki, O. (2001). Growth Models in Microporous Materials *Microporouse Mesoporouse Mater.* 48, 1-9.

Agger, JR; Hanif, N; Anderson, MW; Terasaki, O. (2001). Crystal growth in framework materials. *Solid State Science.* 3. 809-819.

Aghabozorg, H. R.; Ghassemi, M. R.; Salehirad, F. et al. (2001). Investigation of parameters affecting zeolite NaA crystal size and morphology. *Iranian J. Cryst. Min.* 8, 2, 107-116.

Alibouri, M.; Ghoreishi, S. M. & Aghabozorg, H .R. (2009a). Hydrodesulfurization activity of using NiMo/Al-HMS nanocatalyst synthesized by supercritical impregnation. *Ind. Eng. Chem. Res.* 48, 4283-4292.

Alibouri, M.; Ghoreishi, S. M. & Aghabozorg, H. R. (2009b). Hydrodesulfurization of dibenzothiophene using CoMo/Al-HMS nanocatalyst synthesized by supercritical deposition. *The Journal of Supercritical Fluids* 49, 239-248.

Anderson, M. W.; Agger, J. R.; Thornton, J. T. & Forsyth, N. (1996). Crystal Growth in Zeolite Y Revealed by Atomic Force Microscopy. *Angewandte Chemie, International Edition in English.* 35, (11), 1210-1213.

Anderson, MW; Agger, JR; Pervaiz, N; Weigel, SJ; Cheetham, AK. (1998). Zeolite crystallization and transformation determined by atomic force microscopy. In: Treacy et al.; editor. Proceedings of the 12th International Zeolite Conference. Baltimore. MRS. 1487-94.

Anderson, M. W. (2001). Surface microscopy of porous materials. *Curr. Opinion Solid State Mat. Sci.* 5, 407 -415.

Anderson M. W. & Agger, J. A. (1998). Proc. Int. Zeolite 12th Conf., Vol. 3 (Materials Research Society, Warrendale, PA).

Anderson, MW; TErasaki, O. (2000). Understanding and Utilising novel microporous and mesoporous catalysts. In: Orafao, JM; Faria, JL; Figueiredo, JL. Editor. Actas do XVII Simposio Ibero-americano de Catalise. Porto. FEUP. 37-8.

Baerlocher, C.; Meier, W. M. & Olson, D. H. (2001) Atlas of Zeolite Framework Types (Elsevier, Amsterdam).

Barrer, R. M. & Villiger, H. Z. (1969). Crystal structure of synthetic zeolite L. *Z. Kristallogr.* 128, 352 -370.

Binder, G; Scandella, L; Schumacher, A; Kruse, N; Prins, R. (1996) Microtopographic and molecular scale observations of zeolite surface structures: Atomic force microscopy on natural heulandites. *Zeolites*, 16, 2-6.

Binnig, G.; Quate, CF.; Gerber, C. (1986). Atomic force microscopy. *Phys Rev Lett.* 56. 930-3.

Bosnar, S.; Subotic, B. (1999). Microporous Mesoporous Mater. 28. 483-493.

Bosnar, S.; Antonic, T.; Bronic, J.; Subotic, B. (2004). *Microporous Mesoporous Mater.* 76. 157-165.

Breck, D. W. & Acara, N. A. (1965). crystalline zeolite L. *U. S. Patent*: 3,216,789.

Brent, R. & Anderson, M. W. (2008). Fundamental Crystal Growth Mechanism in Zeolite L Revealed by Atomic Force Microscopy. *Angew. Chem.* 120, 5407 –5410.

Bruhwiler, D. & Calzaferri, G. (2004). Molecular sieves as host materials for supramolecular organization. *Micropor. Mesopor. Mat.* 72, 1-23.

Bruhwiler, D. & Calzaferri, G. (2005). Selective functionalization of the external surface of zeolite L. *Comptes Rendus Chimie.* 8, 391-398.

Cadby, A.; Dean, R.; Fox, A. M.; Jones, R. A. L. & Lidzey, D. G. (2005). Mapping the fluorescence decay lifetime of a conjugated polymer in a phase-separated blend using a scanning near-field optical microscope. *Nanoletters* . 5(11), 2232-7.

Caputo, D.; Gennaro, B. D.; Liguori, B.; Testa, F.; Carotenuto, I.; Piccolo, C. (2000). *Mater. Chem. Phys.* 66. 120-125.

Celik, F. E.; Kim, T. J. & Bell, A. T. (2010). Effect of zeolite framework type and Si/Al ratio on dimethoxymethane carbonylation. *Journal of Catalysis.* 270, 185–195.

Cleveland, J. P.; Anczykowski, B.; Schmid, A. E.; & Elings V. B. (1998). Energy dissipation in tapping-mode atomic force microscopy. *Appl. Phys. Letts.* 72(20), 2613-2615.

Cora, I.; Streletzky, K.; Thompson, R. W.; Phillies, G. D. J.(1997). *Zeolites.* 18. 119-131.

Cundy, C. S.; Henty, M. S.; Plaisted, R. J. (1995). Zeolite Synthesis Using a semi-continuous Reactor (part I). *Zeolites.* 15. 353-372.

Cundy, C. S.; Henty, M. S.; Plaisted, R. J. (1995). Zeolite Synthesis Using a Semi-continuous Reactor (part II). *Zeolites.* 15. 400-407.

Cundy, C. S.; Cox, P. A. (2005). The Hydrothermal Synthesis of Zeolites: Precursors, Intermediates and Reaction Mechanism. *Microporous Mesoporous Mater.* 82, 1-78.

Duan, G.; Zhang, Ch.; Li, A.; Yang, X.; Lu, L. & Wang, X. (2008). Preparation and Characterization of Mesoporous Zirconia Made by Using a Poly (methyl methacrylate). *Nanoscale Res. Lett.* 3, 118 -122.

Ebrahimpoor Ziaie, E.; Rachtchian, D. & Sadegh Hassani, S. (2008). Atomic force microscopy as a tool for comparing lubrication behavior of lubricants. *Materials Science: An Indian Journal.* 4, 2, 111-115.

Fonseca Filho, H. D.; Mauricio, M. H. P.; Ponciano, C. R. & Prioli, R. (2004). Metal layer mask patterning by force microscopy lithography. *Material Science and Engineering B.*, 112, 194-199.

Franke, M. & Rehse, N. (2008). Three-Dimensional Struc ture Formation of Polypropylene Revealed by in Situ Scanning Force Microscopy and Nanotomography. *Macromolecules* 41, 163-166.

Garcia, R. & Perez, R. (2002). Dynamic atomic force microscopy methods. *Surf. Sci. Rep.* (2002) 47, 197-301.

Giessibl, F. (1995). Atomic Resolution of the Silicon (111)-(7x7) Surface by Atomic Force Microscopy. *Science* 267, 68-71.

Hashimoto, S. (2003). Zeolite photochemistry: impact of zeolites on photochemistry and feedback from photochemistry to zeolite science. *J. Photochem. Photobio. C.* 4, 19 -49.

Hobbs, J. K. McMaster, T.J. Miles, M. J. & Barham, P.J. (1998). Direct observations of the growth of spherulites of poly(hydroxybutyrate-co-valerate) using atomic force microscopy. *Polymer.* 39(12), 2437-2446.

Hobbs, J. K. (1997). PhD Thesis, University of Bristol.

Hobbs, J. K. (2006). Progress in understanding polymer crystallization. Springer-Verlag, Berlin, Lecture Notes in Physics, Eds G Reiter and G Strobl.

Iwasaki, A; Hirata, M.; Kudo, I.; Sano, T.; Sugawara, S.; Ito, M.; Watanabe, M. (1995). In-situ measurement of crystal growth of rate of zeolite, *Zeolites.* 15, 308-314.

Hobbs, J. K.; Mullin, N.; Weber, Ch. H. M.; Farrance O. E. & Vasilev, C. (2009). Watching processes in soft matter with SPM. 12, 7-8, 26-33.

Kalipcilar, H.; Culfaz, A. (2000). Synthesis of Submicron Silicalite-1 Crystals from Clear Solutions. *Cryst. Res. Tecnol.* 35. 933-942.

Klinov, D. & Magonov, S. N. (2004). True molecular resolution in tapping-mode atomic force microscopy with high-resolution probes. *Appl. Phys. Lett.* 84, 2697-2699.

Kosanovic, C.; Bosnar, S.; Subotic, B.; Svetlicic, V.; Misic, T.; Drazic, G. & Havancsak, K. (2008). Study of the microstructure of amorphous aluminosilicate gel before and after its hydrothermal treatment. *Microporous and Mesoporous Materials.* 110, 2-3, 177-185.

Leggett, G. J.; Brewer N. J. & Chong K. S. L. (2005). Friction force microscopy: towards quantitative analysis of molecular organisation with nanometre spatial resolution. *Phys. Chem. Chem. Phys.* 7(6), 1107-1120.

Liu, J. & Xue, D. (2008). Thermal Oxidation Strategy towards Porous Metal Oxide Hollow Architectures. *Adv. Mater.* 20, 2622-2627.

Liu, J.; Liu, F.; Gao, K.; Wu, J. & Xue, D. (2009). Recent developments in the chemical synthesis of inorganic porous capsules. *J. Mat. Chem.* 19, 6073-6084.

Liu, J.; Xia, H.; Xue, D. & Lu, L.(2009b). Double-Shelled Nanocapsules of V2O5-Based Composites as High-Performance Anode and Cathode Materials for Li Ion Batteries. *J. Am. Chem. Soc.* 131, 12086 -12087.

Luo, Q. L.; Xue, Li, Z. & Zhao, D. (2000). Synthesis of nanometer-sized mesoporous oxides. *Studies Surf. Sci. Catal.* 129, 37-41.

MacDougall, J. E.; Cox, Sh. D.; Stucky, G. D.; Weisenhom, A. L.; Hansma, P. K.; Wise, W. S. (1991). Molecular resolution of zeolite surfaces as imaged by atomic force microscopy. *Zeolites.* 11, 426-427.

Magonov, S. N., & Reneker, D. H. (1997). Characterization of polymer surfaces with Atomic force microscopy. *Ann. Rev. Mater. Sci.* 27, 175-222.

Magonov, S.; Elings, N. V. & Papkov, V. S. (1997). AFM Study of Thermotropic Structural Transitions of Poly(diethylsiloxane). *Polymer.* 38, 297-307

Mcpherson, A.; Malkin, AJ.; Kuznetsov, YG. (2000). Atomic force microscopy in the study of macromolecular crystal growth. *Ann Rev Biophys Biomol Struct.* 29.361-410.

McMaster, T. J., Hobbs, J. K., Barham, P. J., Miles, M. J., (1997). AFM study of in situ real time polymer crystallization and spherulite structure, *Probe Microscopy.* 1, 43-56.

Mintova, S.; Valtchev, V.; Onfroy, T.; Marichal, C.; Knozinger, H. & Bein,T. (2006). Variation of the Si/Al ratio in nanosized zeolite Beta crystals. *Microporous and mesoporous materials.* 90, 1-3, 237-245.

Mohanty, P. & Landskron, K. (2008). Periodic Mesoporous Organosilica Nanorice. *Nanoscale Res. Lett.* 4, 169-172.

Neudeck, Ph. G.; Trunek, A. J. & Anthony Powell, J. (2004). Atomic Force Microscope Observation of Growth and Defects on As-Grown (111) 3C-SiC Mesa Surfaces. *Mat. Res. Soc. Symp. Proc.* Vol. 815, J5.32.1-J5.32.6.

Pichat, P.; Frannco Parra, C. & Barthomeuf, D. (1975). Infra-red structural study various type L zeolites. *J. Chem. Soc. Faraday Trans.* 71, 991-996.

Sadegh Hassani, S. & Ebrahimpoor Ziaie, E. (2006). Application of Atomic Force Microscopy for the study of friction properties of surfaces. *material science: An Indian journal,* 2(4-5), 134-141.

Sadegh Hassani, S.; Sobat, Z. & Aghabozorg, H. R. (2008a). Nanometer-Scale Patterning on PMMA Resist by Force Microscopy Lithography. *Iran. J. Chem. Chem. Eng.* Vol. 27, No. 4, 29-34.

Sadegh Hassani, S.; Sobat, Z. & Aghabozorg, H. R. (2008b). Scanning probe lithography as a tool for studying of various surfaces. *Nano Science and Nano Technology: An Indian journal,* Volume 2 Issue (2-3),94-98.

Sadegh Hassani, S.; Salehirad, F.; Aghabozorg, H. R. & Sobat, Z. (2010a). Synthesis and morphology of nanosized zeolite L. *Cryst. Res. Technol.* 45, No. 2, 183-187. (© 2010 WILEY-VCH Verlag GmbH & Co. KGaA, Weinheim)

Sadegh Hassani, S.; Sobat, Z. & Aghabozorg, H. R. (2010b). Force nanolithography on various surfaces by atomic force microscope. *Int. J. Nanomanufacturing,* Vol. 5, Nos. 3/4, 217-224.

Sadegh Hassani, S. & Sobat, Z. (2011). Studying of various nanolithography methods by using Scanning Probe Microscope. *Int .J. Nano .Dim.* 1(3), 159-175.

Sadegh Hassani, S. & Aghabozorg, H. R. (2011). Nanolithography study using scanning probe microscope, *Lithography* book , In TechWeb Publisher, in press.

Salehirad, F. & Anderson, M. W (1996) Solid-State ^{13}C MAS NMR Study of Methanol-to-Hydrocarbon Chemistry over H-SAPO-34. *J. catal.*164, 301-314.

Salehirad, F. & Anderson, M. W. (1998a). Solid-State NMR studies of Adsorption Complexes and Surface Methoxy Groups on Methanol- Sorbed Microporous Materials. *J. Catal.* 177, 189-207.

Salehirad, F. & Anderson, M. W. (1998b). NMR studies of methanol-to-hydrocarbon chemistry-part1Primary products and mechanistic considerations using a wide-pore catalyst. *J. Chem. Soc., Faraday Trans.* 94, 2857-2866.

Salehirad, F. & Anderson, M. W. (1998c). Solid-state NMR study of methanol conversion over ZSM-23, SAPO-11 and SAPO-5 molecular sieves-part 2. *J. Chem. Soc., Faraday Trans.* 94, 2857-2866.

Salehirad, F.; Aghabozorg, H. R.; Manoochehri, M. & Aghabozorg, H. (2004). Synthesis of titamium silicalite-2 (TS-2) from methyamine-tetrabutylammonium hydroxide media. *Catalysis Communication.* 5 , 359-365.

Scandella, L; Kruse, N; Prins, R. (1993). Imaging of zeolite surface-structures by atomic force microscopy. *Surf Sci.* 281. L331-4.

Schoeman, B. J. (1997). Progress in Zeolite and Microporous Materials. Parts A-C. 647-654.

Schulz-Ekloff, G.; Wohrle, D.; Van, B. (2002). Chromophores in porous silicas and minerals: preparation and optical properties. *Micropor. Mesopor. Mat.* 51, 91-138.

Shirazi, L.; Jamshidi, E. & Ghasemi, M. R. (2008). The effect of Si/Al ratio of ZSM-5 zeolite on its morphology, acidity and crystal size. *Cryst. Res. Technol.* 43, 12, 1300-1306.

Sig Ko, Y. & Seung Ahn, W. (1999). Synthesis and characterization of zeolite. *Bull. Korean Chem. Soc.* 20, 2, 1-6.

Strobl, G. (2007). The Physics of Polymers: Concepts for understanding their structures and behaviour, Spinger.

Subotic, B; Bronic, J. (2003). In handbook of Zeolite Science and Technology (eds Auer bach, S. M.; Carrado, KA; Dutta, P. K.). Marcel Dekker New York. 1184.

Sugiyama Ono, S.; Matsuoka, O. & Yamamoto, S. (2001). Surface structures of zeolites studied by atomic force microscopy. *Micropor. Mesopor. Mat.* 48, 103-110.

Sugiyama, S.; Yamamoto, S.; Matsuoka, O.; Nozoye, H.; Yu, J.; Zhu, G.; Qiu,, S. & Terasaki, O. (1999). AFM observation of double 4-rings on zeolite LTA crystal surface. *Microporous Mesoporous Mater.* 28, 1-7.

Sugiyama, S; Yamamoto, S; Matsuoka, O; Nozoye, H; Yu, J; Zhu, G; et al. (1999). AFM observation of Double 4-rings on zeolite LTA crystal surface. *Microporous Mesoporous Mater.* 28. 1-7.

Tamayo, J. & Garcia, R. (1998). Relationship between phase shift and energy dissipation in tapping- mode atomic force microscopy. *Appl. Phys. Lett.* 73(20), 2926-2928.

Umemura, A.; Cubillas, P.; Anderson, M. W. & Agger, J. R. (2008). *Studies in Surface Science and Catalysis.* 705-708.

Wakihara, T.; Sasaki, Y.; Kato, H.; Ikuhara, Y. & Okubo, T. (2005). Investigation of the surface structure of zeolite A. *Phys. Chem. Chem. Phys.* 7, 3416–3418.

Williams, C. C. (1999). Two-Dimentional Dopant Profiling by Scanning Capacitance Microscopy*Ann Rev. Mater. Sci.* 29, 471-504.

Xu, J. & Xue, D. J. (2006). Fabrication of copper hydroxyphosphate with complex architectures. *Phys. Chem. B* 110, 7750 – 7756.

Xu, D. & Xue, D. (2006). Chemical bond analysis of the crystal growth of KDP and ADP. *J. Cryst. Growth* 286, 108-113.

Xue, D.; Yan, X. & Wang, L. (2009). Production of specific $Mg(OH)_2$ granules by modifying crystallization conditions. *Powder Technol.* 191, 98-106.

Yamamoto, S.; Sugiyama, S.; Matsuoka, O.; Honda, T.; Banno, Y.; Nozoye, H. (1998). AFM imaging of the surface of natural heulandite. *Microporous Mesoporous Mater*. 21. 1-6.

Yan, X.; Xu, D. & Xue, D. (2007). SO_4^{2-} ions direct the one- dimensional growth of $5Mg(OH)_2 \cdot MgSO_4 \cdot 2H_2O$ *Acta Materl*. 55, 5747 -5757.

Zabala Ruiz, A.; Bruhwiler, D.; Ban, T. and Calzaferri, G. (2005). Synthesis of zeolite L. Tuning size and morphology. *Monatshefte für Chemie.* 136, 77-89.

Zhang, X.; Zhang, Z.; Suo, J. & Li, Sh. (2000). Synthesis of mesoporous silica molecular sieves via a novel templating scheme Original Research Article; *Studies Surf. Sci. Catal.* 129, 23.

Zhdanov, S. P.; Samulevich, N. N. (1980). Proceedings of the 5th International Conference on zeolites. 75-84.

16

One-Dimensional Meso-Structures: The Growth and the Interfaces

Lisheng Huang[1,2,3], Yinjie Su[2] and Wanchuan Chen[1]
[1]*Department of Physics, National Cheng Kung University, Tainan,*
[2]*National Laboratory of Solid State Microstructures, Nanjing University, Nanjing,*
[3]*College of Sciences & College of Materials Science and Engineering, Nanjing University of Technology, Nanjing,*
[1]*Taiwan*
[2,3]*China*

1. Introduction

One-dimensional (1D) meso-structures have become the focus of intensive research worldwide due to their unique physics and potential to revolutionize broad areas of device applications. They act as the most basic building blocks of nano-electronic systems, nano-optics and nano-sensors, so the controlled growth of these meso-structures is important for applying them in these fields. Materials properties can be tuned through control of micro-structural characteristics such as the physical size, shape, and the surface. Efforts to explore structures with multiple length scales unite the frontiers of materials chemistry, physics, and engineering. It is in the design and characterization of advanced materials that the importance of new interdisciplinary studies may be realized [1-4]. Recent research focused on well-faceted meso-structures has shown that the shape as well as the hetero-[5, 6] or homo-junctions [7, 8] contribute much to the tuning of properties of structured materials. Many significant properties, including optical, chemical, as well as electronic, have been revealed to be shape- or junction-related. For example, the lasing behaviors of nonlinear optical nano-scale wires or belts derive from the resonance cavity effect functioned by the parallel end-faces of the nano-structures [9-11]. Quantitative characterization of optical waveguiding in straight and bent nanowires is achievable in active devices [12]. Such study has shown that the optimization of surfaces, boundaries, and interfaces in materials with well-faceted structures plays an important role in furthering the application of these materials.

For efficient fabrication and assembly of well-faceted meso-structures, the anisotropy of the crystal can be utilized to control the nucleation and manipulate the surface energy [13, 14]. Macroscopically, a crystal has different kinetic parameters for different crystal planes guided by certain growth conditions. After initial nucleation, a crystallite will commonly develop into a three-dimensional entity with well-defined, low index crystallographic facets. Thus, the growth anisotropy can be advantageously utilized to create crystals with specific desired characteristics through control of the growth conditions. It has been extremely successful in different growth systems, such as solution-based route for growing shaped nanocrystals, vapor-phase growth of quasi-one-dimensional meso-structures with well-defined cross sections and surface polarities as well as some other exotic configurations through vapor-liquid-solid (VLS) or vapor-solid (VS) process [15-20].

In this chapter, we have examined the growth mechanisms and the morphology evolutions of one-dimensional meso-structures systematically based on the experimental and theoretical aspects of crystal growth. The 1D ZnO meso-structures will be selected as an example to show the morphologic evolution at multiple length scales. The quasi-one-dimensional SnO_2 meso-structures are studied to describe the morphological multiformity of crystal growth. The outline of the chapter is as follows. In Sect. 2, the growth of ZnO meso-structures is discussed, which includes the controlled growth (Sect. 2.1), structural characterization and crystal models (Sect. 2.2), the growth process and mechanism (Sect. 2.3), and structure-related optical properties (Sect. 2.4). In Sect. 3, the SnO_2 zigzag meso-structures growth mechanism is discussed, which includes the controlled growth (Sect. 3.1), structural characterization and crystal models (Sect. 3.2), the morphological evolution mechanism (Sect. 3.3). Concluding remarks are given in Sect. 4.

2. ZnO meso-structures

2.1 Growth control

ZnO, a wide direct band-gap semiconductor, is piezoelectric and transparent to visible light [21]. It is attracting much attention for application in UV light-emitters, varistors, transparent high power electronics, surface acoustic wave devices, piezoelectric transducers, gas-sensors, photo-catalysts, and as a window material for display and solar cells [22-31]. The wurtzite structure of the ZnO crystal has pronounced anisotropy. It possesses three fast growth directions of $<2\overline{1}\overline{1}0>$, $<01\overline{1}0>$, and $\langle 0001 \rangle$. Currently much effort has been focused on the fabrication of ZnO nano-/micro-scale structures. A number of methods, based on solid reaction, solution based synthesis, and vapor rout have been developed to grow this material. These methods include the reaction of zinc salt with base, thermal decomposition, pulse laser deposition (PLD), thermal evaporation/vapor phase transport (CVD), metal-organic CVD, molecular beam epitaxy (MBE), electrochemical deposition, chemical bath deposition, aqueous solution decomposition, modified micro-emulsion, and sol-gel methods [32-44]. ZnO nano-/micro-structures of varied geometries, exemplified by wires/rods, belts/ribbons, comb-like structures, tetra-pod whiskers and their various assemblages have all been produced by our group (Figs. 1a-d).

We have also reported a new type of modulated and well-faceted ZnO microfibers, which was synthesized via a convenient CVD process [8]. Considering the decomposition of of $Ni(NO_3)_2$ at high temperature, we used nickel oxide (Ni_2O_3) as a catalyst. This proved to be an efficient way for growing the modulated microfibers. Fig. 2 shows a typical SEM morphology of the as-synthesized product. It is evident that the products are composed of microfibers with periodic junctions at a significant percentage (over 95%) of the yield and over 80% reproducibility from run to run.

The fibers with very thin junctions usually grow parallel to each other, and the roots appear to be compressed and broad. The lengths of the fibers typically range in between 200 and 500 μm. the longest one observed was nearly a millimeter. The spacing between two neighboring junctions normally ranges from 5 to 30 μm. The side surfaces of the fibers are well-faceted. Note that the V-shaped junction derives from the concavo-concave morphology, and the angle between the left and the right facets is exactly of 60 or 120 degrees. The fiber is characteristically decorated by periodically prism-like junction arrays. We refer to this structure as a "junction-prism" structure.

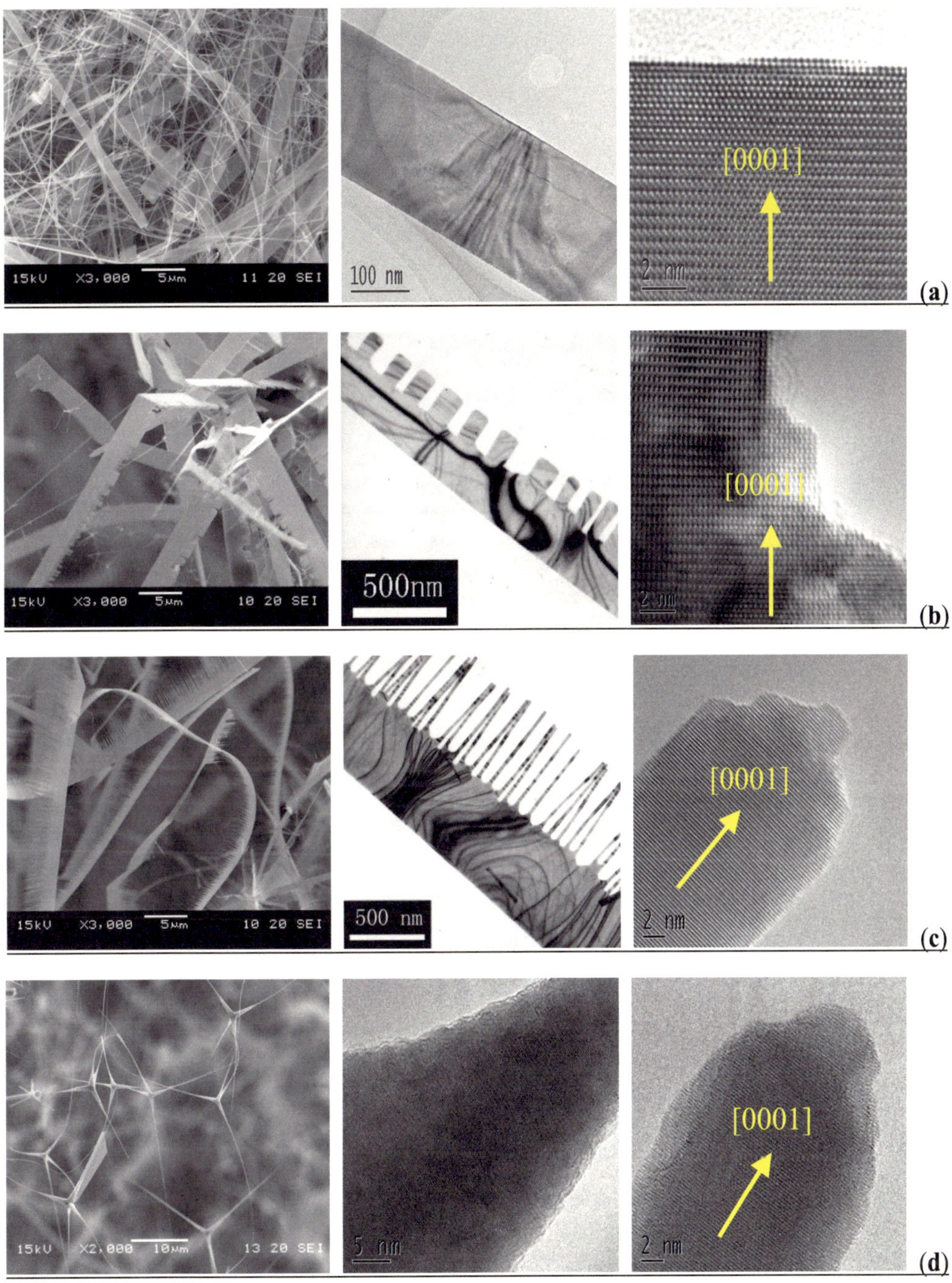

Fig. 1. ZnO meso-structures of belts (a), comb-like structures (b, c) and tetra-pod whiskers (d).

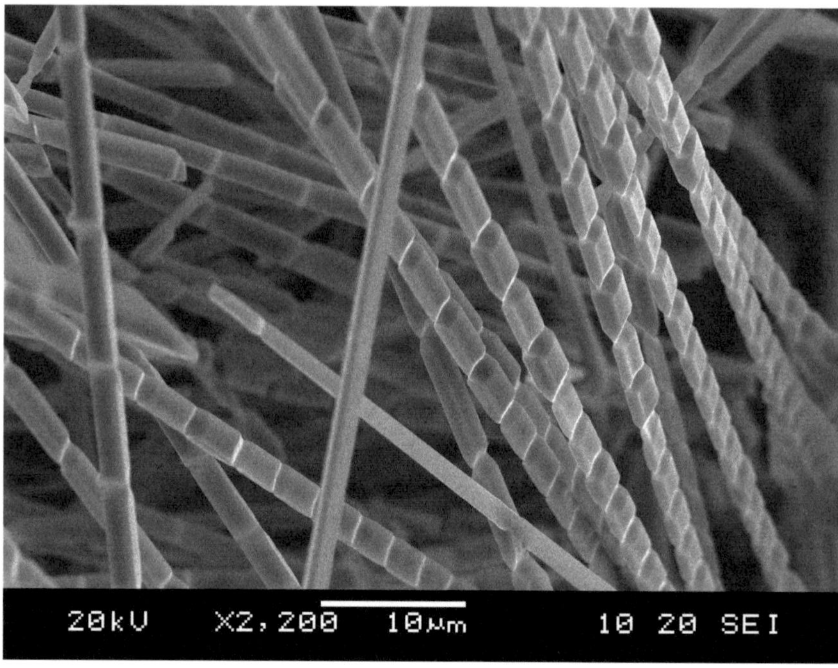

Fig. 2. ZnO junction-prism structure.

2.2 Structural characterization and crystal models

XRD and EDS measurements were performed for element analysis and phase determination. XRD studies show a typical wurtzite structure of ZnO with cell constants of a=0.324 nm and c=0.519 nm (Fig. 3a, JCPDS No. 36-1451). EDS studies (equipped in TEM) at the head, junction and root of a fiber show only peaks belonging to Zn and O without any other impurities (Figs. 3b-d).

In our studies, the presence of a small amount of nickel oxide is critical to synthesizing these modulated and well-faceted ZnO fibers. Although the vapor-liquid-solid (VLS) crystal growth mechanism explains the catalysis growth of some microstructures, no element nickel was observed found in our samples. Of course, the possibility exists that the quantities may be less than can be measured by XRD or EDS analysis. It is likely that the role of the nickel is the same as that of indium oxide and lithium carbonate for nanoring growth [45]. While no catalysts added for their growth, we believe that another important intrinsic factor for growth of the modulated microstructures is the intense anisotropy of the wurtzite-structured ZnO along different axes.

Crystallographic orientations of the fibers were obtained by EBSD. Fig. 4a shows the microfiber with a flat facet upturned, which was automated EBSD mapped for the selected area (Fig. 4b). As EBSD requires a highly tilted surface (near 70° tilt), several microfibers were searched until one was found to give indexable EBSD patterns with illumination corresponding to a flat surface tilted to 70°. The map displayed is corrected for the 70° tilt whereas the SEM image is not tilt-corrected. Pole figures obtained from the EBSD map data show the $[2\bar{1}\bar{1}0]$ direction aligned with the growth direction, the broad surfaces parallel to

{0001} plane and the side surfaces parallel to $\{01\bar{1}0\}$ (Fig. 4c). A schematic unit cell displayed in the orientation was obtained by EBSD (Fig. 4d). The growth direction of the microfiber is $[2\bar{1}\bar{1}0]$ (*a* axis) and the side surfaces are ± (0001). The broad top and bottom surfaces are parallel $\{01\bar{1}0\}$ planes.

It was found there are two types of oriented fibers in the production. Fig. 5a shows the crystal models. This is consistent with the crystal structure of the ZnO. The wurtzite structure of the ZnO crystal has pronounced anisotropy, it possesses three fast growth directions of $<2\bar{1}\bar{1}0>$, $<01\bar{1}0>$, and $\langle 0001 \rangle$. Generally, [0001] is the fastest based on the kinetic mechanism involved. [0001] growth minimizes the area of exposed {0001} faces (Fig. 5b). Under thermodynamic equilibrium conditions, the surface energy of the polar {0001} planes is larger than that of the nonpolar planes of $\{01\bar{1}0\}$ or $\{2\bar{1}\bar{1}0\}$. Moreover, the surface energies differ less between the $\{01\bar{1}0\}$ and $\{2\bar{1}\bar{1}0\}$ planes. Fig. 5c illustrates the basic configurations evolved from ZnO hexagonal unit in Fig. 5b. Changing the growth condition to activate various growth facets, microstructures would be synthesized in shapes with higher complexity than those of the familiar wire, rod, belt, and sphere-like structures. Thus, it is often found that the produced well-faceted ZnO fibers with periodic junction-prisms preferentially grow along $[01\bar{1}0]$ as opposed to the $[2\bar{1}\bar{1}0]$ direction. The structure model shown in Fig. 5a (insert) illustrates a $[01\bar{1}0]$ preferred growth axis of the fiber and the geometric relationships between all its outer facets.

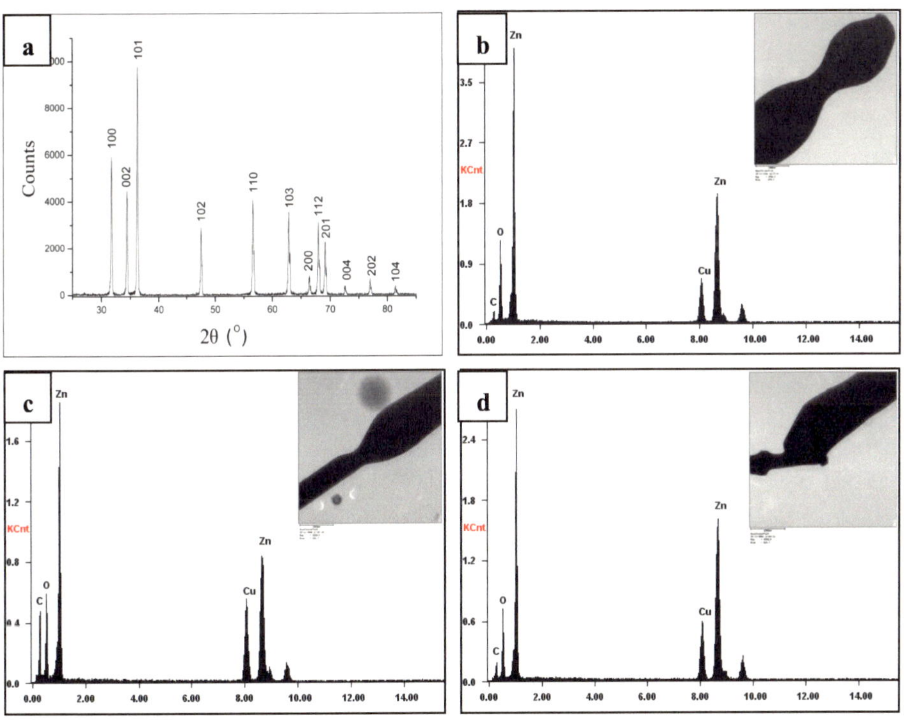

Fig. 3. XRD pattern showing wurtzite structure of ZnO (a), EDS studies at the head, the junction and the root of a fiber (b-d).

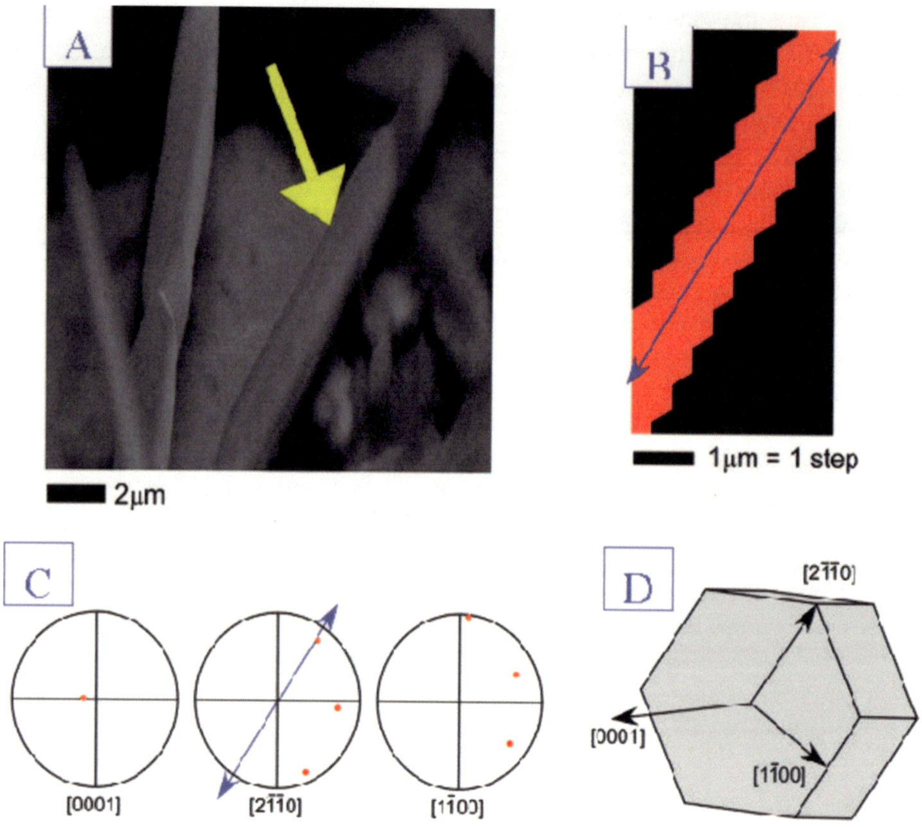

Fig. 4. EBSD measurement for a microfiber.

2.3 Growth process and mechanism

SEM investigations on the microfibers demonstrate growth mechanism of the junction-prisms structures. Fig. 6a shows a newly growing head of a fiber, the growth unit of a nearly hexagonal prism is grown perpendicularly on the nanobelt base. Presumably, the small growth head would develop anisotropically and contact the adjacent unit forming a junction (Fig. 6b). Some similar configurations have been studied by our group [46]. We found that ZnO nano/microcombs are kin to these fibers: every tooth could be considered as a segment unit of the fiber. If the teeth grow short in the [0001] direction and thick in diameter, then they would contact each other and the morphology should be identical to these junction-like fibers (Fig. 6c, d). Additional experiment also showed that large amount necklace-shape structures could be produced. Every microstructure consisted of a row of rhomboids that are equally separated on a straight base of a narrow nanobelt (Fig. 6e, 6f). An anisotropic growth process is shown as follows: a nanobelt base was formed by fast catalysis growth along $[2\bar{1}\bar{1}0]$ or $[01\bar{1}0]$, followed by slow growth along [0001], forming the separate units of nearly hexagonal prisms. Developing on the nanobelt, these small units merged and formed junctions.

Some SEM images of the fibers show segment units quenched at different stages of their growth, and careful examination of the unit's morphology gives insight into the growth process. Fig. 7 (right panel) shows typical units in various stages of growth, along with their schematics (left panel). Although the SEM images are of different fibers, it is presumed that each fiber undergoes a similar sequence of steps during the growth process.

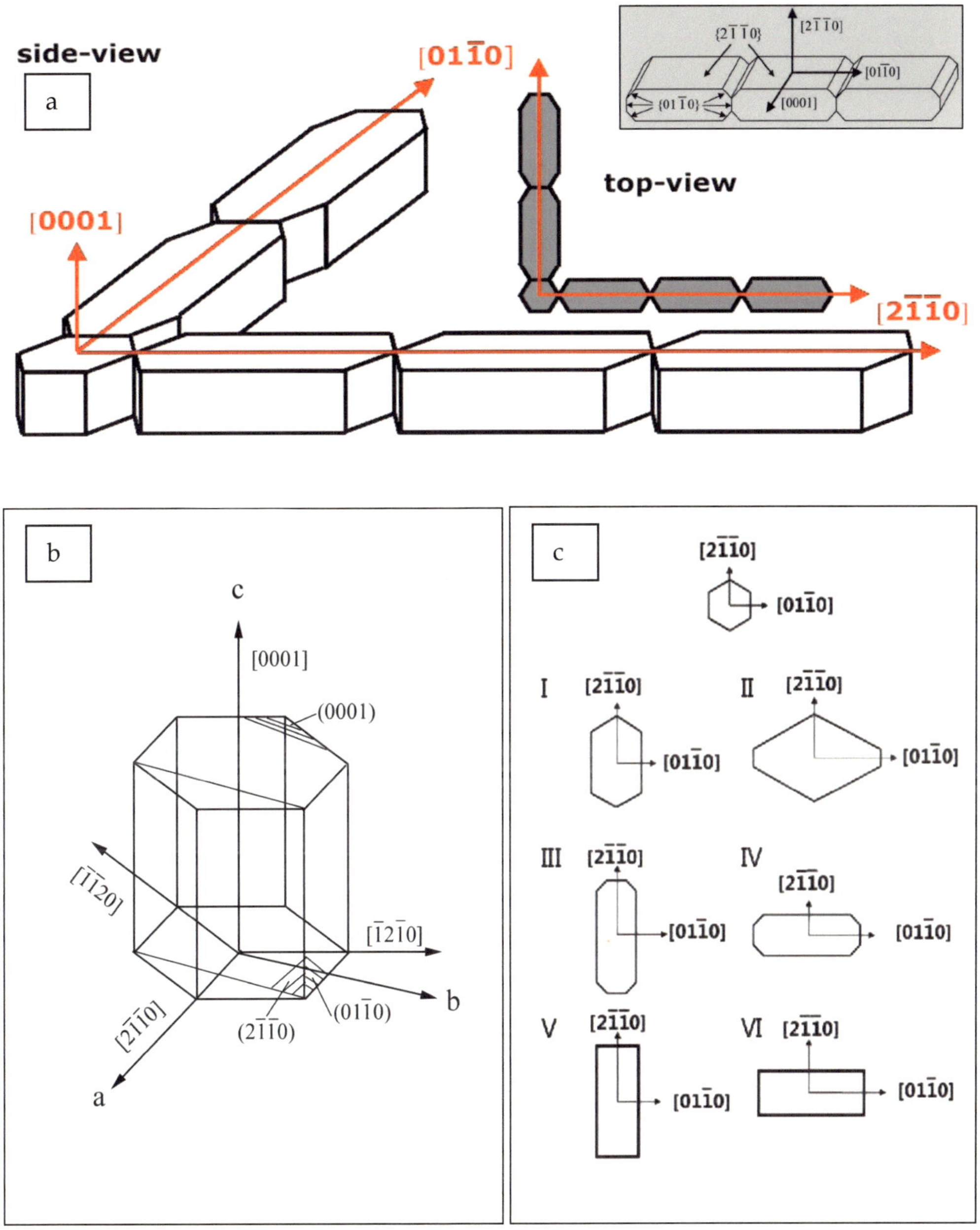

Fig. 5. (a) crystal mode of the two types of oriented ZnO fibers, (b) crystal mode of wurtzite structure of ZnO, (c) the basic configurations evolved from ZnO hexagonal unit.

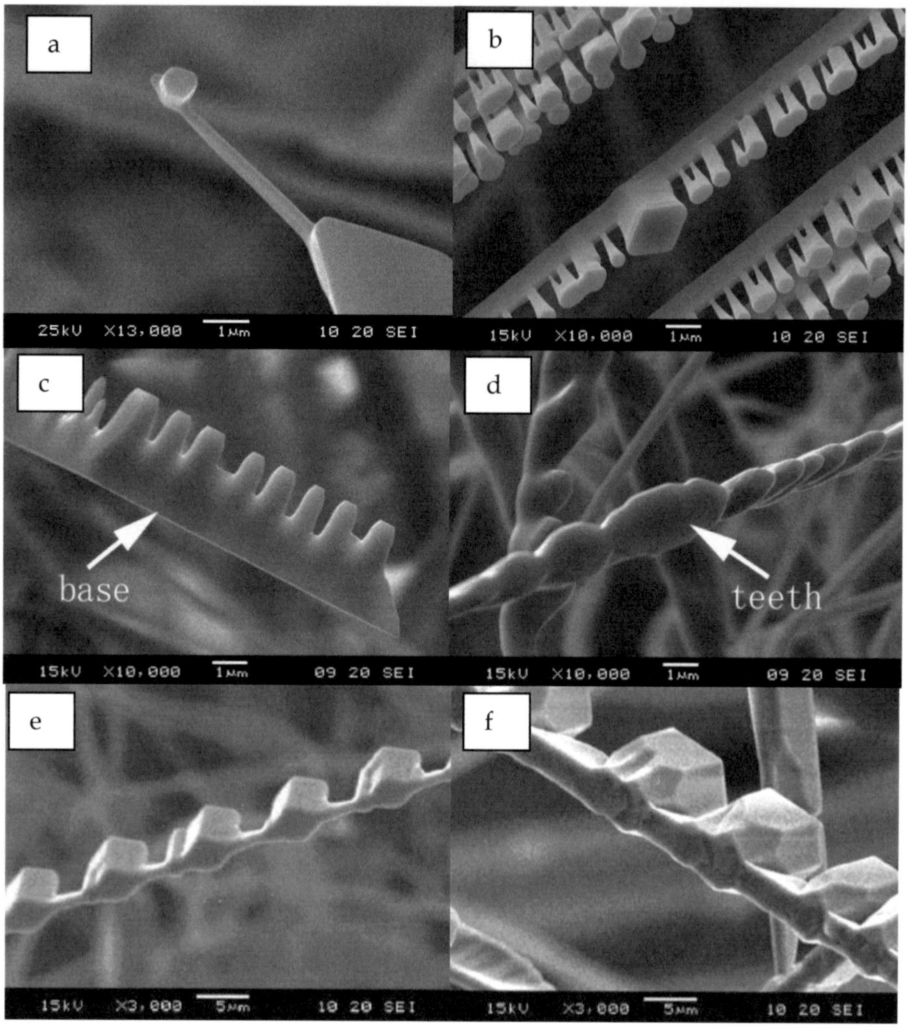

Fig. 6. (a) a newly growing head of a fiber, (b) small growth heads would develop anisotropically and contact the adjacent unit forming a junction, (c, d) ZnO micro-combs are kin to these fibers, (e, f) necklace-shape structures.

For example, consider the growth of $[01\bar{1}0]$ oriented fibers:

1. The first step is to grow a $[01\bar{1}0]$ oriented base with $\{2\bar{1}\bar{1}0\}$ side surfaces and top surfaces of {0001}. Subsequent grow will be self-modulated by nucleation and growth of the epitaxial pyramids on the c-face, (0001), of the base. The separate units growing along the c-axis on the base have hexagonal shapes (Fig. 7a, left panel are the crystal models). The six sided surfaces are equivalent planes of $\{01\bar{1}0\}$.
2. The units constructing a regular fiber with periodic junctions exhibit a prolonged eight-square shape, where the four profile $\{01\bar{1}0\}$ faces are partly exposed, and two broad $\{2\bar{1}\bar{1}0\}$ faces neighbor them (Fig. 7b). This evolution could be explained by an enhanced growth along $[01\bar{1}0]$ (the base growth direction) and a confined epitaxy perpendicular to $[01\bar{1}0]$.

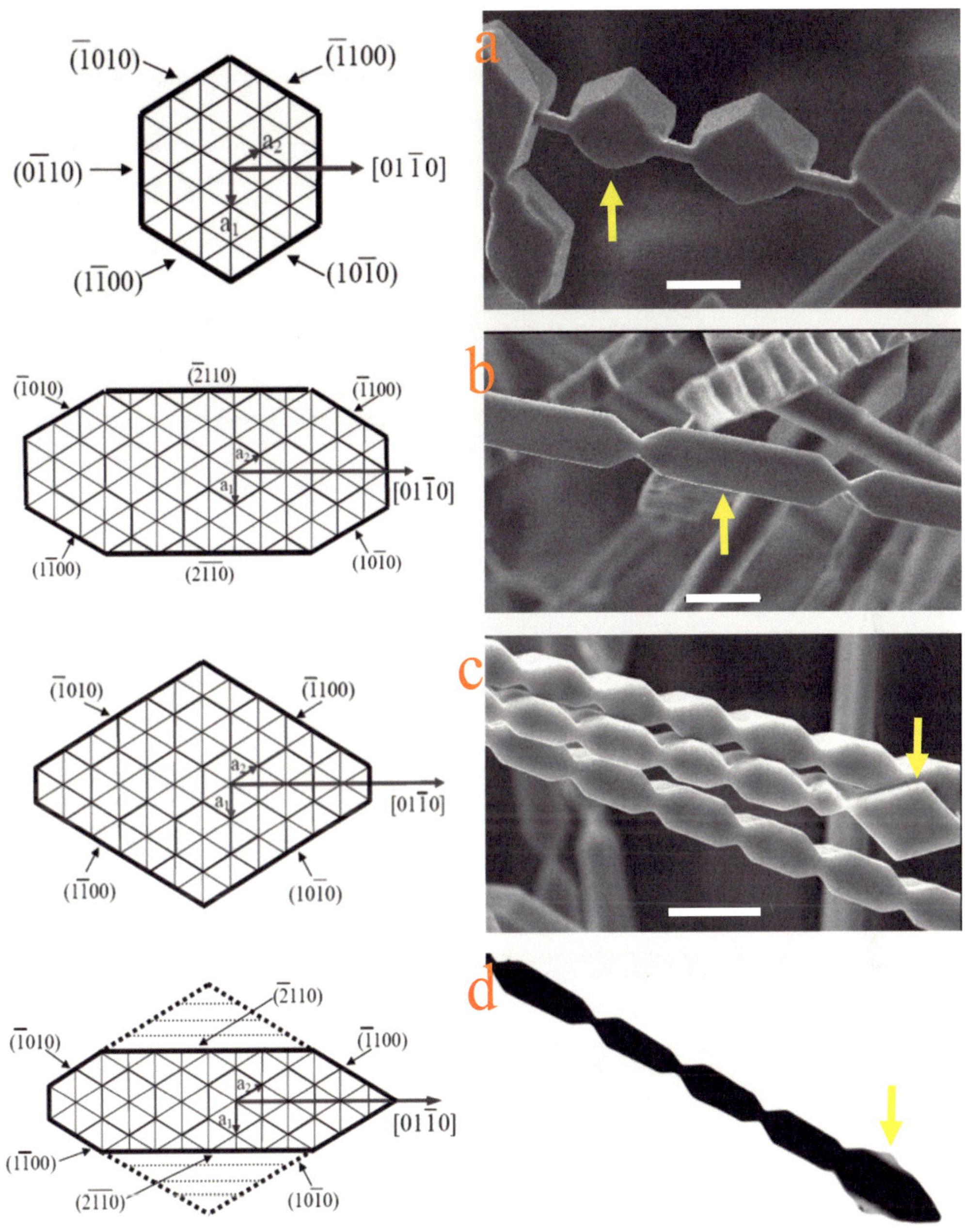

Fig. 7. Typical units in various stages of growth along with their schematics.

3. A prolonged hexagonal head of a fiber shows evidence for enhanced growth along $[01\bar{1}0]$ (Fig. 7c).
4. TEM observations show that the anti-confined epitaxy process is perpendicular to $[01\bar{1}0]$. Thin epitaxial layers are growing on the two broad $\{2\bar{1}\bar{1}0\}$ facets, and the growth would cease once the four $\{01\bar{1}0\}$ side surfaces are completely grown out of existence and the two $\{2\bar{1}\bar{1}0\}$ facets disappear (Fig. 7d).

Thus, we can deduce that two-step anisotropic growth as well as the confinement effect of the base (substrate) could result from these modulated and well-faceted junction-like fibers. A schematic representation of the growth process for the modulated fiber is illustrated in Fig. 8. Note that in the practical growth process, the base of the fiber has been combined into one united body with the segment units, but sometimes one side of the fiber is thicker than that of the other side. This is evident in the contrast in SEM images (Fig. 2). In order to grow very regular segment units, the growth condition should be controlled.

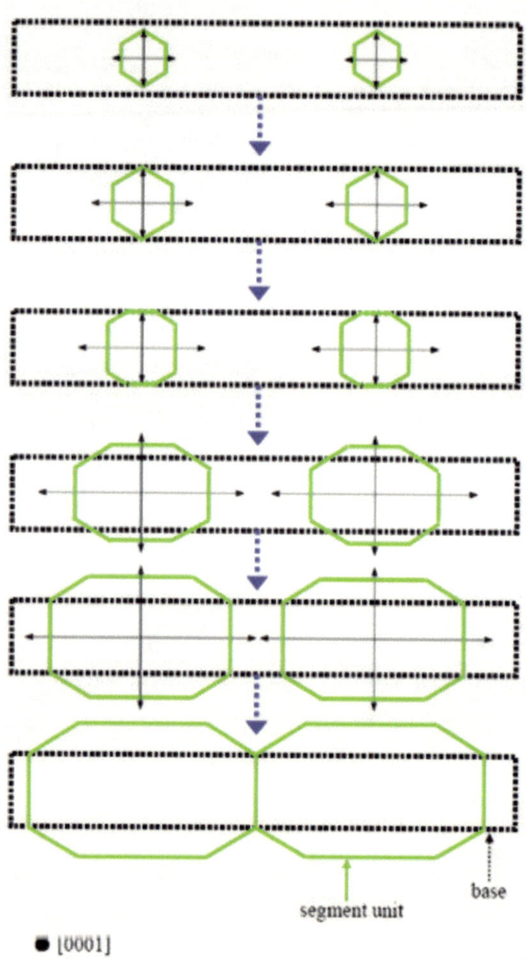

Fig. 8. A schematic representation of the growth process for the modulated fiber.

2.4 Structure-related optical properties

Because it approximately undergoes a thermodynamically equilibrium during the growth process of the fibers, all facets of a fiber are commonly low index crystallographic faces. According to the growth process and the crystal models, all the side surfaces of $\{01\bar{1}0\}$ and $\{2\bar{1}\bar{1}0\}$ should have coordinative crystalline qualities. μ-Raman studies further proved this point of view. The Raman spectra obtained at the $\{01\bar{1}0\}$ facets of a junction and at the $\{2\bar{1}\bar{1}0\}$ facets on the stem are shown in Fig. 9 (the inset shows the sample), respectively. No apparent difference is observed between the spectra from the junction facets and the stem facets.

Since the wurtzite structure of ZnO belongs to the C_{6v}^{4} space group, the zone center optical phonons are: A_1+$2E_2$+E_1 [47]. In the spectra, two Raman active E_2 modes were observed at 101 and 437 cm^{-1}, and four Raman active modes--A_1 and E_1 transverse (TO), at 380 and 407 cm^{-1}, and longitudinal (LO), at 574 and 583 cm^{-1}, with second order vibrations observed at 208, 334 and 1050-1200 cm^{-1}. These results can be entirely explained on the basis of the ZnO crystal [48], and signify the good crystalline properties of the junction stem facets of a fiber.

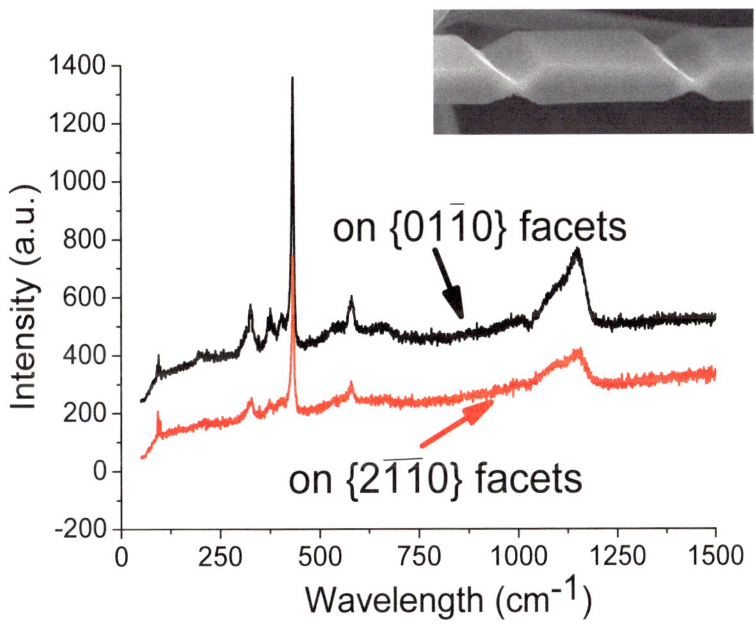

Fig. 9. Raman spectra obtained at the $[01\bar{1}0]$ facets of a junction and at the $\{2\bar{1}\bar{1}0\}$ facets on the stem.

Room temperature micro-PL spectra shown in Fig.10 indicate the enhancement of the green light emission at the junction. The spectrum obtained from the part between the two junctions consists of an intensive UV peak at 383 nm and a weak green band around 510 nm. The spectrum around the junction indicates that the green band is strong. This was further demonstrated using a PL microscope. The PL microscopy images show the fibers emitting strong green light at the junctions (Fig. 11h).

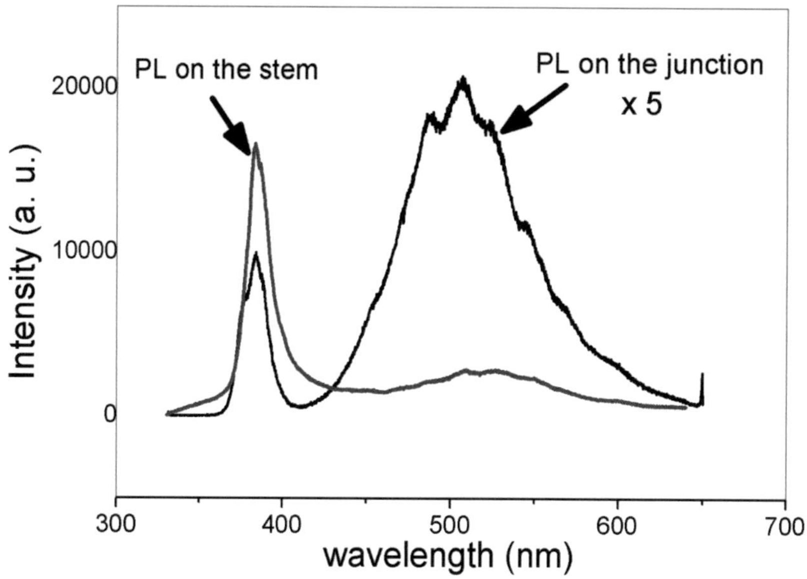

Fig. 10. Room temperature micro-PL spectra on the stem and at the junction.

It is generally accepted that the UV peak at 383 nm resulted from free excitonic emission of ZnO [49], while the green band arises from the recombination of a shallowly trapped electron within a deeply trapped hole [50]. Note that two neighboring units form one thin junction, the V-like slots upon/below the junctions are not suitable as a platform (substrate) for uniform epitaxial growth of the crystals, thus the intrinsic defects such as oxygen vacancies easily develop, resulting in the enhancement of the green light emission [51]. However, the further results of fluorescence microscopy suggest that the inhomogeneous PL emission of green light along the fiber stem, which is characterized by the periodic enhancement at the nearly isometric junctions, should be mainly attributed to the wave-guide property of the well-faceted fibers.

The produced well-faceted ZnO fibers with periodic junctions preferentially grow along $[01\bar{1}0]$ as opposed to the $[2\bar{1}\bar{1}0]$ direction by catalyzing growth. These fibers usually grow broad roots, and the bottom surfaces are $(000\bar{1})$ (Fig. 11a). When the fibers were dispersed onto the quartz substrates by drop-casting, most of the fibers attach to the substrate with broad {0001} facets (Fig. 11c and 11e), the fragments without broad roots (Fig. 11b) attach with $\{2\bar{1}\bar{1}0\}$ facets (Fig. 11d and 11f).

These natural junction-prism arrays as well as the well-faceted surfaces associated with the transparent and homogeneous nature of crystalline ZnO medium offer sharp interfaces between ZnO and air (or other media) for guiding the propagation of light effectively. The optical morphology of the fiber shown in a typical barcode-like black-bright contrast (Fig. 11d) was imaged with a transmission optical microscope. Note in the experimental setup, the parallel light used to illuminate the sample in the microscope came from a lamp underlying the sample, while the camera was located atop the sample. The dark contrast regions correspond to the junction-prisms, while the bright contrast regions correspond to the building blocks of the fiber, which are separated by the junctions. This typical optical phenomenon suggests that refraction and reflection are strongly modulated by the junction-prism arrays within this structural fiber. When parallel light propagates perpendicular to the boundary between the ZnO crystal ($n_{ZnO} \approx 2$) and air medium ($n_{air} \approx 1$), it splits into two parts: light transmitted into ZnO and the light reflected back into air. Considering the reflection and transmission coefficients of ZnO crystal, about 88 % of the incident light was refracted. Moreover, because the ZnO crystal is optically denser than air, no light should enter the air from the V-shape surfaces of the upper junction-prism, where the entire incident light was reflected back due to total reflection.

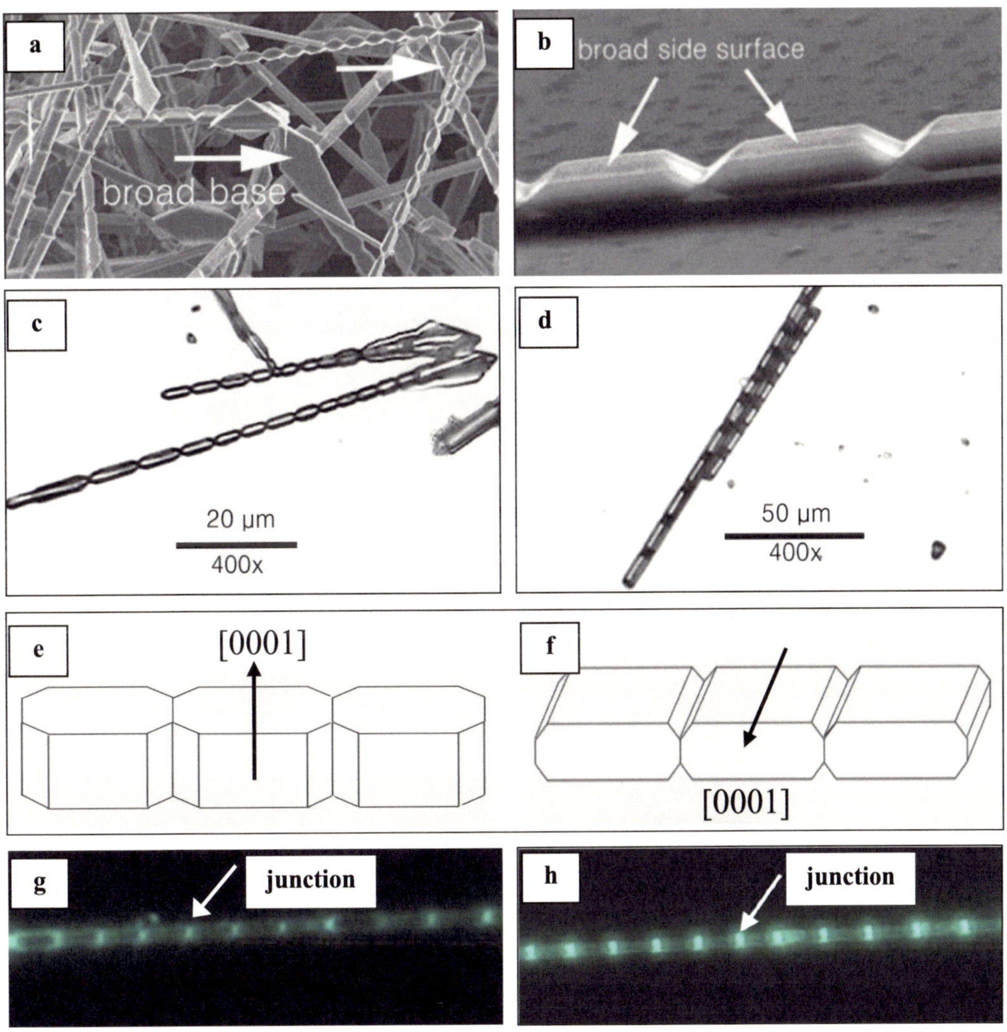

Fig. 11. Most of the fibers attach to the substrate with broad {0001} facets (a, c, e and g), other fragments without broad roots attach with $\{2\bar{1}\bar{1}0\}$ facets (b, d, f and h).

When the fibers were excited by UV light (wavelength: 325-380 nm) with a fluorescence microscope, it is interestingly found that the enhanced green light emits the periodic junctions (Fig. 11h). This result could be explained by the optical waveguide behavior of the well-faceted structure with the junction-prism arrays of ZnO. As to the side surfaces of single building blocks of a fiber, every two parallel broad $\{2\bar{1}\bar{1}0\}$ surfaces and two narrow {0001} surfaces could serve as a natural square cavity/waveguides (Fig. 12a). In general, the Vo* centers contributing to the defect-related green emission should be present at the surface region of a given ZnO crystal [50]. An ideal model elucidates the featured enhancement of the green light emitting at the junction-prisms. Analyzing one of these Vo* centers, its emitting light is easily reflected by the two $\{2\bar{1}\bar{1}0\}$ surfaces along the z-axis (i.e., $[01\bar{1}0]$). Note Fig. 12b-1, if the fiber is uniform and cuboid in shape, it should be an ideal bar-like waveguide and the emitting light from the total-reflection widows would be sent out from the ends of the fiber due to total reflection. In this case, it can be considered as an ideal optical fiber. However, the junction-prism arrays of the present fibers destroy the total-reflection condition (Fig. 12b-2). Thus, the junction-prism arrays change the propagation paths of the emitted light and most of the light from the total-reflection windows is guided out of the junction-prism regions directly, resulting in enhanced illumination at the junctions. Moreover, even if the emitted light goes straight though a junction-prism, it would encounter the next junction, and be sent out at last. The thicker the junction of a fiber, the more easily light is arrested by the junction-prism (Fig. 12b-3). All these observations show that the periodic junction-prisms, which provide emitting windows for intrinsic emissions, naturally tune the guided light in the well-faceted fibers.

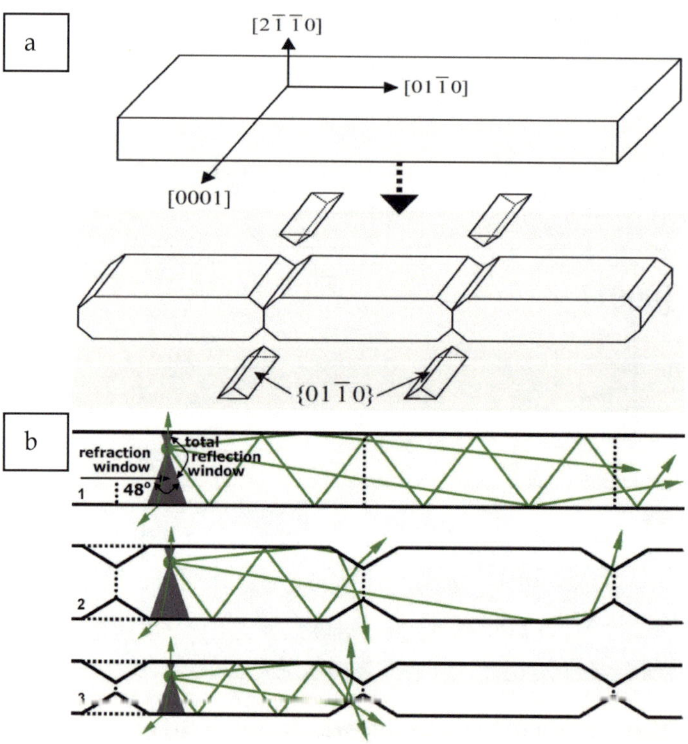

Fig. 12. Schematic illustrations of light reflection at the surface of the junction-prism structure.

3 SnO_2 zigzag shaped meso-structures

3.1 Controlled growth

SnO_2 has been paid attention in a variety of applications in chemical, optical, electronic and mechanical fields, due to its unique high conductivity, chemical stability, gas sensitivity and semiconducting properties [52]. Many syntheses of SnO_2 with different morphologies, such as nano-scale belt, wire, disk and dendrite, have been reported [53-56]. Herein, we report on a kinetics-controlled method to realize selective growth of SnO_2 unconventional zigzag shaped fibers. The morphological evolution process was investigated via SEM and TEM. Previously, the method used to grow SnO_2 single crystals is the high temperature gas phase reaction of evaporating SnO_2 or SnO to lead to SnO in the gas phase, and subsequent re-oxidation [57]. Here we used a lower temperature decomposition of SnO solid powders to produce Sn vapor for deposition, and then to oxidize it to SnO_2. In order to selected deposition of structured products, the growth kinetics was controlled [58].

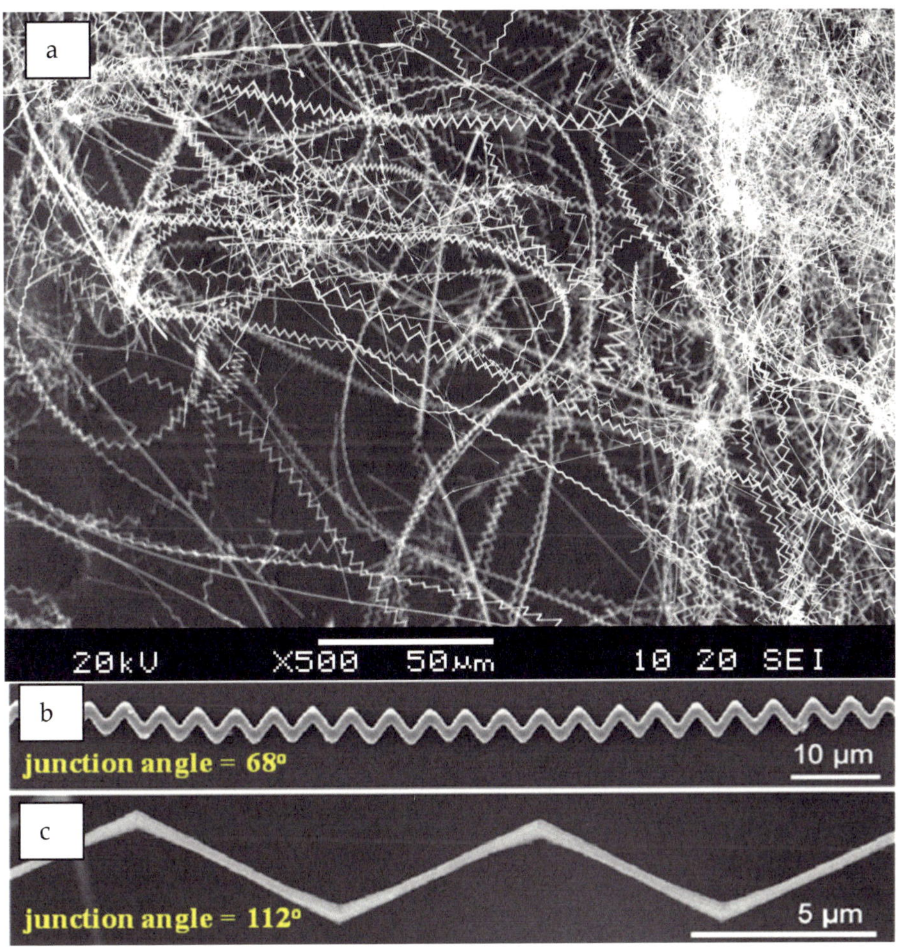

Fig. 13. (a) SEM image of the high yield SnO_2 zigzag fibers, the zigzags with junction angles of about 68° (b) and 112° (c), respectively.

3.2 Structural characterization and crystal models

The SEM image shows the typical growth of zigzag fibers as that shown in Fig. 13a. XRD results show the both structured products are with same crystallography structure: tetragonal rutile SnO_2. The high yield zigzags extend very long and collide with each other. The typical space of one zigzag period ranges from 2 to 10 μm, and the transverse swing is in the range of 5 to 10 μm. The length of the zigzag increases with the growth time, sometimes it can be up to several millimeters. In addition, there are more than three types of the angles of the zigzag junctions. Most of them are about 68° (Fig. 13b), and a few are approximately 112° (Fig. 13c), 90° and 124°, respectively.

The TEM images (Fig. 14a, 14b) give insight of a zigzag angle of 68°. Electron diffractions on the entire junction and on the two blocks reveal that the zigzag is single crystal. The growth directions of the two blocks are parallel to the crystallographic equivalent directions of [101] and $[10\bar{1}]$, respectively. High-resolution TEM images (Fig. 14c, 14d) indicate the entire fiber has same lattice arrangements. The structural models are illustrated in Fig. 14b. Structurally, the ±[101] and $\pm[10\bar{1}]$ in tetragonal SnO_2 are equivalent directions. The angle between the [101] and $[10\bar{1}]$ directions and that between the $[\bar{1}0\bar{1}]$ and $[\bar{1}01]$ is 68°, while the angle between the [101] and $[\bar{1}01]$ directions and that between the $[\bar{1}0\bar{1}]$ and $[10\bar{1}]$ is 112°. The experimental results of about 68° and 112° correlate well with these values. The formation of a zigzag is mainly accomplished though repeated alternation of growth orientations between the ±[101] and $\pm[10\bar{1}]$. Namely, a zigzag could be separated into two types of building blocks, which laterally combine each other periodically. The zigzags with other junction angles should repeatedly shift its growth directions along some other low-index directions, such as from [101] to [001].

3.3 Morphological evolution

Careful examination of the zigzag's morphologies gives insight into the growth habits. As that shown in Fig. 15a, we usually found thin fiber has narrow slab-like morphology with sharp junction corners. The top/down surfaces of the building blocks could be indexed as ±(010) planes, and the side surfaces are $\pm(10\bar{1})$ and ±(101), alternately. After further growth, the morphologies of the sample would become well-faceted with some new ±(100) facets present opposite to the junctions (Fig. 15b, 15c). Although the states are of different fibers, it can be presumed that each fiber undergoes a similar sequence of steps during the morphological evolution. The zigzag fiber would be formed by a two-step growth process. The first step is to fast grow to finalize the zigzag frame; the second step is to laterally grow to thicken its diameter. The evolution process illustrated in Fig. 15d reflects the lateral thickening process. In the beginning, the vapor species favor deposition at the V-like slots and it results in some atom steps (Fig. 15e), and then the new arrived species continuously arrange at the steps parallel to the side surface. The epitaxies would cease once the arrange layers meet the ridges of the junctions, due to the higher energy there. At last, these homo-epitaxies equally thicken a fiber in width and some ±(100) facets are constructed at the same time. Note that the transverse swing of the fiber does not change all along, and the original zigzag frame decides the final frame ($A_0 = A_f$). The longer the growth time, the more (100) surface area is present. Ideally, the final morphology could be predicted to be a rectangular crystal bar with long axis parallel to [001] direction and enclosed by lower energy planes of ±(010) and ±(100). This evolution tents to minimize the surface free energy, so the growth should seek thermodynamical equilibrium and be mainly dominated by surface free energy.

Based on this argument, we can explain why fewer zigzags with 112° angels contrast against the zigzags with 68 ° angels in the products. If the zigzag growth changes from [101] to [10$\bar{1}$] periodically, some higher energy facets of ±(001) would be constructed after the lateral thickening process. Thus it is not favorable from the energy point of view.

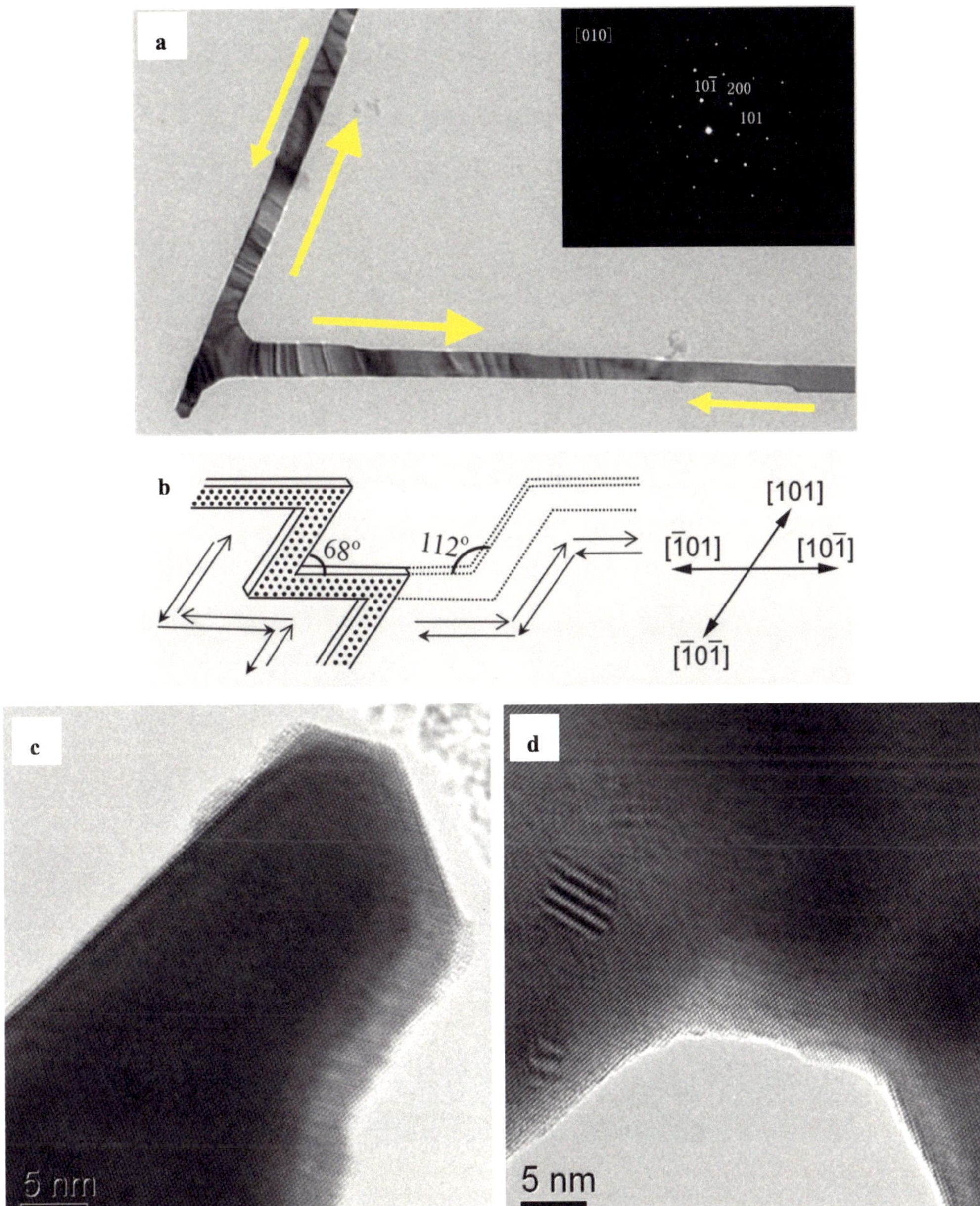

Fig. 14. (a) TEM image showing the junction angles of about 68°, the electron diffractions on the two blocks showing single crystal nature and <101> orientations, (b) crystal structural models for the zigzags, (c, d) high-resolution TEM images taken on the head and V-shape areas.

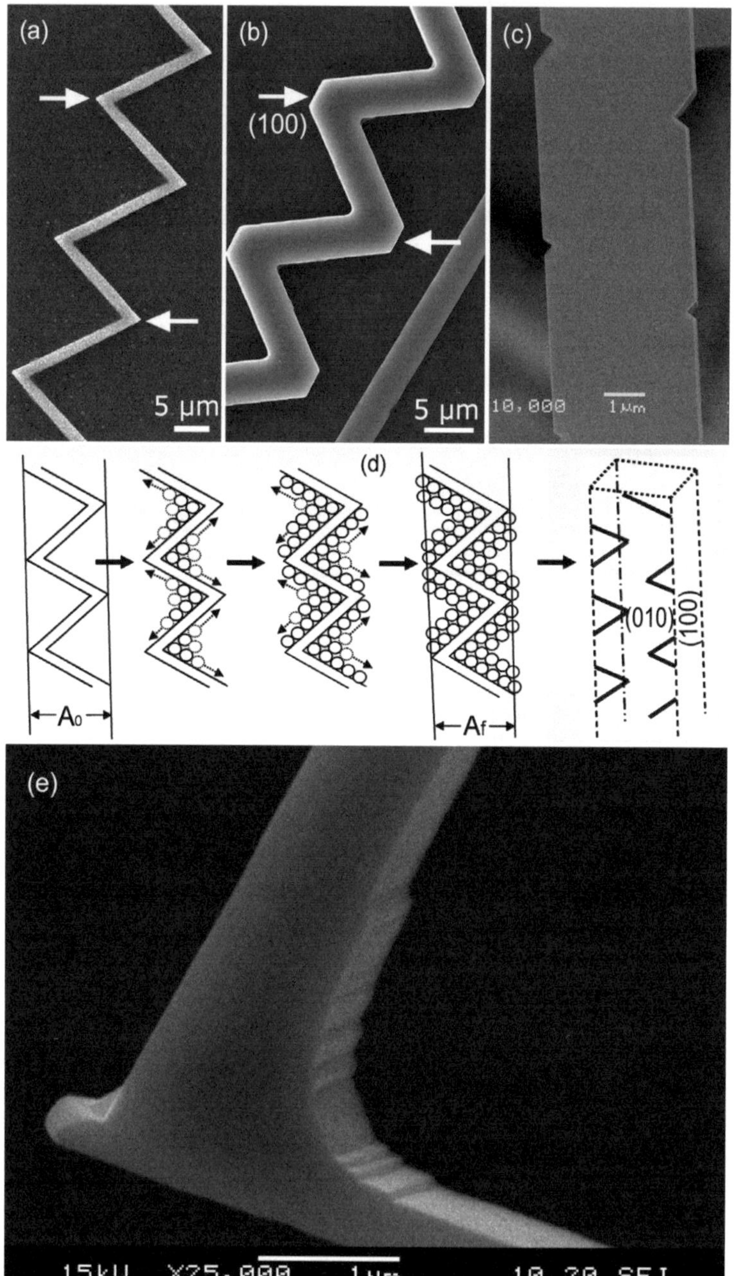

Fig. 15. SEM images of zigzags arranged (a-c) in order to show the evolution process, in (a) the zigzag has narrow slab-like morphology and sharp corners, after epitaxial growth, in (b), (c), the zigzag shows increased width and new ± (100) facets present, schematic illustration for the morphological evolution (d), the vapor species favor deposition at the V-like slots and it results in some atom steps.

4. Concluding remarks

The directional growth of well-faceted ZnO microfibers along different axes could be realized by catalyzing growth. The characterization of the fibers by optical and photoluminescence microscopy showed that the outer facets of the crystalline fibers provide excellent mirror-like surfaces for guiding light propagation along the fiber stem as well as the periodic junction-prisms. The structure-related optical properties of the fibers can be fully explained by a micro-structural model. The model explains several optical properties, such as luminance decreasing at the junction-prisms caused by refraction and total or partial reflection of light, as well as luminance enhancement at the junction-prisms related to waveguiding of the green emission along the ZnO fibers. Further integration of the ZnO junction-prisms into micro-devices could provide micro-scale modulation for light with different wavelengths. Such capability makes such fibers potentially suitable for enhanced light-illumination arrays. Reproducibly high-yield growth of SnO_2 zigzag nanofibers was achieved via controlling the reactant vapor concentration. The formation of the zigzag fiber based on the pre-growing nanobelt is suggested to be in a two-step process: the first is frame growth, which is accomplished through repeated orienting along equivalent directions; the second is lateral epitaxy, which thickens a fiber and results in well-faceted morphology. It is note that the present of intrinsic equivalent directions and the oscillation of external growth kinetics are key roles for producing zigzag structures. The elucidation of the growth mechanism should provide a fully controlled route for reproducibly high-yield growth of zigzag fibers of SnO_2 and give some valuable hints to synthesis other zigzag fibers. This well-faceted zigzag fiber could be studied as optical waveguide in its periodic structure and gas sensor component. These results have shown that the optimization of surfaces, boundaries, and interfaces in 1D meso-scale materials with well-faceted structures plays an important role in furthering the application of these materials.

5. Acknowledgement

We thank Dr. L. Pu, Dr. M. K. Lee, Prof. C. Tien and Prof. L. J. Chang for helpful discussions. L. S. Huang is truly grateful for financial support from NSFC (No. 60606020) and NCKU's "Aim for the Top University Project".

6. References

[1] A. P. Alivisatos, "Semiconductor clusters, nanocrystals, and quantum dots" Science 271, 933-937 (1996).

[2] A. D. Yoffe, "Semiconductor quantum dots and related systems-electronic, optical, luminescence and related properties of low dimensional systems" Advances In Physics 50, 1-208 (2001).

[3] Z. L. Wang, "ZnO nanowire and nanobelt platform for nanotechnology" Mater. Sci. Eng. R. 64, 33-71 (2009).

[4] H. Joyce, Q. Gao, H. Tan, Jagadish, C. "Phase perfection in zinc blende and wurtzite III-V nanowires using basic growth parameters" Nano Lett.10, 908- 915 (2010).

[5] B. M. Wong, F. Lonard, Q. M. Li, T. George "Nanoscale Effects on Heterojunction Electron Gases in GaN/AlGaN Core/Shell Nanowires" Nano Lett. 11, 3074-3079 (2011).

[6] X. F. Qiu, C. Burda, R. L. Fu, L. Pu, H. Y. Chen, J. J. Zhu, "Heterostructured Bi_2Se_3 nanowires with periodic phase boundaries" J. Am. Chem. Soc. 126, 16276-16277 (2004).
[7] L. Manna, D. J. Milliron, A. Meisel, E. C. Scher, A. P. Alivisatos, "Controlled growth of tetrapod-branched inorganic nanocrystals" Nat. Mater. 2, 382-385 (2003).
[8] L. S. Huang, S. Wright, S. Yang, D. Z. Shen, B. X. Gu, Y. W. Du, "ZnO well-faceted fibers with periodic junctions" J. Phys. Chem. B. 108, 19901-19903 (2004).
[9] M. H. Huang, S. Mao, H. Feick, H. Yan, Y. Wu, H. Kind, E. Weber, R. Russo, P. Yang, "Room-temperature ultraviolet nanowire nanolasers" Science 292, 1897-1899 (2001).
[10] J. Johnson, H. J. Choi, K. P. Knutsen, R. D. Schaller, P. Yang, R. J. Saykally, "Single gallium nitride nanowire lasers" Nat. Mater. 1, 106-110 (2002).
[11] M. Law, D. Sirbuly, J. Johnson, J. Goldberger, R. Saykally, P. Yang, "Ultralong nanoribbon waveguides for sub-wavelength photonics integration" Science 305,1269-1271 (2004).
[12] C. J. Barrelet, A. B. Greytak, C. M. Lieber, "Nanowire photonic circuit elements," Nano Lett. 4, 1981-1985 (2004).
[13] X. G. Peng, L. Manna, W. D. Yang, J. Wickham, E. C. Scher, A. Kadavanich, A. P. Alivisatos, "Shape control of CdSe nanocrystals," Nature 404, 59-61 (2000).
[14] R. Jin, Y. C. Cao, E. Hao, G. S. Métraux, G. C. Schatz, C. A. Mirkin, "Controlling anisotropic nanoparticle growth through plasmon excitation," Nature 425, 487-490 (2003).
[15] Z. R. Tian, J. Voigt, J. Liu, B. Mckenzie, M. McDermott, M. Rodriguez, H. Konishi, H. F. Xu, "Complex and oriented ZnO nanostructures," Nat. Mater. 2, 821-826 (2003).
[16] T. Schlli, R. Daudin, G. Renaud, A. Vaysset, O. Geaymond, A. Pasturel, "Substrate enhanced supercooling in AuSi eutectic droplets" Nature 464, 1174- 1177 (2010).
[17] M. Tabuchi, A. M. Y. Ohtake, Y. Takeda, "X-ray CTR scattering measurement to investigate the formation process of InP/GaInAs interface" J. Phys.: Conf. Ser. 83, 012031 (2007).
[18] M. Verheijen, R. Algra, M. Borgstrm, G. Immink, E. Sourty, W. van Enckevort, E. Vlieg, E. Bakkers, "Three dimensional morphology of GaP-GaAs nanowires revealed by transmission electron microscopy tomography" Nano Lett. 7, 3051–3055 (2007).
[19] D. E. Perea, N. Li, R. M. Dickerson, A. Misra, S. T. Picraux, "Controlling Heterojunction Abruptness in VLS-Grown Semiconductor Nanowires via in situ Catalyst Alloying" Nano Lett. 11, 3117–3122 (2011).
[20] O. Shpyrko, R. Streitel, V. Balagurusamy, A. Grigoriev, M. Deutsch, B. Ocko, M. Meron, B. Lin, P. Pershan, "Surface crystallization in a liquid AuSi alloy" Science 313, 77-80 (2006).
[21] D. C. Look, "Recent advances in ZnO materials and devices" Mat. Sci. Eng. B. 80, 383-387 (2001).
[22] D. M. Bagnall, Y. F. Chen, Z. Zhu, S. Koyama, M. Y. Shen, T. Goto, "Optically pumped lasing of ZnO at room temperature" Appl. Phys. Lett. 70, 2230-2232 (1997).
[23] P. M. Martin, M. S. Good, J. W. Johnston, L. J. Bond, S. L. Crawford, "Piezoelectric films for 100-MHz ultrasonic transducers" Thin Solid Films 379, 253–258 (2000).
[24] J. Q. Xu, Q. Y. Pan, Y. A. Shun , Z. Z. Tian, "Grain size control and gas properties of ZnO gas sensors" Sensors Actuators B. 66, 277–279 (2000).

[25] M. H. Huang, S. Mao, H. Feick, H. Q. Yan, Y. Y. Wu, H. Kind, E. Weber, R. Russo, P. D. Yang, "Room-temperature ultraviolet nanowire nanolasers" Science 292, 1897–1899 (2001).
[26] J. H. Song, Y. Zhang, , C. Xu, W. Z. Wu , Z. L. Wang, "Polar Charges Induced Electric Hysteresis of ZnO Nano/Microwire for Fast Data Storage" Nano Lett. 11, 2829–2834 (2011).
[27] H. Kind, H. Yan, B. Messer, M. Law, P. Yang, "Nanowire ultraviolet photodetectors and optical switches" Adv. Mater. 14, 158–160 (2002).
[28] Y. Wu, H. Yan, P. Yang, "Semiconductor nanowire array: potential substrates for photocatalysis and photovoltaics" Top. Catal. 19, 197–202 (2002).
[29] G. Gordillo, "New materials used as optical window in thin film solar cells" Surf. Rev. Lett. 9, 1675–1680 (2002).
[30] W. I. Park, G. C. Yi, J. W. Kim, S. M. Park, "Schottky nanocontacts on ZnO nanorod arrays" Appl. Phys. Lett. 82, 4358–4360 (2003).
[31] A. F. Yu, H. Y. Li, H. Y. Tang, T. J. Liu, P. Jiang, Z. L. Wang, "Vertically integrated nanogenerator based on ZnO nanowire arrays" Physica Status Solidi. 5, 162-164 (2011).
[32] N. Audebrand, J. P. Auffredic, D. Louer, "X-ray diffraction study of the early stages of the growth of nanoscale zinc oxide crystallites obtained from thermal decomposition of four precursors" Chem. Mater. 10, 2450-2461 (1998).
[33] A. M. Morales, C. M. Lieber, "A laser ablation method for the synthesis of crystalline semiconductor nanowires" Science 279, 208–211 (1998).
[34] Z. W. Pan, Z. R. Dai, Z. L. Wang, "Nanobelts of semiconducting oxides" Science 291, 1947-1949 (2001).
[35] M. H. Huang, Y. Y. Wu, H. Feick, N. Tran, E. Weber, P. D. Yang, "Catalytic growth of zinc oxide nanowires by vapor transport" Adv. Mater. 13, 113–116 (2001).
[36] Y. G. Wei, W. Z. Wu, R. D. Guo, J. Yuan, D. Suman, Z. L. Wang, "Wafer-Scale High-Throughput Ordered Growth of Vertically Aligned ZnO Nanowire Arrays" Nano Lett. 11, 2829–2834 (2011).
[37] G. D. Yuan, W. J. Zhang, S. T. Lee "p-Type ZnO Nanowire Arrays" Nano Lett. 8, 2591–2597 (2008).
[38] Y. Li, D. W. Meng, L. D. Zhang, F. Phillip, "Ordered semiconductor ZnO nanowire arrays and their photoluminescence properties" Appl. Phys. Lett. 76, 2011-2013 (2000).
[39] D. S. Boyle, K. Govender, P. O'Brien, "Novel low temperature solution deposition of perpendicularly orientated rods of ZnO: substrate effects and evidence of the importance of counter-ions in the control of crystallite growth" Chem. Commun. 1, 80-81 (2002).
[40] L. Vayssieres, K. Keis, S. Lindquist, A. Hagfeldt, "Purpose-built anisotropic metal oxide material: 3D highly oriented microrod array of ZnO" J. Phys. Chem. B 105, 3350-3352 (2001).
[41] L. Vayssieres, "Growth of arrayed nanorods and nanowires of ZnO from aqueous solutions" Adv. Mater. 15, 464–466 (2003).
[42] L. Guo, S. H. Yang, C. L. Yang, J. N. Wang, W. K. Ge, "Synthesis and Characterization of Poly (vinylpyrrolidone)-Modified Zinc Oxide Nanoparticles" Chem. Mater. 12, 2268-2274, (2000).

[43] Z. R. Tian, J. A. Voigt, J. Liu, B. Mckenzie, M. J. Mcdermott, "Biomimetic arrays of oriented helical ZnO nanorods and columns" J. Am. Chem. Soc. 124, 12954–12955 (2002).
[44] L. E. Greene, M. Law, D. H. Tan, P. D. Yang "General Route to Vertical ZnO Nanowire Arrays Using Textured ZnO Seeds" Nano Lett. 5, 1231–1236 (2005).
[45] X. Y. Kong, Y. Ding, R. S. Yang, Z. L. Wang, "Single-crystal nanorings formed by epitaxial self-coiling of polar nanobelts" Science 303, 1348-1351 (2004).
[46] Y. J. Su, Y. J. Zhou, L. S. Huang, Y. F. Liu, S. Z. Shi, Y. N. Lv, "The morphology evolution mechanism of ZnO quasi-one-dimensional nanostructures" J. Inor. Mater. 24, 6 (2009).
[47] C. A. Arguello, D. L. Rousseau, S. P. S. Porto, "First-Order Raman Effect in Wurtzite-Type Crystals" Phys. Rev. 181, 1351-1363 (1969).
[48] T. C. Damen, S. P. S. Porto, B. Tell, "Raman Effect in Zinc Oxide" Phys. Rev. 142, 570-574 (1966).
[49] L. E. Greene, M. Law, J. Goldberger, F. Kim, J. C. Johnson, Y. F. Zhang, R. J. Saykally, P. D. Yang, "Low-temperature wafer-scale production of ZnO nanowire arrays" Angew. Chem. Int. Ed. 42, 3031-3034 (2003).
[50] A.Van Dijken, E. A. Meulenkamp, D.Vanmaekelbergh, A. Meijerink, "Identification of the transition responsible for the visible emission in ZnO using quantum size effects" J. Lumin. 90, 123-128 (2000).
[51] L. S. Huang, L. Pu, Y. Shi, R. Zhang, B. X. Gu, Y. W. Du, Y. D. Zheng, "Light propagation tuned by periodic junction-prisms within well-faceted ZnO fibers" Opt. Express, 13, 5263-5269, (2005).
[52] Z. M. Jarzebski, J. P. Marton, "Physical properties of SnO2 materials" J. Electrochem. Soc. 123, 199c-205c (1976).
[53] Z. R. Dai, Z. W. Pan, Z. L. Wang, "Ultra-long single crystalline nanoribbons of tin oxide" Solid State Commun. 118, 351-354 (2001).
[54] C. Y. Wen, J. Tersoff, K. Hillerich, M. C. Reuter, J. H. Park, S. Kodambaka, E. A. Stach1,and F. M. Ross "Periodically Changing Morphology of the Growth Interface in Si, Ge, and GaP Nanowires" Phys. Rev. Lett. 107, 025503 (2011).
[55] J. Q. Hu, Y. Bando, D. Golberg, "Self-catalyst growth and optical properties of novel SnO2 fishbone-like nanoribbons" Chem. Phys. Lett. 372, 758-762 (2003).
[56] A. Beltran, J. Andres, E. Longo, E. R. Leite, "Thermodynamic argument about SnO2 nanoribbon growth" Appl. Phys. Lett. 83, 635-637 (2003).
[57] B. Thiel, R. Helbig, " Growth of SnO2 single crystals by a vapour phase reaction method" J. Crystal Growth 32, 259-264 (1976).
[58] L. S. Huang, L. Pu, Y. Shi, R. Zhang, B. X. Gu, Y. W. Du, S. Wright, "Controlled growth of well-faceted zigzag tin oxide mesostructures" Appl. Phys. Lett., 87, 163124 (2005).

17

Green Synthesis of Nanocrystals and Nanocomposites

Mallikarjuna N. Nadagouda
Water Supply and Water Resources Division,
National Risk Management Research Laboratory,
U. S. Environmental Protection Agency, Ohio,
USA

1. Introduction

Metal nanomaterials have attracted considerable attention because of their unique magnetic, optical, electrical, and catalytic properties and their potential applications in nanoelectronics (1-5) as well as in various wet chemical synthesis methods (6-14). There is also great interest in synthesizing metal and semiconductor nanoparticles due to their extraordinary properties—properties which are different than when they are in bulk. Green chemistry principles are also regaining popularity for this type of synthesis (8, 15-25). Green chemistry is the design, development, and implementation of chemical products and processes to reduce or eliminate the use and generation of substances that are hazardous to human health and to the environment (25). An example of a greener application of metal nanoparticles is the use of silver and gold nanoparticles, produced from vegetable oil, that are being used in antibacterial paints (26).

Polymer-inorganic nanocomposites have also attracted a lot of attention recently due to their unique, size-dependent chemical and physical properties (26-30). In response to this, different methods of preparing novel nanocomposites with desired properties and functions have been developed (31-35). Such methods should produce materials in which the unique properties of the nanoparticles are preserved (30). One of the main approaches is the dispersion of the previously prepared nanocrystals in polymers. Another is the generation of nanocrystals in polymers *in situ*. In the latter approach, various nanocables, nanowires and nanoparticulates, generated *in situ*, have been reported (36-45).

2. Production of nanomaterials using greener methods

Three areas of opportunity to engage in green chemistry when synthesizing metal nanoparticles by the reduction of the corresponding metal ion salt solutions are: (i) choice of solvent, (ii) the reducing agent employed, and (iii) the capping agent (or dispersing agent). There has also been growing attention in identifying environmentally friendly materials that are multifunctional in this area. For example, the vitamin B_2 can function as both a reducing and a capping agent for Au and Pt metals (15). In addition to its high water solubility, biodegradability and low toxicity when compared to other reducing agents, such as sodium borohydride ($NaBH_4$), sodium citrate and hydroxylamine hydrochloride, and B_2 is the most widely used, behaviorally-active drug in the world. By using natural, available resources like B_2, it is possible to prepare nanospheres, nanowires, and nanorods by using solvents of varying densities. It is possible to make multiple shape nanostructures by altering the density of the solvents. This green approach can also be extended to silver and palladium noble nanostructures.

Similarly, vitamin B_1 has also been used as a reducing and a capping agent (46). The method is a one-pot method and is greener in nature. By this method, bulk quantities of nanoballs of aligned nanobelts as well as nanoplates of the noble metal palladium in water can be synthesized without the need of any external capping, surfactant agents, and/or large amounts of insoluble templates that have been commonly deployed.

Vitamin C has also been used to fabricate novel core-shell (Fe and Cu), metal (noble metals) nanocrystals. Transition metal salts such as Cu and Fe were reduced using ascorbic acid in solution, a benign, naturally-available antioxidant, and then the simultaneous addition of noble metal salts. This process resulted in the formation of a core-shell structure, depending on the core and shell material used for the preparation (21). Pt yielded a tennis ball-shaped structure, with a Cu core; whereas Pt and Au formed regular spherical nanoparticles. Au, Pt, and Pd formed cube-shaped structures with Fe as the core.

Another interesting route to the synthesis of dendritic Ag structures without the use of any reducing chemical is the transmetallic reaction between copper and silver. The copper–carbon substrate of a transmission electron microscopy (TEM) grid reacted with the aqueous silver nitrate solution within minutes to yield spectacular tree-like silver dendrites. This occurred without using any added capping or reducing reagents (47). These results demonstrate a facile, aqueous, room-temperature synthesis of a range of noble metal nano- and meso-structures (see Figures 1 and 2) that have widespread technological potential in the design and development of next-generation fuel cells, catalysts, and antimicrobial coatings.

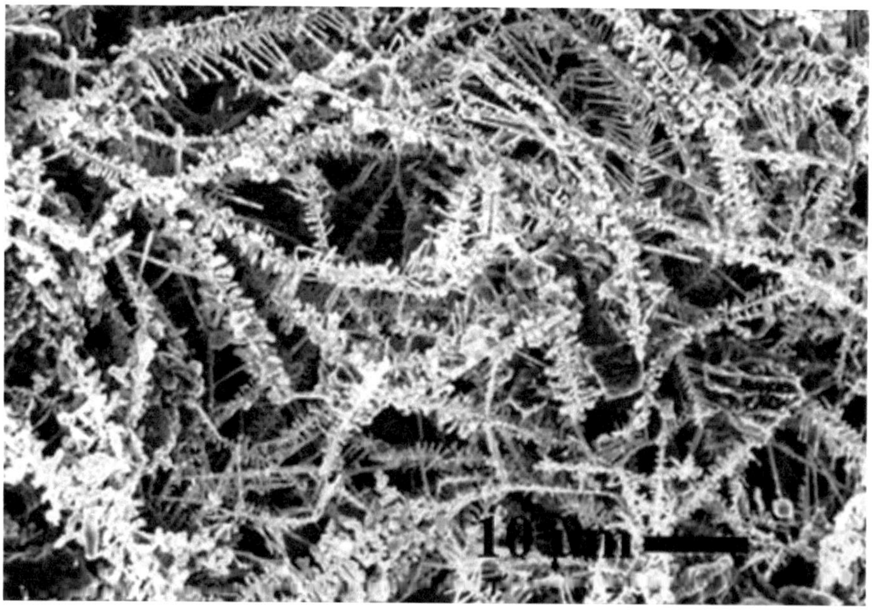

Fig. 1. Scanning electron microscopy image of silver dendrite, formed with copper shavings and activated carbon.

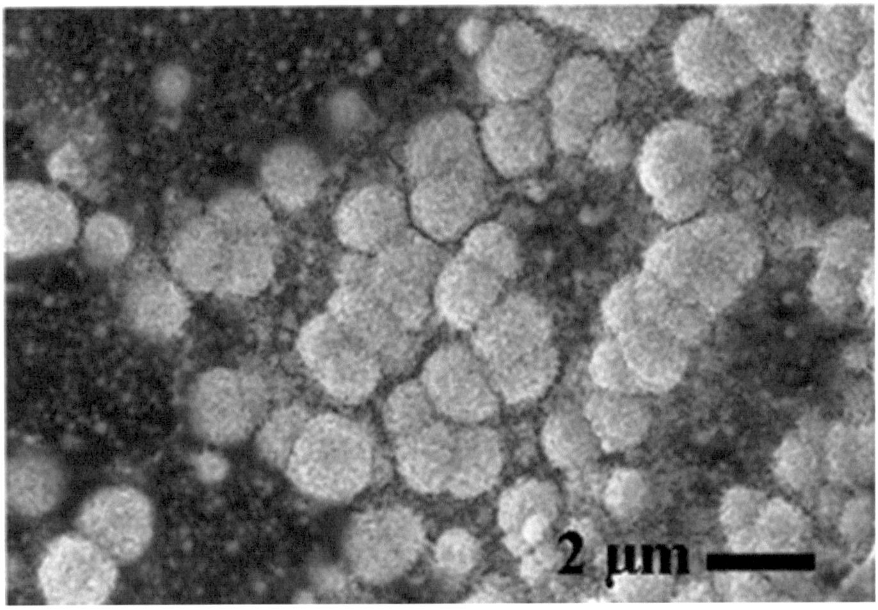

Fig. 2. Scanning electron microscopy image of spongy Pd, formed on a transmission electron microscopy copper grid.

Another material that was investigated in this study was green tea. Green tea has attracted significant attention recently, both in the scientific and consumer communities due to its health benefits for a variety of disorders, ranging from cancer to weight loss. This publicity has led to the increased consumption of green tea by both the general and the patient population, and the inclusion of green tea extract in several nutritional supplements, including multivitamin supplements. There are several polyphenolic catechins in green tea such as viz, (−) epicatechin (EC), (−) epicatechin-3-gallate (ECG), (−) epigallocatechin (EGC), (−) epigallocatechin-3-gallate (EGCG), (+) catechin, and (+) gallocatechin (GC). These compounds are strong antioxidants and hence, can reduce metals salts. One such example is the preparation of noble metals using tea/coffee extract (48). This one-pot method uses no surfactant, capping agent, and/or template. The size of the obtained nanoparticles ranges from 20–60 nm (see Figures 3 and 4) and are crystallized in face-centered cubic symmetry. This method is general and may be extended to other noble metals such as gold (Au) and platinum (Pt).

To prepare the coffee extract, 400 mg of coffee powder (Tata Bru coffee powder 99%) was dissolved in 50 mL of water. Then, 2 ml of 0.1N$AgNO_3$ ($AgNO_3$, Aldrich, 99%) was mixed with 10 ml of coffee extract and shaken to ensure thorough mixing. The 40 reaction mixture was allowed to settle at room temperature. For the tea extract, 1 g of tea powder (Red label from Tata, India Ltd. 99%) was boiled in 50 ml of water and filtered through a 25 μl Teflon filter. A similar procedure was repeated for Pd Q4 nanoparticles (using 0.1 N PdCl2, Aldrich, 99%). To evaluate 45, the source (tea and coffee extract) effect on morphology of the Ag and Pd nanoparticles was prepared and several experiments were performed using the above described procedure using the sources as shown in Table 1.

Sl No.	Item Brand Names
1	Sanka coffee
2	Bigelow tea
3	Luzianne tea
4	Starbucks coffee
5	Folgers coffee
6	Lipton tea

Table 1. Various brands of tea/coffee used to generate nanoparticles.

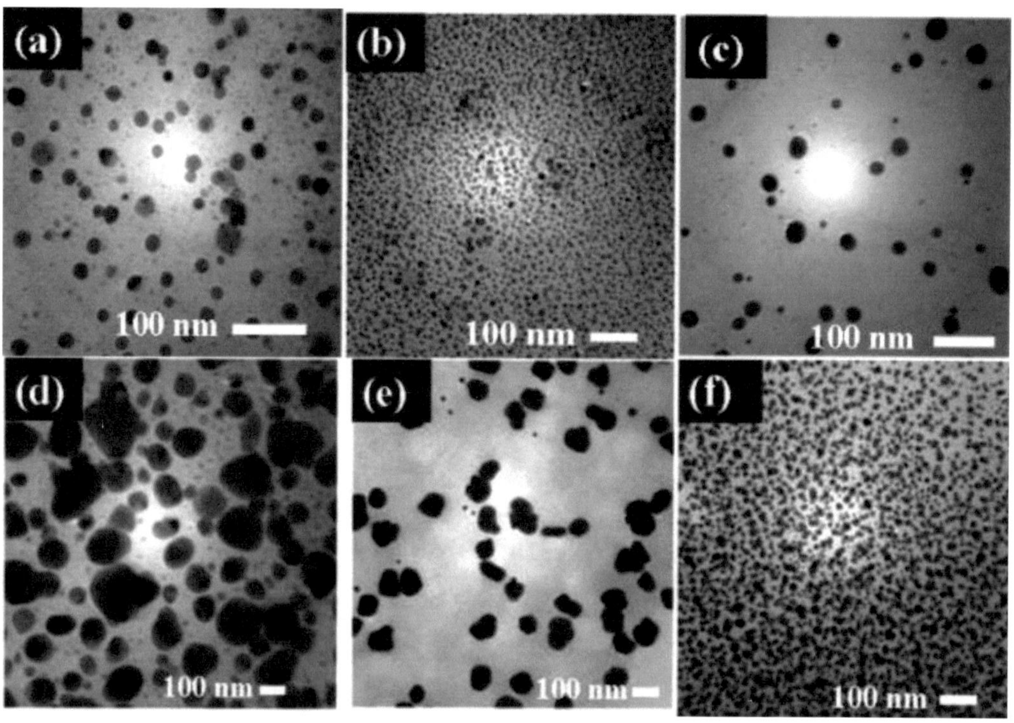

Fig. 3. TEM image of silver nanoparticles, synthesized using (a) Bigelow tea, (b) Folgers coffee, (c) Lipton tea, (d) Luzianne tea, (e) Sanka coffee, and (f) Starbucks coffee extract at room temperature. The process involved one step and did not use any hazardous reducing chemicals or non-degradable capping agents.

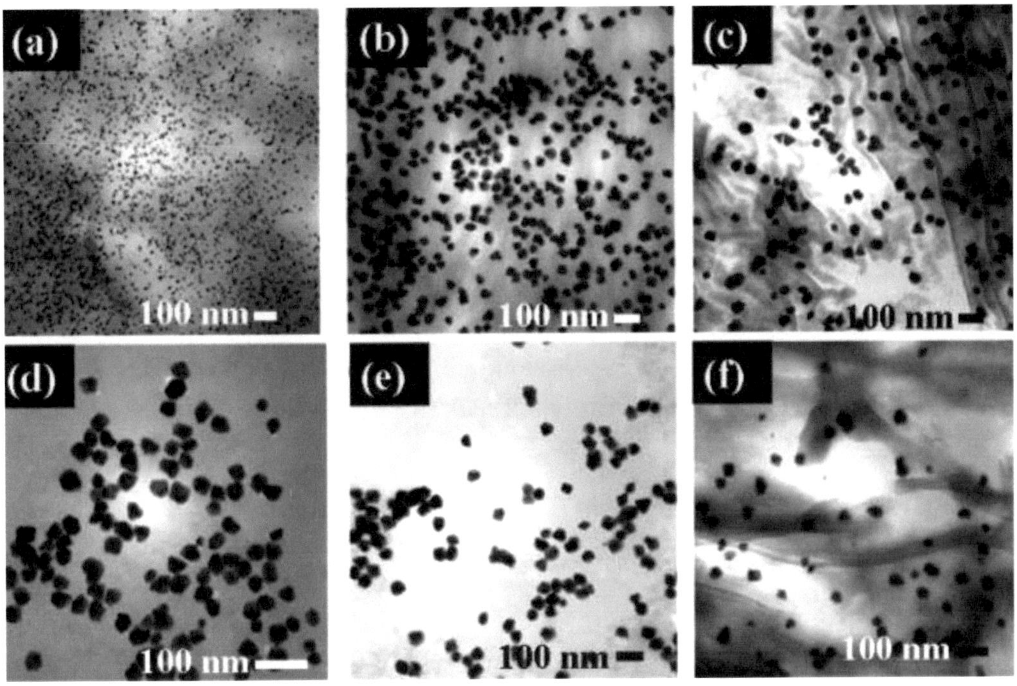

Fig. 4. TEM image of palladium nanoparticles, synthesized using (a) Sanka coffee, (b) Bigelow tea, (c) Luzianne tea, (d) Starbucks coffee, (e) Folgers coffee and (f) Lipton tea extract at room temperature. The process involved one step and did not use any hazardous reducing chemicals or non-degradable capping agents.

Apart from our work with noble nanometals, we have developed a greener, more straight forward, single-step approach for the synthesis of bulk quantities of nanofibers of the electronic polymer, fully-reduced polyaniline (leucoemarldine) without using any reducing agents, surfactants, and/or large amounts of insoluble templates. The nanofibers undergo a spontaneous redox reaction with noble metal ions under mild aqueous conditions, resulting in deposition of various shapes such as leaves, particulates, nanowires, and cauliflower for Ag, Pd, Au, and Pt, respectively. Thus, this approach affords a facile entry into this technologically important class of metal-polymer nanocomposites (49).

3. Microwave assisted synthesis of noble nanostructures and composites

Microwaves play an important role in green chemistry. The use of microwaves can reduce energy consumption and the time used to obtain desired materials. Over the past couple of years, microwave (MW) chemistry has moved from a laboratory curiosity to a well-established, synthetic technique used in many academic and industrial laboratories around the world. Even though the overwhelming number of MW-assisted applications used today are still performed on a laboratory scale, it expected that this technology may be used on a larger, perhaps even production-size, scale in conjunction with radio frequency or conventional heating. Microwave chemistry is based on two main principles: the dipolar mechanism and the electrical conductor mechanism.

The dipolar mechanism occurs when, under a very high frequency electric field, a polar molecule attempts to follow the field in the same alignment. When this happens, the molecules release enough heat to drive the reaction forward. In the later mechanism, the irradiated sample is an electrical conductor and the charge carriers, ions and electrons, move through the material under the influence of the electric field and lead to polarization within the sample. These induced currents and any electrical resistance will heat the sample.

Microwave heating has received considerable attention as a promising new method for the one-pot synthesis of metallic nanostructures in solutions. Because of this, the microwave-assisted synthetic approach for producing silver nanostructures has recently been reviewed. In the review process, researchers have successfully demonstrated the application of this method in the preparation of silver (Ag), gold (Au), platinum (Pt), and palladium (Pd) nanostructures. MW heating conditions allow not only for the preparation of spherical nanoparticles within a few minutes, but also for the formation of single crystalline polygonal plates, sheets, rods, wires, tubes, and dendrites. The morphologies and sizes of the nanostructures can be controlled by changing different experimental parameters, such as the concentration of metallic salt precursors, the surfactant polymers, the chain length of the surfactant polymers, the solvents, and the operation reaction temperature. In general, nanostructures with smaller sizes, narrower size distributions, and a higher degree of crystallization have been obtained more consistently via MW heating than by heating with a conventional oil-bath.

The use of microwaves to heat samples is a practical boulevard for the greener synthesis of nanomaterials (50) and provides many desirable features, such as shorter reaction times, reduced energy consumption, and better product yields. For example, Kundu et al. (51) have synthesized electrically conductive gold nanowires within 2-3 min using DNA as a reducing and nonspecific capping agent using a MW irradiation method. Similarly, uniform and stable polymer-stabilized colloidal clusters of Pt, Ir, Rh, Pd, Au, and Ru have been synthesized by MW irradiation with a modified domestic MW oven (52). The resulting colloidal clusters have small average diameters and narrow size distributions. Further, polychrome silver nanoparticles have been prepared using a soft solution approach under MW irradiation from a solution of silver nitrate ($AgNO_3$) in the presence of poly (N-vinyl-2-pyrrolidone) without any other reducing agent. Different morphologies of silver colloids with attractive colors could be obtained using different solvents as the reaction medium (53).

The MW method can find diversified applications; for example, bulk quantities of nanocarbons with pre-selected morphology can be synthesized in a simple and rapid MW heating approach directly from conducting polymers (54). On the same grounds, the successful preparation of highly active and dispersed metal nanoparticles on a mesoporous material has been accomplished in a conventional MW oven using an eco-friendly protocol in which ethanol and acetone–water were employed as both solvents and reducing agents. The materials exhibited different particle sizes, depending on the metal and the time of MW irradiation and the ensuing nanoparticles were found to be very active and selective in the oxidation of styrene (55).

Recently, Nadagouda et al. (23) have accomplished bulk syntheses of Ag and Fe nanorods using polyethylene glycol (PEG) under MW irradiation conditions. Due to tremendous increases in the biological applications of these nanostructures, there is a continued interest in using biodegrable polymers or surfactants to cap these nanoparticles in order to prevent their aggregation. Most of these biodegradable polymers or surfactants have the tendency to be soluble in water and it is of great interest to know that good dispersion or capping can be obtained using these biodegradable polymers or surfactants. The PEG was chosen as a reducing agent and stabilizing agent for several reasons. First, PEG is biodegradable (as well as non-toxic) and has high water solubility at room temperature, unlike other polymers. It can also form complexes with metal ions and, thereafter, reduce to metals. Finally, it contains alcoholic groups that were exploited for the reduction and the stabilization of the nanoparticles. Favorable conditions to make Ag nanorods were established and the process was expanded to make Fe nanorods with uniform size and shape. The nanorods' formation depended upon the concentration of PEG used in the reaction with Ag salt (see schematic diagram 1). Ag and Fe nanorods crystallized in face-centered cubic symmetry. In a typical procedure, aqueous silver nitrate ($AgNO_3$) solution (0.1 M) and different molar ratios of PEG (molecular weight 300) were mixed in a 10 mL test tube at room temperature to form a clear solution. The reaction mixture was irradiated in a CEM Discover focused MW synthesis system maintaining a temperature of 100 °C (monitored by a built-in infrared sensor) for 1 h with a maximum pressure of 280 psi. The resulting precipitated Ag nanorods (see Figures 5-7) were then washed several times with water to remove excess PEG.

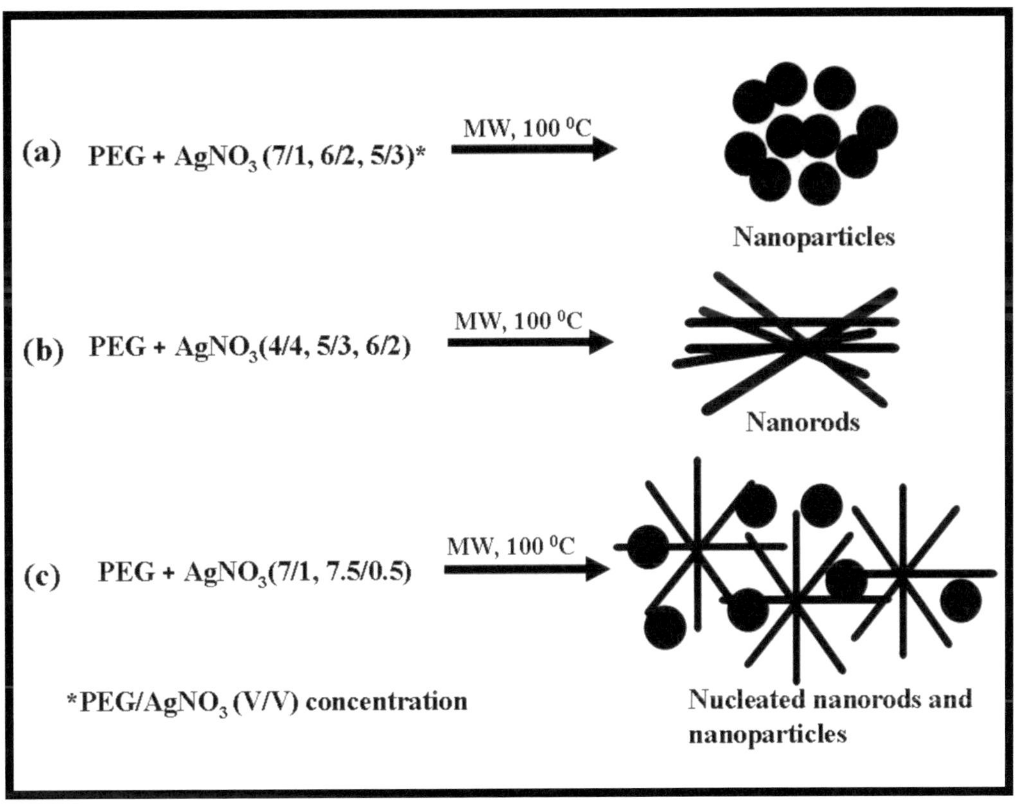

Scheme 1. Schematic illustrations of experimental mechanisms that generated Ag (a) nanoparticles, (b) nanorods, and (c) nucleated nanorods and nanoparticles.

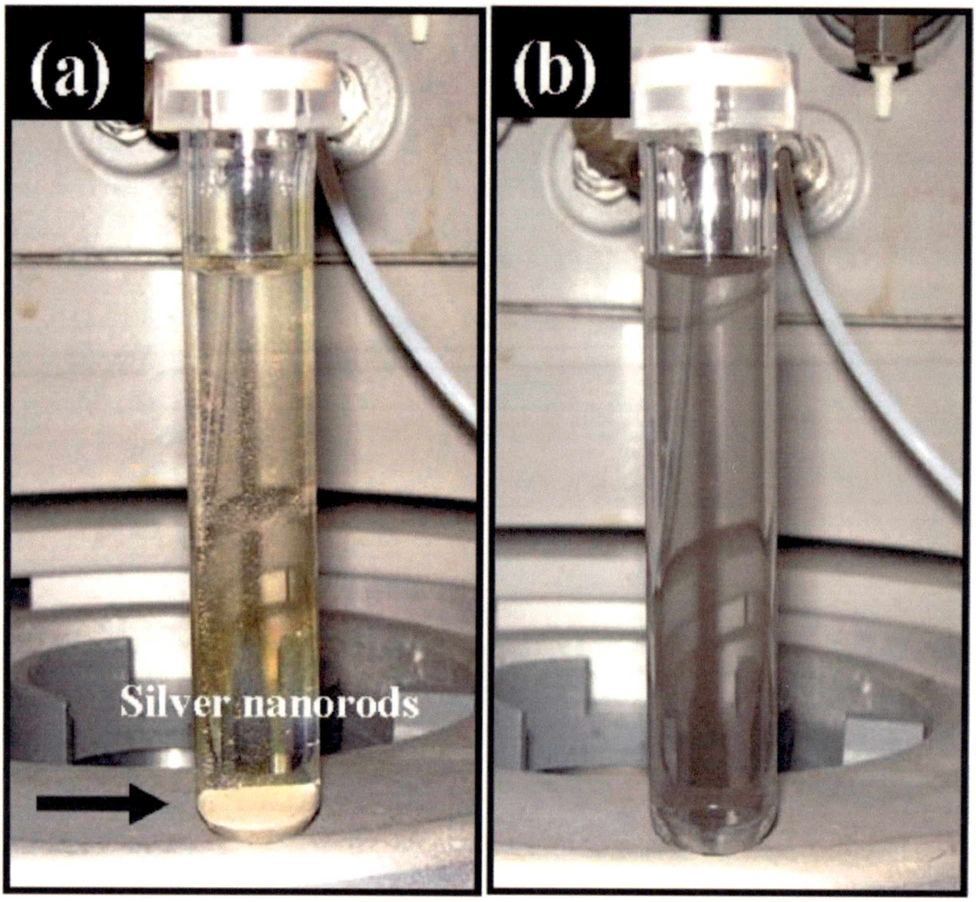

Fig. 5. Photographic image of (a) precipitated Ag nanorods after microwave irradiation for 2 min; and (b) control reaction of the same reaction composition carried out using an oil bath at 100 °C for 1 h.

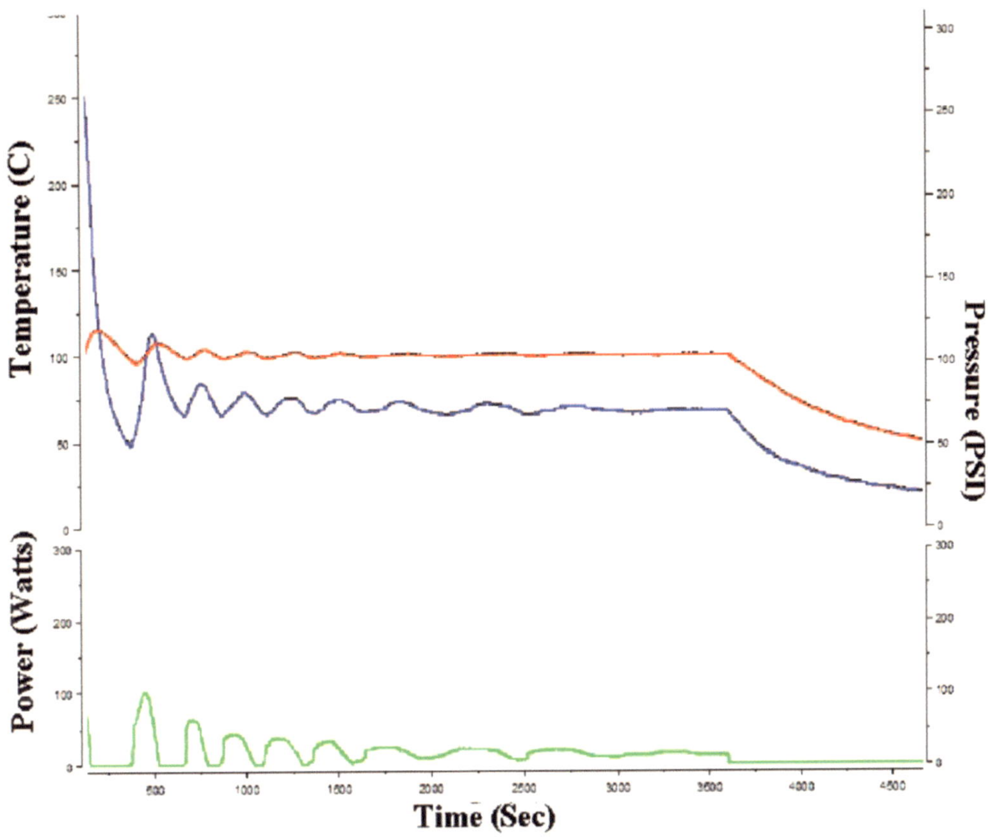

Fig. 6. Reaction profile of 4 mL PEG(300) + 4 mL 0.1 N $AgNO_3$, irradiated at 100 °C for 1 h using MW.

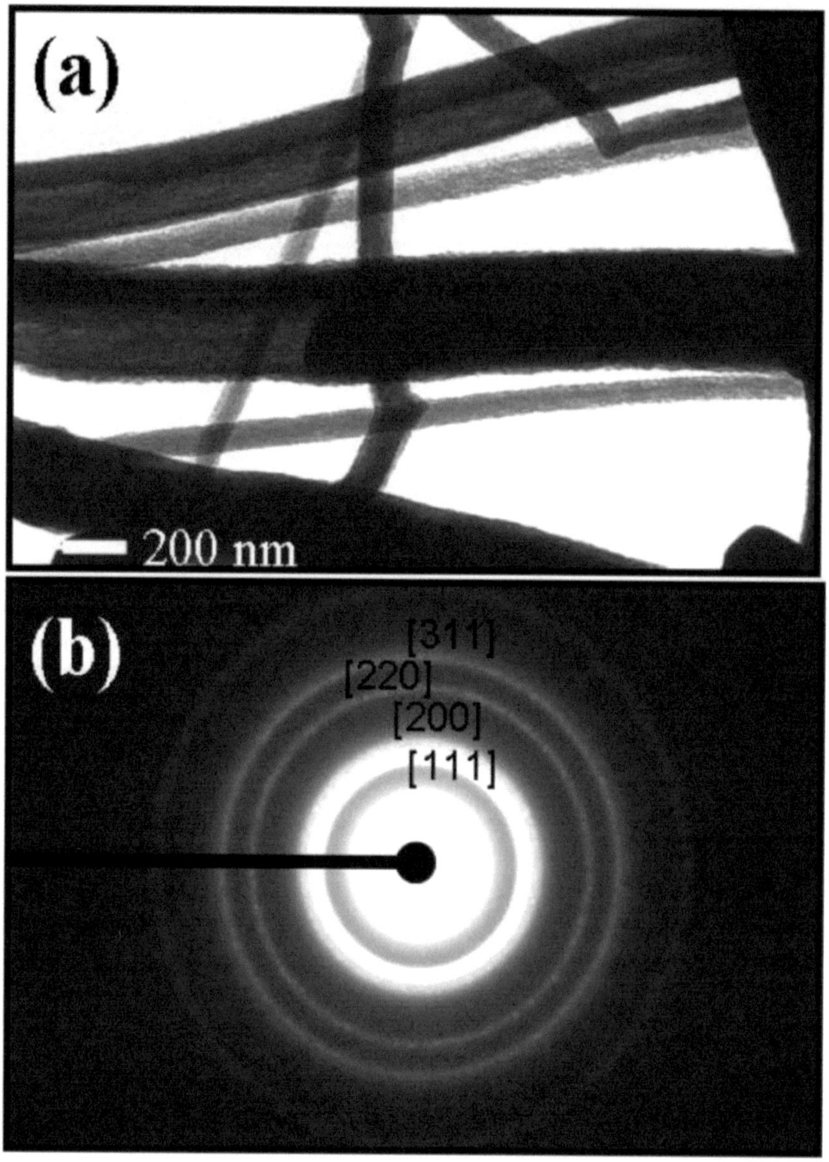

Fig. 7. TEM images of Ag nanorods from (a) 4 mL PEG(300) + 4 mL 0.1 N $AgNO_3$ under MW conditions, and (b) its SAED pattern obtained, from a bundle of Ag nanorods randomly deposited on the TEM grid.

Shape-controlled synthesis of gold (Au) nanostructures with various shapes such as prisms, cubes, and hexagons was accomplished via the MW-assisted spontaneous reduction of noble metal salts using an aqueous solution of varying concentrations of α-D-glucose, sucrose, and maltose (22). The expeditious reaction was completed under MW irradiation in 30–60s with the formation of different shapes and structures (see Figure 8) and potential application to the generation of nanospheres of Ag, Pd, and Pt. The noble nanocrystals underwent catalytic oxidation with monomers such as pyrrole to generate noble nanocomposites, which have potential functions in catalysis, biosensors, energy storage systems, nano-devices, and other ever-expanding technological applications.

In a typical experiment, an aqueous solution of $HAuCl_4$ (5 mL, 0.01 N) was placed in a 20 mL glass vessel and then mixed with 300 mg of R-D-glucose. The reaction mixture was exposed to high-intensity microwave irradiation (1000 W, Panasonic MW oven equipped with inverter technology) for 30-45 s. Similarly, experiments were conducted using 0.01 N PtCl4, 0.01 N $PdCl_2$, and 0.1 N $AgNO_3$. In the cases of $PdCl_2$ and $AgNO_3$, 300 mg of poly (vinyl pyrrolidinone) (PVP) was added to prevent aggregation and the formation of silver mirror (Tollen's process) on the surface of the glass walls.

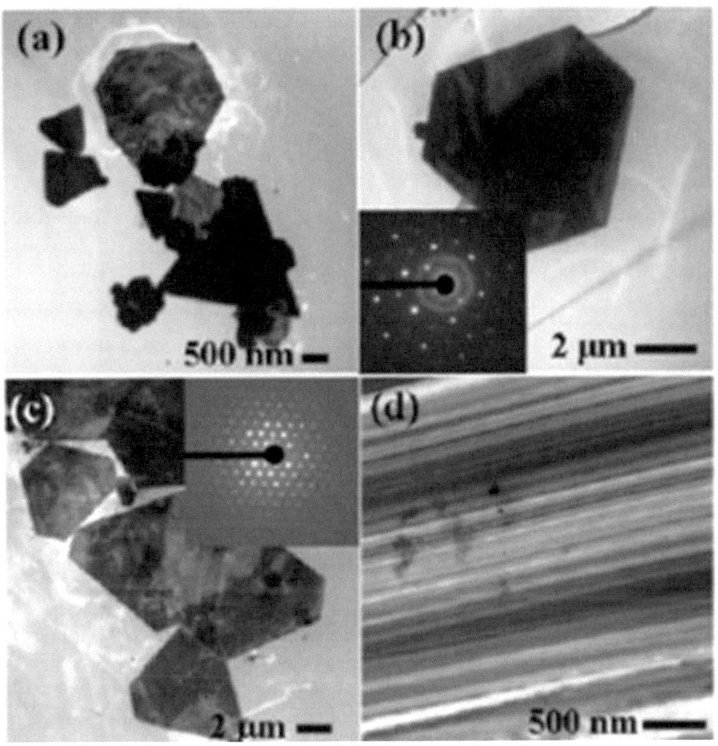

Fig. 8. TEM images of Au nanostructures, synthesized (low concentration of sugar) using MW irradiation with natural polymers such as (a) sucrose, (b) α-D-glucose, or (c, d) maltose. The insets show corresponding electron diffraction patterns.

A green approach was also developed that generated bulk quantities of nanocomposites containing transition metals such as Cu, Ag, In, and Fe at room temperature. A biodegradable polymer, carboxymethyl cellulose (CMC), was reacted with respective metal salts to obtain desired composites (20).

These nanocomposites exhibited broader decomposition temperatures when compared with control CMC and Ag-based CMC nanocomposites, exhibiting a luminescent property at longer wavelengths.

The other noble metals (such as Au, Pt, and Pd) did not react at room temperature with aqueous solutions of CMC, but did react rapidly under MW irradiation (MW) conditions at 100 ^{0}C.

This environmentally-benign approach provides facile entry to the production of multiple-shaped noble nanostructures without using any toxic reducing agents and/or capping/surfactant agents. The method also uses a benign biodegradable polymer, CMC, could find widespread technological and medicinal applications.

Recently, Yu et al. (56) prepared uniform water-soluble silver nanoparticles by reducing silver nitrate with basic amino acids in the presence of soluble starch via MW heating in aqueous medium. Although the fundamental of MW irradiation for this system has yet to be studied completely, the authors believed that MW irradiation plays a major role in the synthesis of the uniform silver nanoparticles. The choice of benign solvent and renewable reacting components and targeted heating approaches amply support the notion that the green chemical synthesis of metal nanoparticles with well-controlled shapes, sizes, and structures is possible.

Microwave irradiation that accomplishes the cross-linking reaction of poly (vinyl alcohol) (PVA) with metallic and bimetallic systems has also been achieved (19).

Nanocomposites of PVA cross-linked metallic systems such as Pt, Cu, and In, and bimetallic systems such as Pt-In, Ag-Pt, Pt-Fe, Cu-Pd, Pt-Pd, and Pd-Fe were prepared expeditiously by reacting the respective metal salts with 3 wt. % PVA under MW irradiation, maintaining the temperature at 100 ^{0}C. This is a radical improvement over the methods used for preparing the cross-linked PVA described in the literature (see Figure 9).

The general preparative procedure is versatile and provides a simple route to manufacturing useful metallic and bimetallic nanocomposites with various shapes, such as nanospheres, nanodendrites, and nanocubes.

Recently, there has been an increasing interest in synthesizing carbon nanotube (CNT)-metal nanoparticle/polymer composites. The larger surface areas and high electric conductivity render them as ideal supporting materials for metal nanoparticle catalysts such as Ag, Au, Pt, and Pd nanoparticles, which have shown great promise in catalysis, surface-enhanced Raman scattering (SERS), and electrochemical and fuel cells. CNTs are also ideal templates for attaching metal nanoparticles and nanoparticle-fused metal nanowires for hydrogen storage and for chemical and biological sensing applications.

Fig. 9. Photographic image of cross-linked PVA with various metallic and bimetallic systems: (a) Pt, (b) Pt-In, (c) Ag-Pt, (d) Cu, (e) Pt-Fe, (f) Pt with higher concentration ratio, (g) Cu-Pd, (h) In, (i) Pt-Pd, and (j) Pd-Fe.

The cross-linking reaction of PVA with single-walled carbon nanotubes (SWNTs), multi-walled carbon nanotubes (MWNTs), and buckminsterfullerene (C-60) using MW irradiation was achieved with 3 wt. % PVA under MW irradiation, maintaining a temperature of 100 ^{0}C, representing a radical improvement over literature methods to prepare such cross-linked PVA composites (Figure 10) (57). This general preparative procedure is versatile and provides a simple route for manufacturing useful SWNT, MWNT, and C-60 cross linked PVA nanocomposites.

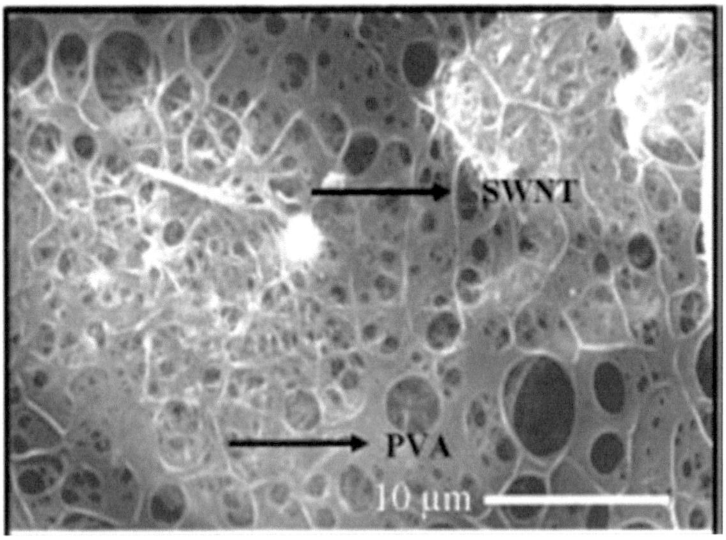

Fig. 10. SEM images of SWNT cross-linked PVA nanocomposites.

Alignment and decoration of noble metals on CNTs wrapped with CMC was also achieved under MW condition. CNTs, such as SWNT, MWNT, and C-60, were well dispersed using the sodium salt of CMC under sonication (58). The addition of respective noble metal salts then generated noble metal-decorated CNT composites at room temperature. However, aligned nanocomposites of CNTs could only be generated by exposing the above nanocomposites to MW irradiation. The general preparative procedure is flexible and provides a straightforward route to manufacturing functional metal coated CNT nanocomposites (Figure 11).

Varma et al. (59) have developed a simple method for the bulk synthesis of monodispersed spinel ferrite nanoparticles with size selectivity using readily available inorganic precursors via a water-organic interface . Hydrothermal as well as MW hydrothermal methods are applicable but the use of MW has the advantage of low temperature, expedient synthesis. The synthesized particles are highly dispersible and are stable in nonpolar organic solvents, which is important in their use as ferrofluids and other magnetic applications. Surface functionalization of the As-synthesized particles with lysine made them water dispersible for possible biological applications (Figure 12).

It has been stated that volumetric and selective heating using MW irradiation may reduce the thermal gradients in the reaction, thereby generating a more homogeneous product with faster consumption of the starting materials. (60)

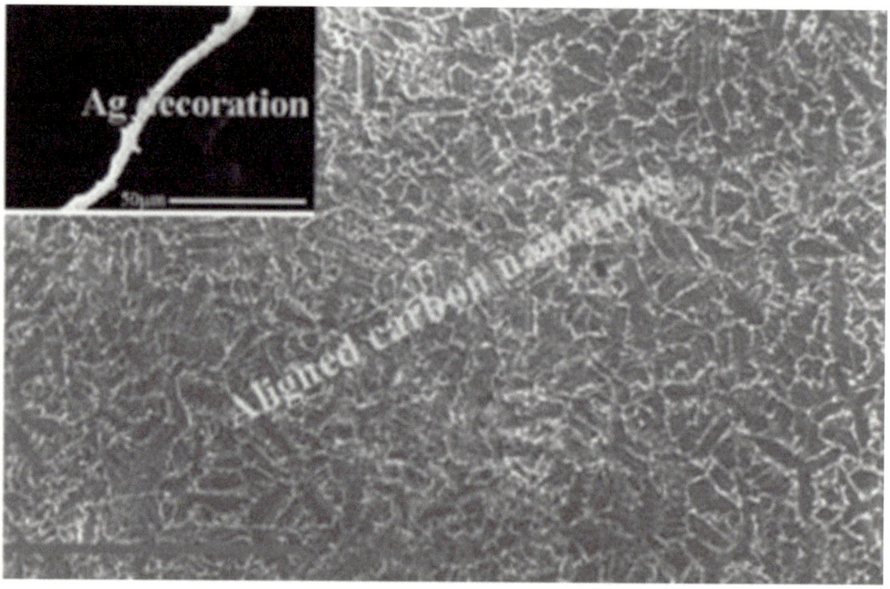

Fig. 11. Aligned CNTs in CMC polymer matrix.

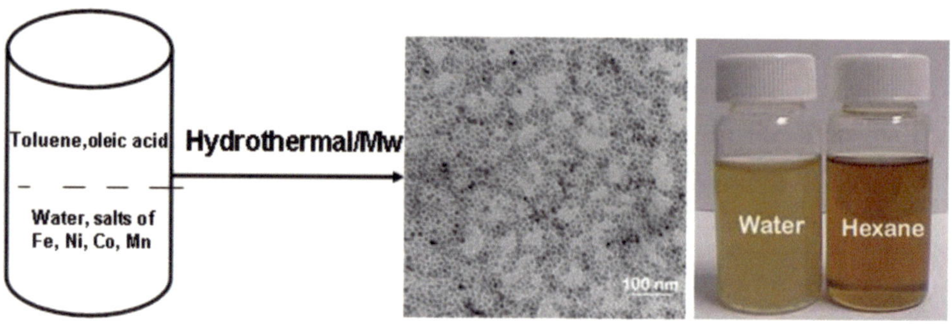

Fig. 12. MW hydrothermal synthesis and fictionalization of nanoferrites.

Varma et al.(61) have synthesized, for the first time under MW irradiation conditions, dendritic ferrites with micro-pine morphology (see Figure 13) without using any reducing or capping reagents. With this adjustment, nano ferrites could then be functionalized (Scheme 2) and coated with Pd metal, which catalyzes various C-C coupling reactions. An assortment of magnetic, nanoparticle-supported metal catalysts have been readily prepared from inexpensive starting materials and shown to catalyze a variety of organic transformations such as oxidation (62), hydration (63), and reduction (hydrogenation) (64). Superior activity and the inherent stability of these catalyst systems coupled with their easy magnetic separation, which eliminates the prerequisite of catalyst filtration after completion of the reaction, are some of the supplementary sustainable attributes of these protocols.

Fig. 13. TEM image of dendritic α-Fe_2O_3.

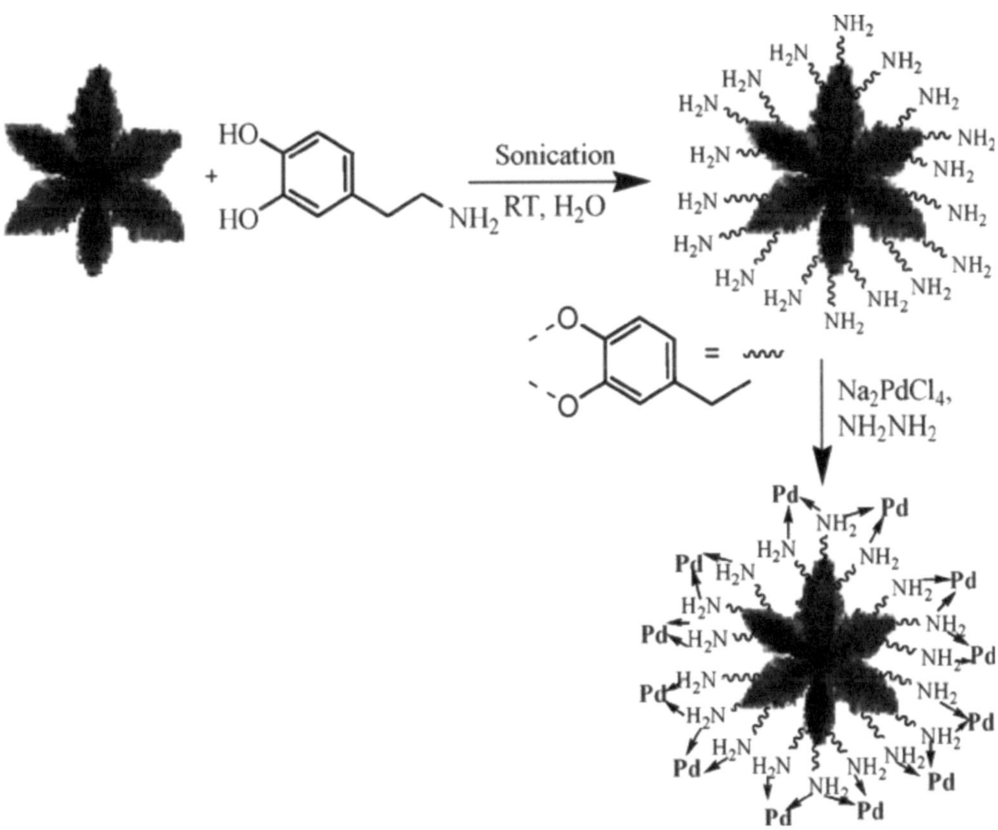

Scheme 2. Schematic diagram of α-Fe_2O_3 functionalization with amino group and Pd.

Microwave strategy can be also be expanded to a solid state reaction. Porous nanocrystalline TiO_2 and carbon coated TiO_2 using sugar dextrose as a template has been achieved through MW and the results were compared with conventional heating furnace (65). Out of three compositions, namely, 1:1, 1:3, and 1:5 (metal: dextrose), 1:3 favors formation of consistent porous structures (see Figure 14). This general and eco-friendly method uses a benign natural polymer, dextrose, to create spongy porous structures and can be extended to other transition metal oxides such as ZrO_2, Al_2O_3, and SiO_2.

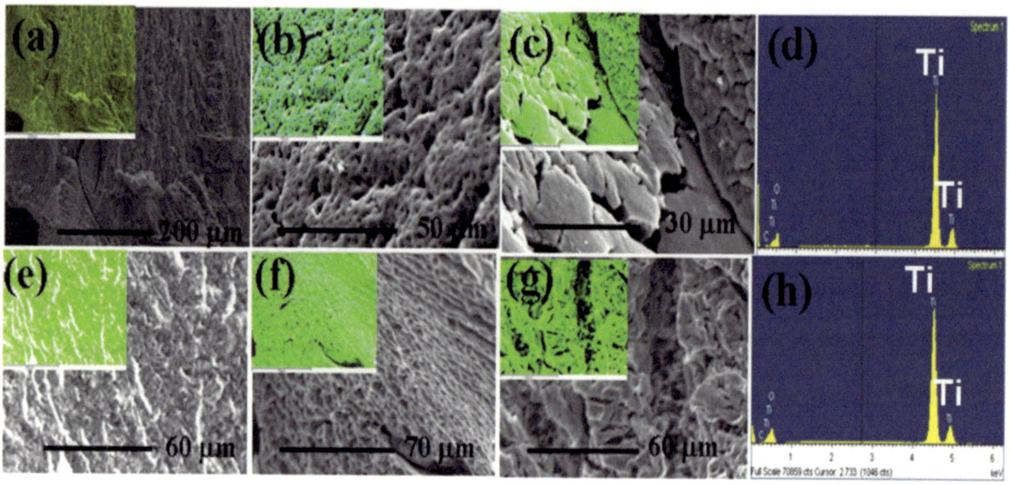

Fig. 14. (a)-(c) SEM images of 1:1, 1:3, and 1:5 (titania: dextrose molar ratio) titania sponges synthesized by microwave combustion and subsequently heated at 850 ^{0}C for 1 h by a conventional furnace (the inset shows the x-ray mapping images of the same with the green region showing titania and the red region showing carbon). (d) Representative (1:3 titania: dextrose molar ratio) sample of energy dispersive x-ray analysis (EDX) showing the presence of titania. (e)-(g) SEM images of 1:1, 1:3, and 1:5 (titania: dextrose molar ratio) titania sponges synthesized by heating at 850 ^{0}C for 1 h by conventional furnace (the inset shows the x-ray mapping images of the same with the green region showing titania and the red region showing carbon). (h) Representative sample (1:3 titania: dextrose molar ratio) of energy dispersive x-ray analysis (EDX) showing the presence of titania.

4. References

[1] X.Wang andY.Li, Chem. Commun., 2007, 2901.
[2] Y. Sun and Y. Xia, Science, 2002, 298, 2176.
[3] J. Chen, J. M. McLellan, A. Siekkinen, Y. Xiong, Z.-Y. Li and Y. Xia, J. Am. Chem. oc., 2006, 128, 14776.
[4] J. W. Stone, P. N. Sisco, E. C. Goldsmith, S. C. Baxter and C. J.Murphy, Nano Lett., 007, 7, 116.
[5] B. Wiley, Y. Sun and Y. Xia, Acc. Chem. Res., 2007, 40, 1067.
[6] J. Du, B. Han, Z. Liu and Y. Liu, Cryst. Growth Design, 2007, 7, 900.
[7] B. Wiley, T. Herricks, Y. Sun and Y. Xia, Nano Lett., 2004, 4, 2057.
[8] C. J. Murphy, A. M. Gole, S. E. Hunyadi and C. J. Orendorff, Inorg. Chem., 2006, 5, 7544.

[9] B. J. Wiley, Y. Chen, J. M. McLellan, Y. Xiong, Z.-Y. Li, D. Ginger and Y. Xia, Nanoletters, 2007, 4, 1032.
[10] Y.Xiong, H. Cai, B. J. Wiley, J.Wang, M. J. Kimand Y. Xia, J. Am. Chem. Soc., 2007, 129, 3665.
[11] J. Fang, H. You, P. Kong, Y. Yi, X. Song and B. Ding, Cryst. Growth Design, 2007, 7, 864.
[12] A. Narayan, L. Landstrom and M. Boman, Appl. Surf. Sci., 2003, 137, 208.
[13] H. Song, R. M. Rioux, J. D. Hoefelmeyer, R. Komor, K. Niesz, M. Grass, P. Yang and G. A. Somorjai, J. Am. Chem. Soc., 2006, 128, 3027.
[14] C. C. Wang, D. H. Chen and T. C. Huang, Colloids Surf. A, 2001, 189, 145.
[15] M. N. Nadagouda and R. S. Varma, Green Chem., 2006, 8, 516.
[16] P. Raveendran, J. Fu and S. L.Wallen, J. Am. Chem. Soc., 2003, 125,
[17] M. N. Nadagouda and R. S. Varma, Green Chem., 2007, 9, 632.
[18] R. R. Naik, S. J. Stringer, G. Agarwal, S. E. Jones and M. O. Stone, Nat. Mater., 2002, 1, 169.
[19] M. N. Nadagouda and R. S. Varma, Macromol. Rapid Commun., 2007, 28, 465.
[20] M. N. Nadagouda and R. S. Varma, Biomacromolecules, 2007, 8, 2762–2767.
[21] M. N. Nadagouda and R. S. Varma, Cryst. Growth Design, 2007, 7((12)), 2582–2587.
[22] M. N. Nadagouda and R. S. Varma, Cryst. Growth Design, 2007, 7((4)), 686–690.
[23] M. N. Nadagouda and R. S. Varma, Cryst. Growth Design, 2008, 8((1)), 291–295.
[24] J. A. Dahl, L. S.Maddux and J. E. Hutchison, Chem. Rev., 2007, 107, 2228.
[25] P. T. Anastas and J. C. Warner, Green, Chemistry: Theory and Practice, Oxford University Press, Inc., New York, 1998.
[26] V. L. Colvin, M. C. Schlamp, A. P. Alivisatos, Nature 1994, 370, 354.
[27] S. Maeda, S. P. Armes, Chem. Mater. 1995, 7, 171.
[28] P. G. Hill, P. J. S. Foot, R. Davis, Mater. Sci. Forum 1995, 191, 43.
[29] R. E. Schwerzel, K. B. Spahr, J. P. Kurmer, V. E. Wood, J. A. Jenkins, J. Phys. Chem. A. 1998, 102, 5622.
[30] T. Trindade, M. C. Neves, A. M. V. Barros, Scr. Mater. 2000, 43, 567.
[31] J. J. Tunney, C. Detellier, Chem. Mater. 1996, 8, 927.
[32] C. O. Oriakhi, M. M. Lerner, Chem. Mater. 1996, 8, 2016.
[33] L. Ouahab, Chem. Mater. 1997, 9, 1909.
[34] J. H. Choy, S. J. Kwon, S. J. Hwang, Y. H. Kim, W. Lee, J. Mater. Chem. 1999, 9, 129.
[35] C. Sanchez, F. Ribot, B. Lebeau, J. Mater. Chem. 1999, 9, 35.
[36] L. Luo, S. Yu, H. Qian, T. Zhou, J. Am. Chem. Soc. 2005, 127, 2822.
[37] S. Xiong, L. Fei, Z. Wang, H. Y. Zhou, W. Wang, Y. Qian, Eur. J. Inorg. Chem. 2005, 2006, 207.
[38] H. Qian, L. Luo, J. Gong, S. Yu, T. Li, L. Fei, Cryst. Growth Design 2005, 6, 607.
[39] J. Gong, L. Luo, S. Yu, H. Qian, L. Fei, J. Mater. Chem. 2006, 16, 101.
[40] W. Wu, Y. Wang, L. Shi, Q. Zhu, W. Pang, G. Xu, F. Lu, Nanotechnology 2005, 16, 3017.
[41] H. Kong, J. Jang, Chem. Commun. 2006, 3010.
[42] Z. Li, H. Huang, C. Wang, Macromol. Rapid Commun. 2006, 27, 152.
[43] S. Porel, S. Singh, T. P. Radhakrishnan, Chem. Commun. 2005, 2387.
[44] G. A. Gaddy, A. S. Korchev, J. L. McLain, B. L. Slaten, E. S. Steigerwalt, G. Mills, J. Phys. Chem. B 2004, 108, 14850.
[45] Z. H. Mbhele, M. G. Salemane, C. G. C. E. van Sittert, J. M. Nedeljkovic, V. Djokovic, A. S. Luyt, Chem. Mater. 2003, 15, 5019.

[46] M. N.Nadagouda, V. Polshettiwar and R. S. Varma, "J. Mater. Chem., 9, 2026 – 2031, (2009).
[47] N. N. Mallikarjuna, and R. S. Varma Aust. J. Chem., 62, 260–264(2009).
[48] M. N. Nadagouda, R. S. Varma, Green Chem. 2008, 10, 859.
[49] M. N. Nadagouda and R. S. Varma "Bulk and template-free synthesis of narrow diameter reduced polyaniline nanofibers at room temperature" Green Chemistry, 9, 632-637(2007).
[50] J.A. Dahl, B. L. S. Maddux, J. E. Hutchison, *Chem. Rev.* 2007, *107,* 2228.
[51] S. Kundu, H. Liang, Langmuir 2008, 24, 9668.
[52] W. Tu, H. Liu, J. Mater. Chem. 2000, 10, 2207.
[53] R. He, X. Qian, J. Yin, Z. Zhu, *J. Mater. Chem.* 2002, *12,* 3783.
[54] X. Zhang, S. K. Manohar, Chem. Commun. 2006, 2477.
[55] J. M. Campelo, T. D. Conesa, M. J. Gracia, M. J. Jurado, R. Luque, J. M. Marinasa, A. A. Romeroa, Green Chem. 2008, 10, 853.
[56] B. Hu, S. -B. Wang, K. Wang, M. Zhang, S. -H. Yu, J. Phys. Chem. C 2008, 112, 11169.
[57] M. N. Nadagouda, R. S. Varma, Macromol. Rapid Commun. 2007, 28, 842.
[58] M. N. Nadagouda, R. S. Varma, Macromol. Rapid Commun. 2008, 29, 155.
[59] B. Baruwati, M. N. Nadagouda, R. S. Varma, J. Phys. Chem. C. 2008, 112, 18399.
[60] A. Gerbec, D. Magana, A. Washington, G. F. Strouse, J. Am. Chem. Soc. 2005, 127, 15791.
[61] V. Polshettiwar, M. N. Nadagouda, R.S. Varma: Chem. Commun., 2008, DOI: 10.1039/B814715A.
[62] V. Polshettiwar, R. S. Varma, Org. Bio. Chem. *Org. Biomol. Chem.,* 2009, 7, 37–40
[63] V. Polshettiwar, R. S. Varma, Chem. Eur. J. 2009, 15, 1582-1586.
[64] V. Polshettiwar, B. Baruwati, R. S. Varma, Green. Chem. 2009, 11, 127-131.
[65] M. N.Nadagouda and R. S. Varma, Smart Materials and Structures, 5, 1-6 (2006).

18

Crystal Habit Modification Using Habit Modifiers

Satyawati S. Joshi
University of Pune,
India

1. Introduction

The synthesis of inorganic materials with a specific size and morphology has recently received much attention in the material science research area. Morphology control or morphogenesis is more important for the chemical industry than size control. Many routes have been reported to control the crystal growth and eventually modify the morphology of the crystals. For crystal-habit modification, crystals are grown in the presence of naturally occurring soluble additives, which usually adsorb or bind to the crystal faces and influence the crystal growth or morphology. A number of recent investigations show that such type of crystal-habit modifiers can be used to obtain inorganic crystals with organized assemblies. (Xu, et al. 2007, Yu & Colfen 2004, & Colfen, 2001).

The crystal-habit modifiers may be of a very diverse nature, such as multivalent cations, complexes, surface active agents, soluble polymers, biologically active macromolecules, fine particles of sparingly soluble salts, and so on. (Sarig et al.,1980) These crystal modifiers often adsorb selectively on to different crystal faces and retard their growth rates, thereby influencing the final morphology of the crystals. (Yu & Colfen, 2004) The strategy that uses organic additives and/or templates with complex functionalization patterns to control the nucleation, growth, and alignment of inorganic crystals has been universally applied for the biomimetic synthesis of inorganic materials with complex forms. (Qi et al., 2000) The biomimetic process uses an organized supramolecular matrix and produces inorganic crystals with characteristic morphologies. (Xu et al., 2007& Loste &Meldrum, 2001) Understanding the mechanism involved in such a matrix-mediated synthesis has a great potential in the production of engineering materials. Thus, catalyst particles of controlled size and morphology, magnetic materials with appropriate anisotropy, highly porous materials, composites, and well-organized crystallite assemblies can be produced by this synthesis method. (Sinha etal.,2000) Using water-soluble polymers as crystal modifiers for controlled crystallization is widely expanding and becoming a benign route for controlling and designing the architectures of inorganic materials. (Yu & Colfen, 2004) Investigators have used different double hydrophilic block copolymers, such as poly(ethylene glycol)-block-poly(methacrylic acid), to control the morphology of a number of inorganic salts, namely, $CaCO_3$, (Sedlak & Colfen, 2001, Rudolff et al., 2002, Meng et al., 2007, Guo et al., 2006, He et al., 2006, Wang et al., 2005, Meldrum et al., 2007, Gorna et al.,2007, & Colfen &Qi, 2001) $BaCO_3$, $CdCO_3$, $MnCO_3$, $PbCO_3$, (Yu et al.,2003) $BaCrO_4$, (Liu et al.2005& Yu et al.,

2002) $BaSO_4$, (Qi et al., 2000, Robinson et al.,2002, Wang et al., 2005,& Yu et al., 2005) tolazamide, Pb- WO_4, (Kuldipkumar et al., 2005) and so forth. In the early stages, gel matrices have been used for the control of nucleation and morphology in aqueous solution-based crystal growth. (Yu et al., 2007 Oaki &Imai, 2003) Investigators have used poly(vinyl alcohol) (PVA)-, agar-, gelatin-, and pectin-based gel matrices to control the morphology of inorganic crystals such as PbI_2, AgI, $Ag_2Cr_2O_7$, $PbSO_4$, $PbCl_2$, and so forth. (Henisch, 1988) The advantage of a gel medium is believed to be the reduction of the nucleation rate and suppression of convection. (Yu & Colfen, 2004) The functional groups, such as amine, amide, carboxylic acid and so forth, are known to significantly influence the mineralization process. Among the reported common gel matrices used as crystal-habit modifiers, PVA is a water soluble synthetic polymer with excellent film-forming and emulsifying properties. PVA is a crystalline polymer with a monoclinic structure and is known for its biological activities. (Merrill & Bassett, 1975) Also, PVA is reported to have been used for the morphology control of $K_2Cr_2O_7$, AgBr, and $CaCO_3$, (Sinha, 2001) and even for the selective nucleation of $CaCO_3$ polymorphs. (Lakshminarayanan, 2003) In this chapter, the results on morphological changes using polymers as habit modifiers are discussed on the basis of nucleation theory and growth process.

2. Crystal habit

Although crystals can be classified according to seven crystal systems, the relative sizes of the faces of a particular crystal can vary considerably. This variation is called a modification of habit.

2.1 Crystal habit modifications

2.1.1 Crystal morphology and structure

The morphology of a crystal depends on the growth rates of the different crystallographic faces. Some faces grow very fast and have little or no effect on the growth form; while slow growing faces have more influence. The growth of a given face is governed by the crystal structure and defects on one hand and by the environmental conditions on the other. (Mullin 2002)

A number of proposed mechanisms and theories have been put forth to predict the equilibrium form of a crystal. According to the Bravais rule, the important faces governing the crystal morphology are those with the highest reticular densities and greatest interplanar distances, d_{hkl}. Or in simpler terms, the slowest growing and most influential faces are the closest packed and have the lowest Miller indices. The surface theories suggest that the equilibrium form should be such that the crystal has a minimum total surface free energy per unit volume.

The crystals may grow rapidly, or be stunted, in one direction; thus an elongated growth of a prismatic habit gives a needle shaped crystal (acicular habit) and a stunted growth gives a flat plate-like crystal (tabular, platy or flaky habit). The relative growths of the faces of a crystal can be altered and often controlled by a number of factors. Rapid crystallization, produced by the sudden cooling or seeding of a supersaturated solution, may result in the formation of needle crystals. The growth of a crystal may be stunted in certain directions due to presence of impurities in the crystallizing solution. A change of solvent often changes the crystal habit.

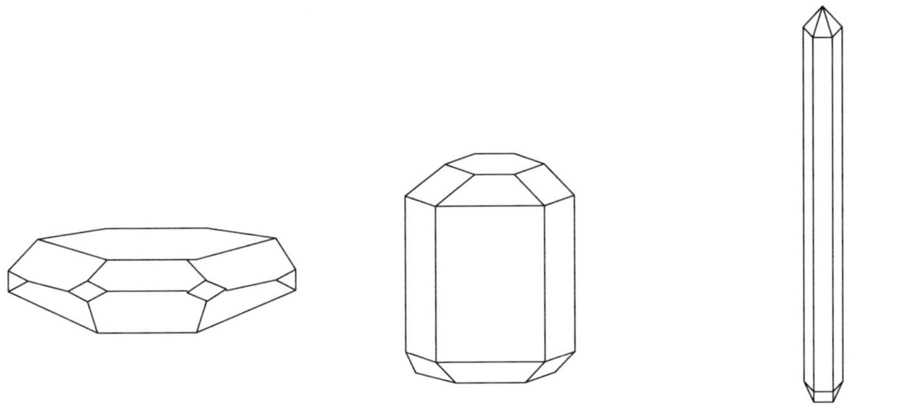

Fig. 1. Crystal habit illustrated on a hexagonal crystal

2.1.2 Crystal surface structure

The structure of a growing crystal face at its interface with the growth medium has been characterized by a quantity as surface roughness or surface entropy factor or the alpha factor defined by

$$\alpha = \xi \, \Delta H / kT \qquad (1)$$

Where ξ is an anisotropy factor related to the bonding energies in the crystal surface layers, ΔH is the enthalpy of fusion and **k** is the Boltzmann constant. Values of α less than 2 are indicative of rough crystal surface which will allow continuous growth to proceed. The growth will be diffusion controlled and the face growth rates, v, will be linear with respect to the supersaturation, σ, i.e.

$$v \; \alpha \; \sigma \qquad (2)$$

For α >5, a smooth surface is indicated

2.2 Effect of crystal size

In order for crystallization to occur, there must exist in a solution a number of minute solid bodies, nuclei or seeds that act as a centre of crystallization, the classical theory of nucleation stemming from the work of Gibbs (1948) Volmer (1939) and others is based on the condensation of vapor to liquid and this treatment may be extended to crystallization from melts and solutions. Crystallization process can be explained on the basis of nucleation and growth process.

2.2.1 Nucleation

Schematically the nucleation steps are as shown below:

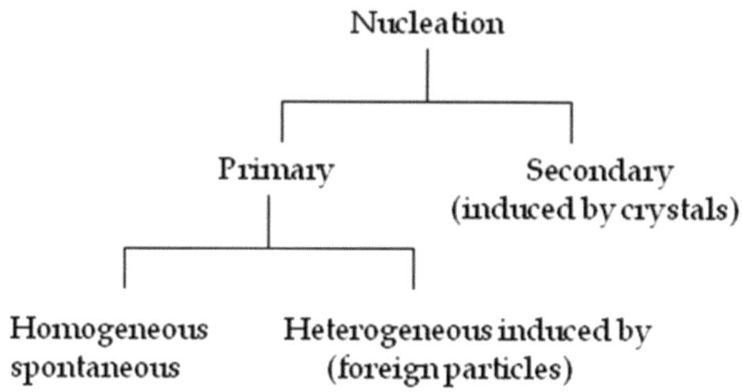

Let us consider the free energy changes associated with the process of homogenous nucleation. The overall excess free energy, ΔG, between a small solid particles of solute (assume here a sphere of radius r for simplicity) and the solute in solution is equal to the sum of surface excess free energy ΔGs i.e. excess free energy between the surface of the particles and the bulk of the particles, and the volume excess free energy, ΔGv, i.e. the excess free energy between very large particles (r = ∞) and the solute in the solution , ΔGs is a positive quantity, the magnitude of which is proportional to r^2. In a supersaturated solution Gv is a negative quantity proportional to r^3. Thus

$$\Delta G = \Delta G_s + \Delta G_v \quad (3)$$

$$= 4\pi r^2\gamma + \frac{4}{3}\pi r^3 \Delta G_v \quad (4)$$

ΔGv also can be understood as free energy change of formation per unit volume
γ is the interfacial tension or surface energy.
ΔGs and ΔGv are opposite in sign and depend differently on r. the free energy of formation, ΔG, passes through a maximum (ΔG_{crit}) corresponds to critical nucleus, r_c. For a spherical cluster, it is obtained by setting dΔG/dr = 0

$$\frac{d\Delta G}{dr} = 8\pi r\gamma + 4\pi r^2 \Delta G_v = 0 \quad (5)$$

$$r_c = \frac{-2\gamma}{\Delta G_v} \quad (6)$$

From equations 1 & 3 we get

$$\Delta G_{crit} = \frac{16\pi\gamma^3}{3(\Delta G_v)^2} = \frac{4\pi\gamma r_c^2}{3} \quad (7)$$

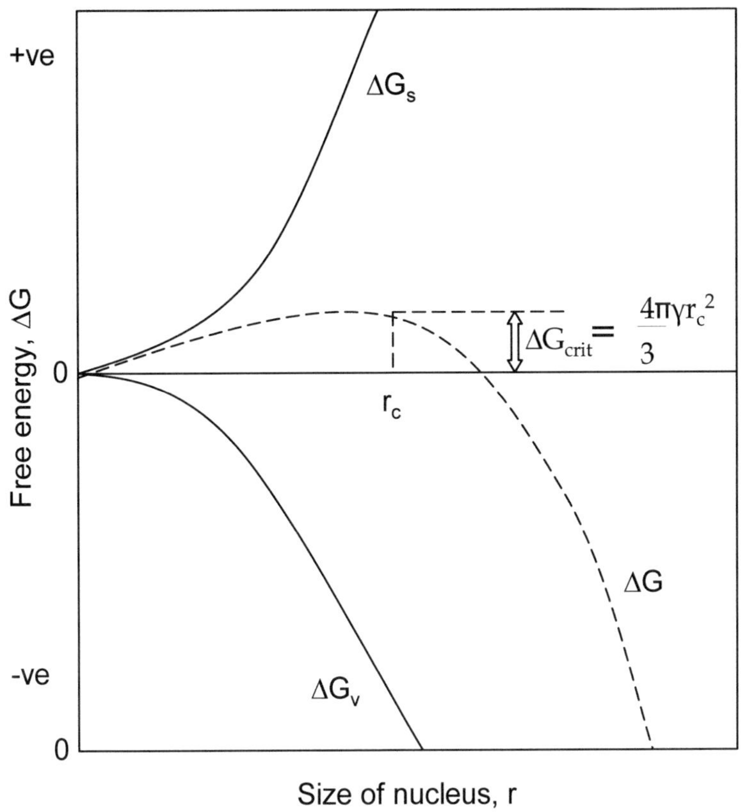

Fig. 2. Free energy diagram for nucleation explaining the existence of a 'critical nucleus'.

The behavior of a newly crystalline lattice structure in a supersaturated solution depends on its size, the crystal may grow or redissolve and it undergoes decrease in free energy of the particle. Particles smaller than r_c will dissolve if present in a liquid in order to achieve reduction in free energy. Similar particles larger than r_c will continue to grow.

There will be fluctuations in the energy about the constant mean value i.e. there will be a statistical distribution of energy, or molecular velocity, in the molecules constituting the system, and in those supersaturated regions where the energy level rises temporarily to a high value, nucleation will be favored.

The rate of nucleation J e.g. the number of nuclei formed per unit time per unit volume can be expressed in the form of Arrhenius reaction velocity equation

$$J = A \exp(-\Delta G/kT) \tag{8}$$

k = Boltzmann constant, the gas constant per molecule

The basic Gibbs-Thomson relationship for a non-electrolyte may be written as

$$\ln S = \frac{2\gamma v}{kTr} \tag{9}$$

Where S is defined by equation 10

$$S = \frac{C}{C^*} \tag{10}$$

Where C is the solution concentration and C* is equilibrium saturation at given temperature and v is the molecular volume; this gives

$$-\Delta G_v = \frac{2\gamma}{r} = \frac{kT \ln S}{v} \tag{11}$$

Hence, from equation 7

$$\Delta G_{crit} = \frac{16\pi \gamma^3 v^2}{3 (kT \ln S)^2} \tag{12}$$

And from equation 8

$$J = A \exp\left(\frac{16\pi \gamma^3 v^2}{3k^3T^3(\ln S)^2}\right) \tag{13}$$

This equation indicates that three main variables govern the rate of nucleation: temperature, *T*; degree of supersaturation, *S*; and interfacial tension, γ. Equation 13 may be rearranged to give

$$\ln S = \left(\frac{16\pi \gamma^3 v^2}{3k^3T^3\ln(A/J)}\right)^{1/2} \tag{14}$$

And if, arbitrarily, the critical supersaturation, $S_{crit,}$ is chosen to correspond to a nucleation rate, J, of say one nucleus per second per unit volume, then equation 14 becomes

$$\ln S_{crit} = \left(\frac{16\pi \gamma^3 v^2}{k^3T^3 \ln A}\right)^{1/2} \tag{15}$$

From equation 6 and 11, the radius of a spherical critical nucleus at a given supersaturation can be expressed as

$$r_c = \frac{2\gamma v}{kT\ lnS} \tag{16}$$

For the case of non-spherical nuclei, the geometrical factor 16π/3 in equations 7 and 12-14 must be replaced by an appropriate (e.g. 32 for cube).
Critical reviews of nucleation mechanism have been made by Nancollos and Purdie (1964), Nielsen (1964), Walton (1967), Strickland-Constable (1968), Zettlemoyer (1969), Nyvlt et al (1985) and Sohnel, Garside (1992) and Kashchiev (2000).

2.2.2 Mechanism of growth

Nucleation occurs over some time with constant precursor concentration. Eventually surface growth of clusters begins to occur which depletes the initial supply. When the initial concentration falls below the critical level for nucleation (critical supersaturation level), nucleation ends. A general analysis of the growth process is then important to understand nanocrystal synthesis. In general, the surface to volume ratio in smaller particles is quite high. As a result of the large surface area present, it is observed that surface excess energy becomes more important in very small particles, constituting a non-negligible percentage of the total energy. Hence, for a solution that is initially not in thermodynamic equilibrium, a mechanism that allows the formation of larger particles at the cost of smaller particles reduces the surface energy and hence plays a key role in the growth of nanocrystals. A colloidal particle grows by a sequence of monomer diffusion towards the surface followed by reaction of the monomers at the surface of the nanocrystal. Coarsening effects, controlled either by mass transport or diffusion, are often termed the Ostwald ripening process. This diffusion limited Ostwald ripening process is the most predominant growth mechanism and was first quantified by Lifshitz and Slyozov [Lifshitz and Slyozov, 1961], followed by a related work by Wagner [Wagner and Elektrochem, (1961)] known as the LSW theory.
The diffusion process is dominated by the surface energy of the nanoparticle. The interfacial energy is the energy associated with an interface due to differences between the chemical potential of atoms in an interfacial region and atoms in neighboring bulk phases. For a solid species present at a solid/liquid interface, the chemical potential of a particle increases with decreasing particle size, the equilibrium solute concentration for a small particle is much higher than for a large particle, as described by the Gibbs–Thompson equation. The resulting concentration gradients lead to transport of the solute from the small particles to the larger particles. The equilibrium concentration of the nanocrystal in the liquid phase is dependent on the local curvature of the solid phase. Differences in the local equilibrium concentrations, due to variations in curvature, set up concentration gradients and provide the driving force for the growth of larger particles at the expense of smaller particles [Sugimoto, (1987)].

2.3 Habit modification by polymers

2.3.1 Habit modification by polymers of inorganic materials

All crystal growth rates are particle size dependent and size range. For microscopic, submicroscopic particles, the size effect becomes significant.
The morphology-controlling effect of PVA is long known and utilized as a capping agent during the synthesis of nanoparticles. Using radiation chemical reduction, we have

successfully synthesized morphology controlled copper and silver metal nanoparticles by using PVA as a capping agent. (Joshi et al.,1998, & Temgire & Joshi,2004) In the presence of crystal habit- modifying polymers, the crystal growth or nucleation is diverted from the non-uniform to a uniform shape. In most of the earlier studies using PVA as a crystal-habit modifier, a gel matrix made out of PVA has been used for the control of nucleation and morphology in aqueous-solution-based crystal growth. (Merrill & Bassett, 1975, Sinha et.al., 2001, Lakshminarayanan et al., 2003, Joshi et al.,1998 & Temigre & Joshi,2004) Ammonium perchlorate (AP) is one of the most extensively used solid propellant oxidizers in the propellant industry. The percentage of oxidizer in the propellant formulation varies from 70 to 80% by weight, depending on the energetic requirements and compatibility with the other ingredients. Because of the high percentage in the propellant formulation, the performance of the propellants (specific impulse and burning rate) varies with the oxidizer properties, and in turn, the performance of the oxidizer varies with the particles' size and morphology. (Sutton & Oscar, 2001) Hence, in the present investigation, PVA has been used as a supermolecular matrix to control the morphology of AP.

AP is the most commonly used rocket propellant oxidizer and one of the extensively studied ammonium compounds. The morphology of the oxidizer has an important role in the formulation and performance of solid propellants, and the AP crystallized from its saturated solution gives needle-shaped crystals. The nucleation of the crystals was observed immediately after drying began. The crystals grown in the PVA showed entirely different morphologies, such as rectangular prism and rectangular wedge, in comparison to the morphologies of the AP crystals grown in the absence of PVA. The SEM images obtained are shown in Figures 3–8. Three different sets of SEM images were chosen for different concentrations, such as a low salt concentration, equal salt-polymer concentration, and high salt concentration (crystals grown from mixtures A, C, and E) (Vargeese et al.2008)

The images of the crystals grown immediately after mixing the solution and after 24 h of reaction time are shown in panels a and b of each figure, respectively. Figures 3–5 show the images of the crystals grown from PVA 14000. Figure 3 shows that the crystals have an irregular morphology and do not have any growth orientation toward a particular plane. The images also indicate that the crystals have an irregular shape, although they tend to grow in an organized manner. This could be due to the polymer-substrate interaction that prevents the crystals from growing in an organized manner. At low salt concentrations, there is too much hydrogen bonding between the hydroxyl groups of the polymer and the hydrogen of the ammonium ion. Adsorption characteristics of polymers are different from those of other systems because of the polymers' flexibility. In addition to the usual adsorption factors, such as adsorbate–adsorbent and adsorbate-solvent interactions, a major aspect is the conformation of molecules at the interface and its role in dispersion. PVA is a flexible linear molecule with no charge and which can potentially adsorb on the surface. Bridging is considered to be a consequence of the adsorption of individual intermolecular polymer molecules on the surface. This happens through hydrogen bonding. Because of the high polymer concentration, not all the segments of the polymer are in direct contact with the surface. Also, the diffusion of ions is slow at high polymer concentrations. The possibility that the solution does not contain enough AP to grow in an organized manner cannot be ruled out. The viscous nature of the solutions containing a large quantity of PVA polymer leads to rectangular wedge- shaped crystals because diffusion is predominant and convection is suppressed for the transformation of solutes.

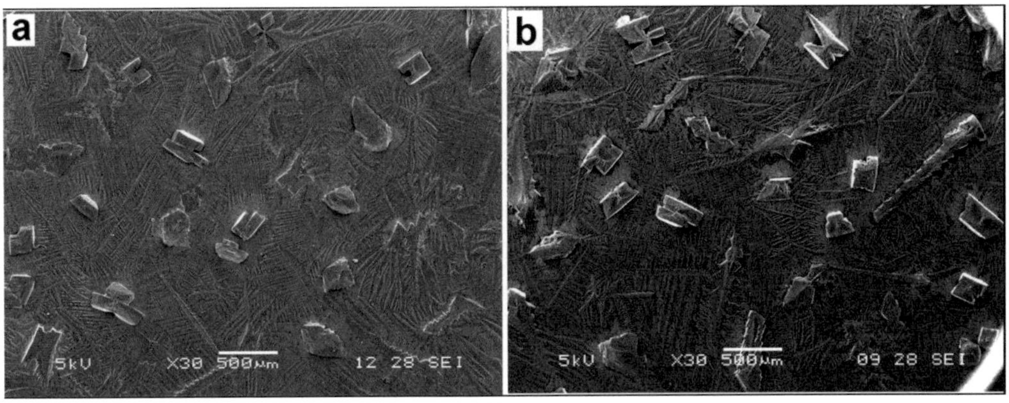

Fig. 3. SEM images of crystals grown from mixture A (PVA 14000) after (a) 0 h and (b) 24 h.

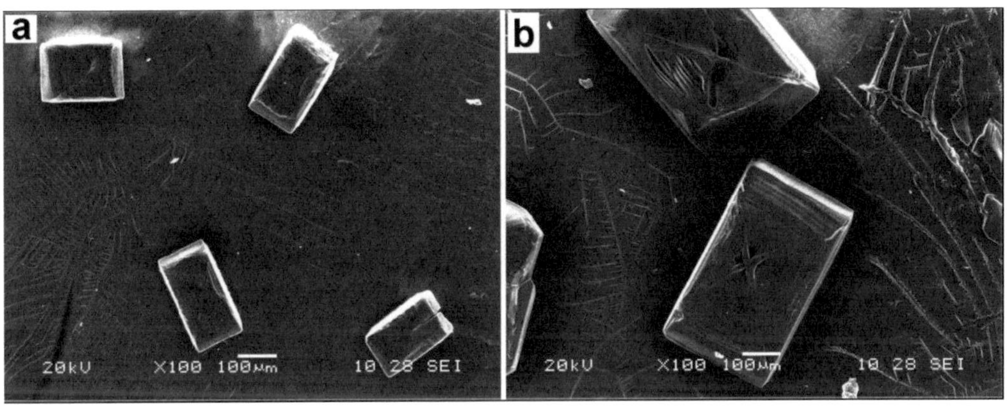

Fig. 4. SEM images of crystals grown from mixture C (PVA 14000) after (a) 0 h and (b) 24 h.

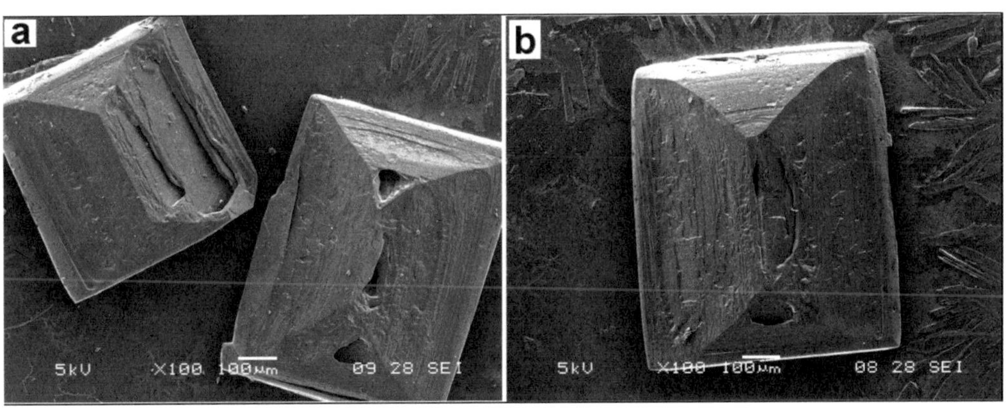

Fig. 5. SEM images of crystals grown from mixture E (PVA 14000) after (a) 0 h and (b) 24 h.

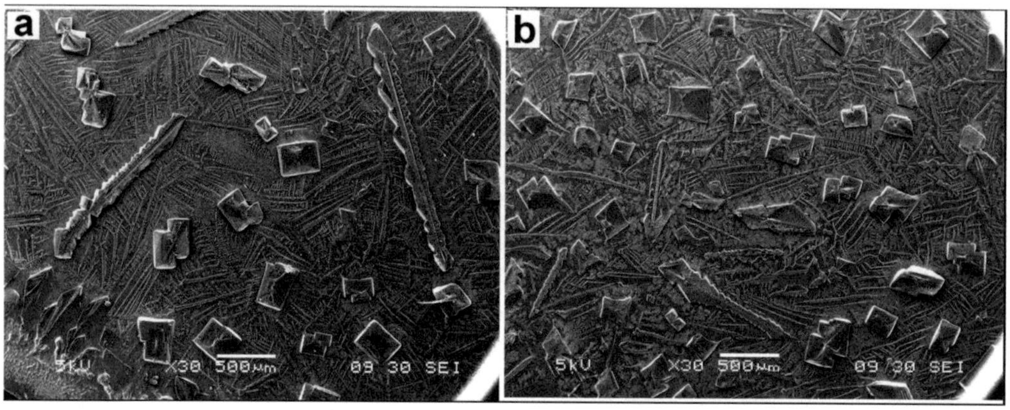

Fig. 6. SEM images of crystals grown from mixture A (PVA 125000) after (a) 0 h and (b) 24 h.

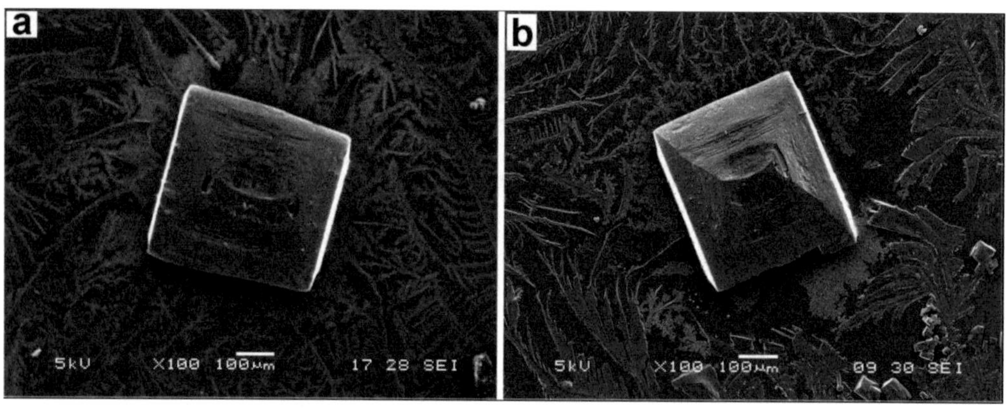

Fig. 7. SEM images of crystals grown from mixture C (PVA 125000) after (a) 0 h and (b) 24 h.

Fig. 8. SEM images of crystals grown from mixture E (PVA 125000) after (a) 0 h and (b) 24 h.

As seen from the SEM images, the salt-to-polymer solution ratio change is reflected in the morphology of the crystals. Although the crystals grown from mixture A have an irregular morphology, the particles grown from the other compositions contain individual crystals with a specific morphology. The crystals grown from mixture C have a rectangular prism shape, whereas the crystals grown from the solution containing a high salt concentration (mixture E) have a rectangular wedge shape. The crystals grown from mixtures B and D also show a more or less similar morphology, comparable to the crystals grown from mixtures C and E, respectively. Hence, the crystals grown from mixtures B and D are not discussed here. The crystals obtained from mixture E have grown in another plane, leading to a rectangular wedge-like morphology in response to the change in polymer-to-salt ratio. This outgrowth to another plane of crystals obtained from mixture E is one of the observed modifications from the crystals grown from mixture C. The morphological evolution of the crystals from rectangular prism to wedge-like shape clearly shows the dependency of the particles' morphology on the polymer concentration and the minimum salt concentration requirement.

The advantage of using PVA as a habit modifier is a reduced agglomeration, leading to samples containing only individual particles. Studies show that, for a chemical system in which PVA and AP molecules are involved, the PVA induces the crystallization of individual AP crystals irrespectively of the polymer-substrate concentration variations. Here, the colloidal action of the PVA or the surfactant activity of the polymer could prevent the particles' agglomeration. The PVA could possibly get adsorbed on the surface, thereby preventing the agglomeration and leading to the formation of individual particles of AP. That is, immediately after the nucleation has started, the polymer might isolate the salt solution into packets and force them to grow as individual particles.

It has also been speculated that the densification of the gel medium increases the random noise for crystal growth because the polymer gel matrices disturb the progress of the growing surface. (Oaki & Imai, 2003) The PVA gel forms an organized matrix under the optimum concentration of components, which ensures the homogeneous distribution of the cations in the polymeric network structures. The chain alignment and interchain separation of PVA, which depend on temperature and concentration, lead to the formation of a polymeric matrix with complex structures. These structures chelate the cations through a process of weak chemical bonding (such as van der Waals, hydrogen bonding) and steric entrapment. (Sinha, 2001) The polymer matrix not only provides an organized surface of mineralization, but also induces a vector growth on the polymeric surface, the direction of which differs from the characteristically preferred direction of the unit cell. (Lee et al., 1999, Addadi &Weiner, 1992, Walsh & Mann, 1995, Mann et al,1988) This results in the formation of extend and nonequilibrium morphologies, as well as metastable phases with lattice parameters on the order of the spacing available in the polymeric matrix. However, the nucleation of a particular space group on a charged polymeric surface not only depends on the lattice geometry, but also includes spatial charge distribution, hydration, defect sites, and surface relaxation. (Mann, 1988) These factors affect the collision frequency and in turn the activation energy for nucleation; hence, the transition state theory might be considered to explain the nucleation of biominerals. (Sinha et.al., 2001)

The shape of inorganic crystals is normally related to the intrinsic unit cell structure, and the crystal shape is usually the outside embodiment of the unit cell replication and amplification. (Yu and Colfen, 2004, Colfen & Mann, 2003, & Jongen et al., 2000)The diverse crystal morphologies that a mineral, identical to that for calcium carbonate, can have are due to the different surface energies and external growth environments of the crystal faces. (Wulff *&Kristallogr 1901)* The morphological evolutions (from irregular to organized crystal assemblies) of the AP crystals are seen in the SEM images. The polymer-substrate interaction is clearly seen in the polymer pattern observed near the crystals. The polymer seems to have grown in the form of dendrites surrounding the crystals with primary and secondary branches. Usually, the rate of nucleation is governed by the temperature, the degree of supersaturation, and the interfacial tension. Crystals often grow from the center of the face and spread outward toward the edges in layers, and these layers may have a thickness of several 1000 Å. During this growth, dissolved impurities may affect the thickness and shape of the layers, which in turn change the morphology of the crystals. Usually, the effect of these impurities is highly specific and depends upon a number of parameters. The growth rate of a crystal face is usually related to its surface energy, if the same growth mechanism acts on each face. The fast growing faces have high surface energies, and they will vanish in the final morphology, and vice versa. This treatment assumes that the equilibrium morphology of a crystal is defined by the minimum energy resulting from the sum of the products of the surface energy and the surface area of all exposed faces (Wulff rule). (Yu & Colfen 2004) The driving force for this spontaneous oriented attachment is that the elimination of the pairs with a high surface energy will lead to the substantial reduction of the surface free energy from a thermodynamic viewpoint. (Banfield et al., 2000 & Alivisatos, 2000) The surface roughness on the molecular level is governed by energetic factors arising from fluid-solid interactions at the interface between the crystal and its growth environment. A change in the solvent often changes the crystal habit, and this may sometimes be explained in terms of interface structure changes. The structure of the growing crystal surface at its interface with the growth medium has an important effect on the particular mode of crystal growth adopted. A functional group with a high affinity ensures the anchoring of the molecule on a particular phase, and the polymeric chain protects the surface from coalescing with the next one through electrostatic repulsion or steric hindrance. (Joshi et al.1998) This result suggests a significant interaction between the polymeric hydroxyl groups and the crystallizing AP, resulting in the considerable influence on both the primary crystallization and the superstructure. (Qi et. al.2000)

2.3.2 Habit modification by polymers of nanomaterials

2.3.2.1 Zinc oxide nanoparticles

In the studies of nanomaterials, it has been observed that the size and shape of a nanomaterial depends on nature of stabilizer i.e. surfactant, ligand, polymer to salt ratio, reaction temperature and time. The synthetic method applied also plays a role. The systematic adjustment of the reaction parameters can be used to control the size and shape, a quality of nanocrystals and inorganic crystals as well. Nanoparticles are small and thermodynamically unstable. After the primary nucleation, the particles grow via molecular addition. Particles can grow by aggregation with other particles called secondary growth. Their growth rates may be arrested during the reaction either by adding surface protecting agents. Nanocrystal dispersions are stable if interaction between the capping groups and solvent is favorable providing an energetic barrier to counter act van der Waals' forces.

In our recent studies, we have synthesized flower-like ZnO nanostructures comprising of nanobelts of 20 nm width by template and surfactant free low-temperature (4 ^{0}C) aqueous solution route. The ZnO nanostructures exhibit flower-like morphology, having crystalline hexagonal wurtzite structure with (001) orientation. The flowers with size between 600 and 700 nm consist of ZnO units having crystallite size of 40 nm. Chemical and structural characterization reveals a significant role of precursor: ligand molar ratio, pH, and temperature in the formation of single-step flower-likeZnO at low temperature. Plausible growth mechanism for the formation of flower like structure has been discussed in detail. Photoluminescence studies confirm formation of ZnO with the defects in crystal structure. The flower-like ZnO nanostructures exhibit enhanced photochemical degradation of methyleneblue (MB) with the increased concentration of ligand, indicating attribution of structural features in the photocatalytic properties. (Vaishampayan et.al.2011)

ZnO exhibits a varied range of novel structures. The relative surface activities of various growth facets under given conditions determine the surface morphology of the grown structure. Macroscopically, a crystal has different kinetic parameters for different crystal planes, which are emphasized under controlled growth conditions. Thus, after an initial period of nucleation and incubation, a crystallite will commonly develop into a three-dimensional object with well-defined, low-index crystallographic faces.

Wurtzite ZnO being a polar crystal, Zn forms a positive polar plane and O forms a negative polar plane. Zn^{2+} and O^{2-} ions are tetrahedrally coordinated and stack alternatively along the c-axis thus, ZnO grows along the c-axis. When EA is added in the aqueous solution, it gets hydrolyzed and forms EAH+ molecule. Thus, by Coulomb interaction EAH+ molecule gets adsorbed on the negative polar plane retarding the growth of ZnO along the negative polar plane. Therefore, when appropriate amount of EA is used, it covers the side surfaces of ZnO crystal, enhancing growth along the (0 0 1) direction. When EA concentration is lower, i.e. not enough to cover the whole surface, the Oswald ripening takes place and thereby role of EAH+ in the growth of ZnO crystal results in the formation of flower-like structure where individual petal is formed by the overlay of nanobelts.

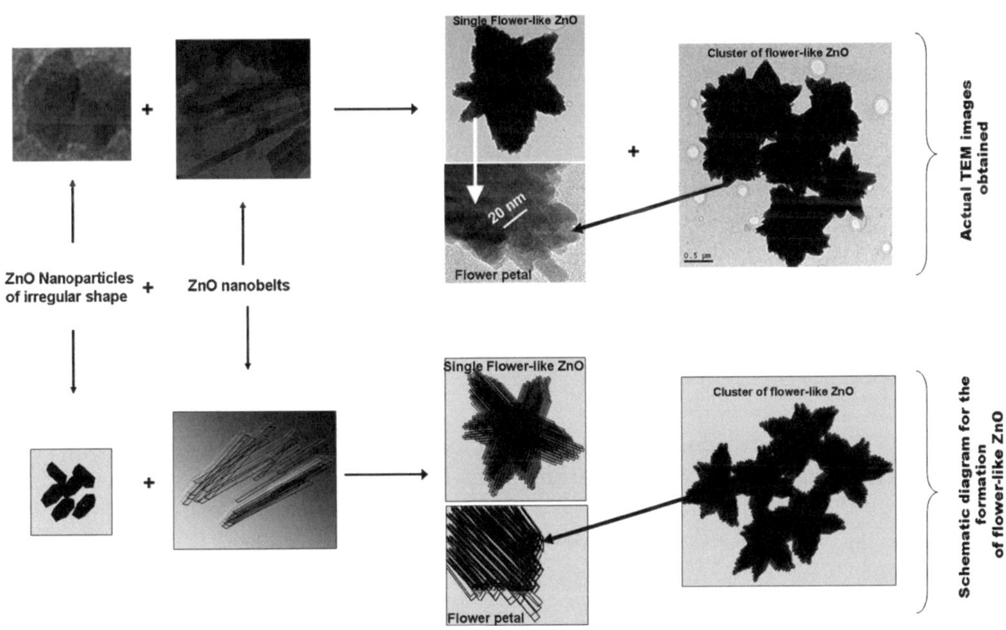

Fig. 9. Diagrammatic representation of formation of flowerlike structure.

Also, some particles of crystallite size 30 nm are seen on the nanobelts. The nanobelts and nanoparticles are formed by conventional nucleation followed by crystal growth process. The thermal energy released during the hydrolysis of EA facilitates nanoparticles to arrange themselves in between the nanobelts so as to form a compact flower-like structure. Also, as the concentration of EA increases, the nanobelts appear to have tapering feature. Thus, the ZA:EA molar ratio plays an important role in framing the morphology of the final product. The EA chelates the cations through a process of weak interactions such as van der Waals forces, hydrogen bonding, and steric entrapment. The ligand not only provides organized surface for structure formation but also induces a vector growth on the surface, the direction of which differs from characteristically preferred direction of the unit cell. This results in non-equilibrium morphologies as well as metastable phases.

We have also synthesized ZnO by aqueous thermolysis method. (Patil & Joshi, 2007) PVA of two different molecular weights was used as a capping agent and as a fuel. A TEM study of ZnO nanoparticles was undertaken to highlight the shape, size and size distribution as well as the crystallinity of the particles. Fig. 10 shows the TEM image of zinc oxide nanocrystals after TGA. The micrograph of sample A (PVA 14,000) showed uniform distribution with nearly spherical morphology (Fig. 10a). All the particles are separated from each other. While sample B (PVA 125,000) synthesized with higher molecular weight exhibits cuboid like morphology and few particles appear to be close to spherical shape (Fig. 10b). In this micrograph the crystals are structurally perfect and attached like beads due to cross-linking of the polymer. This may be due to migration of defects to the surface of crystal during the calcination and growth of particles (Gu et.al, 2004) Fig. 10c shows very small particles of ZnO mostly of spherical shape for sample C (PVP 40,000). PVP does not form gel at room temperature. Therefore, it directly gets solidified while heating the precursor. As the solution is not viscous, particles formed are not cross-linked in the polymer matrix. Hence due to lack of steric interactions as compared to PVA, particles synthesized by PVP are dense but quite separated from each other. The particle sizes of XRD and TEM are comparable. The surface morphology of zinc oxide nanoparticles changes greatly with an increase in oxidation temperature. This can be clearly observed in scanning electron micrographs (SEM). Fig. 10 also shows SEM micrographs for samples A and B after TGA (500 °C) in air. At annealing temperature of 500 °C the ZnO nanoparticles consist of fine grains. Since the grains are agglomerated together, grain boundaries cannot be distinguished clearly. In sample A (Fig.10d) numerous micropores were observed compared to sample B (Fig.10e). This porosity may be generated by the evolution of gases and removal of organic matter, which was loaded in the polymer network and also due to heat generated during combustion of polymers. A significant change in surface morphology is observed in the ZnO annealed at 1000 °C (Fig.10 f&g), well facet grains are observed acquiring dumbbell morphology; their size becomes larger, with a wide range of distribution. At such high temperature, migration of grain boundaries occurs causing the coalescence of small grains and the formation of large grains. In order to understand the process occurring during thermolysis, we have to consider the 'cross-linking' of the polymeric network, which depends on the average molecular weight, degree of polymerization and solubility of the polymer in water. Our observation shows that PVA of molecular weight 14,000 takes almost 24 h to dissolve in water while higher molecular weight PVA, 125,000 dissolves in 2 h. Both have a very good gelling property. Poly-vinyl alcohol is linear flexible molecule with no charge. Therefore, it adsorbs non-specifically on the surface of oxides. The interaction with the surface takes place through hydrogen bonds between polar functional groups of the polymer chain and hydroxylated and protonated groups on the surface. Though the interaction energy between surface and each chain segment is smaller than kT, chains adsorb very well because of large number of contact points. The affinity of the macromolecule for the surface usually increases with its molecular weight. The conformation of the adsorbed polymer remains similar to that of free macromolecule and exhibit tails and loops between contact points. The adsorbed layer provides an excellent steric protection against aggregation (Gennes, 1987, Dickson & Ericsson, 1991). This can be schematically depicted as below.

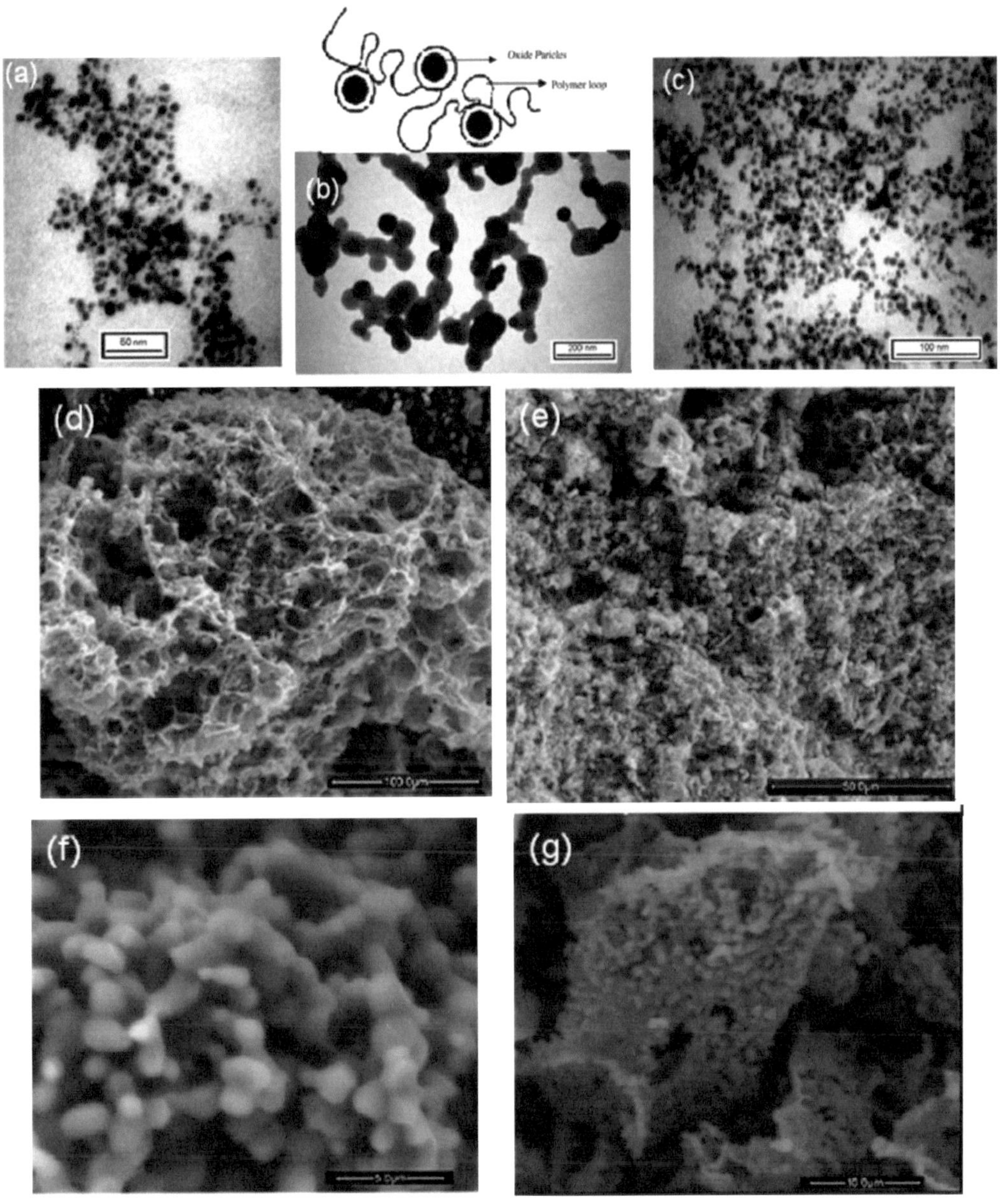

Fig. 10. TEM of ZnO for: (a) sample A, (b) sample B and (c) sample C and SEM of ZnO after TGA in air: (d) sample A at 800× magnification; (e) sample B at 2000× magnification and SEM of ZnO after annealing at 1000 ◦C: (f) sample B at 12000× magnification; (g) sample C at 6000× magnification.

Here polymer is adsorbed and acting as a bridge between particles. The linear chains of PVA can be cross-linked in aqueous medium, i.e. water (Kirk-Othmer, 1983). The cross-linking between the chains may provide small cages wherein the 'sol' of the reactant mixture gets trapped. During thermolysis, the 'sol' trapped in the cages may get converted to ultrafine particles of zinc oxide. Thus, the cages formed by the cross-linking may offer resistance to the agglomeration of the particles and the particle growth. The degree of polymerization can also affect the formation and morphology (Temgire & Joshi, 2004).

2.3.2.2 Palladium nanoparticles

Stable palladium nanocluster catalysts prepared by chemical and -radiolytic reduction methods were found to give very high turn-over frequency numbers in hydrogenation of styrene oxide and 2-butyne-1,4-diol (B3D) as compared to the conventional catalysts. (Telkar et al.2004) A systematic study was carried out on the effects of different transition metals, their reduction methods, types of polymer used as a capping agent, and the concentration and composition of solvent used during catalyst preparation on the size and shape of nanoparticles. The reduction method of metal precursor directly influenced the morphology of the nanoparticles, affecting the catalyst activity considerably. The cubic-shaped nanoparticles (5–7 nm) were obtained in chemical reduction, while radiolytic reduction method gave spherical nanoparticles (1–5 nm).

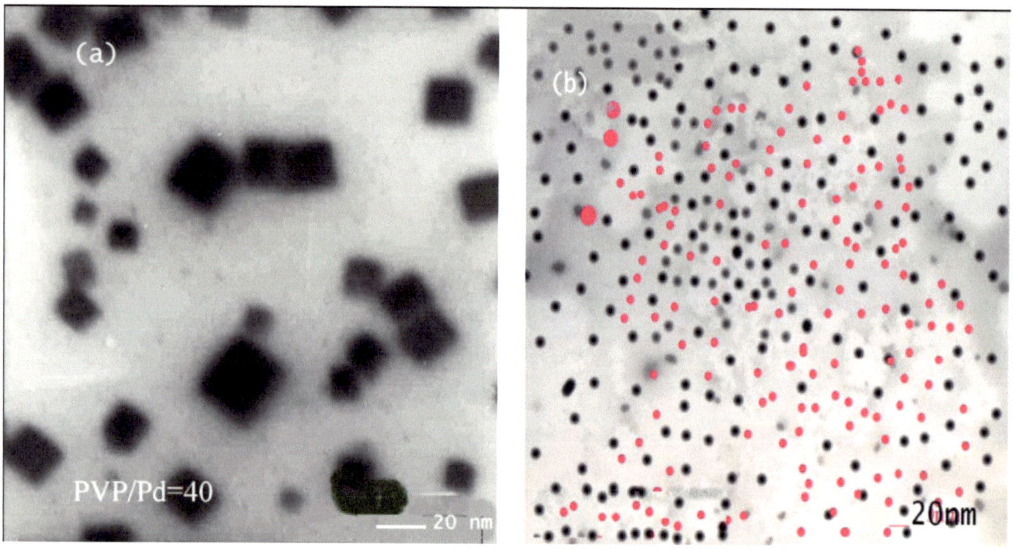

Fig. 11. TEM photograph for Pd nanoparticles prepared by (a) chemical method (AV = 80 kV; magnification = 40,000×) and (b) radiolytic method (PVP/Pd: 40) (AV = 80 kV; magnification = 80,000×). (Applied Catalysis A: General 273 (2004) 11–19)

The activity results, along with the particle sizes of various nanocluster catalysts, are presented in Table 1. The catalyst activity of RRPd (Radiolytic Reduction) catalyst was higher than that of CRPd (Chemical Reduction) catalyst when PVP/Pd ratio was 1, which was in accordance with the fact that the particle size of RRPd was less (5.2 nm) than that of CRPd sample (7.1 nm). As the PVP/Pd ratio was increased from 10 to 40, the activity trend was reversed with respect to particle sizes, thus CRPd samples showed higher catalyst activities than that of RRPd samples. The catalyst activity of CRPd was maximum for PVP/PD ratio of 40, in spite of the fact that particle size of RRPd sample was one-fifth of that of CRPd sample. Such a trend was consistent for both 2-butyne-1, 4-diol and styrene oxide hydrogenation. However, the extent of activity enhancement was dramatically higher for styrene oxide hydrogenation. Besides the reduction in the size, the polymer concentration seems to have a significant effect on the adsorption of reactants. It is reported that, for polymer concentration higher than 50 mg/l, fully developed steric layers are formed around the particle; these act as an effective diffusion barrier that blocks further growth of the metal particles. Pd particles are considered to adsorb onto the polymer. At higher PVP/Pd ratios, beads on string-type of complexes may thus be formed, adsorbing multi-particle complexes (Boonekamp&Kelly, 1994), which is also one of the reasons for higher activity of nanoparticles at higher PVP concentration. The selectivity to B_2D obtained was more than 98% for the Pd catalysts having PVP/Pd ratio in the range of 1-30. A further increase in PVP/Pd ratio to 40 caused a marginal decrease in 2-butene-1,4-diol selectivity from 98 to 91% due to the formation of butane-1,4-diol. In the case of catalysts prepared by the radiolytic method, the particle size reduced to 1 nm with increase in concentration of PVP. Surprisingly, the activity was found to reduce with decrease in particle size. This observation was consistent for both (2-butyne-1, 4-diol (B3D)) as well as styrene oxide hydrogenation (Table 1). This trend can be attributed to differences of shape of the Pd nanoparticles formed by chemical and radiolytic reduction methods. TEM photographs of Pd nanoparticles (Fig.11) showed a distinct morphological change depending on the method of preparation of nanoparticles. Pd particles prepared by ethanol reduction showed particles with a square outline, from which the three-dimensional shapes determined, were found to be cubic. Similar morphology was observed for Pt colloids prepared by H_2 gas reduction. (Ahmadi et al.1996) However, they also obtained a mixture of tetrahedral, polyhedra and irregular-prismatic particles. Milligan and Morris also observed cubical gold nanoparticles for hydroxylamine hydrochloride as the reducing agent (Milligan & Morriss, 1964). In contrast to this, the radiolytic reduction of $PdCl_2$ gave colloidal Pd particles of spherical and oval shapes, 1-5 nm diameter. The final structure and size of the clusters depend on the mechanism of growth process. In the case of radiolytic reduction, the solvated electrons and H• are strong reducing agents and with high rate of reduction, the free metal ions are generally reduced at each encounter (Belloni et al., 1998). In the chemical reduction method, an adsorption of excess of metal ions on the reduced metal clusters, get reduced at a slower rate. This difference in reduction mechanism of radiolytic and chemical reduction may give rise to two distinct shapes of the nanoparticles. It is known that the active sites are more concentrated on the edges of the catalyst these sites may be formed in chemically reduced Pd nanoparticles, leading to higher catalyst activity for these samples. This clearly indicates that not only the particle size but also the shape of the nanoparticles influences the activity of the catalyst (Chen et al., 2000). Nanoparticles of other metals reduced by ethanol also showed cubic shapes while the radiolytic reduction gave spherical particles. As mentioned earlier, the concentration of a stabilizer influenced the nanoparticle size dramatically in case of radiolytically reduced Pd colloids. However, the shape remained spherical, thus confirming that the stabilizer concentration did not contribute to the shape of the nanoparticles. In order to further understand why the catalyst activity decreased in spite of considerable size reduction of radiolytically reduced Pd catalysts; stabilizing polymer alone was irradiated in a separate experiment. There was an increase in the viscosity of irradiated PVP, which indicates increased cross-linking of the polymer (Wang et al.1997). With increase in concentration of polymer, (PVP/Pd = 40), the polymer cross-linking may hinder the access of substrate to the Pd metal particles thereby decreasing the activity of radiolytic nanosize particles.

Method of preparation	PVP/Pd	Particle size (nm)	B_3D hydrogenation			Styrene oxide hydrogenation	
			TOF ($x10^{-5}h^{-1}$)^{a}	Selectivity (%)		TOF ($x10^{-4}h^{-1}$)a	Selectivity PEA (%)
				B_1D	B_2D		
Chemical method (CRPd)	1	7.1	3.0	1.0	99	1.5	99.9
	10	6.1	3.2	1.8	98.2	1.4	99.8
	20	6.0	4.3	1.6	98.4	3.4	99.6
	30	5.5	4.6	1.6	98.4	7.0	99.8
	40	5	5.7	8.8	91.2	10.2	99.9
Radiolytic method (RRPd)	1	5.2	3.6	1.0	99.0	1.9	99.5
	10	5.0	3.2	1.7	98.3	1.8	99.5
	20	4.0	3.0	1.6	98.4	1.5	99.5
	30	3.0	2.8	1.5	98.5	1.3	99.4
	40	1	2.4	1.4	98.6	1.0	99.8

Table 1. Effect of polymer to Pd ratio prepared by chemical and radiolytic methods for hydrogenation reactions.

2.3.2.3 Copper chromite nanoparticles

Amorphous and monodispersed copper chromite nanoparticles were prepared by aqueous thermolysis method using PVA and different ratios of urea-PVA as fuel in air (Hrishikeshi and Joshi, unpublished results, 2011). Morphology and size of nanoparticles were measured by SEM and TEM analysis. Copper chromite ($CuCr_2O_4$) is a tetragonally distorted normal spinel; this distortion is due to Jahn Teller effect of Cu^{+2} (d^9) ions in tetrahedral sites. It is a p-type semiconductor which is widely used as a catalyst for the oxidation of CO (Hertl et al., 1973), hydrocarbons (Mc Cabe & Mitchell, 1983) alcohols (Solymosi & Krix 1962) and as a burn rate catalyst in composite solid propellants, (Prince, 1957, Solymosi & Krix 1962, Patil et al., 2008) Well resolved square bipyramidal morphology was seen in all copper chromite samples using PVA alone. The habit modification of copper chromite was observed due to presence of urea. The urea molecule is planar in the crystal structure, but the geometry around the nitrogens is pyramidal in the gas-phase minimum-energy structure. In solid urea, the oxygen center is engaged in two N-H-O hydrogen bonds. The resulting dense and energetically favorable hydrogen-bond network probably changes the morphology after combustion process.

Figure 12 shows scanning electron micrograph of Copper chromite (a) using only PVA and (b) using Urea and PVA after annealing at 800°C. As obtained as well as annealed samples show uniform and compact distribution of copper chromite $CuCr_2O_4$ nanoparticles. There is almost no porosity in the as obtained as well as in annealed samples. Polymer is adsorbed and acting as bridge between particles. The linear chains of PVA can be cross linked in aqueous medium (Kirk-Othmer, 1983). The cross linking between the chains may provide small cages wherein the "sol" of the reactant mixture gets trapped. During combustion, the "sol" trapped in the cages may get converted to ultrafine particles of copper chromite. Thus cages formed by the cross linking may offer resistance to the agglomeration of the particles and particle growth. Perfect square bipyramidal morphology is seen in PVA capped and orthorhombic in annealed samples. Sharpness of edges decreases gradually with increase in urea content in the fuel mixture.

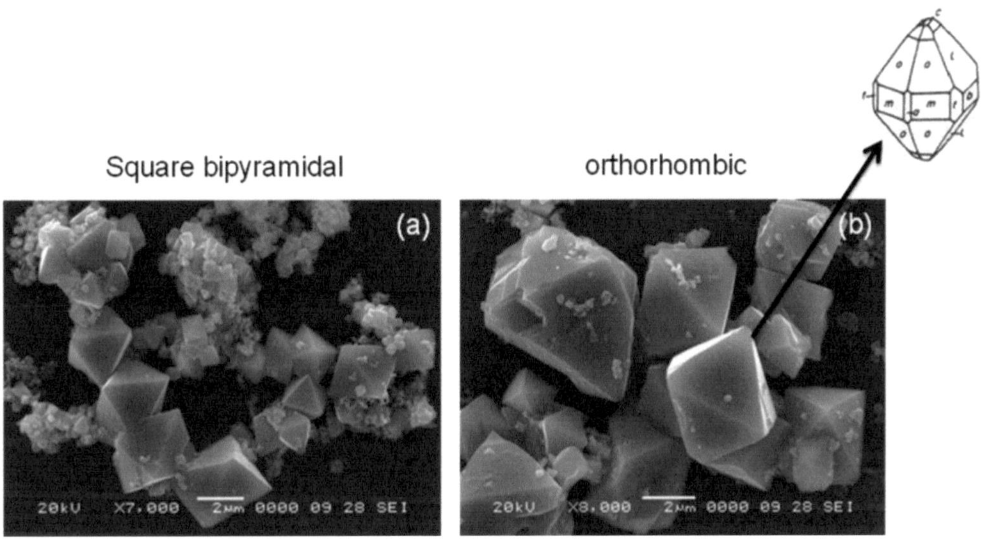

Fig. 12. SEM micrographs of Copper chromite (a) using only PVA and (b) using Urea and PVA after annealing at 800°C.

I have tried to discuss the morphological changes and habit modification of some of the materials studied in our group, on the basis of the theories put forth and the literature.

3. Conclusion

In this Chapter, the habit modification and morphological changes of some inorganic materials in microsize and nanosize are discussed. In most of these studies polymers play multiple roles as a fuel in combustion synthesis, encapsulating agent and as a habit modifier in other synthesis method applied. We have observed that the size, shape, morphology of the synthesized material depends on various factors like nature of polymer, its degree of polymerization, molecular weight, reaction time, synthetic method applied and also on heat of reaction. In the methods applied at high temperature, rapid nucleation time gives rise to short burst of nuclei which might react with intermediate species and the reactions are more kinetically controlled. When the synthesis was carried out at low temperatures, nucleation process is slow and thermodynamically driven process. With aging, growth process becomes more favorable. Final morphology of the material depends on equilibrium conditions related to minimum surface energy, rate of nucleation and growth.

4. Acknowledgement

My sincere thanks are due to Ms. Tajana Jevtic, Publishing Process Manager, In Tech, for inviting me to write a chapter on the work related to crystal growth. It is rewarding to be a contributory of the book, "Crystal Growth". I take this opportunity to thank all my research students, who have worked hard and contributed to the field of Nanoscience and related area. I need to mention the effort of Ms. Shubhangi Borse, in editing the manuscript, as per the requirements of the prescribed format. I acknowledge my family members, for their co-operation, wholehearted support and constant encouragement during the preparation of this chapter.

5. References

Addadi, L., Weiner, S. (1992) Preparation of portland cement components by poly(vinyl alcohol) solution polymerization *Angew.Chem., Int. Ed., 31,* 153–169.

Ahmadi T.S., Wang Z.L., Green T.C., Henglein A., El-Sayed M.A. (1996) Shape-Controlled Synthesis of Colloidal Platinum Nanoparticles, *Science* 272, 1924.

Alivisatos, A. P. (2000) Naturally Aligned Nanocrystals *Science,* 289, 736–737.

Banfield, F., Welch, S. A., Zhang, H., Ebert, T. T. Penn, R. L. (2000) Aggregation-Based Crystal Growth and Microstructure Development in Natural Iron Oxyhydroxide Biomineralization Products Banfield, *Science* 2000, *289,* 751–754.

Boonekamp E.P, Kelly J.J, Fokkink L.G.J (1994) Adsorption of nanometer-sized palladium particles on Si(100) surfaces, *Langmuir,* 10, 4089–4094.

Colfen, H. (2001) Double-Hydrophilic Block Copolymers: Synthesis and Application as Novel Surfactants and Crystal Growth Modifiers, *Macromol. Rapid Commun., 22,* 219–252.

Colfen, H., Qi, L. A. (2001) Systematic Examination of the Morphogenesis of Calcium Carbonate in the Presence of a Double-Hydrophilic Block Copolymer, *Chem. Eur. J., 7,* 106–116.

Colfen, H., Mann, S. (2003) Higher-Order Organization by Mesoscale Self-Assembly and Transformation of Hybrid Nanostructures, *Angew. Chem., Int. Ed., 42,* 2350–2365.

De Gennes P.G. (1987) Polymers at an interface; a simplified view, *Adv. Colloid Interface Sci.,* 27 189

Dickson E., Ericsson, L. (1991) *Adv. Colloid Interface Sci.,* 143 9.

Gibbs,J.W (1948) *Collected works,* Vol.1, *Thermodynamics,* Yale University Press,New Haven.

Gorna, K., Munoz-Espi, R., Grohn, F., Wegner, G. (2007) Bioinspired Mineralization of Inorganics from Aqueous Media Controlled by Synthetic Polymers, *Macromol. Biosci., 7,* 163–173.

Gu, F., Wang,S.F., Lu, M.K.,. Xu, D, Yuan. D.R. (2004) Structure evalution & highly enhanced luminescence of DY^{+3} -doped ZnO nanocrystals by Li doping via combustion, *Langmuir* 20, 3528.

Guo, X., Yu, S., Cai, G. (2006) Crystallization in a Mixture of Solvents by Using a Crystal Modifier: Morphology Control in the Synthesis of Highly Monodisperse $CaCO_3$ Microspheres, *Angew. Chem., 118*, 4081–4085.

He, L., Zhang, Y., Ren, L., Chen, Y., Wei, H., Wang, D. (2006) Double-Hydrophilic Polymer Brushes: Synthesis and Application for Crystallization Modification of Calcium Carbonate *Macromol., Chem. Phys. 207*, 684–693.

Henisch, H. K. (1988) In *Crystals in Gels and Liesegang Rings*; Cambridge University Press: Cambridge,; Chapter 2, 29–47.

Jongen, N., Bowen, P., Lemaitre, J., Valmalette, J. C.; Hofmann, H. J. (2000) Precipitation of Self-Organized Copper Oxalate Polycrystalline Particles in the Presence of Hydroxypropylmethylcellulose (HPMC): Control of Morphology, *Colloid Interface Sci., 226*, 189–198.

Joshi, S. S., Patil, S. F., Iyer, V., Mahamuni, S. (1998) Radiation induced synthesis and characterization of copper nanoparticles, *Nanostruct. Mater.*, 10(7) 1135–1144.

Kashchiev, D. (2000) Nucleation, Butterworth-Heinemann Oxford.

Kirk-Othmer Encyclopedia of Chemical Technology, vol. 23, John Wiley and Sons, New York, 1983, p. 856.

Kuldipkumar, A., Tan, Y. T. F., Goldstein, M., Nagasaki, Y., Zhang, G. G. Z., Kwon, G. S. (2005) Amphiphilic Block Copolymer as a Crystal Habit ModifierKuldipkumar, *Cryst. Growth Des., 5*, 1781–1785.

Lakshminarayanan, R., Valiyaveettil, S., Loy, G. L., (2003) Selective Nucleation of Calcium Carbonate Polymorphs: Role of Surface Functionalization and Poly(Vinyl Alcohol) Additive, *Cryst. Growth Des., 3*, 953–958.

Lee, S. J., Benson, E. A., Kriven, W. M. (1999) Preparation of portland cement components by poly(vinyl alcohol) solution polymerization, *J. Am. Ceram. Soc., 82*, 2049–2055.

Lifshitz, I.M., Slyozov, V. V. (1961) *J. Phys Chem. Solids*, 19, 35

Liu, S., Yu, J., Cheng, B., Zhang, Q. (2005) Controlled Synthesis of Novel Flower-shaped $BaCrO_4$ Crystals, *Chem. Lett., 34*, 564–565.

Loste, E., Meldrum, F. C. (2001) Control of calcium carbonate morphology bytransformation of an amorphous precursor in a constrained volume, *Chem. Commun.*, 901–902.

Makoto, K. (2001) *Mem. Nat. Def. Acad., Math., Phys., Chem. Eng., 1*, 1–8.

Mann, S. (1988) Molecular recognition in biomineralization, *Nature, 332*, 119–124.

Mann, S., Heywood, R., Rajam, S., Birchall, J. D. (1988) Controlled crystallization of $CaCO_3$ under stearic acid monolayers, *Nature, 334*, 692–695.

Meldrum, F. C., Ludwigs, S. (2007) Template-Directed Control of Crystal Morphologies *Macromol. Biosci., 7*, 152–162.

Meng, Q., Chen, D., Yue, L., Fang, J., Zhao, H., Wang, L. (2007) Hyperbranched Polyesters with Carboxylic or Sulfonic Acid Functional Groups for Crystallization Modification of Calcium Carbonate, *Macromol. Chem. Phy., 208*, 474–484.

Merrill, L. Bassett, W. A. (1975) The crystal structure of $CaCO_3$ (II), a high-pressure metastable phase of calcium carbonate, *Acta Crystallogr. B., 31*, 343–349.

Mullin, J. W. (2002) In *Crystallization*; Butterworth-Heinemann: Oxford,; Chapter 6, 216–314.

Mullin. J.W., (1961) *Crystallization* (4th Edition Reprinted) Butterworth-Heinemann, ISBN O 7506 4833 3, Oxford.

Nanocallas, G.H. and Purdie, N. (1964) The kinetics of crystal growth. Quarterly Reviews of the chemical Society, 18,1-20.

Nielsen, A.E (1964) Kinetics of precipitation, Pregamon, Oxford.

Nyvlt, J. (1995) The ostwale Rule of Stages. Crystal Research and Technology, 30, 445-451.

Oaki, Y., Imai, H., (2003) Experimental Demonstration for the Morphological Evolution of Crystals Grown in Gel Media, *Cryst. Growth Des,*. 2003, *3*, 711–716.

Patil P. R., Joshi S. S., (2007) Polymerized organic–inorganic synthesis of nanocrystalline zinc oxide, *Materials Chemistry and Physics*,105, 354–361.

Patil P. R., Krishnamurty V. N., Joshi S. S. (2008) Effect of Nano-Copper Oxide and Copper Chromite on the Thermal Decomposition of Ammonium Perchlorate, *Propellants, Explosives, Pyrotechnics*, 4, 33, 266 - 270.

Prince, E., (1957)*Acta.Crystallogr*. 10, 554

Qi, L., Colfen, H., Antonietti, M. (2000) Control of Barite Morphology by Double-Hydrophilic Block Copolymers, *Chem. Mater., 12*, 2392–2403.

Qi, L., Colfen, H., Antonietti, M. (2000) Crystal Design of Barium Sulfate using Double-Hydrophilic Block Copolymers, *Angew. Chem., Int. Ed., 39*, 604–607.

Robinson, K. L., Weaver, J. V. M., Armes, S. P., Marti, E. D., Meldrum, F. C. (2002) Synthesis of controlled-structure sulfate-based copolymers *via* atom transfer radical polymerisation and their use as crystal habit modifiers for $BaSO_4$, *J. Mater. Chem., 12*, 890–896.

Rudolff, J., Antonietti, M., Colfen, H., Pretula, J., Kaluzynski, K., Penczek, S. (2002) Double-Hydrophilic Block Copolymers with Monophosphate Ester Moieties as Crystal Growth Modifiers of $CaCO_3$, *Macromol. Chem. Phys., 203*, 627–635.

Said A.A., (1991) The role of Copper-chromium oxide catalyst in the thermal decomposition of ammonium perchlorate, *J. Therm. Anal.*, 37,959.

Sarig, S., Mullin, J. W. (1980) Size Reduction of Crystals in Slurries by the Use of Crystal Habit Modifiers, *Ind. Eng. Chem. Process Des. DeV., 19*, 490–494.

Sedlak, M., Colfen, H. (2001) Synthesis of Double-Hydrophilic Block Copolymers with Hydrophobic Moieties for the Controlled Crystallization of Minerals, *Macromol. Chem. Phys., 202*, 587–597.

Sinha, A., Agrawal, A., Das, S. K., Ravi Kumar, B., Rao, V., Ramachandrarao, P. (2001) On the growth of monoclinic calcium carbonate in poly(vinyl alcohol), *J. Mater. Sci. Lett., 20*, 1569–1572.

Sinha, A., Kumar Das, S., Rao, V., Ramachandrarao, P. (2000) Synthesis of organized inorganic crystal assemblies, *Curr. Sci., 79*, 646–648.

Sohnel, O. and Garside, J. (1992) Precipitation: Basic Principle and Industrial Applications, Butterworth-Heinemann Oxford.

Solymosi F., Krix E., (1962) Catalysis of solid phase reactions effect of doping of cupric oxide catalyst on the thermal decomposition and explosion of ammonium perchlorate, *J.Catal.*, 1, 468.

Strickland-Constable, R.F (1968) Kinetics and Mechanism of crystallization, Academic Press London.

Sugimoto, T. (1987) *Adv. Colloid Interface Sci.*, 28, 165.

Sutton, G. P., Oscar, B. (2001) In *Rocket Propulsion Elements*; John Wiley & Sons: New York,; Chapter 12. 474–519.

Telkar M.M., Rode C.V., Chaudhari R.V., Joshi S.S., Nalawade A.M., (2004) Shape-controlled preparation and catalytic activity of metal nanoparticles for hydrogenation of 2-butyne-1,4-diol and styrene oxide, *Applied Catalysis A: General,* 273, 11–19.

Temigre, M. K., Joshi, S. S. (2004) Optical and structural studies of silver nanoparticles, *Radiat. Phys. Chem.*, *71*, 1039–1044.

University Press: Cambridge, (1988) Chapter 2, 29–47.

Vaishampayan M., Joshi. S. S., Mulla I. S., (2011) Low temperature pH dependent synthesis of flower-like ZnO nanostructures with enhanced photocatalytic activity *Material Research Bulletin*, 46(5), 771-778.

Vargeese. A. A., Joshi. S. S., Krishnamurthy V. N. (2008) Role of Poly(vinyl alcohol) in the Crystal Growth of Ammonium Perchlorate, *Crystal Growth & Design, 8, 3,* 1060-66.

Volmer,M. (1939) *Kinetic der Phasenbildung*, Steinkopff, Leipzig.

Wagner, C. (1961) *Elektrochem.Z*, 65, 581

Walsh, D., Mann, S. (1995) Molecular recognition in biomineralization, *Nature, 377,* 320–323.

Walton.A.G. (1967) The formation and properties of precipitates, Interscience, New York.

Wang, F., Xu, G., Zhang, Z., Xin, X. (2005) Morphology control of barium sulfate by PEO-PPO-PEO as crystal growth modifier, *Colloids Surf., A, 259*,151–154.

Wang, T., Rother, G., Colfen, H. (2005) A New Method to Purify Highly Phosphonated Block Copolymers and Their Effect on Calcium Carbonate Mineralization, *Macromol. Chem. Phys., 206,* 1619–1629.

Wulff, G. Z. (1901) On the question of the rate of growth and dissolution of crystal surfaces, *Kristallogr. Mineral., 34,* 449–530.

Xu, A., Ma, Y., Colfen, H. (2007) Biomimetic mineralization, *J. Mater. Chem.,* 17, 415-449

Yu, J. G., Zhao, X. F., Liu, S. W., Li, M.; Mann, S., Ng, D. H. L. (2007) Poly(methacrylic acid)-mediated morphosynthesis of $PbWO_4$ micro-crystals, *Appl. Phys. A: Mater. Sci. Process., 87,* 113–120.

Yu, J., Liu, S., Cheng, B. (2005) Effects of PSMA additive on morphology of barite particles, *J. Cryst. Growth., 275,* 572–579.

Yu, S., Colfen, H. (2004) Bio-inspired crystal morphogenesis by hydrophilic polymers, *J. Mater. Chem., 14,* 2124–2147.

Yu, S., Colfen, H., Antonietti, M. (2002) Control of the Morphogenesis of Barium Chromate by Using Double-Hydrophilic Block Copolymers (DHBCs) as Crystal Growth Modifiers, Chem. *Eur. J.*, 8, 2937–2945.

Yu, S., Colfen, H., Antonietti, M. (2003) Polymer-Controlled Morphosynthesis and Mineralization of Metal Carbonate Superstructures, *J. Phys. Chem. B, 107*, 7396–7405.

Zettlemoyer, A.C (ed.) (1969) Nucleation, Dickker, New York.

Part 3

Growth of Organic Crystals

19

Protein Crystal Growth Under High Pressure

Yoshihisa Suzuki
Institute of Technology and Science,
The University of Tokushima,
Japan

1. Introduction

In this chapter, I would like to describe following two main roles of high pressure (up to 250 MPa) on protein crystal growth.

1. High pressure as a tool for enhancing crystallization of a protein
2. High pressure as a tool for modifying a three-dimensional (3D) structure of a protein molecule

For the first role, Visuri et al. reported that the total amount of obtained crystals of glucose isomerase (GI) was drastically increased with increasing pressure, for the first time (Visuri et al., 1990). Such drastic enhancements probably play an important role in increasing the success rate of 3D structure analysis of protein molecules, since crystallization is still the rate limiting step in the structure analysis process. Although they were the pioneers of this field, they did not do further studies on the growth mechanisms of GI crystals. After their pioneer work, many studies have been done on solubility (Section 2), nucleation (Section 3), and growth kinetics (Section 4) under high pressure. Here I would like to review and classify these studies, and present the potential of high pressure as a tool for enhancing protein crystallization.

For the second role, Kundrot & Richards were the pioneers of this field. They analyzed 3D structure of hen egg-white lysozyme under high pressure at the atomic level, for the first time (Kundrot & Richards, 1987). High-pressure protein crystallography is a prerequisite to understanding effects of pressure on an enzymatic activity of a protein at the atomic level (Makimoto et al., 1984). The structural information also plays an important role in the studies of deep-sea organisms (Yayanos, 1986). Pressure probably influences the protein structure through the structure of surrounding water molecules. Thus the protein structure under high pressure has to be solved with water of hydration at ambient temperatures, since a flash cooling method obviously influences the crystal structure (Charron et al., 2002); a freezing process of the method probably influences the structure of the surrounding water molecules, and the process prevents us from an *"in situ"* analysis of the protein structure with water of hydration. In addition, mainly due to the technical difficulties, total number of studies on high-pressure protein crystallography is not so many at this stage. In Section 5, I would like to review the studies on *"in situ"* protein structure analysis under high pressure, and present a new methodology for an ideal *"in situ"* structure analysis.

2. Solubility of protein crystals under high pressure

Solubility is generally important and indispensable to study an equilibrium state between a solution and crystal. Solubility is usually measured at the beginning of the crystallization research to determine the supersaturation σ ($\sigma = \ln(C/C_e)$, C: concentration of the solution, C_e: solubility), because the crystallization phenomena are often well described by using σ. Supersaturation is also named as the *driving force for crystallization*. For the studies at atmospheric pressure, σ has been useful for the discussion on the protein crystallization (Rosenberger et al., 1996). Thus, for a high-pressure study, the high-pressure σ is also useful to discuss the mechanisms of protein crystallization. To determine the high-pressure σ, the high-pressure solubility is indispensable.

2.1 Methodology

Many researchers have measured the solubility under high pressure. However, the solubility varies according to the method, even though the composition of the solution is almost the same (Suzuki et al., 2000b). The variation prevents us from the quantitative discussion. Therefore, the more sophisticated method is expected to measure the more accurate solubility. Here I would like to present several methods for the measurement of solubility under high pressure with their merits and demerits.

2.1.1 Change in the concentration of a supernatant solution with time (*ex situ*)

In general, this method is the most popular one for the solubility of crystals of small molecules. Groß et al. and Lorber et al. reported the solubility of tetragonal lysozyme crystals under high pressure by *ex situ* measurement (Groß & Jaenicke, 1991; Lorber et al., 1996). They incubated the supersaturated solution under high pressure for a certain period, then reduced the pressure and measured the concentration in the supernatant solution. They measured the solubility only from the supersaturated state, where the solubility is always uncertain from 0 mgmL^{-1} to the asymptotic concentration. In addition, Suzuki et al. showed that an asymptotic concentration from a supersaturated state did not correspond to that from an undersaturated state in a realistic time scale (Suzuki et al., 2000b). Although this method provides concentration data directly, it takes very long time to attain an equilibrium condition.

2.1.2 Change in the concentration of a supernatant solution with time (*in situ*)

We have designed an *in situ* precise method of solubility measurement using a Mach-Zehnder interferometer (Suzuki et al., 1998) and measured *in situ* the solubility of lysozyme (Suzuki et al., 2000b). Using the method, the relative change in the concentration with time during equilibration was measured accurately and continuously starting from a supersaturated state (growth relaxation) and an undersaturated state (dissolution relaxation). The asymptotic concentration for the dissolution relaxation was regarded as the solubility. This method is more precise than *ex situ* one (described in 2.1.1). However, a long time period is still required to establish a solubility curve by this method, and this remains as one of the most serious barriers to further high-pressure studies.

2.1.3 Change in the crystal size with pressure

Takano et al. provided a much-improved technique based on in situ observation (Takano et al., 1997). They measured the equilibrium pressure in situ from supersaturation and undersaturation. They gradually increased the pressure of the sample over a period of few days, and continuously recorded the images of the lysozyme crystal. The solubility was determined from changes in both the size of the crystal and the amount of the transmitted light through the crystal. Although they could shorten the period for one plot, a long time period was still required to establish a solubility curve.

We also measured the solubility of triclinic lysozyme crystals by observing the crystals before and after pressurization for two hours with high precision, although the measurements had been conducted *ex situ* (Suzuki et al., 2011).

2.1.4 Change in the concentration distribution around a crystal (*in situ*)

In order to decrease the time necessary for the solubility measurement under high pressure, another interferometric technique has been developed, which can determine the solubility of lysozyme within at most 3 hours (Sazaki et al., 1999). In the interferometric method, an equilibrium temperature of a given concentration is determined by observing the concentration distribution around a crystal. The distribution can be visualized by using a Michelson interferometer (Sazaki et al., 1999).

Under high pressure, the concentration distribution around the crystals was observed in situ with the Michelson interferometer. Figure 1 shows interferograms of the solution around a GI crystal under 100 MPa (Suzuki et al., 2002b). Here, the concentration of glucose isomerase was 35.4 mgmL^{-1}. If the temperature of the sample was set lower than its equilibrium temperature, the crystal grew (24.7 °C), and the fringes were bent in the vicinity of the crystal (Fig. 1(a)), because of the decrease in the concentration around the crystal. On the other hand, when the temperature was raised higher than its equilibrium temperature (44.1 °C), the crystal dissolved and the fringes bent in the opposite direction (Fig. 1(b)). From observation of the fringes around the crystal, we determined the equilibrium temperature of the crystal and solution of a given concentration.

Fig. 1. Interferograms around the glucose isomerase crystal under 100 MPa (Suzuki et al., 2002b). Concentration of glucose isomerase in bulk solution is 35.4 mgmL^{-1}. (a) Growth (24.7 °C), (b) dissolution (44.1 °C). The scale bar represents 1 mm.

Although this technique reduced the measurement time for one data point drastically (within 3 hours), the error of the data points was generally larger than the method of 2.1.3.

2.1.5 Change in the position of steps or the morphology of ledges of crystals (*in situ*)

Among the many studies on protein solubilities so far, *in situ* observation of steps on crystal faces using a laser confocal microscope combined with a differential interference contrast microscope (LCM-DIM (Sazaki et al., 2004)) has been the most powerful method (step-observation method) for measuring the equilibrium temperatures T_e of protein crystals (Van Driessche et al., 2009; Fujiwara et al., 2010). Van Driessche et al. reported that this method yielded the highest precision in measurements of T_e of tetragonal hen egg-white lysozyme crystals (Van Driessche et al., 2009), and we found it produced the fastest results (Fujiwara et al., 2010). For high-pressure solubility, Fujiwara et al. measured that fastest and with highest precision, at this stage.

To tell the truth, we applied this method to measure high-pressure solubility of GI crystals for the first time (Suzuki et al., 2009, 2010a). In these papers, we also use the changes in the morphology of a ledge of a crystal, while the precision was not so high as that of the data measured by Fujiwara et al.

2.2 Solubility data

In addition to the above studies, many studies on the solubility of proteins under high pressure have been reported. The solubility, C_e, of tetragonal (Groß & Jaenicke, 2001; Lorber et al., 1996; Takano et al., 1997; Sazaki et al., 1999; Suzuki et al., 2000a, 2000b, 2002a; Kadri et al., 2002; Fujiwara et al., 2010) and monoclinic (Asai et al., 2004) hen lysozyme, turkey lysozyme (Kadri et al., 2002), and subtilisin (Webb et al., 1999; Waghmare et al., 2000a) crystal increased with increasing pressure, while that of orthorhombic hen lysozyme (Sazaki et al., 1999; Suzuki et al., 2002a), glucose isomerase (Suzuki et al, 2002b, 2005, 2009), and thaumatin (Kadri et al., 2002) crystal decreased with increasing pressure.

Protein crystals	Pressure dependence	Measurement time for one plot	Accuracy
Hen Lysozyme			
Tetragonal	↑ (Positive)	< 70 minutes	< ± 0.7 K in T_e
		(Fujiwara et al., 2010)	
Orthorhombic	↓ (Negative)	< 3 hours	< ± 4.8 K in T_e
		(Sazaki et al., 1999)	
Monoclinic	↑	< 6 hours	< ± 0.4 mgmL^{-1} in C_e
		(Asai et al, 2004)	
Triclinic	↑	< 6 hours	< ± 1.0 mgmL^{-1} in C_e
		(Suzuki et al., 2011)	
Turkey Lysozyme	↑	9 days	-
		(Kadri et al., 2002)	
Purafect Subtilisin	↑	7 days	-
		(Webb et al., 1999)	
Glucose Isomerase	↓	< 90 minutes	< ± 2.5 K in T_e
		(Suzuki et al., 2009)	
Thaumatin	↓	9 days	-
		(Kadri et al., 2002)	

Table 1. Effects of pressure on the solubility of proteins.

The above results are listed in Table 1. In general, the decrease in solubility with pressure results in the increase in nucleation rates and growth rates of crystals. From Table 1, three of eight crystals exhibit the decrease in solubility with pressure. Thus, application of high pressure to a protein solution would be useful for crystallizing previously uncrystallized proteins.

2.3 Thermodynamic analyses

From solubility data, thermodynamic parameters are often calculated using van't Hoff plots. If we assume that the effect of the activity coefficient is negligible, we can estimate the partial molar enthalpy of dissolution, ΔH, from Eq. (1) and the partial molar entropy of dissolution, ΔS, from Eq. (2).

$$\frac{\partial \ln C_e}{\partial (1/T)} = \frac{-\Delta H}{R}, \tag{1}$$

$$\frac{\partial RT \ln C_e}{\partial T} = \Delta S, \tag{2}$$

where C_e: solubility (mg mL^{-1}); R: gas constant.
To estimate ΔH $\ln C_e$ is plotted against T^{-1}. If we assume that ΔH does not depend on temperature, ΔH is estimated from the slope by linear fitting of the plot. In this section, weighted fitting was done, because the temperature error was large at lower concentration region. ΔS is estimated from T-$RT\ln C_e$ plots.
From the dependence of pressure on solubility, if we assume that the effect of the activity coefficient is negligible, the volume change accompanying the dissolution, ΔV ($\Delta V \equiv \bar{V} - V_c$, $\bar{V}$: the partial molar volume of the solute, V_c: molar volume of the crystal), is expressed as,

$$\Delta V = -RT[\frac{\partial \ln C_e}{\partial P}]_T. \tag{3}$$

If ΔV does not depend on pressure up to P MPa, the molar volume change accompanying the dissolution at 0.1 MPa is expressed as,

$$\Delta V = -RT\frac{\ln C_{e,P} - \ln C_{e,0.1}}{P\text{-}0.1}, \tag{4}$$

where $C_{e,P}$ and $C_{e,0.1}$ indicate the solubility at P MPa and 0.1 MPa, respectively.
All the above thermodynamic functions (ΔH, ΔS and ΔV) reported so far are listed in Table 2.
Negative value of ΔV indicates that the partial molar volume of a protein, $\bar{V}$, is smaller than the molar volume of a crystal, V_c, and *vice versa*. Figure 2 represents a simplified model of the states of the protein in the crystal and in solution.
Consider now the change from crystalline to the solution state. If we neglect any change in volume of the protein molecule, the bulk water, and the waters of hydration 2 (those around parts of the protein exposed in both crystal and solution), then ΔV is given by the volume of the waters of hydration 1 (around the contact surfaces of the protein) minus the volume occupied by these same water molecules as "free" water when the protein is in the crystalline state. For this volume change to be negative, as found for tetragonal lysozyme

crystals, the water molecules must be more tightly packed when hydrating the contact regions than when "free" in the bulk water. This, in fact, is expected to be the case for contacts containing a large number of hydrophilic residues. Correspondingly, the positive volume change on dissolution of glucose isomerase crystals implies that the contact surfaces tend to structure the waters of hydration such that they occupy a larger volume than in the bulk. It predicts that the contacts in glucose isomerase crystals should be more hydrophobic than in tetragonal lysozyme crystals.

Protein crystals	ΔV / cm^{3}mol^{-1}		ΔH / kJmol^{-1}		ΔS / Jmol^{-1}K^{-1}	
	0.1 MPa	Authors	0.1 MPa	100 MPa	0.1 MPa	100 MPa
Hen Lysozyme						
Tetragonal	-18±46	S&S	130±10	70±10	460±40	280±40
	-11.6	L				
	-5	We				
	-3.0±0.5	K				
Orthorhombic	5±18	S&S	35±3	35±5	140±10	140±20
Monoclinic		A	102±6	79±2		
Triclinic		S2011	113±4	97±4		
Turkey Lysozyme	-15±1	K				
Purafect Subtilisin	-21±1	We				
	-30±7	Wa				
Glucose Isomerase	54±31	S2002	160±40	210±60	420±100	580±180
Thaumatin	11±1	K				

Table 2. Thermodynamic functions of dissolution obtained by solubility data of protein crystals. Characters listed in Authors column indicate references as follows. S&S: (Sazaki et al., 1999; Suzuki et al., 2002a); L: (Lorber et al., 1996); We: (Webb et al., 1999); K: (Kadri et al., 2002); A: (Asai et al., 2004); S2011: (Suzuki et al., 2011); Wa: (Waghmare et al., 2000a); S2002: (Suzuki et al., 2002b).

The decrease in ΔS of tetragonal lysozyme crystal with pressure can be explained by a decrease in ΔV with pressure. Since the solution is more compressible than the crystal, the magnitude of ΔV is smaller under high pressure than under atmospheric pressure. Smaller $|\Delta V|$ under high pressure can lead to a smaller change in a degree of freedom. In the case of GI crystals, on the other hand, the increase in ΔS can be explained by the increase in ΔV with pressure. The change in ΔH is still not clear. Further crystallographic study on the hydration of the intermolecular contact regions or the other independent measurements of ΔV, ΔH and ΔS may explain these phenomena.

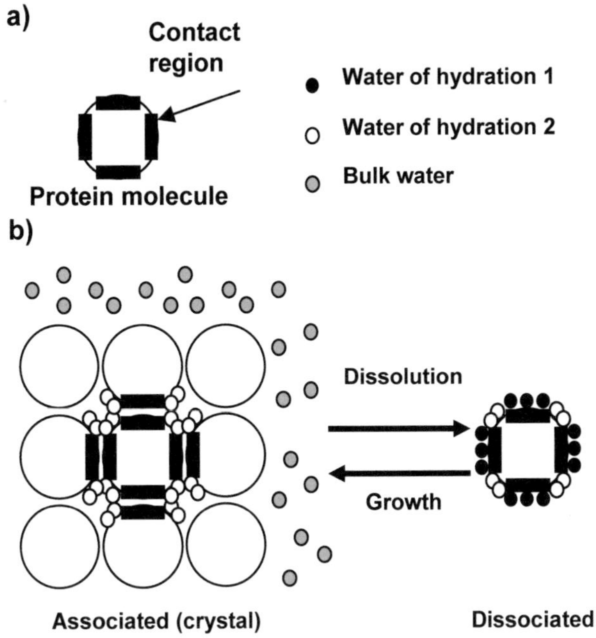

Fig. 2. Schematic diagrams of protein molecules in the crystal and solution (Suzuki et al., 2002a). In the crystal, the molecules are in contact with each other at the regions indicated in black. Hydrated water 1 and 2 represent water molecules hydrating the contact and non-contact regions of the protein, respectively. The volume change on dissolution is given mainly by the difference in volume between hydrated water 1 and bulk water.

3. Nucleation of protein crystals under high pressure

The mechanisms of high-pressure acceleration of 3D nucleation will play the most important role in the improvement of the success rate of crystallization, since the success rate of the 3D nucleation corresponds to that of the crystallization. The precise analyses of the supersaturation dependencies of 3D nucleation rate, J, will clarify the mechanisms.

Except for the data presented in our studies on GI crystals (Maruoka et al., 2010; Suzuki et al., 2010c), the effects of pressure on J have not been reported yet. Although Groß et al. discussed the effects of pressure on the nucleation kinetics using the Oosawa theory of protein self-assembly (Groß et al., 1993), and the group of Glatz discussed the activation volume of the nucleation using the number of crystals (Saikumar et al., 1998; Webb et al., 1999; Waghmare et al., 2000b; Pan & Glatz, 2002), neither group measured J under high pressure directly.

Thus, in this section, I would like to focus on our studies on GI crystals (Suzuki et al., 2009, 2010c; Maruoka et al., 2010).

3.1 Classical nucleation theory

The 3D nucleation rate (Volmer & Weber, 1926), J , is modified and expressed as follows (Suzuki et al., 1994):

$$J = \nu s n_t \exp(\frac{-\Delta G^*}{kT}), \tag{5}$$

where ν, s, n_t, ΔG^*, k, and T represent the collision rate of GI tetramers with critical nuclei, the sticking parameter for the addition of a GI tetramer to a critical nucleus, the number of GI tetramers in the unit volume of a solution, the Gibbs free energy for the formation of a critical nucleus of a GI crystal, the Boltzmann constant, and the absolute temperature, respectively. The variables s and ΔG^* can be expressed as follows (Boistelle & Lopez-Valero, 1990):

$$s = n\exp(\frac{-\varepsilon}{kT}), \tag{6}$$

and

$$\Delta G^* = \frac{f\Omega^2\gamma^3}{(kT\sigma)^2}, \tag{7}$$

where n, ε, f, Ω and γ represent the total number of tetramers adjacent to the surface of a nucleus, the activation energy for the addition of a GI tetramer to a critical nucleus, the shape factor, the average volume occupied by a GI tetramer, and the surface free energy of the GI crystal, respectively. Substituting equation (6) and (7) for (5) and taking the natural log of both sides, we obtain the following expression:

$$\ln J = \ln(\nu n n t) - \frac{\varepsilon}{kT} - \frac{f\Omega^2\gamma^3}{(kT)^3} \times \frac{1}{\sigma^2}. \tag{8}$$

3.2 Methodology

A high-pressure vessel with transparent sapphire windows was used (Maruoka et al., 2010; Suzuki et al., 2010c). An inner cell (inner volume = 1 × 6 × 20 mm^3) for *in situ* observation was made of glass slides, and equipped with soft silicone tubes for sample loading. The cell was set in the vessel, and crystals in the cell under high pressure were observed through the sapphire windows using a stereoscopic microscope (Nikon, SMZ800, objective: EDPlan×2 (N. A. = 0.2)). The solution and pressure medium were separated by soft silicone tubes of the cell. The solution around the crystals was pressurized *via* the tubes. The pressure in the vessel was well controlled automatically by a feedback system with a pressure sensor (accuracy of pressure: ± 0.5 MPa) and could be kept constant for a long time. The temperature of the cell was directly controlled using a Cu jacket with a Peltier element. We could control the temperature from 15.0 to 35.0 °C with the accuracy of ± 0.2 °C.

A supersaturated solution of a given GI concentration was transferred into an inner cell. The number of the observable crystals per unit volume N was counted with time t using a stereoscopic microscope. The nucleation rate J is defined as the slope of the tangent line of the t - N plots at the point of inflection. In practice, we fit Gompertz function, which is a sigmoid function, to the t - N plots, since Foubert et al. fit the Gompertz function to their data of released crystallization heat of fat crystals, and the fit of the Gompertz model seemed to be better than that of the mostly used Avrami model (Foubert et al., 2003). The Gompertz function we used is expressed as,

$$N = a\exp\{-\exp[-k(t - t_c)]\} \tag{9}$$

where t represents time, and a, k, and t_c are fitting parameters. We assume that J is defined as the slope of the tangent line of the Gompertz function at the point of inflection, since the slope provides the maximum value. From eq. (9), J is expressed as,

$$J = ak/e \tag{10}$$

Induction time τ is calculated by substituting $N = 1$ into eq. (9), and expressed as,

$$\tau = t_c - (1/k)\ln(-\ln(1/a)) \tag{11}$$

3.3 Three-dimensional nucleation rates

N increased with time in a sigmoidal-like fashion (Fig. 3). The Gompertz function fitted well all the t - N plots. From the fitting parameters and eq. (10), J was calculated and plotted against σ (Fig. 4). J increased with increasing pressure at the same σ. We also determined τ^{-1} using eq. (11). τ^{-1} also increased with increasing pressure at the same σ (Fig. 4). The increase in J and τ^{-1} with pressure at the same σ indicates that they are kinetically accelerated under high pressure.

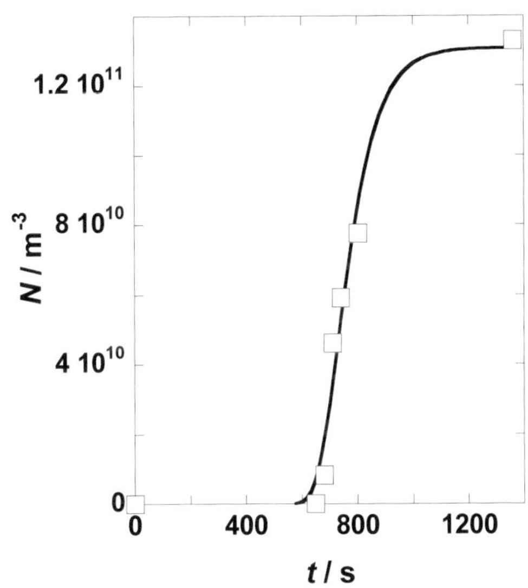

Fig. 3. Time course of the number of observed microcrystals at T = 20 °C, C = 27.07 mgmL^{-1}, and P = 100 MPa (Maruoka et al., 2010). Solid curve indicates a Gompertz function.

Although nucleation of protein crystals under high pressure has been already studied by a few researchers (Suzuki et al., 1994; Waghmare et al., 2000b; Pan & Glatz, 2002), no one has succeeded in measuring J directly and discussing the dependence of J on σ. We previously measured J of tetragonal lysozyme crystals under high pressure by *in situ* observation of the number of crystals using a diamond anvil cell (Suzuki et al., 1994). J decreased with increasing pressure at a constant concentration. However, since the solubility was not measured at that time, we could not separate the effects of solubility change under high pressure. Waghmare et al. and Pan et al. assumed that the final number of subtilisin crystals was proportional to J

(Waghmare et al., 2000b; Pan & Glatz, 2002), and they did not observe the transient number of the crystals. Thus, our results shown in Fig. 4 successfully clarified, for the first time, that the 3D nucleation of GI crystals was kinetically accelerated under high pressure.

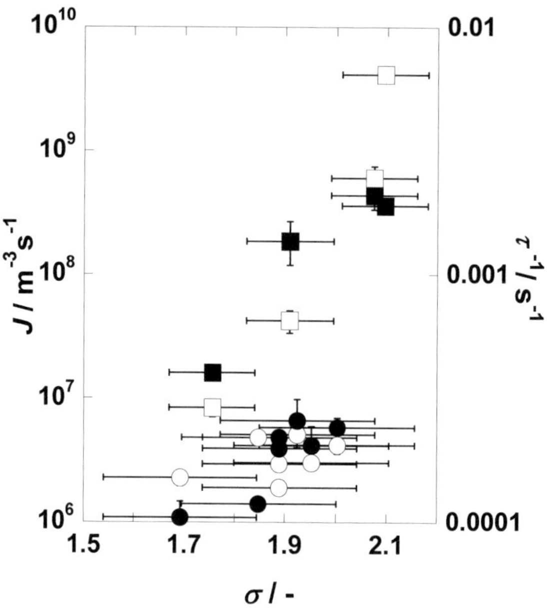

Fig. 4. J and τ^{-1} with supersaturation σ (Maruoka et al., 2010). Open and closed symbols indicate J and τ^{-1}, respectively. Circles and squares indicate the data measured under 0.1 and 100 MPa, respectively.

3.4 Kinetic analyses

We plotted the natural logarithm of J against $1 / \sigma^2$. ln$J$ increased with increasing pressure at the same value, $1 / \sigma^2$. Using eq. (10) and tentatively assuming that f equals 25 (Boistelle & Lopez-Valero, 1990), the average surface free energies γ at 20°C are calculated to be $(8 \pm 3) \times 10^{-5}$ and $(9 \pm 2) \times 10^{-5}$ Jm^{-2} at 0.10 and 100 MPa, respectively. γ does not change with pressure within experimental errors. This result does not correspond with our previous results, in which γ decreased with increasing pressure (Suzuki et al., 2005). This inconsistency is mainly due to the experimental errors in J. To confirm the pressure dependency of γ in detail, we will need to measure the two-dimensional (2D) nucleation rates with σ under high pressure (Suzuki et al., 2009, 2010a; Van Driessche et al., 2007).

On the other hand, the intercept of the linear fitting function shown in Fig. 2, $\ln(\nu n n_t) - \varepsilon / kT$, at 100 MPa is much larger than that at 0.10 MPa. This indicates that the activation energy for the addition of a GI tetramer to a critical nucleus, ε, decreases drastically with increasing pressure, since ν should not change so much, and n and n_t of 100 MPa are less than those of 0.10 MPa at the same σ.

3.5 Two-dimensional (2D) nucleation rates

We preliminarily measured *in situ* 2D nucleation rates J_s of 2D islands on the {011} face of glucose isomerase crystals at 0.1, 25 and 50 MPa. J_s increased with increasing pressure. For these plots, GI concentration in the bulk solution C (= 5.6 mg mL^{-1}) and temperature (T = 26.4 °C) were constant throughout the measurement. Thus, the increases in J_s are completely due to the increase in pressure.

4. Growth kinetics of protein crystals under high pressure

To understand the mechanisms of crystal growth precisely, growth kinetics should be clarified. In this section, I would like to describe mainly following two topics.

1. Effects of high pressure on growth rates of crystal faces, R
2. Effects of high pressure on step velocities, V

4.1 Effects of high pressure on growth rates of crystal faces, *R*

Kinetic analyses of R provide useful information about growth mechanisms. Pressure effects on the kinetics of R of protein crystals are listed in Table 3.

Protein crystals		Pressure effects	Authors
Hen Lysozyme			
	Tetragonal	Inhibition	Suzuki et al., 2000a
	Orthorhombic	Inhibition	Nagatoshi et al., 2003
	Monoclinic	Acceleration	Asai et al., 2004
Purafect Subtilisin		Inhibition	Waghmare et al., 2000a
Glucose Isomerase		Acceleration	Suzuki et al., 2005

Table 3. Effects of pressure on the growth kinetics of protein crystals.

4.1.1 Growth theory

How does pressure affects R kinetically? The following three hypotheses are conceivable. (1) An increasing pressure reduces the volume of the system, and thus elevates the protein concentration. (2) The rising pressure brings about changes in the crystals' growth mode. (3) Changes in growth parameters such as an activation energy, surface free energy, etc. occur with elevations in pressure.

(1) Elevation of the protein concentration through a reduction in the system volume

Let us first consider hypothesis (1). How much does the concentration change with increasing pressure? Kundrot & Richards reported that from 0.1 to 100 MPa the volume contraction of a tetragonal hen lysozyme (t-HEWL) crystal, the solvent in the crystal and the bulk solution were 1.1, 3.7 and 3.7 % (Kundrot & Richards, 1987, 1988), respectively. For a t-HEWL crystal, regardless of the increase in the protein concentration resulting from the volume contraction, the growth kinetics decelerates with increasing pressure (Suzuki et al., 2000a). In the case of the other protein crystals listed in Table 3, the volume contraction of the system is probably of the same order as that of the t-HEWL crystal (i. e. several percent). Thus, this hypothesis (1) can not explain the inhibition of the growth kinetics of protein crystals. In addition, it hardly explains the significant acceleration of the growth kinetics of monoclinic hen lysozyme and GI crystals.

(2) Changes in the crystals' growth mode
To evaluate the second hypothesis, we should first confirm the crystal growth mode under all the growth conditions. Since all the crystals described in this review had clear facets, the crystals formed via a layer-by-layer growth mechanism. Therefore, they must have grown in a spiral growth mode with screw dislocations or in a 2D nucleation growth mode (Fig. 5).

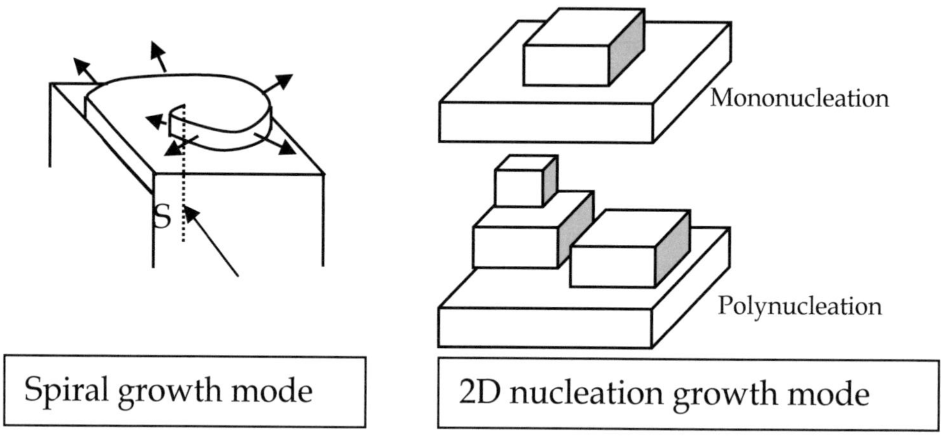

Fig. 5. Schematic illustrations of typical growth modes of crystals with clear faces.

If the density of the screw dislocations is sufficiently low, the growth rate R of the spiral growth mode is expressed as (Burton et al., 1951; Cabrera & Levine, 1956),

$$R = \frac{K_s h}{19 f_0 \kappa} \sigma^2 . \qquad (12)$$

Here, K_s is a step kinetic coefficient, h a step height, f_0 the area which is occupied by a molecule on the crystal face, and κ a ledge free energy. In Table 3, only for t-HEWL and orthorhombic lysozyme (o-HEWL) and GI crystals, R of specific faces of the crystals were precisely measured. Supersaturation dependencies of R of the above three crystals were not fitted well using the equation (12).
Next, we take into account the 2D nucleation growth model. Actually, there are two models that can represent the 2D nucleation growth mode: one operating by mononucleation and the other by polynucleation. Through the following reasoning, we judged that the latter is the growth mode in the present cases. The growth rate in the mononucleation mode is proportional to the surface area of the relevant face (Markov, 1995). However, all R referred in the present review did not depend on the surface area, although different crystals of different size were used. In addition, to confirm the growth mode directly, we observed the surface topography *in situ* using a reflection type laser confocal microscope combined with a differential interference contrast microscope (LCM-DIM system) (Sazaki et al., 2004) for GI crystals. The 2D nucleation and subsequent lateral growth of the 2D islands were clearly observable. Thus, we concluded that the GI crystal grew in the polynucleation mode. In the case of t-HEWL, we also confirmed polynucleation growth under high pressure.
R in the polynucleation mode is expressed as (Suzuki et al., 2000a, 2005; Nagatoshi et al., 2003),

$$R = k_1 \exp(\frac{2\sigma}{3})(\exp\sigma - 1)^{2/3}\sigma^{1/6}\exp(\frac{-k_2}{\sigma}), \tag{13}$$

where k_1 and k_2 are expressed as,

$$k_1 = (\frac{\pi}{3})^{1/3} a^{13/3} h^{4/3} \nu \lambda_0^{-2/3} C_e^{4/3} \exp(-\frac{\varepsilon + \varepsilon_{ad} + 2\varepsilon_{kink}}{3kT}), \tag{14}$$

and

$$k_2 = \frac{\pi\gamma^2}{3k^2T^2}. \tag{15}$$

In Eqs. (13), (14) and (15), the following symbols are used: a is the distance between the molecules in the crystal; h is the step height; ν is the thermal frequency of a solute; λ_0 is the average distance between the kinks on a step; ε is the activation energy for a solute molecule to be incorporated into a critical nucleus; ε_{ad} is the activation energy for a solute molecule to be adsorbed on the crystal surface; ε_{kink} is the activation energy for a solute molecule to be incorporated into a kink site; γ is the molecular surface energy that represents the excess free energy due to unsatisfied bonds of a molecule at a step edge; and k is the Boltzmann constant. By nonlinear least squares fitting, eq. (13) reproduces the experimental data well. All experimental data were best fitted to the 2D nucleation growth mode of the polynucleation type. Hence we concluded that there was no change in the growth mode with increasing pressure. Thus pressure effects are mainly due to (3) Changes in growth parameters such as an activation energy, surface free energy, etc. with pressure.

4.1.2 Summary of kinetic analysis of *R*

For t-HEWL, o-HEWL, and GI crystals, R were measured with σ under high pressure. By fitting these data with the equation (13), kinetic constants k_1 and k_2 were calculated as shown in Table 4.

The value of k_1 of the {110} face of t-HEWL crystals at 100 MPA for the best fit ($k_1 = 7.1 \times 10^6$ nms^{-1}) is extraordinarily large and this value has less physical meaning. This is owing to the lack of data and error of R. As in eq. (14), there are too many factors to determine which dominate the increase in k_1 with pressure. Thus, I also showed the result in which I fixed k_1 value. The increase in k_1 and the decrease in k_2 result in the acceleration of growth kinetics. In Table 4, the dependence of k_1 and k_2 on pressure is generally complicated.

First, in the case of t-HEWL crystal, for both faces, k_1 increases with increasing pressure, while k_2 increases. This shows that the effect of the increase in surface free energy dominates the overall inhibition of the growth kinetics. Second, results of o-HEWL crystals show different pressure dependencies. Both k_1 and k_2 decrease with an increase of pressure. The decrease in k_1 dominates the inhibition of the growth kinetics under high pressure. Third, the acceleration of growth kinetics of GI crystals is mainly due to the decrease in surface free energy (k_2).

To study the effects of pressure on each parameter precisely, further accumulation of the data of R is needed.

Protein crystals and constants	Authors	Pressure / MPa		
		0.1	50	100
Hen Lysozyme				
Tetragonal	Suzuki et al., 2000a			
{110} face				
k_1 / nms^{-1}		0.84	3.5	7.1×10^6
k_2 / -		3.6	8.7	45
k_2 (k_1 = 0.84 nms^{-1} fixed) / -		3.6	5.4	9.6
{101} face				
k_1 / nms^{-1}		0.14	0.33	0.86
k_2 / -		1.0	3.8	6.5
k_2 (k_1 = 0.14 nms^{-1} fixed) / -		1.0	2.0	2.6
Orthorhombic	Nagatoshi et al., 2003			
k_1 / nms^{-1}		4.7 ± 1.3		1.7 ± 0.6
k_2 / -		2.0 ± 0.4		1.5 ± 0.5
Glucose Isomerase	Suzuki et al., 2005			
{101} face				
k_1 / nms^{-1}		7 ± 6		0.60 ± 0.08
k_2 / -		18 ± 2		3 ± 1

Table 4. Kinetic constants of 2D polynucleation growth.

4.2 Effects of high pressure on step velocities, *V*

In the case of a 2D nucleation growth model, since R is proportional to $J_s^{1/3}V^{2/3}$ (J_s: 2D nucleation rate, V: step velocity) (Markov, 1995), the model analyses of R are indirect and prevent further detailed analysis. *In situ* observation of the steps enables us to directly and separately measure J_s and V, which are necessary to elucidate the causes of high-pressure acceleration of the nucleation and growth. Hence, direct observation of individual elementary steps plays a crucial role in studies of crystallization under high pressure.

Direct observations of individual elementary steps on protein crystal surfaces have been carried out mainly by atomic force microscopy (AFM) (Durbin & Carlson, 1992; Durbin et al., 1993; McPherson et al., 2000). However, AFM does not work under high pressure (> 6 atm) at the present stage (Higgins et al., 1998). Besides, the scan of a cantilever would potentially affect the soft surfaces of protein crystals. Advanced optical microscopy is a promising alternative to directly and noninvasively observe individual elementary steps even under high pressure. Among various kinds of advanced optical microscopy, we adopted laser confocal optical microscopy combined with differential interference contrast microscopy (LCM-DIM), by which we have already succeeded in observing the elementary steps of GI crystals under atmospheric pressure (Suzuki et al., 2005). The development of a high-pressure vessel with an optical window thin enough to suppress optical aberration, also played a crucial role in the LCM-DIM system.

At this stage, V of GI crystals under high pressure are only available data (Suzuki et al, 2009, 2010a).

4.2.1 Theory of step velocities, *V*

Assuming a direct incorporation process, step velocity on a specific face of a GI crystal V is expressed as follows (Suzuki et al., 2009, 2010a):

$$V = \beta_{\mathrm{step}}\Omega(C - C_{\mathrm{e}}) \tag{16}$$

where β_{step} and Ω represent the step kinetic coefficient of the incorporation process of GI tetramers, which are the growth units of GI crystals, at kink sites on steps of GI crystals and the average volume occupied by a GI tetramer in the crystal, respectively. We used bulk concentration C instead of the concentration adjacent to a crystal surface, C_{surf}. β_{step} is expressed as follows (Chernov, 1984):

$$\beta_{step} = \nu \frac{p}{\lambda} a \exp(-\frac{\varepsilon_{kink}}{kT}), \tag{17}$$

where ν, p, λ, a, and $\varepsilon_{\mathrm{kink}}$ represent the vibrational frequency of a GI tetramer, unit cell length parallel to a step, kink spacing, unit cell length perpendicular to the step, and activation energy of the incorporation of a GI tetramer into a kink site on the GI crystal surface, respectively. The variables ν, p, and a seldom change with increasing pressure. λ probably does not change with increasing pressure too much, since the shape of a step does not change with increasing pressure (Suzuki et al., 2009, 2010a). Most of the steps on GI crystal surfaces were straight ones and the shape of the steps did not change with increasing pressure. This means that λ did not change significantly; the change in β_{step} was mainly due to the change in $\varepsilon_{\mathrm{kink}}$.

Based on the dependence of pressure on β_{step}, the volume change in going to the activation state in the incorporation process of GI molecules at the kink site on the step, $\Delta V^{\ddagger}$ ($\Delta V^{\ddagger} \equiv V^{\ddagger} - \bar{V}$, $V^{\ddagger}$: partial molar volume of the activated GI tetramer in the solution), is expressed as follows (Laidler, 1987):

$$\Delta V^{\ddagger} = -RT[\frac{\partial \ln \beta_{\mathrm{step}}}{\partial P}]_T . \tag{18}$$

If $\Delta V^{\ddagger}$ does not depend on pressure up to P, $\Delta V^{\ddagger}$ at 0.10 MPa is expressed as follows:

$$\Delta V^{\ddagger} = -RT\frac{\ln \beta_{step,P} - \ln \beta_{step,0.10}}{P - 0.10}, \tag{19}$$

where $\beta_{\mathrm{step},P}$ and $\beta_{\mathrm{step},0.10}$ indicate the step kinetic coefficients at P and 0.10 MPa, respectively.

4.2.2 Experimental

This study made use of an LCM-DIM system (Olympus Optical Co., Ltd.). To measure V of GI crystals precisely, a super-luminescent diode (SLD) laser (Amonics Ltd., model ASLD68-050-B-FA: λ = 680 nm), whose coherence length is about 10 μm, was adopted as a light source.

Figure 6 shows a schematic illustration of a high-pressure vessel and a GI crystal. A high-pressure vessel with a 1-mm-thick sapphire window (Syn-corporation, Ltd., PC-100-MS) was specially designed and used for the *in situ* observation of crystal surfaces under high pressure. We used an O-ring to provide a seal between the sapphire window and a stainless steel support. The surface of the support attached to the sapphire window was processed by mirror polishing to increase the pressure that the O-ring could withstand. In this study, achieving a balance in the thickness of the sapphire window was particularly important, since a thinner window decreases optical aberration, while a thicker one raises the withstand pressure. To our knowledge, the vessel used in this study provides top performance in *in situ* observation of crystal surfaces. The volume of sample space in this vessel is 8.3 mm^3 (4.3 mm in height and 1.6 mm in diameter).

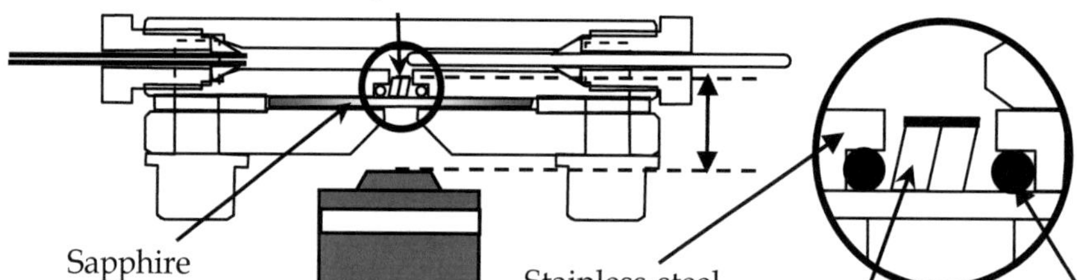

Fig. 6. Schematic illustration of an experimental setup.

To compensate for the optical aberration caused by the light transmitted through the sapphire window, an objective with a compensation ring for a cover glass with a thickness of 0 - 2 mm (Olympus Optical Co., Ltd., SLCPlanFl 40X) also played an important role. Precise adjustments of the compensation ring of the objective and the shear amount of the Nomarski prism were indispensable for obtaining a high contrast level of elementary steps.

For *in situ* observation of elementary steps under high pressure, seed crystals were placed directly on the sapphire window of the high-pressure vessel to minimize the optical aberration. The seed crystals were prepared as follows. A suspension of GI crystals (Hampton Research Co., Ltd., HR7-100), containing 0.91 M ammonium sulfate and 1 mM magnesium sulfate in 6 mM tris hydrochloride buffer (pH = 7.0) (Tris-HCl is known as the most insensitive buffer to pressure) (Neuman et al., 1973), was dissolved (~ 33 mg mL^{-1}) and filtered (Suzuki et al., 2002b). Then the filtrate was transferred onto the sapphire window and sealed with an o-ring and a glass slide.

After a few small crystals appeared on the window at 10 °C, the crystals were grown at room temperature (~ 22 °C) until they reached an appropriate size for the observation (typically ≥ 100 μm). The crystals placed on the window were rinsed with a GI solution of 5.6 mg mL^{-1}, and then the window with the crystals was fitted into the high-pressure vessel filled with a GI solution of 5.6 mg mL^{-1}. In this study we prepared ≤ 10 crystals (size ~ 150 μm) in the 1.6 mmφ O-ring on the sapphire window. Thus, the average separation between the crystals was ~ 300 μm.

4.2.3 Step velocities

Step velocities V on the {011} faces at 0.1 and 50 MPa were measured in the range of protein concentrations C = 5.3 - 8.9 mg mL^{-1}. As shown in Figure 7(a), V increased with increasing pressure. The increase in V is attributed to both kinetic and thermodynamic contributions as shown in eq. (16).

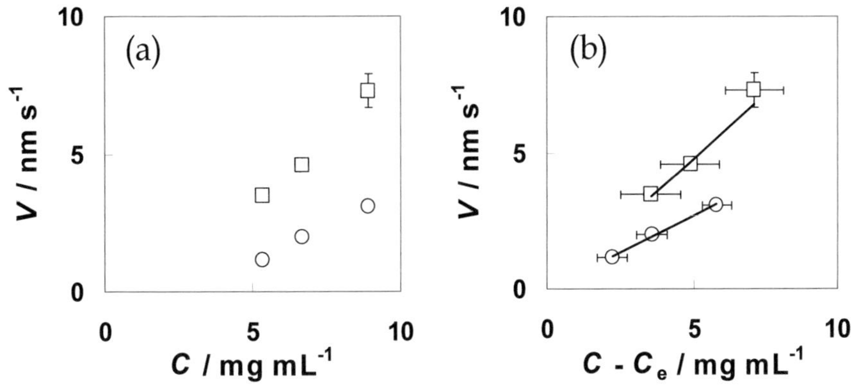

Fig. 7. Step velocities V on the {011} faces of GI crystals as a function of C (a) and $C - C_e$ (b) (Suzuki et al., 2009, 2010a). V was measured at 0.1 MPa (○) and 50 MPa (□). Temperature was 25.0 °C. The lines shown in (b) indicate the results of weighed linear fitting.

To separate the kinetic contribution (β_{step}) from the thermodynamic one ($C - C_e$), we replotted V as a function of $C - C_e$ (Figure 7 (b)). The slopes of the straight lines shown in Figure 7 (b) correspond to $\beta_{step}\Omega$ in eq. (16) at 0.1 and 50 MPa. We have measured Ω under 100 MPa by X-ray crystallography, and found that Ω decreased by only 1.1% with increasing pressure: Ω were (4.79 ± 0.03) × 10^{-25} and (4.74 ± 0.08) × 10^{-25} m^3 at 0.1 and 100 MPa, respectively (Tsukamoto, 2009). Thus, we concluded that β_{step} increased with increasing pressure kinetically. β_{step} thus obtained were (3.2 ± 0.2) × 10^{-7} and (5.7 ± 0.9) × 10^{-7} m s^{-1} at 0.1 and 50 MPa (here we assume that Ω at 50 MPa is same as that at 0.1 MPa), respectively. From these data, we calculated $\Delta V^{\ddagger}$ to be –28 ± 8 cm^{3}mol^{-1} using equation (19).
C_e values at 25°C (2.6 ± 1.4) and (0.8 ± 0.4) mgmL^{-1} at 0.10 and 50 MPa, respectively (Suzuki et al., 2009, 2010a). From these data, we calculated ΔV to be -60 ± 40 cm^{3}mol^{-1} using equation (4). The absolute value of $\Delta V^{\ddagger}$ is almost half that of ΔV.
Such volumetric parameters are strongly related to the dehydration process during the incorporation of a growth unit into a kink site. Thus, further data accumulation will be useful for understanding the dehydration process, which should be one of the most important mechanisms of protein crystallization.

5. X-ray crystallography of protein crystals under high pressure

From the viewpoint of "in situ" high-pressure protein crystallography at the atomic level, five reports have been published so far (Kundrot & Richards, 1987; Urayama et al., 2002; Collins et al., 2005, 2007; Colloc'h et al., 2006). Kundrot et al. reported a three-dimensional structure of a protein (lysozyme) molecule under 100 MPa for the first time (Kundrot &

Richards, 1987). They used a Beryllium (Be) vessel with a stainless steel capillary tube. They revealed anisotropic contraction of the molecule and the increase in the number of ordered water in the crystal with increasing pressure. Urayama et al. and Collins et al. also used Be vessels (Urayama et al., 2002; Collins et al., 2005, 2007). Colloc'h et al. used a diamond anvil cell (DAC), and they could collect 2.3 Å resolution data of urate oxidase (Colloc'h et al., 2006).

However, each method has some problems. All the Be vessels equipped capillary tubes, and the tubes were obstacles to the free rotation of the vessels on goniometers during data collection processes. In the case of a DAC, the accuracy of pressure measurements with ruby fluorescence is low, although the DAC can generate much higher pressures than the Be vessels can do. The error of the pressure measurements in a DAC is usually larger than 10 MPa. In addition, there are usually geometrical constraints on data collection processes with a DAC.

A stand-alone type Be vessel solves all the above problems. Without connecting to the capillary tube, the vessel can freely rotate. With a simple free-piston type pressure indicator, we can monitor pressure in the vessel. In this section, I would like to present our most recent work on high-pressure x-ray protein crystallography.

5.1 Methodology

A stand-alone type high-pressure Be vessel (Syn-corporation Ltd.) was constructed for "in situ" high-pressure protein crystallography at the atomic level (Suzuki et al., 2010b). The vessel equips a Be tube, a stainless steel base, a pressure valve, a coupler joint and a free-piston type pressure indicator. The pressure indicator was composed of a free piston and two springs, and pressure was monitored within ± 1 MPa. From calibration plots of the indicator, we obtained the following equation.

$$h = (0.46 \pm 0.05) + (0.0983 \pm 0.0009)\, P. \tag{20}$$

Here h is the displacement of the piston in mm, and P shows pressure in MPa. The Be tube contains 1% BeO, which reduces X-ray transmittance, and this BeO content is less than that in Urayama's Be tube (2.5%) (Urayama et al., 2002). The thickness of the tube wall is 1.08 mm, and it is also less than that of Kundrot's tube (2.25 mm) (Kundrot & Richards, 1987).

5.2 High pressure X-ray analyses of crystals grown at ambient pressure

Glucose isomerase from *Streptomyces rubiginosus* (Hampton Research, HR7-100) was used without further purification. A GI crystal (~ 0.5 mm) was prepared at atmospheric pressure on the inner wall of a glass capillary (Hampton Research, HR6-164) with its growing solution. The solution contains 0.91 M ammonium sulfate, 1 mM magnesium sulfate, and these are dissolved in 6 mM tris hydrochloride buffer (pH = 7.0). The details of the preparation of the crystal are as follows. First, smaller seed crystals were prepared as described elsewhere (Suzuki et al, 2002b). Second, one of the smaller seed crystals was transferred into a growth solution (the GI concentration of the solution was 28 mgmL^{-1}) in a glass capillary. Third, the seed crystal was incubated for 3 days at 26 °C. Last, the capillary with the crystal and solution was transferred into our high-pressure vessel without replacing the solution. We did not remove the solution, since hydrostatic pressure should be transmitted *via* the solution. A diffraction data set was collected at room temperature on an imaging-plate detector (Rigaku Co., R-AXIS VII) using a rotating copper-anode in-house generator operating at 40 kV and 20 mA (0.8 kW). Such a relatively mild condition is suitable for "in situ" structure analyses, since a high-intensity X-ray radiation easily increases the temperature of a crystal and water molecules around the crystal; the crystal easily dissolves and deteriorates.

Rotation diffraction spots of a GI crystal and powder diffraction rings of a Be tube are shown in Fig. 8. The data were processed using Crystal Clear (Rigaku Corporation, Tokyo).

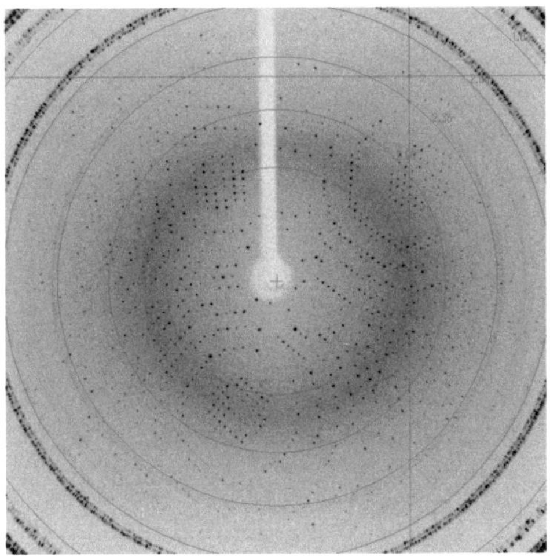

Fig. 8. Rotation diffraction spots of a GI crystal and diffraction rings of a polycrystalline Be tube (Suzuki et al., 2010b).

We successfully collected a 2.0 Å resolution data set of a GI crystal. Pressure could be kept constant at 100 ± 1 MPa for > 24 hours in stand-alone conditions (without connecting to a pressure generator). Although the crystal dissolved a little after the data collection process (> 3 hours irradiation with X-rays), we confirmed that this vessel is sufficiently useful for "in situ" high-pressure protein crystallography.

Strictly speaking, no one has done true "in situ" high-pressure protein crystallography, and a direct result has not been achieved yet. Kundrot et al., Urayama et al., Collins et al., and Colloc'h et al. prepared their crystals at atmospheric pressure, and then pressurized and analyzed the crystals. In such cases, proteins in a crystal shrink with keeping their bonding structure in the crystal during the pressurization. Thus, the pressurized structure in the crystal can be different from that in a solution. In contrast, Charron et al. prepared thaumatin crystals under high pressure, and analyzed the crystals at 100 K after depressurization (Charron et al., 2002). The 3D molecular structures of thaumatin in the depressurized crystals were essentially same as those of unpressurized control crystals. The result suggested that the crystal lattice of thaumatin is elastic. Our group analyzed depressurized, pressurized, and unpressurized GI crystals (Tsukamoto, 2009). A GI structure of the depressurized crystal was essentially same as that of thee unpressurized crystal. Only a GI structure of the pressurized crystal shrinked a little under 100 MPa. This result seems to support Charron's conclusion. However, all the above results are indirect.

To achieve direct results, we should collect a high-resolution, "in situ", and high-pressure data set of a crystal that has nucleated and grown under high pressure. Our setup will achieve the direct results. We can incubate a sufficiently supersaturated protein solution in our vessel under a pressure as long as possible with connecting to a pressure-generating apparatus. After the appropriate nucleation and sufficient growth of crystals from the solution, we can separate the vessel from the capillary tube and directly collect a high-resolution diffraction data set of the crystals with keeping the pressure in the vessel constant.

6. Conclusions

In this chapter, I have presented the great potentialities of high pressure for the promotion of studies on the fundamental growth mechanisms of protein crystals and correlation between the function and 3D structures of protein molecules. Key points in this review are described shortly as follows.

1. As a tool for enhancing the crystal growth, three of eight proteins show the decrease in its solubility under high pressure. Application of high pressure during the screening processes would be useful because of such high probability.
2. Acceleration of growth and nucleation kinetics of glucose isomerase crystals occurred under high pressure.
3. Step velocities under high pressure provided us direct information on activation volume. Activation volume was negative in the case of glucose isomerase crystals. Precise discussion on the activation volume will be useful for understanding dehydration mechanisms during an incorporation process of a protein molecule into a kink site.
4. Usefulness of our standalone-type Be vessel for high-pressure protein crystallography was confirmed. With the vessel, precise high-pressure 3D structure analysis of protein crystals which are also grown under high pressure.

7. Acknowledgments

In this chapter, our studies were supported mainly by two research and education programs and two grants. Studies on solubility, nucleation rates, face growth rates, and step velocities under high pressure were partially supported by "Program for an improvement of education" promoted by the University of Tokushima, the inter-university cooperative research program of the Institute for Materials Research, Tohoku University, and Grants-in Aid (Nos. 16760014 and 19760009 (Y.S.)) for Scientific Research from the Ministry of Education, Culture, Sports, Science and Technology of Japan.

8. References

Asai, T., Suzuki, Y., Sazaki, G., Tamura, K., Sawada, T., & Nakajima, K. (2004). Effects of High Pressure on the Solubility and Growth Kinetics of Monoclinic Lysozyme Crystals. *Cellular and Molecular Biology*, Vol. 50, No. 4, (June 2004), pp. 329-334, ISSN 0145-5680

Boistelle, R., & Lopez-Valero, I. (1990). Growth Units and Nucleation: the Case of Calcium Phosphates. *Journal of Crystal Growth*, Vol. 102, No. 3, (May 1990), pp. 609-617, ISSN 0022-0248

Burton, W.K., Cabrera, N., & Frank, F.C. (1951). The Growth of Crystals and the Equilibrium Structure of Their Surfaces. Philosophical Transactions of the Royal Society of London A., Vol. 243, No. 866, (June 1951), pp. 299-358, ISSN 1471-2962

Cabrera, N., & Levine, M.M. (1956). On the Dislocation Theory of Evaporation of Crystals. *Philosophical Magazine*, Vol. 1, No. 5, (May 1956), pp. 450-458, ISSN 0031-8086

Charron, C., Robert, M.-C., Capelle, B., Kadri, A., Jenner, G., Giegé, R., & Lorber, B. (2002). X-ray Diffraction Properties of Protein Crystals Prepared in Agarose Gel under Hydrostatic Pressure. *Journal of Crystal Growth*, Vol. 245, No. 3-4, (November 2002), pp. 321-333, ISSN 0022-0248

Chernov, A.A. (1984). *Modern Crystallography III Crystal Growth*, Springer-Verlag, ISBN 3-540-11516-1, Berlin Heidelberg New York Tokyo

Collins, M.D., Hummer, G., Quillin, M.L., Matthews, B.W., & Gruner, S.M. (2005). Cooperative Water Filling of a Nonpolar Protein Cavity Observed by High-Pressure Crystallography and Simulation. *Proceedings of the National Academy of Sciences of the United States of America*, Vol. 102, No. 46, (November 2005), pp. 16668-16671, ISSN 0027-8424

Collins, M.D., Quillin, M.L., Hummer, G., Matthews, B.W., & Gruner, S.M. (2007). Structural Rigidity of a Large Cavity-Containing Protein Revealed by High-Pressure Crystallography. *Journal of Molecular Biology*, Vol. 367, No.3, (March 2007), pp. 752-763, ISSN 0022-2836

Colloc'h, N., Girard, E., Dhaussy, A.-C., Kahn, R., Ascone, I., Mezouar, M., & Fourme, R. (2006). High Pressure Macromolecular Crystallography: the 140-MPa Crystal Structure at 2.3 Å Resolution of Urate Oxidase, a 135-kDa Tetrameric Assembly. *Biochimica et Biophysica Acta*, Vol. 1764 (2006), No. 3, (March 2006), pp 391-397, ISSN 1570-9639

Durbin, S.D., & Carlson, W.E. (1992). Lysozyme Crystal Growth Studied by Atomic Force Microscopy. *Journal of Crystal Growth*, Vol. 122, No. 1-4, (August 1992), pp. 71-79, ISSN 0022-0248

Durbin, S.D., Carlson, W.E., & Saros, M.T. (1993). In Situ Studies of Protein Crystal Growth by Atomic Force Microscopy. *Journal of Physics D: Applied Physics*, Vol. 26, No. 8B, (August 1993), pp. B128-B132, ISSN 0022-3727

Foubert, I., Dewettinck, K., & Vanrolleghem, P.A. (2003). Modelling of the Crystallization Kinetics of Fats. *Trends in Food Science & Technology*, Vol. 14, No. 3, (March 2003), pp. 79-92, ISSN 0924-2244

Fujiwara, T., Suzuki, Y., Sazaki, G., & Tamura, K. (2010). Solubility Measurements of Protein Crystals under High Pressure by *In Situ* Observation of Steps on Crystal Surfaces. *Journal of Physics: Conference Series*, Vol. 215, (March 2010), pp. 012159-1-5, ISSN 1742-6588

Groß, M., & Jaenicke, R. (1991). Growth Inhibition of Lysozyme Crystals at High Hydrostatic Pressure. *FEBS Letters*, Vol. 284, No. 1, (June 1991), pp. 87-90, ISSN 0014-5793

Groß, M., & Jaenicke, R. (1993). A Kinetic Model Explaining the Effects of Hydrostatic Pressure on Nucleation and Growth of Lysozyme Crystals. *Biophysical Chemistry*, Vol. 45, No. 3, (January 1993), pp. 245-252, ISSN 0301-4622

Higgins, S.R., Eggleston, C.M., Knauss, K.G., Boro, C.O. (1998). A Hydrothermal Atomic Force Microscope for Imaging in Aqueous Solution up to 150 °C. *Review of Scientific Instruments*, Vol. 69, No. 8, (August 1998), pp. 2994-2998, ISSN 0034-6748

Kadri, A., Lorber, B., Jenner, G., & Giegé, R. (2002). Effects of Pressure on the Crystallization and the Solubility of Proteins in Agarose Gel. *Journal of Crystal Growth*, Vol. 245, No. 1-2, (November 2002), pp. 109-120, ISSN 0022-0248

Kundrot, C.E., & Richards, F.M. (1987). Crystal Structure of Hen Egg-White Lysozyme at a Hydrostatic Pressure of 1000 Atmospheres. *Journal of Molecular Biology*, Vol. 193, No. 1, (January 1987), pp. 157-170, ISSN 0022-2836

Kundrot, C.E., & Richards, F.M. (1988). Effect of Hydrostatic Pressure on the Solvent in Crystals of Hen Egg-White Lysozyme. *Journal of Molecular Biology*, Vol. 200, No. 2, (March 1988), pp. 401-410, ISSN 0022-2836

Laidler, K.J. (1987). *Chemical Kinetics, 3rd ed.*, Harper & Row, ISBN 0060438622, New York, USA

Lorber, B., Jenner, G., & Giegé, R. (1996). Effect of High Hydrostatic Pressure on Nucleation and Growth of Protein Crystals. *Journal of Crystal Growth*, Vol. 158, No. 1-2, (January 1996), pp. 103-117, ISSN 0022-0248

Makimoto, S., Suzuki, K., & Taniguchi, Y. (1984). Pressure Dependence of the a-Chymotrypsion-Catalyzed Hydrolysis of Amide and Anilides. Evidence for the single-proton-transfer mechanism. *The Journal of Physical Chemistry*, Vol. 88, No. 24, (November 1984), pp. 6021-6024, ISSN 0022-3654

Markov, I.V. (1995). *Crystal Growth for Beginners : Fundamentals of Nucleation, Crystal Growth, and Epitaxy*, World Scientific, ISBN 9810215312, Singapore, Singapore

Maruoka, T., Suzuki, Y., & Tamura, K. (2010). Effects of High Pressure on the Three-Dimensional Nucleation Rates of Glucose Isomerase Crystals. *Journal of Physics: Conference Series*, Vol. 215, (March 2010), pp. 012158-1-5, ISSN 1742-6588

McPherson, A., Malkin, A.J., Kuznetsov, Y.G. (2000). Atomic Force Microscopy in the Study of Macromolecular Crystal Growth. Annual Review of Biophysics and Biomolecular Structure, Vol. 29, pp. 361-410, ISSN 1056-8700

Nagatoshi, Y., Sazaki, G., Suzuki, Y., Miyashita, S., Matsui, T., Ujihara, T., Fujiwara, K., Usami, N., & Nakajima, K. (2003). Effects of High Pressure on the Growth Kinetics of Orthorhombic Lysozyme Crystals. *Journal of Crystal Growth*, Vol. 254, No. 1-2, (June 2003), pp 188-195, ISSN 0022-0248

Neuman, R.C., Jr., Kauzmann, W., & Zipp, A. (1973). *The Journal of Physical Chemistry*, Vol. 77, No. 22, (October 1973), pp. 2687-2691, ISSN 0022-3654

Pan, X., & Glatz, C.E. (2002). Solvent Role in Protein Crystallization as Determined by Pressure Dependence of Nucleation Rate and Solubility. *Crystal Growth & Design*, Vol. 2, No. 1, (January 2002), pp. 45-50, ISSN 1528-7483

Rosenberger, F., Vekilov, P.G., Muschol, M., & Thomas, B.R. (1996). Nucleation and Crystallization of Globular Proteins – What We Know and What is Missing. *Journal of Crystal Growth*, Vol. 168, No. 1-4, (October 1996), pp. 1-27, ISSN 0022-0248

Saikumar, M.V., Glatz, C.E., & Larson, M.A. (1998). Lysozyme Crystal Growth and Nucleation Kinetics. *Journal of Crystal Growth*, Vol. 187, No. 2, (May 1998), pp. 277-288, ISSN 0022-0248

Sazaki, G., Matsui, T., Tsukamoto, K., Usami, N., Ujihara, T., Fujiwara, K., & Nakajima, K. (2004). In Situ Observation of Elementary Growth Steps on the Surface of Protein Crystals by Laser Confocal Microscopy. *Journal of Crystal Growth*, Vol. 262, No. 1-4, (February 2004), pp. 536-542, ISSN 0022-0248

Sazaki, G., Nagatoshi, Y., Suzuki, Y., Durbin, S.D., Miyashita, S., Nakada, T., & Komatsu, H. (1999). Solubility of Tetragonal and Orthorhombic Lysozyme Crystals under High Pressure. *Journal of Crystal Growth*, Vol. 196, No. 2-4, (January 1999), pp. 204-209, ISSN 0022-0248

Suzuki, Y., Miyashita, S., Komatsu, H., Sato, K., & Yagi, T. (1994). Crystal Growth of Hen Egg White Lysozyme under High Pressure. Japanese Journal of Applied Physics, Vol. 33, No. 11A, (November 1994), pp. L1568-L1570, ISSN 0021-4922

Suzuki, Y., Sawada, T., Miyashita, S., Komatsu, H. (1998). *In Situ* Measurements of the Solubility of Crystals under High Pressure by an Interferometric Method. *Review of Scientific Instruments*, Vol. 69, No. 7, (July 1998), pp. 2720-2724, ISSN 0034-6748

Suzuki, Y., Miyashita, S., Sazaki, G., Nakada, T., Sawada, T., & Komatsu, H. (2000a). Effects of Pressure on Growth Kinetics of Tetragonal Lysozyme Crystals. *Journal of Crystal Growth*, Vol. 208, No. 1-4, (January 2000), pp. 638-644, ISSN 0022-0248

Suzuki, Y., Sawada, T., Miyashita, S., Komatsu, H., Sazaki, G., & Nakada, T. (2000b). An Interferometric Study of the Solubility of Lysozyme Crystals under High Pressure. *Journal of Crystal Growth*, Vol. 209, No. 4, (February 2000), pp. 1018-1022, ISSN 0022-0248

Suzuki, Y., Sazaki, G., Miyashita, S., Sawada, T., Tamura, K., & Komatsu, H. (2002a). Protein Crystallization under High Pressure. *Biochimica et Biophysica Acta*, Vol. 1595, No. 1-2, (March 2002), pp. 345-356, ISSN 0167-4838

Suzuki, Y., Sazaki, G., Visuri, K., Tamura, K., Nakajima, K., & Yanagiya, S. (2002b). Significant Decrease in the Solubility of Glucose Isomerase Crystals under High Pressure. *Crystal Growth & Design*, Vol. 2, No. 5, (September 2002), pp 321-324, ISSN 1528-7483

Suzuki, Y., Sazaki, G., Matsui, T., Nakajima, K., & Tamura, K. (2005). High-Pressure Acceleration of the Growth Kinetics of Glucose Isomerase Crystals. *The Journal of Physical Chemistry B*, Vol. 109, No. 8, (March 2005), pp. 3222-3226, ISSN 1089-5647

Suzuki, Y., Sazaki, G., Matsumoto, M., Nagasawa, M., Nakajima, K., & Tamura, K. (2009). First Direct Observation of Elementary Steps on the Surfaces of Glucose Isomerase Crystals under High Pressure. *Crystal Growth & Design*, Vol. 9, No. 10, (October 2009), pp. 4289-4295, ISSN 1528-7483

Suzuki, Y., Sazaki, G., Matsumoto, M., Nagasawa, M., Nakajima, K., & Tamura, K. (2010a). First Direct Observation of Elementary Steps on the Surfaces of Glucose Isomerase Crystals under High Pressure (Additions and Corrections). *Crystal Growth & Design*, Vol. 10, No. 4, (April 2010), pp. 2020-2020, ISSN 1528-7483

Suzuki, Y., Tsukamoto, M., Sakuraba, H., Matsumoto, M., Nagasawa, M., & Tamura, K. (2010b). Design of a standalone-type beryllium vessel for high-pressure protein crystallography. *Review of Scientific Instruments*, Vol. 81, No. 8, (August 2010), pp. 084302-1-3, ISSN 0034-6748

Suzuki, Y., Maruoka, T., & Tamura, K. (2010c). Activation Volume of Crystallization and Effects of Pressure on the Three-Dimensional Nucleation Rate of Glucose Isomerase. *High Pressure Research*, Vol. 30, No. 4, (December 2010), pp. 483-489, ISSN 0895-7959

Suzuki, Y., Konda, E., Hondoh, H., & Tamura, K. (2011). Effects of Temperature, Pressure, and pH on the Solubility of Triclinic Lysozyme Crystals. *Journal of Crystal Growth*, Vol. 318, No. 1, (March 2011), pp. 1085-1088, ISSN 0022-0248

Takano, K.J., Harigae, H., Kawamura, Y. & Ataka, M. (1997). Effect of Hydrostatic Pressure on the Crystallization of Lysozyme Based on In Situ Observations. *Journal of Crystal Growth*, Vol. 171, No. 3-4, (February 1997), pp. 554-558, ISSN 0022-0248

Tsukamoto, M. (2009). *Master Thesis*, Graduate School of Advanced Technology and Science, The University of Tokushima

Urayama, P., Phillips, Jr., G.N., & Gruner, S.M. (2002). Probing Substrates in Sperm Whale Myoglobin Using High-Pressure Crystallography. *Structure*, Vol. 10, No. 1, (January 2002), pp. 51-60, ISSN 0969-2126

Van Driessche, A.E.S., Sazaki, G., Otalora, F., Gonzalez-Rico, F.M., Dold, P., Tsukamoto, K., & Nakajima, K. (2007). Direct and Noninvasive Observation of Two-Dimensional Nucleation Behavior of Protein Crystals by Advanced Optical Microscopy. *Crystal Growth & Design*, Vol. 7, No. 7, (September 2007), pp. 1980-1987, ISSN 1528-7483

Van Driessche, A.E.S., Gavira, J.A., Patiño Lopez, L.D., & Otalora, F. (2009). Precise Protein Solubility Determination by Laser Confocal Differential Interference Contrast Microscopy. *Journal of Crystal Growth*, Vol. 311, No. 13, (June 2009), pp. 3479-3484, ISSN 0022-0248

Visuri, K., Kaipainen, E., Kivimaki, J., Niemi, H., Leisola, M., & Palosaari, S. (1990). A New Method for Protein Crystallization Using High Pressure. *Bio/Technology*, Vol. 8, No. 6, (June 1990), pp. 547-549, ISSN 0733-222X

Volmer, M., & Weber, A. (1926). Keimbildung in Übersättigten Gebilden. *Zeitschrift für Physikalische Chemie*, Vol. 199, pp. 277-301, ISSN 0044-3336

Waghmare, R.Y., Webb, J.N., Randolph, T.W., Larson, M.A., & Glatz, C.E. (2000a). Pressure Dependence of Subtilisin Crystallization Kinetics. *Journal of Crystal Growth*, Vol. 208, No. 1-4, (January 2000), pp. 678-686, ISSN 0022-0248

Waghmare, R.Y., Pan, X.J., & Glatz, C.E. (2000b). Pressure and Concentration Dependence of Nucleation Kinetics for Crystallization of Subtilisin. *Journal of Crystal Growth*, Vol. 210, No. 4, (March 2000), pp. 746-752, ISSN 0022-0248

Webb, J.N., Waghmare, R.Y., Carpenter, J.F., Glatz, C.E., & Randolph, T.W. (1999). Pressure Effect on Subtilisin Crystallization and Solubility, *Journal of Crystal Growth*, Vol. 205, No. 4, (September 1999), pp. 563-574, ISSN 0022-0248

Yayanos, A.A. (1986). Evolutional and Ecological Implications of the Properties of Deep-Sea Barophilic Bacteria. *Proceedings of the National Academy of Sciences of the United States of America*, Vol. 83, No. 24, (December 1986), pp. 9542-9546, ISSN 0027-8424

20

Protein Crystal Growth

Igor Nederlof[1], Eric van Genderen[1], Flip Hoedemaeker[2], Jan Pieter Abrahams[1] and Dilyana Georgieva[1]
[1]*Leiden University,*
[2]*Kabta Consultancy Ltd.,*
The Netherlands

1. Introduction

The biological activity of most proteins is determined by their 3D structure. For instance, a substantial number of molecular diseases are caused by protein structural alterations, which are genetically encoded. Drugs operate by binding to proteins, inducing alteration of their functional structure and thereby affecting their biological activity. Hence the design and improvement of drugs is greatly facilitated by knowledge of the 3D structures of their macromolecular targets. In the light of these considerations, it is clear that elucidation of the 3D structure of proteins is of prime importance for understanding the underlying mechanisms of molecular diseases. It was initially believed that any protein that could be made soluble and could be purified would be relatively easy to crystallize. However, the results have indicated that solubility and purity of proteins, although being important factors, do not secure a yield of useful crystals. The crystallization behavior of proteins turns out to be very complex.

In an effort to identify the naturally occurring protein folds, large structural genomics consortia were set up. The somewhat disappointing outcome of these efforts is that only about 3% of all proteins that were targeted by these consortia yielded a crystal structure (http://targetdb.pdb.org/statistics/TargetStatistics.html), despite massive investments in high-throughput, automated protein production, purification and crystallization. It is clear that in order to improve the current situation, better strategies for protein crystallization are required, combined with techniques that allow the use of smaller nano-crystalline material.

2. Crystallization of bio-macromolecules

Biocrystallization involves the three classical steps of nucleation, growth, and cessation of growth, even though the protein crystals contain on average 50% of disordered solvent (Figure 1) (Matthews et al., 1968). However, crystal growth of biological molecules differs substantially from small molecule crystalogenesis. The reason is the much larger number of parameters involved in biocrystallization, as well as the specific physico-chemical properties of the biological compounds. The main difference from small molecule crystal growth is the conformational flexibility and chemical versatility of macromolecules and their greater sensitivity to external factors. An overview of different parameters affecting the crystallization of biomacromolecules is presented in table 1 (Bergfors T, 2009).

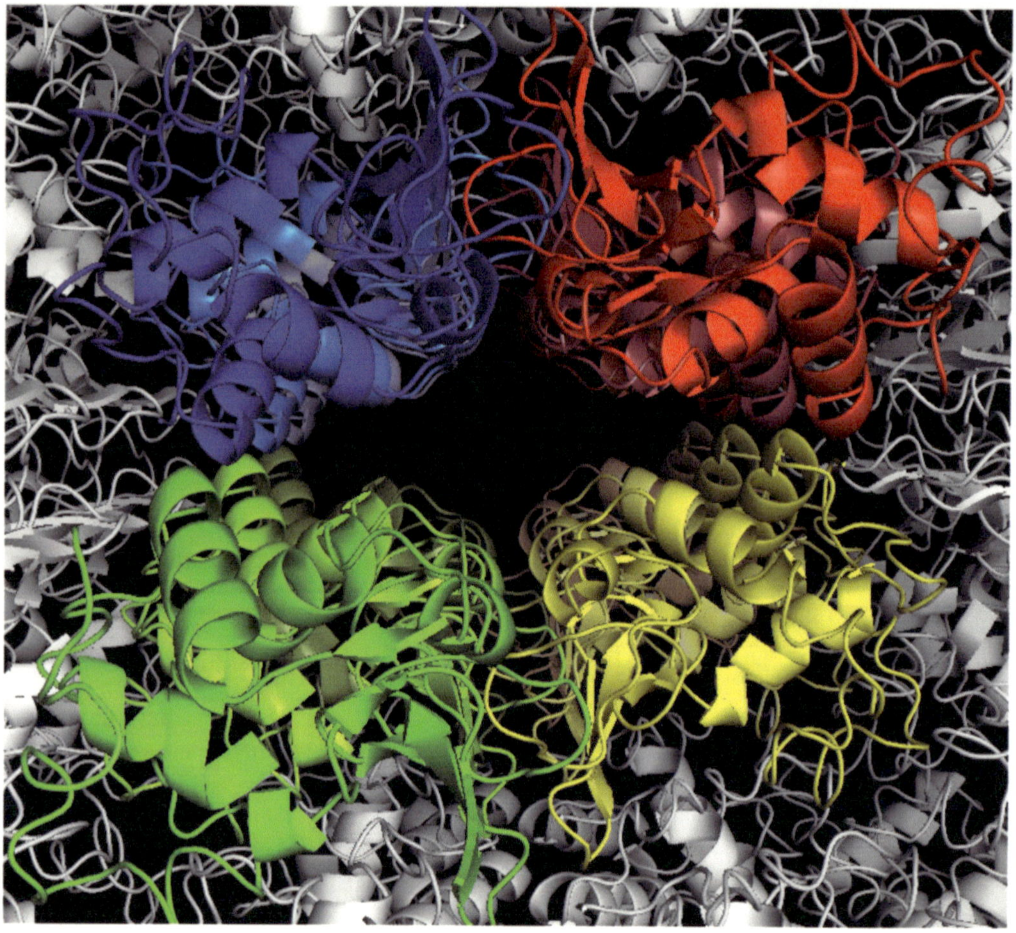

Fig. 1. Crystal packing in lysozyme crystals (pdb:1Lyz) shows large cavities. These cavities are filled with disordered solvent (not shown).

Intrinsic physico-chemical properties • Supersaturation • Temperature, pH • Ionic strength and purity of chemicals • Pressure, electric and magnetic fields • Vibration and sound	**Biochemical and biophysical parameters** • Sensitivity of conformation to physical parameters • Binding of ligands • Specific additives • Aging of samples
Biological parameters • Rarity of biological macromolecules • Bacterial contaminants • Biological sources of organisms and cells	**Purity of macromolecules** • Macromolecular contaminants • Sequence (micro) heterogeneity • Conformational (micro) heterogeneity • Batch effects

Table 1. Overview of parameters affecting bio-macromolecular crystallization.

Another important prerequisite for successful crystallization is the quality of the macromolecular samples. Bio-macromolecules are extracted from living cells or synthesized *in vitro* and they are frequently difficult to prepare at a high degree of purity and homogeneity. Besides traces of impurities, the different treatments proteins are subjected to may decrease their stability and activity through different kinds of alterations. As a general rule, purity and homogeneity are regarded as conditions of prime importance. Accordingly, purification, stabilization, storage and handling of macromolecules are other essential steps prior to crystallization.

3. Purity of bio-macromolecular samples

The concept of purity has a special meaning when biological crystallogenesis is concerned. Molecular samples need to be not only chemically pure, but they must also be conformationally uniform (Giege et al., 1986). This concept is based on the fact that the best crystals are grown from solutions containing well-defined entities with identical physico-chemical properties. For X-ray crystallographic studies, the aim is to grow 'single crystals' diffracting to high resolution with a low mosaicity and prolonged stability in the X-ray beam. It is therefore understandable that contaminants may compete for sites on the growing crystals and generate lattice errors leading to internal disorder, dislocations, poor diffraction or early cessation of growth (Vekilov et al., 1996). Because of the high molecular weight of molecules in a single crystal (up to millions of daltons), and hence low molarity of their solutions even relatively small amounts of contaminant may induce formation of non-specific aggregates, alter macromolecular solubility, or interfere with nucleation and crystal growth (Skouri et al., 1996; McPherson et al., 1996). Successful crystallization of rare proteins and nucleic acids support the importance of purity and homogeneity (Wierenga et al., 1987; Thegesen et al., 1996; Aoyama et al., 1996; Douna et al.,: 1993). Usually most of the contaminants are eliminated during the different purification steps, however traces of polysaccharides, lipids or proteases may still be present and hinder crystallization. Small molecules, like peptides, oligonucleotides, amino acids, as well as uncontrolled ions should also be considered as contaminants. Buffering molecules remaining from a purification step can be responsible for irreproducible crystallization. For instance, phosphate ions are relatively difficult to remove and may crystallize in the presence of divalent cations (Ca^{2+}, Mg^{2+}). Counterions play a critical role in the packing of biomolecules. Often macromolecules do not crystallize or yield different habits in the presence of various buffers adjusted at the same pH.

Bio-molecular samples containing traces of contaminants can further be subjected to purification through recrystallization, column chromatography, ultra-centrifugation, fractionated precipitation, affinity purification and other techniques. Microheterogeneity in pure macromolecules can only be revealed by very sensitive methods. The most common causes for heterogeneity are uncontrolled fragmentation and post-synthetic modification.

Proteolysis normally takes place in many physiological processes and represents a major difficulty that needs to be overcome during protein extraction from the living cells that produce the desired protein (Achstetter et al., 1985; Barrett et al., 1986; Dalling et al., 1986; Bond et al., 1987; Arfin et al., 1988; Wandersman et al., 1989). The reason is that proteases are localized in various cellular compartments or excreted in the extracellular medium. Upon cell disruption, cellular compartments are mixed with extracellular proteases and control over proteolysis is lost. Decrease of protein size and stability, modification of their charge or hydrophobicity, partial or total loss of activity are usually signs of proteolysis. Traces of protease may not be detectable even when overloading electrophoresis gels, but they can cause damage during concentration or storage of samples.

Co- or post-translational enzymatic modifications generate microheterogeneity in proteins when different groups, for instance oligosaccharide chains, occupy specific modification sites on the protein, or when correct modifications are unevenly distributed. Only certain modifications are reversible, for instance phosphorylation, but others like glycosylation or methylation are not. Microheterogeneity can also appear during storage, for instance by deamidation of asparagines or glutamine residues is a well-documented phenomenon.

Pure, chemically uniform macromolecules can be fully functional in a biochemical activity assay even though they are microheterogeneous. Conformational heterogeneity may have several origins: binding of ligands, intrinsic flexibility of molecular backbones, oxidation of cysteine residues or partial denaturation. In the first case, macromolecules should be prepared in both forms, the one deprived of and the other saturated with ligands. In the second case, controlled fragmentation may be helpful. In the last one, oxidation of a single cysteine residue leads to complex mixtures of molecule species for which the chances of growing good crystals are low (Van der Laan et al., 1989).

Although macromolecules may crystallize readily in an impure state (Holley et al., 1961), this is an exception and it is always preferable to achieve a high level of purity before starting crystallization experiments. In order to gain more information about the quality of the protein samples, different techniques can be used. For instance, spectrophotometry and fluorometry give information about the quality of samples if macromolecules or their contaminants have special absorbance or emission properties. SDS-PAGE indicates the size of protein contaminants, but not that of non-protein contaminants. Isoelectric focusing gives an estimate of the pI of protein components in a mixture and electrophoretic titration shows the mobility of individual proteins as a function of pH. The latter method can also suggest the type of chromatography suitable for further purification. Capillary electrophoresis is well adapted for purity analysis (Karger et al., 1996). Amino acid composition and sequencing of N- and C- termini verify in part the integrity of primary structure.

Electrospray ionization and matrix-assisted laser desorption/ionisation mass spectrometry are also powerful tools in the analysis of recombinant protein chemistry. Nuclear magnetic resonance can detect small size contaminants and gives structural information on biomolecules (Wuthrich , 1995).

It is widely believed that the success of crystal trials is largely dependent on various, not very well identified, properties of the protein. For example, a positive correlation has been established between the degree of protein monodispersity in solution and the ability of the protein to crystallize. On the other hand, it's thought to be a negative correlation between the degree of disorder in the protein and its ability to crystallize (Mikol et al., 1989).

A number of biophysical techniques and methods are employed to evaluate the quality and stability of protein solutions. Dynamic light scattering is a useful tool for non-invasive *in situ* monitoring of crystallization trials because it detects the formation of aggregates or nuclei before they become visible under a light microscope (Berne et al., 1976). Fluorescence and light scattering are helpful to rapidly identity stabilizing conditions compromising simple agents (salts, co-factors etc.). Emission fluorescence is used to measure changes if the protein unfolds or undergoes other conformational changes (Konev et al. 1967).

4. Solubility, supersaturation and phase transition

Biological macromolecules follow the same thermodynamic rules as inorganic or organic small molecules concerning supersaturation, nucleation and crystal growth. However, protein macromolecules are organized in tertiary and quaternary structures. The intra-molecular interactions responsible for their tertiary structure, the intermolecular interactions involved in the crystal contacts, and the interactions necessary to solubilise them in a solvent are similar.

To crystallize a biological macromolecule, its solution must have reached supersaturation, which is the driving force for crystal growth. The under- and supersaturated states are defined by the solubility of the macromolecules. When the concentrations of the crystallization agent and the macromolecules correspond to the solubility condition, the saturated macromolecule solution is in equilibrium with the crystallized macromolecules. Below the solubility curve (fig. 2) the solution is under saturated and the system is thermodynamically stable. In this case, phase transition (crystallization) will not occur. Above the solubility curve, the concentration of the biological macromolecules is higher than the concentration at equilibrium. A supersaturated macromolecular solution contains an excess of macromolecules that will appear as a solid phase until the macromolecular concentration reaches the solubility value in the solution. The higher the supersaturation, the faster this solid phase appears. However, at very high supersaturation precipitation, not crystallization occurs, but insoluble macromolecules rapidly separate from the solution in an amorphous state.

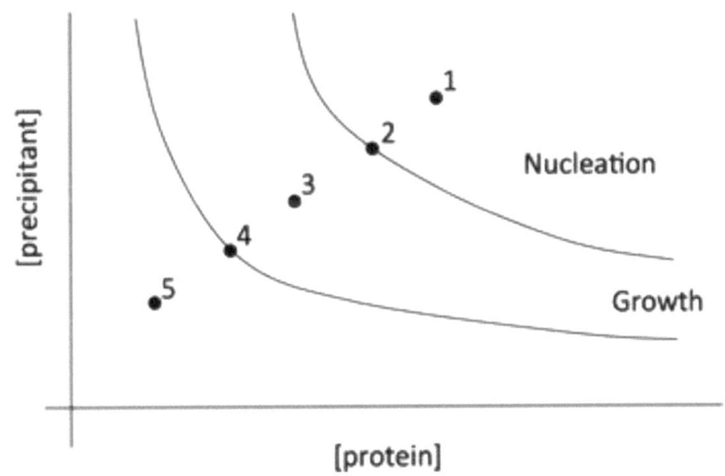

Fig. 2. Solubility curve of a protein, where the phase state of the protein is plotted against the concentration of both protein and precipitant. At the point (1), the protein may precipitate so fast that an amorphous precipitate or at best shower of microcrystals is formed. At (2) the conditions are just right for the protein to form a stable crystal nucleus, which will start to grow – passing (3) – into a stable protein at equilibrium with the mother liquor (4). At (5), the concentration of protein and precipitant are too low for crystal nucleation or growth, and the solution will remain clear. Note that the true solubility curve of any protein is highly multidimensional, with every parameter affecting protein solubility (cf. Table 1) representing a different independent axis.

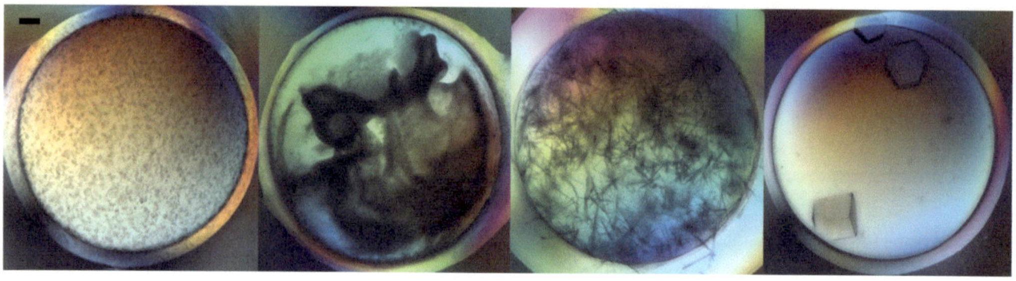

Fig. 3. Glucose Isomerase crystallization condition yielding phase separation (far left) amorphous precipitation (near left) micro-crystals (near right) and macro-crystals (far right) bar on the top left represents 200 micrometer.

5. Crystallization strategies

5.1 Crystallization screens

Finding crystallization conditions for a new protein target is largely based on a trial and error method. The first step is to set up screening trials, exposing the protein to a variety of agents in order to find useful "leads", which can be crystals, crystalline precipitates and phase separation that point to conditions that are conductive to crystallization.
The most popular screens to perform the initial screening step are called sparse-matrix screens. These screens rely on a compilation of conditions that had previously led to successful crystallization. Systematic screens sample the crystallization parameter space in a balanced, rational way using information on the protein properties. Systematic screens are usually used as second remedy or in order to optimize the crystallization conditions.

5.2 Choosing the crystallization method

There are different methods to crystallize biological macromolecules. However, all of them aim at bringing the solution of macromolecules to a supersaturation state (McPherson, 1985; Giege, 1987). It's important to keep in mind that not only the various chemical and physical parameters influence protein nucleation and crystallization, but also the method of crystallization. Therefore, it's wise to try different methods when searching for optimal crystallization conditions. As solubility is dependent on temperature (it could increase or decrease depending on the protein), it's highly recommended to perform crystallization trials at constant temperature unless temperature variation is part of the experiment. Solubility of most chemicals is given in Merck Index. The chemical nature of the buffer is an important parameter for protein crystal growth. It must be kept in mind that the pH of buffers is often temperature dependent, this is particularly significant for Tris buffers. Buffers, which must be used within one unit from their pK value, are well described in textbooks (Perin et al., 1974).
Protein samples often contain large amount of salts of unknown composition when first obtained. Thus it's wise to dialyse a new batch of a macromolecule against a large volume of well-characterised buffer of given pH, to remove unwanted salts and to adjust the pH. Starting from known conditions helps to increase the reproducibility.
Whatever the crystallization method used, it requires high concentration of biological macromolecules as compared to normal biochemistry conditions. Before starting a crystallization experiment, a concentration step is generally needed. It's also important to

keep pH and ionic strength at desired value, since pH may vary when the concentration of macromolecules increases. Also, low ionic strength could lead to early precipitation. Many commercial devices are available based on Different concentration principles such as concentration under pressure, using centrifugation, or lyophilisation. The choice of method for concentration depends on the quantity and the stability of the macromolecules.

Before a crystallization experiment, solid particles such as dust, denatured proteins, and solids coming from purification columns or lyophilization should be removed. This could be achieved by centrifugation or filtration, depending on the available quantity.

The most common method to measure macromolecular concentration is to sample an aliquot, dilute it with buffer, and measure absorbance at 280 nm for proteins within the linear range of a spectrophotometer. Proper subtraction with the reference cell should be made especially when working with additives absorbing in the 260-300 nm wavelength range. When working with enzymes, an alternative method to measure the concentration of protein is to perform activity test, otherwise colorimetric methods can be performed.

5.2.1 Vapour diffusion

The most widely used method of crystallization is vapour diffusion. The protein solution is a hanging, sitting or sandwich drop that equilibrates against a reservoir containing crystallizing agents at either higher or lower concentration than in the drop. Equilibration proceeds by diffusion of the volatile species (water or organic solvent) until vapour pressure in the droplet equals the one of the reservoir. If equilibration occurs by water exchange from the drop to the reservoir, it leads to a droplet volume decrease. Consequently, the concentration of all constituents in the drop will increase. For species with a vapour pressure higher than water, the exchange occurs from the reservoir to the drop. In such a 'reverse' system, the drop volume will increase and the concentration of the drop constituents will decrease. The same principle applies for hanging drops, sitting drops and sandwich drops. Most people use a ratio of 1:1 between the concentration of the crystallizing agent in the reservoir and in the droplet. This is achieved by mixing a droplet of protein at twice the desired final concentration. When no crystal or precipitate is observed, either supersaturation is not reached or one has reached the metastable region. In the latter case changing the temperature by a few degrees is generally sufficient to initiate nucleation. Although unique in this respect, vapour diffusion permit easy variation of physical parameters during crystallization, and many successes were obtained by modifying supersaturation by temperature or pH changes. With ammonium sulphate as the crystallizing agent, it has been shown that the pH in the droplets is imposed by that of the reservoir. Consequently, varying the pH of the reservoir permits gentle adjustments of that in the droplets. From another point of view sitting drops are well suited for attempting epitaxial growth of macromolecule crystals on appropriate mineral matrices. In other words vapour diffusion provides a way to sample the crystallization space with the conditions continuously varying, as the equilibration proceeds. The kinetics of water evaporation determines the kinetics of supersaturating and accordingly affects nucleation rates. Evaporation rates from hanging drops have been determined experimentally in the presence of ammonium sulphate, PEG, MPD and NaCl as crystallizing agents. The main parameters that determine the rate of water equilibration are temperature, initial drop volume, water pressure of the reservoir, and the chemical nature of the crystallization agent. Theoretical modelling has shown in addition the pivotal role of the drop to reservoir distance. It was shown that the effect of this parameter is negligible in classical set-ups and becomes only noticeable when drop to reservoir distance is more than 2 cm. From practical point of view, the time for water evaporation to reach 90% completion can vary from about 25 hours to more than 25 days. The fastest equilibration occurs in the presence of ammonium sulphate and the slowest in the presence of PEG. Equilibration rates are significantly slowed down by increasingly appropriately the distance between the drop and the reservoir. An alternative solution to decrease equilibration rates is to apply a layer of oil over the reservoir.

Fig. 4. Schematic drawing of sitting drop (left), hanging drop (middle), and batch crystallization (right). Well solution is blue, protein mixed with well solution is brown and oil is green.

5.2.2 Batch crystallization methods

Another routinely used method for crystallization is the batch method. The biological macromolecule to be crystallized is mixed with the crystallizing agent at a concentration such that supersaturation is instantaneously reached. Crystallization trials are dispensed and incubated under low-density paraffin oil. The crystallization drops remain under oil, where they are protected from evaporation, contamination and shock. Since supersaturation is reached at the start of the experiment, nucleation tends to be higher, if compared to the vapour diffusion method. However, in some cases fairly large crystals can be obtained when working close to the metastable region. Although the microbatch method has not been compared in a statistically significant scale against hanging drop-vapour diffusion method, a comparison on a small scale has been performed (Baldok et al., *1999)*. The study demonstrated that the methods are not entirely identical, but are equally effective. The results suggest that vapour diffusion method and the microbatch technique will probably produce similar numbers of crystals, but may not produce crystals for the same conditions. Microbatch and vapour diffusion methods are both suitable for high throughput crystallization experiments where all the steps of dispensing, mixing and sealing are automated and performed by a robot. Other crystallization methods worth mentioning, although with more limited success and use are crystallization in gel, dialysis, microfluidics, free interface diffusion. Microfluidic chips are also being used for high throughput crystallization screening.

5.2.3 Crystallization in gels

Special attention has been paid to crystallization in gels (Robert at al., 1987). The protein crystallization process consists of two main steps - the transport of growth units towards the surface of the crystals and second, the incorporation of the growth units into a crystal surface position of high bond strength. The whole growth process is dominated by the slower of these two steps and is either transport controlled or surface controlled. The ratio between transport to surface kinetics, which can be tuned by either enhancing or reducing transport processes in solution, was shown to control the amplitude of growth rate fluctuations. These are the reasons why gels if properly designed are expected to enhance the quality of crystals. It's worth mentioning that crystals growing in gel do not sediment as they do in free solution. They develop at the nucleation site, sustained by the gel network. For small molecule crystals grown in silica gel, the gel often forms cusp-like cavities around the crystal and a thin liquid film that reduces contamination risk, separates the crystal from the gel. Such cavities have not been seen in macromolecular crystals. Recent studies have shown that silica gel can be incorporated in the crystal network almost without disturbing the crystal lattice. Such crystals that still diffract to a high resolution, are mechanically reinforced and are more resistant to dehydration, because the silica gel framework embedded in the crystal slows down water loss due to its hygroscopic properties. Although seeding can be used, it appears that most of the gel-grown crystals are obtained by spontaneous nucleation inside a macroscopically homogeneous gel. When the gel adheres to the walls of the container, no nucleation occurs on the cell walls, neither on dust. So, heterogeneous nucleation is strongly reduced, if not suppressed. Another type of nucleation, namely secondary nucleation, is due to attrition of a previous crystal by the solution flux. When nucleation occurs inside the gel, one observes that all the crystals appear at the same time and have about the same size. They are homogenously distributed in the whole volume.

5.2.4 Dialysis methods

Crystallization by dialysis methods allow for an easy variation of the different parameters that influence the crystallization of biological macromolecules. Different types of dialysis cells are used, but follow the same principle. The macromolecule is separated from a large volume of solvent by a semi-permeable membrane that gives small molecules free passage, but prevents macromolecules from circulating. The kinetics of equilibration will depend on the membrane cut-off, the ratio of the concentration, the temperature and the geometry of the cell.

The method of crystallization by interface diffusion was developed (Salemme, 1972) and used to crystallize several proteins. In the liquid/liquid diffusion method, equilibration occurs by diffusion of the crystallization agent into the biological macromolecule volume. To avoid rapid mixing, the less dense solution is poured gently on the most dense (salt in general) solution. Sometimes, the crystallizing agent is frozen and the protein layered above to avoid rapid mixing.

5.3 The role of heterogeneous substrates in the process of protein nucleation and crystallization

In general, additives play an important role in protein crystallization. Heterogeneous substrates are usually regarded as additives when they are purposefully added to the solution in order to obtain a desired effect (inhibition of nucleation, habit change of crystals). However, impurities of foreign substances may also exist in the solution originating from other sources (the solvent, crystallization agent, etc.). Heterogeneous crystallization which is induced by a properly chosen additive may allow better control of nucleation and growth. The first report of a nucleant inducing nucleation of macromolecules was the epitaxial growth of protein crystals on minerals (McPherson et al., 1988). Other candidate nucleants followed like zeolites, silicates, charged surfaces, porous materials etc. and have been tested for multiple proteins (Sugahara et al., 2008, Takehara et al., 2008). Previous results showed that horsehair and dried seaweed showed increased hits when added to sparse-matrix crystallization trials. The increase in crystallization was 35% when horsehair was added to 10 test proteins (Thakur et al., 2008). The underlying mechanism is explained with epitaxial nucleation in the case of minerals, electrostatic interactions if the nucleants contain charged surfaces, nucleation through specific favourable protein-protein interactions or physical entrapment in the caves of porous materials.

Seeding techniques can be advantageous in both screening of crystallization conditions to obtain crystals as well in the later optimisation steps. The streak seeding technique may provide a fast and effective way to facilitate the optimization of growth conditions without the uncertainty that is intrinsic in the process of spontaneous nucleation (Bergfors, 2003). A probe for analytical seeding is easily made with an animal whisker mounted with wax to the end of a pipette tip. The end of the fibre is then used to touch an existing crystal and dislodge seeds from it. Gentle friction against the crystal is normally sufficient. The probe is then used to introduce seeds into pre-equilibrated drop by rapidly running the fibre in a straight line across the middle of the drop containing protein and precipitant. Sitting drop set-ups are preferable since hanging drops tend to evaporate more quickly.

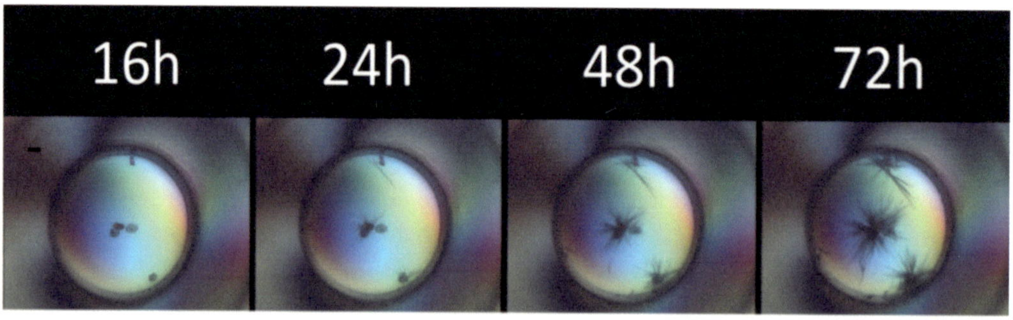

Fig. 5. Lysozyme needle crystals growing on sliced human hair as a nucleant, the black bar in the left picture represents 200 micrometer.

6. Combining heterogeneous crystallization and high throughput methods

A method for the introduction of heterogeneous nucleants in high throughput crystallization experiments has recently been developed (Nederlof et al., 2011). The method includes preparing of crystallization plates that are locally coated with fragments of human hair, allowing automated, high throughput crystallization trials in a fashion entirely compatible with standard vapour diffusion crystallization techniques. The effect of the nucleants was assessed on the crystallization of 11 different proteins in more than 4000 trials. Additional crystallization conditions were found for 10 out of 11 proteins when using the standard JCSG+ screen. In total, 34 additional conditions could be identified. The increase in crystallization conditions ranged between 33.3% (two additional conditions were identified for myoglobin on top of four homogeneous crystallizations) to 1.2% (we identified a single additional condition for insulin, which crystallized in 85 out of 96 conditions); the median increase in crystallization hits was 14%. The method is straightforward, inexpensive and uses materials available in every crystallization lab.

7. Lab automation

In recent years, setting up protein crystallisation trials and analysis of the results has become largely automated. More and more of the crystallisation methods mentioned in section 5 have been made amenable to automation, with the sitting drop method still the most popular experiment type in this respect. Lab automation includes the use of dispensing robots, imaging robots, in situ crystal analysis as well as automated diffraction analysis (Stevens et al., 2000, Berry et al., 2006).

7.1 Automation in dispensing

Dispensing robots that are used routinely are either specialized for dispensing well recipes (e.g. Formulator, MatrixMaker) or drop-setting (NT8, Phoenix, Mosquito), but there are also more generic robots that can do both (Hamilton Star, Tecan Evo). In general, the experimenter will start crystallisation trials with a set of pre-defined conditions, contained in one or more screens. Over 150 of these screens can be bought from commercial vendors in a wide range of formats. A number of these are designed on the basis of statistical analysis of results obtained at structural genomics initiatives. When initial hits are found with screens like these, secondary optimisation experiments need to be performed to produce diffracting crystals. In this stage, interaction with a Lab Information Management System (LIMS), where experiment design can be coupled to experiment preparation and analysis, greatly enhances the potential throughput in a lab and thereby the success rate. There are a number of these software packages that can be used to create grid experiments around an initial hit condition, as well as randomized sparse-matrix screens based on initial successes.

7.2 Automated experiment imaging

Automated experiment analysis is an essential part of the lab setup. Due to the increase in throughput obtained by using dispensing robots it is impossible to routinely scan the results manually under a microscope. The dynamic nature of these experiments can cause the crystallographer to miss events, even crystals. Imaging robots vary from semi-automated microscopes with a moving plate stage and camera to fully automated incubators that are capable of following all lab experiments from start to finish without human intervention. Ideally, images are displayed to the user in the context of the experiment design, so that the results are easily interpreted. If this functionality is integrated with the experiment design and dispensing the optimisation circle is complete. Such LIMS systems (Bard et al., 2004) can be further expanded to follow up on harvested crystals, to assess their diffraction quality and finally the structural data derived. (see 7.4)

7.3 *In situ* crystal analysis

When crystals are found an assessment needs to be made whether the crystals are indeed protein crystals or just salt crystals. And the quality of the crystal needs to be established as well as their usefulness for collecting diffraction data. It has always been difficult to distinguish protein crystals from salt crystals without actually collecting diffraction data. Historically, destructive methods have been used like the "crunch" method and protein dyes, the idea being that crystals similar to the ones destroyed will have the same properties. These methods have not always been conclusive and often the true nature of the crystal was only revealed on the X-ray beam. In recent years, three new techniques have been developed in this field; in situ diffraction analysis, UV detection and second harmonic microscopy.

7.3.1 *In situ* diffraction analysis

A number of years ago, Oxford Diffraction has come up with a device for X-ray diffraction analysis of crystals in the plate where they were grown (Skarzynski 2009, le Maire, et al., 2011). The idea is fairly simple, you center a crystal in the X-ray beam using a visual alignment tool and you subsequently take a single or a small number of X-ray diffraction images to assess whether a crystal is indeed protein, and to get some idea about the diffraction quality (mosaicity, resolution). An advantage is that the method is non-invasive (bearing in mind potential radiation damage) and fast. The method is not suitable for complete diffraction analysis, as the sample can only be rotated by 6°. It is also possible to automatically screen a complete crystallisation plate for potential diffraction. When suitably diffracting crystals are found they will still needed to be harvested and frozen for complete diffraction analysis.

7.3.2 UV detection of protein crystals

An increasing number of imaging devices (see 7.2) make use of a secondary light path in the UV range to detect protein crystals. These imagers make use of the fluorescence in UV by proteins, mostly caused by tryptophan (Judge et al., 2005). Since the protein concentration in the crystal will be much larger than in solution, any protein crystal will light up under UV, provided that the protein contains tryptophan. This is a relatively fast and non-invasive method, UV illumination can cause some ionisation in the drop, but this effect is much less than with X-ray illumination. In order to maximize its use, the experiment media (plates, seals) have to be chosen with care, some plastics are not sufficiently translucent in UV, or fluoresce themselves, adding noise to the image. One also has to bear in mind that some non-protein crystals (ATP, other co-factors added), might also fluoresce in UV. Having visible light and UV cameras integrated in a single imaging device greatly enhances its usefulness to distinguish protein from salt crystals.

7.3.3 Second harmonic microscopy

Fairly recently, a new development in the field of in situ crystal analysis has been reported. The technology makes use of a phenomenon called second harmonic generation (SHG), more often referred to as "frequency doubling" (Wampler et al., 2008). When an intense laser pulse travels through a highly polarizing, non-centrosymmetric material, light emerges with exactly half of the wavelength of the incident beam. The explanation is that two photons of the incident beam merge, creating a single photon with twice the energy. If the incident beam is in the near-infrared, the emerging beam, will be in the visible range. As mentioned, the technology requires intense laser light, delivered in femtosecond pulses. Most chiral crystal classes, with the exception of octahedral and icosahedral crystals, allow for SHG, thus encompassing over 99% of all protein crystals grown so far. A first commercial device using this technology, called SONICC, is available since 2011. When combined with a LIMS and a visible light imaging station, SHG can be used to automatically score and pre-sort the results for the experimenter.

7.4 Automated diffraction analysis

In parallel with automation taking hold of many crystallisation labs, the last part of the protein structure analysis pipeline, diffraction analysis in the X-ray beam, is increasingly automated as well. Not so long ago, a crystallographer would either measure his/her crystals at a home X-ray source or would travel to a synchrotron facility to do so. The process involved manually harvesting of the crystals, preparing them for the X-ray beam (mounting in a capillary or in a cryoloop, freezing), mounting them manually in the X-ray beam, gathering a few trial images to determine optimal settings for exposure, distance etc. and finally recording a set of diffraction images to solve the structure. The most time consuming steps have now been automated (Cork et al., 2006, Song et al 2007). Most notably, crystal mounting robots will now automatically take samples out of a liquid nitrogen dewar and place them in the X-ray beam, eliminating the need for the user to enter the X-ray hub of the synchrotron after every crystal. At the home lab, the mounted crystals are packed in specific dewar compatible with the robot arms at the beamline, and they are mailed to the synchrotron. In many synchrotrons, the user now has a choice of having a local operator collecting the data, or to drive the computers at the facility remotely from their own lab, there is no need to travel to a remote synchrotron anymore. In the near future, the automation can be improved by automatic crystal centering routines (Vernede et al., 2006). With e.g. the use of SHG (see 7.3) crystals can automatically be located inside the cryo loops, and this information can be used to automatically center the crystal in the beam.

8. Conclusions

The chapter covers some of the main aspects of protein nucleation and crystallization. Different diagnostic tools, crystallization techniques and strategies are explained. New tendencies in the field such as combining heterogeneous nucleants and high throughput methods are also presented.

9. References

Abbot, A. (2000) Nature 408, 130-132
Aoyama, K. (1996). J. Cryst. Growth, 168, 198
Achstetter, T. and Wolf, D. H. (1985). Yeast, 1, 139
Arfin, S. M. and Brandshaw, R. A. (1988). Biochemistry, 27, 7979
Baldock, P., Mills, V., Stewart, P.S. (1996). J. Cryst.Growth 168, 170-174
Bard, J., Ercolani, K., Svenson, K., Olland, A., Somers, W. (2004) Methods 34, 329-347
Barrett, A. J. and Salvesen, G. (1986) Research monographs in cell and tissue physiology, Vol 12, pp. 1-661. Elsevier. Amsterdam
Berry, I. M., Dym, O., Esnouf, R. M., Harlos, K., Meged, R., Perrakis, A., Sussman, J. L., Walter, T. S., Wilson, J. & Messerschmidt, A. (2006). Acta Cryst. D62, 1137-1149.
Bergfors T. 2003, J Struct Biol; vol. 142(1): pp 66-76.
Bergfors, T. 2009 Protein Crystallization: Second Edition. International University Line, La Jolla, California, 500 pp.

Berne, B.J.; Pecora, R., 1976, Dynamic Light Scattering; Wiley: New York,
Bond, J. S. and Butler, P.E. (1987). Annu. Rev. Biochem., 56, 333
Cork, C., O'Neill, J., Taylor, J. & Earnest, T. (2006). Acta Cryst. D62, 852-858.
Dalling, M.J. (1986) Plant proteolytic enzymes, 2 Vol. CRC Press, Boca Raton, FL
Doudna, J. A., Grosshans, C., Gooding, A., and Kundrot, C.E. (1993).Proc. Natl.Acad. Sci. USA, 90, 7829
Giege, R., Dock, A.C., Kern, D., Loeber, B., Thierrry, J. and Moras, D. (1986). J. Cryst. Growth, 76, 554
Giege, R. (1987). In Crystallography in molecular biology , NATO ASI Series, Vol. 126, pp. 15-26. Plenum Press, New York and London
Judge, R. A., Swift, K. & Gonzalez, C. (2005). Acta Cryst. D61, 60-66.
Karger, B. L. and Hancock, W. S. (1996). Methods in enzymology, Vol. 270, pp. 1-611 and Vol. 271, pp. 1-543. Academic press London
Konev S. V. (1967), Fluorescence and Phosphorescence of Proteins and Nucleic Acids, Plenum Press, New York
le Maire, A., Gelin, M., Pochet, S., Hoh, F., Pirocchi, M., Guichou, J.-F., Ferrer, J.-L. & Labesse, G. (2011). Acta Cryst. D67, 747-755.
Matthews, B.W. (1968), Journal of Molecular Biology, Volume 33, Issue 2, pp. 491-497
Mikol, V., Hirsche E., Giege R. (1989), FEBS letters Vol. 258, Issue 1, pp. 63-68
McPherson , A. (1985). In Methodsin enzymology, Academic Press, London. Vol. 114, p. 112
McPherson, A., Malkin, A. J., Kuznetsov, Y.G., and Koszelak, S. (1996), J.Cryst. growth, 168, 74
McPherson, A., Shlichta, P. (1988), Science, Vol. 239 Issue 4838, pp 385-387.
Nederlof, I., Hosseini, R., Georgieva, D., Juo, J., Li, Dianfan, Abrahams, J.P. (2011) Crystal Growth & Design, 11 (4), pp 1170-1176
Perrin, D. D. and Dempsey, B. (1974). Buffer for pH and metal ion control. Chapman and Hall Ltd., London and New York.
Takehara, M.; Ino, K.; Takakusagi, Y.; Oshikane, H.; Nureki, O.; Ebina, T.; Mizukami, F.; Sakaguchi, K. (2008), Analytical Biochemistry, 373 (2), pp 322-329.
Thygesen., J., Krumholz, S., Levin, L., Zaytsev-Bashan, A., Harms., J., Barrels, H., (1996). J. Cryst. Growth, 168, 308
Robert M.C., Lefaucheux F. 1988, Journal of Crystal Growth, Vol.90, Issues 1-3, pp 358-367,
Salemme, F.R. (1972), Archives of Biochemistry and Biophysics, Vol. 151, pp 533-539
Skarzynski, T. (2009). Acta Cryst. A, 65, s159
Skouri, M., Lorber, B., Giege, R., Munch, J.P. and Candau, S.J. (1996), J. Cryst Growth, 152, 209
Song J, Mathew D, Jacob SA, Corbett L, Moorhead P, Soltis SM (2007) J Synchrotron Radiation 14, 191-195
Stevens, R.C (2000) Curr. Op. Struct. Biol. 10, 558-563
Sugahara, M.; Asada, Y.; Morikawa, Y.; Kageyama, Y.; Kunishima, N. (2008), Acta Cryst. D 64 (6), pp 686-695.
Thakur, A. S.; Newman, J.; Martin, J. L.; Kobe, B. (2008), Structural Proteomics, Vol. 426, pp 403-409.
Van der Laan, J. M., Swarte, M. B. A., Groendijk, H., Hol, W. G. J., and Drenth , J. (1989). Eur. J. Biochem., 179, 715
Vekilov, P.G. and Rosenberger, F. (1996). J. Cryst. Growth, 158, 540
Wierenga, R. K., Lalk, K. H., and Hol, W.G.J (1987)., J.Mol.Biol., 198, 109
Wampler, R.D, Kissick, D.J.,Dehen, C.J., Gualtieri E.J., Grey, J.L., Wang, H.F., Thompson, D.H., Cheng, J.X. and Simpson, D.J. (2008) J. Am. Chem. Soc, 130, 14076-14077
Wandersman, C. (1989). Mol. Microbiol., 3. 1825
Wuthrich, K. (1995). Acta Cryst. D, pp 51, 249

21

Crystallization of Membrane Proteins: Merohedral Twinning of Crystals

V. Borshchevskiy[2,3] and V. Gordeliy[1,2,3]
[1]*Laboratoire des Protéines Membranaires, Institut de Biology Structurale J.-P.,*
[2]*Research-educational Centre "Bionanophysics", Moscow Institute of Physics and Technology,*
[3]*Institute of Complex Systems (ICS), ICS-5: Molecular Biophysics, Research Centre Juelich,*
[1]*France*
[2]*Russia*
[3]*Germany*

1. Introduction

Membrane proteins are the main functional units of biological membranes. They represent roughly one-third of the proteins encoded in the genome and about 70% of drugs are targeted to membrane proteins. X-ray protein crystallography is one of the most powerful tools to determine protein structure and to provide a basis for understanding molecular mechanisms of protein function. Despite an obvious importance of membrane protein only about 1% of structures in the Protein Data Bank (PDB) are of this type. Moreover, although the number of membrane protein structures deposited to PDB since 1985, date of the first membrane protein structure [1], is increasing it is not yet comparable with the rate achieved for soluble proteins [2]. Currently, the PDB contains more than 70,000 structures, and the structures of membrane proteins do not exceed 500 [3]. Considerable effort made in several laboratories in the last years towards extension of high-throughput crystallography to membrane proteins open a hope of correcting this imbalance. Nevertheless significant challenges must be overcome to achieve this goal. Two major problems toward the determination of membrane proteins structures are: the production of pure, stable and functional protein solubilized in detergents, and the growth of crystals suitable for X-ray crystallography. The latter is often defined as major bottleneck of structural biology of membrane proteins. For a long time, the vapor diffusion method has been the only method which was used to crystallize membrane proteins. This method, which is based on a well-developed approach of crystallization of water soluble proteins, led to relative success, however, it failed to produce crystals of some important membrane proteins. Quite recently new methods were introduced. One of the most promising new method to overcome this problem is the so called *in meso* crystallization approach where lipid systems (e.g. the lipid cubic phase (LCP)) are used as a crystallization matrix. It has been demonstrated that these methods are applicable to different membrane proteins including G-protein-coupled receptors (GPCR), membrane protein complexes and others. One of the first important breakthroughs was bacteriorhodopsin (bR) which for a long time failed to be crystallized by the *in surfo* methods and was solved to resolution about 1.55 Å from the crystals obtained by LCP crystallization. Thanks to the *in meso* method crystallographic structures of almost all functional states of bR are now available with atomic resolution (see [4] for review). Despite this fact the detailed mechanism of bR proton pumping is still to be elucidated. It appeared that a severe problem originates from the tendency of the best (in the sense of resolution) bR crystals to be perfectly twinned. Being a general problem of protein crystallography, twinning may result in controversial structural models of intermediate states in the case of bR. The chapter presented here is aimed to summarize the present knowledge on twinning formation during *in meso* crystallization and the methods to overcome it.

2. *In meso* crystallization

2.1 Crystallization from lipidic cubic phase

A principally new crystallization method - crystallization of membrane proteins in lipidic cubic phases was developed by Rosenbusch and Landau in 1996 [5]. A fundamental difference between methods of standard crystallization and crystallization in the LCP is that in the latter, the solubilized protein is reconstituted back in the native lipid bilayer and after that the crystallization is induced by the addition of a precipitant. Liquid crystalline systems formed by lipids in aqueous media can form infinite bicontinuous periodic minimal surfaces, which have a zero mean curvature and a periodicity in all the three dimensions characterized by a cubic lattice [6-8]. The system consists of two compartments: a continuous curved lipid bilayer forming a three-dimensional well-ordered structure, interwoven with continuous aqueous channels. Macroscopically the phase is very viscous, isotropic, and optically transparent. Membrane cubic phases are found in the cells [9], and they are used in food industry [10] as well as for drug delivery [11]. Practical aspects of crystallization in the lipidic cubic phase look very simple and an example - crystallization of bR - can be described as the following procedure [12]:

1. Weigh into the PCR tube (200 mL) approximately 5 mg of dry MO, incubate tubes with monooleoyl (MO) at 40°C, and spin the lipid down for 10 min at 13,000 × *g* at room temperature.
2. Keep MO at 40°C during an additional 20 min to gain the isotropic fluid lipidic phase and then let the lipid phase cool to room temperature.
3. Mix 1 mL of prepared 10 mg/mL BR solution comprising about 1.2 w/w% of *n*-octyl-β-D-glucopyranoside (OG) with 1 mg of MO. To gain the cubic phase, centrifuge the PCR tubes with the sample at 10,000 rpm for at least 1 h at 22°C (rotating tubes within the rotor every 15 min by 90°). Incubate the samples during 1-2days in the dark at 20-22°C. An alternative way to prepare the cubic phase is described in [13].
4. Add a precipitant to induce crystallization– a ground powder of KH2PO4 mixed with Na2HPO4 (95/5 w/w) with a final concentration of the salt mixture 1–2.5 M (pH 5.6). Repeat homogenizing centrifugation of samples as described in the previous item. Leave the crystallization batch in the dark at 22°C. bR microcrystals (10–20 mm in diameter) usually appear within 1 weak after induction of crystallization (Fig. 1). This protocol of crystallization is close to the original one provided by Rosenbusch and Landau. An alternative way to do such crystallization (it is used in nanovolume high throughput approach) is to add liquid precipitant to the top of the lipidic phase [13].
5. To separate the crystals from the lipidic phase directly from LCP use mechanical manipulation with microtools or, alternatively, add lipase or detergent to the lipidic phase to destroy the lipid phase at room temperature during several hours or days [14].

LCP approach remains most efficient among all other *in meso* approaches introduced later. Nevertheless, it is not yet clear whether other new methods were properly optimized. In other words it is not yet clear what is the real potential of these methods. Therefore, we will describe briefly three more new approaches

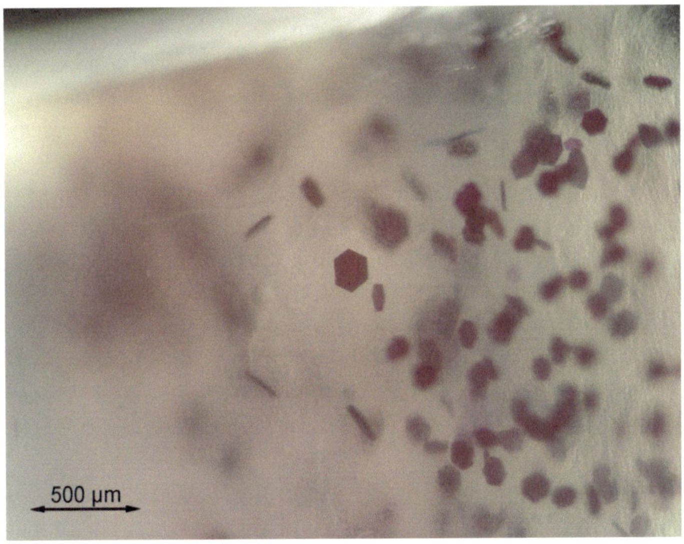

Fig. 1. A crystallization well (a PCR tube) with bR crystals.

2.2 Crystallization from vesicles

An interesting and unusual approach to membrane protein crystallization was proposed in 1998 [15,16]. The authors observed that purple membranes (two-dimensional hexagonal native crystals of bR) treated with the neutral detergent under certain conditions lead to the creation of spherical protein clusters (~50 nm in diameter). Using a standard vapor diffusion method for crystallization from bR vesicles with a high protein/lipid ratio, well diffracting hexagonal crystals were obtained [15-17]. This new crystal belongs to the space group *P622* with unit cell dimensions of $a = b = 104.7$ Å and $c = 114.1$ Å. The highest announced structural resolution achieved by this method is 2.0 Å. It is not compared to the LCP results obtained with the same protein. Until now there is no evidence that a specific case of bR crystallization from vesicles can be extended to other membrane proteins. However, it is not yet clear whether this approach is limited to some specific cases, like bR, or has a more general application.

2.3 Crystallization from bicelles

Just after the second *in meso* method was published another approach - crystallization from bicelles - was proposed. This method was first applied to obtain well diffracting bR crystals [18,19]. Bicelles, known for quite a long time, are a liquid crystal phase consisting of disc-shaped lipid-rich bilayer particles formed from mixtures of dimyristoyl phosphatidylcholine (DMPC) with certain detergents. The detergents mostly used for such a type of crystallization are either dihexanoyl phosphatidylcholine (DHPC) or zwitterionic bile salt derivative, CHAPSO. The bicelle sizes at a 1:3 DMPC/DHPC molar ratio are: the bilayer thickness - 40 Å and the diameter - 400 Å. The lipid detergent ratios present in the bicellar systems are relatively high compared to standard micellar systems [20,21].

The procedure of crystallization of membrane proteins from bicelles is as follows. The first step is preparation of bicelles. Then, solubilized protein is mixed with bicelles. It is considered, but not directly proven, that at this stage, the protein molecules are reconstituted into bicelles. After that the protein is crystallized by a standard vapor diffusion method. bR crystals grown at room temperature are identical to the previously obtained at 37°C twinned crystals: space group $P2_1$ (2.0 Å resolution) with unit cell dimensions of a = 44.7 Å, b = 108.7 Å, c = 55.8 Å, β = 113.6°. The other room-temperature crystals were not-twinned and belong to space group $C222_1$ (2.2 Å resolution) with the following unit cell dimensions: a = 44.7 Å, b = 102.5 Å, c = 128.2 Å. It is important to note that the crystals of the human β_2-adrenergic GPCR were obtained by this method [22]. The structure was solved to 3.5/3.7 Å resolution. It is considerably lower than what was obtained by protein crystallization in the cubic phase [23]. Taking into account the long and dramatic attempts to crystallize a ligand binding GPCR, there is no doubt it was a new considerable success of the method under discussion. The 2.3 Å resolution structure of the murine voltage dependent anion channel (mVDAC) that reveals a high-resolution presentation of membrane protein architecture was also obtained due to bicelles method [24]. Very recent success of the bicelle-like approach is the crystallization of the membrane part of the respiratory complex I [25].

2.4 Crystallization from sponge phases (L_3-phase)

It is interesting that historically crystallization from the sponge phase was described about 10 years after discovering the LCP approach. This is despite the fact that the sponge phase (L_3-phase) is the liquid analogue of the lipidic cubic phase with the reduced bending rigidity of membranes and without a long-range order. When the bending rigidity of the membrane becomes comparable with a thermal energy the ordered cubic phase structure is perturbed by thermally excited collective out-of-plane fluctuations of membranes. The transformation of the cubic to the sponge phase can be induced by different factors, for instance, via adding a solvent such as polyethyleneglycol (Mw = 400), dimethyl sulfoxide, 2-methyl-2,4-pentanediol (MPD), propylene glycol, or Jeffamine M600 to a lipid/ water system [26]. The diameter of aqueous pores in the MO cubic phase is relatively narrow (ca. 3-6 nm) compared to that of the sponge phase (10–15 nm and more) [27]. Evidently the size of the pores of the L_3-phase is compatible with membrane proteins with large hydrophilic parts and lets them diffuse freely within the plane of the membrane surface [26]. Well diffracting crystals of the reaction center from *Rhodobactersphaeroides* were grown in the L_3 by a conventional hanging-drop scheme of the experiment, and were harvested directly without the addition of lipase or cryoprotectant, and the structure was refined to 2.2 Å resolution. The authors of the work claimed that in contrast to the earlier LCP reaction center structure [28], the mobile ubiquinone could be built and refined. In these experiments, the only additional component (relative to the components of the cubic phase crystallization – the MO/membrane protein/detergent/buffer) was a small amphiphilic molecule 1,2,3-heptanetriol or Jeffamine M600. The structure was solved to resolution 2.35 Å [28]. In another work [29], crystals of the light harvesting II complex suitable for X-ray crystallography were obtained with structural 2.45 Å resolution. In this study, the additives used were KSCN, butanediol, pentaerythritolpropoxylate (PPO), *t*-butanol, Jeffamine, and 2-methyl-2,4-pentanediol (MPD). An advantage of the L_3 approach is that the liquid properties of the sponge phase at room temperature can be used directly in hanging- or sitting-drop vapor-diffusion crystallization by commercially available robots. Recently, a sponge phase sparse matrix crystallization screen consisting of different conditions became available [30]. However, unlike the LCP method, this one has not led to a breakthrough in structural biology of membrane protein. There was no structure of a new membrane protein or a principal improvement in structural resolution achieved by this method. Does it mean that the sponge phase approach does not have the same (or higher) power as the LCP method? We would speculate that this approach can be at least considered as a complementary one to the LCP.

3. Overcoming twinning formation

3.1 Introduction to the merohedral twinning of bR $P6_3$ crystals

Although bR can be crystallized by many methods and in different types of symmetries [5,16,18,31], only $P6_3$ crystal grown by *in meso* crystallization diffracts to the highest resolution. At the same time, these crystals often suffer from perfect merohedral twinning [32].

Twinning is one of the most common crystalline defects. A twin crystal consists of several domains oriented in such a way that their reciprocal lattices are superimposed at least in one dimension [33]. There are two forms of twinning: merohedral and non-merohedral. Only part of reflections of individual crystal domains superimpose in non-merohedral twinning, whereas all reflections are superimposed in three space dimensions in the merohedral form [34]. If only two orientations of twin domains are present the merohedral twinning is called hemihedral. It is the most widespread type of merohedral twinning [33]. The hemihedral twinning is intrinsic for hexagonal $P6_3$ crystals of bR grown in the cubic phase of MO [32,35].

Twinning of bR crystals implies the imposition of reflections with Miller indexes *hkl* and *kh-l*, so that the observed crystal reflections is a weighted sum of two different crystallographic reflections:

$$\begin{aligned} I_{hkl}^{obs} &= (1-\alpha)I_{hkl} + \alpha I_{kh-l} \\ I_{kh-l}^{obs} &= (1-\alpha)I_{kh-l} + \alpha I_{hkl} \end{aligned} \tag{1}$$

Where I_{hkl}^{obs} are crystallographic intensities observed in the X-ray experiment, I_{hkl} are crystallographic intensities of the twin domains and α is the twinning ratio, i.e. the volume fraction of equally oriented domains. Twinning is called perfect when α is close to 50 %. The shape and optical properties of twinned crystals are identical to those without twinning. The presence of twinning and estimation of the twinning ration are only possible by using special analysis methods of the diffraction data [36].

Twinning of the crystals complicates the obtaining of a crystallographic structure of the protein. If the twinning ration is $\alpha \neq 50\,\%$, then the system (1) can be solved:

$$\begin{aligned} I_{hkl} &= \frac{(1-\alpha)I_{hkl}^{obs} - \alpha I_{kh-l}^{obs}}{1-2\alpha} \\ I_{kh-l} &= \frac{(1-\alpha)I_{kh-l}^{obs} - \alpha I_{hkl}^{obs}}{1-2\alpha} \end{aligned} \tag{2}$$

After that, the usual tools can be applied for crystallographic analysis. However, as follows from (2), the error in intensity calculation increases and tends to infinity as a tends to 50 % [37]. For this reason the presence of crystal twinning worsens the electron density maps and reduces the reliability of protein models.

The perfect hemihedral twinning of bR crystals shows up in the presence of additional two-fold symmetry since $I_{hkl}^{obs} = I_{kh-l}^{obs}$ (see (1) when $\alpha = 50\,\%$). In this case, the number of independent observations (crystallographic intensities) is two times fewer. The equation system (1) is confluent and the crystallographic intensities cannot be extracted from the X-ray data. In this case, the intensities calculated from the protein model are used to obtain the desired crystallographic intensities according to the equation:

$$
\begin{aligned}
I_{hkl} &= \frac{I_{hkl}^{obs} + I_{hkl}^{cal} - I_{kh-l}^{cal}}{2} \\
I_{kh-l} &= \frac{I_{hkl}^{obs} + I_{kh-l}^{cal} - I_{hkl}^{cal}}{2}
\end{aligned}
\qquad (3)
$$

where I_{hkl}^{cal} are intensities calculated from the protein model. R-factors of protein models obtained from the perfect twinned data overestimate the model reliability, since the difference between the observed and calculated structural factors is undervalued due to the averaging over the reflections related by the twinning law. Hence, the refined crystallographic R-factors from perfectly twinned data are typically a factor of $1/\sqrt{2}$ lower than for low (or un-)twinned data [36,38,39]. In addition, the use for refinement of the intensities calculated according to (3) introduces additional model bias due to the explicit dependence of the detwinned data on the model itself.

An additional problem for X-ray analysis caused by perfect twinning is the inability to use the experimental difference Fourier map. Basing on the mathematical consideration it was shown about 40 years ago that the difference Fourier electron density maps are most sensitive, accurate and less susceptible to model bias method for observing limited structural changes [40]. The difference map is simply the Fourier transform of the amplitudes $(F_{exc} - F_{gr})$ (where F_{gr} and F_{exc} are the structural factors of the ground and excited state of the protein) and phases are taken from the model of the ground state. This type of maps visualizes the changes in the electron density between the first and second crystallographic datasets. If structural changes are visible at a reasonable significance level within a difference Fourier map, then it is a plausible feature of the experimental data. On the opposite side, if changes arise during crystallographic refinement and are not confirmed by the difference Fourier map, then they are probably artifacts. For this reason, the difference Fourier maps are the main criterion for detecting small structural changes in the macromolecular systems and were used in many studies, for instance: myoglobin-CO complex [41-44], photoactive yellow protein [45-48], sensory rhodopsin II [49] and bR [50-57]. In the case of perfect twinning of protein crystals, structural factors F_{gr} and F_{exc} cannot be obtained, and Fourier difference maps cannot be constructed.

Despite the fact that twinning creates problems for protein crystallography, currently there are no rational effective methods of obtaining untwinned crystals. Similarly there are only a few works published on the systematic study of interrelation between twinning formation and crystallization conditions. Description of the phenomenon of twinning is even poorer for the crystals of membrane proteins and particularly for those obtained by *in meso* crystallization.

However, the twinning problem is of particular importance for the case of bR. Among 28 bR structures obtained from $P6_3$ crystals, 19 are from crystals with perfect twinning [32]. The best resolution of bR crystallographic model is 1.43 Å [58]. However, all the structures with the resolution better than 1.9 Å were obtained from crystals with perfect twinning. The only exception is the structure with a resolution 1.55 Å from the crystal with a twinning ratio of 25 %. All the currently published crystallographic studies devoted to the K, L and M bR intermediate states either have a relatively low resolution (> 2.1 Å) [50-56] or were obtained from perfectly twinned crystals [58-63]. The intermediate state structures built using these data are not consistent with each other [53,56,64]. One of the most probable reasons for this is the twinning problem.

It is well known that the changes in bR structure during the transition from the ground state to intermediates are relatively small [50-55,58,60,62]. Thus, X-ray data of very high quality are required to obtain the structures of intermediate states. In particular, crystals should be untwinned as twinning reduces the quality of the electron density maps and the reliability of protein models, as well as suppresses the utilization of the Fourier difference maps. To elucidate the molecular mechanism of bR proton transport, it is crucial to obtain highly ordered crystals without twinning.

3.2 Physical detwinning of bR crystals

In 2004 [35] it was shown that the twinned crystals of bR consist of large scale domains. Each of the domains is a hexagonal plate with the size in the hexagonal plane equal to that of the whole crystal and the thickness comparable to that of the crystall (as it is shown in Fig.2). In most cases the crystals were split in two plates with no twinning. However in some cases the crystals were split in three and more plates. Thus it may be supposed that most of bR $P6_3$ crystals consist of only two twinning domains. However the presence of three and even more domains is also possible. But the size of these domains is always comparable to the size of the twinned crystal. The attempts to mechanically separate the twin domains had no effect. However it was noted that the slow decrease of mother liquid molarity may result in crystal slicing. Basing on this idea the approach for physical detwinning of bR crystals was proposed. According to the procedure the molarity of salt in mother liquid was slowly reduced from 3 to 1 M which induces splitting of agglutinated plates. Some of the split crystals diffracted well enough to determine the twin ratio which in all cases was equal to zero within the experimental error.

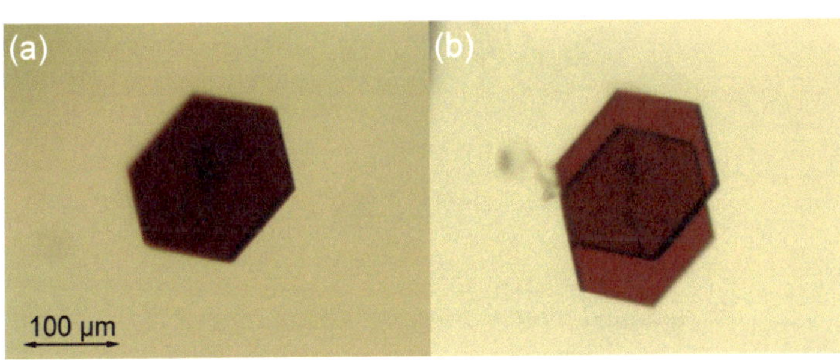

Fig. 2. bR crystal splits into two parts: (a) initial crystal, (b) two parts of the crystal separated by gradual decrease of the salt concentration.

Unfortunately, it turned out that the procedure of physical separation of the crystals often leads to a significant drop in the diffraction quality of the crystals, and therefore is not applicable in practice for obtaining high-resolution X-ray analysis.

3.3 Direct observation of twin domains

As it was mentioned before the twinning fraction of the crystal can only be estimated by the analysis of the statistical distribution of its crystallographic intensities. This implies that to determine the twinning, one has to fulfill the whole procedure of obtaining the crystallographic data, including the dissolution of the crystallographic sample, the separation of crystals from the crystallization matrix and X-ray data collection. Meanwhile, this resource- and time-consuming procedure has a small useful output: nine out of ten crystals have the twinning ratio close to 50 %.

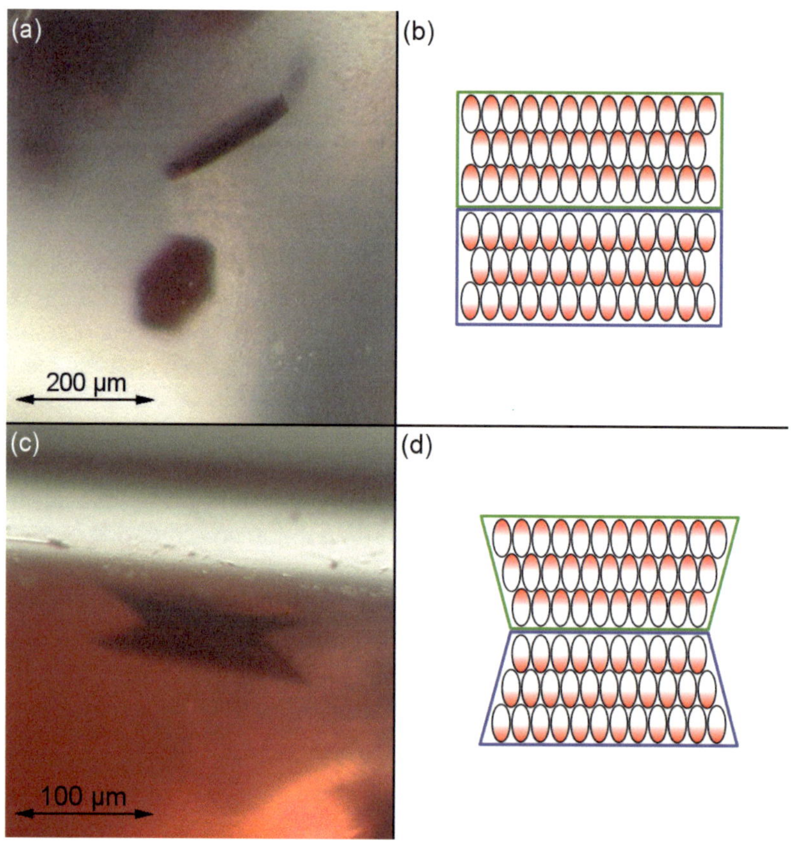

Fig. 3. bR crystals usually obtained by *in meso* crystallization in OG (a) and their schematic representation (b). bR crystals obtained by *in meso* crystallization in CYMAL-5 (c) with their schematic representation (d). Two different twin domains are shown in blue and green color. Red color represents the negative charge of CP side of bR.

One of the ways to simplify this procedure was found during crystallization trials with different detergent types [32]. It was observed that the crystals grown in the presence of 5-cyclohexyl-1-pentyl-β- D-maltoside (CYMAL-5) at concentrations of about 10 % have a shape of two truncated pyramids stuck together along the smaller of the hexagonal sides (Figure 3 c). Crystals in one crystallization probe had all the possible values of relative volumes of domains (from 0 when one domain was missing; to 0.5 when the domains had equal volume). The twinning ratio was surprisingly correlated with the relative domain volumes, which was confirmed by statistical analysis of X-ray intensities. The twinning fraction was close to 0 % when one of the domains was much smaller than the other, and close to 50 % for crystals with approximately equal parts. In addition, some of the crystals were split in two parts during fishing. Each of the domains had no twinning. Thus, it was concluded that the truncated pyramids represent twin domains as shown at Fig.3d. It is possible to select non-twinned crystals by careful inspection of the crystals shape in stereomicroscope, which significantly reduces the time and resources on the procedure for selection of crystals suitable for X-ray diffraction studies and produces additional information about the nature of the twinning formation.

3.4 Interrelation of crystal growth rate and twinning fraction

Additional information on the nature of bR twinning came from the statistical distribution of twinning ratio among several hundreds of crystals [32]. For this purpose bR crystals were grown in a wide range of crystallization conditions: at different concentrations of salt and protein, types and concentrations of detergents. More than 300 crystals were obtained and X-ray data were collected from all of them to determine their twinning ratios.

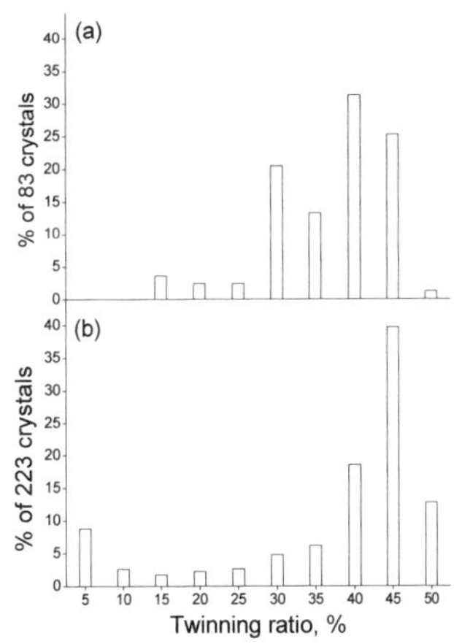

Fig. 4. Distribution of twinning ratios in two groups of crystals with the characteristic growth time less than 1.5 months (empty columns) (a) and more than 1.5 months (hatched columns) (b). The first and second groups consist of 83 and 227 crystals, respectively.

It turns out that regardless of the specific crystallization conditions the crystals with low twinning ratio (< 20 %) were observed with higher probability in samples where the first crystals appeared relatively late (in 2-3 weeks after sample preparation, rather than 2-3 days) and growth proceeds for a longer time period (for ~10 weeks). If the first crystals appeared in the sample relatively early and their growth was rapid then almost all crystals

had a high twinning ratio. The distribution of twinning ratio for 83 crystals grown less than 1.5 months and for 227 crystal with growth time of more than 1.5 months is shown at Figure 4. 11 % of the slowly grown crystals had the twinning ratio smaller than 10 %. Meanwhile all the fast grown crystals had the twinning ratio higher than 10 % and only 5 % had the twinning ratios between 10 % and 20 %.
It was suggested before for the soluble protein plastocyanin that slow growth favors the formation of untwinned crystals [39]. Confirmation of this relationship for a membrane protein, probably indicates the general nature of this phenomenon. It is plausible that in all cases when protein crystals suffer from twinning, one should search for the crystallization conditions of slow crystal growth.

3.5 Crystallisation in β-XylOC$_{16+4}$ mesophase

A presumably new approach to obtaining non-twinned bR crystals unexpectedly comes from the *in meso* crystallization in the "exotic" β-XylOC$_{16+4}$ mesophase.
The crystallization trials with this lipid were excited by the inequality of lipid and detergent libraries used for handling membrane proteins. The library of detergents with different hydrophilic and hydrophobic parts used for solubilization, purification and crystallization of membrane proteins is quite large. The fittest detergent may be found in the library for each specific membrane protein. This fact significantly increases the number of crystallized membrane proteins [65]. On the contrary the library of lipids used for the cubic phase creation is discouragingly small. MO is the most common lipid for *in meso* crystallization. Three other monoglycerols are reported to be suitable for this type of crystallization: monopalmitolein [5], monovaccenin [66], 2,3-dihydroxypropyl-(7Z)-hexadec-7-enoate [67] and 2,3-dihydroxypropyl-(7Z)-tetradec-7-enoate [68]. The library of matrix lipids for *in meso* crystallization should be increased for further success of the method.
Recently we presented the results of bR crystallization in the β-XylOC$_{16+4}$ cubic phase used for this purpose for the first time. β-XylOC$_{16+4}$ (Fig.2 in [69]) represents a recently developed isoprenoid-chained lipid family [70,71].
β-XylOC$_{16+4}$ forms a cubic phase almost at the same conditions as MO. It turns to be possible to crystallize bR in the β-XylOC$_{16+4}$ cubic phase using the standard protocol of *in meso* crystallization [69]. Several dozens of crystals were obtained. Three of them diffracted well enough and the X-ray dataset was collected for them. Two crystals diffracted up to 2 Å . The third one was worse and gave diffraction up to 2.7 Å .
The crystals obtained in the cubic phase of β-XylOC$_{16+4}$ and MO have the same $P6_3$ symmetry. The diffraction quality of bR crystals obtained in β-XylOC$_{16+4}$ is better than that of the first bR crystals obtained in MO [72] (the resolution is 2.0 Å and 2.5 Å, correspondingly). A further search for optimal crystallization conditions will possibly improve the diffraction properties as it was done in the case of MO.
It is important to mention that three studied crystals had a low twinning ratio. The twinning ratio was 37 and 34 % in two cases (for the crystals with diffraction resolution of 2.0 Å), and the third crystal (with resolution of 2.7 Å) had no twinning. As follows from §3.4 and [32], only 28 % of crystals obtained in the MO cubic phase have the twinning ratio smaller than 34 %. Thus the probability to find in one crystallization probe three crystals with small twinning ratios is relatively low which is unlikely to be a coincidence. The β-XylOC$_{16+4}$ cubic phase may favor the formation of low-twinned crystals.

3.6 The nature of the twinning phenomenon

Experiments described in 3.2-3.5 gave enough information to produce untwinned bR crystals for the investigation of the proton transport mechanism. On the other hand they gave some hints to understand the nature of the phenomenon of twinning formation in bR crystal.

bR crystals belong to class I in the nomenclature introduced in [73]. The hexagonal plane of bR crystals is perpendicular to the crystallographic axis *c* which implies that crystal growth occurs trough layer-by-layer two-dimensional nucleation on the *ab* surfaces of the crystal [74]. This assumption is in accordance with the model of *in meso* crystal growth proposed by M. Caffrey [75] and is confirmed by atomic force microscopy [76]. The contact surface between twinning domains is also perpendicular to *c* axis as it is demonstrated in Fig.3. Consequently, this surface also emerges as a result of two-dimensional nucleation on the *ab*-surface.

The contact surface may be formed either by two cytoplasmic (CP) surfaces of bR or two extracellular (EC) ones. The twinning ratio of the majority of crystals is > 30 %, and most of them consist of two domains. This peculiarity may be explained by different energies of interaction for CS-CS and EC-EC contacts in the protein crystal. As follows from the pdb-structure (1C3W [77] for instance) EC surface of bR is almost neutral and CP is negatively charged. On the other hand there is no specific interaction seen in pdb-structures between two adjacent protein layers, they interact by Van-der-Waals contacts between only two amino acids [72]. That means that even a weak electrostatic interaction may play an important role in the total energy of layer interaction.

Thus we can imagine the following process of crystal formation: the first twin domain emerges soon after (or even during) nucleation with two twin domains interacting by their EC surfaces. The crystal itself has two CP surfaces at its external faces. The probability to form a new twin domain on the CP surface is relatively low due to unfavourable electrostatic interaction. Consequently, the crystal continues to grow without formation of new twinning domains.

It may be noted that the distribution of the twinning fraction of slowly growing crystals has a sinuous pattern: there are local maxima with the twinning ratio < 10 % and> 35 %, and a minimum is located in between them. This non-obvious behavior may be explained by computer modelling of the growth of twin crystals.

As it was mentioned before crystal growth occurs through the two-dimensional nucleation at the surface of the crystal (slow step) and a relatively fast growth of the new layer in two dimensions. Thus one can use a one-dimensional model to simulate crystal growth in the direction perpendicular to the *ab* crystallographic plane. Crystal growth begins from a single layer and proceeds by consecutive addition of new layers to each surface of the crystal alternatively. When a new layer is added three different types of contacts may be formed.

1. CP-EC contacts which corresponds to normal crystal growth. Let us assign to this event a relative probability of 1.
2. EC-EC contacts which corresponds to the formation of the twinning domain. We will assign the probability P_1 to this event.
3. CP-CP contacts which also gives rise to a twinning domain as probability P_2 is assigned to this event.

The usual thickness of $P6_3$ bR crystal is about 20 μm that corresponds to about 4000 protein layers. This number of layers was used in the simulation of the crystal growth.

There are two variables which will dictate the number of formed twinning domains and the twinning ratio of the crystal: the probabilities P_1 and P_2. These probabilities may be varied to fit the experimental dependencies shown at Fig.4.

The first feature noted while exploring this model was that the symmetrical conditions ($P_1 = P_2$) cannot reproduce the experimental data. Under relatively low probabilities of twinning formation the distribution shows a peak at zero twinning ratio. The height of the peak decreases as the probability of twin formation increases and the distribution over the nonzero range remains quite flat until the peak at zero vanishes (Fig. 5a).

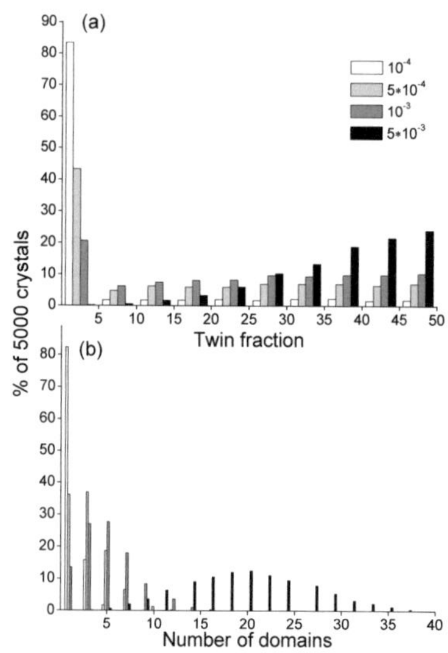

Fig. 5. Distribution of twinning fraction (a) and (b) the number of twinning domains calculated for 5000 crystals under conditions of symmetric domain nucleation (P_1=P_2) and for probabilities in the range 10^{-4} - 5×10^{-3}.

When an asymmetry in the probabilities is introduced to the model with $P_1=10^{-3}$, the peak at zero value changes very little, while the rest of the distribution has low values at low twinning ratios which gradually increase towards higher twinning ratios (Fig. 6a). Two important things are worth noting here. Firstly, when P_2 is smaller than P_1 it has almost no influence on the distribution. P_2 is the probability of forming CP-CP. As it was described above this event is quite improbable because the two negatively charged CP surfaces are pushing apart. Thus P_2 can be fixed at 0 at the following consideration. Second, the introduction of asymmetry leads to a shift in the peak of the number of twin domains distribution (compare Fig. 5b and 6b) from six domains (for $P_1 = P_2 = 10^{-3}$) to two domains ($P_1 = 10^{-3}$, $P_2 = 0$) which is in accordance with the experimental results.

Under the asymmetrical conditions the model resembles the experimentally observed distributions. Small changes in P_1 lead to dramatic changes in the fractions of non-twinned and perfectly twinned crystals, while the fraction of crystals with an intermediate twinning ratio changes much more slowly. The best fit of the experimentally observed distributions corresponds to a probability P_1 of 3×10^{-3} for fast crystal growth, where less than 1 % of crystals grow without twinning, and of 1.25×10^{-3} for slow growth, where 10 % of crystals have no twinning (Fig.7).

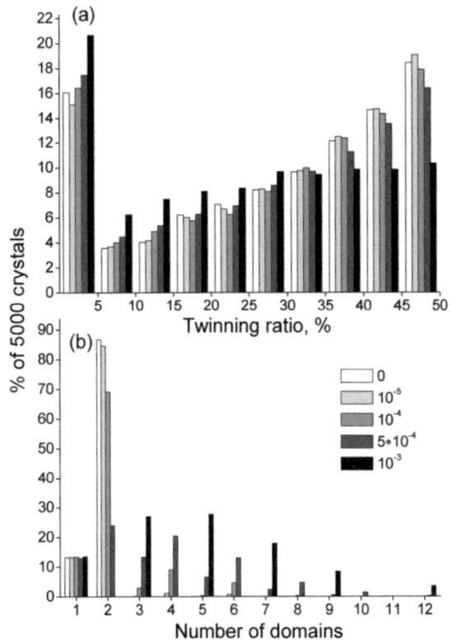

Fig. 6. Distribution of twinning ratio (a) and twinning domains (b) calculated for 5000 crystals under conditions of asymmetric domain nucleation $P_1=10^{-3}$, P_2 in the range between 0 and 10^{-3}.

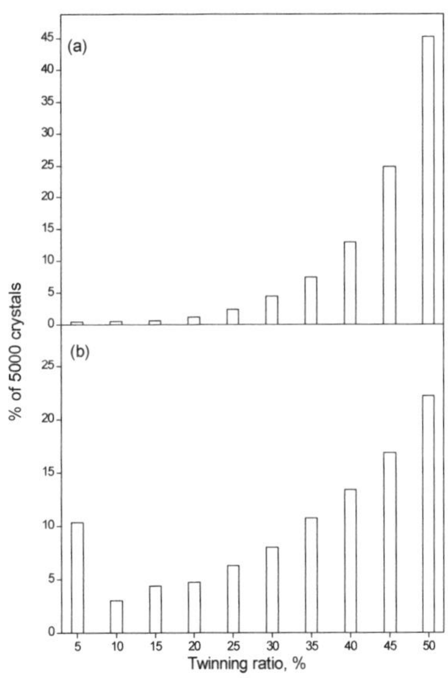

Fig. 7. Modelled distributions of twinning ratios simulating experimental distributions for slow (a) and fast (b) crystal growth. The P_1 probabilities for the models are 1.25×10^{-3} and 3×10^{-3}, correspondingly; $P_2 = 0$.

The model resembles the principal features of the experimentally observed distributions having quite a bad fit at the region of the high twinning ratio. This feature may be explained either by the underestimation of the twinning ratio by computational procedures owing to noise in the diffraction intensities or by the heterogeneity of the crystallization medium, which is responsible for the inevitable differences in the growth rates of different crystal surfaces.

The described model of twinning formation explains how the probabilities P_1 and P_2 determine the type of twinning fraction distribution. However, this model does not explain what is the relation between the rate of crystal growth and probabilities P_1 and P_2. Unfortunately, the theory of *in meso* crystallization is quite poorly understood at the moment and it cannot be used to explain this dependence. However, we can imagine the following thermodynamic explanation:

The limiting step of the crystal growth is the two-dimensional nucleation on the crystal surface. According to the classical two-dimensional theory of crystallization the thermodynamic potential of nucleus formation is [78]:

$$\Delta G = -n(\mu_v - \mu_c) + \sum_i l_i \kappa_i \tag{4}$$

where μ_v and μ_c are chemical potentials of the protein molecule in the volume and on the crystal surface, n is the number of molecules in a nucleus. The second term describes the surface energy, where κ_i is a specific surface energy and l_i is the length of *i-th* edge.

Basing on (4) we can write down the general expression for the free energy:

$$\Delta G^* = -A\Delta\mu + B \tag{5}$$

where A and B are the values which are not dependent on $\Delta\mu = \mu_v - \mu_c$. A depends on the specific surface energy and is virtually equal for normal and twinning nucleation. B depends on the interaction of the molecules in different layers and is significantly different for normal and twinning nucleation.

The rate of two-dimentional nucleation represents as:

$$J \approx \Gamma e^{\frac{-\Delta G^*}{kT}} \tag{6}$$

where Γ poorly depends on $\Delta\mu$. The experimental fact that twinning domains have a macroscopic size results in the condition that the probability of normal layer nucleation is significantly higher than that of the twinning formation. Consequently:

$$J_0 >> J_1; J_2 => \Delta G_0^* < \Delta G_1^*; \Delta G_2^* \tag{7}$$

where J_0, J_1 and J_2 are the rates for normal and two twinning (CS-CS and EC-EC) nucleations, and ΔG_0^*, ΔG_1^* и ΔG_2^* are the corresponding free energies. The rate of crystal growth is regulated by supersaturation $\Delta\mu$ and at very high values of supersaturation the difference between J_0, J_1 and J_2 vanishes (see (5)). When $\Delta\mu$ decreases, the absolute value of ΔG^* also diminishes and the growth rate drops down. As follows from (5) , ΔG_1^* and ΔG_2^* approache 0 faster than ΔG_0^* and under a certain value of $\Delta\mu$ become positive (the formation of twin crystals ceases). Simultaneously, due to exponential dependency (6) the difference between probabilities of normal and twinning nucleations grows.

The described explanation is applicable for any crystals where twinning is formed by two-dimensional nucleation. For this type of crystals the correlation between the growth rate and the probability of twinning formation may be a common feature. Taking into consideration the presence of the lipidic cubic phase may give better understanding of the mechanism of the twinning formation.

The most important feature of the *in meso* crystallization mechanism for this consideration is the presence of lamellar lipid environment around the growing protein crystal (see Fig.1a in [75]). It is obvious that the highly curved transitional lipid phase should be present between the bulky cubic and lamellar phases. The changes in the cubic phase curvature will simultaneously cause the corresponding changes in the curvature of the transitional phase.
It was proposed in [79] that the protein *in meso* crystallization is provoked by excess of elastic energy in the curved lipid bilayer. This type of energy is accumulated by the crystallization system due to hydrophilic-hydrophobic mismatch between the lipid bilayer and the protein molecule and the value of this energy is strongly dependent on the bilayer curvature radius and the length of protein hydrophobic-hydrophilic boarder. The rate of crystal growth is regulated through the changes of the elastic energy caused by variations in the bilayer curvature. The decrease of the curvature radius results in the slowdown of crystal growth.
On the other hand the variations in the length of the hydrophilic-hydrophobic boarder also influence the crystallization rate. There are two substantially different hydrophilic-hydrophobic boarders of the protein molecule (one is at the EC side of the protein and the other is at the CP one). The curved bilayer is also asymmetrical relative to the perpendicular to its surface. Consequently, the elastic energy of deformation is dependent on the orientation of the protein in the curved bilayer.
The protein molecule has to cross the highly curved transitional bilayer during the crystallization and the corresponding energy barrier of this process is different for different orientations of protein molecules. And the character of the elastic energy dependence on the bilayer curvature is also different for the two possible protein orientations.
The decrease of the bilayer curvature during crystallization results in a slowdown of crystal growth and simultaneously reduces the curvature of the transitional region. The energy barriers for two different protein orientations change differently and this fact results in different probabilities of the formation of the normal or twinned protein layer in the crystal.

4. Conclusions

Twinning of protein crystals is an unwelcome phenomenon for crystallographers and may be a barrier, like in the case of bR crystals, on the way to elucidating protein function. For this reason the efforts were applied to understand and overcome it. Nowadays the twinning of bR $P6_3$ crystals is one of the most studied and characterised twinning phenomena of protein crystals.
First of all it was directly shown that the LCP grown twinned crystals of bR consist of large scale domains. Each of the domains is a hexagonal plate with the size equal to that of the whole crystal [35]. It is important the crystals may be split into several non-twinned domains by slow changes of salt concentration in the mother liquid so that the split parts preserved high diffraction quality. Further systematic investigation showed that the rate of crystal growth strongly affects the twinning-ratio distribution of the crystals. Searching for crystallization conditions leading to slow crystal growth, it is possible to select crystallization trials that contained up to 10% non-twinned crystals [32]. In addition, the conditions were found allowing selection of crystals with low twinning by visual inspection of their shape with no need for analysis of the diffraction intensity distribution. This discovery further facilitates the process of selection of non-twinned crystals. The experimental data obtained so far allow the formulation of a theory of twinning formation

which in particular sheds some light on the general question of the process of *in meso* crystallization. Most recently some hints were found that the usage of different crystallization matrixes may allow to improve the yield of non-twinned crystals in crystallization [69].

5. Acknowledgements

Authors are grateful to Georg Büldt, Rouslan Efremov and Ekaterina Round for their contribution to the chapter. Authors are supported by the program "Chairesd'excellence" édition 2008 of ANR France, CEA(IBS)-HGF(FZJ) STC 5.1 specific agreement, the German Federal Ministry for Education and Research (PhoNa - Photonic Nanomaterials), the MC grant for training and career development of researchers (Marie Curie, FP7-PEOPLE-2007-1-1-ITN, project SBMPs), an EC FP7 grant for the EDICT consortium (HEALTH-201924), Russian State Contracts No. 02.740.11.0299, 02.740.11.5010, P974 of activity 1.2.2, and No. P211 of activity 1.3.2 of the Federal Target Program "Scientific and Academic Research Cadres of Innovative Russia" for 2009–2013.

6. References

[1] Deisenhofer J. et al., Structure of the protein subunits in the photosynthetic reaction centre of Rhodopseudomonas viridis at 3[angst] resolution, *Nature* 318 (1985) 618-624.

[2] White S.H., The progress of membrane protein structure determination, *Protein Sci.* 13 (2004) 1948-1949.

[3] The Protein Data Bank, http://www.pdb.org/ (2011)

[4] Hirai T. et al., Structural snapshots of conformational changes in a seven-helix membrane protein: lessons from bacteriorhodopsin, *Current Opinion in Structural Biology* 19 (2009) 433-439.

[5] Landau E.M. et Rosenbusch J.P., Lipidic cubic phases: A novel concept for the crystallization of membrane proteins, *Proc.Natl.Acad.Sci.USA* 93 (1996) 14532-14535.

[6] Mariani P. et al., Cubic phases of lipid-containing systems. Structure analysis and biological implications, *J.Mol.Biol.* 204 (1988) 165-189.

[7] Luzzati V. et al., Structure of the cubic phases of lipid-water systems, *Nature* 220 (1968) 485-488.

[8] Scriven L.E., Equilibrium Bicontinuous Structure, *Nature* 263 (1976) 123-125.

[9] Landh T., From entangled membranes to eclectic morphologies: cubic membranes as subcellular space organizers, *FEBS Lett.* 369 (1995) 13-17.

[10] Fontell K., Cubic phases in surfactant and surfactant-like lipid systems, *Colloid & Polymer Science* 268 (1990) 264-285.

[11] B.Ericsson et al., Cubic Phases as Delivery Systems for Peptide Drugs, в: *Polymeric Drugs and Drug Delivery Systems*, American Chemical Society, 1991) 251-265.

[12] V.I.Gordeliy et al., Crystallization in lipidic cubic phases: A case study with Bacteriorhodopsin, *Membrane Protein Protocols: Expression, Purification, and Crystallization*, ed. B.Selinsky, publ.: Humana Press, Totowa NJ, (2003) 305-316.

[13] Caffrey M. et Cherezov V., Crystallizing membrane proteins using lipidic mesophases, *Nature Protocols* 4 (2009) 706-731.

[14] Nollert P. et Landau E.M., Enzymic release of crystals from lipidic cubic phases, *Biochem.Soc.Trans.* 26 (1998) 709-713.

[15] Kouyama T. et al., Polyhedral assembly of a membrane protein in its three-dimensional crystal, *J.Mol.Biol.* 236 (1994) 990-994.

[16] Takeda K. et al., A novel three-dimensional crystal of bacteriorhodopsin obtained by successive fusion of the vesicular assemblies, *J.Mol.Biol.* 283 (1998) 463-474.

[17] Denkov N.D. et al., Electron cryomicroscopy of bacteriorhodopsin vesicles: mechanism of vesicle formation, *Biophys.J.* 74 (1998) 1409-1420.

[18] Faham S. et Bowie J.U., Bicelle crystallization: a new method for crystallizing membrane proteins yields a monomeric bacteriorhodopsin structure, *J.Mol.Biol.* 316 (2002) 1-6.

[19] Faham S. et al., Crystallization of bacteriorhodopsin from bicelle formulations at room temperature, *Protein Sci.* 14 (2005) 836-840.

[20] Sanders C.R. et Schwonek J.P., Characterization of magnetically orientable bilayers in mixtures of dihexanoylphosphatidylcholine and dimyristoylphosphatidylcholine by solid-state NMR, *Biochemistry* 31 (1992) 8898-8905.

[21] Sanders C.R. et Prestegard J.H., Magnetically orientable phospholipid bilayers containing small amounts of a bile salt analogue, CHAPSO, *Biophys.J.* 58 (1990) 447-460.

[22] Rasmussen S.G. et al., Crystal structure of the human beta2 adrenergic G-protein-coupled receptor, *Nature* 450 (2007) 383-387.

[23] Cherezov V. et al., High-resolution crystal structure of an engineered human beta2-adrenergic G protein-coupled receptor, *Science* 318 (2007) 1258-1265.

[24] Ujwal R. et al., The crystal structure of mouse VDAC1 at 2.3 A resolution reveals mechanistic insights into metabolite gating, *Proc.Natl.Acad.Sci.U.S.A* 105 (2008) 17742-17747.

[25] Efremov R.G. et al., The architecture of respiratory complex I, *Nature* 465 (2010) 441-445.

[26] S.Engstrom et al., Solvent-induced sponge (*L3*) phases in the solvent-monoolein-water system, *The Colloid Science of Lipids*, ed. B.Lindman, B.Ninham, publ.: Springer Berlin / Heidelberg,1998) 93-98.

[27] Ridell A. et al., On the water content of the solvent/monoolein/water sponge (L3) phase, *Colloids and Surfaces A: Physicochemical and Engineering Aspects* 228 (2003) 17-24.

[28] Katona G. et al., Lipidic cubic phase crystal structure of the photosynthetic reaction centre from Rhodobacter sphaeroides at 2.35A resolution, *J.Mol.Biol.* 331 (2003) 681-692.

[29] Cherezov V. et al., Room to move: crystallizing membrane proteins in swollen lipidic mesophases, *J.Mol.Biol.* 357 (2006) 1605-1618.

[30] Wohri A.B. et al., A lipidic-sponge phase screen for membrane protein crystallization, *Structure.* 16 (2008) 1003-1009.

[31] Schertler G.F. et al., Orthorhombic crystal form of bacteriorhodopsin nucleated on benzamidine diffracting to 3.6 A resolution, *J.Mol.Biol.* 234 (1993) 156-164.

[32] Borshchevskiy V. et al., Overcoming merohedral twinning in crystals of bacteriorhodopsin grown in lipidic mesophase, *Acta Crystallogr.D.Biol.Crystallogr.* 66 (2010) 26-32.

[33] Parsons S., Introduction to twinning, *Acta Crystallogr.D Biol.Crystallogr.* 59 (2003) 1995-2003.

[34] Dauter Z., Twinned crystals and anomalous phasing, *Acta Cryst. D*59 (2003) 2004-2016.

[35] Efremov R. et al., Physical detwinning of hemihedrally twinned hexagonal crystals of bacteriorhodopsin, *Biophys.J.* 87 (2004) 3608-3613.

[36] Yeates T.O., Detecting and overcoming crystal twinning, *Methods Enzymol.* 276 (1997) 344-358.

[37] Royant A. et al., Detection and characterization of merohedral twinning in two protein crystals: bacteriorhodopsin and p67(phox), *Acta Crystallogr.D Biol.Crystallogr.* 58 (2002) 784-791.

[38] Royant A. et al., Detection and characterization of merohedral twinning in two protein crystals: bacteriorhodopsin and p67(phox), *Acta Crystallogr.D Biol.Crystallogr.* 58 (2002) 784-791.

[39] Redinbo M.R. et al., The 1.5-A crystal structure of plastocyanin from the green alga Chlamydomonas reinhardtii, *Biochemistry* 32 (1993) 10560-10567.

[40] Henderson R. et Moffat J.K., The difference Fourier technique in protein crystallography: errors and their treatment, *Acta Cryst. B*27 (1971) 1414-1420.

[41] Schlichting I. et al., Crystal-Structure of Photolyzed Carbonmonoxy-Myoglobin, *Nature* 371 (1994) 808-812.

[42] Srajer V. et al., Photolysis of the carbon monoxide complex of myoglobin: nanosecond time-resolved crystallography, *Science* 274 (1996) 1726-1729.

[43] Srajer V. et al., Protein conformational relaxation and ligand migration in myoglobin: a nanosecond to millisecond molecular movie from time-resolved Laue X-ray diffraction, *Biochemistry* 40 (2001) 13802-13815.

[44] Schotte F. et al., Watching a protein as it functions with 150-ps time-resolved x-ray crystallography, *Science* 300 (2003) 1944-1947.

[45] Genick U.K. et al., Structure of a protein photocycle intermediate by millisecond time-resolved crystallography, *Science* 275 (1997) 1471-1475.

[46] Genick U.K. et al., Structure at 0.85 A resolution of an early protein photocycle intermediate, *Nature* 392 (1998) 206-209.

[47] Perman B. et al., Energy transduction on the nanosecond time scale: early structural events in a xanthopsin photocycle, *Science* 279 (1998) 1946-1950.

[48] Ren Z. et al., A molecular movie at 1.8 A resolution displays the photocycle of photoactive yellow protein, a eubacterial blue-light receptor, from nanoseconds to seconds, *Biochemistry* 40 (2001) 13788-13801.

[49] Moukhametzianov R. et al., Development of the signal in sensory rhodopsin and its transfer to the cognate transducer, *Nature* 440 (2006) 115-119.

[50] Edman K. et al., High-resolution X-ray structure of an early intermediate in the bacteriorhodopsin photocycle, *Nature* 401 (1999) 822-826.

[51] Sass H.J. et al., Structural alterations for proton translocation in the M state of wild-type bacteriorhodopsin, *Nature* 406 (2000) 649-653.

[52] Royant A. et al., Helix deformation is coupled to vectorial proton transport in the photocycle of bacteriorhodopsin, *Nature* 406 (2000) 645-648.

[53] Matsui Y. et al., Specific damage induced by X-ray radiation and structural changes in the primary photoreaction of bacteriorhodopsin, *J.Mol.Biol.* 324 (2002) 469-481.

[54] Kouyama T. et al., Crystal structure of the L intermediate of bacteriorhodopsin: Evidence for vertical translocation of a water molecule during the proton pumping cycle, *J.Mol.Biol.* 335 (2004) 531-546.

[55] Edman K. et al., Deformation of helix C in the low temperature L-intermediate of bacteriorhodopsin, *J.Biol.Chem.* 279 (2004) 2147-2158.
[56] Takeda K. et al., Crystal structure of the M intermediate of bacteriorhodopsin: allosteric structural changes mediated by sliding movement of a transmembrane helix, *J.Mol.Biol.* 341 (2004) 1023-1037.
[57] Yamamoto M. et al., Crystal structures of different substates of bacteriorhodopsin's M intermediate at various pH levels, *J.Mol.Biol.* 393 (2009) 559-573.
[58] Schobert B. et al., Crystallographic structure of the K intermediate of bacteriorhodopsin: Conservation of free energy after photoisomerization of the retinal, *J.Mol.Biol.* 321 (2002) 715-726.
[59] Lanyi J. et Schobert B., Crystallographic structure of the retinal and the protein after deprotonation of the Schiff base: the switch in the bacteriorhodopsin photocycle, *J.Mol.Biol.* 321 (2002) 727-737.
[60] Lanyi J.K. et Schobert B., Mechanism of proton transport in bacteriorhodopsin from crystallographic structures of the K, L, M-1, M-2, and M-2 ' intermediates of the photocycle, *J.Mol.Biol.* 328 (2003) 439-450.
[61] Schobert B. et al., Crystallographic structures of the M and N intermediates of bacteriorhodopsin: assembly of a hydrogen-bonded chain of water molecules between Asp-96 and the retinal Schiff base, *J.Mol.Biol.* 330 (2003) 553-570.
[62] Lanyi J.K. et Schobert B., Structural changes in the L photointermediate of bacteriorhodopsin, *J.Mol.Biol.* 365 (2007) 1379-1392.
[63] Facciotti M.T. et al., Structure of an early intermediate in the M-state phase of the bacteriorhodopsin photocycle, *Biophys.J.* 81 (2001) 3442-3455.
[64] Lanyi J.K., What is the real crystallographic structure of the L photointermediate of bacteriorhodopsin?, *Biochim.Biophys.Acta* 1658 (2004) 14-22.
[65] Seddon A.M. et al., Membrane proteins, lipids and detergents: not just a soap opera, *Biochim.Biophys.Acta* 1666 (2004) 105-117.
[66] Gordeliy V.I. et al., Molecular basis of transmembrane signalling by sensory rhodopsin II-transducer complex, *Nature* 419 (2002) 484-487.
[67] Misquitta Y. et al., Rational design of lipid for membrane protein crystallization, *J.Struct.Biol.* 148 (2004) 169-175.
[68] Misquitta L.V. et al., Membrane protein crystallization in lipidic mesophases with tailored bilayers, *Structure.* 12 (2004) 2113-2124.
[69] Borshchevskiy V. et al., Isoprenoid-chained lipid [beta]-XylOC16+4--A novel molecule for in meso membrane protein crystallization, *Journal of Crystal.Growth* 312 (2010) 3326-3330.
[70] Yamashita J. et al., New lipid family that forms inverted cubic phases in equilibrium with excess water: molecular structure-aqueous phase structure relationship for lipids with 5,9,13,17-tetramethyloctadecyl and 5,9,13,17-tetramethyloctadecanoyl chains, *J.Phys.Chem. B*112 (2008) 12286-12296.
[71] Hato M. et al., Aqueous phase behavior of lipids with isoprenoid type hydrophobic chains, *J.Phys.Chem. B*113 (2009) 10196-10209.
[72] Pebay-Peyroula E. et al., X-ray structure of bacteriorhodopsin at 2.5 angstroms from microcrystals grown in lipidic cubic phases, *Science* 277 (1997) 1676-1681.
[73] Michel H., Crystallization of Membrane Proteins, (1991)
[74] McPherson A., Crystallization of biological macromolecules, (1999)

[75] Caffrey M., Crystallizing membrane proteins for structure determination: use of lipidic mesophases, *Annu.Rev.Biophys.*38 (2009) 29-51.

[76] Qutub Y. et al., Crystallization of transmembrane proteins in cubo: mechanisms of crystal growth and defect formation, *J.Mol.Biol.*343 (2004) 1243-1254.

[77] Luecke H. et al., Structure of bacteriorhodopsin at 1.55 angstrom resolution, *J.Mol.Biol.*291 (1999) 899-911.

[78] Markov I.V., Crystal growth for the begginers: Fundamentals of Nucleation, Crystal Growth and Epitaxy, second (2003)

[79] Grabe M. et al., Protein interactions and membrane geometry, *Biophys.J.*84 (2003) 854-868.

22

Rational and Irrational Approaches to Convince a Protein to Crystallize

André Abts, Christian K. W. Schwarz, Britta Tschapek, Sander H. J. Smits and Lutz Schmitt
Institute of Biochemistry, Heinrich-Heine University, Düsseldorf, Germany

1. Introduction

The importance of structural biology has been highlighted in the past few years not only as part of drug discovery programs in the pharmaceutical industry but also by structural genomics programs. Mutations of human proteins have been long recognized as the source of severe diseases and a structural knowledge of the consequences of a mutation might open up new approaches of drugs and cure. Although the function of a protein can be studied by several biochemical and/or biophysical techniques, a detailed molecular understanding of the protein of interest can only be obtained by combining functional data with the knowledge of the three-dimensional structure. In principle three techniques exist to determine a protein structure, namely X-ray crystallography, nuclear magnetic resonance spectroscopy (NMR) and electron microscopy (EM). According to the protein data bank (pdb; http://www.rcsb.org) that provides a general and open-access platform for structures of biomolecules, X-ray crystallography contributes more than 90% of all structures in the pdb, a clear emphasis of the importance of this technique.
To perform X-ray crystallography it is essential to have large amounts of pure and homogenous protein to perform an even today still "trail and error"-based screening matrix to obtain well diffracting protein crystals. Therefore, successful protein crystallization requires three major and crucial steps, all of them associate with specific problems and challenges that need to be overcome and solved. These steps are (I) protein expression, (II) protein purification and (III) the empirical search for crystallization conditions. As summarized in Figure 1, every single step needs to be optimized along the long and stoney road to obtain protein crystals suitable for structure determination of your "most-beloved" protein via X-ray crystallography. This chapter will focus on these three steps and suggests strategies how to perform and optimize each of these three steps on the road of protein structure determination.

2. Protein expression (I)

To crystallize a protein, the first requirement is the expression of your protein in high amounts and most importantly on a regular basis. This implies that it is possible to obtain a freshly purified protein at least weekly. In general, it is possible to express a protein either homologously or heterologously (see Figure 1 - (I) expression). Especially for large proteins,

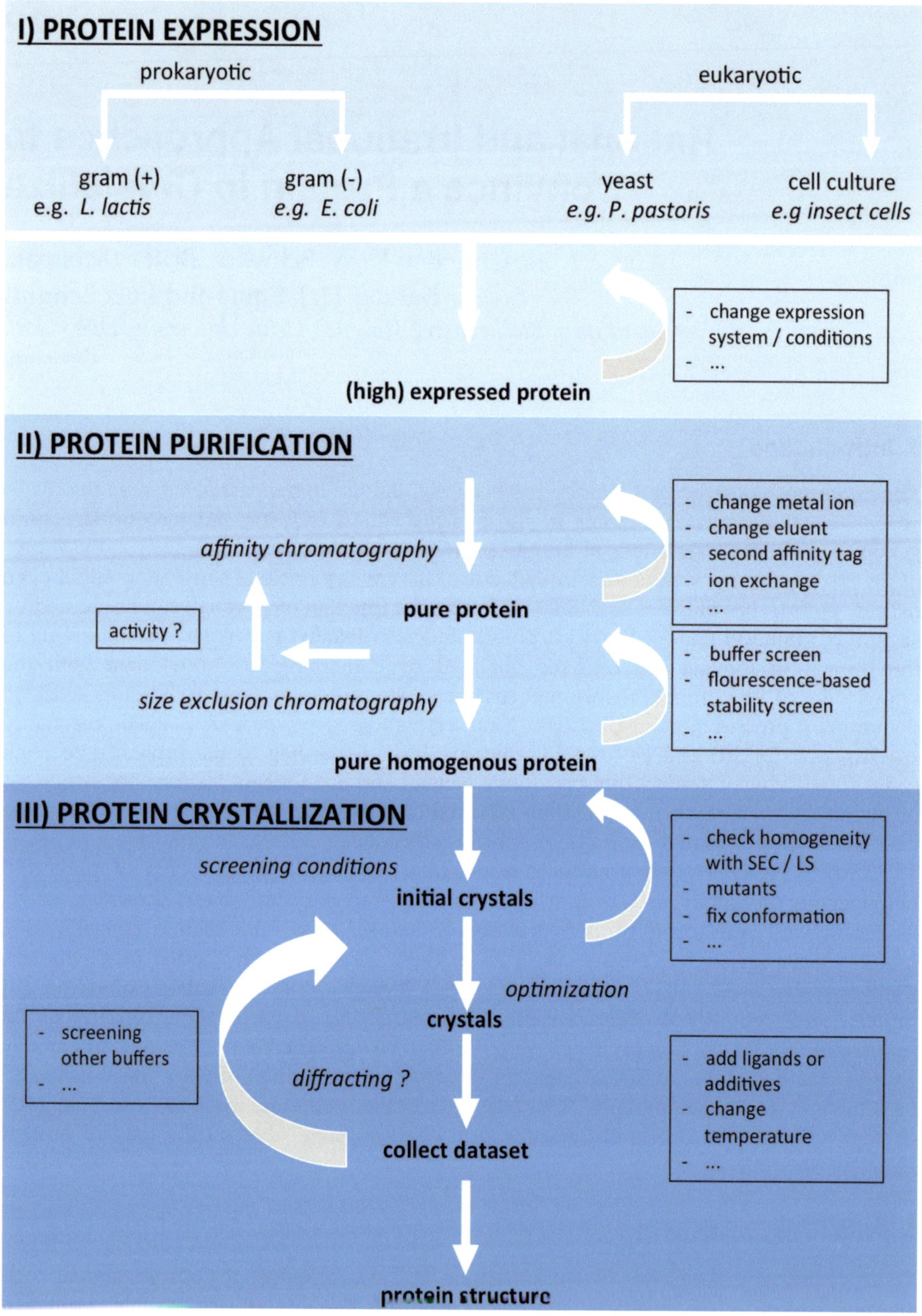

Fig. 1. Schema highlighting the three steps towards a protein crystal. (I) Expression (II) Purification (III) Protein crystallization.

proteins containing a co-factor or a ligand, the natural habitat is likely the best choice to express the protein. However, often the natural host, for example humans, produce only low amounts of protein and suitable overexpression protocolls are not available. To circumvent this problem, several expression strains, cell lines as well as a large number of expression vectors have been developed to allow expression of any protein in a different host (heterologous expression). In general, the used organisms for protein expression can be divided into two different groups: prokaryotic and eukaryotic expression hosts. The natural organism of the protein of interest mainly dictates the choice, which expression system to use. If working with a bacterial protein, it is very likely that also a prokaryotic host is able to express the protein in high amounts. The same holds true for proteins originating from a eukaryotic host, which is likely best overexpressed in a eukaryotic host. These proteins often require posttranslational modification such as glycosylation or disulfide bond formation, which are possible in eukaryotic expression hosts. The most common used heterologous expression host is the gram negative bacteria *Escherichia coli* since it is commercially available and a large number of expression cassettes have been developed. Thus, it is the most widely used expression system with expression rates of several mg/L of culture. The best characterised and understood expression hosts are described in more detail below and the commecially available systems are listed in table 1.

2.1 Expression hosts – Prokaryotic

2.1.1 Gram negative - *E. coli*

As mentioned above, *E. coli* is the most common used expression system (Figure 1 - (I) expression, left side). This is further highlighted by the fact that 80% of all protein structures deposited in the protein data bank were overexpressed in *E. coli* (Sorensen and Mortensen 2005). There are several advantages promoting *E. coli* as expression host: (A) Cultivation of *E. coli* is simple and a doubling time of 30 minutes is rather quick allowing the fast generation of biomass, (B) genetics are well understood and any genetical manipulation is well established, (C) expression levels of up to 60 % of the total protein mass within the cell make the next step, protein purification rather straight forward and finally (D) the cultivation does require only standard equipment normally present in every biochemical laboratory and therefore expression using *E.coli* is relatively cheap. In the last decades, many different plasmid based expression systems have been developed such as the pET vector systems, which contain several different expression plasmids with a choice for the affinity tag on either termini of the protein as well as the possiblity to use a dual cassette when expressing two or more proteins at once. The selection pressure derived from different antibiotics, and the resistance genes encoded on these plasmids further simplify laboratory practice. Only cells harbouring the right plasmids are able to grow and therefore expenditure on sterility is low.

The typical *E. coli* expression system is plasmid-based, which can be transferred to different *E. coli* strains ((Sorensen and Mortensen 2005), Novagen pET vector table). An *E. coli* expression vector consists mainly of five important parts: the replicon, a resistance marker, a promotor and a so-called multiple cloning site (MCS) (Baneyx 1999; Jonasson, Liljeqvist et al. 2002). The replicon is the crucial part of a plasmid to maintain it inside a cell. It is recognized and duplicated by the replication machinery (Baneyx 1999). The selection marker allows the identification of cells carrying a plasmid as it encodes for a resistance, e.g. against antibiotics (see above)(Sorensen and Mortensen 2005). The promotor sequence is the recognition site for the RNA polymerase, however it is inactive under initial cultivation conditions. The addition of an inducer (sometimes also a temperature change) switches the promotor from 'off' to 'on' whereby the expression is initiated (Jana and Deb 2005). Common inducers are isopropyl-β-D-thiogalactopyranosid (IPTG) in the pET system or arabinose for pBAD vectors (Invitrogen™). The multiple cloning site (MCS) is a short DNA segment combining many (up to 20) restriction sites. This feature simplified the insertion of genes into the plasmid enormously and made cloning procedures very convenient. However, new cloning strategies, which are independent of restriction enzymes and ligases, are emerging and will replace the standard approaches some day (see for example Li and Elledge 2007).

2.1.2 Gram positive - *L. lactis*

Within the prokaryotic expression system also some gram (+) bacteria are used for protein overexpression (Figure 1 – (I) expression left side). Here, lactic acid bacteria play a privotal role. They are used in food industry and are known since 1873 when Joseph Lister isolated the first strain (Teuber 1995; Mierau and Kleerebezem 2005). For the overexpression of recombinant proteins, there are many lactic acid bacteria around, *lactococci, lactobacilli, streptococci* and *leuconostocs*. (Gasson 1983; van de Guchte, Kok et al. 1992; de Vos and Vaughan 1994). The best characterized and most widely used host is *Lactococcus lactis*, which is famous for its usage within food fermentation and like for *E. coli* the genome, metabolisms and molecular modifications are well known and established (Bolotin, Wincker et al. 2001; Guillot, Gitton et al. 2003). Thus, it has been called the 'bug of the next millennium'(Konings, Kok et al. 2000). All established gram (+) bacteria expression hosts are able to overexpress proteins homologously or heterologously. Since only one, namely the cyctoplasmic membrane is present (Kunji, Slotboom et al. 2003), this host is in comparison to the two-membrane system of gram (-) bacteria, a good choice to express eukaryotic as well as prokaryotic membrane proteins or proteins with membrane anchors (Kunji, Slotboom et al. 2003). The promoter used for expression in *L. lactis* is induced by the external addition of nisin. Nisin is an antimicrobial active peptide, which interacts with lipid II in the cytoplasmic membrane of gram (+) bacteria and causes cell lysis. Interestingly, nisin is produced by *L. lactis* itself. The expression strain is deleted of the nisin producing genes and therefore external nisin can be used as inductor. Nisin binds to NisK, which as part of a two-component system, phosphorylates NisR, which in turn binds to the promotor P_{nisA} thereby allowing synthesis of the protein located downstream on the plasmid. Since nisin is also active against *L. lactis* itself, the concentration range of nisin used in such expression studies is relatively narrow to circumvent killing of the *L. lactis* expression strain. This is clearly a draw back of this expression system since expression can basically only turned on with a certain nisin concentration. Between the inducer concentration and the expressed protein a linear behaviour is observed. Unfortunately the nisin concentration range between the minimal and maximum nisin concentration is very small, nisin concentration higher than 25 ng per liter of cells, cause cell death.

However, *L. lactis* has been proven to be a very efficient expression system. Kuipers et al. created many expression hosts and plasmids to produce any protein of interest by cloning it downstream of the P_{nisA} promotor. With this nisin inducible (NICE) expression system, it is now possible to induce the protein production with minimal concentration (0.1 – 5 ng) of nisin (de Ruyter, Kuipers et al. 1996; Kuipers, de Ruyter et al. 1998). The amount of produced recombinant protein can reach up to 50 % of the total intracellular proteins (Kuipers, Beerthuyzen et al. 1995; de Ruyter, Kuipers et al. 1996). Following a few examples for expressed proteins in *L. lactis* whose structures have been solved: an ECF-type ABC transporter (PDB:3RLB)(Erkens, Berntsson et al. 2011), a peptide binding protein OppA

(PDB:3RYA)(Berntsson, Doeven et al. 2009) and the multidrug binding transcriptional regulator LmrR (PDB:3F8B)(Madoori, Agustiandari et al. 2009).

2.2 Eukaryotic expression hosts

The great benefit of choosing a eukaryotic host for overexpression of a protein of interest are the availability of a posttranslational modification system as well as the frequently enhanced protein folding (Midgett and Madden 2007). Eukaryotic proteins tend to misfold or lack biological activity when expressed in prokaryotic expression systems such as *E. coli* (Cregg, Cereghino et al. 2000; Midgett and Madden 2007). To overexpress these proteins, different yeast strains, insect cells or even mammalian cell lines have been developed as expression hosts (Figure 1 - (I) expression, right side). Eukaryotic expression systems are often more expensive, provide low expression levels and are sometimes hard to handle, when compared to bacterial systems. However, the genetic and cellular contexts are more similar to the original protein-expressing organism (Midgett and Madden 2007). In the following sections, some of the commonly employed eukaryotic expression systems will be described.

2.2.1 Yeast expression systems - *Saccharomyces cerevisiae* and *Pichia pastoris*

The most widely used yeast strains to express protein are *Saccharomyces cerevisiae* and *Pichia pastoris*, which offer the major advantage of a posttranslational modification system for glycosylation, proteolytic processing as well as disulfide bond formation, which for some proteins are essential for the function and/or correct folding (Cregg, Cereghino et al. 2000; Midgett and Madden 2007). The handling of yeast expression systems is similar to prokaryotic systems with respect to the genetic background and cultivation. Similar to the bacterial vector systems, expression in yeast starts with a plasmid-based cloning part which can be performed in *E. coli* (Cregg 2007). Afterwards the expression cassette gets integrated into the genome by simple homologous recombination in the yeast. One major advantage in *P. pastoris* is the insertion of multiple copies of the protein DNA-sequence into genomic DNA, which increases expression yield.

The biggest advantage of yeast as expression system is that well established protocols for fermentation are available. Optimal fermentation of *P. pastoris* can end up with more than 130 gram of cells per liter of culture. Even if expression levels in the cell are not that high the mass of cells easily compensates for this disadvantage (Wegner 1990; Cregg, Cereghino et al. 2000; Hunt 2005; Cregg 2007; Midgett and Madden 2007). Examples of crystal structures from proteins expressed in *P. pastoris* are a human monoamine oxidase B (PDB:3PO7) (Binda, Aldeco et al. 2010) and a protein involved in cell adhesion NCAM2 IG3-4 (PDB:2XY1)(Kulahin, Kristensen et al. 2011).

2.2.2 Insect cells

The expression system in insect cells is beside yeast a well-characterised alternative to express eukaryotic proteins (Midgett and Madden 2007). As insect cells are higher eukaryotic systems their posttranslational modification machinery can carry out more complex alterations than yeast strains. They also have a machinery for the folding of mammalian proteins. The most commonly used vector system for recombinant protein expression in insect cells is baculovirus, which can also be used for gene transfer and expression in mammalian cells (Smith, Summers et al. 1983; D., L.K. et al. 1992; Altmann, Staudacher et al. 1999). A few examples of proteins expressed in insect cells that resulted in crystal structures are the transferase Ack1 (PDB:3EQP)(Kopecky, Hao et al. 2008), a human hydrolase (PDB: 2PMS)(Senkovich, Cook et al. 2007) and myosin VI (PDB:2BKI)(Menetrey, Bahloul et al. 2005).

2.2.3 Mammalian cell lines

The expression of proteins in mammalian cell lines is the most expensive and complex alternative. Especially for human membrane proteins this expression system has been proven to express the most active protein (Tate, Haase et al. 2003; Lundstrom 2006; Lundstrom, Wagner et al. 2006; Eifler, Duckely et al. 2007). The resulting protein amount, however, obtained from mammalian cell lines is mostly only sufficient for functional studies. Using mammalian cells lines is the most challenging variant of protein overexpression and therefore only choosen if any of the other expression system described failed. Some examples of protein structures expressed in mammalian cell lines are the hydrolase PCSK9 (PDB:2QTW)(Hampton, Knuth et al. 2007) and the acetylcholine receptor AChBP (PDB:2BYQ)(Hansen, Sulzenbacher et al. 2006).
Table 1 sums up advantages and disadvantages of the above mentioned overexpression systems used for protein crystallography.

Expression system	Pros	Cons
Prokaryotic		
Gram negative *E. coli*	- costs - simplicity (genetic/culture) - yield	- inclusion bodies - protein folding - posttranslational modifications - protein secretion
Gram positive *L.lactis*	- costs - protein secretion - one membrane	- posttranslational modifications
Eukaryotic		
Yeast	- high cell densities - costs - simple cultivation - posttranslational modifications	- hyper glycosylation, non-native - lipid composition is different to mammalian cells
Insect cells	- more native lipid environment - good track record of functional proteins	- costs - non-native lipid environment - glycosylation pattern different to mammalian cells - protein amount
Mammalian cells	- native conditions for human membrane proteins to investigate diseases - posttranslational modifications - lipid environment - good track record of functional protein	- costs - difficult to establish - protein amount

Table 1. Overview of expression systems. Summarized are the advantages and disadvantages.

3. Purification

After having expressed your protein of interest, the race for crystals is by no means finished. The next step on the long road to structure determination is to isolate the protein or - phrasing it differently - to remove all other proteins present in the cell (Figure 1 - (II) purification). An elegant method to do so is the genetic attachment of an affinity tag on either site of the protein or in some cases on both sides (Waugh 2005). This affinity tag has the possibility to bind high affine to a immobilized ligand on a matrix, while all other proteins have a much more reduced binding affinity and therefore flow through the matrix (Figure 1 - (II) purification 1st step). This allows a one-step purification, which in almost all cases is relatively harmless for the protein and likely does not interfere with folding and/or overall structure of the protein. There are a lot of affinity tags available as well as matrix materials (Terpe 2003). The well known and most often used affinity tag is the poly-histidine tag (Porath, Carlsson et al. 1975; Gaberc-Porekar and Menart 2001), which can vary in length as well as in position but the overall purification strategy is the same. From all the structures solved nowadays, almost 60 % of the proteins are purified via a histidine tag; mainly due to the great purification efficiency, which can be as large as 90% after a single purification step (Gaberc-Porekar and Menart 2001; Arnau, Lauritzen et al. 2006). Therefore, most commercially available expression systems and methods contain a his-tag encoded on the plasmid. Besides the his-tag, there are other tags avaible and used for protein purification, of which the Strep-, CBP-, GST-, MBP-tag are described below.

3.1 Choice of the right tag

3.1.1 Polyhistidine-tag (his-tag)

As mentioned above the polyhistidine-tag is the most common affinity tag and the required affinity resins and chemicals are relatively inexpensive. The purification step is a so-called immobilized metal ion affinity chromatography (IMAC) (Porath, Carlsson et al. 1975). Here, a matrix is able to bind bivalent metal ions. For example nitrilotriactetic acid (NTA), which is a chelator and binds metal ions like Ni^{2+}, Zn^{2+}, Co^{2+} or Cu^{2+} (Hochuli, Dobeli et al. 1987). These metal ions have a high affinity to the imidazole group of the amino acid histidine. A stretch of histindines in a row with for example an *E. coli* protein is very unusual. Thus, the genetical introduction of several, in most cases 6- 10 histidines in a row selects for specific binding of this protein. As eluant very elegantly imidazole can be used, which competes with the histidine tag and elutes the protein of interest. When used in low concentrations, imidazole can also be used to remove unspecifically bound proteins, which bind with low affinity to the matrix (Hefti, Van Vugt-Van der Toorn et al. 2001). Normally, a protein with a 6-10 histidine tag should be bound to the matrix relative strongly and 100-250 mM imidazole in the buffer is required to elute the protein from the resin. In contrast, proteins with a low affinity to the matrix can already be eluted with 10-50 mM imidazole (the "impurities" of *E. coli*). Therefore, a linear imidazol gradient, for example, separates the protein of interest and impurities (Hochuli, Dobeli et al. 1987; Gaberc-Porekar and Menart 2001). Although the polyhistidine-tag is the most common and mostly an efficient variant, there are a few applications where the his-tag can cause problems. Metalloproteins can interact either directly with the his-tag or with the ions immobilized on the matrix. In comparison to some other affinity-tags, the specificity of the his-tag is not that high and in some cases this results in the co-purification of other proteins (Waugh 2005).

3.1.2 Strep-tag

In comparison to the his-tag, which binds to immobilized metal ions, the strep-tag II constists of a small octapeptide (WSHPQFEK), which binds to the protein streptavidin (Schmidt, Koepke et al. 1996). The commercial available matrix is a streptavidin variant and is called Strep-Tactin. This variant is able to bind the Strep-tag II octapeptide under mild buffer conditions and can be gently eluted with biotin derivates such as desthiobiotin (Schmidt, Koepke et al. 1996; Voss and Skerra 1997). Especially for metal-ion containing enzymes it is a promising alternative to the his-tag (Groß, Pisa et al. 2002). However, as chemicals are more expensive and the matrix has a lower binding capacity, compared to NTA resins, it is often not the first option choosen. Moreover, it cannot be used under denaturating conditions since Strep-Tactin denatures and will not bind the tag anymore (Terpe 2003; Waugh 2005). Examples of proteins crystallized after a Strep-tag purification are OpuBC (PDB:3R6U)(Pittelkow, Tschapek et al. 2011) and *Af*ProX (PDB:3MAM)(Tschapek, Pittelkow et al. 2011) as well as the sodium dependent glycine betain transporter BetP from *Corynebacterium glutamicum* (PDB:2WIT)(Ressl, Terwisscha van Scheltinga et al. 2009).

3.1.3 CBP-tag

Another peptide tag, is the calmodulin binding peptide, first described in 1992 (Stofko-Hahn, Carr et al. 1992). This peptide is prolonged compared to the Strep-tag II, consisting of 26 amino acids and binds with nanomolar affinity to calmodulin in the presence of Ca^{2+} (Blumenthal, Takio et al. 1985). It is derived from the C-terminus of the skeletal-muscle myosin light-chain kinase, which makes the system an excellent choice for proteins expressed using a prokaryotic expression system, since in prokaryotic systems nearly no protein interacts with calmodulin. This allows extensive washing to remove impurities and elution with EGTA, which complexes specifically Ca^{2+}, and a protein recovery around 90 % can be achieved (Terpe 2003). A drawback of this tag however is that the CBP tag can only be fused to the C-terminus of the protein since it has been shown that CBP on the N-terminus negatively influences the translation and thereby the expression rate (Zheng, Simcox et al. 1997).

3.1.4 GST-tag

With respect to the length of the tags, the his-tag contains only a few amino acids, the Strep-tag II and the CBP-tag already contain 8 – 26 amino acids, but it is possible to fuse whole proteins with 26 – 40 kDa to a recombinant protein. Here, the high affinity binding of the protein to their substrate is used to purify the protein of interest (Smith and Johnson 1988). In the case of the glutathione S-transferase (GST, 26 kDa) the protein specifically binds to immobilized glutathione. To elute the fusion protein from the resin, non-denaturating buffer conditions employing reduced glutathione are used (Terpe 2003). The tag can help to protect the recombinant protein from degradation by cellular proteases. It is recommended to cleave off the GST-tag after purification with a specific protease like thrombin or TEV (Tobacco Etch Virus) protease (Terpe 2003).

3.1.5 MBP-tag

Another affnitiy tag, which can be fused to the protein of interest, is the maltose binding protein (MBP) from *E. coli*. This protein has a molecular weight of 40 kDa and has the ability to bind to a cross-linked amylose matrix. The binding affinity is in the micro molar range and the tag can be used in a pH range from 7.0 - 8.0, however, denaturating buffer conditions are not possible (di Guan, Li et al. 1988). The elution of the recombinant protein is recommended with 10 mM maltose. A great opportunity of the MBP-tag is the increasing solubility effect of the recombinant protein in prokaryotic expression systems and even more pronounced in eukaryotic systems (Sachdev and Chirgwin 1999). Like the CBP-tag, a fusion at the N-terminal side might influence translation and expression rates (Sachdev and Chirgwin 1999).

3.1.6 Tag position and double tags

As described above, the position of the tag either at the N- or C-terminus has a considerable influence on translation and expression rate as well as on the biological function (Arnau, Lauritzen et al. 2006). If information regarding activity of the protein is already available especially about the location of interaction sites, this should be included in the protein design, meaning tag position etc. In general, the tag should be placed at the position of the protein, which is less important for interactions and/or expression. To minimize the influence of the tag on folding and/or activity in some cases it helps to create a linker region of a few amino acids between the tag and the protein (Gingras, Aebersold et al. 2005). A very efficient and sophisticated solution is, the addition of amino acids between tag and protein of interest, which functions not only as an accessibility increasing factor, but, also encodes for a recognition site for proteases like thrombin or TEV. Due to this arrangement the tag - protein interaction is minimized and the tag can be cleaved off if necessary (Arnau, Lauritzen et al. 2006). In some special cases a combination of two affinity tags results in enhanced solubility and more efficient purification. To enhance the purity of a protein, often a construct of two different short affinity tags like his-tag and Strep-tag or CBP-tag can be engineered (Rubio, Shen et al. 2005). Also a combination of two his-tag or two strep-tag kept apart by a linker region enhances the binding affinity extremely. This allows more stringent washing steps prior to elution of the protein (Fischer, Leech et al. 2011).

3.2 Size exclusion chromatography and ion exchange chromatography

Despite the usage of affinity tags a second purification step is sometimes required (Figure 1 - (II) purification). Which kind of purification procedure is required depends on the nature of impurities. If these impurities differ in molar mass compared to the protein of interest, a method based on size separation can be applied. Size exclusion chromatography (SEC) also separates different oligomeric species of the protein from each other, which otherwise would strongly inhibit crystallization and also allows analysis of stability and monodispersity of the protein (Regnier 1983a; Regnier 1983b).

However, in many cases, SEC is not sufficient to remove all impurities. Then separation by overall charge of the protein might be an option. Depending on the isoelectric point of the protein either anion or cation exchange chromatography can be performed. The protein binds to a matrix under very low ionic strength and is eluted afterwards either by increasing the ionic strength or by pH variation. Similar results can be achieved by hydrophobic interaction chromatography. Here, proteins with different surface properties show differences in their binding strength and binding of the protein is done inversely as during ion exchange chromatography. High ionic strength favors protein binding to a hydrophobic matrix and elution takes place when reducing the ionic strength. Although there are many other possibilities to increase the purity of a protein, the above mentioned techniques are without any doubt the most widely used and general applicable methods.

3.3 How to get a homogenous protein solution

In some cases isolated proteins are stable and homogenous at high concentrations after the purification and can be directly used for crystallization experiments. Often, however, the protein does not behave ideal and precipitates at high concentrations or forms aggregates or inhomogenous, oligomeric species; all of them prohibit crystal growth. SEC is a very elegant method to visualize the stability and oligomeric state of a protein. If the stability or the homogeneity of a protein sample is critical, you need to adapt your purification protocol and search for an optimized procedure. Different approaches are summarized below, for example a buffer screen to enhance protein solubility, multi-angle light scattering experiments to determine the absolute mass and the oligomeric state of the protein sample or fluorescence-based experiments to investigate the stability of the protein of interest.

3.3.1 Purified proteins – An *in vitro* system

After a protein is expressed in a soluble form, the subsequent purification procedure changes the environment of the protein dramatically. The cytoplasm of the cells, where the overexpression takes place, is packed with macromolecules. In *E. coli*, for example, the concentrations of proteins, RNAs and DNAs are about 320 mg/mL, 120 mg/mL, and 18 mg/mL, respectively (Cayley, Lewis et al. 1991; Zimmerman and Trach 1991; Elowitz, Surette et al. 1999) resulting in an overall concentration of macromolecules of above 450 mg/mL. During cell lysis and the first purification step, likely an IMAC (see above), the protein is separated from almost all other cell components. This rigorous procedure is accompanied with a severe change of the environment into an *in vitro* system. As a result proteins often tend to aggregate, precipitate or form inhomogeneous oligomeric states that prevent the formation of crystals in further experiments. Therefore one of the biggest challenges in structural studies is the preparation of protein solutions with high concentrations (as a rule of thumb 10-20 mg/mL) in a homogenous state. To fulfill these requirements, the *in vitro* system needs to be optimized with respect to different parameters as highlighted in *Figure 1 – (II) purification*. If a sufficient protein sample cannot be obtained, different strategies are available to increase the important characteristics of the protein: purity and homogeneity. As mentioned above, the usage of different metal ions during IMAC, ion exchange, a second affinity chromatography etc. can be sufficient to enhance purity. This might also lead to an increased stability. However, if the stability and/or homogeneity of a protein is still a problem, screening for a new buffer composition is essential to succeed during crystallization trials.

3.3.2 Buffer composition

Many examples illustrate the importance of an adequate buffer composition for protein stability, homogeneity, conformation, and activity (Urh, York et al. 1995; Holm and Hansen 2001; Jancarik, Pufan et al. 2004; Collins, Stevens et al. 2005). Some buffers are very frequently used and recommended by manufactures (see for example Qiagen, Roche, New England BioLabs, Fermentas, etc.). All of them contain a buffer reagent that keeps the pH constant in a well-defined range. Well-known examples are phosphate, tris (hydroxymethyl) aminomethane (Tris), or HEPES (4-(2-hydroxyethyl)-1-piperazineethanesulfonic acid) that buffer at the physiological relevant pH range of 6- 9 (Durst and Staples 1972; Chagnon and Corbeil 1973; Tornquist, Paallysaho et al. 1995). In recent years, the development of other buffer systems has been quite successful (Taha 2005) (for a list of buffers and corresponding pH ranges, see for example: http://delloyd.50megs.com/moreinfo/buffers2.html). Next to the well-defined pH, the stability and homogeneity of proteins depend on many other parameters, for example ionic strength, the presence of ligands and/or co-factors, divalent ions, glycerol, etc. The appropriate buffer composition cannot be predicted so far and needs to be identified by trial-and-error approaches.

3.4 Protein purification – How to overcome problems

In this part we would like to present some pitfalls that might occur during protein purification and provide some ‚rationales' to overcome these problems. As usual, the crucial step of solving a problem is its identification. Here, we are trying to sensitize the reader to indications, which might point towards problems related to instability and/or inhomogeneity of the protein sample. Moreover, such problems cannot always be recognized without the adequate technique(s). Therefore, we are introducing techniques that are capable to visualize the state of proteins.

3.4.1 Visible protein precipitations during IMAC

A very obvious stability problem is the formation of precipitations in the elution fractions of a chromatography step (see Figure 2). In this example, the his-tagged protein was eluted with a linear imidazole gradient from 10 to 500 mM imidazole and eluted at about 250 mM imidazole. Protein precipitation occured immediately after elution (Figure 2A and B) and continued (Figure 2C) resulting in a low amount of soluble protein. This aggregation can be reduced by dilution with a IMAC buffer (typically lacking imidazole) immediatly after the elution. Thereby, dilution hinders the concentration-dependent aggregation. In many cases, this rational is not sufficient to prevent precipitation. After applying, for example, a *buffer screen* (see Figure 1 - (II) purification) the new defined buffer is used for the chromatography or the eluting protein is diluted into the new buffer (see Figure 2D).
Other elution strategies of his-tagged proteins from an IMAC column are available. As described before, competing the poly-histidine from the IMAC column by imidazole is the most common elution strategy, however, for some proteins other strategies are superior, for example, replacing imidazole by histidine. Imidazole is only a mimic for histidine. If one uses histidine instead of imidazole aggregation can be avoided as concentration of the eluent can be reduced by a factor of ten. An example for a protein sensitive to imidazole concentration is shown in Figure 3B. Here a comparative SEC chromatogram is shown. After elution from the IMAC column with imidazole only a very small amount of the protein elutes at the volume corresponding to the size of a monomer or the dimer, respectively (Figure 3B, continuous line). Most of the protein passes the column very fast and elutes at the void volume indicating large radii meaning aggregated protein. Yields of dimeric (at about 150 mL) and monomeric (at about 180 mL) proteins are strongly increased after an elution with histidine (dashed line) compared to an elution with imidazole (continous line) and only the monomeric species could be crystallized (data not shown). The choice of the eluent in IMAC might therefore be an important step in a purification protocol. Another elution strategy of his-tagged proteins is a pH change from 8 to 4. In an acidic environment, histidines become positively charged and are therefore released from the column matrix. This strategy results in a sharp elution from the matrix and the protein is eluted highly concentrated. Although this strategy is recommended by the manufacturers (see GE Healthcare, Qiagen, etc.) the desired protein needs to retain activity at acidic pHs. The

bivalent metal ions (Ni^{2+}, Co^{2+}, Zn^{2+},..., see above) that complex the his-tag can be removed from the matrix by chelating reagents as ethylenediaminetetraacetic acid (EDTA) as another elution strategy (Muller, Arndt et al. 1998)

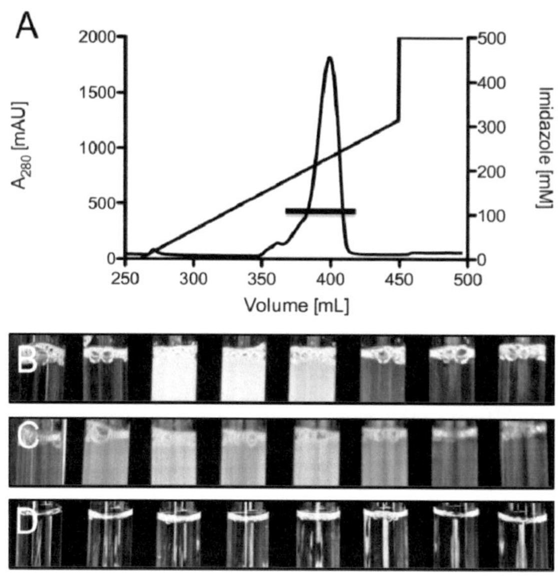

Fig. 2. Elution fractions of an IMAC. The protein was eluted via a linear imidazole gradient from 10 to 500 mM and the absorption at 280 nm was recorded. The elution fractions were collected and photographed. A: IMAC chromatogram of the his-tagged protein. Elution fractions containing the desired protein (indicated by a bar) are collected and shown in B – D. B and C: Elution fractions of the protein in 50 mM Tris-HCl, 150 mM NaCl, pH 8.0 immediately and 10 min after the elution, respectively. D: The elution fractions were immediately mixed in a 1 to 1 ratio with a buffer that enhances the protein stability (50 mM citrate, 50 mM LiCl, pH 6.00) evaluated during a solubility screen.

3.4.2 Invisible aggregations

Sometimes aggregation of proteins in solution can not be detected directly by eye. This inhomogeneity of protein samples can be visualized SEC, a method that separates proteins by their hydrodynamical radius (see above). Protein aggregates are eluting at the void volume, since they are clumbed together resulting in a big hydrodynamical radius (see Figure 3A and B). If invisible aggregation is detected the buffer composition needs to be adjusted. In one case we applied this technique to visualize the state of a protein after an IMAC, and the resulting elution profile is shown in Figure 3A (continous line). Comparable to the imidazole-induced precipitation described above, the protein aggregated and elutes within the void volume of the column (about 40 mL). Moreover, several other protein species elute from 55 to 80 mL indicating a highly inhomogeneous protein sample. The running buffer of the SEC was 50 mM Tris-HCl, pH 8.0 and 150 mM NaCl. Remarkably, a simple change to a new buffer (20 mM HEPES, 150 mM NaCl, pH 7.0) Resulted in a stable and homogenous protein sample (Figure 3A, dotted line), which was suitable for crystallization trials. Next to the rigorous change in the homogeneity of the protein, the biological activity of the protein could only be determined in the new buffer system. The influence of the buffer composition for the protein activity is a well-known phenomenon (Urh, York et al. 1995; Holm and Hansen 2001; Zaitseva, Jenewein et al. 2005) and in many cases the activity goes hand in hand with an optimal buffer for the purification. Mentionable, the new buffer was not found by trial-and-error approaches. We searched for literature dealing with homologous proteins, especially for established purification protocols. This literature search revealed the new buffer, illustrating that not every step towards a protein structure determination must be a trial-and-error process.

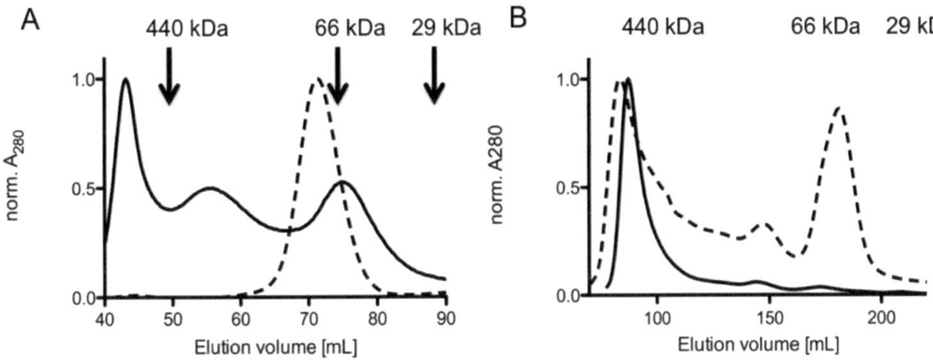

Fig. 3. Size exclusion chromatograms (UV 280nm) of proteins in different buffers. A: The homogeneity of a protein was analyzed in two different buffers; continuous line: 50 mM Tris-HCl, 150 mM NaCl, pH 8.00; dotted line: 20 mM Hepes, 150 mM NaCl, pH 7.00. B: The protein was eluted of the IMAC column either with imidazole (continous line) or with histidine (dotted line), concentrated and applied to the SEC.

Another example for the influence of the buffer composition was published bei Mavaro et al. (Mavaro, Abts et al. 2011). Instead of the buffer agent, the ionic strength of the buffer was the crucial determinant. Purification of the protein in low-salt buffer resulted in an inhomogenous protein sample containing a mixture of aggregates, dimers and monomers without biological activity. However, a simple change to high-salt buffer allowed the purification of homogenous dimeric protein, that was able to bind its substrate.

3.4.3 Overcoming protein instability

In the previous sections different strategies were mentioned to enhance the stability and the homogeneity of purified proteins and in all cases the buffer composition was the solution. Still, the essential question how to determine the optimal buffer to make a protein feel happy *in solution* is not answered? Some rationales and experiences are listed above: different elution strategies for IMAC purifications, the usage of frequently used buffer agents and a literature research for established purification protocols of related proteins. However, in many cases these approaches do not solve the problems occuring during the purification. But, is there a general methodology to overcome the problems? Unfortunately, the answer is as frustrating as challenging - there is not a general panacea around for the right buffer composition of a protein. If a new buffer needs to be found, trial-and-error approaches have to be applied. A lot of different parameters are influencing the state of a protein, i. e. the buffer agent, the salt concentration, presence of metal ions with different valences, the hydrophobicity, and even the temperature of the buffer. The analysis of the protein in different buffers can be done by SEC and/or light scattering experiments. However, screening of all the different variables is very labor- and cost-intensive, and time-consuming, moreover only combinations of two or more additives might be sufficient to enhance the solubility and homogeneity of the protein. Therefore high-throughput methods are needed that handle a lot of different conditions simultaneously using as few as possible protein sample.

3.4.4 Buffer screen – Enhancing the solubility

Many publications are available suggesting methods for a solubility screening to allow the crystallization of initially inhomogeneous, aggregating protein samples (Jancarik, Pufan et al. 2004; Zaitseva, Holland et al. 2004; Collins, Stevens et al. 2005; Sala and de Marco 2010; Schwarz, Tschapek et al. 2011). In all of these methods aggregating protein samples are mixed with commercially available crystallization screens incubated for a period of time, and analyzed for precipitation visually using a light microscope. Screening conditions resulting in no precipitations are analyzed upon their composition, and protein samples are further examined with respect to their solubility and homogeneity under these conditions by SEC or light scattering experiments. This technique allows high-throughput screening in a 96-well format, where an automated pipetting system mixes only 50-200 nL of protein solution with 50-200 nL of buffer solutions to minimize the needed protein sample and increase the screening efficiency. Several buffer screens are commercially available that cover many different buffer agents, salt concentrations and other buffer parameters (i. e. from Hampton Research, Molecular Dimensions, Sigma, Jena Bioscience, Qiagen). After a solubility screening was applied, we were able to stabilize a previously unstable protein at concentrations above 3 mg/mL (see above "Protein precipitations during IMAC" and Figure 2D) at concentrations of up to 100 mg/mL for weeks (Schwarz, Tschapek et al. 2011). Typically, the new buffer (50 mM citrate, 50 mM $LiCl_2$, pH 6,00) should be used during the entire purification procedure starting with cell lysis. In the described case, the new buffer contains citrate, which is incompatible with an IMAC purification. Therefore the protein was immediately mixed with the new buffer after the elution of the IMAC column.

3.4.5 Size-exclusion chromatography versus light scattering experiments

Size-exclusion chromatography (SEC) and light scattering experiments (LS) are very helpful tools to analyze the homogeneity (Collins, Stevens et al. 2005) and the molecular mass of proteins; however both of them have advantages and disadvantages compared to each other. In SEC experiments proteins are separated based on their hydrodynamic radius by partitioning between a mobile phase and a stationary liquid within the pores of a matrix. All SEC columns are characterized by the volumes V_0, the liquid volume in the interstitial space between particles, V_i, the volume contained in the matrix pores and V_T, the total diffusion volume (V_0 + V_i) (Regnier 1983a; Regnier 1983b). In dependency of the hydrodynamic radius molecules are eluting at specific retention volumes in between V_0 and V_T with big molecules eluting first. After a calibration of a SEC column with proteins of known

molecular weight (i. e. Sigma-Aldrich, "Kit for Molecular Weights") the molecular mass of the protein of interest can be roughly estimated; the elution volume is correlated to the $\log_{10}$ of the molecular weight (therefore, the hydrodynamic radius is considered to be proportional to the molecular weight). However, many extraneous mechanisms such as adsorptive, hydrophobic and ionic effects are further limiting the correlation between the retention volume and the molecular mass giving sometimes rise to wrong estimations.
Light scattering (LS) experiments can be applied to overcome these disadvantages and investigate the exact molecular weight of the protein sample. The rayleigh scattering of particles of monochromatic light depends directly on the molar mass of the particle. If you know the exact number of particles you can calculate the average molar mass of these particles. This technique is very powerful when used online after separation of the protein depending on their hydrodynamic radius, meaning SEC. This technique is always superior to normal SEC but requires special equipment and especially more time. However, if the protein fold is not really globular or other effects occur (see above: ionic, hydrophobic, etc.) assumption on size and oligomeric state based on SEC is not possible at all. For protein crystallization information about monodispersity, which can be provided by such an experiment, is an additional benefit.

3.4.6 Analysis of the homogeneity – High-throughput methods

Despite the development of various sophisticated methods, a bottleneck of homogeneity screening is high-throughput analysis. As mentioned above, proteins need to be analyzed by SEC and/or LS experiments after visual read-out of the protein-buffer droplets. Therefore, fluorescence-based solubility screens were developed that allow the high-throughput analyzes of many samples in a 96-well format (Ericsson, Hallberg et al. 2006; Alexandrov, Mileni et al. 2008; Kean, Cleverley et al. 2008). All these assays use fluorophores as reporters of the protein state. A suitable fluorophore is, for example, Sypro Orange, which exhibits different fluorescence properties as a function of its environment. This dye is almost dark in hydrophilic environment, however, after binding to hydrophobic molecules, it emits light at 570 nm. In inhomogenous and unfolded protein samples hydrophobic amino acids are exposed to the surface of proteins (Murphy, Privalov et al. 1990). An increase in the fluorescence signal of Sypro Orange correlates therefore with unfolding events of proteins. The homogeneity screening can be performed in basically two ways: temperature- or time-dependent. For the first setup the protein sample is heated gradually in distinct steps (i. e. 1 °C) and the emission is monitored at 570 nm. Hereby, a "melting" temperature is determined, which is characterized by 50% fluorescence of the maximal fluorescence at the highest temperature; the higher the melting temperature, the higher the stability of the protein (Ericsson, Hallberg et al. 2006). Secondly, the protein sample is incubated at a specific temperature (i. e. 40°C) and the fluorescence is measured for a period of time. The "half-life" time, at which 50 % fluorescence of the maximum fluorescence in one sample is detected, can be compared to all buffer conditions. In Figure 4 an example of the time-dependent approach is shown. Here, the protein is incubated in different buffers with various salt concentrations. The emission of Sypro Orange is recorded each minute at 570 nm. An analysis of all time-dependent fluorescence plots indicates that the protein is most stable in buffers containing 125 mM NaCl but unfolds fast in 1 M ammonium sulfate. These assays result in qualitative indications about a favourable environment of proteins that enhance the stability. Ericsson et al. proved the concept of this method by applying it to different proteins (Ericsson, Hallberg et al. 2006). The stability optimization yielded a twofold increase in initial crystallization leads. Moreover these assays enable the search for putative ligands of the protein. Upon binding of a substrate in the binding pocket or an inhibitor, the stability of the protein increases, which can be detected experimentally.

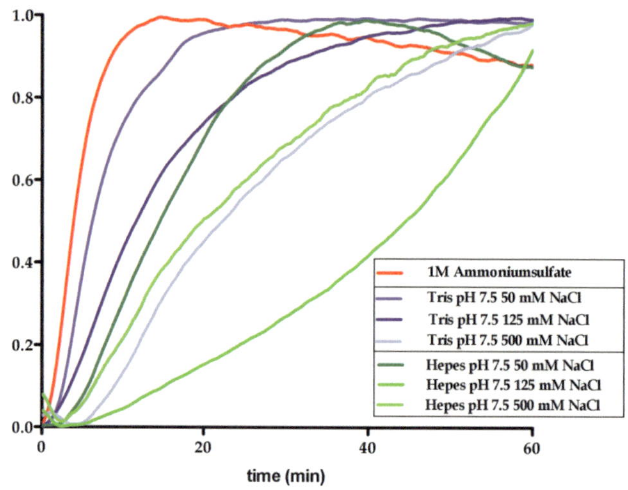

Fig. 4. Time-dependent stability optimization screen using Sypro Orange as reporter. The protein is diluted 1:50 into each test buffer containing Sypro Orange, excited with 490 nm and the fluorescence at 570 nm is measured for 60 minutes automatically with a PLATE READER (Fluorostar, BMG Labtech). Normalized fluorescence is plotted against the time.

4. Protein crystallization: Introduction

Protein crystals suitable for X-ray diffraction experiments and usable for subsequent structure determination are normally relatively large with a size of at least 10 to 100 µm. In contrast to crystals of mineral compounds, protein crystals are rather soft and sensitive to mechanical stress and temperature fluctuations. These properties are due to weak interactions between single proteins within the crystal, their high flexibility as well as the size of the macromolecules. The periodic network of building blocks is held together by dipole-dipole interactions, hydrogen bonds, salt bridges, van der Waals contacts or hydrophobic interactions. All of them have binding energies in the low kcal/mol range.

Especially the limited number of crystal contacts and their directionality are the largest difference to the high interactions generally observed in salt crystals. An example of the interactions within a protein crystal is shown in Figure 5. This picture highlights the main pitfalls in protein crystallization. A protein is a highly irregular shaped and flexible macromolecule which allows weak and stinted interactions at very specific locations of its surface. All vacuity is filled with buffer, in general not contributing to any kind of interactions between the protein molecules. Figure 5A shows a protein of around 30 kDa, which crystallizes in a rather small unit cell (shown in black). Only one protein monomer is located in the asymmetric unit of the unit cell, the other shown monomers represent symmetry related proteins. Figure 5B highlights the three-dimensional packing of protein molecules within a crystal.

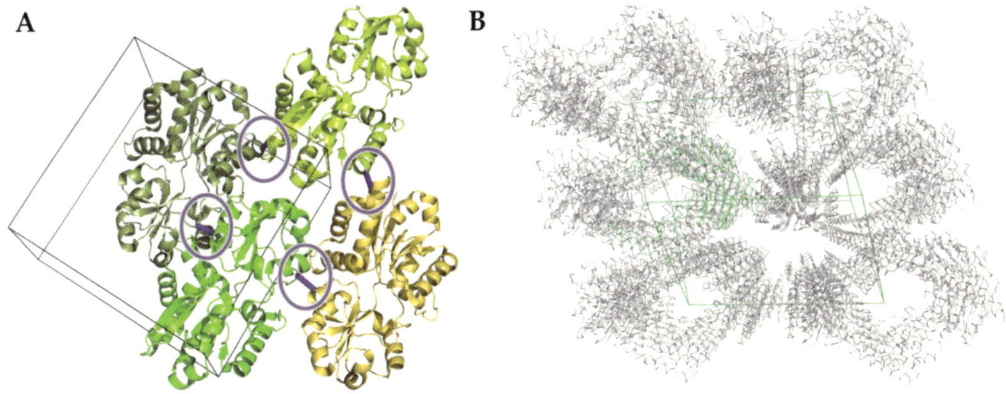

Fig. 5. Example of the packing within a crystal. A: The unit cell is shown in black, crystal contacts are highlighted with purple circles and lines. B: Three-dimensional crystal packing of a different protein. The unit cell as well as one protein monomer are depicted in green.

The flexibility as well as the other mentioned characteristics of proteins are responsible for the problems occuring during crystallization trials and despite extensive efforts not every protein is suitable for crystallization. If one cannot generate crystals one has to move back several steps and change the properties of the protein, e. g. surface properties by mutation of single amino acids, truncation of the protein or sometimes only changing buffer compositions that result in a more suitable protein for crystallization (see Figure 1 and also below). There are several prediction servers available that help choosing the 'right' protein and modification (Linding, Jensen et al. 2003; Goldschmidt, Cooper et al. 2007). However, protein crystallization still remains an empirical approach, sometimes called voodoo, while crystallography is science.

4.1 Phase diagram

The conditions or protocols for obtaining good crystals are still poorly understood and despite all progress and efforts protein crystallization is a trail-and-error approach. However, a step towards a better understanding of crystal growth can be achieved by analyzing the phase diagram of a protein-water mixture. The phase diagram is a simple illustration to help understanding how protein crystals are formed. Mostly, it is shown as a function of two ambient conditions that can be manipulated, i. e. the temperature and the concentration. Three-dimensional diagrams (two dependent parameters) have also been reported (Sauter, Lorber et al. 1999) and even a few more complex ones have been determined as well (Ewing, Forsythe et al. 1994). Figure 6 shows a schematic phase diagram for a protein solution as a function of protein concentration and precipitant concentration. The phase diagram is broken down into four distinct zones (Rosenbaum and Zukoski 1996; Haas and Drenth 1999; Asherie 2004):

1. Undersaturated zone: Under these condition the protein will stay in solution as neither the concentration of the protein nor of the precipitant is high enough to reach supersaturation.
2. Precipitation zone: Is the protein concentration or the precipitant concentration too high, the protein precipitates out of solution; this kind of solid material is not useful for crystallographic studies.

3. Labil zone: This is the most important configuration of the two parameters, as nucleation and initial crystal growth take place under these conditions.
4. Metastable zone: After initial crystals are formed and start growing in the labil zone, protein concentration decreases in the drop and the metastable zone will be reached. Here the crystal can grow further to its final maximum size.

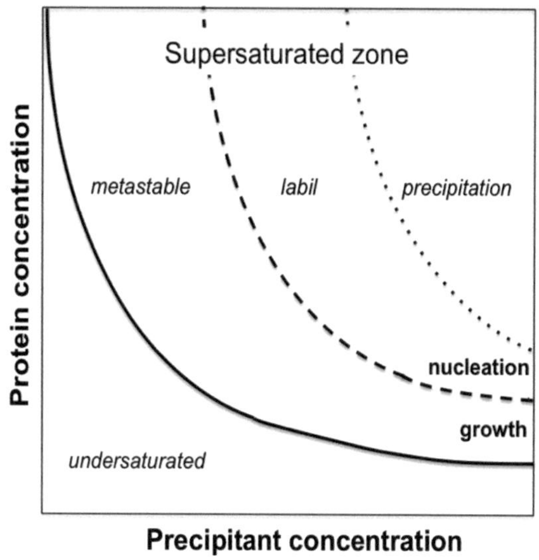

Fig. 6. A basic solubility phase diagram for a given temperature (adapted from (Rupp 2007).

The curve separating the undersaturated zone from the supersaturated one is called *solubility curve*. If conditions are chosen below the solubility curve, the protein will stay in solution and never crystallize. This means when a protein crystal is placed in a solvent, which is free of protein, it will start to dissolve. If the volume of the droplet is small enough it will not dissolve completely: it will stop dissolving when the concentration of the protein in the droplet reaches a certain level. At this concentration the crystal loses protein molecules at the same rate at which protein associate to the crystal – the system is at equilibrium. Determination of the solubility of the protein of interest might be a helpful information at the beginning of crystallization experiments. This can be done in a two-dimensional screen varying for example ammonium sulfate concentrations as well as the protein concentration.

4.2 Crystallization techniques

Crystallization is a phase transition phenomenon. Protein crystals grow from a supersaturated aqueous protein solution. Varying the concentration of precipitant, protein and additives, pH, temperature and other parameters induce the supersaturation. However, as mentioned before, prediction of this kind of phase diagrams is *a priori* impossible.
Protein crystallization can be divided into two main steps:

1. Generating initial crystals: 'Searching the needle in a haystack'
2. Empirical optimization of these crystallization condition

The first step is mostly based on experiences from other crystallization trials with different proteins. Nowadays several supplier offer crystallization screens that contain solutions for initial experiments that were used successfully in the past for crystallization trials (Jancarik and Kim 1991), so-called "sparse matrix screens". There are also some trials around to use more systematic approaches (Brzozowski and Walton 2000) to get more information about solubility prior and simultaneous to crystallization (incomplete factorials, solubility assays). Both kinds of screens can be applied to different crystallization techniques.

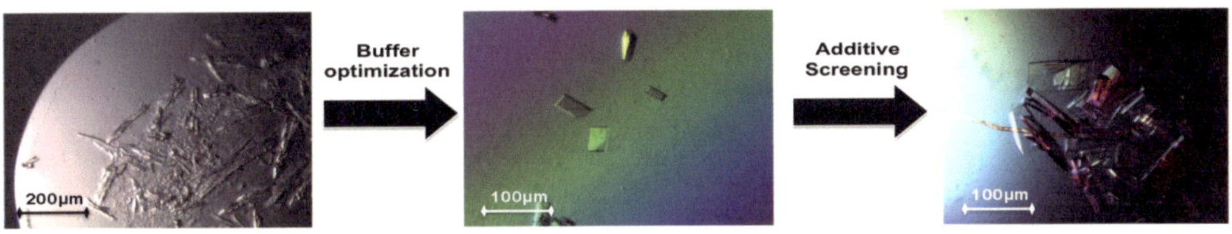

Fig. 7. Crystal optimization. First steps in crystal optimization are shown. Initial protein crystals look weak and fragile, after screening around this initial buffer composition crystal evaluation by eye results in less fragile, homogeneous looking crystals. However, diffraction quality was poor. Therefore an additive screening was performed that resulted in a different crystal form. These crystals finally were able to diffract X-rays to a reasonable resolution.

A lead/hit in that initial step might not be a 'real' crystal rather than a crystalline precipitate or just phase separation. In the next step, fine-tuning the buffer composition further optimizes this hit. Varying pH, salt concentration, type and concentration of precipitant and protein concentration are expected to yield larger and hopefully also better-diffracting crystals. In this step, the chemicals used are much more defined and therefore it is a more systematic than empirical screening (see Figure 7).

4.2.1 Vapor diffusion

The most popular and simplest technique to obtain protein crystals is the vapor diffusion method either in the sitting or hanging drop variant (see Figure 8). For both a defined volume (mostly < 1µl) of protein solution is mixed with an equivalent volume of screening solution and then equilibrated against the original precipitant/screening concentration. During this equilibration, the vapor pressure of the solution rises as the protein crystallizes (protein in solution lowers water activity) while the water evaporates to maintain equilibrium, which causes the precipitant concentration to rise. Therefore, if the crystal growth is sensitive to the precipitant concentration, vapor diffusion can rapidly force the mixture to unstable conditions where growth and nucleation are too rapid. This is the main disadvantage of vapor diffusion: Growing large crystals might be problematic!

4.2.2 Micro batch method

In this set-up the protein solution is mixed with screening solution at concentrations required for supersaturation right at the beginning of the experiment. Typical drop sizes of micro batch experiments ranges from 1-2 µl. The drop is then covered with oil, which acts as an inert sealing to protect the drops during incubation from evaporation (see Figure 8).

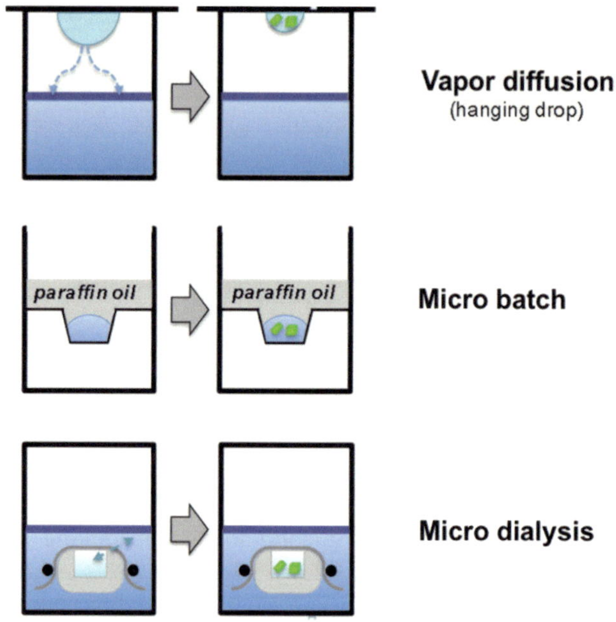

Fig. 8. Protein crystallization techniques. Schematic representation of a) vapor diffusion, b) micro batch and c) micro dialysis crystallization techniques widely used for crystal growth (adapted from (Drenth 2006)).

4.2.3 Micro dialysis

Dialysis is another way to change the buffer composition and increase its concentration in the crystallization experiment gradually (see Figure 8). Micro-dialysis buttons are exposed to different screening buffers. This method requires rather high amounts of protein but might yield large crystals.

After obtaining initial crystal hits in a commercial screen the tough part of crystal optimization starts. By varying pH, salt concentration, temperature, precipitant concentration or protein concentration these initial crystals should be reproduced and become larger, more regular shaped or are simply growing faster. A further improvement of crystal quality might be achieved by the addition of small amounts of so called 'additives'. At this point basically each chemical compound might be sufficient to improve the crystal quality. Luckily, there are some preferable working additives, which have been proven to produce better crystal in more than one case. Especially compounds that are known to reduce undirected interactions in proteins like organic solvents, i. e. DMSO or phenol, or detergents and reducing agents are very often used at this stage and helpful to force more homogeneous well diffracting crystals.

4.3 Crystal nucleation

There are two fundamental steps during protein crystallization: Nucleation and crystal growth. If one cannot obtain single crystals of adequate quality for analysis, this is generally a consequence of problems associated with the growth phase (see above). But failure to obtain any crystals at all or failure to obtain single, supportable nuclei reflects difficulties in the nucleation step. Therefore control of nucleation is a powerful tool to optimize protein crystals

or sometimes it is the only way to get crystals at all. Nucleation can take place either homogeneous meaning in the bulk of the solution, when the supersaturation is high enough for the free-energy barrier to nucleus formation to be overcome or heterogeneous mostly by solid material in the crystallization solution. This can also occur even when the supersaturation is not achieved. Therefore in order to control nucleation one has to work with highly clean solutions to avoid nucleation by the second mentioned possibility. The nucleation zone can be bypassed by insertion of crystals, crystal seeds or other nucleants to the protein/precipitant mixture. Addition of crystals or tiny fragments of crystals is called seeding. This method is then subdivided into macro- and micro-seeding dependent on the size of the nucleant added. In macro-seeding experiments one single, already well-formed but small crystal is placed into a new crystallization solution at lower saturation. Microseeding in contrast requires small fragments of a crystal or almost invisible microcrystalline precipitate. These 'seeds' are then transferred into a fresh crystallization solution either by a seeding wand which is dipped into the microseed mixture to pick up seeds and then touched across the surface of the new drop or by a animal whisker or hair that is stroked over the surface of the parent crystal to trap the nuclei and then is drawn through the new drop. As this method also enhances the speed of crystal growth it can be used with sensitive substrate that undergo decomposition over time. Oswald et al. proved this in 2008 by solving the structure of ChoX from *Sinorhizobium meliloti* in complex with a highly hydrolyzing substrate, acetylcholine (Oswald, Smits et al. 2008). In classical vapor diffusion experiments crystals appear after four weeks but data showed only little electron density in the ligand-binding site and turned out to result from a choline bound instead of acetylcholine. Hydroxylation was favored due to the relatively long time for crystal growth but also because of an acetic pH in the crystallization set-up. To circumvent these problems accelerated crystal growth was required. In this case micro-seeding results in crystals suitable for data collection in less than 24 hours.
Recent years more effort in nucleation control yielded in fancy materials that can be used as nuclei for crystals. These methods use the second way of nuclei formation, as a solid material is introduce into the crystallization solution as an 'universal' nucleant (Chayen, Saridakis et al. 2006). There have been several substances that have been tried more or less successful. Some have been useful for individual proteins, but mostly they were not applicable in general (McPherson and Shlichta 1988; Chayen, Radcliffe et al. 1993; Blow, Chayen et al. 1994). In 2001, Chayen et al. proposed the idea of using porous silicon whose pore size is comparable with the size of a protein molecule. In theory such pores may confine and concentrate the protein molecules at the surface of the silicone and thereby encourage them to form crystal nuclei (Chayen, Saridakis et al. 2001). These nucleants have made it to commercial availability (www.moleculardimensions.com) and have proven to be suitable for different kinds of proteins and even membrane proteins that have not been possible to crystallize before formed nice crystals in the presence of these nucleants.

4.4 Cryoprotection

Exposure of a protein crystal at room temperature results in dramatic radiation damage due to radicals formed by the ionizing X-ray photons. To reduce that harmful disintegration of the protein crystal the crystal is cooled to ≈100K with the help of liquid nitrogen (Low, Chen et al. 1966; Hope 1988; Rodgers 1994; Garman 1999). However, it is common for the cooling process to disrupt the crystal order and decrease diffraction quality. Thus, the crystal must be cooled fast so that the water in the solvent channels is in the vitreous rather than in the crystalline state at the end of this procedure. As for pure water this cooling has to take place very quick (10^{-5}s, (Johari, Hallbrucker et al. 1987), some water molecules can be replaced by a cryoprotective solution prior to cooling (Juers and Matthews 2004). This exceeds the time window for freezing up to 1-2s (Garman and Owen 2006) however, finding a good 'cryoprotectant' for a special protein crystal again involves substantial screening. Once flash frozen in liquid nitrogen, the crystal must be kept below the glass transition temperature of the cryobuffer at or below 155K at all times (Weik, Kryger et al. 2001).

4.5 What can you do when all efforts did not succeed in crystals?

4.5.1 Buffer composition – Again!

The choice of the right buffer used for crystallization experiments is very crucial. As shown above, every protein needs its own buffer composition to feel kind of happy in this aqueous artificial environment. Especially as high protein concentrations (>10mg/ml) are required for crystallization, one might has to test several buffer compositions again (see also Figure 1). As a rule of thumb you should obtain around 50% of clear drops immediately after mixing protein and buffer solution. If you detect drastically more precipitation in your drops you should think first about lower protein concentration but of course secondly about changing your buffer system again.

4.5.2 How to obtain a rigid protein suitable for crystallization?

To overcome the problem of flexibility of some regions in the protein addition of ligands is often a very powerful tool to fix the protein in a single conformation that is more favorable for crystallization. A good example for this strategy is the crystallization of so-called substrate binding proteins (for a recent review see (Berntsson, Smits et al. 2010)). These proteins catch their substrate in the periplasm of bacteria or on the outer membrane of archaea and then deliver it to their cognate transport system located in the membrane. The mechanism of substrate binding is quite well understood. These binding proteins all consist of two domains, which rotate towards each other during the binding event. In solution without substrate they are quite flexible and NMR-studies proved a equilibrium between open and closed conformation (Tang, Schwieters et al. 2007). Analysis of all available structures for this class of proteins showed that more than 95% were crystallized with a ligand bound (Berntsson, Smits et al. 2010). Thus, a stabilization of the two domains seems to simplify crystal contact formation dramatically. Although people always want to obtain a functional conformation of the protein in their crystal structure, it is sometimes helpful to think about how to stop the protein from doing its job. A non-functional protein is in general less flexible and fixed in one conformation. One example for successful implementation of this strategy is the crystal structure of NhaA from *Escherichia coli* solved in 2005 (Hunte, Screpanti et al. 2005). Here, Hunte et al., downregulated the protein activity by working at an acidic pH of 4. Although the protein shows almost no activity at this pH the structure reveals the basis for mechanism of Na+/H+ exchange and also its regulation by pH could be understood.

4.5.3 Rational protein design for crystallization. Surface engineering

The first example of rational protein design that yielded a good diffracting protein crystal is given by Lawson et al. in 1991 (Lawson, Artymiuk et al. 1991). They compared amino acids involved in crystal contact formation of the rat ferritin protein L. (which is highly homologous to human ferritin H, the target protein) with the amino acids present at that position in human ferritin H. A replacement of Lys86, found in the human sequence, with Glu, which occurs in rat, recreated a Ca^{2+} binding bridge that mediates crystal contacts in the rat ortholog. As this method was successful for several other proteins (McElroy, Sissom et al. 1992; Braig, Otwinowski et al. 1994; Horwich 2000), a general protocol was required. The concept Derewenda et al. proposed in 2004 is based on the general equation for the free energy that drives protein crystallization:

$$\Delta G = \Delta H - T(\Delta S_{protein} + \Delta S_{solvent})$$

As the enthalpy values of intermolecular interactions in a crystal lattice are rather small (see above), crystallization is very sensitive to entropy changes of both protein and solvent. The formation of ordered protein aggregates carries a negative entropy term. This can only be overcome by positive entropy from the release of water bound to the protein. However, large hydrophilic residues (e.g. lysines, arginies, glutamates, glutamines) exposed on the protein surface need to be ordered. Since they are rather flexible this can cause problems. This can be overcome by mutating large amino acids into smaller ones, for example alanines. Among these large amino acids lysines and glutamates play a particular role, as they are always (with only very few exeptions) located on the protein surface (Baud and Karlin 1999). Both lysines and glutamates are typically disfavored at interfaces in protein protein complexes (Lo Conte, Chothia et al. 1999), therefore it is rather straight forward to assume that lysine and glutamate to alanine mutants are good targets for protein crystallization if wildtype protein hardly forms crystals. However this also means that you have to go several steps backwards on road to a protein structure determination (see Figure 1).

4.5.4 Affinity tag removal: Philosophic question???

Another variant in protein crystallization nowadays is the affinity tag used for purification of the desired protein. The decision about position and choice of the affinity tag are mostly made at the beginning of the long way to a crystal structure (see Figure 1). However, it becomes crucial again at the crystallization step. In general most people like to remove the tag before crystallization to prove a physiological conformation. But, there are examples where the tag played a pivotal role in crystallization (Smits, Mueller et al. 2008a). The crystal structure of the octopine dehydrogenase from *Pecten maximus* is shown in Figure 9 (Smits, Mueller et al. 2008b), with the interactions sides/crystal contacts highlighted in green.
In Figure 9A contacts look quite similar to that presented in Figure 5. However when having a closer look on the his-tag, you recognize that it is located in a cavity formed by another monomer of that protein. In that cavity it can perform several hydrogen bonds with amino acids from the other monomer resulting in a very strong interaction which yields good quality crystals.

4.5.5 Crystallization using antibody fragments

A number of ways to stabilize proteins for crystallography have been developed, for example genetic engineering, co-crystallization with natural ligands and reducing surface entropy (see above). Recently, crystallization mediated by antibody fragments has moved into the focus of crystallographers especially to obtain crystals of membrane proteins (Ostermeier, Iwata et al. 1995; Hunte and Michel 2002). Membrane protein crystallization is even tougher compared to soluble proteins, because of the amphipathic surface of the molecules. As they are located in the lipid bilayer most of their surface is hydrophobic and must be covered to keep them in solution. This is maintained by detergents. The detergent micelles cover the hydrophobic surface and therefore this area is no longer available to form crystal contacts. Crystal contacts can only be formed by the polar surfaces of these proteins. As many membrane proteins contain only relatively small hydrophilic domains, a strategy to increase the probability of getting well-ordered crystals is required. Antibody fragments can play this role. They can be designed for binding at specific regions in the protein and then function as additional polar domain in the membrane protein complex (for example see (Ostermeier, Iwata et al. 1995; Huber, Steiner et al. 2007).

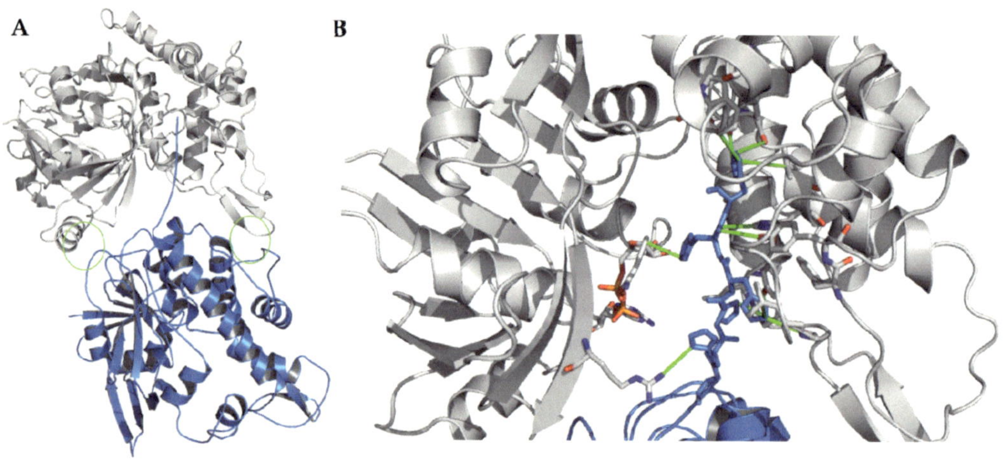

Fig. 9. Crystal contacts in OcDH protein. A: Overall view on two monomers. Surface Crystal contacts are highlighted in the green circles. B: Zoom in on the His-tag of one monomer. The his-tag of one monomer in the crystal structure is located near the binding site in a deep cavity formed by the other monomer. Therefore it is able to form several hydrogen bonds (highlighted in green) with side and main chains of the other protein but also with the ligand bound in this binding site (orange).

5. Conclusion

For what reason do we effort so much work on good quality crystals?

Single good quality crystals constitute an essential prerequisite for structural investigations of biological macromolecules using X-ray diffraction. The harder one works on crystal quality the easier the determination of a reasonable atomic model of the molecule of interest becomes. The vast majority of problems encountered in crystal structure determination can typically be traced back to data quality issues caused by crystal imperfections. Consequently, although primary focus of structural biology is on the macromolecule that makes up a crystal, there is also considerable interest in the physical properties, nucleation and growth of the crystals themselves. Statistics of various Structural Genomics Centers proved that protein crystallization is still despite all the progress in the technology of crystallization robotics is still a rather tough field in biological science. Success rate ranges for small prokaryotic proteins from 10-30% and decreases dramatically to a few percent for human proteins. The struggle obtaining crystals for protein structure determination is justified. After all efforst looking at electron density and subsequent the protein structure is still one of the most intriguing as well auspicious parts in structural biology

6. References

Alexandrov, A. I., M. Mileni, et al. (2008). "Microscale fluorescent thermal stability assay for membrane proteins." Structure 16(3): 351-359.

Altmann, F., E. Staudacher, et al. (1999). "Insect cells as hosts for the expression of recombinant glycoproteins." Glycoconj J 16(2): 109-123.

Arnau, J., C. Lauritzen, et al. (2006). "Current strategies for the use of affinity tags and tag removal for the purification of recombinant proteins." Protein expression and purification 48(1): 1-13.

Asherie, N. (2004). "Protein crystallization and phase diagrams." Methods 34(3): 266-272.

Baneyx, F. (1999). "Recombinant protein expression in *Escherichia coli*." Current opinion in biotechnology 10(5): 411-421.

Baud, F. and S. Karlin (1999). "Measures of residue density in protein structures." Proc Natl Acad Sci U S A 96(22): 12494-12499.

Berntsson, R. P., M. K. Doeven, et al. (2009). "The structural basis for peptide selection by the transport receptor OppA." EMBO J 28(9): 1332-1340.

Berntsson, R. P., S. H. Smits, et al. (2010). "A structural classification of substrate-binding proteins." FEBS Lett 584(12): 2606-2617.

Binda, C., M. Aldeco, et al. (2010). "Interactions of Monoamine Oxidases with the Antiepileptic Drug Zonisamide: Specificity of Inhibition and Structure of the Human Monoamine Oxidase B Complex." J Med Chem.

Blow, D. M., N. E. Chayen, et al. (1994). "Control of nucleation of protein crystals." Protein Sci 3(10): 1638-1643.

Blumenthal, D. K., K. Takio, et al. (1985). "Identification of the calmodulin-binding domain of skeletal muscle myosin light chain kinase." Proc Natl Acad Sci U S A 82(10): 3187-3191.

Bolotin, A., P. Wincker, et al. (2001). "The complete genome sequence of the lactic acid bacterium *Lactococcus lactis* ssp. lactis IL1403." Genome research 11(5): 731-753.

Braig, K., Z. Otwinowski, et al. (1994). "The crystal structure of the bacterial chaperonin GroEL at 2.8 A." Nature 371(6498): 578-586.

Brzozowski, A. M. and J. Walton (2000). "Clear strategy screens for macromolecular crystallization." J Appl Cryst 34: 97-101.

Cayley, S., B. A. Lewis, et al. (1991). "Characterization of the cytoplasm of *Escherichia coli* K-12 as a function of external osmolarity. Implications for protein-DNA interactions in vivo." J Mol Biol 222(2): 281-300.

Chagnon, A. and M. Corbeil (1973). "Use of an organic buffer (HEPES) in human lymphocytoid cell line cultures." In Vitro 8(4): 283-287.

Chayen, N. E., J. W. Radcliffe, et al. (1993). "Control of nucleation in the crystallization of lysozyme." Protein Sci 2(1): 113-118.

Chayen, N. E., E. Saridakis, et al. (2001). "Porous silicon: an effective nucleation-inducing material for protein crystallization." J Mol Biol 312(4): 591-595.

Chayen, N. E., E. Saridakis, et al. (2006). "Experiment and theory for heterogeneous nucleation of protein crystals in a porous medium." Proc Natl Acad Sci U S A 103(3): 597-601.

Collins, B., R. C. Stevens, et al. (2005). "Crystallization Optimum Solubility Screening: using crystallization results to identify the optimal buffer for protein crystal formation." Acta Crystallogr Sect F Struct Biol Cryst Commun 61(Pt 12): 1035-1038.

Cregg, J. M. (2007). Pichia Protocols, Humana Press.

Cregg, J. M., J. L. Cereghino, et al. (2000). "Recombinant protein expression in Pichia pastoris." Molecular biotechnology 16(1): 23-52.

D., O. R., M. L.K., et al. (1992). Baculovirus Expression Vectors A Laboratory Manual. New York.

de Ruyter, P. G., O. P. Kuipers, et al. (1996). "Controlled gene expression systems for Lactococcus lactis with the food-grade inducer nisin." Applied and environmental microbiology 62(10): 3662-3667.

de Vos, W. M. and E. E. Vaughan (1994). "Genetics of lactose utilization in lactic acid bacteria." FEMS microbiology reviews 15(2-3): 217-237.

di Guan, C., P. Li, et al. (1988). "Vectors that facilitate the expression and purification of foreign peptides in *Escherichia coli* by fusion to maltose-binding protein." Gene 67(1): 21-30.

Drenth, J. (2006). "Principles of Protein X-Ray Crystallography." Springer 3.

Durst, R. A. and B. R. Staples (1972). "Tris-tris-HCl: a standard buffer for use in the physiologic pH range." Clin Chem 18(3): 206-208.

Eifler, N., M. Duckely, et al. (2007). "Functional expression of mammalian receptors and membrane channels in different cells." Journal of structural biology 159(2): 179-193.

Elowitz, M. B., M. G. Surette, et al. (1999). "Protein mobility in the cytoplasm of *Escherichia coli*." J Bacteriol 181(1): 197-203.

Ericsson, U. B., B. M. Hallberg, et al. (2006). "Thermofluor-based high-throughput stability optimization of proteins for structural studies." Anal Biochem 357(2): 289-298.

Erkens, G. B., R. P. Berntsson, et al. (2011). "The structural basis of modularity in ECF-type ABC transporters." Nat Struct Mol Biol 18(7): 755-760.

Ewing, F., E. Forsythe, et al. (1994). "Orthorhombic lysozyme solubility." Acta Crystallogr D Biol Crystallogr 50(Pt 4): 424-428.

Fischer, M., A. P. Leech, et al. (2011). "Comparative Assessment of Different Histidine-Tags for Immobilization of Protein onto Surface Plasmon Resonance Sensorchips." Analytical chemistry.

Gaberc-Porekar, V. and V. Menart (2001). "Perspectives of immobilized-metal affinity chromatography." Journal of biochemical and biophysical methods 49(1-3): 335-360.

Garman, E. (1999). "Cool data: quantity AND quality." Acta Crystallogr D Biol Crystallogr 55(Pt 10): 1641-1653.

Garman, E. F. and R. L. Owen (2006). "Cryocooling and radiation damage in macromolecular crystallography." Acta Crystallogr D Biol Crystallogr 62(Pt 1): 32-47.

Gasson, M. J. (1983). "Plasmid complements of *Streptococcus lactis* NCDO 712 and other lactic streptococci after protoplast-induced curing." Journal of bacteriology 154(1): 1-9.

Gingras, A. C., R. Aebersold, et al. (2005). "Advances in protein complex analysis using mass spectrometry." The Journal of physiology 563(Pt 1): 11-21.

Goldschmidt, L., D. R. Cooper, et al. (2007). "Toward rational protein crystallization: A Web server for the design of crystallizable protein variants." Protein Sci 16(8): 1569-1576.

Groß, R., R. Pisa, et al. (2002). "Isolierung der trimeren Hydrgenase aus Wollinella succinogenes durch StrepTactin-Affinitätschromatographie." Biospektrum 1: 101.

Guillot, A., C. Gitton, et al. (2003). "Proteomic analysis of *Lactococcus lactis*, a lactic acid bacterium." Proteomics 3(3): 337-354.

Haas, C. and J. Drenth (1999). "Understanding protein crystallization on the basis of the phase diagram." Journal of Crystal Growth 196(2-4): 388-394.

Hampton, E. N., M. W. Knuth, et al. (2007). "The self-inhibited structure of full-length PCSK9 at 1.9 A reveals structural homology with resistin within the C-terminal domain." Proc Natl Acad Sci U S A 104(37): 14604-14609.

Hansen, S. B., G. Sulzenbacher, et al. (2006). "Structural characterization of agonist and antagonist-bound acetylcholine-binding protein from *Aplysia californica*." J Mol Neurosci 30(1-2): 101-102.

Hefti, M. H., C. J. Van Vugt-Van der Toorn, et al. (2001). "A novel purification method for histidine-tagged proteins containing a thrombin cleavage site." Analytical biochemistry 295(2): 180-185.

Hochuli, E., H. Dobeli, et al. (1987). "New metal chelate adsorbent selective for proteins and peptides containing neighbouring histidine residues." Journal of chromatography 411: 177-184.

Holm, J. and S. I. Hansen (2001). "Effect of hydrogen ion concentration and buffer composition on ligand binding characteristics and polymerization of cow's milk folate binding protein." Biosci Rep 21(6): 745-753.

Hope, H. (1988). "Cryocrystallography of biological macromolecules: a generally applicable method." Acta Crystallogr B 44 (Pt 1): 22-26.

Horwich, A. (2000). "Working with Paul Sigler." Nat Struct Biol 7(4): 269-270.

Huber, T., D. Steiner, et al. (2007). "In vitro selection and characterization of DARPins and Fab fragments for the co-crystallization of membrane proteins: The Na(+)-citrate symporter CitS as an example." J Struct Biol 159(2): 206-221.

Hunt, I. (2005). "From gene to protein: a review of new and enabling technologies for multi-parallel protein expression." Protein expression and purification 40(1): 1-22.

Hunte, C. and H. Michel (2002). "Crystallisation of membrane proteins mediated by antibody fragments." Curr Opin Struct Biol 12(4): 503-508.

Hunte, C., E. Screpanti, et al. (2005). "Structure of a Na+/H+ antiporter and insights into mechanism of action and regulation by pH." Nature 435(7046): 1197-1202.

Jana, S. and J. K. Deb (2005). "Strategies for efficient production of heterologous proteins in Escherichia coli." Applied microbiology and biotechnology 67(3): 289-298.

Jancarik, J. and S. H. Kim (1991). "Sparse matrix sampling: a screening method for crystallization of proteins." J Appl Cryst 24: 409-411.

Jancarik, J., R. Pufan, et al. (2004). "Optimum solubility (OS) screening: an efficient method to optimize buffer conditions for homogeneity and crystallization of proteins." Acta Crystallogr D Biol Crystallogr 60(Pt 9): 1670-1673.

Johari, G. P., A. Hallbrucker, et al. (1987). "The glass-liquid transition of hyperquenched water." Nature 330: 552-553.

Jonasson, P., S. Liljeqvist, et al. (2002). "Genetic design for facilitated production and recovery of recombinant proteins in Escherichia coli." Biotechnology and applied biochemistry 35(Pt 2): 91-105.

Juers, D. H. and B. W. Matthews (2004). "Cryo-cooling in macromolecular crystallography: advantages, disadvantages and optimization." Q Rev Biophys 37(2): 105-119.

Kean, J., R. M. Cleverley, et al. (2008). "Characterization of a CorA Mg2+ transport channel from Methanococcus jannaschii using a Thermofluor-based stability assay." Mol Membr Biol 25(8): 653-663.

Konings, W. N., J. Kok, et al. (2000). "Lactic acid bacteria: the bugs of the new millennium." Curr Opin Microbiol 3(3): 276-282.

Kopecky, D. J., X. Hao, et al. (2008). "Identification and optimization of N3,N6-diaryl-1H-pyrazolo[3,4-d]pyrimidine-3,6-diamines as a novel class of ACK1 inhibitors." Bioorg Med Chem Lett 18(24): 6352-6356.

Kuipers, O. P., M. M. Beerthuyzen, et al. (1995). "Autoregulation of nisin biosynthesis in *Lactococcus lactis* by signal transduction." The Journal of biological chemistry 270(45): 27299-27304.

Kuipers, O. P., P. G. de Ruyter, et al. (1998). "Quorum sensing-controlled gene expression in lactic acid bacteria." J Biotechnol 64: 15-21.

Kulahin, N., O. Kristensen, et al. (2011). "Structural model and trans-interaction of the entire ectodomain of the olfactory cell adhesion molecule." Structure 19(2): 203-211.

Kunji, E. R., D. J. Slotboom, et al. (2003). "*Lactococcus lactis* as host for overproduction of functional membrane proteins." Biochimica et biophysica acta 1610(1): 97-108.

Lawson, D. M., P. J. Artymiuk, et al. (1991). "Solving the structure of human H ferritin by genetically engineering intermolecular crystal contacts." Nature 349(6309): 541-544.

Li, M. Z. and S. J. Elledge (2007). "Harnessing homologous recombination in vitro to generate recombinant DNA via SLIC." Nat Methods 4(3): 251-256.

Linding, R., L. J. Jensen, et al. (2003). "Protein disorder prediction: implications for structural proteomics." Structure 11(11): 1453-1459.

Lo Conte, L., C. Chothia, et al. (1999). "The atomic structure of protein-protein recognition sites." J Mol Biol 285(5): 2177-2198.

Low, B. W., C. C. Chen, et al. (1966). "Studies of insulin crystals at low temperatures: effects on mosaic character and radiation sensitivity." Proc Natl Acad Sci U S A 56(6): 1746-1750.

Lundstrom, K. (2006a). "Structural genomics for membrane proteins." Cellular and molecular life sciences : CMLS 63(22): 2597-2607.

Lundstrom, K., R. Wagner, et al. (2006b). "Structural genomics on membrane proteins: comparison of more than 100 GPCRs in 3 expression systems." Journal of structural and functional genomics 7(2): 77-91.

Madoori, P. K., H. Agustiandari, et al. (2009). "Structure of the transcriptional regulator LmrR and its mechanism of multidrug recognition." EMBO J 28(2): 156-166.

Mavaro, A., A. Abts, et al. (2011). "Substrate recognition and specificity of NISB, the lantibiotic dehydratase involved in nisin biosynthesis." J Biol Chem.

McElroy, H. H., G. W. Sissom, et al. (1992). "Studies on engineering crystallizability by muttion of surface residues of human thymidylate synthase." J. Cryst. Growth 122: 265-272.

McPherson, A. and P. Shlichta (1988). "Heterogeneous and epitaxial nucleation of protein crystals on mineral surfaces." Science 239(4838): 385-387.

Menetrey, J., A. Bahloul, et al. (2005). "The structure of the myosin VI motor reveals the mechanism of directionality reversal." Nature 435(7043): 779-785.

Midgett, C. R. and D. R. Madden (2007). "Breaking the bottleneck: eukaryotic membrane protein expression for high-resolution structural studies." Journal of structural biology 160(3): 265-274.

Mierau, I. and M. Kleerebezem (2005). "10 years of the nisin-controlled gene expression system (NICE) in Lactococcus lactis." Applied microbiology and biotechnology 68(6): 705-717.

Muller, K. M., K. M. Arndt, et al. (1998). "Tandem immobilized metal-ion affinity chromatography/immunoaffinity purification of His-tagged proteins--evaluation of two anti-His-tag monoclonal antibodies." Anal Biochem 259(1): 54-61.

Murphy, K. P., P. L. Privalov, et al. (1990). "Common features of protein unfolding and dissolution of hydrophobic compounds." Science 247(4942): 559-561.

Ostermeier, C., S. Iwata, et al. (1995). "Fv fragment-mediated crystallization of the membrane protein bacterial cytochrome c oxidase." Nat Struct Biol 2(10): 842-846.

Oswald, C., S. H. Smits, et al. (2008). "Microseeding - a powerful tool for crystallizing proteins complexed with hydrolyzable substrates." Int J Mol Sci 9(7): 1131-1141.

Pittelkow, M., B. Tschapek, et al. (2011). "The Crystal Structure of the Substrate-Binding Protein OpuBC from *Bacillus subtilis* in Complex with Choline." Journal of molecular biology 411(1): 53-67.

Porath, J., J. Carlsson, et al. (1975). "Metal chelate affinity chromatography, a new approach to protein fractionation." Nature 258(5536): 598-599.

Regnier, F. E. (1983a). "High-performance liquid chromatography of biopolymers." Science 222(4621): 245-252.

Regnier, F. E. (1983b). "High-performance liquid chromatography of proteins." Methods Enzymol 91: 137-190.

Ressl, S., A. C. Terwisscha van Scheltinga, et al. (2009). "Molecular basis of transport and regulation in the Na(+)/betaine symporter BetP." Nature 458(7234): 47-52.

Rodgers, D. W. (1994). "Cryocrystallography." Structure 2(12): 1135-1140.

Rosenbaum, D. F. and C. F. Zukoski (1996). "Protein interactions and crystallization." Journal of Crystal Growth 169(4): 752-758.

Rubio, V., Y. Shen, et al. (2005). "An alternative tandem affinity purification strategy applied to Arabidopsis protein complex isolation." The Plant journal : for cell and molecular biology 41(5): 767-778.

Rupp, B. (2007). "Biomolecular Crystallography: Principles, Practice, and Application to Structural Biology." Taylor & Francis Ltd 1.

Sachdev, D. and J. M. Chirgwin (1999). "Properties of soluble fusions between mammalian aspartic proteinases and bacterial maltose-binding protein." Journal of protein chemistry 18(1): 127-136.

Sala, E. and A. de Marco (2010). "Screening optimized protein purification protocols by coupling small-scale expression and mini-size exclusion chromatography." Protein Expr Purif 74(2): 231-235.

Sauter, C., B. Lorber, et al. (1999). "Crystallogenesis studies on yeast aspartyl-tRNA synthetase: use of phase diagram to improve crystal quality." Acta Crystallogr D Biol Crystallogr 55(Pt 1): 149-156.

Schmidt, T. G., J. Koepke, et al. (1996). "Molecular interaction between the Strep-tag affinity peptide and its cognate target, streptavidin." Journal of molecular biology 255(5): 753-766.

Schwarz, C. K., B. Tschapek, et al. (2011). "Crystallization and preliminary X-ray crystallographic studies of an oligomeric species of a refolded C39 peptidase-like domain of the *Escherichia coli* ABC transporter haemolysin B." Acta Crystallogr Sect F Struct Biol Cryst Commun 67(Pt 5): 630-633.

Senkovich, O., W. J. Cook, et al. (2007). "Structure of a complex of human lactoferrin N-lobe with pneumococcal surface protein a provides insight into microbial defense mechanism." Journal of molecular biology 370(4): 701-713.

Smith, D. B. and K. S. Johnson (1988). "Single-step purification of polypeptides expressed in *Escherichia coli* as fusions with glutathione S-transferase." Gene 67(1): 31-40.

Smith, G. E., M. D. Summers, et al. (1983). "Production of human beta interferon in insect cells infected with a baculovirus expression vector." Molecular and cellular biology 3(12): 2156-2165.

Smits, S. H., A. Mueller, et al. (2008a). "Coenzyme- and His-tag-induced crystallization of octopine dehydrogenase." Acta Crystallogr Sect F Struct Biol Cryst Commun 64(Pt 9): 836-839.

Smits, S. H., A. Mueller, et al. (2008b). "A structural basis for substrate selectivity and stereoselectivity in octopine dehydrogenase from *Pecten maximus*." J Mol Biol 381(1): 200-211.

Sorensen, H. P. and K. K. Mortensen (2005). "Advanced genetic strategies for recombinant protein expression in *Escherichia coli*." Journal of biotechnology 115(2): 113-128.

Stofko-Hahn, R. E., D. W. Carr, et al. (1992). "A single step purification for recombinant proteins. Characterization of a microtubule associated protein (MAP 2) fragment which associates with the type II cAMP-dependent protein kinase." FEBS letters 302(3): 274-278.

Taha, M. (2005). "Buffers for the physiological pH range: acidic dissociation constants of zwitterionic compounds in various hydroorganic media." Ann Chim 95(1-2): 105-109.

Tang, C., C. D. Schwieters, et al. (2007). "Open-to-closed transition in apo maltose-binding protein observed by paramagnetic NMR." Nature 449(7165): 1078-1082.

Tate, C. G., J. Haase, et al. (2003). "Comparison of seven different heterologous protein expression systems for the production of the serotonin transporter." Biochimica et biophysica acta 1610(1): 141-153.

Terpe, K. (2003). "Overview of tag protein fusions: from molecular and biochemical fundamentals to commercial systems." Applied microbiology and biotechnology 60(5): 523-533.

Teuber, M. (1995). "The genus Lactococcus." The genera of lactic acid bacteria 2.

Tornquist, K., J. Paallysaho, et al. (1995). "Influence of Hepes- and CO2/HCO(3-)-buffer on Ca2+ transients induced by TRH and elevated K+ in rat pituitary GH4C1 cells." Mol Cell Endocrinol 112(1): 77-82.

Tschapek, B., M. Pittelkow, et al. (2011). "Arg149 Is Involved in Switching the Low Affinity, Open State of the Binding Protein AfProX into Its High Affinity, Closed State." Journal of molecular biology 411(1): 36-52.

Urh, M., D. York, et al. (1995). "Buffer composition mediates a switch between cooperative and independent binding of an initiator protein to DNA." Gene 164(1): 1-7.

van de Guchte, M., J. Kok, et al. (1992). "Gene expression in *Lactococcus lactis*." FEMS microbiology reviews 8(2): 73-92.

Voss, S. and A. Skerra (1997). "Mutagenesis of a flexible loop in streptavidin leads to higher affinity for the Strep-tag II peptide and improved performance in recombinant protein purification." Protein engineering 10(8): 975-982.

Waugh, D. S. (2005). "Making the most of affinity tags." Trends in biotechnology 23(6): 316-320.

Wegner, G. H. (1990). "Emerging applications of the methylotrophic yeasts." FEMS microbiology reviews 7(3-4): 279-283.

Weik, M., G. Kryger, et al. (2001). "Solvent behaviour in flash-cooled protein crystals at cryogenic temperatures." Acta Crystallogr D Biol Crystallogr 57(Pt 4): 566-573.

Zaitseva, J., I. B. Holland, et al. (2004). "The role of CAPS buffer in expanding the crystallization space of the nucleotide-binding domain of the ABC transporter haemolysin B from *Escherichia coli*." Acta Crystallogr D Biol Crystallogr 60(Pt 6): 1076-1084.

Zaitseva, J., S. Jenewein, et al. (2005). "Functional characterization and ATP-induced dimerization of the isolated ABC-domain of the haemolysin B transporter." Biochemistry 44(28): 9680-9690.

Zheng, C. F., T. Simcox, et al. (1997). "A new expression vector for high level protein production, one step purification and direct isotopic labeling of calmodulin-binding peptide fusion proteins." Gene 186(1): 55-60.

Zimmerman, S. B. and S. O. Trach (1991). "Estimation of macromolecule concentrations and excluded volume effects for the cytoplasm of *Escherichia coli*." J Mol Biol 222(3): 599-620.

23

Growth of Organic Nonlinear Optical Crystals from Solution

A. Antony Joseph and C. Ramachandra Raja
*Department of Physics, Government Arts College (Autonomous),
Kumbakonam,Tamil Nadu,
India*

1. Introduction

Investigations on the growth of good quality single crystals play an important role in the development of modern scientific world with advanced technology. Behind the development in every new solid state device and the explosion in solid state device, there stands a single crystal. Crystal growth is an important field of materials science, which involves controlled phase transformation. In the past few decades, there has been a growing interest in crystal growth process, particularly in view of the increasing demand for materials for technological applications (Brice, 1986). Researchers worldwide have always been in the search of new materials and their single crystal growth.
Solids exist in two forms namely single crystals, polycrystalline and amorphous materials depending upon the arrangement of constituent molecules, atoms or ions. An ideal crystal is one, in which the surroundings of any atom would be exactly the same as the surroundings of every similar atom. Real crystals are finite and contain defects. However, single crystals are solids in the most uniform condition that can be attained and this is the basis for most of the uses of these crystals. The uniformity of single crystals can allow the transmission without the scattering of electro magnetic waves. The methods of growing crystals are very wide and mainly dictated by the characteristics of the material and its size (Buckley, 1951).

2. Nucleation

Comprehensive study on the growth of crystals should start from an understanding of nucleation process. Nucleation is the process of generating within a metastable phase, initial fragments of a new and more stable phase. In a supersaturated or super-cooled system when a few atoms or molecules join together, a change in energy takes place in the process of formation of the cluster. The cluster of such atoms or molecules is termed as "embryo". An embryo may grow or disintegrate and disappear completely. If the embryo grows to a particular size, critical size known as 'critical-nucleus' then greater is the possibility for the nucleus to grow into a crystal. Thus nucleation is an important event in crystal growth.

2.1 Kinds of nucleation

Nucleation is broadly classified into two types. These two types are frequently reserved to as primary and secondary nucleation. The former occurs either spontaneously or induced artificially. The spontaneous formation of crystalline nuclei within the interior of parent phase is called homogeneous nucleation. On the other hand if the nuclei form heterogeneously around ions, impurity molecules or on dust particles or on surface of the container or at structural singularities such as dislocation or imperfection, it is called heterogeneous nucleation. If the nuclei are generated in the vicinity of crystals present in supersaturated system then this phenomenon is often referred to as "secondary" nucleation. Nucleation can often be induced by external influence like agitation, mechanical shock, friction, spark, extreme pressure, electric and magnetic fields, UV -rays, X-rays, gamma rays and so on.

2.2 Classical theory of nucleation

The formation of the crystal nuclei is a difficult and a complex process, because the constituent atoms or molecules have to be oriented into a fixed lattice. In practice, a number of atoms or molecules may come together as a result of statistical incidence to form an ordinary cluster of molecules known as embryo.

2.3 Kinetic theory of nucleation

The main aim of the nucleation theory is to calculate the rate of nucleation. Rate of nucleation is nothing but the number of critical nuclei formed per unit time per unit volume. In kinetic theory, nucleation is treated as the chain reaction of monomolecular addition to the cluster and ultimately reaching macroscopic dimensions is represented as follows:

$$A_1 + A_1 \rightleftharpoons A_2$$

$$A_2 + A_1 \rightleftharpoons A_3$$

- -

$$A_{i-1} + A_1 \rightleftharpoons A_i$$

$$A_i + A_1 \rightleftharpoons A_{i+1}$$

- -

$$A_{j-1} + A_1 \rightleftharpoons A_j^*$$

Two monomers collide with one another to form a dimer. A monomer joins with a dimer to form a trimer. This reaction builds a cluster having i-molecules known as i-mer. As the time increases the size distribution in the embryos changes and larger ones have increase in size. As the size attains a critical size Aj*, the further growth into macroscopic size is guaranteed, as there is a possibility for the reverse reaction i.e., the decay of a cluster into monomers.

3. Stability of nucleus

An isolated droplet of a fluid is most stable when its surface free energy and therefore its area is a minimum. According to Gibbs (Gibbs & Longmans, 1982), the total free energy of a crystal in equilibrium would be minimum for a given volume if its surrounding is at constant temperature and pressure.
If the volume free energy per unit volume is considered to be constant then,

$$\sum a_i \sigma_i = \text{minimum} \tag{1}$$

Where, a_i is the area of ith face and σ_i is the corresponding surface energy per unit area.

4. Energy formation of spherical nucleus

The total free energy change associated with the process of homogenous nucleation shall be considered as follows. Let ΔG be the overall excess free energy of an embryo between the two phases mentioned above. ΔG can be represented as a combination of volume and surface energies since an embryo possesses both these energies.

$$\Delta G = \Delta G_s + \Delta G_v \tag{2}$$

Where, ΔG_s is the surface excess free energy and ΔGv is the volume excess free energy. Assuming the second phase to be spherical.

$$\Delta G = 4 \pi r^2 \sigma + 4/3 \pi r^3 \Delta G_v \tag{3}$$

Where ΔG_v is the free energy change per unit volume is a negative quantity and 'σ 'is the free energy change per unit area. The quantities ΔG, ΔG_s and ΔG_v are represented in Fig 1. The surface excess free energy increases with r^2 and the volume excess free energy decreases with r^3 so the total free energy change increases with increase in size of the nucleus and attains a maximum and then decreases for further increase in the size of nucleus. The size corresponding to the nucleus in which the free energy change is maximum is known as the critical size and can be obtained mathematically by maximizing the equation (3).

i.e., $$\frac{d\Delta G}{dr} = 0$$

Or

$$\frac{d\Delta G}{dr} = 8\pi r\sigma + 4\pi r^2 \, \Delta Gv = 0$$

When r = r* (radius of critical nucleus), simplifying we have

$$r^* = -2\sigma \;/\; \Delta G_V \tag{4}$$

The free energy change associated with the formation of critical nucleus can be estimated by substituting equation (4) in equation (3)

$$\Delta G^* = 16 \pi \sigma^3 \;/\; 3 \Delta G_V^2 \tag{5}$$

$$G^* = 4/3 \pi r^{*2} \sigma$$

$$\Delta G^* = 1/3\, S.\sigma \tag{6}$$

Where 'S' is the surface area of the critical nucleus. Though the present phase is at constant temperature and pressure, there will be variation in the energies of the molecules. The molecules having higher energies temporarily favour the formation of the nucleus. The rate of nucleation can be given by Arhenius reaction which is a velocity equation since the nucleation process is basically a thermally activated process. The nucleation rate J is given by

$$J = A\exp\left\{\frac{-\Delta G^*}{KT}\right\} \qquad (7)$$

Where, 'A' is the pre-exponential constant, 'K' is the Boltzman constant and 'T' is the absolute temperature.

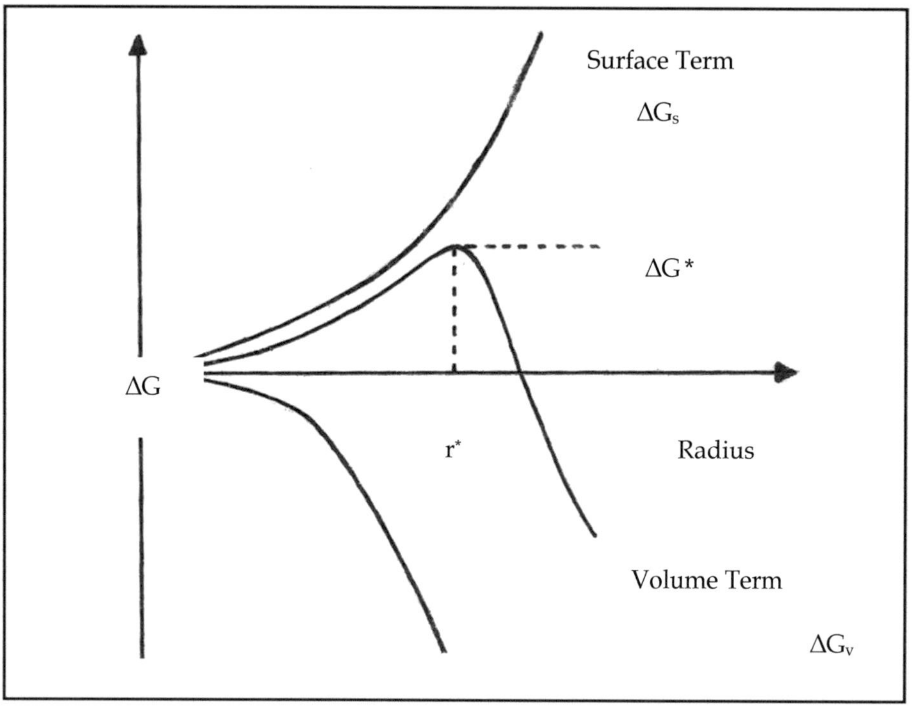

Fig. 1. Free energy diagram.

5. Classification of crystal growth

The growth of the single crystal developed over the years to satisfy the needs of modern technology. The free energy of the growing crystal must be lower than initial stage of the system. It is the common condition for all crystal growth process. The crystal growth method is classified into three types namely growth from melt, from vapour, from the solution. The selection method of crystal growth depending upon the physical properties of material

- Growth from solid ------> solid-solid phase transformation
- Growth from liquid ------> liquid-solid phase transformation
- Growth from vapour ------> vapour-solid phase transformation

We can consider the conversion of the polycrystalline piece of a material into a single crystal by causing the grain boundaries to sweep through and pushed out of the crystal in the solid-solid growth of crystals. The crystal growth from liquid falls into four categories namely,

i. Gel growth,
ii. Flux growth,
iii. Hydrothermal growth and
iv. Low temperature solution growth.

Low temperature solution growth is the most widely practiced next to growth from melt. Crystal growth from solution always occurs under condition in which the solvent and crystallizing substance interact. The expression "solution" is most commonly used to describe the liquid which is the result of dissolving a quantity of given substance in a pure liquid known as solvent. Usually water is used as the solvent rarely other liquid is also used as solvent.

6. Solvent selection

The solution is a homogeneous mixture of a solute in a solvent. Solute is the component present in a smaller quantity. For a given solute, there may be different solvents. Apart from high purity starting materials, solution growth requires a good solvent. The solvent must be chosen taking into account the following factors:

1. A good solubility for the given solution.
2. A positive temperature co-efficient of solubility.
3. A small vapour pressure.
4. Non-corrosiveness.
5. Non-flammability.
6. Less viscosity.
7. Low price in the pure state.

7. Solution preparation and crystal growth

After selecting the desirable solvent with high purity solute to be crystallized, the next important part is preparation of the saturated solution. To prepare a saturated solution, it is necessary to have an accurate solubility-temperature data of the material. The saturated solution at a given temperature is placed in the constant temperature bath. Whatman filter papers are used for solution filtration. The filtered solution is taken in a growth vessel and the vessel is sealed by polythene paper in which 10–15 holes were made for slow evaporation. This solution was transferred to crystal growth vessels and crystallization is allowed to take place by slow evaporation at room temperature or at a higher temperature in a constant temperature bath. As a result of slow evaporation of solvent, the excess of solute which got deposited in the crystal growth vessel results in the formation of seed crystals.

8. Low temperature solution growth methods

Solution growth is the most widely used method for the growth of crystals, when the starting materials are unstable at high temperatures. In general, this method involves seeded growth from a saturated solution. The driving force i.e., the supersaturation is achieved either by temperature lowering or by solvent evaporation. This method is widely used to grow bulk crystals, which have high solubility and have variation in solubility with temperature (James & Kell, 1975).

Low temperature solution growth (LTSG) can be subdivided into the following categories:

i. Slow cooling method
ii. Slow evaporation method
iii. Temperature gradient method

8.1 Slow cooling method

In this process, supersaturated solution is prepared by keeping quantity of the solution same as that of the initial stage and temperature of the solution is reduced in small step. By doing so, solution which is just saturated at initial temperature will become supersaturated solution. Once supersaturation is achieved, growth of single crystal is possible. The main disadvantage of slow cooling method is the need to use a range of temperature. Wide range of temperature may not be desirable because the properties of the grown crystal may vary with temperature. Even though this method has technical difficulty of requiring a programmable temperature control, it is widely used with great success.

8.2 Slow evaporation method

In this process the temperature of the solution is not changed, but the solution is allowed to evaporate slowly. Since the solvent evaporates, concentration of solute increased and therefore supersaturation is achieved. For example 40 g of solute in 100 ml solvent is considered as saturated solution at 50°C. Now the solution is allowed to evaporate at the same temperature. The 100 ml of the solution is reduced to some lower level say 70 ml. Then 40 g in 70 ml at 50°C is supersaturated solution. The evaporation technique has an advantage that the crystals grow at a fixed temperature. But inadequacies of the temperature control system still have a major effect on the growth rate. This method can effectively be used for materials having very low temperature coefficient of solubility.

8.3 Temperature gradient method

This method involves the transport of the materials from hot region containing the source materials to be grown to a cooler region, where the solution is supersaturated and the crystal grows. The advantages of this method are that [a] the crystal is grown at fixed temperature, [b] this method is insensitive to changes in temperature provided both the source and the growing crystal under go the same change. [c] economy of the solvent. On the other hand, small changes in temperature difference between the source and the crystal zones have a large effect on the growth rate.

In general, crystal growth from solution is mainly influenced by super saturation. Super saturation may be achieved by any methods (described above) which are based on the principle that solution which is saturated at a particular temperature will behave as unsaturated at high temperature. The disadvantages are the slow growth rate and in many cases inclusion of the solvent in to the growing crystal. Materials having moderate to high solubility in temperature range, ambient to 100°C at atmospheric pressure can be grown by LTSG method. This method is well suitable for those materials which suffer from decomposition in the melt and which undergo structural transformation while cooling from the melting point. The other advantages of LTSG method are the low working temperature, easy operation and feasible growth condition.

9. Nonlinear optical crystals

Non Linear optics deals with the interaction of intense electromagnetic fields in suitable medium producing magnified fields different from the input field in frequency, phase or amplitude Nonlinear optics is now established as an alternative field to electronics for the future photonic technologies. The fast-growing development in optical fiber communication systems has stimulated the search for new highly nonlinear materials capable of fast and efficient processing of optical signals. Organic nonlinear optical (NLO) materials have been intensely investigated due to their potentially high nonlinearities and rapid response in electro-optic effect compared to inorganic NLO materials. In recent years, there has been considerable interest in the study of organic NLO crystals with good nonlinear properties because of their wide applications in the area of laser technology, optical communication, optical information processing and optical data storage technology (Chenthamarai et al., 2000). Among the organic crystals for NLO applications, amino acids display specific features of interest such as (i) molecular chirality, which secures acentric crystallographic structures, (ii) absence of strongly conjugated bonds, leading to wide transparency ranges in the visible and UV spectral regions, (iii) Zwitterionic nature of the molecule, which favours crystal hardness. Further they can be used as a basis for synthesizing organic compounds and derivatives (Eimerl et al., 1990). In our laboratory, we have grown NLO crystals such as L-Alaninium Succinate (LAS), L-Valinium Succinate (LVS), L-Alaninium Fumarate (LAF), L-Valinium Fumarate (LVF) and reported in the journal of repute (Ramachandra Raja, 2009a, 2009b, 2009c, 2010).

In this chapter, we have discussed the growth of organic nonlinear optical crystal. L-Alaninium Succinate (LAS) and L- Valinium Succinate (LVS) which have been grown by slow evaporation solution growth technique in detail. The characteristic studies such as single crystal and powder X-ray Diffraction (XRD) analysis, UV–Vis–NIR spectrum, FT-IR, nuclear magnetic resonance studies, TGA/DTA studies and SHG are also discussed.

10. Growth and characterization of L- Alaninium Succinate (LAS)

10.1 Crystal growth

LAS have been grown from aqueous solution by slow evaporation. The starting material was synthesized from commercially available L-Alanine (AR grade) and Succinic acid (AR grade), taken in the equimolar ratio 1:1. In deionized water, L-Alanine and Succinic acids were allowed to react by the following reaction to produce LAS.

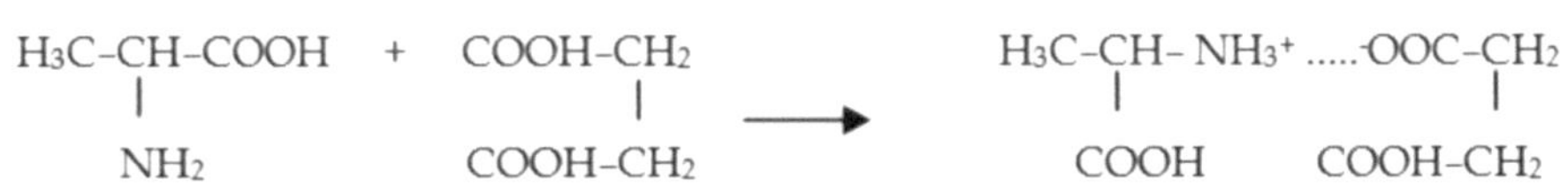

L- Alanine Succinic Acid L-Alaninium Succinate

Calculated amount of the reactants were thoroughly dissolved in deionized water and stirred well for about 3 hours using a magnetic stirrer to obtain a homogenous mixture. Then the solution was allowed to evaporate slowly until the solvent was completely dried. The purity of the synthesized salt was further increased by successive recrystallization process. The synthesized powder of LAS was dissolved thoroughly in double distilled water to form a saturated solution. The solution was then filtered twice to remove any insoluble impurities. Growth was carried out by low-temperature solution growth technique by slow evaporation in a constant temperature bath controlled to an accuracy of ±0.01°C. Crystals begin to grow inside the solution and were removed from the solution after 3 weeks, washed and dried in air.

10.2 Characterization studies

10.2.1 Single crystal XRD analysis

In order to estimate the crystal data, the single crystal XRD analysis of grown LAS crystal have been carried out using ENRAF NONIUS CAD-4 X-ray diffractometer equipped with MoKα (λ = 0.71069 Å) radiation. The X-ray diffraction study on grown crystal reveals that the grown crystal belongs to orthorhombic system with the following unit cell parameters: a = 5.77 Å, b = 6.02 Å, c = 12.32 Å and $\alpha = \beta = \gamma = 90^{\circ}$, the cell volume = 428 $Å^3$. These lattice parameters are tabulated in the Table 1.

10.2.2 Powder XRD analysis

The structural property of the single crystals of LAS has been studied by X-ray powder diffraction technique. Powder X-ray diffraction studies of LAS crystal is carried out, using Rich Seifert diffractometer with CuK_α (λ =1.54060 Å) radiation. The sample is scanned for 2θ values from 10° to 90° at a rate of 2°/min. Figure 2 shows the Powder XRD pattern of the pure LAS crystal. The diffraction pattern of LAS crystal has been indexed by Reitveld index software package. The lattice parameter values of LAS crystal has been calculated by Reitveld unit cell software package and are matched with single crystal XRD data. The comparison of lattice parameters between single crystal and powder XRD is shown in Table 1.

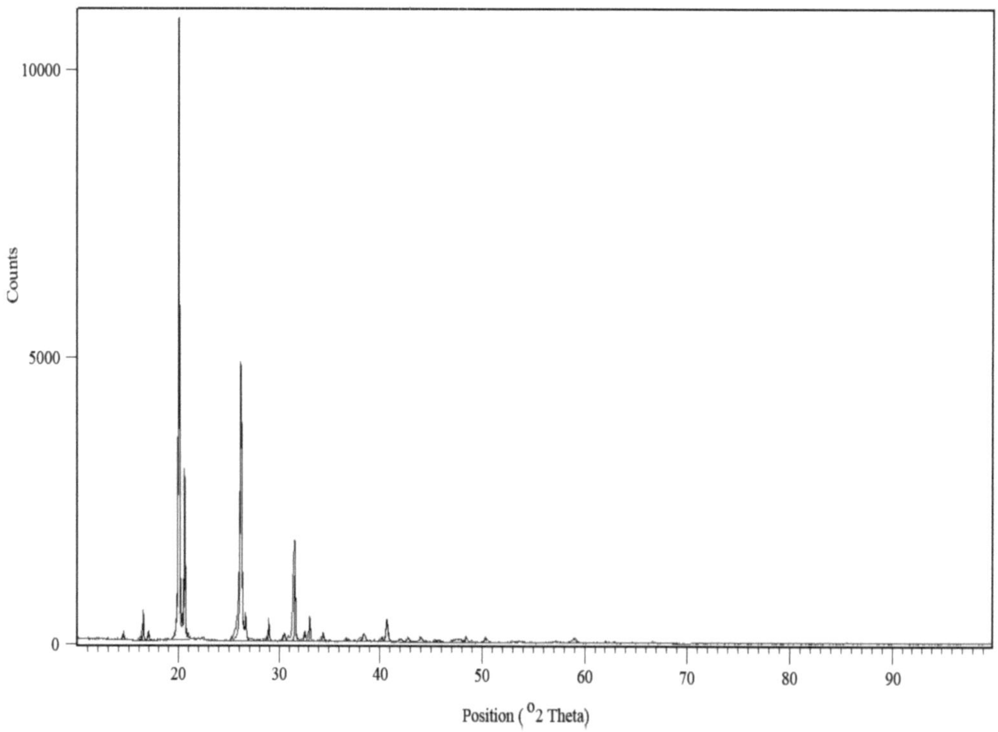

Fig. 2. Powder XRD pattern of LAS crystal.

XRD	a Å	b Å	c Å	α deg	β deg	γ deg	Volume $Å^3$
Single crystal	5.77	6.02	12.32	90	90	90	428
Powder	5.74	5.98	12.53	90	90	90	430

Table 1. The cell parameters of LAS crystal.

Position °2 θ	d- spacing Å	(h k l)
14.4335	6.13689	(0 0 2)
16.4368	5.39317	(0 1 1)
16.9673	5.22574	(1 0 1)
20.0604	4.42642	(0 1 2)
20.6379	4.30384	(1 0 2)
26.1583	3.40675	(0 1 3)
26.6351	3.34684	(1 0 3)
28.9666	3.08254	(0 0 4)
30.5183	2.92926	(1 1 3)
31.5955	2.83181	(2 0 1)
32.5805	2.74841	(0 1 4)
33.0863	2.70753	(0 2 2)
34.4077	2.60652	(2 1 0)
38.4281	2.34257	(2 0 3)
40.6932	2.21726	(1 2 3)
42.0137	2.15057	(0 2 4)
42.789	2.11339	(2 0 4)
43.9448	2.06045	(2 2 1)
45.5981	1.98951	(2 1 4)
47.5815	1.91111	(0 3 2)
48.4142	1.88017	(2 0 5)
50.3422	1.81259	(1 2 5)
58.9467	1.56559	(3 2 2)

Table 2. Powder XRD data of LAS crystal.

It is observed that LAS belongs to orthorhombic system and cell parameters values are in good agreement with the single crystal XRD data. The h, k, l values, d-spacing and 2θ values are tabulated in Table 2.

10.2.3 UV-Vis-NIR analysis

The UV-Vis-NIR transmittance spectrum of grown LAS crystal has been recorded with a Lambda 35 double-beam spectrophotometer in the range 190–1100 nm to find the suitability of LAS crystal for optical applications. The recorded spectrum is shown in Fig. 3. The crystal shows a good transmittance in the visible region which enables it to be a good material for optoelectronic applications. As observed in the spectrum, there is no significant absorption in the entire range tested. A good optical transmittance from ultraviolet to infrared region is very useful for nonlinear optical applications. Most of the nonlinear optical effects are studied using Nd:YAG laser operating at a fundamental wavelength of 1064 nm. Absorption, if any, near the fundamental or the second harmonic at 532 nm, will lead to a loss of conversion efficiency of second harmonic generation (SHG). From the UV-Vis-NIR spectrum, it is clear that the transparency of the grown crystals extends up to UV region.

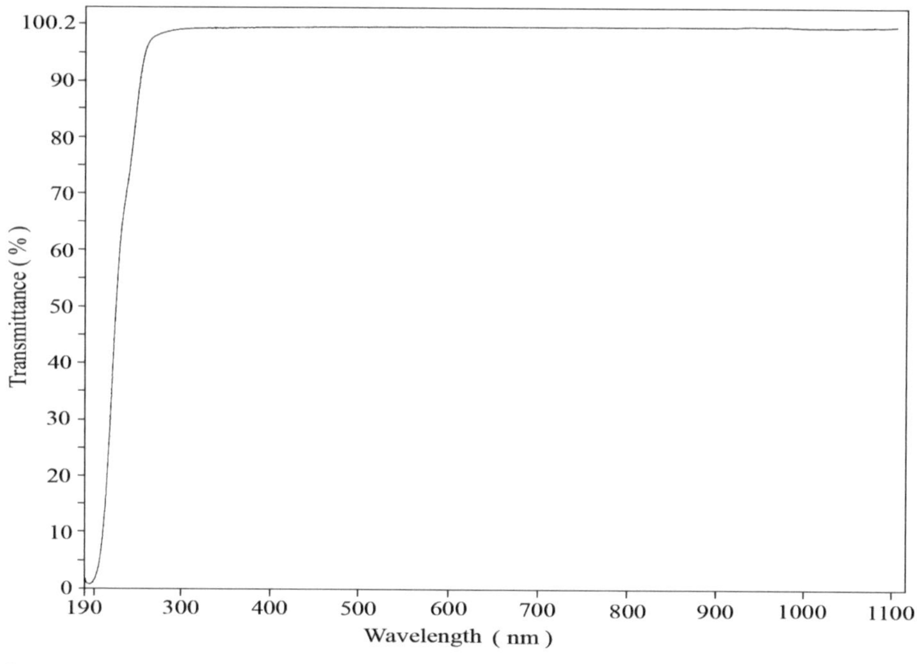

Fig. 3. Transmission spectrum of LAS crystal.

The lower cut-off wavelength is as low at 190 nm. The wide range of transparency suggests that the crystal is a good candidate for nonlinear optical applications (Aravindan et al., 2007). This transmittance window (190–1100 nm) is sufficient for the generation of second harmonic light (λ = 532 nm) from the Nd:YAG laser (λ=1064nm) (Natarajan et al, 2008). The lower cut-off near 190 nm combined with the very good transparency, makes the usefulness of this material for optoelectronic and nonlinear optical applications.

10.2.4 FT-IR analysis

The infrared spectrum of LAS has been carried out to analyse the chemical bonding and molecular structure of the compound. The FT-IR spectrum of the crystal has recorded in the frequency region from 400 cm^{-1} to 4000 cm^{-1} with Perkin–Elmer FT-IR spectrometer model SPECTRUMRX1 using KBr pellets containing LAS powder obtained from the grown single crystals. The observed FT–IR spectrum of LAS is as shown in Fig. 4.

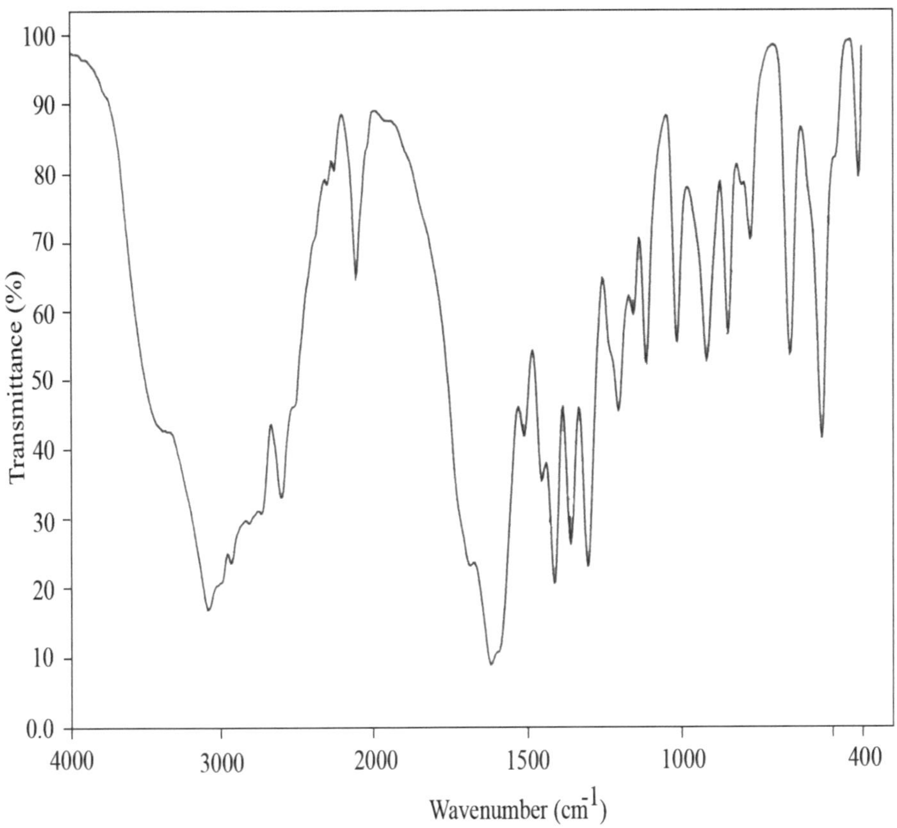

Fig. 4. FT-IR spectrum of LAS crystal.

The characteristics vibration of LAS has been compared with L-alaninium alanine nitrate (Aravindan et al., 2007) and L-alanine cadmium chloride (Dhanuskodi et al., 2007) as shown in Table 3. The asymmetric stretching vibration of NH_3^+ is observed at 3086 cm^{-1} of LAS is confirming the presence of NH_3^+ in compound. The NH_3^+ absorption range of amino acids (3130–3100 cm^{-1}) is shifted to lower wave number, due to formation of amino salts, and in LAS, it is observed at 3086cm^{-1}.

Amino group absorption bands are noted at 2604 cm^{-1} (symmetric stretching), 1620cm^{-1} (bending), and 1111 cm^{-1} (rocking). These bands are due to NH_3^+ ions. During the formation of amino salts, the NH_2 group in amino acids is converted in to NH_3^+ ion. The strong absorption at 1413 cm^{-1} indicates the symmetric stretching vibration frequency of carbonyl group. The bending and rocking vibrations of COO− are observed at 772 cm^{-1} and 539 cm^{-1}, respectively. CH_2 wagging (1304cm^{-1}) and CH_3 stretching (1204 cm^{-1}) vibrations are also observed (Ramachandran & Natarajan, 2007).

Wavenumber (cm^{-1})	Assignment
3086	NH_3^+ asymmetric stretching
2604	NH_3^+ symmetric stretching
1620	NH_3^+ bending
1453	CH_3 bending
1413	COO^- symmetric stretching
1360	CH_3 symmetric bending
1304	CH_2 wagging
1204	CH_3 symmetric stretching
1111	NH_3^+ rocking
1012	CH_3 rocking
917	CCN symmetric stretching
848	C-CH_3 bending
772	COO^- bending
539	COO^- rocking
412	COO^- rocking

Table 3. Assignments of FT-IR bands observed for LAS crystal.

10.2.5 NMR studies

The ^{1}H- and ^{13}C-NMR spectra of LAS have been recorded using D_2O as solvent on a Bruker 300MHz (Ultrashield) TM instrument at 23°C (300 MHz for ^{1}H-NMR and 75 MHz for ^{13}C-NMR) to confirm the molecular structure. The spectra are shown in Figures 5 and 6 respectively and the chemical shifts are tabulated with the assignments in Table 4.

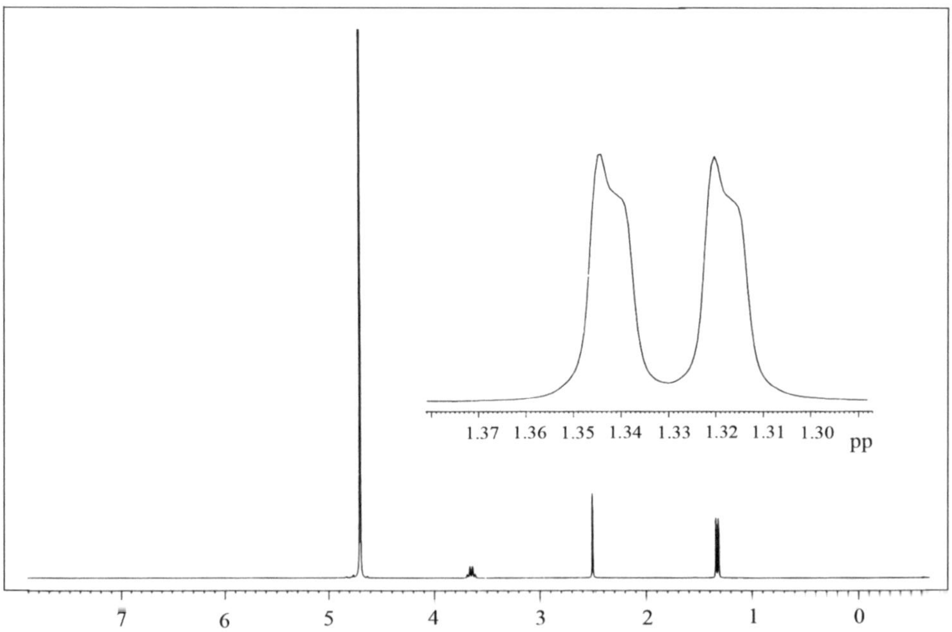

Fig. 5. ^{1}H-NMR spectrum of LAS crystal.

The resonance peaks at δ = 1.33 ppm of ^{1}H-NMR spectrum is due to the CH_3 group and peaks observed at δ = 3.65 ppm is due to the CH group of L-Alanine. The methyl proton signal at δ = 1.33 ppm is split into a proton doublet due to the coupling of the neighboring proton (CH) and the signal at δ = 3.65 ppm is split into a proton quartet due to the coupling of three neighboring protons (CH_3). The resonance peak observed as a singlet at δ = 2.57 ppm exhibits the presence of methylene (CH_2) proton of succinic acid.

The signal at δ = 4.69 ppm is due to the solvent (D_2O). The signals due to NH and COOH do not show up because of fast deuterium exchange reactions takes place in these two groups, with D_2O being used as solvent (Bruice, 2002). Because of the presence of the methylene (CH_2) groups of LAS, electron contributions towards the rest of the compound get enhanced, so that the protons are more protected in LAS. Such property is not noticed in L-alaninium oxalate (LAO), due to the absence of methylene groups so that the proton groups in LAS absorbs signals at the values lesser than the value of LAO (Dhanuskodi & Vasantha, 2004). The ^{13}C NMR spectrum of LAS contains five signals. The resonance peaks observed at δ = 16.00 ppm and at δ = 50.33 ppm are due to the carbon environments of CH_3 and CH groups of L-alanine respectively. The signal at δ = 29.06 ppm is due to the presence of two methylene (CH_2) groups of succinic acid. The resonance signal observed at δ = 175.52 ppm is due to the free carboxylic acid from L -alanine. In solution, the two carboxyl groups of succinic acid are equivalent due to the fast exchange of H^+ between them and give rise to a single signal at δ = 177.41 ppm.

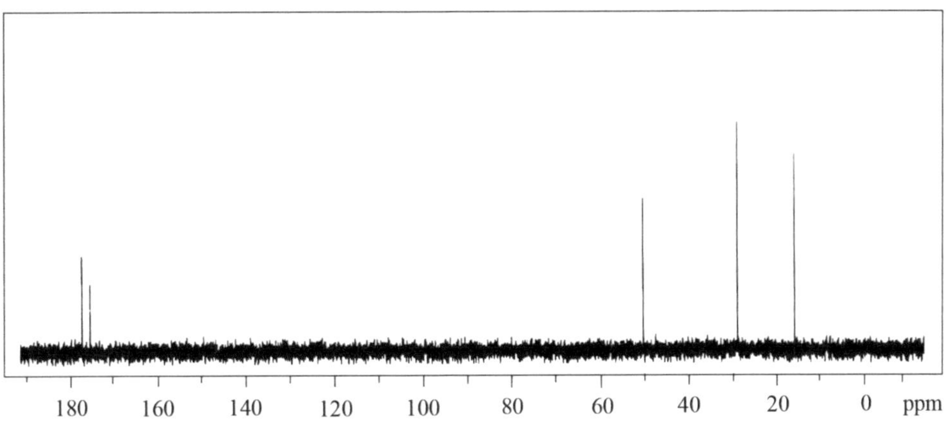

Fig. 6. ^{13}C-NMR spectrum of LAS crystal.

Spectra	Chemical Shift (ppm)	Group identification
^{1}H NMR	1.33	- CH_3-
	2.57	- CH_2 -
	3.65	- CH -
	4.69	- D_2O
^{13}C-NMR	16.00	- CH_3
	29.06	- CH_2 -
	50.33	- CH -
	175.52	COOH of L-Alanine
	177.41	COOH of Succinic Acid

Table 4. The chemical shifts in ^{1}H-NMR and ^{13}C-NMR spectra of LAS.

10.2.6 Thermal analysis

The Thermo Gravimetric Analysis (TGA), Differential Thermal Analysis (DTA) spectra of grown LAS crystal have been obtained using the instrument NETSZCH SDT Q 600 V8.3 Build 101. The TGA and DTA have been carried out in nitrogen atmosphere at a heating rate of 20°C/min from 0°C to 1000°C. The TGA curve is presented in Fig. 7.
The initial mass of the materials to analysis was 2.5720 mg and the final mass left out after the experiment was only 1.729 % of initial mass. The TGA trace shows that the material exhibit very small weight loss of about 1.17 % in the temperature up to 155°C due to loss of water. TGA curve shows that there is the weight loss (85%) between 178°C and 274°C indicating that the decomposition of LAS crystals. From the Fig. 7, the appearance of endothermic in the DTA at 178°C corresponds to TGA results. From the TGA, DTA analyses, it is clearly understood that the LAS is thermally stable upto 178°C.

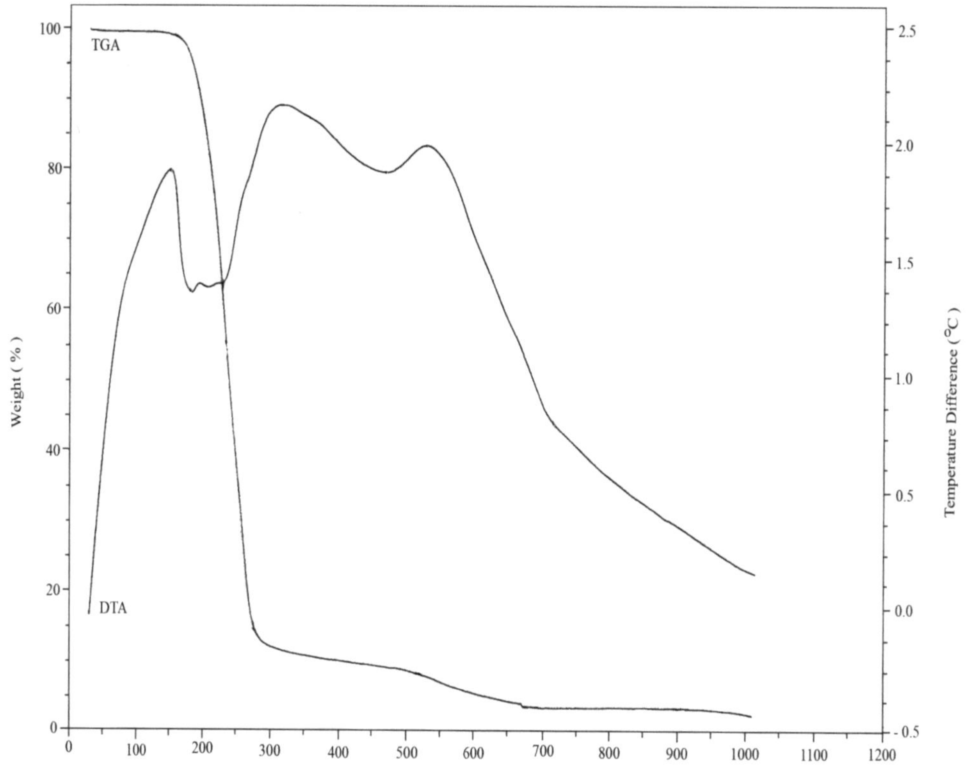

Fig. 7. TGA / DTA curve of LAS crystal.

10.2.7 Second harmonic generation analysis

A preliminary study of the powder SHG measurement of LAS has been performed using Kurtz powder technique (Kurtz & Perry, 1968) with 1064nm laser radiations. An Nd:YAG laser producing pulses with a width of 8 ns and a repetition rate of 10 Hz was used. The crystal was crushed into powder and densely packed in a capillary tube. It is observed that the crystal converts the 1064 nm radiation into green (532 nm) while passing the Nd:YAG laser output into the sample which confirms the SHG. The observed intensity of output light is obtained as 12 mV and for the same incident radiation, the output of KDP is observed as 52 mV. It was found that the efficiency of SHG is 23% when compared with that of the standard KDP (Ramachandra Raja , 2009c).

11. Growth and characterization of L- Valinium Succinate (LVS)

11.1 Experimental procedure

The nonlinear optical crystal L-Valinium Succinate (LVS) were grown by slow evaporation solution growth method. The LVS was synthesized from analar grade L-Valine and Succinic acid which were taken in equimolar ratio 1:1 using following reaction:

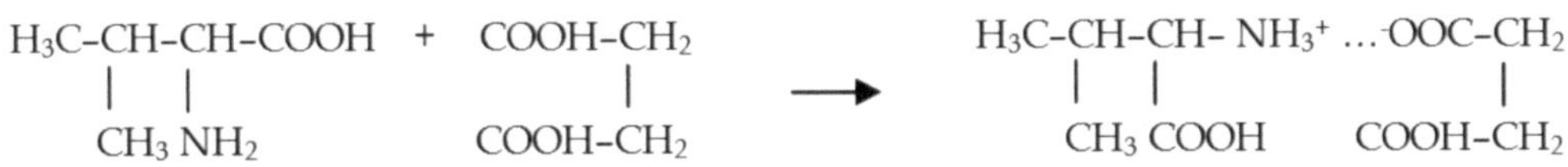

L- Valine Succinic Acid L-Valinium Succinate

The calculated amounts of reactants were thoroughly dissolved in double distilled water and stirred continuously using magnetic stirrer. The saturated solution may contain impurities such as solid and dust particles and therefore it was filtered using filter paper. Then the filtered solution was covered by polythene paper in which 10 to 15 holes were made for slow evaporation. This solution was transferred to crystal growth vessels and crystallization was allowed to take place by slow evaporation at a temperature range of 35°C in a constant temperature bath of accuracy ±0.01^{o}C. As a result of slow evaporation of water, the excess of solute has grown as LVS crystals in the period of two weeks.

11.2 Characterization studies

11.2.1 Single crystal XRD analysis

The X-Ray diffraction pattern of LVS crystals have been studied by ENRAF NONIUS CAD4 single crystal X-Ray diffractometer with MoKα radiation (λ=0.71069 Å). The single crystal X-ray diffraction study of crystals is used to identify the cell parameters. It is observed that the LVS crystal belongs to orthorhombic system with following cell parameters: a = 9.85 Å, b = 5.35 Å, c = 12.26 Å and α = β = γ = 90 ^{o}, the cell volume = 646 Å^3. From the lattice parameters it is clear that for grown crystal a ≠ b ≠ c and α = β = γ = 90 ^{o} and they are compared with powder XRD data and tabulated in Table 5.

XRD	a Å	b Å	c Å	α deg	β deg	γ deg	Volume Å^3
Single crystal	9.85	5.35	12.26	90	90	90	646
Powder	9.99	5.36	12.19	90	90	90	652

Table 5. The cell parameters of LVS crystal.

11.2.2 Powder XRD analysis

The powder X-ray diffraction (XRD) pattern of LVS crystals has been obtained using Rich Seifert X-ray diffractometer. The crushed powder sample was subjected to intense X-rays of wavelength 1.54060 Å (CuK$_α$) at a scan speed of 1°/minute. The powder X-ray pattern of LVS is shown in Fig. 8. The observed powder XRD pattern has been indexed by Rietveld Index software package. The lattice parameters have been calculated by Rietveld Unit Cell software package and they are shown in Table 5. It is observed that LVS belongs to orthorhombic system and cell parameters values are agreed with the single crystal XRD data. The h, k, l values, d-spacing and 2θ values are tabulated in Table 6.

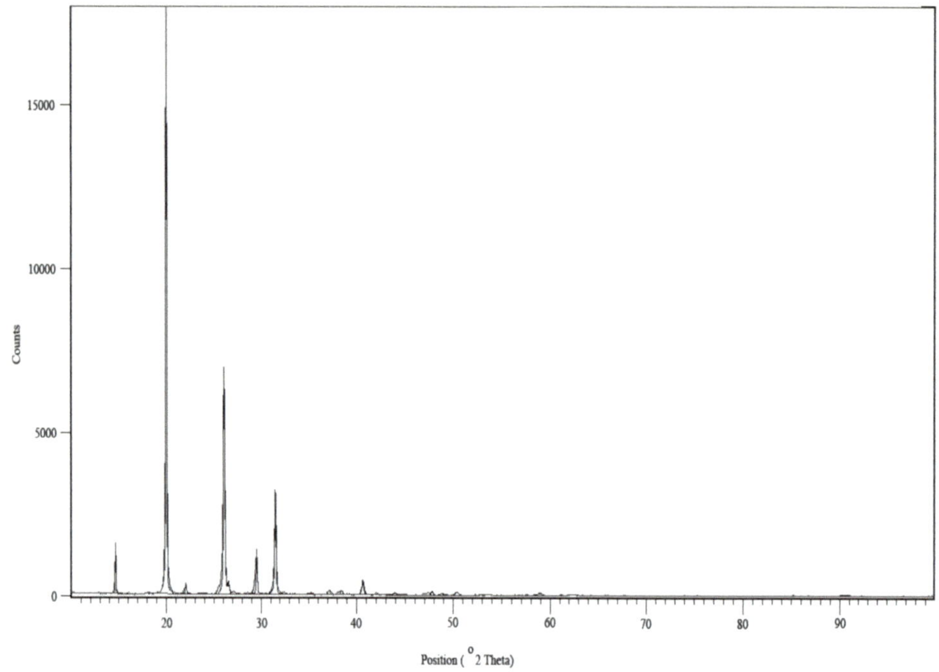

Fig. 8. Powder XRD pattern of LVS crystal.

Position°2 θ	d- spacing Å	(h k l)
14.6613	6.04207	(0 0 2)
20.0051	4.43853	(0 2 1)
22.0352	4.03399	(1 0 2)
23.6856	3.75650	(0 1 3)
26.1127	3.41259	(1 2 1)
29.5373	3.02427	(0 0 4)
31.4841	2.84157	(0 3 2)
35.0908	2.55733	(0 3 3)
37.1320	2.42131	(2 0 2)
38.3887	2.34489	(2 2 0)
40.6179	2.22120	(1 0 5)
43.8867	2.06304	(1 3 4)
47.7577	1.90447	(1 0 6)
48.8085	1.86590	(2 2 4)
50.3581	1.81206	(1 4 4)
58.9110	1.56775	(3 3 0)
90.4585	1.08503	(2 2 10)

Table 6. Powder XRD data of LVS crystal.

11.2.3 UV-Vis-NIR analysis

To find the optical transmission range of LVS crystals, the UV-Vis-NIR spectrum has been recorded using Lambda 35 double beam spectrophotometer in the range between 190 nm and 1100 nm and it is shown in Fig. 9. When the transmittance is monitored from longer to shorter wavelengths, LVS is transparent from 190 nm to 1100 nm. Optical absorption with lower cut-off wavelength near 190 nm makes the crystal suitable for UV tunable laser and SHG device applications.

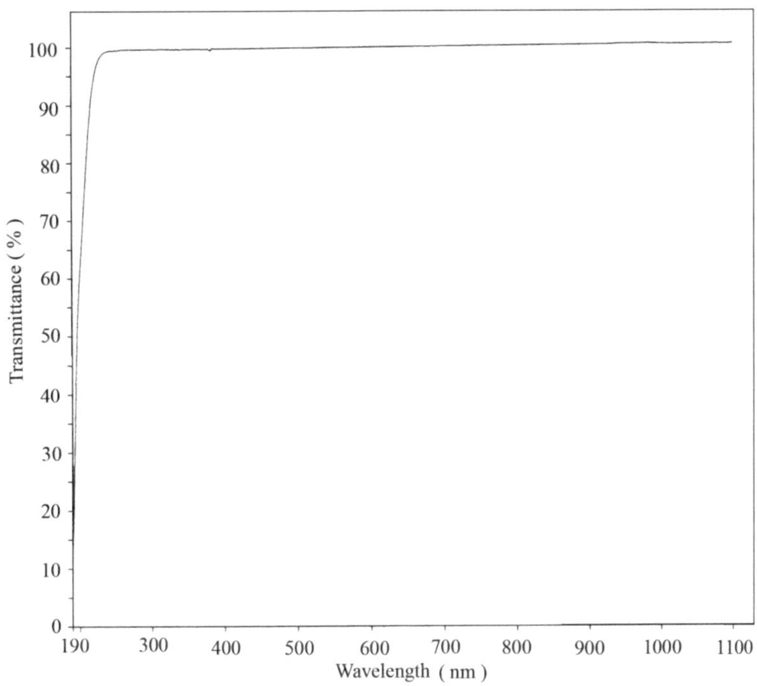

Fig. 9. UV-Vis-NIR spectrum of LVS crystal.

11.2.4 FT-IR analysis

The Fourier transform infrared spectrum of LVS have been recorded in between the region 400 - 4000 cm^{-1} using Perkin Elmer Fourier transforms infrared spectrometer (model SPECTRUM RX1) with the help of KBr pellets as shown in Fig. 10. The presence of functional groups was identified and they are stacked in Table 7. The presence of NH_3^+ group in LVS has confirmed by peaks at 3429 cm^{-1} and 3156 cm^{-1}. It is due to protonation of NH_2 group by the COOH group of succinic acids (Nakamo, 1978; Sajan et al., 2004). The symmetric and asymmetric bending of NH_3^+ was obtained at 1587 and 1508 cm^{-1} respectively. The strong absorption at 1393 cm^{-1} indicates that the symmetric bending of CH_2. The CH_2 wagging vibration produces a sharp peak at 1327 cm^{-1}. The C-CH bending vibration produced its characteristic peak at 1270 cm^{-1}. The rocking vibration at 1177 cm^{-1} establishes the presence NH_3^+ group. The peak at around 1137 cm^{-1} is assigned to NH_3^+ wagging. The stretching vibration of C-O-C, C-C-N and C-C are positioned at 2108 cm^{-1}, 1063 cm^{-1} and 1029 cm^{-1} respectively. Meanwhile, for the peaks at 945 cm^{-1} is due to CH_2

rocking. The bending vibration of COO^- is observed at 662 cm^{-1}. The bending and rocking vibration of COO^- are observed at 714 cm^{-1} and 430 cm^{-1} respectively.

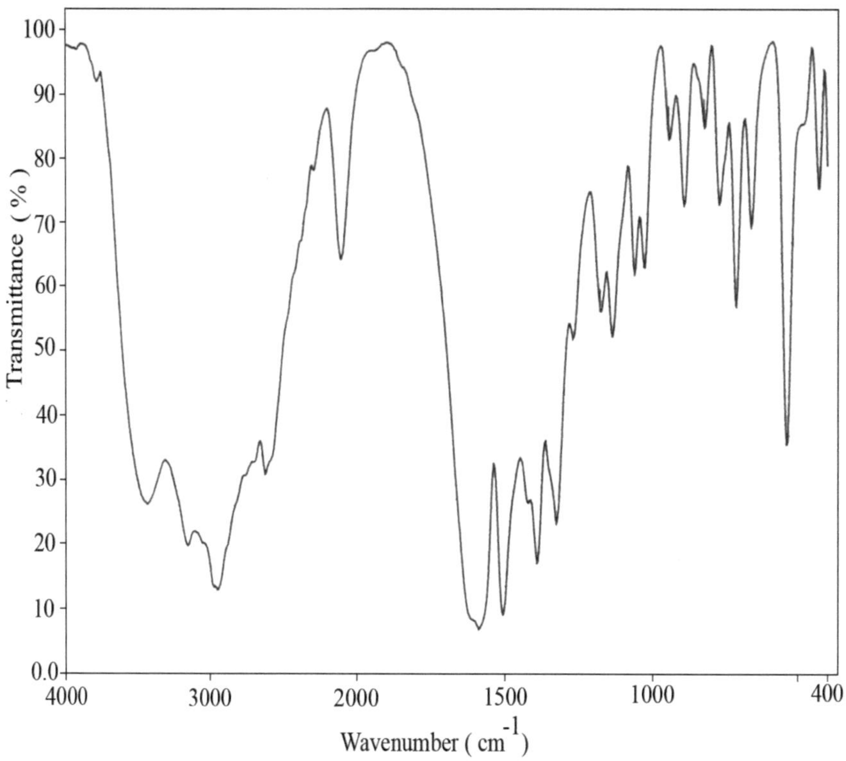

Fig. 10. FT-IR spectrum of LVS crystal.

11.2.5 NMR studies

The ^1H- and ^{13}C-NMR spectra of LVS have been recorded for the crystals dissolved in water (D_2O) using BRUKER 300 MHz (Ultrashield)™ instrument at 23°C (300 MHz for ^1H-NMR and 75 MHz for ^{13}C-NMR) for the confirmation of molecular structure. The ^1H - NMR spectrum of LVS is shown in the Fig. 11. The deuterium exchange proton NMR spectrum of LVS crystal is found to contain resonance signals integrated for a total of 12 protons. From the spectrum, it is observed that the two methyl proton signal is split into two doublets due to the coupling of neighbouring (-CH) proton which is confirmed from the signal at δ = 0.84 ppm, δ = 0.89 ppm respectively. The –CH group signal is split into a multiplet due to the hyperfine splitting of neighbouring three (-CH_3) protons is confirmed from the signal of LVS crystal centered at = δ 2.13 ppm. The doublet signal observed at δ 3.48 ppm is attributed to a (-CH) proton next to carboxylic acid. There is one peak found at δ = 2.51 ppm due to the -CH_2- group of succinic acid. The signal at δ = 4.69 ppm is due to the solvent D_2O. The signals due to N-H and COOH do not show up because of fast deuterium exchanges which took place in those two groups, where the D_2O was used as the solvent (Bruice 2002). The chemical shift values of LVS with assignments are tabulated in Table 8.

Wavenumber (cm^{-1})	Assignment
3429	NH_3^+ symmetric stretching
3156	NH_3^+ asymmetric stretching
2946	CH_2 asymmetric stretching
2626	NH_3^+ symmetric stretching
2108	C-O-C stretching
1587	NH_3^+ symmetric bending
1508	NH_3^+ asymmetric bending
1393	CH_2 symmetric bending
1327	CH_2 wagging
1270	C-CH bending
1177	NH_3^+ rocking
1137	NH_3^+ wagging
1063	C-C-N stretching
1029	C-C stretching
945	CH_2 rocking
893	C-C-N stretching
823	COO^- rocking
773	NH wagging
714	COO^- bending
662	COO^- bending
541	C-C=O wagging
430	COO^- rocking

Table 7. Assignments of FT-IR bands observed for LVS crystal.

The ^{13}C-NMR spectrum is shown in Fig. 12. The characteristic absorption peaks of ^{13}C-NMR spectrum of LVS are explained as follows. The signals at $\delta = 17.82$ ppm and $\delta = 16.55$ ppm are attributed to the two methyl group of LVS. An intense signal is observed at $\delta = 28.93$ ppm is due to presence of two methylene groups of succinic acid. The signal of (CH) at $\delta = 29.00$ ppm is integrated for one carbon due to presence of carboxylic acid isopropyl carbon. The peaks at $\delta = 60.12$ ppm is due to tertiary carbon connected to amino group. The peak at $\delta = 173.97$ ppm and $\delta = 177.28$ ppm are due to deuterium exchange of carbon in carbonyl group. A peak with higher intensity at $\delta = 177.28$ ppm can be safely attributed to carbonyl carbons of two COOH groups of succinic acid present in the same chemical environment.

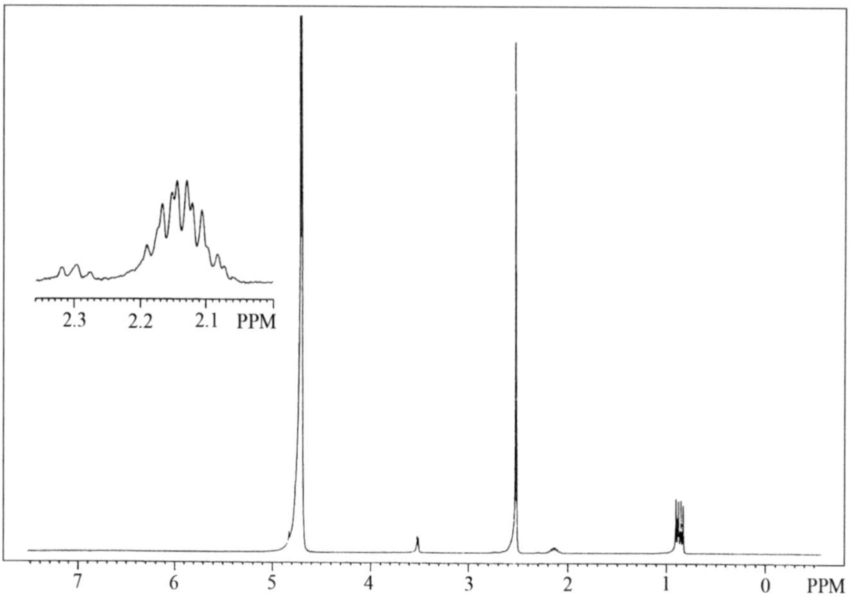

Fig. 11. ^{1}H-NMR spectrum of LVS crystal.

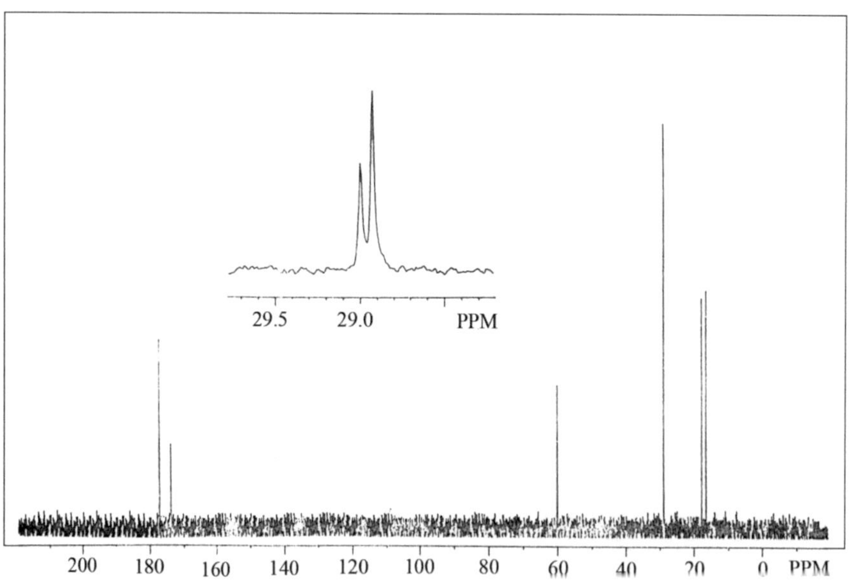

Fig. 12. ^{13}C-NMR spectrum of LVS crystal.

Spectra	Chemical Shift (ppm)	Group identification
^{1}H NMR	0.84 & 0.89	- (CH_3) -
	2.13	- CH -
	2.51	- CH_2 -
	3.48	- CH -
	4.69	D_2O
^{13}C-NMR	16.55 & 17.82	- (CH_3) -
	29.00	- CH –
	28.93	- CH_2 -
	60.12	- CH -
	173.97	COOH of L-Valine
	177.28	COOH of Succinic Acid

Table 8. The chemical shifts in ^{1}H-NMR and ^{13}C-NMR spectra of LVS.

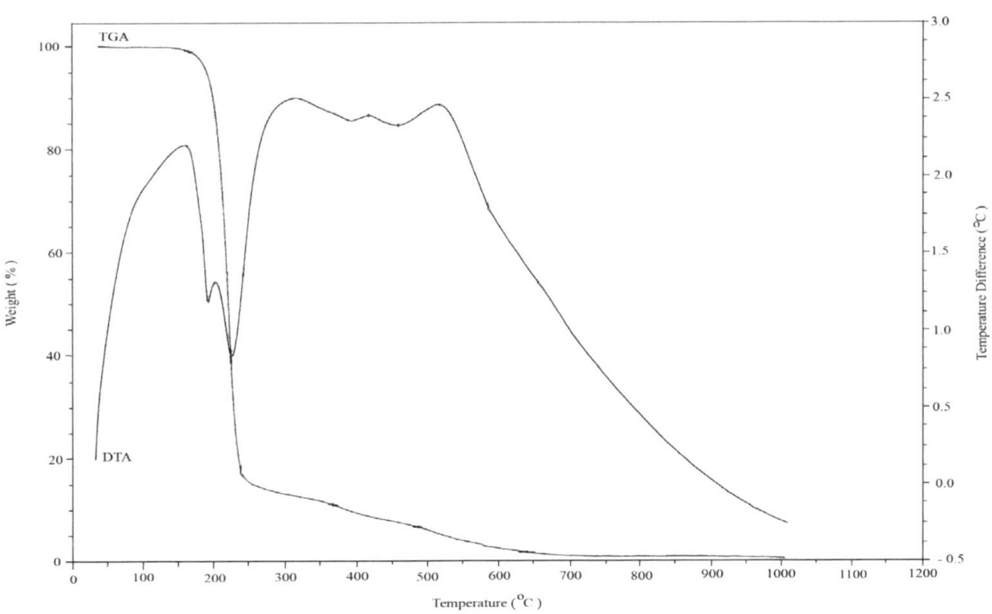

Fig. 13. TGA / DTA curve of LVS crystal.

11.2.6 Thermal analysis

The Thermo Gravimetric Analysis (TGA), Differential Thermal Analysis (DTA) spectra of grown LVS crystal have been obtained using the instrument NETSZCH SDT Q 600 V8.3 Build 101. The TGA and DTA have been carried out in nitrogen atmosphere at a heating rate of 20°C/min from 0°C to 1000°C. The TGA curve is presented in Fig. 13. The initial mass of the materials to analysis was 3.0160 mg and the final mass left out after the experiment was only 0.8631 % of initial mass.

The TGA trace shows that the material exhibit very small weight loss of about 1.04 % in the temperature up to 160°C due to loss of water. TGA curve shows that there is a weight loss of about 92 % between 160°C to 500°C indicating that the decomposition of LVS crystals. From the Fig. 13, the appearance of sharp endothermic in the DTA at 222°C corresponds to TGA results. From the TGA, DTA analyses, it is clearly understood that the LVS is thermally stable upto 222°C.

11.2.7 Second harmonic generation analysis

The nonlinear optical susceptibility of grown LVS crystals have been measured through second harmonic generation using standard Kurtz and Perry method (Kurtz & Perry, 1968). The output of laser beam having the bright green emission of wavelength 532 nm confirms the second harmonic generation output. The observed intensity of output light is 31 mV. For the same incident radiation, the output of KDP was observed as 55 mV. The second harmonic efficiency of LVS is 0.56 times that of KDP.

12. Conclusions

Thus the chapter fully discussed about solution crystal growth methods and nucleation. Then the growth and characterization of Single crystals of L- Alaninium Succinate (LAS) and L-Valinium Succinate (LVS) have been grown by slow evaporation method from saturated solution are also discussed. From X-Ray diffraction, it is observed that LAS and LVS crystal belongs to orthorhombic system. The UV-Vis-NIR spectral studies confirm that the grown crystals have wider transparency range in the visible and UV spectral regions and both LAS and LVS crystals have lower cut-off at 190 nm. The good transparency shows that LAS and LVS crystal can be used for nonlinear optical applications. The modes of vibration of the molecules and the presence of functional groups have been identified using FT-IR technique. The chemical structure of the grown crystals is established by ^{1}H and ^{13}C NMR techniques. From Thermal analysis, the melting point of LAS and LVS are identified is 178 °C and 222°C respectively. The SHG output proves that LAS and LVS crystals can be used as nonlinear optical materials.

13. Acknowledgment

The authors thank Dr. P.K. Das, Indian Institute of Science, Bangalore for the measurement of powder SHG efficiency. The authors are thankful to St. Joseph's College, Trichy, India, and SASTRA University, Thanjavur, India and Central Electro Chemical Research Institute (CECRI), Karaikudi, India for spectral facilities. The authors also express their gratitude to, Indian Institute of Technology, Chennai, India for XRD facilities

14. References

Aravindan, A., Srinivasan, P., Vijayan, N., Gopalakrishnan, R., & Ramasamy P. (2007). Investigations on the growth, optical behaviour and factor group of an NLO crystal: L-alanine alaninium nitrate. *Cryst. Res. Technol.,* Vol. 42, No.11, (November 2007), pp. 1097–1103, ISSN 0232-1300

Brice, J.C. (January 1987). *Crystal Growth Processes,* John Wiley and Sons, ISBN 0-470-20268-8, New York.

Bruice P.Y. (2002). *Organic Chemistry,* Pearson Education Pvt. Ltd, Singapore.

Buckley H.E. (4 May 1951). *Crystal Growth,* Wiley, New York.

Chenthamarai, S., Jayaraman, D., Ushasree, P.M., Meera, K., Subramanian, C., & Ramasamy P. (2000). Experimental determination of induction period and interfacial energies of pure and nitro doped 4-hydroxyacetophenone single crystals. *Mater. Chem. Phys.,* Vol. 64, No. 3, (May 2000), pp. 179–183, ISSN 0254-0584

Dhanuskodi, S., & Vasantha, K. (2004). Structural, thermal and optical characterizations of a NLO material: L-alaninium oxalate. *Cryst. Res. Technol.,*Vol. 39, No. 3, (March 2004), pp. 259–265, ISSN 0232-1300

Dhanuskodi, S., Vasantha, K., & Angeli Mary, P.A. (2007). Structural and thermal characterization of a semiorganic NLO material: l-alanine cadmium chloride. *Spectrochim. Acta A* , Vol. 66, No. 3, (March 2007) pp. 637-642, ISSN 1386-1425

Eimerl, D., Velsko, S., Davis, L., & Wang, F. (1990). Progress in nonlinear optical materials for high power lasers. *Prog. Cryst. Growth Charact.,* Vol. 20, No. 1, pp. 59-113, ISSN 0960-8974

Kurtz, S.K., & Perry, T.T. (1968). A Powder Technique for the Evaluation of Nonlinear Optical Materials. *J. Appl. Phys.,* Vol. 39, pp.3798 – 3813, ISSN 0021-8979

Milton , B., Boaz, Leyo Rajesh, A., Xavier Jesu Raja, S., & Jeromedas S. (2004). Growth and characterization of a new nonlinear optical semiorganic lithium paranitrophenolate trihydrate (NO_2–C_6H_4–OLi ·3H_2O) single crystal. *J. Cryst. Growth,* Vol. 262, pp. 531-535, ISSN 0022-0248

Natarajan, S., Shanmugam, G., and Martin Britto Dhas, S.A. (2008), Growth and characterization of a new semi organic NLO material: L-tyrosine hydrochloride', *Cryst. Res. Technol.,* Vol. 43, pp. 561–564.

Ramachandra Raja, C. & Antony Joseph, A. (2009) Crystal growth and characterization of new non linear optical single crystals of L- alaninium fumarate. *Materials Letters,* Vol. 63, No. 28 (November 2009) 2507- 2509, ISSN 0167-577X

Ramachandra Raja, C. & Antony Joseph, A. (2009). Crystal growth and comparative studies of XRD, spectral studies on new NLO crystals: L- valine and L- valininium succinate. *Spectrochim. Acta. A,* Vol. 74, No. 3 (October 2009), pp. 825-828, ISSN1386-1425

Ramachandra Raja, C., Gokila, G. & Antony Joseph, A. Growth and spectroscopic characterization of a new organic nonlinear optical crystal: L-alaninium succinate. *Spectrochim. Acta. A* Vol. 72, No. 4, (May 2009) pp. 753-756, ISSN1386-1425

Ramachandra Raja, C, & A. Antony Joseph (2010). Synthesis, spectral and thermal studies of new nonlinear optical crystal: L-valinium fumarate. *Materials Letters* Vol. 64, No.2. (January 2010), pp. 108-110, ISSN 0167-577X

Part 4

Theory of Crystal Growth

24

Simulation of $CaCO_3$ Crystal Growth in Multiphase Reaction

Pawel Gierycz
Faculty of Chemical and Process Engineering, Warsaw University of Technology, Institute of Physical Chemistry, Polish Academy of Sciences, Warsaw, Poland

1. Introduction

Calcium carbonate formation and aggregation processes have been studied from many years and there are widely described in the literature (i.e. Kitano et al., 1962, Montes-Hemandez et al., 2007, Reddy & Nancollas, 1976). However, the mechanism of the process, which depends on the way of reaction conducting (i.e. Dindore et al., 2005, Feng et al., 2007, Jung et al., 2005, Schlomach et al., 2006) is till now not fully understood and investigated due to increasing application of $CaCO_3$ in commercial production of new materials, pharmaceuticals and many others.

Calcium carbonate occurs in nature in three polymorphic modifications: rhombohedral calcite, orthorhombic aragonite usually with needle-like morphology and hexagonal vaterite with spherical morphology. The most needed from the practical point of view is the most stable thermodynamically calcite. One of its important applications area is connected with fabrication of functional solids where the fully controlled precipitation process must be applied. The big interest in this field is due to the fact that application of produced materials is determined by many strictly defined parameters.

In recent years many researchers deal with application of organic additives (Bandyopadhyaya, 2001) as a template to produce inorganic materials and conduction of reaction in a macro- or microemulsions or in sol-gel matrixes. Such methods give an opportunity to control of precipitation process or to modify product properties but unsolved problem remains purity of the obtained powder. There are also many ways of $CaCO_3$ precipitation conducting without any additives (i.e. Cafiero et al., 2002, Sohnel & Mullin, 1982, Rigopoulos & Jones, 2003a). Although a lot of investigations described in the literature (i.e. Chakraborty & Bhatia, 1995, Chen et al., 2000, Cheng et al., 2004) there are still many questions about the full mechanism of crystals nucleation and growth of freshly precipitated particles.

Generally, crystallization is a particle formation process by which molecules in solution or vapor are transformed into a solid phase of regular lattice structure, which is reflected on the external faces. Crystallization may be further described as a self-assembly molecular building process. So, crystallographic and molecular factors are thus very important in affecting the shape, purity and structure of crystals (Colfen & Antonietti, 2005, Collier &

Hounslow, 1999, Mullin, 2001). There are two established mechanisms of crystals growth described in literature (Jones et al., 2005, Judat & Kind, 2004, Spanos & Koutsoukos, 1998). The Ostwald ripening involves the larger crystals formation from smaller crystals which have higher solubility than larger ones (the smaller crystals act as fuel for the growth of bigger crystals). Another important growth mechanism revealed in recent years is nonclassical crystallization mechanism by aggregation, i.e. coalescence of initially stabilized nanocrystals which grow together and form one bigger particle (Judat & Kind, 2004, Myerson, 1999).
There are some papers dealing with calcium carbonate formation through oriented aggregation of nanocrystals (Collier & Hounslow, 1999, Myerson, 1999, Wang et al., 2006) or through self assembled aggregation of nanometric crystallites followed by a fast recrystalization process (Judat & Kind, 2004). This way of particles formation control to fabricate ordered structures is inspired by processes observed in biological systems and is one of top topics of modern colloid and materials chemistry (Judat & Kind, 2004, Myerson, 1999, Wang et al., 2006).
Each way of reaction conduction needs its own modeling. In the literature there are many different models and simulations (i.e. Bandyopadhyaya et al., 2001, Hostomsky & Jones, 1991, Malkaj et al., 2004,) done for different particular reactions.
Each model describing the crystal formation (i.e. Quigley & Roger, 2008, Tobias & Klein, 1996, Wachi & Jones, 1991) has to take into account both particulate crystal characteristics and fluid-particle transport processes. The crystals formation and further solid–liquid separation of particulate crystals from solution involves suspension and sedimentation. During these processes solid matter may change phase from liquid to solid or vice versa. New particles may be generated and existing ones can be lost. Thus, both the liquid and solid phases are subject to the physical laws of change: conservation of mass and flow. The crystals may be also separated from fluids by flow through reactor. So, any model well-describing the particular crystallization process has to take into account the conditions of reaction leading in the reactor.

2. Modelling and simulation

The behaviour of real crystallization processes is determined by the interaction of multiple process phenomena, which all have to be modelled to fully describe the process. Over the past decades simulation has become a standard tool for solution of these model equations. Different tools have been developed to solve many typical chemical engineering problems particularly for standard fluid phase processes. Also for more complex processes, as population balance models for crystallization processes (Randolph & Larson, 1988) and computational fluid dynamics (Ferziger & Perić, 1996) problems based on the Navier–Stokes equations, commercial simulation tools are available.
However, still not every kind of crystallization process models can be solved with the available tools. Moreover, every particular crystallization process needs a specific treatment taking into account process parameters, hydrodynamic conditions, crystallizer construction, etc. A separate problem, which also has to be considered is connected with the accuracy of the simulation. The accurate calculations are time consuming and the accuracy is strongly connected with the way in which the simulation is performed. So, for each accurate simulation of a particular crystallization process it is necessary to elaborate both the appropriate physico-chemical description of the process and the proper way of simulation performance.

2.1 Conservation equations - Computational fluid dynamics

Conservation relates to accounting for flows of heat, mass or momentum (mainly fluid flow) through control volumes within vessels and pipes. This leads to the formation of conservation equations which enables to predict results of operation performed in defined equipment.
In continuum mechanics the general equation for all conservation laws can be expressed in the following form (Spiegelman, 2004):

$$\frac{\partial \Phi}{\partial t} + \nabla(\mathbf{F} + \Phi \mathbf{V}) - H = 0 \tag{1}$$

where: Φ - any quantity (in units of stuff per unit volume), $\mathbf{F}$ - the flux of Φ in the absence of fluid transport, $\mathbf{V}$ – velocity of transport, H - a source or sink of Φ.
To derive an equation for *conservation of mass* it is necessary to substitute $\Phi = \rho$ (density - the amount of mass per unit volume), $\mathbf{F} = 0$ (mass flux can be only change due to transport) and $H = 0$ (mass cannot be created or destroyed). After that we get the following equation, called, *the continuity equation*:

$$\frac{\partial \rho}{\partial t} + \nabla(\rho \mathbf{V}) = 0 \tag{2}$$

In the case of *conservation of energy (heat)* in a single phase material (Spiegelman, 2004), the amount of heat per unit volume is $\Phi = \rho\, c_P\, T$ where c_P is the specific heat at constant pressure (energy per unit mass per degree Kelvin) and T is the temperature. The heat flux consists of two components due to conduction (the heat flux is $\mathbf{F} = -k\nabla T$ where k is the thermal conductivity, "-" because heat flows from hot to cold) and transport ($\rho\, c_P\, T\, \mathbf{V}$). Heat can be also created in a investigated volume due to viscous dissipation, radioactive decay, shear heating etc. (H – all the terms creating heat). Thus the conservation of heat equation assumes the form:

$$\frac{\partial \rho c p T}{\partial t} + \nabla(\mathbf{F} + \rho c_p T\mathbf{V}) = \nabla \cdot k\nabla T + H \tag{3}$$

For constant c_P and k, after introducing thermal diffusivity $\kappa = k/(\rho\, c_P)$, the equation can be rewritten in the form:

$$\frac{\partial T}{\partial t} + \mathbf{V} \cdot \nabla T = \kappa\nabla^2 T + H/(\rho c_p) \tag{4}$$

Conservation of momentum (or force balance) can be derived (Spiegelman, 2004) assuming that the amount of momentum per unit volume is $\Phi = \rho\mathbf{V}$ and the forces which can change the momentum are connected with the stress that acts on the surface ($\mathbf{F} = -\sigma$) and gravity ($H = \rho g$, where g – terrestrial acceleration). So, the equation has the following form:

$$\frac{\partial \rho \mathbf{V}}{\partial t} + \nabla(\rho \mathbf{V}\mathbf{V}) = \nabla\sigma + \rho g \tag{5}$$

which can rewritten in the following way:

$$\frac{\partial \mathbf{V}}{\partial t} + (\mathbf{V} \cdot \nabla)\mathbf{V} = \frac{1}{\rho}\nabla\sigma + g \tag{6}$$

For an isotropic incompressible fluids the above equation can be rewritten into Navier-Stokes equation:

$$\frac{\partial \mathbf{V}}{\partial t} + (\mathbf{V} \cdot \nabla)\mathbf{V} = -\frac{1}{\rho}\nabla P + \nu \bullet \nabla^2 \mathbf{V} + g \tag{7}$$

where: ν is the dynamic viscosity, P – fluid pressure.
In the case of crystallization a further conservation equation is required to account for particle numbers. This is *the population balance*. It is another transport equation which, based on particle formation (nucleation, growth, agglomeration, breakage, etc.), allows for prediction of particle size distribution, $n(L, t)$ in the defined crystallizers. Generally the population balance equation (PBE) can be written in the following form (Jones et al., 2005)

$$\frac{\partial n}{\partial t} + \nabla(\mathbf{v_i} n) + \nabla(\mathbf{v_e} n) = B_0 + B - D \tag{8}$$

where: B_0 is nucleation rate, B and D are the "Birth" and "Death" functions for agglomeration and breakage of crystals, where $\mathbf{v_i}$ is the "internal" velocity and $\mathbf{v_e}$ is the "external" velocity.
The "internal" velocity describes the change of particle characteristic, e.g. its size, volume or composition, and the "external" velocity, the fluid velocity, in the crystallizer. The "internal" velocity, for well-mixed systems, is approximated by the crystal growth rate (G):

$$\nabla(\mathbf{v_i} n) = \nabla(Gn) \tag{9}$$

and usually assumes the following form:

$$\nabla(\mathbf{v_i} n) = \nabla(Gn) = \frac{\partial(Gn)}{\partial L} \tag{10}$$

where: L is a crystal size.
For non well-mixed systems, the velocity derivatives, in addition to crystal growth, have to be included to the equation.
The population balance is a partial integro-differential equation that can be normally solved by numerical methods, except for some simplified cases. Different numerical discretization schemes for solution of the population balance (Kumar & Ramkrishna, 1996, Nicmanis & Hounslow, 1998, Ramkrishna, 2000) and compute correction factors in order to preserve total mass are widely described in the literature (Hostomsky & Jones, 1991, Rigopoulos & Jones, 2003a, Wojcik & Jones, 1998, Wuklow et al., 2001).
Computation Fluid Dynamics, (CFD) is the numerical analysis and solution of system involving all transport processes via computer simulation (Jones et al., 2005,). It is strongly dependent on the development of computer related technologies and on the advancement of our understanding and solving of ordinary and partial differential equations. Direct numerical solving of complex flows in real conditions requires a huge amount of computational power and is very much dependent on the physical models applied. That is why, an ideal model applied for such calculations should introduce minimum amount of complexity into the model equations, while capturing the essence of the relevant physics.

One of the most important flow phenomena is turbulence. If it is present in a certain flow it appears to be the dominant over all other flow phenomena. That is why successful modelling of turbulence greatly increases the quality of numerical simulations. Although, all analytical and semi-analytical solutions of simple flow cases were solved at the end of 1940s, there are still many open questions on modelling of turbulence and properties of turbulence it-self. Till now, no universal turbulence model exists yet.

In the case of crystallization, CFD involves the numerical solution of conservation continuity, momentum and energy equations coupled with constitutive laws of rate (kinetic) processes together with the population balance accounting the solid particles formed and destroyed during crystallization. So, the CFD model solution comprises both the flow properties and a particle size distribution what leads to the formation of conservation equations which enables to predict results of operation performed in defined equipment. Attempts to generate a theoretical model-based description of the interaction of fluid dynamics and crystallization face the multi-scale nature of this interaction.

Usually, the population balance is represented by a partial differential equation of particle size and time and the mass balance, in most cases, is expressed as ordinary differential equations. On the other hand, the growth and nucleation kinetics of particles are often based on empirical correlations.

The main problem connected with a numerical simulation is a problem of discretization of the all coordinates (Euclidean space, particle size, time). Discretization significantly affects the accuracy, the computational costs and even convergence properties of numerical algorithms. Therefore, the selection of the proper discretization grids has to be carefully considered in the context of the characteristic scales of the modelled phenomena.

Usually, for the fluid flow calculation, the Euclidean space is divided into a number of CFD grid cells with elementary volumes. The size of these cells is above the Kolmogorov turbulence scale (order of magnitude 10^{-4} m) but small enough to well resolve the convective flows and energy transport within the unit (Ferziger & Perić, 1996). Such discretization is sufficient to resolve the most of the phenomena occurring in mass crystallization but needs to be improved in the case of reactive crystallization processes where micromixing phenomena play the significant role. The time coordinate of the CFD problem is also discretized using small time steps (seconds) to resolve fast fluctuations.

The particle size coordinate (the population balance equation) has to be also discretized. There are many methods available to perform this discretization (Ramkrishna, 2000, Hounslow, 1990, Hounslow et al., 1988, Hill & Ng, 1995) but in all cases, the most important in the proper evaluation of the size of CFD cell with the appropriate number of particles. If the CFD cells are too small or have too low number of particle the statistical requirements of the population balance is not fulfilled. This may result in an incorrect solution. The next problem connected with the discretization is necessity of solution of several dozens of the equations in each CFD grid cell what would certainly result in prohibitive computational cost and possibly introduce convergence problems. Therefore, some means of model reduction must be employed to allow a numerical simulation. Moreover, all these methods have been developed with a focus on the way in which the systems are mixed.

Generally, CFD models can be implemented to "well-mixed" and "non well-mixed" systems. Assumption of well-mixing is commonly used for the modelling of crystallization processes, what simplifies the simulation and reduces its time. Such approach can be accepted in the case of theoretical calculations and small, laboratory scale, cristallizers. However, even in a stirred tank with impellor (Rielly & Marquis, 2001) we deal with very inhomogeneous fluid mechanical environment. The turbulence quantities and the relevant mean-flow may vary by orders of magnitude throughout the vessel, especially around the impellor. Therefore it is clear, that the 'well-mixed' assumption will lead to significant errors on the rates of growth, nucleation and agglomeration, and consequently, on the crystal size distribution. In these cases, information of the solid concentration distribution, as well as local velocities, shear rates and energy dissipation rates would be needed for the proper design of the process. Crystallization systems frequently show also high levels of supersaturation around the points where it is generated (cooling surfaces, evaporation interfaces, etc.) causing as well as suspension significant local density variations (Sha & Palosaari, 2000). Consequently, crystallization rates locally vary throughout the crystallizer even in case when no reactive crystallization occurs. Therefore, assumption of uniform conditions throughout the reactor volume can not be accepted.

Moreover, many crystallization processes are directly affected by the local fluid dynamic state. One of the most important factors is the shear rate which strongly influences both the frequency and the efficiency of particle collisions (agglomeration (Hounslow et al., 2001)) as well as particle-impeller collisions (Gahn & Mersmann, 1999) which are depended on the relative velocity of the particle and local streamlines around the impeller blade.

The mixing problem increases with increasing of the scale of operation. Typically, fluid dynamics phenomena act on 'micro-scale' (CDF grid), a much smaller scale compared to crystallization phenomena which are usually considered on the 'macroscale' (unit). To solve the problem one can compute the population balance in each CFD grid, accounting for the full locality of the crystallization kinetics or use the scale or spatial resolution for the population balance. The first approach is not recommended because it can violate the statistical assumptions used for the formulation of the population balance equation and needs a long, tome consuming calculations (computational costs). The second method enables for selection of some compartments, representing a certain region in the crystallizer, which can be treated as homogenous (well-mixed) and well described by CFD. Such approach can be a compromise between one single, well mixed unit and the over-detailed system. The use of this model requires the exchange of information between the two scales of calculations: "inner" inside the compartment and "outer" between the compartments. Taking into account these assumptions some compartmental mixing models (Wei & Garside, 1997) for modelling precipitation processes based on the engulfment theory (Baldyga & Bourne, 1984a, 1984b, 1984c) has been elaborated. Also, several mesomixing and micromixing models have been proposed to describe the influence of mixing on chemical reactions on the meso- and molecular scale (Villermaux & Falk, 1994, Baldyga et al., 1995).

2.2 Batch reactor

Batch (or semi-batch) reactor is one of the most popular reactors widely used in chemical and pharmaceutical (especially batch crystallizers) industry. Batch crystallization processes, commonly investigated, are still not well understood because the process is strongly influenced by fluid mixing, particle aggregation and particle breakage. For a batch crystallizer with nucleation, aggregation, breakage and growth occurring (Wan & Ring, 2006). the population balance equation is given by (Randolph & Larson, 1988) as:

$$\frac{\partial n(v,t)}{\partial t} + \frac{\partial [G(v)n(v,t)]}{\partial v} = b(v) - d(v) \tag{11}$$

where: $n(v)$ is the number-based population of particles in the crystallizer being a function of the particle volume v, $G(v)$ is the volume dependent growth rate, $b(v)$ is the volume dependent birth rate and $d(v)$ is the volume dependent death rate. For the initial condition, $n(v,t=0) = n_o(v)$.

For aggregation, the birth $b_a(v)$ and death $d_a(v)$ rate terms can be given by (Hulburt & Katz, 1964):

$$b_a(v) - d_a(v) = \int_0^{1/v} \beta(v-u,u)n(v-u)n(u)du - n(v)\int_0^{\infty} \beta(v,u)n(u)du \tag{12}$$

where $\beta(v,u)$ is the aggregation rate constant (a measure of the frequency of collision of particles of size v with those of size u).

In the case of breakage, the birth $b_b(v)$ and death $d_b(v)$ rate terms can be given by (Prasher, 1987):

$$b_b(v) - d_b(v) = \int_v^{\infty} S(w)\rho(v,w)n(w)dw - S(v)n(v) \tag{13}$$

where $S(v)$ is the breakage rate that is a function of particle size v, $\rho(v,w)$ is the daughter distribution function defined as the probability that a fragment of a particle of size w will appear at size v. The population balance equations can be solved by the use of the standard method of moments (SMOM) and the quadrature method of moments (QMOM) (Wan & Ring, 2006). Using these methods the population balance can be simplified into a series of a few discrete moment equations (some of them as number of particles (${}_vm_0$), volume of particles (${}_vm_1$), etc. have physical significance) defined, for k-th volume-dependent moment, in the following way:

$${}_v m_k = \int_v^{\infty} v^k n(v)\,dv \tag{14}$$

Such calculated ${}_vm_0$ and ${}_vm_1$ represent the total number and total volume of particles in the system

Because in the CFD code, the particle density function is described as a function of particle size x, instead of particle volume v and the population balance is written in terms of $n(x)$ instead of $n(v)$ the population balance, eq. [11], can be rewritten as:

$$\frac{\partial n(x,t)}{\partial t} + \frac{\partial[G(x)n(x,t)]}{\partial x} = b(x) - d(x). \tag{15}$$

and the k-th length-dependent moment as:

$${}_L m_k = \int_v^{\infty} x^k n(x)\,dx \tag{16}$$

The both SMOM and QMOM models has been tested (Wan & Ring, 2006) using numerical cases with nucleation, growth, aggregation and breakage and the obtained results have been compared with the analytical measurements. For all cases the OMOM model gave the very good (the accuracy < 1 %) description of the particle size distribution in the batch reactor.
The particle size distribution in a batch crystallizer can be also simulated in different way. As an example can be given a process of obtaining of calcium carbonate (Kangwook et al., 2002) when we deal with the following overall precipitation reaction:

$$Ca(OH)_2 + Na_2CO_3 = CaCO_3 + 2NaOH \tag{17}$$

where the feeds are a solution of sodium carbonate and a solution of calcium hydroxide at certain, defined concentrations, and the main product is calcium carbonate. The main variable which is to be estimated is particle size distribution of precipitated $CaCO_3$.
The precipitation occurs, when the calcium ions and carbonate ions are present at supersaturated concentration levels. Supersaturation implies that the ionized species are present in the solution where the solubility of the species is exceeded. If we assume that the ionization reactions are fast compared to the precipitation i.e. the ionization reactions reach equilibrium instantaneously and that the perfect mixing in the reactor is obtained we can write the mass balance of the precipitation reactor as follows:

$$\frac{d(VC_i)}{dt} = q_j C_j^F - qC_j - k_q V \int_0^\infty G(L,t)n(L,t)L^2 dL \tag{18}$$

$$\frac{dV}{dt} = q^F - q \tag{19}$$

where: C_j^F is the concentration of species j in the j-th feed stream, C_j is the reactor concentration of species j, q_j^F is the feed flow rate of stream j, q^F is the total feed flow rate, q is the total outlet flow rate, V is the volume of contents in the reactor, k_a is the area factor, L is the characteristic particle size, $G(L, t)$ is the growth rate of particle, $n(L, t)$ is the particle size distribution (number of particles per volume of solvent per particle size), t is the time of reaction, j is equal to 1 for $Ca(OH)_2$ or 2 for Na_2CO_3.
The mass balance equation should be solved together with the population balance equation:

$$\frac{d[Vn(L,t)]}{dt} + V\frac{\partial[G(L,t)n(L,t)](Vn(L,t))}{\partial L} = VP(L, n, t) - qn(L.t) \tag{20}$$

where: $P(L.n.t)$ is a number of density
with corresponding initial condition $n(L,t_0) = n_{t0}(L)$ and boundary condition $n(L_0,t) = n_{L0}(t)$ (i.e. the number of nucleated particles), where L_0 is the nucleated particle size.
Next important equation needed is an equation describing nucleation rate. Typically, nucleation and growth rates of precipitation and crystallization processes are represented by semi-empirical power laws. A proper, nucleation model has to take into account the both primary nucleation induced by supersaturation without particles and secondary nucleation related to the existing particles in the reactor. Growth rate is a function of supersaturation and particle size and can be calculated from the following equations (Eek et al., 1995):

$$n_{L0}(t) = \frac{1}{G(0,t)} \int_0^{\infty} a_n C_s^{b_n} n L^{2,5} dL \tag{21}$$

$$G_L(t) = a_t C_s^{b_t} \frac{1}{1-exp[-a_L(L-b_L)]} \tag{22}$$

where: C_s is the supersaturation of the solute, which is defined as: $[(C_1C_2)^{0.5}-1]$ (expressed in terms of the normalized concentration) and an a_n, b_n, a_t, b_t, a_L, b_L are the parameters.
The kinetic equations have strong nonlinearity due to the power terms what combined with the mass balance equation makes the problem difficult. However, the computationally demanding part of the precipitation reactor model is the population balance equation. In general, the population balance equation can be converting into a set of ordinary differential equations. Many, various forms of the finite element method and the finite difference method can be applied for this purpose. The details on the solution techniques can be found in a (Ramkrishna, 2000) book.
The population balance equation can be simplified in the case when the right-hand side of eq. (20) is a linear or an independent function of the density number. Then a closed-form of the solution can be obtained using the method of characteristics (Varma & Morbidelli, 1997). We can further simplify the model equations assuming that the aggregation and breakage are negligible and the growth rate takes a separable form of G_tG_L in eq. (22) (where: G_t is the time-dependent part of the growth rate and G_L is the size-dependent part). In this case we get the following equations for the particle size distribution:

$$n(L,t) = \frac{V(t_0)}{V(t)} n_{t_0}(L_b) \frac{G(L_b)}{G(L)} \quad \text{for } L_b = L(L,t) \geq 0 \tag{23}$$

$$n(L,t) = \frac{V(t_b)}{V(t)} n_{L0}(t_b) \frac{G(0)}{G(L)} \quad \text{for } t_b = L(L,t) > 0 \tag{24}$$

where t_0 is the start time of growth reaction, L_b in the birth size and t_b is the birth time of the L size particle at time t, which can be obtained by solving the following equations:

$$\int_{L_b}^{L} \frac{1}{G(l)} dl = \int_{t_0}^{t} G_t(\tau) d\tau \tag{25}$$

$$\int_{0}^{L} \frac{1}{G(l)} dl = \int_{t_b}^{t} G_t(\tau) d\tau \tag{26}$$

In the case of size dependent growth, there are no general theoretical kinetics and the separable form is the exclusively used empirical form.
In order to simulate the precipitation reactor, the mass balance and the population balance equation should be solved together. They can be solved used an explicit integration method in which the algebraic equations are solved just once at the beginning of each integration step and held constant or finite element method (Kangwook et al., 2002).

The usefulness of this model for the calcium carbonate precipitation (both an explicit integration and finite element method gave almost the same results) has been checked successfully by (Kangwook et al., 2002) but is necessary to remember, that the assumption of negligible agglomeration and breakage (limits of the model) can be applied only for the reactor where the particle density is maintained on the low level (Kataki & Tsuge, 1990).
For the calcium carbonate precipitation in the batch reactor, breakage can be treated as a negligible phenomenon but the agglomeration is usually significant according to the high particle density in the reactor (Collier & Hounslow, 1999). So, if we want to avoid the aggregation and breakage phenomena in this reactor we have to operate the process in a special way, maintaining the low particle density.
Generally, the presented approach can be implemented for simple precipitation reaction and is, especially, very useful in the case of "run-to-run" or "on-line" controlling of the particle size distribution in a batch (or semi-batch) reactors (Kangwook et al., 2002).

2.3 Crystallization in tube

Every model describing crystallization in a tube has to take into account the fluid dynamics, the fluid flow through the tube and crystallization processes acting simultaneously. The simplest model describing crystallization from solution with feed concentration c_0, in a wall-cooled tube with a defined length and radius, where the supersaturation is generated by cooling of the solution by means of an energy withdrawal at the wall, can be derived making the following assumptions (Kulikov et al., 2005):

- the system is considered to be quasi-homogeneous - it is assumed that the flow through the tube causes very well mixing of the fluid and solid (very small crystals) phases. So, instead of writing separate transport equations for the fluid and the solid phases, a single equation for the whole suspension is formulated. This results in assuming no slip and no particle drag which also implies no segregation of the particles,
- mixture properties (density ρ, molecular viscosity ν, specific heat capacity c_p, thermal conductivity λ) are assumed to be constant,
- no heat of crystallization is released,
- agglomeration and particle breakage are not considered.

The fluid dynamics of the homogeneous mixture can be described by the Reynolds-averaged Navier–Stokes equations consisting of the equations for mass and momentum conservation:

$$\nabla \tilde{\mathbf{V}} = 0 \tag{27}$$

$$\frac{\partial \tilde{\mathbf{V}}}{\partial t} + (\tilde{\mathbf{V}} \cdot \nabla)\tilde{\mathbf{V}} = -\frac{1}{\rho}\nabla p + (\nu + \nu_t) + \nabla^2 \tilde{\mathbf{V}} + g \tag{28}$$

where: $\tilde{\mathbf{V}}$ is the vector of Reynolds-averaged velocities, p is the static pressure, g is the gravitational acceleration, ν and ν_t are viscosity and the turbulent viscosity, respectively.
Using the introduced assumptions, a boundary condition was set for the flow at the tube inlet, no-slip condition was used at the wall and the standard k–ε model (Ferziger & Perić, 1996) has been used, as a turbulence model for a closure of the system. So, the energy balance can be expressed as:

$$\frac{\partial T}{\partial t} + \tilde{\mathbf{V}} \cdot \nabla T = \frac{\lambda}{c_p \rho} \nabla^2 T \tag{29}$$

where: T is the temperature and a boundary temperature condition is specified at the walls of the tube.

The population balance equation used in this model has been taken from (Marchisio et al., 2003). It contains (Kulikov et al., 2005) the accumulation term, the particle growth term, the convective transport term, terms reflecting molecular and turbulent diffusion of particles with the molecular diffusion coefficient D_m and the turbulent diffusion coefficient D_t, respectively, as well as particle birth b and death d terms and can be written in the following form:

$$\frac{\partial n}{\partial t} + \frac{\partial (Gn)}{\partial L} + \nabla (\tilde{\mathbf{V}} n) - (D_m + D_t) \nabla^2 n = b - d \tag{30}$$

where: $n(L, x, t)$ is a particle size distribution and L is a characteristic particle size.

In this case, it is assumed a simple kinetics with the growth term G obeying McCabe's law (size-independent growth) and being first order dependent on supersaturation:

$$G = k_1 [c - c_s(T)] \tag{31}$$

where: c and c_s is the solution concentration and the equilibrium concentration at saturation, respectively. k_1 is a constant.

As it assumed both birth term B and death term D are set to zero:

$$B = 0 \qquad D = 0 \tag{32}$$

and nucleation B_0 is accounted for as a left boundary condition as follows:

$$n(L = 0, x, t) = \frac{B_0}{G} \tag{33}$$

and expressed by a power law equation:

$$B_0 = (1 - \alpha) k_2 \exp\left(-\frac{k_3}{[c/c_s(T) - 1]^2} \right) \tag{34}$$

where: a is the volume fraction of solids and k_2 and k_3 are constants.

The initial condition for nucleation are given by the following equations;

$$n(L, t = 0) = 0 \qquad c(t = 0) = c_0 \tag{35}$$

and mass balance for the solute in the liquid phase by the following:

$$\frac{\partial n}{\partial t} + \nabla (\tilde{\mathbf{V}} c) - D_c \nabla^2 c = -3 \rho_{cr} k_v G \int_0^\infty n L^3 dL \tag{36}$$

where: D_c is the solute diffusivity, ρ_{cr} is the density of the crystals and k_v is are the shape factor of the particles.

The presented model specified in eqs (27)-(36) is a multidimensional dynamic problem containing partial differential equations formulated in spatial coordinates x, one internal particle size distribution coordinate L and the time coordinate t. The locally distributed velocities, temperatures, and particle size distribution are the unknown variables which cannot be calculated analytically and have to be obtained by a numerical simulation.
As it was mentioned before the numerical simulation can be done using two approaches. The first aims at the reduction of the complexity of the population balance discretization by selection a small number of variables characterizing the particle size distribution. It causes some loss of accuracy in the solution of the population balance, which is reformulated in terms of these variables. Transport equations are also reformulated for these variables and solved along with the CFD problem on the proper spatial grid. Usually, these variables are the moments of the distribution function i.e. the Quadrature Method of Moments (Marchisio et al., 2003). A main disadvantage of this approach is the inaccurate reconstruction of the particle size distribution when no a-priori information about its shape is available.
The second approach is based on the reduction of the spatial resolution for the population balance only. Most crystallization phenomena like growth, agglomeration, etc. do not change significantly on the resolution of the CFD grid and can be considered to act on larger scales. This allows for the representation of the population balance by collecting a set of CFD cells in an 'ideally-mixed' compartment. The population balance equations can then be solved in this compartment by a highly accurate discretization scheme. Set of such ideally-mixed compartments represents different regions of the crystallizer. This approach has been well described in the literature (Kramer et al., 2000).
It is difficult to claim the superiority of one of these approaches over the other. The proper selection of the approach very much depends on the application to which it is addressed. The compartmental approach better describes the major crystallization phenomena in a cooling crystallizer with complex breakage and aggregation behaviour while the reduced population balance approach better describes a high spatial fluctuation of supersaturation, e.g., in reactive crystallization.

2.4 Bubble column reactor

Bubble column reactor is an apparatus in which simplicity of design gives rise to extraordinary complexity in the physical and chemical phenomena. That is why modeling of the precipitation process in this reactor needs an integration of reaction kinetics, population balance and hydrodynamic principles.
Such successful modeling of the bubble column reactor applied for the precipitation of calcium carbonate by carbon dioxide absorption into lime has been done by (Rigopoulos. & Jones, 2003a). They used their own (Rigopoulos. & Jones, 2003b) finite element method for solving the time-dependent population balance equation with combined nucleation, growth, agglomeration, and breakage. The previous studies of gas-liquid precipitation (Rigopoulos. & Jones, 2001) which used the method of moments, took into account only a nucleation growth. However, experiment in both gas-liquid (Wachi & Jones, 1991) and liquid-liquid (Tai & Chen, 1995, Collier & Hounslow, 1999) precipitation of $CaCO_3$ have evidenced the presence of agglomeration and demonstrated its importance in determining of the product crystal size distribution.
The time-dependent population balance equation (Rigopoulos. & Jones, 2003b) that describes the evolution of the particle size distribution in a finite, spatially uniform domain,

with particle volume as the "internal" coordinate, and including nucleation, growth, and agglomeration, can be written as follows:

$$\frac{dn(V,t)}{dt} = \frac{n_{in}(V,t)-n(V,t)}{\tau} - \frac{\partial}{\partial V}[G(V)n(V,t)] + B_0\delta(V - V_0) +$$

$$+\frac{1}{2}\int_0^V \beta_a(V-V',V')n(V-V',t)n(V',t)dV' - n(V,t)\int_0^\infty \beta_a(V,V')n(V',t)dV' \qquad (37)$$

where: $n(v,t)$ and $n_{in}(v,t)$ is the population density at the reactor and at the inlet, respectively; $G(v)$ is the volumetric growth rate; and B_0, β_a, and V_0 are the nucleation rate, agglomeration kernel, and volume of the nuclei, respectively, δ is a width of boundary layer.
The equation should be solved with the following initial and boundary conditions:

$$n(V,0) = n_0(V) \qquad \text{(initial distribution)} \qquad (38)$$

$$n(0,t) = 0 \qquad \text{(no crystals zero of size)} \qquad (39)$$

The mass balance equation is derived from the concept of penetration theory (Astarita, 1967) where mass transfer and chemical reactions at the interface are treated simultaneously. The interface and the bulk are considered as two separate dynamic reactors which operate independently and interact at discrete time intervals. Thus, the diffusion and reaction of chemical components is described by the following equations (c_i is the concentration of component i, superscripts I and B denotes variables at the interface and bulk, respectively):

$$\frac{\partial c_i^I}{\partial t} = D\frac{\partial^2 c_i^I}{\partial x^2} + \sum_{k=1}^{K} r_k(c_1^I, c_2^I, \ldots, c_n^I) \qquad (40)$$

with initial and boundary conditions:

$$t = 0, \quad x > 0 \rightarrow c_i^I = c_i^B \qquad (41)$$

$$x = 0, \quad t > 0 \rightarrow c_i^I = c_i^* \quad \text{(volatile species)} \qquad (42)$$

$$x = \delta, \quad t > 0 \rightarrow \frac{dc_i^I}{dx} = 0 \quad \text{(non-volatile species)} \qquad (43)$$

In the most cases of the bubble column the precipitation phenomena at the interface can be neglected because of the very short contact time between the reagents compared to the bulk. Generally. nonideal mixing should be considered in the column but for the relatively short height of the column and intense recirculation a full mixing in the bulk can be assumed. In this case the mass balance in the bulk can be described in the following way:

$$\frac{\partial c_i^B}{\partial t} = \sum_{k=1}^{K} r_k(c_1^B, c_2^B, \ldots, c_n^B, n_1, n_2, \ldots, n_m) \qquad (44)$$

$$\frac{\partial n_j}{\partial t} = f_j\,(c_1^B, c_2^B, \ldots, c_n^B, n_1, n_2, \ldots, n_m) \tag{45}$$

where f_j (c_i^B,n_j) is a function into which the original population balance is transformed via the finite element discretization. Initial conditions are calculated from the mixing of bulk and interface at the end of the previous contact time. The solution of the interface equations is obtained numerically with an implicit iterative scheme, while the bulk equations are calculated according to Adams method (Hindmarsh, 1983).
The reactions occurring during the process can be described in the following way:

$$CO_{2(g)} = CO_{2(l)} \tag{46}$$

$$CO_{2(g)} + OH^- = HCO_3^- \tag{47}$$

$$HCO_3^- + OH^- = CO_3^{2-} + H_2O \tag{48}$$

$$Ca^{2+} + CO_3^{2-} = CaCO_{3(s)} \tag{49}$$

The first step (eg. (46)) is a CO_2 absorption in water at the gas-liquid equilibrium. The equilibrium can be described by Henry's law, taking into account that we deal with an ionic system, as follows:

$$\log\left(\frac{H}{H_0}\right) = -\sum_i I_i h_i \tag{50}$$

$$h_i = h_+ + h_- + h_g \tag{51}$$

where: H and H_0 is a Henry's constant for an ionic and nonionic system, respectively, I_i is an ionic strength of component "*i*" and h_i, h_-, h_+, h_g is a component "*i*", anions, cations and gas contribution, respectively.
The kinetics of carbon dioxide absorption into alkali solutions are determined by the conversion of CO_2(aq) into HCO_3^- (eq. (47)), which proceeds at a great, but finite rate. This reaction is followed by an instantaneous ionic reaction eq. (48). and the precipitation reaction eq. (49).
The rate of $CaCO_3$(s) production is determined by a crystallization mechanism but always volumetric crystal growth is size-dependent even when the linear growth is size-independent (McCabe's law). To obtain the rate of change for the whole crystal mass is necessary to integrate the volumetric growth function over the whole range of crystal volumes:

$$\frac{dc_{CaCO3}}{dt} = \int_{V_0}^{V_\infty} G(V)\,n(V)\,dV\,\frac{\rho_{CaCo3}}{M_{CaCO3}} \tag{52}$$

where: ρ_{CaCO3} and M_{CaCO3} is calcium carbonate density and molar mass, respectively.
To estimate the rate of crystal mass production, which is coupled with the population balance, it is necessary to derive a complete kinetic model of precipitation taking into account the whole information concerning the crystal formation i.e. nucleation, crystal growth as well as agglomeration and breakage.

The growth rate kinetics is usually described by the linear growth rate (the increase in particle diameter or radius) G_l, by the following expression:

$$G_l = k_g (\lambda_s - 1)^2 \tag{53}$$

where: k_g is a kinetic constant and λ_s is the saturation ratio (Rigopoulos. & Jones, 2003b).
Nucleation process can have a variety of mechanisms (homogeneous, heterogeneous, secondary, etc). In the bubble column it can be assumed (Rigopoulos. & Jones, 2003b) that in the beginning of the process high supersaturation levels induce primary nucleation, but later, secondary nucleation causes the rise of crystal growth. So, the overall nucleation model consists of the sum of the two models: primary and secondary. The primary nucleation which depends mainly on supersaturation is usually described by a power law:

$$B_0 = k_{n1} (\lambda_s - 1)^{k_{n2}} \tag{54}$$

Secondary nucleation is induced by the existing crystals (Garside & Davey, 1980) and is a function of the crystal mass (M_c):

$$B_0 = k_{ns} (\lambda_s - 1)^{k_{n3}} M_c^{k_{n4}} \tag{55}$$

where: k_{n1}, k_{n2}, k_{n3}, k_{n4} and k_{ns} are the appropriate constants (Rigopoulos. & Jones, 2003b).
Thus, the overall nucleation model can be expressed as follows:

$$B_0 = k_{n1} (\lambda_s - 1)^{k_{n2}} + k_{ns} (\lambda_s - 1)^{k_{n3}} M_c^{k_{n4}} \tag{56}$$

It is necessary to point out that calcium carbonate can appear, during the precipitation process, in three different polymorphs where the most prevailing polymorph appears to be calcite. That is why, a kinetic model should account for their simultaneous presence in the solution (Chakraborty & Bhatia, 1996). Usually, because of the complexity and difficulty of such calculations, the considerations are limited only to calcite.
Agglomeration of crystals is a very complex and system-dependent process. Usually, it can be simplified and treated as a two-step process. The first step of agglomeration, i.e. the formation of flocculates through collisions and interparticle attraction, is similar to the phenomena occurring in colloids and aerosols. The second step is the growth of crystalline material between the clusters at so-called cementing sites (Hounslow et al., 2001). In the case of the bubble column (Rigopoulos. & Jones, 2003b) the agglomeration can be assumed to be roughly proportional to growth (Hounslow et al., 2001) and described as the second-order dependence on supersaturation, in the following way:

$$\beta_a (V', V - V') = k_a (\lambda_s - 1)^2 [(V')^{1/3} + (V - V')^{1/3}]^3 \tag{57}$$

where: k_a is the agglomeration constant (Rigopoulos. & Jones, 2003b).
Hydrodynamics of the gas-liquid precipitation strongly depends on the gas holdup which determines the rates of the chemical phenomena. The Eulerian-Eulerian multiphase CFD model (Rigopoulos, & Jones, 2001), where the turbulence in the liquid phase is calculated with k-ε model (Schwarz & Turner, 1988), can be used for its description. This model can be

successfully used for modeling large-scale equipment because it gives sufficiently accurate results with respect to averaged properties. However, it is less successful in reproducing fine details i.e. the radial phase distribution.

The above model very well (good agreement with the experiment) described the precipitation of $CaCO_3$ by CO_2 absorption into lime, in the bubble column. The conjunction of penetration theory and CFD predictions of the gas holdup seems to yield an adequate description of the reactor performance. Such integration of the population balance, reaction kinetics and hydrodynamic principles allowed for proper formulation of modeling approach for the gas-liquid precipitation process and the model can be used as a tool for the analysis and scale-up of industrial-class equipment.

2.5 Thin film reactor

The another approach (Kędra-Królik & Gierycz, 2010) is necessary when the precipitation goes in the thin film. It happens in the Rotating Disc Precipitation Reactor (Kędra-Królik & Gierycz, 2006, Kędra-Królik & Gierycz, 2009) used for calcium carbonate production. The reaction in liquid phase goes in contact with continuously flowing gaseous carbon dioxide in the thin film formed on the surface of the rotating disc (Kędra-Królik & Gierycz, 2006, Kędra-Królik & Gierycz, 2009). This creates a constant surface area of gas-liquid interface and the carbonation reaction of lime water involves gas, liquid and solid phase. The reactions occurring during the process are described by eqs. (46-49).

The model (Kędra-Królik & Gierycz, 2010) has taken into account not only kinetics of the multiphase reaction but also crystal growth rate. The film theory (Wachi & Jones, 1991, Danckwerts, 1970) describes the mass balance of reactants in these reactions as follows:

$$\frac{\partial c_{CO2}}{\partial t} = D_{CO2}\left(\frac{\partial^2 c_{CO2}}{\partial x^2}\right) - kc_{CO2}c_{OH} \tag{58}$$

$$\frac{\partial c_{OH}}{\partial t} = D_{OH}\left(\frac{\partial^2 c_{OH}}{\partial x^2}\right) - kc_{CO2}c_{OH} \tag{59}$$

$$\frac{\partial c_{CO3}}{\partial t} = D_{CO3}\left(\frac{\partial^2 c_{CO3}}{\partial x^2}\right) + kc_{CO2}c_{OH} - G' - B' \tag{60}$$

where: t is time, c_{CO2}, c_{OH}, c_{CO3} are the concentrations of gas reactant ($CO_{2(g)}$), liquid reactant (OH^-) and the product (CO_3^{2-}), respectively, G', B' are rate of nucleation and crystal growth, respectively; k is second order chemical reaction constant; D_{CO2}, D_{OH}, D_{CO3} are diffusivity of ($CO_{2(g)}$), (OH^-) and (CO_3^{2-}), respectively.

The component (CO_3^{2-}) is formed by reaction (48) and consumed by the precipitation reaction (49). It is assumed also that the concentration of (CO_3^{2-}) is constant across the diffusion layer. Thus the population balance of the precipitated particles is given by the following equation (Hill & Ng, 1995):

$$\frac{\partial N}{\partial t} + G\frac{\partial N}{\partial L} = D_P\left(\frac{\partial^2 N}{\partial x^2}\right) \tag{61}$$

where: N is a population density of particles, G is linear growth rate; L is a coordinate of particle dimension; D_P is the diffusivity of particles. Substituting: $N = P/L$, we get:

$$\frac{\partial P}{\partial t} + G\frac{\partial P}{\partial L} = D_P\left(\frac{\partial^2 P}{\partial x^2}\right) + \frac{G}{L}P \tag{62}$$

where: $P(x,L_i,t)$ is a number of density discretized in L_i, L_i is a particle size coordinate, L_0 is an effective nucleic size, for newly nucleated particles.
In the case of the precipitation of $CaCO_3$ in the thin film the small crystals are obtained due to the very high nucleation rate compared to the crystal growth rate (Kędra-Królik & Gierycz, 2010). For such very small particles, the diffusivity of the crystals (D_P) within the liquid film can be described by the Stokes-Einstein equation (Hostomsky & Jones, 1991):

$$D_P = k_B T / (6\pi\mu r) \tag{63}$$

where: k_B is the Boltzmann constant, T is temperature, μ is viscosity and r is radius of particle
The number rate of nucleation (J_n) and linear crystal growth (G) can be expressed by the Nielsen equations (Hounslow, 1990):

$$J_n = k_n (c - c^*)^n \tag{64}$$

$$G = k_g (c - c^*)^g \tag{65}$$

where: n, g – the orders of nucleation and growth, respectively; c, c^* - the concentration and equilibrium saturation concentration, respectively; k_n, k_g – nucleation and growth rate constants, respectively.
The equations can be rewritten to the following forms:

$$J_n = k_n \left(\sqrt{c_{Ca} c_{CO3}} - \sqrt{K_{sp}}\right)^n \tag{66}$$

$$G = k_g \left(\sqrt{c_{Ca} c_{CO3}} - \sqrt{K_{sp}}\right)^g \tag{67}$$

where: c_{Ca} is the concentrations of (Ca^{2+}), K_{sp} is solubility of the product (calcium carbonate).
The corresponding mass based rate equations both of nucleation and growth can be expressed by the following equations (Wachi & Jones, 1991, Danckwerts, 1970):

$$B' = a\rho J_n L_0^3 \quad (a = \pi/6 \text{ for the sphere}) \tag{68}$$

$$G' = \sum_{i=0}^{\infty} \beta\rho P(x, L_i) G L_i^2 \quad (\beta = \pi \text{ for the sphere}) \tag{69}$$

where: ρ is crystal density.

The boundary conditions for the gas-liquid interface, assuming that except for the gaseous reactant ($CO_{2(g)}$) every component is non-volatile, are as follows:

$$\text{at } x = 0;\ t > 0 \rightarrow c_{CO2} = c_{CO2}^0,\ dc_{OH}/dx = 0,\ dc_{CO3}/dx = 0,\ dP/dx = 0 \tag{70}$$

and for the film formed on the disk surface, assuming that newly nucleated particles have an effective nucleic size equal to L_0, as follows:

$$\text{at } 0 < x < \delta;\ t > 0; \rightarrow L = L_0;\ \frac{\partial P}{\partial t} + G\frac{\partial P}{\partial L} = D_P\left(\frac{\partial^2 P}{\partial x^2}\right) + J_n;\ L = \infty;\ P = 0 \tag{71}$$

Solving the mass balance equations (eqs. 58-60) and population equation (eq. 62) with the boundary conditions we can calculate both the discretized density number of particles ($P(t,x,L_i)$) and discretized diameter L_i. The model describes properly the change of precipitation rate in the liquid film. The $Ca(OH)_2$ concentration decreases because of the very high nucleation rate. Higher supersaturation leads to smaller mean crystal size, since the nucleation rate is much more sensitive to the level of supersaturation than the growth rate. It agrees very well with the experiment (Kędra-Królik & Gierycz, 2006, Kędra-Królik & Gierycz, 2009) and is caused by the fact that the high level of supersaturation is accumulated within the liquid film due to the large diffusion resistance.

The proposed model very well describes the $CaCO_3$ crystals formation in the rotating disc reactor and can be used and recommended for accurate calculation of the particle size and distribution obtained by gas-liquid precipitation in the reactor. However, it is necessary to remember that the model has not taken into account agglomeration of the obtained crystals and cannot be used for calculation of the aggregation process in the reactor.

3. Conclusion

The aim of this paper was to present the different approaches to the proper and accurate modeling and simulation of $CaCO_3$ formation and growth in multiphase reaction. This very complex problem has been presented for most popular, different types of reactors, i.e. batch, tube and thin film reactor as well as bubble column.

The batch (or semi-batch) precipitation process has been described by closed-form solution of population balance equation, which has not taken into account aggregation and breakage, what simplifies the simulation. However, the presented strategy is general and can be applied to batch or semi-batch processes described by more complex types of population balance equations.

In the case of tube reactor integration of simulation of crystallization and fluid dynamics was successfully applied by means of the Method of Moments. The used method allowed for reconstructing the solids fraction profiles on the fine CFD grid, while preserving the full information on particle size distribution on the coarser compartment scale. The technique is well established and has moderate computational costs.

The thin film reactor has been described by the model which takes into account both kinetics of the multiphase reaction and crystals growth rate. Results of calculation agreed very well with the experiment and the model described properly the change of precipitation rate from bulk liquid to the film region and showed that the higher supersaturation leads to smaller mean crystal size, since the nucleation rate is more sensitive to the level of supersaturation than the growth rate.

The gas-liquid precipitation process in the bubble column was modeled by integration the population balance, reaction kinetics and hydrodynamic principles. The used model well-described the precipitation of $CaCO_3$ by CO_2 absorption into lime and can be recommended the analysis and scale-up of industrial-class equipment. It gave also some explanations for the experimental results. It showed that the crystal mean size increase after the pH drop is due to the disappearance of the smaller crystals by dissolution, the secondary nucleation take place because a new wave of nucleation-growth is induced by the existing crystals and crystal agglomeration starts to take place at relatively high pH and proceeds to a considerable extent because the aggregates are less frequently disrupted than in stirred tanks.

Moreover, a wide review of different methods and approaches to the accurate description of crystallization processes as well as main CFD problems has been presented in this chapter. It can serve as a basic material for formulation and implementation of new, accurate models describing not only multiphase crystallization processes but also any processes taking place in different chemical reactors.

Combined population balance and kinetic models, computational fluid dynamics and mixing theory enable well prediction and scale-up of crystallization and precipitation systems but it is necessary to remember that each process (performed in the well defined reactor) needs always its own modeling.

4. References

Astarita, G. (1967). *Mass Transfer with Chemical Reaction*; Elsevier Publishing, ISBN 66-25758,: Amsterdam, The Netherlands

Baldyga, J. & Bourne, J.R. (1984a). A fluid mechanical approach to turbulent mixing and chemical reaction. Part I: Inadequacies of available methods. *Chemical Engineering Communications*, Vol.28, No.4-6, pp. 231-241, ISSN 0098-6445

Baldyga, J. & Bourne, J.R. (1984b). A fluid mechanical approach to turbulent mixing and chemical reaction. Part II: Micromixing in the light of turbulence theory. *Chemical Engineering Communications*, Vol.28, No.4-6, pp. 243-258, ISSN 0098-6445

Baldyga, J. & Bourne, J.R. (1984c). A fluid mechanical approach to turbulent mixing and chemical reaction. Part III: Computational and experimental results for the new micromixing model. *Chemical Engineering Communications*, Vol.28, No.4-6, pp. 259-281, ISSN 0098-6445

Baldyga, J., Podgorska, W. & Pohorecki R. (1995), Mixing-precipitation model with application to double feed semibatch precipitation. *Chemical Engineering Science*, Vol.50, No.8, (April 1995), pp. 1281-1300, ISSN 0009-2509

Bandyopadhyaya, R.; Kumar, R. & Gandhi K.S. (2001). Modelling of $CaCO_3$ nanoparticle formation during overbasing of lubricating oil additive. *Langmuir*, Vol.17, No.4, (February 2001), pp. 1015-1029, ISSN 0743-7463

Cafiero, L.M.; Baffi, G.; Chianese, A. & Jachuck, R.J.J. (2002). Process intensification: precipitation of barium sulfate using a spinning disk reactor. *Industrial & Engineering Chemistry Research*, Vol.41, No.21, (October 2002), pp. 5240-5246, ISSN 0888-5885

Chakraborty, D. & Bhatia, S.K. (1996). Formation and aggregation of polymorphs in continuous precipitation. 2. Kinetics of $CaCO_3$ precipitation. *Industrial & Engineering Chemistry Research*, Vol.35, No.6, (June 1996), pp. 1995-2006, ISSN 0888-5885

Chen, J.F.; Wang, Y.H.; Guo, F.; Wang, X.M. & Zheng, Ch. (2000). Synthesis of nanoparticles with novel technology: High-gravity reactive precipitation. *Industrial & Engineering Chemistry Research*, Vol.39, No.4, (April 2000), pp. 948-954, ISSN 0888-5885

Cheng, B.; Lei, M.; Yu, J. & Zhao, X. (2004). Preparation of monodispersed cubic calcium carbonate particles via precipitation reaction. *Materials Letters*, Vol.58, No.10, (April 2004), pp. 1565-1570, ISSN 0167-577X

Colfen, H. & Antonietti, M. (2005). Mesocrystals: Inorganic superstructures made by highly parallel crystallization and controlled alignment. *Angewandte Chemie International Edition*, Vol.44, No.35, (September 2005), pp. 5576-5591, ISSN 1433-7851

Collier, A.P. & Hounslow, M. J. (1999). Growth and aggregation rates for calcite and calcium oxalate monohydrate. *American Institute of Chemical Engineering Journal*, Vol.45, No.11, (November 1999), pp. 2298–2305, ISSN 0001-1541

Danckwerts, P.V. (1970). *Gas-Liquid Reaction*, McGraw-Hill, ISBN 007015287X, New York, US

Dindore, V.Y.; Brilman, D.W.F. & Versteeg, G.F. (2005). Hollow fiber membrane contactor as a gas–liquid model contactor. *Chemical Engineering Science*, Vol.60, No.2, (January 2005), pp. 467-479, ISSN 0009-2509

Feng, B.; Yonga, A.K. & Ana, H. (2007). Effect of various factors on the particle size of calcium carbonate formed in a precipitation process. *Materials Science and Engineering A*, Vol.445-446, (February 2007), pp. 170-179, ISSN 0921-5093.

Ferziger, J.H. & Perić, M. (1996). *Computational Methods for Fluid Dynamics*, Springer-Verlag, ISBN 3-540-59434-5, Berlin, Germany

Gahn, C. & Mersmann, A. (1999). Brittle fracture in crystallization processes. Part A. Attrition and abrasion of brittle solids. *Chemical Engineering Science*, Vol.54, No.9, (May 1999), pp. 1273–1282, ISSN 0009-2509

Garside, J. & Davey, R. J. (1980). Invited review secondary contact nucleation: kinetics, growth and scale-up. *Chemical Engineering Communications*, Vol.4, No.4&5, pp. 393-424, ISSN 0098-6445

Hill, P.J. & Ng, K.M. (1995). New discretization procedure for the breakage equation. *American Institution of Chemical Engineers Journal*, Vol.41, No.5, (May 1995), pp. 1204–1217, ISSN 0001-1541

Hindmarsh, A.C. (1983). Odepack, a systematized collection of ode solvers. In: *Scientific Computing. Vol.1 of IMACS Transactions on Scientific Computation*, R. S. Stepleman, M. Carver, R. Peskin, W.F. Ames & R. Vichnevetsky, (Eds.), 55-64, IMACS/North-Holland, Amsterdam, The Netherlands

Hostomsky, J. & Jones, A.G. (1991). Calcium carbonate crystallization, agglomeration and form during continuous precipitation from solution. *Journal of Physics D: Applied Physics*, Vol.24, No.2, (February 1991), pp. 165-170, ISSN 0022-3727

Hounslow, M.J. (1990). A discretized population balance for continuous systems at steady state. *American Institution of Chemical Engineers Journal*, Vol.36, No.1, (January 1990), pp. 106-116, ISSN 0001-1541

Hounslow, M.J.; Ryall, R.L. & Marshall, V.R. (1988). A discretized population balance for nucleation, growth, and aggregation. *American Institution of Chemical Engineers Journal*, Vol.34, No.11, (November 1988), pp. 1821-1832, ISSN 0001-1541

Hounslow, M.J.; Mumtaz, H.S.; Collier, A.P.; Barrick, J.P. and Bramley, A.S. (2001). A micro-mechanical model for the rate of aggregation during precipitation from solution. *Chemical Engineering Science*, Vol.56, No.7, (April 2001), pp. 2543–2552, ISSN 0009-2509

Hulburt, H.M. & Katz, S. (1964). Some Problems in Particle Technology - Statistical Mechanical Formulation. *Chemical Engineering Science*, Vol.19, No.8, (August 1964), pp. 555-574, ISSN 0009-2509

Jones, A.G.; Rigopoulos, S. & Zauner, R. (2005). Crystallization and precipitation engineering. *Computers & Chemical Engineering*, Vol.29, No.6, (May 2005), pp. 1159-1166, ISSN 0098-1354

Judat, B. & Kind, M. (2004). Morphology and internal structure of barium sulfate—derivation of a new growth mechanism. *Journal of Colloid and Interface Science*, Vol.269, No.2, (January 2004), pp. 341-353, ISSN 0021-9797

Jung, T.; Kim, W.S. & Choi, Ch.K. (2005). Effect of monovalent salts on morphology of calcium carbonate crystallized in Couette-Taylor reactor. *Crystal Research and Technology*, Vol.40, No.6, (June 2005), pp. 586-592, ISSN 0232-1300

Kangwook L.; Jay, H.L.; Dae R.Y. & Mahoney, A.W. (2002). Integrated run-to-run and on-line model-based control of particle size distribution for a semi-batch precipitation reactor. *Computers and Chemical Engineering*, Vol.26, No.7-8, (August 2002), pp. 1117–1131, ISSN 0098-1354

Kataki, Y. & Tsuge, H. (1990). Reactive crystallization of calcium carbonate by gas–liquid and liquid–liquid reactions. *Canadian Journal of Chemical Engineering*, Vol.68, No.3, (June 1990), pp. 435–442, ISSN 0008-4034

Kędra-Królik, K. & Gierycz P. (2006). Obtaining calcium carbonate in a multiphase system by the use of new rotating disc precipitation reactor. *Journal of Thermal Analysis and Calorymetry*, Vol.83, No.3, pp. 579-582, ISSN 1388-6150

Kędra-Królik, K. & Gierycz P. (2009). Precipitation of nanostructured calcite in a controlled multiphase process. *Journal of Crystal Growth*, Vol.311, No.14, (July 2009), pp. 3674-3681, ISSN 0022-0248

Kędra-Królik, K. & Gierycz P. (2010). Simulation of nucleation and growing of $CaCO_3$ nanoparticles obtained in the rotating disk reactor. *Journal of Crystal Growth*, Vol.312, No.12-13, (June 2010), pp. 1945-1952, ISSN 0022-0248

Kitano, Y.; Park, K. & Hood, D.W. (1962). Pure aragonite synthesis. *Journal of Geophysical Research*, Vol.67, No.12, pp. 4873-4874, ISSN 0148-0227

Kramer, H.J.M.; Dijkstra, J.W.; Verheijen, P.J.T. & van Rosmalen, G.M. (2000). Modeling of industrial crystallizers for control and design purposes, *Powder Technology*, Vol.108, No.2-3, (March 2000), pp. 185–191, ISSN 0032-5910

Kulikov, V.; Briesen, H. & Marquardt, W. (2005). Scale integration for the coupled simulation of crystallization and fluid dynamics. *Chemical Engineering Research and Design,* Vol.83, No.6, (June 2005), pp. 706–717, ISSN 0263-8762

Kumar, S. & Ramkrishna, D. (1996). On the solution of population balance equations by discretization—II. A moving pivot technique. *Chemical Engineering Science,* Vol.51, No.8, (April 1996), pp. 1333-1342, ISSN 0009-2509

Malkaj, P.; Chrissanthopoulos, A. & Dalas, E. (2004). Understanding nucleation of calcium carbonate on gallium oxide using computer simulation. *Journal of Crystal Growth,* Vol.264, No.1-3, (March 2004), pp. 430-437, ISSN 0022-0248

Marchisio, D.L.; Vigil, R.D. and Fox, R.O. (2003). Implementation of quadrature method of moments in CFD codes for aggregation-breakage problems. *Chemical Engineering Science,* Vol.58, No.15, (August 2003), pp. 3337–3351, ISSN 0009-2509

Montes-Hernandez, G.; Renard, F.; Geoffroy, N.; Charlet, L. & Pironon, J. (2007). Calcite precipitation from CO_2–H_2O–$Ca(OH)_2$ slurry under high pressure of CO_2. *Journal of Crystal Growth,* Vol.308, No.1, (October 2007), pp. 228-236, ISSN 0022-0248

Mullin, J.W. (2001). *Crystallization,* Butterworth-Heinemann, ISBN 978-075-0648-33-2, Oxford, UK

Myerson, A.S. (1999). *Molecular modelling applications in crystallization,* Cambridge University Press, ISBN 0 521 55297 4, Cambridge, UK

Nicmanis, N. & Hounslow, M.J. (1998). Finite-element methods for steady-state population balance equations. *American Institution of Chemical Engineers Journal,* Vol.44, No.10, (October 1998), pp. 2258-2272, ISSN 0001-1541

Prasher, C.L. (1987). *Crushing & Grinding Process Handbook,* Wiley, ISBN 047110535X, New York, US

Quigley, D. & Roger, P.M. (2008). Free energy and structure of calcium carbonate nanoparticles during early stages of crystallization. *Journal of Chemical Physics,* Vol.128, No.22, (June 2008) pp. 2211011-2211014, ISSN 0021-9606

Ramkrishna, D. (2000). *Population Balances. Theory and Applications to Particulate Systems in Engineering,* Academic Press, ISBN 0-12-576970-9, San Diego, US

Randolph, A.D. & Larson, M.A. (1988). *Theory of Particulate Processes,* Academic Press, ISBN 0125796528, New York, US

Reddy, M.M. & Nancollas, G.H. (1976). The crystallization of calcium carbonate : IV. The effect of magnesium, strontium and sulfate ions. *Journal of Crystal Growth,* Vol.35, No.1, (August 1976), pp. 33-38, ISSN 0022-0248

Rigopoulos, S. & Jones, A.G. (2001). Dynamic modelling of a bubble column for particle formation via a gas-liquid reaction. *Chemical Engineering Science,* Vol.56, No.21-22, (November 2001), pp. 6177-6183, ISSN 0009-2509

Rigopoulos, S. & Jones, A.G. (2003a). Modeling of semibatch agglomerative gas−liquid precipitation of $CaCO_3$ in a bubble column reactor. *Industrial & Engineering Chemistry Research,* Vol.42, No.25, (December 2003), pp. 6567- 6575, ISSN 0888-5885

Rigopoulos, S. & Jones, A.G. (2003b) Finite element scheme for solution of the dynamic population balance. *American Institute of Chemical Engineering Journal,* Vol.49, No.5, (May 2003), pp. 1127-1139, ISSN 0001-1541

Schlomach, J.; Quarch, K. & Kind, M. (2006). Investigation of precipitation of calcium carbonate at high supersaturations. *Chemical Engineering & Technology*, Vol.29, No.2, (February 2006), pp. 215-220, ISSN 1521-4125

Schwarz, M.P. & Turner, W.J. (1988). Applicability of the standard *k*-ε turbulence model to gas-stirred baths. *Applied Mathematical Modelling*, Vol.12, No.3, (June 1988), pp. 273-279, ISSN 0307-904X

Sha, Z. & Palosaari, S. (2000). Mixing and crystallization in suspensions. *Chemical Engineering Science*, Vol.55, No.10, (May 2000), pp. 1797–1806, ISSN 0009-2509

Sohnel, O. & Mullin, J.W. (1982). Precipitation of calcium carbonate. *Journal of Crystal Growth*, Vol.60, No.2, (December 1982), pp. 239-250, ISSN 0022-0248

Spanos, N. & Koutsoukos, P.G. (1998). Kinetics of precipitation of calcium carbonate in alkaline pH at constant supersaturation. spontaneous and seeded growth. *Journal of Physical Chemistry B*, Vol.102, No.34, (August 1998), pp. 6679-6684, ISSN 1520-6106

Spiegelman M. (2004). *Myths & Methods in Modeling*, LDEO, Columbia University, New York, US

Tai, C.Y. & Chen, P.-C. (1995). Nucleation, agglomeration and crystal morphology of calcium carbonate. *American Institute of Chemical Engineering Journal*, Vol.41, No.1, (January 1995), pp. 68-77, ISSN 0001-1541

Tobias, J. & Klein, M.L. (1996). Molecular dynamics simulations of a calcium carbonate/calcium sulfonate reverse micelle. *Journal of Physical Chemistry B*, Vol.100, No.16, (April 1996), pp. 6637-6648, ISSN 1520-6106

Varma, A. & Morbidelli, M. (1997). *Mathematical Methods in Chemical Engineering*, Oxford University Press, ISBN 0-19-509821-8, New York, US

Villermaux, J. & Falk, L. (1994). A generalized mixing model for initial contacting of reactive fluids. *Chemical Engineering Science*, Vol.49, No.24(2), (December 1994), pp. 5127-5140, ISSN 0009-2509

Wachi, S. & Jones, A.G. (1991). Mass transfer with chemical reaction and precipitation. *Chemical Engineering Science*, Vol.46, No.4, pp. 1027-1033, ISSN 0009-2509

Wan, B. & Ring, T.A. (2006). Verification of SMOM and QMOM population balance modeling in CFD code using analytical solutions for batch particulate processes. *China Particuology*, Vol.4, No.5, (October 2006), pp. 243-249, ISSN 1672-2515

Wang, T.; Antonietti, M. & Colfen, H. (2006). Calcite mesocrystals: "Morphing" crystals by a polyelectrolyte. *Chemistry – A European Journal*, Vol.12, No.22, (July 2006), pp. 5722-5730, ISSN 0947-6539

Wei, H.Y. & Garside, J. (1997). Application of CFD modelling to precipitation systems. *Chemical Engineering Research and Design*, Vol.75, No.2, (February 1997), pp. 219-227, ISSN 0263-8762

Wojcik, J. & Jones, A.G. (1998). Dynamics and stability of continuous MSMPR agglomerative precipitation: Numerical analysis of the dual particle coordinate model. *Computers and Chemical Engineering*, Vol.22, No.4-5, pp. 535-545, ISSN 0098-1354

Wuklow, M.; Gerstlauer, A. & Nieken, U. (2001). Modeling and simulation of crystallization processes using parsival. *Chemical Engineering Science*, Vo.56., No.7, (April 2001), pp. 2575-2588, ISSN 0009-2509

25

Colloidal Crystals

E. C. H. Ng[1], Y. K. Koh[2] and C. C. Wong[1]
[1]*Singapore-MIT Alliance, Nanyang Technological University,*
[2]*DSO National Laboratories,*
Singapore

1. Introduction

A colloidal system consists of insoluble particles well-dispersed in a continuous solvent phase, with dimensions (generally less than 1μm in at least one important dimension) that are relatively larger than the molecules of the solvent. When the particles in this system are arranged in periodic arrays, analogous to a standard atomic crystal with repeating subunits of atoms or molecules, they form colloidal crystals (Pieranski, 1983). Gem opal (silica particles in close packed arrangement), iridescent butterfly wings made of periodic and spongelike pepper-pot structure (Biró et al, 2003) and sea mouse with hexagonal close packed structure of holes (McPhedran et al, 2003) are typical examples of colloidal crystals found in nature.

In 1935, discovery of tobacco and tomato virus by Stanley (Kay, 1986) provided excellent examples of naturally occurring monodisperse colloidal crystals. By centrifuging the dilute suspension of virus particles, crystals formed at the bottom of centrifuge tube can be examined by X-ray or light diffraction. The ease to obtain close-packed arrangement of colloidal particles has fascinated many, especially researchers working in chemical sensors, photonic band gap (PBG[1]) crystals, nanopatterning and sensors. 3D Colloidal crystals with periodicity ranging from 100nm to 1μm diffract visible wavelength according to Bragg's law, serving as waveguide and reflective surfaces for many useful devices in communication (Avrutsky et al, 2000; Liu, 2005) and solar harvesting (Mihi et al, 2011). Monolayers of colloidal arrays can serve as lithography mask (Lee et al, 2010) and physical mask for nanoimprinting. Periodicity can then be manipulated with the right particle size.

In this chapter, we discuss briefly the concepts of a colloidal system and types of interaction that give rise to crystallization, growth techniques and characterization tools available to help readers who are interested and new to colloidal systems. However, we limit the scope to simple colloidal particles which are spherical and of identical size, despite the possibility to self-assemble colloidal structures of complicated dimensions (Chen et al, 2006; Lu et al, 2001). Of the many growth methods, we place particular emphasis on capillary growth and its dependence on interparticle interactions, the substrate, and the manipulation of the solvent meniscus.

---

[1]Photonic band gap crystals are periodic structures which are able to block light propagation through the crystal in one or more directions. (Yablonovitch, 1987)

2. General concepts of a colloidal system

2.1 Colloidal system as a model for condensed matter

Colloidal systems share several similar cooperative phenomena with condensed matter (atomic crystal): ordering, phase transitions and stability of the resulting phases. As the ratio of colloidal particle size to atomic dimension is huge (~10^3), parameters like the time scales of diffusion and lattice distances are also scaled up appropriately to milliseconds and micrometers respectively. This allows one to probe into real time processes that are otherwise inaccessible in atomic systems (Arora & Tata, 1996). As a result, a suspension of colloidal particles has provided fascinating models to investigate the physics of nucleation and growth (Gasser et al, 2001; Habdas & Weeks, 2002), phase transitions (Bartlett et al, 1992; Gast & Russel, 1998; Sirota et al, 1989), and fundamental problems of crystallization kinetics (Auer & Frenkel, 2001; Yethiraj et al, 2004).

A stable suspension of monodisperse colloidal particles normally appears milky white and will only become iridescent when interparticle distance shrinks to submicron range and satisfy Bragg's diffraction. Extensive study of equilibrium phase behavior of monodisperse colloidal system has shown that face-centered cubic (fcc) ordering is the equilibrium phase (Hales, 1997), above a threshold volume fraction (Pusey & Megan, 1986). Interestingly, body centered cubic (bcc) system was also discovered at ionic strength lower than 2.7 x 10^{-6} M KCl and volume fraction less than 0.008 (Monovoukas & Yiannis, 1989). The ability to change interparticle forces using electrolytes soon becomes a major advantage of colloidal system in modeling condensed matter.

Understanding all the interaction forces involved for particles in close proximity is nontrivial, as their magnitude or strength will decide their stability in suspension (no aggregation), ease of crystallization and final packing arrangement. Since many practical applications in interface science and colloidal science revolve around the problems of controlling forces between colloidal particles and between surfaces of different curvatures, many have devoted considerable effort to model surface forces and engineer their interactions in either short range or long range. It is impossible to give a detail perspective for all proposed models regarding colloids, only a few important concepts are introduced here, chosen based on the authors' preference for convenience. A more comprehensive review of all the interactions involved can be found in "Ordering and Phase transitions in Charged Colloids" (Arora & Tata, 1996).

2.2 Hard sphere model

One can first treat colloidal particles as hard and electrically neutral particles. Besides van der Waals forces, they only interact by steric repulsion when they are brought into physical contact. For monodisperse spherical particles, the close-packing limit is 0.74, which is equivalent to atomic packing factor of fcc atomic crystal. In another words, fcc packing arrangement would be the most stable form of colloidal crystal at equilibrium.

In practice, high concentration of colloidal suspension tends to flocculate before maximum packing limit is reached. Brownian motion allows particles to gain thermal energy easily and collide with each other to form clusters. If many aggregates form upon collision (especially at high temperature and high concentration), there is a high chance that amorphous colloidal aggregates will form, preventing further packing. In order to obtain FCC crystal structure, slow sedimentation of large particles in less concentrated suspension (volume fraction, v.f. << 0.5) by gravitational forces was explored (Crandall & Williams,

1977; Mayoral et al, 1997; Míguez et al, 1997). A phase transition from fluid to crystal can be observed as the local volume fraction increases beyond 0.49 (Pussey & Megan, 1986). This process is rather slow and may take up to a few days to weeks, depending on the initial concentration and desired volume of crystal grown.

In 1968, Hoover and Francis confirmed the existence of first-order melting transition for colloidal hard spheres and reported the densities of coexisting phases via Monte Carlo simulation (Hoover & Ree, 1968). It was found that melting and crystallization occur at reduced pressure of p = 8.27. The corresponding volume fractions, Φ is thus in the range of 0.50 to 0.55.

2.3 Charged spheres

Almost all colloidal particles are not electrically neutral. They may contain a large number of acid, base or other functional groups which are susceptible to dissociation in a polar solvent. Their functional groups are normally determined by the synthesis path or catalysts used. For instance, polystyrene beads synthesized using KPS (Potassium persulfate) catalyst will give rise to negatively charged particles. Hence, hard sphere model using exclusively steric repulsion is only a good approximation, but not an accurate approach to address real colloidal system.

Fig. 1 depicts the charge distribution surrounding negatively charged particles in water suspension. Given a neutral system of suspension, each colloidal particle of radius R carries negative charges of Ze. The water solvent contains an equivalent amount of counterions as well as possible stray ions such as salt (Na^+ and Cl^-). In a pure colloidal suspension with no salt present, the negative charges on particle surfaces will form thick double layer (~10^1 nm) and exert repulsion on each other, preventing formation of particle clusters. Part of the counterions will be attracted to near surface of negatively charged particles, while the remaining can move around in bulk suspension. If a small amount of salt (e.g. NaCl) is added into the colloidal system, positive ions (Na^+) will be attracted to particle surface, screening the repulsions between negatively charged particles. As a result, interparticle distance decreases and the attraction between particles could be enhanced. Super high salt concentration will collapse the double layer, and if the particles happen to be in physical contact by chance (Brownian motion), they will form irreversible aggregates immediately, giving no chance of forming ordered structures.

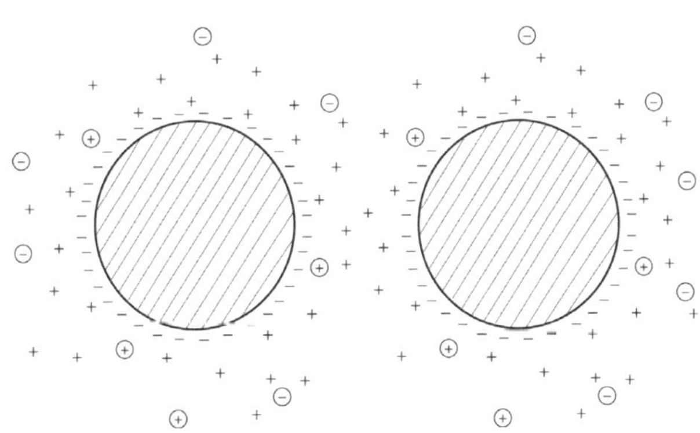

Fig. 1. Colloidal particles of radius R carrying negative charge -, with counterions + present in vicinity. + and – signs which are enclosed in circles represent stray ions.

2.4 DLVO model

To further assist reader, a model based on DLVO theory is introduced. It is named after Derjaquin, Landau, Verwey and Overbeek. The theory describes the interaction forces between charged surfaces in a liquid medium. It combines both the van der Waals forces and electrostatic interaction of charged surface, computed in mean field approximation in the limit of low surface potentials.

Here, it is assumed that:

a. Each colloidal particle is surrounded by its own double layer and behaves as if it was electrically neutral unless it approaches another particle closely enough
b. Only small and very mobile ions are first allowed to move. The mean field must be consistent with the boundary conditions on the surface of colloidal particles in suspension and the distribution obeys Boltzmann statistics (Brownian motion).
c. The potential energy of an elementary charge on a surface is much smaller than the thermal energy, kBT.
d. The particles are allowed to change position and find their own equilibrium states.

For DLVO potential,

$$V_{DLVO} = \frac{64\pi R c_o \Gamma_o^2 kT}{k^2} \exp(-\kappa r) - \frac{AR}{12r} \tag{1}$$

R is particle radius, c_o is ion concentration, k is Boltzmann constant, T is temperature, κ is inverse double layer thickness, r is separation between particles, and A is Hamaker constant. Surface potential of particles can be described by

$$\Gamma_o = \tanh(\frac{ze\Phi_o}{4kT}) \tag{2}$$

where z is valence of counter ions, e is the electronic charge, and Φ is the surface potential. It is obvious that the first part of equation (1) is related to electric double layer interaction (repulsion by similar charge of particles) where as the second part can be referred to attractive van der Waals interaction. As discussed earlier, electric double layer provides an energy barrier against irreversible agglomeration. This energy barrier, as indicated in Fig. 2 (right) as V_{max}, normally has a magnitude ranging from 0 ~ 100 kT, depending on the suspension parameters. Unlike the atomic model, two minima exist between two approaching colloidal particles, which are irreversible (primary) and reversible (secondary) respectively. In atomic crystals, atoms are held strongly in the single equilibrium position (as indicated in Fig. 2 (left)). High energy barrier or chemical bonds must be broken to disrupt the ordering. A brief comparison of interaction potential between atoms and colloidal particles in close proximity is best illustrated in Fig. 2 for clarity.

Surprisingly, the existence of secondary minimum allows scientists to "anneal" or remove defects in ordered arrays of wet colloidal crystal at this metastable state, before reaching the final irreversible dry state of colloidal crystal. The related works of particulate mobility in wet colloidal crystal will be further discussed in section 3.2.

For crystallization to occur, van der Waals forces must exceed double layer repulsion. This could only happen when particles are so close enough, overcoming the energy barrier V_{max}. This is possible as double-layer interaction energy is finite or increases slowly when r approaches zero, while V_{vdw} decreases exponentially when r → 0. Unlike electrostatic interaction, van der Waals interaction is highly insensitive to the change of pH and electrolyte concentration. Small addition of salt may change the magnitude of V_{max} significantly and remove the secondary minimum. Thus, no ordering in long range is ever possible if this situation happens.

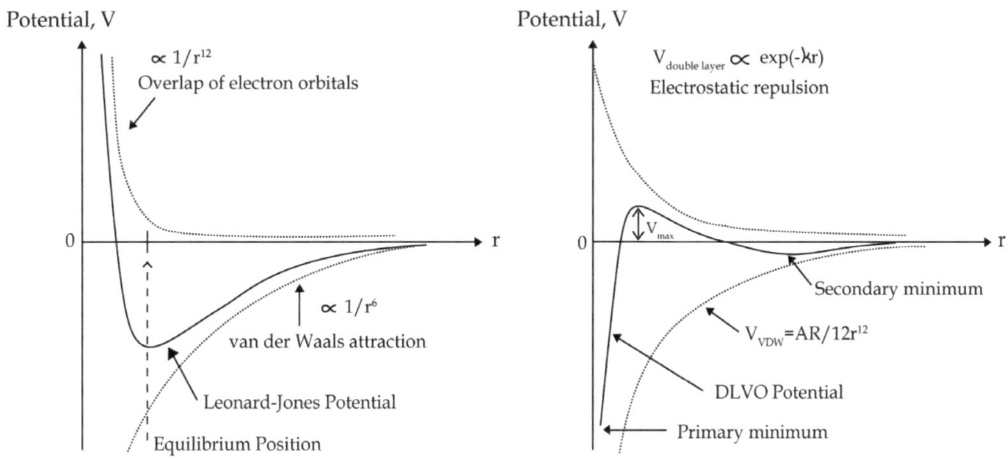

Fig. 2. A comparison of interaction potentials between atoms (Left) and colloidal particles (Right) in close proximity, described by Leonard-Jones potential and DLVO potential respectively.

3. Ordering and kinetics of colloidal particles

The ordering and kinetics of colloidal particles could be quite different from atomic crystals. The medium of colloidal dispersion gives rise to the dynamic motion of particles (Brownian) and exerts capillary forces on the particles near the interfaces (Kralchevsky & Nagayama, 1994). Fluid instability at interface (e.g. dewetting) further complicates the kinetics of crystallization process (Davis, 1980; Troian, 1989). All these contribute to several technical issues in achieving perfect crystal growth.

Hence, it is important to address the relevant critical forces involved in particle ordering. First, the proposed mechanism will be presented here with supporting research works from literature review and authors' publications. However, the discussion will be mostly focused on assembly process induced by capillary forces. Other assembly methods like e-field induced assembly (Holgado et al, 1999; Prieve et al, 2010; Yethiraj and Blaaderen, 2003; Zhang & Liu, 2004) and spin-coating assembly (Jiang & McFarland, 2004) can be referred to the recommendations in the reference list.

As mentioned in section 2, self-assembly via gravitation force is rather slow, and limits its application. Besides, the particles will only settle if their size and density is sufficiently high. This sedimentation tendency is best characterized by the Peclet number (see equation 3). Hence, this practical issue led to the studies of filtration and centrifugation (Park et al, 1999; Velev & Lenhoff, 2000), which however do not necessarily offer good control over packing quality and thickness of crystal grown. The formation of polycrystalline domains with different lattice orientation in colloidal crystals is commonly found, possibly due to nucleation at different locations of the specimen surface and the subsequent growth of crystal domains in different directions (Pusey et al, 1989).

Here we describe colloidal self-assembly under the effect of evaporation and capillary force, a phenomenon observed in the "coffee-ring" experiment [Deegan et al, 1997]. When a drop of coffee is spilled onto a solid substrate, a dense ring of stain will be formed upon drying, leaving the center of the initial droplet empty. The dense deposition of coffee solids near the outer ring signifies an intense movement of solid particles from the droplet interior to the drying perimeter. This driving force is attributed to evaporation, causing outward capillary flow (both solvent and solids) to the pinned contact line of the drying droplet, in order to replenish solvent loss from the edge (Dushkin et al, 1993). Deegan and his group then confirmed a power-law growth of ring mass with time in coffee-ring deposition, a mechanism which is independent of substrate type, carrier fluid and deposited solids. In short, the deposition process could be very fast, depending only significantly on the evaporation rate.

Dating back to 1992, the mechanism and stages of evaporation-induced assembly was first investigated in detail using an experimental cell containing a thin well of monodisperse micrometer-size latex particles, which allowed in-situ microscopic observation (Denkov et al, 1992). Similar to coffee-ring experiment, the particles were brought convectively to the edge of evaporation front, and the ordering of particles was found to be initiated when the thickness of water layer approached the underlying diameter of particles. Examples of particles moving to the drying edge of water film are shown in Fig. 3 and Fig. 4. Self-assembly starts when water thickness is about one particle diameter (860 nm) and it is normally monolayer near the perimeter of drying droplet.

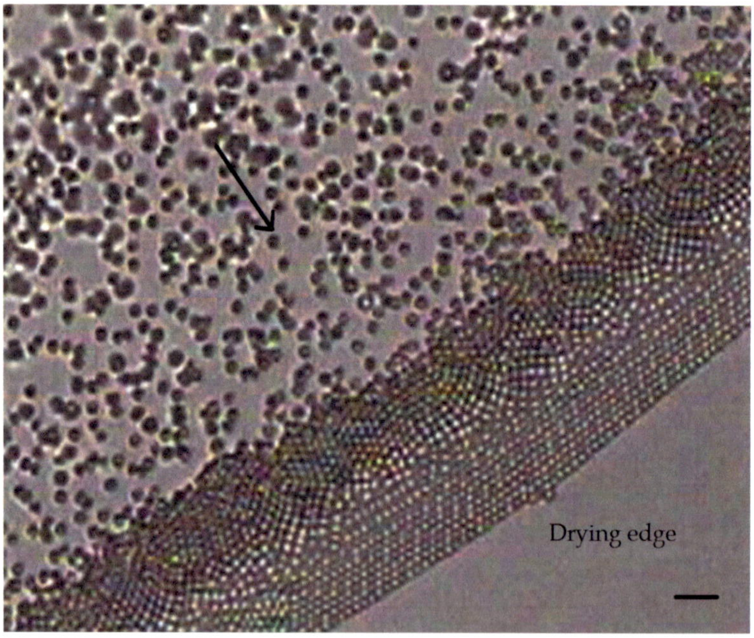

Fig. 3. Evaporation driven self-assembly in sessile droplet. Polystyrene particles (860 nm) move convectively to the drying edge, forming monolayer crystal at the outer perimeter. Transitional phases (e.g. buckling, square) are observed few particle diameters away from the perimeter (copyright reserved). Arrow indicates the direction of particle movement.

In evaporation-induced self-assembly, much attention is paid to the drying stage when the thickness of water layer is close to particle diameter. As evaporation proceeds, thinning of water layer causes deformation of menisci (shown in Fig. 4) between particles at drying front. Further evaporation from the concave menisci increases the curvature and local capillary pressure, driving more water influx from thicker water layer to thinner region. Partially immersed particles also experience capillary attraction, leading to further packing of particles to form close-packed crystal. It is also worth noting that large and partially submerged particles can be immobilized by the thinning water layer as the vertical component of surface tension force pressing the particles against the horizontal substrate is huge. This phenomenon also serves as a condition for contact line pinning, which we will discuss later in section 4.2. Besides, formation of multilayer colloidal crystal is possible if wetting angle is large or thickness of water layer is large (Jiang et al, 1999).

Due to the almost fixed contact line formed by colloidal droplet on horizontal substrate, a variation of colloidal crystal thickness is commonly observed, from one single layer near the edge to multilayer further away from the contact line (Fig. 3). In spite of evaporation, slight reduction of water layer thickness over time does not help in the uniformity. Hence, an approach to let contact line move along deposition direction was introduced. Fig. 5 illustrates an experimental setup of vertical deposition, where a substrate is submerged vertically or at an inclined angle inside a colloidal suspension of dilute concentration (< 0.1 volume fraction). The temperature of the suspension can be controlled by a water bath surrounding the suspension, driving the convective flow of particles to contact line for controlled deposition. As long periods of evaporation may change suspension concentration over time, the temperature is normally adjusted to be very low, close to room temperature. Then the moving contact line is driven accordingly by withdrawing the substrate or pumping out the colloidal suspension at a controlled speed (Zhou & Zhao, 2004). The speed of the moving contact line will then determine the thickness of colloidal crystal obtained.

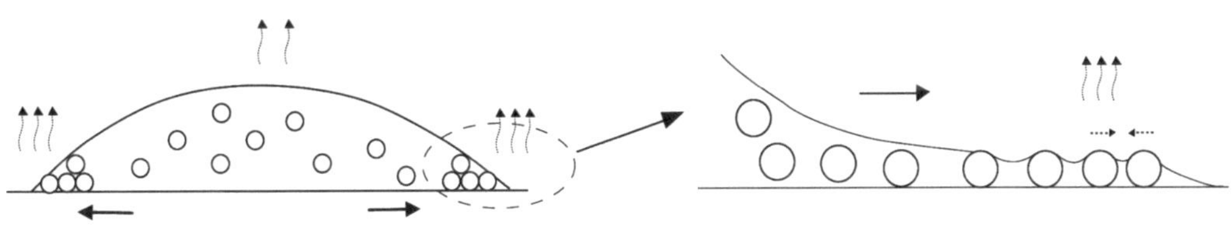

Fig. 4. Sessile droplet of colloidal suspension is shown on the left and a magnified version of thinning water layer of colloidal particles is depicted on the right. The scaling is for illustration purpose only. Curvy arrows indicate evaporation or water loss, which is faster at the edge. Straight arrows indicate convective flow of particles to the edge of sessile droplet. And dashed arrows on the right figure indicate capillary attraction between two partially submerged particles.

Other parameters such as incline angle of submerging substrate, surfactant addition, evaporation rate, ambient temperature, solvent volatility, particle concentration and substrate hydrophilicity have been explored to control the thickness of deposition (Denkov et al, 1992; Dimitrov & Nagayama, 1996; Im et al, 2003; Jiang et al, 1999; Kralchevsky & Denkov, 2001; Mclachlan, 2004). All these parameters must be optimized together with the speed of moving contact line, in order to obtain large area of uniform colloidal crystal.

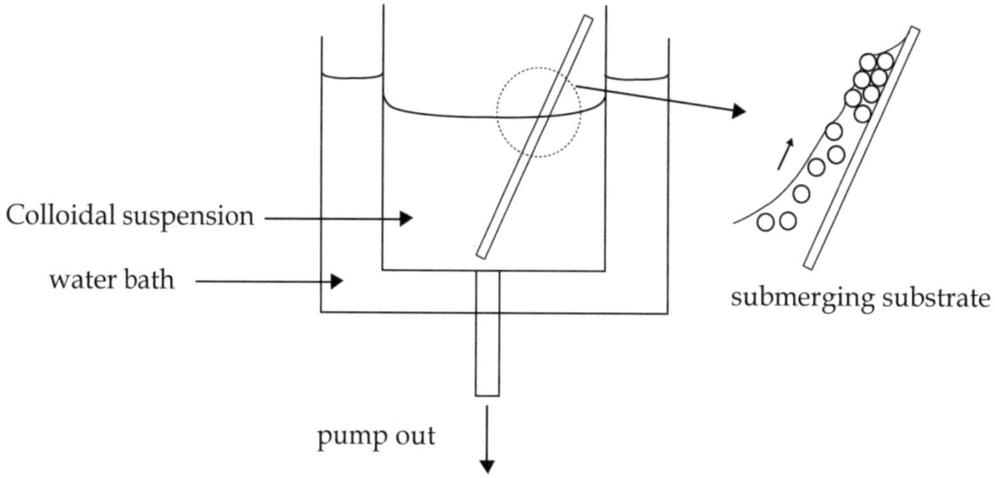

Fig. 5. Figure depicts an experimental setup for flow-controlled vertical deposition, an assembly method to control movement of contact line along crystal growth direction (Zhou & Zhao, 2004). Inset on the right shows capillary rise region near the submerged substrate.

3.1 Kinetic stages of colloidal crystallization

Since capillary forces can be significant in influencing the final assembly during drying process, systems with continuous environmental changes must be taken into account if one aims to design a process to obtain long-range ordered structures. There was also a desire to correlate the microscopic interaction and macroscopic boundary forces to provide a more complete model of colloidal self-assembly. The macroscopic forces are primarily capillary forces experienced by partially immersed particles under the thinning of water layer. The microscopic forces are contributed by electrostatic, steric interactions and van der Waals forces between closely arranged particles. In the case of larger particles, they will experience stronger effect of macroscopic boundaries before they have a chance to interact by microscopic forces. One can assume dominant function of capillary forces in this type of system. On the contrary, particles which are relatively smaller than the volume in which they are confined will be driven by microscopic interaction forces before experiencing capillary forces.

In order to understand the transition dynamics from well-dispersed colloids in suspension to dry colloidal crystalline film by evaporation-induced self-assembly, in situ transmission spectroscopy was introduced to monitor the assembly stages (Koh & Wong, 2006; Koh et al, 2008). Treating the assembly process as an emergent photonic crystal, several sequential stages of colloidal ordering with different lattice parameters in wet suspension could be interpreted based on Bragg's law (equation 4). Fig. 6 depicts an experimental setup by Koh et al, to investigate the self-assembly stages at the meniscus near the contact line formed on vertical substrate in vertical deposition.

Due to the good wettability of the hydrophilic glass substrate, thin meniscus formed near the substrate was higher than the bulk meniscus, providing a clear line of sight for transmitted beam. As evaporation proceeds, consistent arrival flux of particles at the thin meniscus could either stay or sediment back into bulk suspension. The tendency to sediment can be described by Peclet number, Pe:

$$Pe = m_B g R / kT \tag{3}$$

where m_B is buoyancy mass of a particle with radius R and g is the gravitational acceleration. This number is also a good indicator to determine the stability of a colloid suspension. For Pe >> 1, the particles will tend to sediment and form agglomerates. In vertical deposition, commonly used particles are small, having Pe of unity or smaller. For instance, polystyrene spheres of 0.1 μm and 10 μm in water solvent at a temperature of 23°C give Peclet number of 5×10^{-5} and 5030 respectively.

A sequence of transmission spectra were taken from time t_1 to t_5, as shown in Fig. 7. The diffraction features shown in Fig. 7 can be correlated to photonic band gap structures (Ho et al, 1990; Koh et al, 2008). Under slow evaporation, it can be assumed that ordering in the suspension will lead to the equilibrium FCC structure (Monovoukas & Yiannis, 1989). Besides, (111) plane of colloidal crystal obtained is confirmed to be parallel to the substrate, providing an important reference for photonic band gap calculation

The spectra was at first featureless as light was hardly transmitted due to random scattering by the disordered structure. As local volume fraction at thin meniscus increases with evaporation, first feature A was observed at t_1 (300 min). This indicates the onset of order in the colloidal suspension: the first kinetic stage of ordering. Here, the rising local concentration shrinks the average distance between particles. Since particles are fully submerged in solvent, it is believed that the interaction forces between the particles are the only driving force for ordering, overwhelming the randomizing Brownian forces. It is also found that feature A corresponds to a larger lattice parameter (368nm) of fcc colloidal crystal, compared to 276 nm of the expected equilibrium colloidal crystal formed by hard-sphere packing of the polystyrene particles with diameter of 195 nm. This transition structure is stabilized by the interparticle forces, in which the existence of DLVO potential barriers prevents the particles from coming into direct contact. Here, we treat the interaction potential between two particles in a solvent using DLVO theory (see section 2.4) where the total potential is taken as the sum of the repulsive and attractive forces.

With continued growth of the colloidal crystal, a second feature (B) appears at a shorter wavelength (from t_2 onwards in Fig. 7). This corresponds to wet FCC colloidal crystal with the particles in direct physical contact (zero separation) with water in the interstices. Water-retaining capillary pores are normally formed in the interstices of deposited colloidal crystal as the meniscus recedes below the self-ordered crystal. The subsequent loss of water upon drying of wet colloidal crystal immediately gives rise to a blue shift, feature C at t_4 (770 min) and t_5 (800 min), which can be correlated to the change of dielectric contrast as water in the interstices is replaced by air. In the transitions from t_1 to t_3 and t_3 to t_5, coexistences of the two features A and B are observed at t_2 and t_4 simultaneously. These are likely due to the simultaneous existence of the corresponding transition structures where not all region of wet colloidal crystal shrink and dry at the same time. The growth of intensity for feature B from t_2 to t_3 indicates the area increase of double-layer collapse across the studied area of colloidal crystal (area of the incident light beam). Also, the intensity increase from t_4 to t_5 explains the continual evaporation of water from the wet colloidal crystal to form dry crystal, revealing feature C.

The essence of this work is the demonstration of three distinctive stages in colloidal self-assembly, as shown in Fig. 8. This model is further supported with sequential changes of lattice parameters derived from the transmission spectra in Fig. 7. It was then confirmed that feature A had a lattice parameter of 368 nm, which was larger than the equilibrium colloidal crystal with lattice parameter of 276 nm. The larger interparticle distance in wet suspension during self-assembly indicates the existence of DLVO potential barrier, which prevents particles from coming into direct contact. This finding highlights the important role of interaction forces for small particles, despite the earlier understanding that colloidal self-assembly at liquid menisci is driven solely by capillary forces. This could only be true for the sizes of the particles that are small relative to the volume that is confined by the macroscopic boundaries. It is believed that the capillary forces are only brought into action from the second stage onwards, where the forces collapse the electric double layer, thus bringing

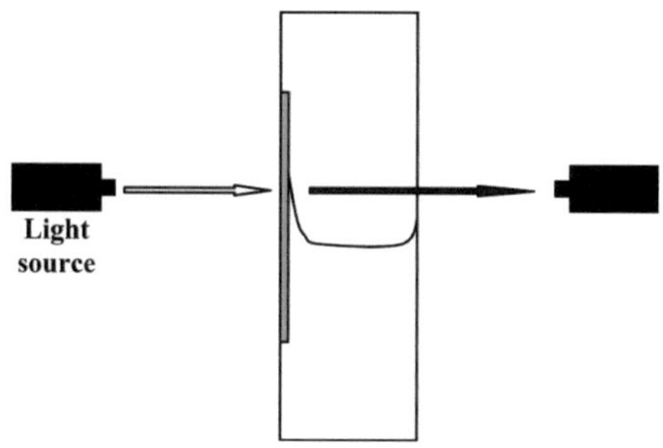

Fig. 6. Schematic shows an in situ transmission spectroscopy of colloidal self-assembly. A glass substrate is located in a plastic cuvette of colloidal suspension. This whole apparatus is kept in a temperature-controlled chamber, which controls evaporation from the suspension. Reprinted with permission from *Langmuir,* Vol. 24, No. 10 (May 2008), pp. 5245-5248. Copyright 2008 American Chemical Society.

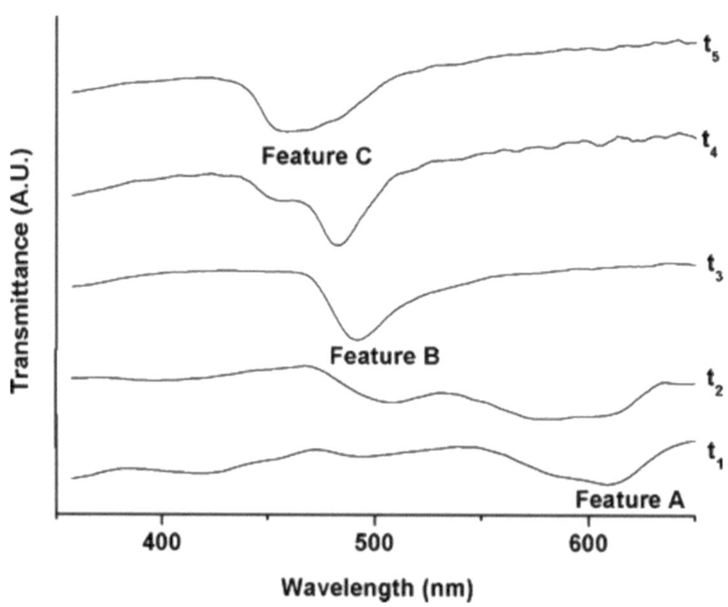

Fig. 7. Transmission spectra show emerging features A (608 nm), B (491 nm), and C (462 nm) at different time interval. Two features are observed simultaneously at times t_2 and t_4. The spectra are offset vertically for clarity. Spectra are taken at t_1 (300 min), t_2 (320 min), t_3 (400 min), t_4 (770 min) and t_5 (800 min). Reprinted with permission from *Langmuir,* Vol. 24, No. 10 (May 2008), pp. 5245-5248. Copyright 2008 American Chemical Society.

particles into direct contact, with solvent trapped in the interstices (Fig. 8, middle). The assembly process then ends with the final replacement of water with air, bringing a change in periodic dielectric contrast as water evaporates from wet colloidal crystal (Fig. 8, Top). Once this stage is achieved, the colloidal crystal obtained will be stable and strong enough to resist structural changes against liquid infiltration process. This robustness enables infiltration with other functional materials to obtain inverse opal structures or opals with different material properties via double templating (Yan et al, 2009).
Understanding the dynamic transition of colloidal crystal may provide some insights towards improving long-range quality of colloidal crystal. First, the electric double layer around each particle must be as thin as possible, and yet prevent premature aggregation. This is because large double layer thickness will give rise to large shrinkage stress during the final collapse of double layer (A to B), resulting in macroscopic cracks (Jiang et al, 1999). In the next section, the manipulation of DLVO potential to study particulate mobility of self-assembly process will be discussed.

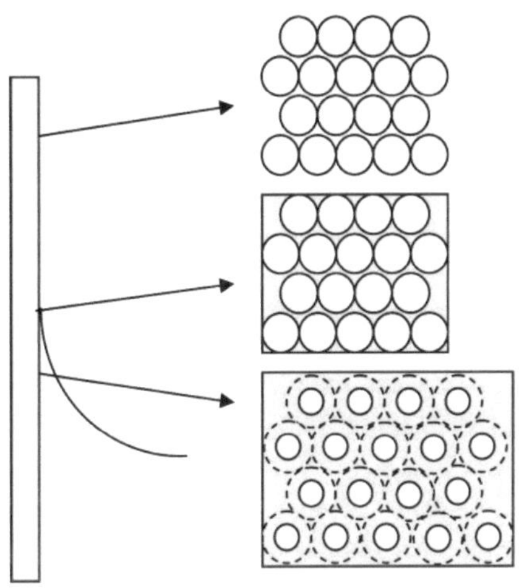

Fig. 8. Three distinct stages in colloidal self-assembly process. First stage (bottom) shows a transition structure with a large lattice parameter, corresponding to feature A in Fig. 7. As the meniscus moves, this structure collapses to a smaller lattice parameter (middle), with water retained in its interstices, giving rise to feature B in the transmission spectra. Finally, a dried colloidal crystal (top) corresponding to feature C is obtained. Reprinted with permission from *Langmuir,* Vol. 24, No. 10 (May 2008), pp. 5245-5248. Copyright 2008 American Chemical Society.

3.2 Particulate mobility in vertical deposition

The absence of repulsion forces between charged particles and oppositely charged surfaces often lead to disordered clustering (Yan et al, 2008). At high particle concentration in a confined volume (e.g. thin meniscus), repulsion force is the key to prevent irreversible clustering and the prerequisite for sufficient particulate mobility to obtain higher packing quality, if sufficient time of ordering is given. Since understanding the collective behavior of the particles in an environment of high mobility is indispensable, the mobility and electrostatic interactions of negatively charged substrates and positive colloids were studied, with optimized parameters of ionic strength, volume fraction, and solvent evaporation temperature in vertical deposition (Tan et al, 2010).

3.2.1 Stick-slip behavior of colloidal deposition

When the positive polystyrene colloids are self-assembled on a negatively charged substrate, a uniform array of alternating linear patterns is usually obtained with limited widths (Ray et al 2005; Tan et al, 2010). As shown in Fig. 9, these 1D particulate bands are deposited at relatively regular intervals across the entire substrate. A magnified view of SEM photos reveals close-packed ordering within each band. However, the ordering quality is poor with abundant point and planar defects.

This alternating band is no different to the case of negatively charged particles being self-assembled on glass surface of same charge (Teh et al, 2004). The only difference is the presence of tiny clusters scattered within the so called "empty band" region. As discussed by Teh, the alternating bands can be attributed to the stick-slip motion of meniscus growth front during deposition, while the presence of scattered clusters in the "empty bands" could be caused by electrostatic attractions between colloids and substrate. Since the surface charge density of silica glass was determined to be -0.32mC/m^2 (Behrens & Grier, 2001), strong electrostatic attraction will immobilize positive particles if they happen to come close to the glass surface. This could explain the disordered random ordering of positive colloids in the "empty-band" region.

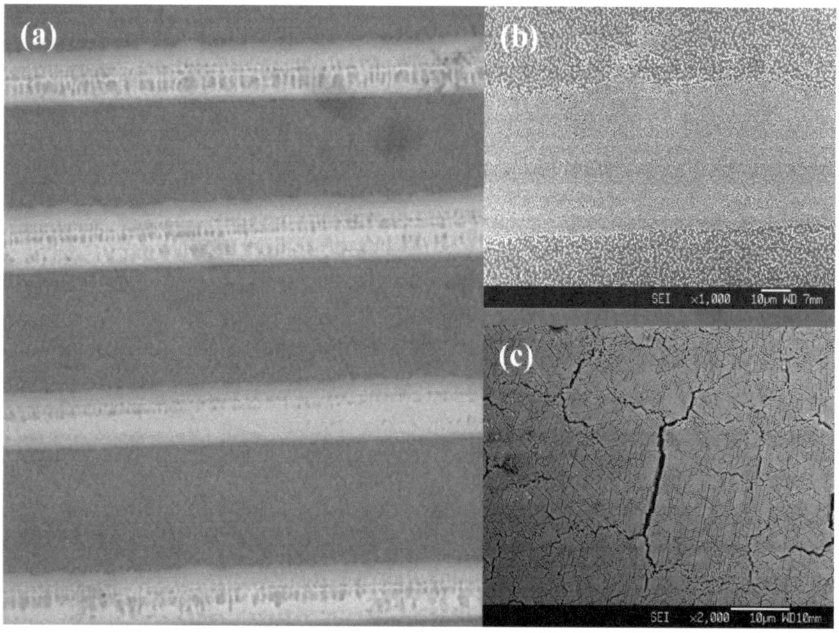

Fig. 9. (a) Microscopic picture shows interplay of stick-slip deposition and attractive deposition of positive colloids on negatively charged glass surface. (b) and (c) are magnified SEM photos showing alternating bands of ordered region and scattered clusters of random ordering. Particle concentration = 0.5 vol%, temperature = 35ºC. Reprinted with permission from *Langmuir*, Vol. 26, No. 10 (May 2010), pp. 7093-7100. Copyright 2010 American Chemical Society.

At the pinned contact line of meniscus, meniscus will first recede and deform with the consistent withdrawal of solvent in flow-controlled vertical deposition. Then the continual flux of particles to the thinning region will increase the local volume fraction and decrease interparticle distance, causing the ordering to nucleate in the confined meniscus. If the meniscus thickness is larger than one particle diameter and recedes much slower than particle deposition, multilayer band with hcp orientation will be obtained. When the solution recedes further with contact angle reaching the minimum receding angle, the meniscus contact line will slip rapidly to a lower pinning level with a new contact angle. This process is then repeated with dynamic change of angle and alternating bands of deposition. More details can be found in Teh's published work (Teh et al, 2004).

3.2.2 Effect of volume fraction and ionic strength

Multilayer colloidal crystal shown in Fig. 9 was obtained with 0.5 vol% of colloidal suspension at 35°C. When a more dilute concentration of 0.1 vol% was used (not shown), multilayered bands can be replaced by monolayers with locally ordered configurations. This can be explained by the lower particle flux to the drying edge during deposition, at the same withdrawal rate of suspension. At the same time, longer time is required for colloids to reach threshold concentration in the confined thin meniscus, implying that particles will have sufficient time to self-assemble in an orderly manner. Larger interparticle distance and lower frequency of collision due to Brownian motion are believed to slow down irreversible aggregation and random electrostatic adsorption of particles onto charged substrate. Hence, lower volume fraction of colloidal suspension is normally used to obtain long-range ordered crystal as slow increment of local volume fraction at thin meniscus is expected to impart greater inplane colloidal mobility during assembly process. Another comparison of concentration effect (0.05 and 0.01 vol%) on monolayer crystalline quality is given in Fig. 10. Dilution leads to better ordering quality, in agreement with other work (Zhou & Zhao, 2004).

Besides volume fraction, inplane mobility is also affected by electrostatic interaction between ordering particles in thin meniscus layer (Maskaly, 2006; Tan et al, 2008). For positive colloids, addition of salt (ionic strength increases) will reduce Debye screening length of the electric double layer. The result of this is twofold. First, positive colloids of similar charge will approach each other closer, and can be configured into stable in-plane ordered array with minimal cracks upon drying. Second, shorter Debye length will give positive colloids extra time to form ordered array, before being adsorbed onto negatively charged surface by electrostatic attraction. For example, the addition of 10 μM KCl was reported to give highest density of hcp domains, further supported by the distinctive 6-fold coordinated diffraction spots of a hexagonal lattice (Fig. 10 b). Detailed evidence can be referred to the relevant publication (Tan et al, 2010).

Besides, Tan also postulated that the assembly of charged colloids may achieve an intricate balance between particle-particle repulsion and particle-substrate attraction, when a colloidal suspension of low volume fraction and low ionic strength is used. This is a condition where the particles are sufficiently far apart to reorient themselves into geometrically and thermodynamically favored close-packed arrangement.

Fig. 11 shows the phase behavior of positive polystyrene colloids assembled on a negative silica glass substrate at 25°C, at various ionic strength and volume fraction. It indicates that aggregates are likely to occur at high ionic strength across the whole studied range of volume fractions (0 to 0.1 vol%). For low ionic strength (< 10μM), long-range hcp ordering could be obtained with the use of low volume fraction (< 0.6 vf%).

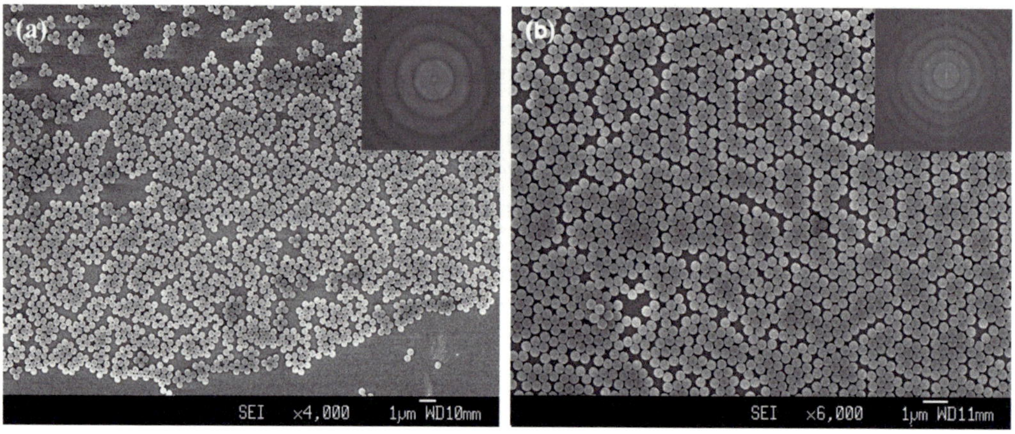

Fig. 10. Hexagonal close-packed domains and corresponding FFT inset obtained from vertical depositon at 25°C and 10 μM KCl. (a) Particle concentration = 0.05 vol % (b) Particle concentration = 0.01 vol%. Reprinted with permission from *Langmuir*, Vol. 26, No. 10 (May 2010), pp. 7093-7100. Copyright 2010 American Chemical Society.

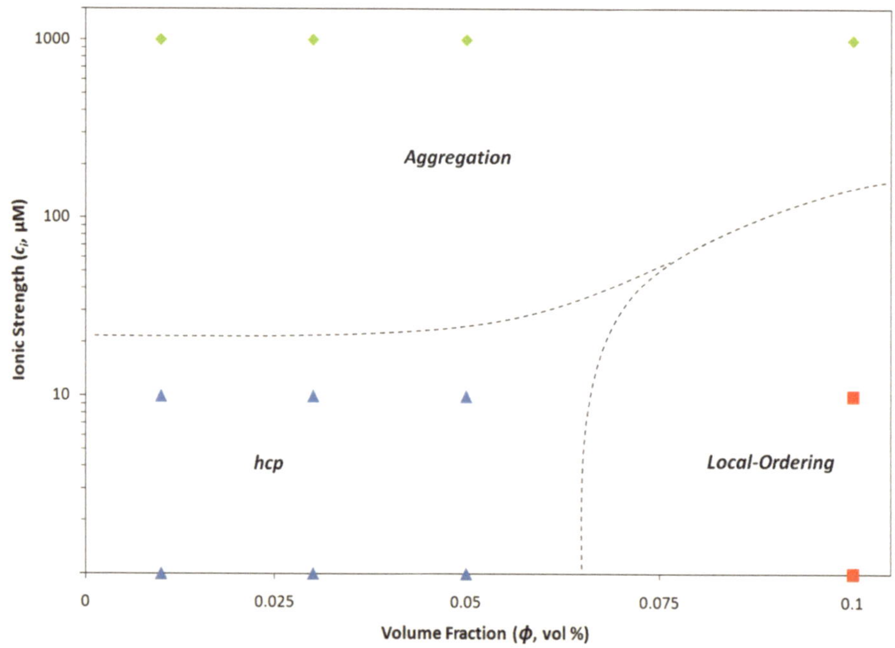

Fig. 11. Phase behavior of self-assembled colloidal structure, by deposition of positively charged polystyrene colloids on a negative borosilicate glass substrate at 25°C. Approximate boundaries between the ordered and glassy phases at different conditions are indicated by dashed lines. The axis of ionic strength is plotted in logarithm scale. Reprinted with permission from *Langmuir*, Vol. 26, No. 10 (May 2010), pp. 7093-7100. Copyright 2010 American Chemical Society.

3.2.3 Effect of temperature

There are two major effects brought by temperature increase (Dimitrov & Nagayama, 1996; McLachlan et al, 2004). First, high evaporation rate drives higher flux of particles to the drying edge, leading to faster crystal growth rate. Larger thickness of colloidal deposition is possible if solution withdrawal and meniscus deformation rate (thickness reduction) is much slower than particle flux. Second, kinetically active particles will bump into each other with high frequency. If they are of similar charge with substrate, higher ordering could be obtained since mobility is sufficiently large for further packing. However, substrate of opposite charge will likely immobilize colliding particles, giving no chance of good ordering (irreversible clustering).

3.2.4 Mobility in binary colloid

Earlier discussion has been devoted to the fine control of PS colloidal mobility in single-component crystallization. However, close-packed lattices have limited available symmetries (FCC or HCP) and associated properties which are too restrictive for diverse potential applications, especially in photonics. Using a mixed suspension of two particle types, colloidal crystals with lower symmetries can be made possible to provide novel properties in photonics, sensing and filtering (Bartlett et al, 1992; Kitaev & Ozin, 2003; Sharma et al, 2009; Tan et al, 2008).

Layer-by-layer growth is commonly used to self-assemble these structures, with right conditions of surfactants, temperature and ionic strength. Unlike the uniform negatively charged glass substrate used in self-assembly of positive colloids (see section 3.2.2), an underlying negatively charged L-colloidal template can be used to provide a periodic potential landscape to assist ordering of the next layer of colloids (named S-colloids) with opposite charge. These steps could be repeated to get additional layers of LS_n structures.

In LS_2 structures (see inset in Fig. 12a), each interstice in the first layer of hexagonally close packed (hcp) particles (L) is filled by one small particle (S). LS_6 structure is also possible where each interstice of first layer is filled by three particles (S) instead (Sharma et al, 2009). Unfortunately, both lower and higher densities of binary structures are usually observed together in LBL experiments. This led to an extended work to study the particulate mobility of colloids over an ordered potential landscape of an assembled hcp monolayer of opposite charge (Tan et al, 2010).

In previous work, Tan reported that a low ionic strength of 10 μM KCl could vastly improve the ordering of attractive binary colloidal structures in layer-by-layer (LbL) growth, presumably because the S-colloids possess sufficient mobility to self-assemble into a highly symmetrical LS_2 2D-superlattice. Besides ionic strength, crystalline quality also depends strongly on evaporation temperature and the most uniform LS_2 was obtained at low temperature (25°C). Fig. 13 shows a comparison of temperature effects (25°C vs 35°C) over LS_2 assembly using 10μM of KCl electrolyte. Similar to the case of single-component crystallization on glass surface of opposite charge, lower evaporation rate and slow crystal growth rate provide additional time for S-particles to reorient and stabilize into a thermodynamically favorable in-plane LS_2-superlattice (in suspension), before settling onto the oppositely charged template (L-layer). On the other hand, a slight increase of temperature to 35°C disrupts the inplane ordering structure, resulting in disordered binary arrays. This can be explained by greater Brownian motion of particles and faster loss of water due to evaporation, in which water is the key to mobility during ordering.

Thus a consensus can be drawn that an intricate balance of the particulate mobility by all three parameters, (1) evaporation temperature, (2) volume fraction, and (3) ionic strength, is critical for the quality of self-assembled colloidal crystals. Other than these, the delivery speed of the particles to crystal growth front should equal the crystal growth rate. These can be optimized by evaporation control and meniscus receding rate. Regardless of the type of substrates and the charge of particles used, particulate mobility must be assured to guarantee sufficient time of reordering to achieve final irreversible crystallization. Next we will discuss various templating efforts to obtain perfect single crystal and crystals with complex symmetry, which are impossible to be achieved by conventional self-assembly on bare substrates.

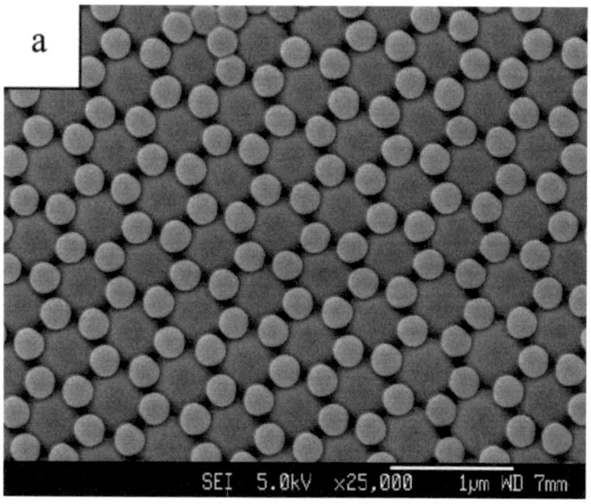

Fig. 12. Layer-by-layer assembly of positive colloids (250 nm) onto a hcp monolayer of negatively charged particles (550 nm) in flow-controlled vertical deposition, revealing LS_2 structure. Reprinted with permission from *Langmuir*, Vol. 26, No. 10 (May 2010), pp. 7093-7100. Copyright 2010 American Chemical Society.

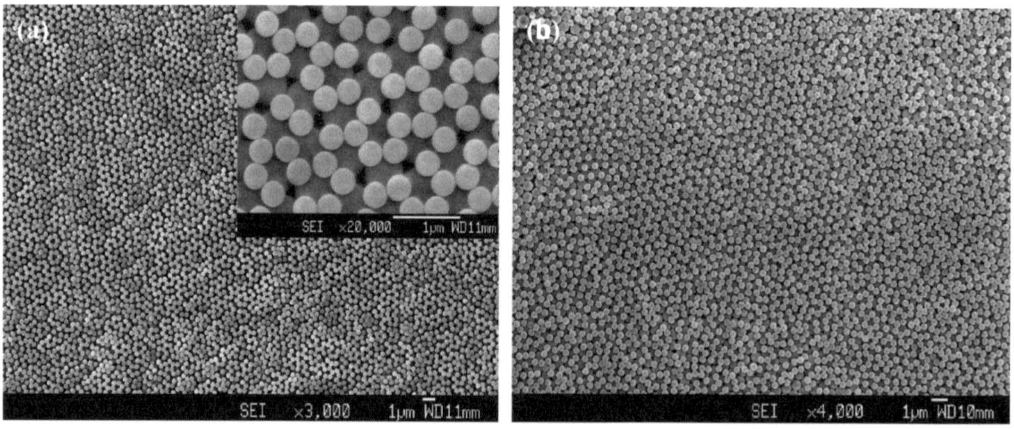

Fig. 13. Layer-by-layer assembly of positive colloids (371 nm) onto a hcp monolayer of negatively charged particles (601 nm) in flow-controlled vertical deposition, at temperature of (a) 25°C and (b) 35°C. Reprinted with permission from *Langmuir*, Vol. 26, No. 10 (May 2010), pp. 7093-7100. Copyright 2010 American Chemical Society.

4. Towards perfect crystallization

4.1 Template-assisted self-assembly

Despite many efforts to grow large-area perfect crystal (> 1000μm^3) that are useful for optical devices, most do not offer sound practicality for large scale integration. Sedimentation is extremely slow, limited by absence of control over number of layers and uniformity over topology. The method based on vertical deposition requires strict control of surface charge density of particles or substrate, particle and electrolyte concentration, temperature, and humidity.

Geometrical confinement has been long studied to affect the phase behavior of colloidal particle ordering (Schmidt & Löwen, 1997; Ramiro-Manzano et al, 2007). By using thin parallel plates or a wedge cell of few particle diameters in gap, crystal transitions (e.g. buckling) can be observed with changes of cell thickness. Similar confinement approach was also extended by Park et al to obtain much better ordering and orientation compared to the colloidal crystals grown from bare substrate (Park et al, 1997). Using pressure, they injected a suspension of colloidal particles into a well-confined rectangular cubic cell, with solvents being drained out through the channels (< particle diameter) built lithographically along the side walls of the cell. This left behind accumulating particles at the bottom of the cell and rapid crystallization of particles over 1 cm^2 with well-controlled number of layers (1 ~ 50 layers) could be easily attained. With sonication (Sasaki & Hane, 1996), the packing quality of close-packed lattice could be further improved under flow. This cell can then be dried off in an oven and dismantled later. The perfect colloidal crystal confined by these physical walls could then be integrated into device-making. Crytals grown this way have 3D domain size of 12 μm x 0.5 cm x 2 cm, which is almost equivalent to the cell size of 12μm x 2 cm x 2cm.

Coupling with the laminar flow of colloidal particles, flow-driven organization of particles in microchannels (rectangular grooves) was studied (Kumacheva et al, 2003). Using template with periodic rectangular grooves, the influence of the width and size of such rectangular grooves on colloidal self-assembly was investigated. Depending on the commensurability of the particle into the grooves, various structures like close-packed hexagonal, rhombic and disordered structures were reported. If a large mismatch (> 15%) exists between the ideal and experimental ratios, defects would be introduced in the crystal structure. In a subsequent study, a much narrower groove was explored, in which only parts of a colloidal sphere could (snugly) fit into. Precise ordering of colloidal lines was thus obtained (Allard et al, 2004).

Fig. 14. (a) Anisotropically etched V-shaped channels into a Si (100) wafer obtained using lithography. (b) SEM photo depicts six-layer (100) single crystal made of silica particles, sitting in a V-grooved channel. (c) Vacancy defects observed in the top layer of micro-spheres. (Yang & Ozin, 2000) – Reproduced by permission of The Royal Society of Chemistry.

In order to enable realization of colloidal crystals in devices, they must be fabricated into planarized microphotonic crystal chips. Ozin and his group pioneered the fabrication of planarized microphotonic structures by the deposition of micro-spheres into V-shaped grooves (apex angle = 70.6°) (Ozin & Yang, 2000 & 2001). The nucleation was believed to first occur along the apex of the V-grooves, followed by the subsequent layers of growth. The depth of such grooves will determine the number of layers of colloidal crystal formed and the close-packing tendency ensures projection of (100) crystal orientation, terminating at the crystal-air interface (see Fig. 14).

Emerging micro- and nanofabrication technologies in template structuring allows one to shrink the pattern size to tens of nanometers (e.g. e-beam lithography). The ability to create nano- or microfeatures consistently enables precise deposition of each particle and control over its packing symmetry, packing efficiency and packing quality of the resulting crystal. Using topologically patterned templates, Van Blaaderen et al showed that slow sedimentation of colloidal particles into the "holes" of topologically patterned templates enables formation of fcc colloidal crystals with crystal orientation of (100) planes parallel to the patterned surface (Blaaderen et al, 1997). This is different from the usual (111) plane orientation obtained, with respect to the surface. They succeeded to achieve large oriented crystals at which the defect structures were tailored by surface graphoepitaxy[2] approach. As the template used has a known orientation, photonic crystals grown could be sliced such that the exposed (001) and (110) facets of the fcc crystal structure could be integrated into specific applications. It was also shown that intentional mismatch of hole pitch and particle diameter can give rise to defect structures, such as randomly stacked (111) planes above the first few layers from the surface of template. This is similar to the case of growing epitaxial layer of CdTe (111) on GaAs (001) substrate, with a mismatch of about 14.6% (Bourret et al, 1993). Other than this, other cubic packing system like body-centered cubic (bcc) and simple cubic (sc) colloidal crystals have also been reported using similar approach (Hoogenboom et al, 2004).

It is obvious that templating offers a remedy to the shortcomings of spontaneous colloidal self-assembly, especially in manufacturing crystals tailored for realistic photonic applications. Other than the potential to obtain defect-free colloidal crystal of fcc structure with different plane orientation, it is also possible to assemble monodisperse colloids into complex structures or subunits (Romano & Sciortino, 2011; Vinothan et al, 2003), and then lead them to complex crystal symmetries of lower packing density.

According to photonic band structures calculated for various crystal symmetries and dielectrics (Ho et al, 1990; Pradhan et al, 1997; Busch & John, 1998; Moroz & Sommers, 1999; Vlasov et al, 2000), it was confirmed that an fcc colloidal structure has a PBG only in the second Brillouin zone (second-order Bragg diffraction), not in the first Brillouin zone (Blanco et al, 2000). Besides, a sufficiently high refractive index contrast (> 2.8) between the building blocks of the fcc crystal (colloidal particles and the interparticle space) is required to obtain a full omnidirectional band gap. Furthermore, photonic properties of commonly found fcc crystals are very sensitive to structural disorder (Vlasov et al, 2000). In this regard, nonspherical particles also offer immediate advantages in applications that require lattices with lower symmetries and higher complexities.

---

[2] Growth of crystal by substrate topology as opposed to atomic lattice in which a material is crystallized onto an existing crystal of another material, resulting in effective continuation of the crystal structure of the substrate.

Templates used for "nucleation" of complex colloidal clusters or crystal layers are usually engineered by modification of the surfaces via lithography (Chen et al, 2000; Choudhury, 1997; Rijn et al, 1998) and chemical patterning (Bertrand et al, 2000; Delamarche et al, 1998; Ulman, 1996; Xia & Whitesides, 1998). The control of surface chemistry like charge and functional reactivity can be obtained by coating an adhesive monolayer which is specific to the substrate, called self-assembled monolayers (SAM) (Ulman, 1996). For example, one can use thiol functional groups for gold surface, and silanes for silica substrate. The functional groups of this self-assembled molecular layer will then adhere to the corresponding surface with the other desired ends (e.g. charge for particle interactions) projected outwards. Besides, one can also use SAMs to coat different charges (positive, negative or neutral) on each crystal layer grown via layer-by-layer method to obtain binary colloidal crystal in vertical deposition.

Other than direct photolithography or e-beam lithography to create paterns on these SAM layers, SAM patterns can also be transferred to a flat substrate by soft lithography (Xia & Whitesides, 1998) and nanoimprint lithography (Hu & Jonas, 2010; Torres, 2003). Subsequent ordering of colloidal particles on such chemically defined patterns can be achieved in vertical deposition via electrostatic interaction and capillary forces. As discussion in section 3.2, charged microspheres can be self-assembled on the oppositely charged areas of the patterns when the substrate is slowly taken out from the colloidal suspension (Fustin et al, 2004). Depending on the size ratio of colloidal particle and patterned area, complex colloidal aggregates can be grown (Lee et al, 2002).

Despite progressive advancements in lithography systems, high facility cost and maintenance impede practical use of templating in colloidal self assembly. Hence, it is worth exploring simple templating like V-groove, and low-cost templating system like soft lithography and nanoimprinting, to enable innovative ways for template-assisted self-assembly. Next, we will discuss our approach to utilize meniscus pinning to control positional nucleation and inplane-oriented growth of large area monolayer colloidal crystal from one straight surface relief.

4.2 Controlling inplane orientation of large area monolayer colloidal crystal

Among various top-down and bottom-up methods discussed earlier, capillary forces induced convective self-assembly is attractive, requiring only a simple and economical setup. However, common nonidealities like thickness nonuniformity, restricted domain size and empty bands or voids are frequently reported. When a substrate is submerged into a liquid, a wavy contact line is commonly observed at air-liquid-solid interface, due to Rayleigh instability (Davis, 1980). In vertical deposition, it is believed that the trapping of colloidal particles along this wavy contact line will first lead to accumulation of particles, and then multidirectional initiation of colloidal crystal growth (Fig. 15c). Since the domain growth directions tend to be different along the wavy contact line; this eventually limits the final domain size of colloidal crystals obtained. Other than this, dynamic change of receding contact angle of colloidal suspension during liquid or substrate withdrawal will produce colloidal stripes or alternating colloidal and empty bands via the stick-slip mechanism (Adachi et al, 2005; Teh, 2004; Thomson et al, 2008).

Here, we demonstrate the usage of meniscus pinning by surface relief boundaries to control in-plane orientation of monolayer colloidal crystals without the interruption of grain disorientation. By printing a straight surface relief which has a strong affinity to water molecules (common solvent for colloidal particles), a straight wetting line of colloidal suspension could be pinned along the surface relief patterned (Fig. 15a). The photoresist SU-8 has been shown to work in this context. As most photoresists do not have good affinity to water (hydrophobic), their surface can be treated with UV ozone to improve wetting. A small addition of surfactant (SDS) and trapping of colloidal particles along wetting line do give enhanced pinning by almost 100%.

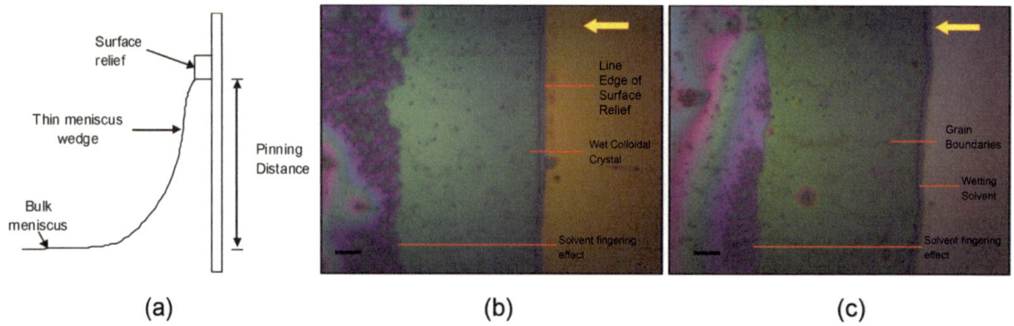

Fig. 15. (a) Schematic shows how water film is pinned and stretched near surface relief when the substrate is being pulled out of colloidal suspension. (b) Optical micrograph shows straight liquid pinning and nucleation along surface relief, and the subsequent inplane oriented growth. (c) Optical micrograph shows that colloidal nucleation on bare substrate always starts with wavy contact line and wavy nucleation line, giving rise to polydomain growth. As indicated, fingering effect of wetting solvent is observed above the wet assembled colloidal crystal. Yellow arrows show liquid receding direction. The scale bar is 8µm. Reprinted with permission from *Langmuir*, Vol. 27, No. 6 (March 2011), pp. 2244-2249. Copyright 2011 American Chemical Society.

The initial establishment of liquid pinning along straight surface relief will allow colloidal particle deposition along the thin meniscus wedge (Fig. 15b). Fast substrate withdrawal or receding bulk meniscus relative to colloidal deposition speed will pull the pinned contact line, either causing depinning or contact-line movement in a fingering pattern (Sharma & Reiter, 1996; Troian et al, 1989), together with the pinned colloidal domains. Hence, depinning of the initial contact line must be avoided.

By optimizing the pinning boundary and withdrawal speed, a well-controlled linear meniscus contact line allows a straight nucleation edge of monolayer crystal growth front, which then acts as a crystal growth seed, permitting the most close-packed direction <11> or <10> (as in 2D hexagonal lattice) to assemble along the surface relief. As a result, this unidirectional growth can give rise to single domain crystals with only twins and vacancies present as residual defects (see Fig. 16). More evidence can be referred to the supporting documents provided at the publication site (Ng et al, 2011).

Conservatively, the domain crystal size obtained can be as large as 1 mm^2, with residual defects of vacancies, twin boundaries and small misoriented domains. Despite these imperfections, the domain orientation of large crystal domain remains similar. More evidence in the form of SEM photos scanned sequentially can be found in the supporting documents published (Ng et al, 2011). To conclude, this novel approach could offer the desired ease of integration for device making, as the inplane-orientation of crystal grown can be easily identified by referring to the engineered surface relief.

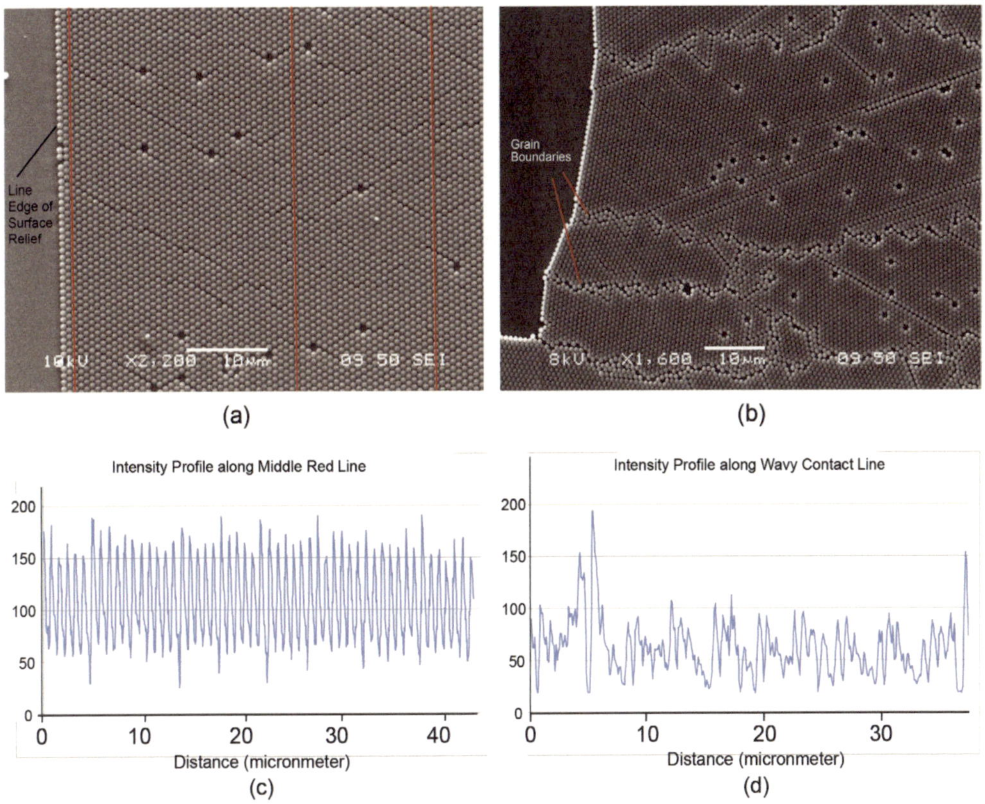

Fig. 16. SEM photo (a) shows inplane domain-oriented growth of colloidal crystal from the edge of straight surface relief, producing high degree of directionality. Particles are lined up in the close-packed direction <10> or <11>, along the surface relief. Red lines drawn serve as a guide to illustrate the perfect orientation under the straight pinning effect. (b) For comparison, colloidal assembly on bare substrate is shown, explaining the effect of wavy lines which result in domain growth of various directions. The desired growth directions are from left to right. (c) and (d) show line-plot profiles generated using ImageJ[3], along middle red line in part a and across central region in part b, respectively. Reprinted with permission from *Langmuir*, Vol. 27, No. 6 (March 2011), pp. 2244-2249. Copyright 2011 American Chemical Society.

5. Characterization of colloidal crystal

The most commonly used modern instruments in imaging dried colloidal crystals are scanning electron microscope (SEM) and transmission electron microscope (TEM). These types of imaging provide a quick view on the periodic structures grown via colloidal self-assembly; it does not however give quantitative data like crystal parameters in three dimensions. They are only good for 2D and topology scanning with smaller field of view

[3] ImageJ is a Java application popular for SEM and TEM image processing and analysis. More on http://rsbweb.nih.gov/ij/

(short range). If the particles are large enough (> 1µm diameter), their motion and ordering can also be observed with optical microscopes (Denkov et al, 1992; Pieranski, 1983; Yan et al, 2008).

Fortunately, the ability of colloidal crystal to diffract light allows one to characterize the crystal structure and quality with ease (Hiltner & Krieger, 1969). For colloidal crystals with lattice spacings in the order of visible-light wavelengths, diffraction method can be used in transmission mode or reflectance mode (Imura et al, 2009; Koh et al, 2006). As discussed earlier, in order for electromagnetic waves to diffract, it must obey the following Bragg's law. Direct reflection peak or transmission peak can then be recorded using UV-vis spectrometer.

$$\lambda = 2n_{eff}d_{111} \tag{4}$$

$$n_{eff} = \sqrt{(\Psi n_{ps}^2 + (1-\Psi)n_{air}^2} \tag{5}$$

$$d_{111} = (\frac{2}{3})^{\frac{1}{2}} D \tag{6}$$

where n_{eff} is effective refractive index of the colloidal suspension, d_{111} is the interlayer spacing between (111) plane, D is particle diameter, n_{ps} and n_{air} are refractive indices of polystyrene particle and air respectively, and ψ is the volume fraction of particles in suspension.

Unfortunately, strong interaction between light and the crystals could result in multiple scattering. It was observed that the Bragg spacings derived from diffraction measurements could deviate strongly from the real lattice spacings (Los et al, 1996). Besides, the available optical spectrum limits the number of Bragg reflections to be observed. These issues could be remedied by small angle X-ray scattering. First, X-ray interacts weakly with colloidal particles, serving as an excellent tool to probe internal structure of photonic crystal. Second, since there is a dramatic difference between X-ray wavelength (~1Å) and the particle diameter (e.g. 1µm), a tiny diffraction angle (narrow focus range) in the order of 10^{-4} rad will be able to supply sufficient information regarding the crystal (Thijssen et al, 2006). To conclude, radius, size distribution and internal structure of particle, crystal structure, lattice parameter and average orientation of colloidal crystal can be investigated via X-ray scattering (Megens et al, 1997; Vos et al, 1997).

Kossel lines, previously used in X-ray diagrams or electron diffraction experiments (Kikuchi lines) had also been used to examine phase transformation of colloidal crystal, crystal structures and their lattice parameters (Clark et al, 1979; Pieranski et al, 1981; Yoshiyama et al, 1984). Fig. 17 demonstrates a simple setup to obtain Kossel diagram, which is either projected on a spherical screen, V or a flat one, F. The colloidal crystal suspension is held in a glass or quartz cuvette, which is immersed in a spherical vessel V (diameter ~ 10 cm) filled with pure water. This water-filled vessel serves to minimize the refraction at the surface of the crystal. A divergent laser beam (normally He-Ne, λ = 632.8 nm) is then focused through a window and the diffracted beam will be projected on the spherical projection screen V or on a flat screen at distance X.

It should be noted that a 2D map of diffraction spots could be printed on the projected screen in Fig. 17, if a collimated white light is used. This offers a distinct advantage in immediate identification of the wavelength diffracted, based on the colour of the projected Laue spots. Other than white light, collimated laser and X-rays can also be used to obtain diffraction spots, by vary the scanning angle of the incoming beam (Williams & Crandall, 1974; Clark & Hurd, 1979).

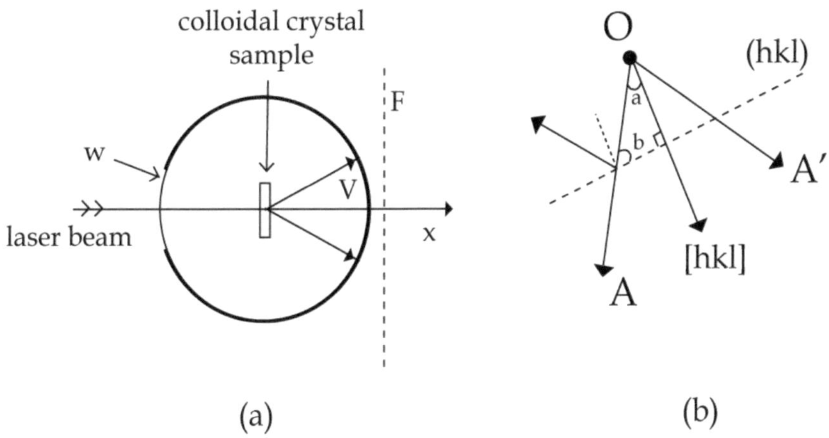

Fig. 17. (a) Experimental set-up of Kossel analysis, where a divergent laser beam is shone through thin colloidal crystal sample and the diffracted beam is projected on the curved screen V or flat screen F. (b) Diagram shows formation of Kossel cone in (a), where the Kossel lines are the intersection between the Kossel cone and the projection screen shown in (a). (Pieranski et al, 1981)

6. Conclusion

In spite of recent progress in colloidal self-assembly, large scale ordering of colloidal crystals with controlled packing symmetry, periodicity, crystal orientation and packing quality in a practical and economical application still remains an active scientific and engineering activity. This interest is intensively fuelled by the many fundamental scientists, application engineers and researchers working in tailored colloidal crystals exhibiting novel functions in applications. This chapter highlights not only our own work in colloidal self-assembly, but also the fundamental concepts and key approaches used to grow colloidal crystals. We think that particle-particle interactions in the process of assembly, mobility and space confinement remain the three crucial keys in fabricating performance-sound colloidal crystals.

For example, one could think of giving directional properties to colloidal particles and control the particle interactions in short range to form tetrahedral clusters. The clusters can then be directed into long range assembly via necessary confinement (e.g. meniscus, template, etc) to achieve equilibrium ordering of a diamond lattice. Simulations can also be deployed to assess the feasibility of related colloidal crystal structures for specific tailored synthesis, assembly and performance. However, the common inherent problem remains over the practicalities: speed, cost, area of crystal grown and ease of integration. In conclusion, the potential is huge, and there is still much room for future research.

7. Acknowledgment

We acknowledge all the works contributed by our group members, including those who have left for good cause. They are Mr. Tan Kwan Wee, Dr. Yan Qing Feng, Dr. Teh Lay Kuan and Dr. Yip Chan Hoe. We also thank Singapore-MIT alliance for the funding support throughout the years.

8. References

Adachi, E., Dimitrov, A. S. & Nagayama, K. (1995). Stripe Patterns Formed on a Glass Surface during Droplet Evaporation. *Langmuir,* Vol.11, No.4, (April 1995), pp. 1057-1060

Allard, M., Sargent, E. H., Lewis, P. C. & Kumacheva, E. (2004). Colloidal Crystals Grown on Patterned Surfaces. *Advanced Materials,* Vol.16, No.15, (August 2004), pp. 1360-1364

Arora, A. K. & Tata, B. V. R. (1996). *Ordering and Phase Transitions in Charged Colloids* (1st ed), Wiley-VCH, ISBN 0471186309

Auer, S. & Frenkel, D. (2001). Prediction of Absolute Crystal-nucleation Rate in Hard-sphere Colloids. *Nature,* Vol.409, No.6823, (February 2001), pp. 1020-1023

Avrutsky, I., Kochergin, V. & Zhao, Y. (2000). Optical Demultiplexing in a Planar Waveguide with Colloidal Crystal. *IEEE Photonics Technology Letters,* Vol.12, No.12, (December 2000), pp. 1647-1649

Bartlett, P., Ottewill, R. H. & Pusey, P. N. (1992). Superlattice Formation in Binary Mixtures of Hard-sphere Colloids. *Physical Review Letters, Vol.*68, No.25, (June 1992), pp. 3801-3805

Behrens, S. H. & Grier, D. G. (2001). The Charge of Glass and Silica Surfaces. *The Journal of Chemical Physics.* Vol.115, No.14, (2001), pp. 6716-6721

Bertrand, P., Jonas, A., Laschewsky, A. & Legras, R. (2000). Ultrathin Polymer Coatings by Complexation of Polyelectrolytes at Interfaces: Suitable Materials, Structure and Properties. *Macromolecular Rapid Communications,* Vol.21, No.7, (April 2000), pp. 319-348

Biró, L. P., Bálint, Z., Kertész, K., Vértesy, Z., Márk, G. I., Horváth, Z. E., Balázs, J., Méhn, D., Kiricsi, I., Lousse, V. & Vigneron, J. P. (2003). Role of Photonic-crystal-type Structures in the Thermal Regulation of a Lycaenid Butterfly Sister Species Pair. *Physical Review E,* Vol.67, No.2, (February 2003), pp. 021907-1 – 021907-6

Blanco, A., Chomski, E., Grabtchak, S, Ibisate, M., John, S., Leonard, S., Lopez, C., Meseguer, F., Miguez, H, Mondian, J. Orzin, G. A., Toader, O. & van Driel, H. M. (2000). Large-scale Synthesis of a Silicon Photonic Crystal with a Complete Three-dimensional Bandgap near 1.5 Micrometres. *Nature,* Vol.405, No.6785, (May 2000), pp. 437-440

Bourret, A., Fuoss, P., Feuillet, G. & Tatarenko, S. (1993). Solving an Interface Structure by Electron Microscopy and X-ray Diffraction: The GaAs(001)-CdTe(111) Interface. *Physical Review Letters,* Vol.70, No.3, (January 1993), pp. 311-315

Busch, K. & John, S. (1998). Photonic Band Gap Formation in Certain Self-organizing Systems. *Physical Review E,* Vol.58, No.3, (September 1998), pp. 3896-3908

Chen, K. M., Jiang, X., Kimerling, L. C. & Hammond, P. T. (2000). Selective Self-Organization of Colloids on Patterned Polyelectrolyte Templates. *Langmuir,* Vol.16, No.20, (October 2000), pp. 7825-7834

Chen, M., Kim, J., Liu, J. P., Fan, H. & Sun, S. (2006). Synthesis of FePt Nanocubes and Their Oriented Self-Assembly. *Journal of the American Chemical Society,* Vol.128, No.22, (June 2006), pp. 7132-7133.

Clark, N. A., Hurd, A. J. & Ackerson, B. J. (1979). Single Colloidal Crystals. *Nature,* Vol.281, No.5726, (September 1979), pp. 57-60

Crandall, R. S., & Williams, R. (1977). Gravitational Compression of Crystallized Suspensions of Polystyrene Spheres. *Science,* Vol.198, No.4314, (October 1977), pp. 293-295

Davis, S. H. (1980). Moving Contact Lines and Rivulet Instabilities. Part 1. The Static Rivulet. *Journal of Fluid Mechanics,* Vol.98, No.2, (1980), pp. 225-242

Delamarche, E., Schmid, H., Bietsch, A., Larsen, N. B., Rothuizen, H., Michel, B. & Biebuyck, H. Transport Mechanisms of Alkanethiols during Microcontact Printing on Gold. *The Journal of Physical Chemistry B,* Vol.102, No.18, (April 1998), pp. 3324-3334

Denkov, N., Velev, O., Kralchevski, P., Ivanov, I., Yoshimura, H. & Nagayama, K. (1992). Mechanism of Formation of Two-dimensional Crystals from Latex Particles on Substrates. *Langmuir,* Vol.8, No.12, (December 1992), pp. 3183-3190

Dushkin, C. D., Yoshimura, H. & Nagayama, K. (1993). Nucleation and Growth of Two-dimensional Colloidal Crystals. *Chemical Physics Letters,* Vol.204, No. 5-6, (March 1993), pp. 455-460

Fustin, C., Glasser, G., Spiess, H. W. & Jonas, U. (2004). Parameters Influencing the Templated Growth of Colloidal Crystals on Chemically Patterned Surfaces. *Langmuir,* Vol.20, No.21, (October 2004), pp. 9114-9123

Gasser, U., Weeks, E. R., Schofield, A., Pusey, P. N. & Weitz, D. A. (2001). Real-Space Imaging of Nucleation and Growth in Colloidal Crystallization. *Science,* Vol.292, No.5515, (April 2001), pp. 258 -262

Gast, A. P. & Russel, W. B. (1998). Simple Ordering in Complex Fluids. *Physics Today,* Vol.51, No.12, (1998), pp. 24-30

Habdas, P. & Weeks, E. R. (2002). Video Microscopy of Colloidal Suspensions and Colloidal Crystals. *Current Opinion in Colloid & Interface Science,* Vol.7, No.3-4, (August 2002), pp. 196-203

Hales, T. C. (1997). Sphere packings, I. *Discrete & Computational Geometry,* Vol.17, No.1, (January 1997), pp. 1-51

Hiltner, P. A. & Krieger, I. M. (1969). Diffraction of Light by Ordered Suspensions. *The Journal of Physical Chemistry,* Vol.73, No.7, (July 1969), pp. 2386-2389

Ho, K. M., Chan, C. T. & Soukoulis, C. M. (1990). Existence of a Photonic Gap in Periodic Dielectric Structures. *Physical Review Letters,* Vol.65, No.25, (December 1990), pp. 3152-3155

Holgado, M., García-Santamaría, F., Blanco, A., Ibisate, M., Cintas, A., Míguez, H., Serna, C. J., Molperceres, C., Requena, J., Mifsud, A., Meseguer, F. & López, C. Electrophoretic Deposition To Control Artificial Opal Growth. *Langmuir,* Vol.15, No.14, (July 1999), pp. 4701-4704

Hoogenboom, J. P., Rétif, C., de Bres, E., van der Boes, M., van Langen-Suurling, A. K. & Romijn, J. (2004). Template-Induced Growth of Close-Packed and Non-Close-Packed Colloidal Crystals during Solvent Evaporation. *Nano Letters,* Vol.4, No.2, (February 2004), pp. 205-208

Hoover, W. & Ree, F. (1968). Melting Transition and Communal Entropy for Hard Spheres. *The Journal of Chemical Physics,* Vol.49, No.8, (October 1968), pp. 3609-3617, ISSN 00219606

Hu, Z. & Jonas, A. M. (2010). Control of Crystal Orientation in Soft Nanostructures by Nanoimprint Lithography. *Soft Matter,* Vol.6, No.1, (2010), pp. 21-28

Im, S. H., Kim, M. H. & Park, O. O. (2003). Thickness Control of Colloidal Crystals with a Substrate Dipped at a Tilted Angle into a Colloidal Suspension. *Chemistry of Materials* Vol.15, No.9 (May 2003), pp. 1797-1802

Imura, Y., Nakazawa, H., Matsushita, E., Morita, C., Kondo, T. & Kawai, T. (2009). Characterization of Colloidal Crystal Film of Polystyrene Particles at the Air-Suspension Interface. *Journal of Colloid and Interface Science,* Vol.336, No.2, (August 2009), pp. 607-611

Jiang, P., Bertone, J. F., Hwang, K. S., Colvin, V. L. (1999). Single-Crystal Colloidal Multilayers of Controlled Thickness. *Chemistry of Materials,* Vol.11, No.8, (1999), pp. 2132-2140

Jiang, P. & McFarland, M. J. (2004). Large-Scale Fabrication of Wafer-Size Colloidal Crystals Macroporous Polymers and Nanocomposites by Spin-Coating. *Journal of the American Chemical Society,* Vol.126, No.42, (October 2004), pp. 13778-13786

Kay, L. E. (1986). W. M. Stanley's Crystallization of the Tobacco Mosaic Virus, 1930-1940. *Isis,* Vol.77, No.3, (September 1986), pp. 450-472

Kitaev, V. & Ozin, G. A. (2003). Self-Assembled Surface Patterns of Binary Colloidal Crystals. *Advanced Materials,* Vol.15, No.1, (January 2003), pp.75-78

Koh, Y. K. & Wong C. C. In Situ Monitoring of Structural Changes during Colloidal Self-Assembly. *Langmuir,* Vol.22, No.3, (January 2006), pp. 897-900

Koh, Y. K., Yip, C. H., Chiang, Y. M. & Wong, C. C. (2008). Kinetic Stages of Single-Component Colloidal Crystallization. *Langmuir* Vol. 24, No. 10 (May 2008), pp. 5245-5248

Kralchevsky, P. A. & Nagayama, K. (1994). Capillary Forces between Colloidal Particles. *Langmuir,* Vol.10, No.1, (January 1994), pp. 23-36

Kralchevsky, P. A. & Denkov, N. D. (2001). Capillary Forces and Structuring in Layers of Colloid Particles. *Current Opinion in Colloid & Interface Science,* Vol.6, No.4, (August 2001), pp. 383-401

Kumacheva, E., Garstecki, P., Wu, H. & Whitesides, G. M. (2003). Two-Dimensional Colloid Crystals Obtained by Coupling of Flow and Confinement. *Physical Review Letters,* Vol.91, No.12, (September 2003), pp. 128301-1 –128301-4

Lee, I., Zheng, H., Rubner, M. F. & Hammond, P. T. (2002). Controlled Cluster Size in Patterned Particle Arrays via Directed Adsorption on Confined Surfaces. *Advanced Materials,* Vol.14, No.8, (April 2002), pp. 572-577

Lee, K. H., Chen, Q. L., Yip, C. H., Yan, Q. & Wong, C. C. (2010). Fabrication of Periodic Square Arrays by Angle-resolved Nanosphere Lithography. *Microelectronic Engineering,* Vol.87, No.10, (October 2010), pp. 1941-1944

Liu, J. (2005). *Photonic devices,* Cambridge University Press, ISBN 0521551951

Lu, Y., Yin, Y. & Xia, Y. (2001). Preparation and Characterization of Micrometer-Sized 'Egg Shells'. *Advanced Materials,* Vol.13, No.4, (February 2001), pp. 271-274

Manoharan, V., Elsesser, M & Pine, D. J. (2003). Dense Packing and Symmetry in Small Clusters of Microspheres. *Science,* Vol.301, No.5632, (July 2003), pp. 483-487

Maskaly, G. R., García, R. E., Carter, W. C. & Chiang, Y. M. (2006). Ionic Colloidal Crystals: Ordered, Multicomponent Structures via Controlled Heterocoagulation. *Physical Review E,* Vol.73, No.1, (January 2006), pp. 011402-1 - 011402-8

Mayoral, R., Requena, J., Moya, J., López, C., Cintas, A., Miguez, H., Meseguer, F., Vásguez, L., Holgado, M. & Blanco, A. (1997). 3D Long-range ordering in an SiO_2 Submicrometer-sphere Sintered Superstructure. *Advanced Materials,* Vol.9, No.3 (March 1997), pp. 257-260

Megens, M., van Kats, C. M., Bösecke, P. & Vos, W. L. (1997). Synchrotron Small-Angle X-ray Scattering of Colloids and Photonic Colloidal Crystals. *Journal of Applied Crystallography,* Vol.30, No.5-2, (October 1997), pp. 637-641

McLachlan, M. A., Johnson, N. P., Rue, R. M. & McComb, D. W. (2004). Thin Film Photonic Crystals: Synthesis and Characterisation. *Journal of Materials Chemistry,* Vol.14, No.2, (2004), pp. 144-150

Míguez, H., Meseguer, F., López, C., Mifsud, A., Moya, J. S. & Vázquez, L. (1997). Evidence of FCC Crystallization of SiO2 Nanospheres. *Langmuir,* Vol.13, No.23, (November 1997), pp. 6009-6011

Mihi, A., Zhang, C. & Braun, P. V. (2011). Transfer of Preformed Three-Dimensional Photonic Crystals onto Dye-Sensitized Solar Cells. *Angewandte Chemie International Edition,* Vol.50, No.25, (June 2011), pp. 5712-5715

Moroz, A & Sommers, C. (1999). Photonic Band Gaps of Three-dimensional Face-centred Cubic Lattices. *Journal of Physics: Condensed Matter,* Vol.11, No.4, (February 1999), pp. 997-1008

Norris, D. J. & Vlasov, Y. A. (2001). Chemical Approaches to Three-Dimensional Semiconductor Photonic Crystals. *Advanced Materials,* Vol.13, No. 6, (March 2001), pp. 371-376

Ozin, G. A. & Yang, S. M. (2001). The Race for the Photonic Chip: Colloidal Crystal Assembly in Silicon Wafers. *Advanced Functional Materials,* Vol.11, No.2, (April 2001), pp. 95-104

Park, S. H., Qin, D. & Xia, Y. (1998). Crystallization of Mesoscale Particles over Large Areas. *Advanced Materials,* Vol.10, No.13, (September 1998), pp. 1028-1032

Park, S. H., Gates, B. & Xia, Y. (1999). A Three-Dimensional Photonic Crystal Operating in the Visible Region. *Advanced Materials,* Vol.11, No.6, (April 1999), pp. 462-466

Pieranski, P. (1983). Colloidal Crystals. *Contemporary Physics,* Vol.24, No.1, (January 1983), pp. 25-73, ISSN 00107514

Pradhan, R. D., Bloodgood, J. A. & Watson, G. H. (1997). Photonic Band Structure of bcc Colloidal Crystals. *Physical Review B,* Vol.55, No.15, (April 1997), pp. 9503-9507

Prieve, D. C., Sides, P. J. & Wirth, C. L. (2010). 2-D Assembly of Colloidal Particles on a Planar Electrode. *Current Opinion in Colloid & Interface Science,* Vol.15, No.3, (June 2010), pp. 160-174

Pusey, P. N. & van Megen, W. (1986). Phase Behaviour of Concentrated Suspensions of Nearly Hard Colloidal Spheres. *Nature,* Vol.320, No.6060, (March 1986), pp. 340-342.

Pusey, P. N., van Megen, W., Bartlett, P., Ackerson, B. J., Rarity, J. G. & Underwood, S. M. (1989). Structure of Crystals of Hard Colloidal Spheres. *Physical Review Letters,* Vol.63, No.25, (December 1989), pp. 2753-2756

Rai-Choudhury, P. (1997). *Handbook of Microlithography, Micromachining, and Microfabrication: Microlithography*, SPIE Press, ISBN 9780819423788

Ray, M. A., Kim, H. & Jia, L. Dynamic Self-Assembly of Polymer Colloids to Form Linear Patterns. *Langmuir,* Vol.21, No.11, (May 2005), pp. 4786-4789

Ramiro-Manzano, F., Bonet, E., Rodriguez, I. & Meseguer, F. (2009). Layering Transitions in Colloidal Crystal Thin Films between 1 and 4 Monolayers. *Soft Matter,* Vol.5, No.21, (2009), pp. 4279-4282

Romano, F. & Sciortino, F. (2011). Colloidal Self-assembly: Patchy from the Bottom Up. *Nat Mater,* Vol.10, No.3, (March 2011), pp. 171-173

Sasaki, M. & Hane, K. (1996). Ultrasonically Facilitated Two-dimensional Crystallization of Colloid Particles. *Journal of Applied Physics,* Vol.80, No.9, (1996), pp. 5427-5431

Schmidt, M. & Löwen, H. (1997). Phase Diagram of Hard Spheres Confined between Two Parallel Plates. *Physical Review E,* Vol.55, No.6, (June 1997), pp. 7228-7241

Sharma, A. & Reiter, G. (1996). Instability of Thin Polymer Films on Coated Substrates: Rupture, Dewetting, and Drop Formation. *Journal of Colloid and Interface Science,* Vol.178, No.2, (March 1996), pp. 383-399

Sharma, V., Yan, Q., Wong, C. C., Carter, W. C. & Chiang, Y. M. (2009). Controlled and Rapid Ordering of Oppositely Charged Colloidal Particles. *Journal of Colloid and Interface Science,* Vol.333, No.1, (May 2009), pp. 230-236

Sirota, E. B., Ou-Yang, H. D., Sinha, S. K., Chaikin, P. M., Axe, J. D. & Fujii, Y. (1989). Complete Phase Diagram of a Charged Colloidal System: A Synchro-tron X-ray Scattering Study. *Physical Review Letters,* Vol.62, No.13, (March 1989), pp. 1524-1527

Tan, K. W., Li, G., Koh, Y.K., Yan, Q. & Wong, C. C. (2008). Layer-by-Layer Growth of Attractive Binary Colloidal Particles. *Langmuir,* Vol.24, No.17, (July 2008), pp. 9273-9278

Tan, K. W., Koh, Y. K., Chiang, Y.M. & Wong, C. C. (2010). Particulate Mobility in Vertical Deposition of Attractive Monolayer Colloidal Crystals. *Langmuir,* Vol.26, No.10 (May 2010), pp. 7093-7100.

Teh, L. K., Tan, N. K., Wong, C. C. & Li, S. Growth imperfections in three-dimensional colloidal self-assembly. *Applied Physics A: Materials Science & Processing,* Vol. 81, No. 7, (November 2005), pp. 1399-1404

Thijssen, J. H. J., Petukhov, A. V., 't Hart, T. C., Imhof, A., van der Werf, C. H. M., Schropp, R. E. I. & van Blaaderen, A. (2006). Characterization of Photonic Colloidal Single Crystals by Microradian X-ray Diffraction. *Advanced Materials,* Vol.18, No.13, (July 2006), pp. 1662-1666

Thomson, N. R., Bower, C. L. & McComb, D. W. (2008). Identification of Mechanisms Competing with Self-assembly during Directed Colloidal Deposition. *Journal of Materials Chemistry.* Vol.18, No.21, (2008), pp. 2500-2505

Tien, J., Terfort, A. & Whitesides, G. M. (1997). Microfabrication through Electrostatic Self-Assembly. *Langmuir,* Vol.13, No.20, (October 1997), pp. 5349-5355

Torres, C. M. S. (2003) *Alternative lithography: unleashing the potentials of nanotechnology,* Springer, ISBN 9780306478581

Troian, S. M., Herbolzheimer, E., Safran, S. A. & Joanny, J. F. (1989). Fingering Instabilities of Driven Spreading Films. *Europhysics Letters (EPL),* Vol.10, No.1, (September 1989), pp. 25-30

Ulman, A. (1996). Formation and Structure of Self-Assembled Monolayers. *Chemical Reviews,* Vol.96, No.4, (January 1996), pp. 1533-1554

van Blaaderen, A., Ruel, R. & Wiltzius, P. (1997). Template-directed Colloidal Crystallization. *Nature,* Vol.385, No.6614, (January 1997), pp. 321-324

van Rijn, C. J. M., Veldhuis, G. J. & Kuiper, S. (1998). Nanosieves with Microsystem Technology for Microfiltration Applications. *Nanotechnology,* Vol.9, No.4, (December 1998), pp. 343-345

Velev, O. D. & Lenhoff, A. M. (2000). Colloidal Crystals as Templates for Porous Materials. *Current Opinion in Colloid & Interface Science,* Vol.5, No.1-2, (March 2000), pp. 56-63

Vlasov, Y. A., Astratov, V. N., Baryshev, A. V., Kaplyanskii, A. A., Karimov, O. Z. & Limonov, A. F. (2000). Manifestation of Intrinsic Defects in Optical Properties of Self-organized Opal Photonic Crystals. *Physical Review E,* Vol.61, No.5, (May 2000), pp. 5784-5793

Vos, W. L., Sprik, R., van Blaaderen, A., Imhof, A., Lagendijk, A. & Wegdam, G. H. (1996). Strong Effects of Photonic Band Structures on the Diffraction of Colloidal Crystals. *Physical Review B,* Vol.53, No.24, (June 1996), pp. 16231-16235

Vos, W. L., Megens, M., van Kats, C. M. & Bösecke, P. (1997). X-ray Diffraction of Photonic Colloidal Single Crystals. *Langmuir,* Vol.13, No.23, (November 1997), pp. 6004-6008

Williams, R. & Crandall, R. S. (1974). The Structure of Crystallized Suspensions of Polystyrene Spheres. *Physics Letters A,* Vol.48, No.3, (June 1974), pp. 225-226

Xia, Y & Whitesides, G. M. (1998). Soft Lithography. *Angewandte Chemie International Edition,* Vol.37, No.5, (March 1998), pp. 550-575

Yablonovitch, E. (1987). Inhibited Spontaneous Emission in Solid-State Physics and Electronics. *Physical Review Letters,* Vol.58, No.20, (May 1987), pp. 2059-2062

Yan, Q., Gao, L., Sharma, V., Chiang, Y.M. & Wong, C. C. (2008). Particle and Substrate Charge Effects on Colloidal Self-Assembly in a Sessile Drop. *Langmuir,* Vol. 24, No.20, (October 2008), pp. 11518-11522

Yan, Q., Pavan, N., Chiang, Y. M. & Wong, C. C. (2009). Three-dimensional Metallic Opals Fabricated by Double Templating. *Thin Solid Films,* Vol.517, No.17, (July 2009), pp. 5166-5171

Yang, S. M. & Ozin, G. A. (2000). Opal Chips: Vectorial Growth of Colloidal Crystal Patterns inside Silicon Wafers. *Chemical Communications,* No.24, (2000), pp. 2507-2508

Yethiraj, A. & van Blaaderen, A. (2003). A Colloidal Model System with an Interaction Tunable from Hard Sphere to Soft and Dipolar. *Nature,* Vol.421, No.6922, (January 2003), pp. 513-517

Yethiraj, A., Wouterse, A., Groh, B. & van Blaaderen, A. (2004), Nature of an Electric-Field-Induced Colloidal Martensitic Transition. *Physical Review Letters,* Vol.92, No.5, (February 2004), pp. 058301-1 – 058301-4

Yoshiyama, T., Sogami, I. & Ise, N. (1984). Kossel Line Analysis on Colloidal Crystals in Semidilute Aqueous Solutions. *Physical Review Letters,* Vol.53, No.22, (November 1984), pp. 2153-2158

Zhang, K. & Liu, X. Y. (2004). In Situ Observation of Colloidal Monolayer Nucleation Driven by an Alternating Electric Field. *Nature,* Vol.429, No.6993, (June 2004), pp. 739-743

Zhou, Z. & Zhao, X. S. (2004). Flow-Controlled Vertical Deposition Method for the Fabrication of Photonic Crystals. *Langmuir* Vol.20, No.4, (February 2004), pp. 1524-1526